Biology
An Introduction

§ The Benjamin/Cummings Series in the Life Sciences

F. J. Ayala
Population and Evolutionary Genetics: A Primer
(1982)

F. J. Ayala and J. A. Kiger, Jr.
Modern Genetics, second edition (1984)

F. J. Ayala and J. W. Valentine
Evolving: The Theory and Processes of Organic Evolution (1979)

M. G. Barbour, J. H. Burk, and W. D. Pitts
Terrestrial Plant Ecology (1980)

C. L. Case and T. R. Johnson
Laboratory Experiments in Microbiology (1984)

R. E. Dickerson and I. Geis
Hemoglobin (1983)

P. B. Hackett, J. A. Fuchs, and J. W. Messing
An Introduction to Recombinant DNA Techniques: Basic Experiments in Gene Manipulation (1984)

L. E. Hood, I. L. Weissman, W. B. Wood, and J. H. Wilson
Immunology, second edition (1984)

J. B. Jenkins
Human Genetics (1983)

K. D. Johnson, D. L. Rayle, and H. L. Wedberg
Biology: An Introduction (1984)

R. J. Lederer
Ecology and Field Biology (1984)

A. L. Lehninger
Bioenergetics: The Molecular Basis of Biological Energy Transformations, second edition (1971)

S. E. Luria, S. J. Gould, and S. Singer
A View of Life (1981)

E. N. Marieb
Human Anatomy and Physiology Lab Manual: Brief Edition (1983)

E. N. Marieb
Human Anatomy and Physiology Lab Manual: Cat and Fetal Pig Versions (1981)

E. B. Mason
Human Physiology (1983)

A. P. Spence
Basic Human Anatomy (1982)

A. P. Spence and E. B. Mason
Human Anatomy and Physiology, second edition (1983)

G. J. Tortora, B. R. Funke, and C. L. Case
Microbiology: An Introduction (1982)

J. D. Watson
Molecular Biology of the Gene, third edition (1976)

W. B. Wood, J. H. Wilson, R. M. Benbow, and L. E. Hood
Biochemistry: A Problems Approach, second edition (1981)

Biology
An Introduction

Kenneth D. Johnson
San Diego State University

David L. Rayle
San Diego State University

Hale L. Wedberg
San Diego State University

The Benjamin/Cummings Publishing Company, Inc.
Menlo Park, California · Reading, Massachusetts
London · Amsterdam · Don Mills, Ontario · Sydney

About the cover: The mandrill, a type of baboon, is an inhabitant of the jungles of West Central Africa. Mandrills are social animals, living in groups of up to 50. The bright colors of the animals aid in communication within the group. For instance, when an animal becomes excited, the colors increase in brilliance on the face and the similarly colored buttocks. To show submission to a dominant male, a subordinate male presents his back side to the dominant male. In these and other ways the social hierarchy of the group is maintained. (See Essay 30–1, "Social Life of Baboons," pp. 546–547.)

Editor-in-Chief: James W. Behnke

Sponsoring Editors: Jane R. Gillen and Andrew Crowley

Developmental Editors: Patricia S. Burner, Carol Verburg, Amy Satran, James Funston, and Deborah Gale

Production Editors: Julie Kranhold and Patricia S. Burner

Photo Researcher: Bonnie Garmus

Interior Designer: John Edeen

Cover Designer: Michael Rogondino

Artists: Barbara Haynes, Kathy Monahan, Wayne Clark, and Carla Simmons

Cover photo © Toni Angermayer
Text photo and art acknowledgments appear after the Glossary.

Library of Congress Cataloging in Publication Data

Johnson, Kenneth D.
  Biology: an introduction.

  Bibliography: p.
  Includes index.
  1. Biology.   I. Rayle, David L. II. Wedberg, Hale L.
III. Title
QH308. 2. J63   1984   574   83-19026
ISBN 0-8053-7887-1

bcdefghij-DO-8987654

The Benjamin/Cummings Publishing Company, Inc.
2727 Sand Hill Road
Menlo Park, CA 94025

Preface

We have written *Biology: An Introduction* for students taking college-level biology for the first and perhaps the only time. With this audience in mind, we have emphasized the most important principles and processes of biology. Many of the concepts are presented from the perspective of their historical development, which we believe gives students a *feel* for biology. As a science, biology is more than a body of knowledge. It is also a means for obtaining knowledge, and understanding how scientists unravel life's mysteries is an important part of understanding biology. We believe that a major strength of this book lies in walking that fine line between providing enough background information to make the concepts understandable and overwhelming our audience with unnecessary detail.

To us, biology is a fascinating subject. It is our goal, both through teaching and writing, to convey some of its fascination to others. We hope this book instills in its reader some of the excitement of biology that made its writing a labor of love.

CONTENT AND ORGANIZATION

Our ideas concerning an introductory biology text have been shaped over many years by our experiences in the classroom. Our teaching experiences, and most importantly, the feedback we have received from thousands of introductory biology students, have made it clear that no single approach, no specific order of topics, is vastly superior to others. However, we have chosen a "levels-of-organization" approach because it works well for us and because it is easily adapted to most other teaching syllabi. We have made every effort to keep each chapter as independent as possible in order to provide some flexibility for instructors who prefer alternative topic orders.

We begin *Biology: An Introduction* with a short Introduction that provides an overview of the physical beginnings of our planet, particularly as they relate to earth as a life-supporting system. This leads naturally into a consideration of what life is, including an introduction to some of the key concepts of biology. We then go on to discuss the scientific method and, finally, the relationship between science and society. The main text follows and is divided into five units.

Unit One, **Molecules and Cells,** begins with two chapters that provide the chemical background necessary to understand modern biology. The chemistry is woven into a discussion of the origin of life. Dovetailing these topics has worked well in our classrooms, for it gives students a biological framework on which to hang the otherwise "dry" principles of chemistry. Besides, the matter of how life began on earth is intriguing for biologists and lay persons alike, and it provides an excellent model to approach what science is and how it works. The remaining chapters in this unit focus on the structure, function, and bioenergetics of cells. The two chapters which cover cellular respiration and photosynthesis contain overviews of these topics, which can stand alone if more complete coverage of the topics is not desired.

Unit Two, **Cell Division, Genetics, and Molecular Biology,** deals with subjects that lend themselves particularly well to an historical approach. We have taken advantage of this in several instances by providing not only the historical background, but also some of the experimental evidence underlying our current state of knowledge. We personally prefer to discuss classical genetics before embarking on the more difficult topic of molecular genetics, but this sequence could certainly be reversed.

Unit Three, **Microorganisms and Plants,** and Unit Four, **Animal Diversity and Physiology,** are the organismal sections of our text. They begin with the principles of classification, then diverge into specific coverage of the main groups of organisms and their physiology. This is probably the part of a biology course where instructors differ the most in terms of their coverage. Therefore, we took special care to make the chapters within these units as self-contained as possible. Instructors not wishing to follow our sequence, or opting for a more limited coverage of diversity or physiology, can design their own paths without much difficulty.

Our text closes with Unit Five, **Evolution, Behavior, and Ecology.** Our review of many course outlines for introductory biology indicates that this is the way most instructors wind up their courses. However, we have designed the chapters so that they can be covered earlier.

SPECIAL FEATURES

• Balanced Coverage

Special efforts have been made to provide balanced coverage of all areas of biology. We have included material on plant diversity and physiology, animal diversity, and the physiology of nonhuman animals—topics often underemphasized in other introductory biology textbooks.

• Special Topic Essays

Topics of special and current interest are explored in these optional essays. Subjects include Agent Orange, the artificial heart, herpes, oncogenes, and many others. These essays highlight important concepts through fascinating examples, and explore material of particular interest to students.

• Clear, Concise Writing Style

In an effort to make the book as readable as possible, we have tried to keep the technical terminology to a minimum. Clear, step-by-step explanations with appropriate examples help students understand even the most difficult topics.

• Artwork

Over 600 figures and photos, many in full color, blend with the text to make this book inviting to study from or just browse through. We have tried to keep the artwork large for ease of reference. Special attention has been given to diagrams illustrating important concepts.

• Integration of Text and Art

Every figure has been worked and reworked with the text so that the two complement each other. In addition, every effort has been made to ensure that each figure appears near its corresponding text. We feel that this careful integration of text and art makes this book a superior teaching and learning tool.

• Pedagogy

To aid the student in mastering the material in *Biology: An Introduction,* each of the 33 chapters ends with a summary, study questions, and suggested readings. A comprehensive glossary and index are included at the back of the book. A **Student Study Guide** written by Dr. Bernice Stewart of Prince Georges Community College in Largo, Maryland, is available through campus bookstores. For each chapter in the text, the Study Guide provides an overview, learning objectives, a detailed chapter outline, a list of new terms, a programmed self-test, a sample exam, and answers to the sample exam.

Dr. H. L. Wedberg has written an **Instructor's Guide** which is available to adopters of the text. For each text chapter, the Instructor's Guide includes a chapter outline, an overview, teaching suggestions, answers to the study questions in the text, and two chapter tests. A 3-ring binder containing the Instructor's Guide and 65 two-color **overhead transparencies** is available to qualified adopters. **Computerized Testing Service** is also available to qualified adopters.

ACKNOWLEDGMENTS

Writing a textbook today requires the dedicated efforts of many behind-the-scenes individuals. We are deeply indebted to the people at Benjamin/Cummings who made this venture a reality. For their unwavering enthusiasm, support, and professional attitude throughout the project, special recognition must go to Jim Behnke, Editor-in-Chief, and Jane Gillen and Andy Crowley, Sponsoring Editors; to Carol Verburg, who contributed significantly as a developmental editor during the early stages; to Amy Satran and Deborah Gale for their developmental assistance on several chapters; to Barbara Haynes, Kathy Monahan, Wayne Clark, and Carla Simmons for their beautiful drawings and diagrams; to Julie Kranhold and the staff at Ex Libris for their careful attention at the final production stage; to Carl May for providing many excellent photographs; to Bonnie Garmus, who did much of the final photo research; and especially to Pat Burner, production coordinator, developmental editor, copyeditor, and general ramrod. Pat's critical eye to every aspect of the manuscript and her unswerving dedication to detail have been major factors contributing to the clarity of both the text and art.

We are also very grateful to Dr. LeRoy McClenaghan and James Funston for developing and revising several sections in the text, and to Dr. Andrew Smith of Arizona State University, who conceived of the animal behavior chapter and wrote the first draft.

Many reviewers (listed to the right) provided helpful criticisms and suggestions for revision at various stages of manuscript development. We are very thankful for their guidance. In addition, many of our colleagues at San Diego State University read chapters, offered suggestions, and answered incessant questions. For your unselfishness and patience, fellow faculty, we thank you.

Kenneth D. Johnson
David L. Rayle
Hale L. Wedberg

Reviewers

John Alcock, Arizona State University
Adela Baer, San Diego State University
Alan Brush, University of Connecticut
Edwin Burling, De Anza College
Jerry Button, Portland Community College
Christine Case, Skyline College
Robert Cleland, University of Washington
Theodore J. Cohn, San Diego State University
Michael Donoghue, San Diego State University
Dennis Emery, Iowa State University
Gina Erickson, Highline Community College
Abraham Flexer, Boulder, Colorado
Susan Foster, Mount Hood Community College
Lawrence Fulton, American River College
Berdell Funke, North Dakota State University
Bernard Hartman, Texas Tech University
John Jackson, North Hennepin Community College
Norman Kerr, University of Minnesota
Philip Laris, University of California, Santa Barbara
Charles Leavell, Fullerton College
William Leonard, University of Nebraska
Joyce Maxwell, California State University, Northridge
Helen Miller, Oklahoma State University
Brian Myres, Cypress College
Michael Novacek, American Museum of Natural History
William K. Purves, Harvey Mudd College
William Romoser, Ohio University
C.K. James Shen, University of California, Davis
Andrew T. Smith, Arizona State University
David Smith, San Antonio College
Donald Smith, North Carolina State University
Bernice Stewart, Prince Georges Community College
Charles Wayne, County College of Morris
David Sloan Wilson, Kellogg Biological Station
Wilfred J. Wilson, San Diego State University
Lawrence Winship, Hampshire College

Brief
Table of Contents

Preface v
Introduction xvii

UNIT I   Molecules and Cells   1

1   Origins I: From Swirling Dust
    to Simple Organic Molecules   2
2   Origins II: From Organic Molecules to Cells   17
3   The Organization of Cells   34
4   How Materials Move Across Membranes   56
5   Energy Concepts and Respiration   70
6   Photosynthesis   92

UNIT II   Cell Division, Genetics, and Molecular Biology   109

7   Mitosis and Asexual Reproduction   110
8   Meiosis and Genetic Variability   123
9   Mendelian Genetics   136
10 Sex Determination and Gene Linkage   153
11 DNA Structure and Function   167
12 DNA → RNA → Protein:
    Information Flow and Its Regulation   181

UNIT III   Microorganisms and Plants   203

13 Classification Systems and Microorganisms   204
14 The Plant Kingdom   221
15 The Structure and Growth of
    Vascular Plants   243
16 Water and Nutrient Transport in
    Vascular Plants   260
17 Plant Development and the Regu-
    lation of Growth   271

UNIT IV   Animal Diversity and Physiology   287

18 The Animal Kingdom: Invertebrates   288
19 The Animal Kingdom: Lower Chordates
    and Vertebrates   310
20 The Circulatory and Immune Systems   334
21 Nutrition and the Digestive System   357
22 Gas Exchange and the Respiratory System   372
23 Homeostasis: Osmoregulation,
    Excretion, and Temperature Regulation   382
24 Hormones and the Endocrine System   398
25 Human Reproduction and Development   412
26 Neurons, Sensory Reception, and
    Muscle Response   438
27 Nervous Systems and the Brain   467

UNIT V   Evolution, Behavior, and Ecology   487

28 The Idea of Evolution   488
29 Modern Evolutionary Theory   504
30 Animal Behavior   532
31 The Ecology of Populations   556
32 The Ecology of Ecosystems   572
33 The Major Ecosystems   590

Appendix A   The Metric System   614
Appendix B   Acids, Bases, and pH   615
Glossary   G-1
Acknowledgments   A-1
Index   I-1

Detailed Table of Contents

Preface   v

Introduction   xvii

The Origin of Earth   xviii
What is Life?   xxii
Two Themes in Biology   xxiv
    Evolution   xxiv
    Organisms Conform to the Laws of Chemistry
      and Physics   xxvi
The Scientific Method   xxvi
Science and Society   xxvii

UNIT I
Molecules and Cells   1

CHAPTER 1
Origins I: From Swirling Dust to Simple Organic Molecules   2

Theories on the Origin of Life   2
    Special Creation: Outside the Realm of Science   2
    Panspermia   3
    Chemical Evolution   4
Elements and Atoms   4
Chemical Bonds   6
    Covalent Bonds   8
    Ionic Bonds   8
    Hydrogen Bonds   11
Carbon and Its Compounds   11
The Origin of Life Revisited   12
    Conditions on the Prebiotic Earth   12

Chemical Evolution in the Laboratory   13
Summary   15
Study Questions   16
Suggested Readings   16
Essay 1-1   Is Chemical Evolution
          Occurring Today?   14

CHAPTER 2
Origins II: From Organic Molecules to Cells   17

Polymerization: Building Large Molecules   17
Carbohydrates   18
    Simple Sugars, or Monosaccharides   18
    Disaccharides   19
    Polysaccharides   19
Lipids   21
    Triglycerides (Fats and Oils)   21
    Phospholipids   22
    Steroids   22
Proteins   24
    Amino Acids   24
    Protein Structure   26
    Protein Function   27
Nucleotides and Nucleic Acids   28
The Origin of Cells   28
    The Evolution of Metabolism   29
    The Emergence of Photosynthesis   29
    The Consequences of Oxygen Production   30
Summary   32
Study Questions   32
Suggested Readings   33
Essay 2-1   The Earth's Ozone Layer:
          Here Today! Gone Tomorrow?   31

CHAPTER 3
The Organization of Cells 34

The Discovery and Visualization of Cells 34
    Light Microscopy 34
    The Cell Theory 37
    Electron Microscopy 37
Cell Size 38
Cell Membranes 40
Prokaryotic Cell Structure 40
Eukaryotic Cell Structure 43
    Nucleus: Control Center of the Cell 43
    Cytoplasm: Seat of Metabolic Activities 44
    Mitochondria: Energy Factories 44
    Plastids: Storage and Photosynthesis 45
    Endoplasmic Reticulum: Protein Synthesis and
        Transport 45
    Golgi Bodies: Packaging and Secretion 48
    Vacuoles: Storage and Water Balance 49
    Lysosomes: Intracellular Digestion 49
    The Cytoskeleton: Intracellular Movement 49
    Cilia and Flagella: Cell Locomotion 51
    Cell Wall: Protection and Support 53
Summary 54
Study Questions 55
Suggested Readings 55
Essay 3–1 The Origin of Organelles 47

CHAPTER 4
How Materials Move Across Membranes 56

Membrane Structure 56
Diffusion 57
    Osmosis 58
    Facilitated Diffusion 63
Active Transport 65
Vesicular Transport 66
Summary 68
Study Questions 69
Suggested Readings 69
Essay 4–1 Water: The Indispensable Liquid 60

CHAPTER 5
Energy Concepts and Respiration 70

Essential Concepts of Energy Metabolism 71
    ATP and Energy Coupling 71

    Activation Energy and Enzymes 74
The Respiration of Glucose 76
    Overview of Glucose Respiration 76
    Glycolysis 78
    The Krebs Cycle 78
    The Respiratory Electron Transport Chain 80
The Fermentation of Glucose 83
The Mobilization of Polysaccharides
   and Fats for Respiration 88
Summary 88
Study Questions 90
Suggested Readings 91
Essay 5–1 Hot Plants: A Respiratory Oddity 84

CHAPTER 6
Photosynthesis 92

Light and the Photosynthetic Pigments 94
The Light and Dark Reactions 97
    Overview of Photosynthesis 97
    The Light Reactions 97
    The Dark Reactions 100
The Uses of Sugar 104
The Limiting Factors of Photosynthesis 106
    Light Intensity 106
    Temperature 106
    Water Availability 107
    Carbon Dioxide Concentration 107
Summary 107
Study Questions 108
Suggested Readings 108
Essay 6–1 The Unraveling of C_3 Photosynthesis 102

UNIT II
Cell Division, Genetics, and Molecular Biology 109

CHAPTER 7
Mitosis and Asexual Reproduction 110

Cell Division in Bacteria 110
Cell Division in Eukaryotes 111
    Chromosome Structure 112
    Mitosis 112
    Cytokinesis 114

The Cell Cycle   116
Asexual Reproduction   116
Summary   119
Study Questions   122
Suggested Readings   122
Essay 7–1   HeLa Cells in Medical Research:
A Useful Tool or Laboratory Pest?   120

CHAPTER 8
Meiosis and
Genetic Variability   123

Haploid and Diploid Chromosome Number   124
Meiosis   125
The First Meiotic Division   125
The Second Meiotic Division   128
Meiosis and Genetic Variability   128
Crossing Over   130
Independent Assortment of Chromosomes   130
Errors in Meiosis   132
Summary   134
Study Questions   135
Suggested Readings   135

CHAPTER 9
Mendelian Genetics   136

Mendel's Laws of Inheritance   136
The Monohybrid Cross   137
The Testcross   142
The Dihybrid Cross   143
Mendel's Laws Extended   146
Incomplete Dominance   146
Multiple Alleles   147
Pleiotropy   148
Polygenic Inheritance   149
Nature versus Nurture   149
Summary   150
Study Questions   151
Suggested Readings   151
Answers to Study Questions   152

CHAPTER 10
Sex Determination
and Gene Linkage   153

The Inheritance of Sex   153
The Inheritance of Sex-Linked Traits   154

Demonstration of a Sex-Linked
Trait in Fruit Flies   154
Sex-Linked Traits in Humans   157
Autosomal Linkage   159
The Effect of Autosomal
Linkage on Inheritance Patterns   159
Recombination of Linked Genes   159
Summary   164
Study Questions   164
Suggested Readings   165
Answers to Study Questions   165
Essay 10–1   Human Sex
Chromosome Abnormalities   155
Essay 10–2   Hemophilia and the
Russian Revolution   162

CHAPTER 11
DNA Structure
and Function   167

On the Track of DNA   167
Evidence from Bacteria   167
Evidence from Viruses   168
The Structure of DNA   169
DNA Replication   173
Information Storage   174
The Nucleotide Alphabet   174
The Nature of Genetic Information   175
Proteins and Traits   175
Recombinant DNA   176
Summary   179
Study Questions   180
Suggested Readings   180

CHAPTER 12
DNA→RNA→Protein:
Information Flow and Its
Regulation   181

The Triplet Code   181
RNA and Transcription   181
Types of RNA   182
Transcription   184
Protein Synthesis   184
The Genetic Code   184
Translating the Message:
The Mechanics of Protein Synthesis   187
Variations on Information Flow   189

Messenger RNA Editing 189
Gene Rearrangements 189
Mutations 191
Types and Causes of Mutations 192
Mutations, Alleles, and Variation 194
The Regulation of Gene Expression 195
Transcriptional Control
in Prokaryotes: The Lac Operon 195
Transcriptional Control in Eukaryotes 196
Translational Control in
Prokaryotes and Eukaryotes 199
Summary 199
Study Questions 201
Suggested Readings 201
Essay 12–1   Will It Cause Cancer? 198
*Essay 12–2   Oncogenes and Cancer: An Exciting
New Research Development 200*

UNIT III
Microorganisms and Plants 203

CHAPTER 13
Classification Systems and Microorganisms 204

The Early Classification Systems 204
The Modern Classification System 205
Kingdom Monera 208
Bacteria 208
Cyanobacteria 210
Viruses 211
Kingdom Protista 213
Amoebas 214
Sporozoans 214
Ciliates 214
Diatoms 216
Kingdom Fungi 216
Summary 219
Study Questions 219
Suggested Readings 220
*Essay 13–1   Herpes: A Virus on
the Rampage 212*

CHAPTER 14
The Plant Kingdom 221

Alternation of Generations 221
The Nonvascular Plants 222
Algae 222
Bryophytes: Mosses and Liverworts 228
The Origins of Vascular Plants 230
Lower Vascular Plants 231
Seed-Producing Vascular Plants 233
Gymnosperms 233
Angiosperms 235
Pollinating Mechanisms 240
Summary 241
Study Questions 242
Suggested Readings 242
Essay 14–1   Chlorella: Algae to Fuel the World? 226

CHAPTER 15
The Structure and Growth of Vascular Plants 243

The Structure of Vascular Plants 244
Leaves 244
Roots 250
Xylem and Phloem 252
Stems 253
The Growth of Vascular Plants 255
Primary Growth 255
Secondary Growth 256
Summary 258
Study Questions 259
Suggested Readings 259
Essay 15–1   Strategies of Desert Plants 248

CHAPTER 16
Water and Nutrient Transport in Vascular Plants 260

Water Transport 260
The Pathway of Water Transport 261
The Mechanism of Water Transport 263
Mineral Nutrition 265
Sugar Transport 267
Summary 270
Study Questions 270
Suggested Readings 270
Essay 16–1   Mycorrhizae 267
Essay 16–2   Aphids as Lab Technicians 268

CHAPTER 17
**Plant Development
and the Regulation of Growth 271**
Seeds and Germination 271
Hormones and Plant Development 274
　Auxin 274
　Gibberellins 276
　Cytokinins 278
　Ethylene 279
Light and Plant Development 280
　Photodormancy and the
　　Discovery of Phytochrome 280
　Photomorphogenesis 281
　Photoperiodism and Flower Induction 282
Senescence 284
Summary 285
Study Questions 286
Suggested Readings 286
Essay 17–1 Herbicides: Boon or Bane? 277

UNIT IV
**Animal Diversity
and Physiology 287**

CHAPTER 18
**The Animal Kingdom:
Invertebrates 288**
Phylum Porifera: Sponges 290
Phylum Cnidaria: Jellyfish, Corals, and Others 291
Phylum Platyhelminthes: Flatworms 293
Class Nematoda: Roundworms 296
Phylum Mollusca: Snails, Clams, Squids, and
　Others 296
　Class Gastropoda: Snails, Slugs, and
　　Their Relatives 297
　Class Bivalvia: Clams, Oysters, and
　　Their Relatives 299
　Class Cephalopoda: Squids, Octopuses,
　　and Their Relatives 299
Phylum Annelida: Segmented Worms 300
Phylum Arthropoda: Spiders, Insects, and Others 301
　General Characteristics of Arthropods 303
　Insects 304
Phylum Echinodermata: Starfish, Sea Urchins,
　and Others 306

Summary 308
Study Questions 309
Suggested Readings 309
Essay 18–1 Schistosomiasis 294

CHAPTER 19
**The Animal Kingdom: Lower
Chordates and Vertebrates 310**
Characteristics and Origin of Chordates 310
The Lower Chordates: Tunicates and Lancelets 312
Vertebrates 314
　Fishes 315
　Amphibians 317
　Reptiles 318
　Birds 322
　Mammals 323
Primates and Human Evolution 326
　The Characteristics of Primates 326
　Human Evolution 328
Summary 332
Study Questions 333
Suggested Readings 333

CHAPTER 20
**The Circulatory and
Immune Systems 334**
An Overview of Organ Systems 334
Internal Transport Systems in Animals 336
　Open versus Closed Circulatory Systems 336
　Evolution of the Vertebrate Heart 336
The Human Circulatory System 338
　The Pump 338
　Blood Pressure 340
　Blood Vessels 341
　The Lymphatic System 346
　Blood 346
The Immune System 349
　Antibody-Mediated Responses 349
　Cell-Mediated Responses 352
　Recognizing Self 352
Summary 353
Study Questions 356
Suggested Readings 356
*Essay 20–1 Heart Disease and the
　　Onset of the Bionic Era 344*
*Essay 20–2 New Strategies in the
　　War Against Cancer 354*

CHAPTER 21
Nutrition and the Digestive System   357

Human Nutritional Requirements   357
    Energy Foods   358
    Essential Organic Precursors   358
    Vitamins   362
    Minerals   364
Digestive Systems   364
The Human Digestive System   366
    Oral Cavity, Pharynx, and Esophagus   366
    Stomach   367
    Small Intestine   367
    Large Intestine   369
Summary   370
Study Questions   371
Suggested Readings   371
Essay 21-1   The Dieter's Dilemma   360

CHAPTER 22
Gas Exchange and the Respiratory System   372

Types of Respiratory Systems   372
    Gills   373
    Tracheae   373
    Lungs   374
The Human Respiratory System   375
    Anatomy   375
    The Movement of Oxygen and
        Carbon Dioxide   376
Breathing Mechanisms   377
    The Mechanics of Breathing   377
    The Regulation of Breathing   380
Summary   380
Study Questions   381
Suggested Readings   381
Essay 22-1   High Altitude Adventure   379

CHAPTER 23
Homeostasis: Osmoregulation, Excretion, and Temperature Regulation   382

Osmoregulation: Environmental Adaptations   382
    Marine Invertebrates   383
    Freshwater Animals   384
    Marine Vertebrates   384
    Terrestrial Animals   385
Excretion   386

The Human Kidney   387
Temperature Regulation   391
    Ectotherms   391
    Endotherms   395
Summary   396
Study Questions   397
Suggested Readings   397
Essay 23-1   Deferring Death: Dialysis   392

CHAPTER 24
Hormones and the Endocrine System   398

Hormones   398
    Insect Hormones   398
    Vertebrate Hormones   399
The Human Endocrine System   402
    The Anterior Pituitary   402
    The Posterior Pituitary   405
    The Hypothalamus   405
    The Thyroid Gland   406
    The Pancreas   406
    The Adrenal Glands   408
    The Gonads   409
Summary   410
Study Questions   411
Suggested Readings   411

CHAPTER 25
Human Reproduction and Development   412

The Female Reproductive System   413
    The Menstrual Cycle   413
The Male Reproductive System   416
    Spermatogenesis   418
Sexual Intercourse   419
Conception   420
Pregnancy   420
    Embryo Development   420
    Fetal Development   425
Birth   427
    Twins   427
Contraception   429
    Methods Based on the Menstrual Cycle   429
    Physical and Chemical Barriers to
        Sperm Movement   431
    Preventing Ovulation   431
    Prevention of Embryo Implantation or
        Embryo Development   433
    Sterilization   433

Sexually Transmitted Diseases   433
Summary   434
Study Questions   437
Suggested Readings   437
Essay 25–1   Amniocentesis   428
Essay 25–2   DES and Its Tragic Legacy   436

CHAPTER 26
Neurons, Sensory Reception, and Muscle Response   438

Neurons   438
    Types of Neurons   439
    Neuron Structure   439
    The Nature of an Impulse   441
    Propagation of the Impulse   445
    Communication Between Neurons   446
Sensory Reception   448
    Sensory Receptors   448
    Sensory Cell Activation   449
    Photoreception and the Human Eye   449
    Hearing and the Human Ear   453
Muscle Response   454
    Structure and Contraction of Skeletal Muscles   457
    Control of Muscle Contraction   461
Summary   465
Study Questions   466
Suggested Readings   466
Essay 26–1   How Fast? How High? How Far?   462

CHAPTER 27
Nervous Systems and the Brain   467

Nervous Systems of Invertebrates   467
The Vertebrate Nervous System   469
    The Peripheral Nervous System   469
    The Central Nervous System   473
The Human Brain   474
    The Thalamus   474
    The Hypothalamus   474
    The Cerebrum   475
    The Limbic System   477
    The Reticular Formation   477
Chemical Activities of the Brain   480
    Neurotransmitters   480
    Natural Opiates   480
    Drugs   482
Summary   483
Study Questions   484
Suggested Readings   485
Essay 27–1   Acupuncture   481

UNIT V
Evolution, Behavior, and Ecology   487

CHAPTER 28
The Idea of Evolution   488

Evidence for Evolution   488
The Beginnings of Evolutionary Thought   492
    Early Ideas on the Origins of Species   492
    Challenges to Special Creation   493
    Lamarck's Theory   493
Darwin and the Concept of Natural Selection   495
    Voyage of the *Beagle*   495
    Formulating the Theory   496
    Publication of the Theory   497
    Natural Selection in a Nutshell   499
Summary   500
Study Questions   503
Suggested Readings   503
Essay 28–1   Microevolution of the Peppered Moth   498

CHAPTER 29
Modern Evolutionary Theory   504

The Hardy–Weinberg Law   505
The Agents of Evolution   505
    Mutation   505
    Gene Flow   507
    Genetic Drift   507
    Natural Selection   509
Speciation   514
    The Concept of Species   514
    Reproductive Isolating Mechanisms   515
    Allopatric Speciation   516
    Sympatric Speciation   518
Evolutionary Trends   523
Summary   530
Study Questions   531
Suggested Readings   531
Essay 29–1   The Hardy–Weinberg Equilibrium   506
Essay 29–2   The Late Cretaceous Cataclysm   528

CHAPTER 30
Animal Behavior   532

Genes and Behavior   534
Instinctive and Learned Behaviors   535

Types of Learning   535
Animal Communication   538
    Communication by Visual Displays   538
    Communication by Sounds   540
    Communication by Chemicals (Odors)   541
Orientation Behavior   541
    Echolocation   541
    Celestial Navigation in Birds   542
Social Behavior   544
    Animal Societies   544
    Evolution and Social Behavior   545
    Territoriality   548
    Mating Behavior in Mammals   550
    Altruism   551
Summary   554
Study Questions   555
Suggested Readings   555
Essay 30–1   Social Life of Baboons   546

CHAPTER   31
The Ecology of Populations   556

Characteristics of Populations   556
Dynamics of Population Growth   557
Regulation of Population Size   558
Interactions Among Populations   559
    Predation and Parasitism   559
    Competition   563
    Symbiosis   565
The Human Population   567
    Human Population Growth   567
    Future Prospects   568
Summary   570
Study Questions   570
Suggested Readings   571
Essay 31–1   Lemming Cycles:
                An Ecological "Whodunit"   560

CHAPTER   32
The Ecology of Ecosystems   572

Trophic Levels   573
Energy Flow Through Ecosystems   577
Nutrient Flow and Biogeochemical
    Cycles in Ecosystems   580
        The Carbon Cycle   583
        The Nitrogen Cycle   585
Summary   588
Study Questions   589
Suggested Readings   589
Essay 32–1   Off-Road Vehicles and Ecosystems   574

CHAPTER   33
The Major Ecosystems   590

Ecological Succession   590
Climax Communities   591
Terrestrial Biomes   594
    Tropical Forests   596
    Grasslands   598
    Deciduous Forests   598
    Coniferous Forests   600
    Chaparral and Brushland   602
    Tundra   603
    Desert   604
Marine Ecosystems   606
    Coastal   606
    Pelagic   608
Freshwater Ecosystems   610
Human Influences on Ecosystems   612
Summary   613
Study Questions   613
Suggested Readings   613

Appendix A   The Metric System   614
Appendix B   Acids, Bases, and pH   615
Glossary   G-1
Acknowledgments   A-1
Index   I-1

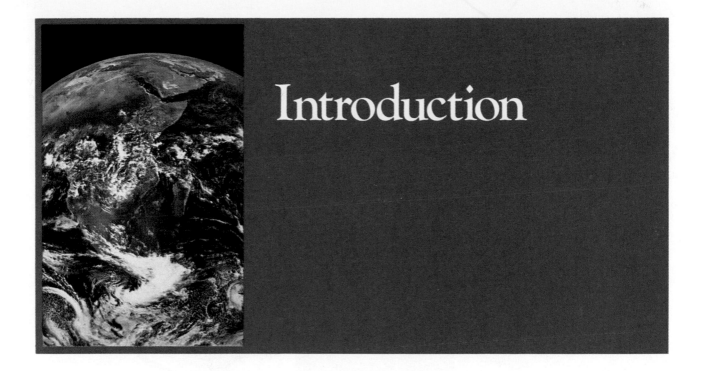

Introduction

We humans are an intelligent and curious lot. We study our world with a fervent zeal, unraveling mysteries that other creatures can't even conceptualize. No wonder we think of ourselves as special. Ironically, though, the more we learn about ourselves and our surroundings, the less special we seem to become. For example, biologists tell us that we are animals, albeit intelligent ones, linked by threads of common ancestry to all other creatures, past and present—that in the giant tree of life we are no more than a tiny branch of new growth. And astronomers have been chipping away at our egocentric views for centuries, giving us an even more humble perspective.

Recent findings in astronomy indicate that the earth is probably not the only life-bearing planet in the universe. Although the possibility of living creatures inhabiting Mars or other planets within our solar system now seems remote, the probability of life existing on planets in other solar systems within the Milky Way galaxy appears to be quite high. And the Milky Way with its hundreds of billions of stars (each potentially a solar system) is but one galaxy in a sea of over 10 billion known galaxies! Thus, there could be untold trillions of earth-like planets in the cosmos, any of which could be inhabited by creatures with a comparable or even higher intelligence and technology than ours.

If there is intelligent life out there, then why not try to communicate with our distant neighbors? This thought has occurred to many scientists, and several attempts have been made to communicate our existence to other parts of the universe. For example, in November, 1974, a coded radio message was transmitted from the Arecibo Observatory in Puerto Rico toward a distant star cluster. This radio pulse took a mere five hours to reach the outer limits of our own solar system. The message is simple enough to be deciphered by any beings who possess an intelligence and technology at least equivalent to ours (Figure I–1). If communication with an extraterrestrial civilization was successful, the cultural shock would leave no human institution untouched. Imagine what might go through the "minds" of any unearthly beings receiving our radio message!

Perhaps other stars formed in much the same way as our sun (Figure I–2). If so, then these stars may also have orbiting planetary systems. Most of these planets would not have the proper physical environments to support life as we know it, just as earth is the only planet among the nine in our solar system that appears to bear life. If you were to embark on a "star trek" mission in search of extraterrestrial life, what type of planet would you seek? Knowing that earth supports life, you would probably look for a planet that is similar

FIGURE I–1

To Whom It May Concern! Graphic representation of a 3-minute radio message beamed to the stars from the Arecibo Observatory in November, 1974. If intelligent, extraterrestrial life intercepted this message, they would presumably plot the series of sound characters in a form like that shown on the left. For the rest of us, the explanation of the plotted message appears on the right. The message has already traveled more than twice the distance to the nearest star.

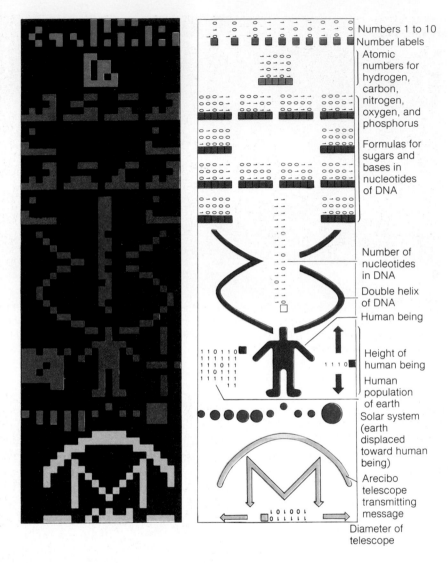

Numbers 1 to 10

Number labels

Atomic numbers for hydrogen, carbon, nitrogen, oxygen, and phosphorus

Formulas for sugars and bases in nucleotides of DNA

Number of nucleotides in DNA

Double helix of DNA

Human being

Height of human being

Human population of earth

Solar system (earth displaced toward human being)

Arecibo telescope transmitting message

Diameter of telescope

to our own. It might be useful, then, to look at the special physical characteristics that the earth acquired during its origin, and how this planet changed through time to become the "living Earth" (Figure I–3).

THE ORIGIN OF EARTH

According to one currently popular idea, the forerunner of our solar system was a large, swirling cloud of dust and gases. Gravitational attraction within the cloud pulled the particles inward, forming a flattened disc that became smaller and denser with time. In accordance with a law of physics, our swirling cosmic cloud must have turned faster and faster as it became smaller, just as a tether ball's angular speed (revolutions per minute) increases as it gets closer to the pole. Eventually, the cloud was spinning so fast that small patches of it were cast into nearby space by centrifugal force, just as you would be thrown to one side of a speeding automobile making a sharp turn. These cast-off dust patches also became smaller and denser as a result of their own gravities, ultimately becoming the nine planets. The central mass of debris condensed into a medium-sized star, our sun.

When the earth first became a solid ball of matter, it had an atmosphere consisting of the lightest, most abundant elements in the universe: hydrogen and helium (Figure I–4a). However, because the primor-

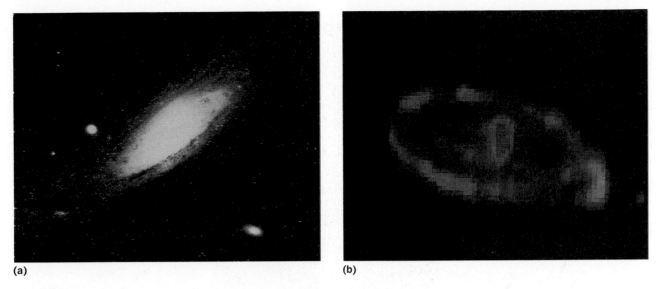

(a) **(b)**

FIGURE I–2
Star Formation. (a) The Andromeda Galaxy as seen with a conventional optical telescope. (b) A computer-processed image of the Andromeda Galaxy recently obtained by the Infrared Astronomical Satellite. The red, orange, and yellow areas indicate regions where young stars are probably forming, each potentially a solar system.

FIGURE I–3
Earth. This photograph was taken by the Apollo 17 astronauts from a distance of 80,000 kilometers (about 60,000 miles).

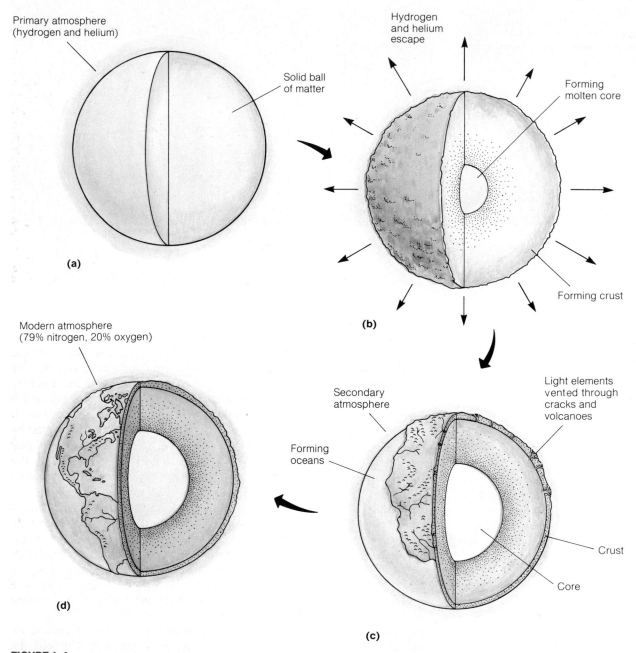

Primary atmosphere
(hydrogen and helium)

Solid ball
of matter

(a)

Hydrogen
and helium
escape

Forming
molten core

Forming crust

(b)

Modern atmosphere
(79% nitrogen, 20% oxygen)

(d)

Secondary
atmosphere

Light elements
vented through
cracks and
volcanoes

Forming
oceans

Crust

Core

(c)

FIGURE I–4
The Shaping of Earth. (a) According to one theory, the dust patch forerunner of earth condensed
into a solid ball with a primary atmosphere of hydrogen and helium. (b) Too light to be retained by
earth's gravitational pull, the primary atmosphere gases gradually escaped into space. Meanwhile,
rocks were liquefied by heat produced by radioactive decay processes occurring deep inside the ball.
This permitted the heavier molten elements to flow toward the earth's core and the lighter elements
to float toward the surface. (c) The internal heat also created pressure, which was periodically vented
through cracks, fissures, and volcanoes in the crust. These vents spewed forth gases that became the
secondary atmosphere, the chemical cauldron of life's beginnings. (d) The earth today has an atmosphere composed largely of nitrogen and oxygen gases.

FIGURE I–5
The Mount Saint Helens Volcano.
Volcanic eruptions are a major source of atmospheric gases today, and they undoubtedly contributed to the formation of the secondary atmosphere some 4.5 billion years ago.

dial earth was relatively small and close to the sun, its gravity was too slight and its temperature too high to retain this primary atmosphere. Thus, most of the hydrogen and helium escaped earth's gravity and floated off into space.

In the meantime, events were occurring in the earth's interior that would eventually lead to the formation of a new, secondary atmosphere. Deep below the surface, radioactive decay processes (similar in principle to those occurring in a nuclear reactor) were generating tremendous amounts of heat. Since this heat was unable to escape through the solid ball of matter, it accumulated and melted much of the subsurface rock. In the molten state, the heavier, denser elements (mainly nickel and iron) sank toward the center, forming a molten core. The lighter elements (such as carbon, oxygen, and nitrogen) "floated" toward the surface to become the earth's outer crust (Figure I–4b). The buildup of heat also produced pressure below the crust. This pressure was vented periodically through cracks and volcanoes, which spewed out lava and gases composed largely of the light crust elements (Figure I–5). These elements reacted with each other and with the hydrogen still present in the atmosphere, forming various gases and, significantly, vast amounts of water. The gases—methane, ammonia, hydrogen sulfide, carbon monoxide, carbon dioxide, and nitrogen—became the earth's secondary atmosphere (Figure I–4c).

The formation of the earth's secondary atmosphere set the chemical stage for events that would transform the earth into a life-bearing planet. The generous amount of water formed was too much to remain suspended as vapor in the atmosphere, and it condensed into liquid form to become the oceans. It rained a lot back then, and the lightning generated by the frequent storms provided part of the "spark" that encouraged chemical reactions to occur among the atmospheric gases. According to the chemical origin of life theory, these simple gases reacted with one another to form larger, more complex substances, which in turn reacted to produce even larger substances. After millions of years of increasing chemical complexity, the surface waters of earth became dotted with the first tiny life forms, probably something resembling the simple bacteria of today. From that moment forward, the earth has had an uninterrupted biological history.

As the early organisms grew in numbers and increased in complexity, they gradually changed the very atmosphere from which they were "born." In using up some of the atmospheric gases to support their growth, and with the emergence of plantlike creatures that produced oxygen gas, the secondary atmosphere ultimately gave way to our present atmosphere. The air we breathe contains about 79% nitrogen gas, 20% oxygen gas, and trace amounts of other gases.

If during your imaginary search for other life-supporting planets you found one with an atmosphere similar to earth's secondary or present one, that would indeed be very encouraging. To be a good candidate for life support, however, your planet should also have plenty of water and a moderate temperature. All three of these conditions hinge primarily on the planet's size and its distance from its sun.

Let us further imagine that your spaceship does happen upon a planet that meets these physical criteria, and you decide to have a closer look. Your mission, you recall, is to find life. If life does exist there, it may not take the forms familiar to us on earth. For example, the adage, "If it quivers, it's alive," may not help you discover life on a planet with no animals. What, then, are the most basic characteristics of all living things?

WHAT IS LIFE?

Asking a biologist to define "life" is like asking a geologist to define "rock." Since both terms refer to natural states of matter, they really defy definition. The best we can do is to *characterize* a living organism in terms of its physical attributes, such as size, form, chemical composition, activities, and other observable features. And by identifying the characteristics that are shared by all organisms, perhaps we can put into words what we all intuitively know to be life.

One of the characteristics of living things is their highly **complex organization** (Figure I–6a). Even very simple organisms, such as microscopic bacteria, are much more complex than the most sophisticated computers. This complexity is built from special types of chemical substances that are found only in biological structures. And from these structures arise the activities that are uniquely life. Included among these activities are the other major characteristics of life: metabolism, growth, sensitivity, and reproduction.

The most fundamental of all biological activities is cellular metabolism. Every organism is a chemical "factory," carrying out thousands of different chemical reactions collectively known as **metabolism.** A major part of metabolism involves the formation of complex molecules from simpler ones obtained from the environment. For example, animals obtain sugars and other small molecules from the food they eat (Figure I–6b). They then transform these building blocks into the more complex substances that make up their bodies. But this process, like most biological activities, requires energy. The energy is provided by another part of metabolism, which breaks down food molecules like sugars, converting their inherent energy to other forms of useful energy. So, organisms use food both as chemical building blocks and as energy sources; but where does the food come from? The green plants of the earth are the ultimate source of energy-laden food—they capture light energy from the sun and use it to build food molecules from simpler substances, namely carbon dioxide and water.

Another characteristic common to all living things is growth. **Growth** is an increase in size or mass, and it results from the various metabolic processes that build complex molecules within organisms. Growth is generally accompanied by **development,** an orderly, progressive series of events that gives form to the body and results in specialization of activities within the organism. For example, your life began as a microscopic fertilized egg, a single cell. This cell divided to produce two cells, these divided to produce four cells, and so on. But this growth process did not culminate in a giant blob of cells, but rather a form that is distinctly human (Figure I–6c–e). During this development, the billions of cells in your growing body took on a variety of highly specialized functions—skin cells protect, eye cells see, blood cells transport, and so forth. Each different type of organism has its own specific pattern of growth and development, and this fact enables us to distinguish an ant from a goat, a tiger from a tulip, and so on.

All organisms display **sensitivity**—the ability to detect and respond to changes in the environment (Figure I–6f). A sunflower bends toward the sun, Canadian geese fly south for the winter, and a frog snatches a fly with its tongue—these are all examples of sensitivity. Sensitivity does not always imply the existence of sophisticated sense organs and a well-developed nervous system. Even single-celled bacteria can sense food nearby and move toward it.

Finally, the capacity for **reproduction**—the act of producing new individuals that are like the parent—is undoubtedly the most distinctive characteristic of living things. You can find examples of chemical processes occurring outside of living systems (such as the reaction between iron and oxygen to form rust), or inanimate objects growing in size (such as a snowball rolling down a snowy bank), or even machines that exhibit sensitivity (such as a signal-triggered garage door opener), but you will never see any nonliving thing produce copies of itself. Reproduction provides for the continuity of life through the generations (Figure I–6g). Indeed, reproduction is the thread that links all present life forms to their ultimate ancestors—the simple creatures that started the experiment of life on earth some 3.6 billion years ago.

FIGURE I–6
The Characteristics of Life. (a) The internal structure of a plant stem illustrates the complex organ-ization of organisms. (b) This grasshopper is munching on a tasty morsel of metabolic fuel. (c–e) Growth and development in humans begins with a fertilized egg, then passes through (c) an embryonic stage (9 weeks) and (d) a fetal stage (4 months) to become (e) a baby. (f) A river otter and crayfish detect each other's presence—a clear example of sensitivity. (g) Different generations of cedar waxwings illustrate that reproduction provides continuity to life on earth.

TWO THEMES IN BIOLOGY

Two major themes in biology weave throughout this book, both so fundamental to the study of modern biology that we will introduce them here: (1) organisms have evolved, and (2) life and its processes conform to the laws of chemistry and physics.

Evolution

Living things have changed, are changing, and will continue to change. This is the basic tenet of biological evolution, and it is so ingrained in the minds of biologists that few of them argue the basic idea anymore, at least not with each other. No one can ignore the fossils, relics of past flora and fauna that represent distinctly different life forms than presently exist. Yet, there are obvious resemblances between modern organisms and fossils of the recent past, and between recent fossils and those slightly older, that makes evolutionary descent an inescapable conclusion (Figure I–7). Evolution is a unifying principle in biology, and it will undoubtedly continue to be a major framework that binds diverse specializations within the life sciences.

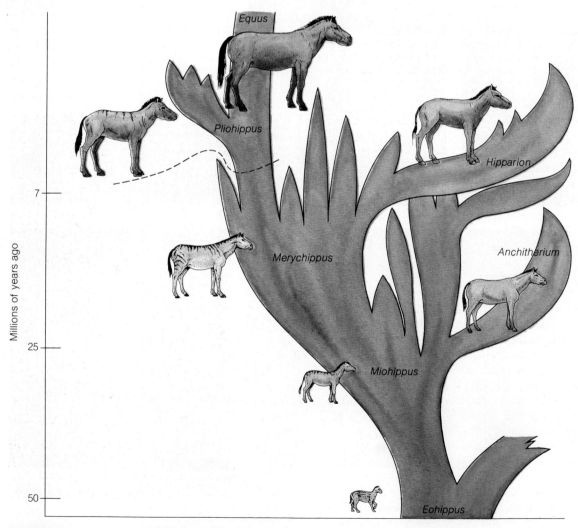

FIGURE I–7
The Evolutionary Tree of Horses. Based on a rather good fossil record, scientists have been able to construct this evolutionary history of horses. Note that only one kind of horse (*Equus*) has survived to modern times.

(a)

(b)

FIGURE I–8
Adaptation. (a) Lurking among the sea anemones is a clown fish, which is unaffected by the paralyzing stings of the anemones' tentacles. Other fish, including predators of the clown fish, are susceptible, however. Thus, the clown fish finds protection among the anemones' tentacles. (b) The ''Question Mark'' butterfly cannot be seen easily because the undersides of its wings match the surrounding dead leaves.

The theory of evolution says more than the fact that living things have changed with time. As Charles Darwin (1809–1882) realized, the idea of evolution is of little value without a reasonable mechanism to explain how it works. To fill this gap, Darwin offered the powerful principle of **natural selection.** He realized that all creatures wage a constant struggle against environmental constraints to survive and reproduce more of their kind. In this struggle, those individuals better suited to the environment survive and reproduce in greater proportion than those less fit, leaving more offspring with their specific characteristics.

Darwin's concept of natural selection, so simple and yet so elegant, has two important implications. The first is **adaptation,** the process of evolutionary change in which organisms become increasingly suited to their specific environmental circumstances (Figure I–8). A fish with gills and fins is adapted to living in water; birds have light bones and feathered wings adapted for flight. But changing circumstances can render a previously well-adapted organism less fit to survive and reproduce, jeopardizing its continued existence as a species. It is estimated that over 95% of all species that have ever lived are now extinct.

The second implication of natural selection is that physical characteristics, such as fins, wings, and hollow bones, are passed from one generation to the next via reproduction; that is, **traits are inheritable.** We know today that traits are passed along in the form of DNA, the genetic material (Figure I–9). Occasionally,

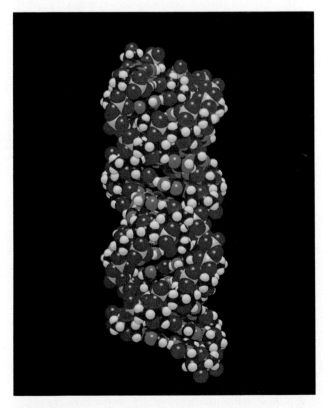

FIGURE I–9
DNA. Computer-generated model of a short segment of DNA. DNA houses the genetic instructions that determine the specific characteristics of organisms.

changes occur in DNA that can alter specific traits (such as fin length in a fish). If the change has adaptive value (if it renders the organism more fit), then the altered trait will probably appear more frequently in subsequent generations because the individuals having it may be more successful than others in the struggle to survive and reproduce. And because the earth offers such a wide assortment of environments, natural selection operating over billions of years has produced an enormous diversity of living things, each type adapted to its own particular environment.

Organisms Conform to the Laws of Chemistry and Physics

Prior to the twentieth century, the distinction between living and nonliving things appeared to be obvious. Organisms were said to have a "vital force," an indescribable "spark of life" found only among the living. The adherents of this theory, the so-called **vitalists,** argued that living things were not only composed of unique types of substances, but they carried out processes that were unmatched in the inanimate world—indeed, even the chemist could not duplicate them. Louis Pasteur (1822–1895), a renowned vitalist, contended that only *living* yeasts could carry out the fermentation of sugar to alcohol; that is, he considered fermentation to be a vital process. In 1897, Eduard Buchner proved Pasteur wrong by demonstrating that a "nonliving" extract prepared from broken yeast cells could change grape juice into wine. With this demonstration of a biological process occurring in the absence of any possible vital force, vitalism as an acceptable scientific idea was dead.

Vitalism has been supplanted by **reductionism,** the theory (or philosophy) that life has a purely chemical basis and its operation can be explained entirely in terms of the physical laws that pertain to all natural phenomena. Reductionism opened up a new era in biology. It meant that life and its processes could be studied using the powerful analytical tools of the chemist and physicist.

Over the past 50 years or so, inquiries into the "anatomy of life" have dispensed with the scalpel in favor of the biochemist's mortar and pestle. All of the basic types of molecules that make up living things have now been identified and synthesized in laboratories. And with the appropriate molecules present, all of the chemical processes that take place in cells can be duplicated in test tubes. Given the natural properties of the molecules found in cells, there is nothing mystical or "vital" about the chemical processes in which they participate.

The phenomenal advances in biochemical research have brought us to the point where we can manipulate the genetic constitution of certain types of cells almost at will. For example, some "genetically-engineered" bacteria produce *human* hormones. With new advances in molecular biology coming every week, we can look forward to solving some very old problems, such as cancer, in the not-too-distant future.

THE SCIENTIFIC METHOD

Biology is a natural science, and that has certain implications with regard to how biologists study life. In the broadest sense, science refers to a body of systemized knowledge. But "science" also implies a means for obtaining that knowledge. There are three generally recognized ways to obtain new knowledge. One is *intuition,* a "mental flash" of insight. Intuition is an important process in our everyday lives, as it is in science. It can provide instant insights into problems that otherwise might take years to solve. However, intuition is not always reliable and often leads us down the wrong track. Moreover, even if an intuitive insight seems perfectly clear to you, you are still left with the task of convincing others that it is correct. The second avenue to knowledge is by way of an authority, such as a guru in spiritual matters or a book for factual information. Depending on their sources of facts, however, authorities differ and often contradict one another. In order to decide among differing authoritative views, you may want to know how the authorities obtained their knowledge. You will probably find that the authoritative knowledge you accept, or the intuitive notions that prove correct, will have observed or experiment-based facts to support them. Observation and experimentation are the tools of the third means for gaining knowledge—the **scientific method.**

The scientific method entails a series of logical steps that, in actual practice, are not always taken in sequence. In fact, scientific discoveries are often made by mistakes in the application of the method, but that does not make them any less scientific, as long as they are still subject to observed or experimental justification. In most instances, however, scientific knowledge starts with some casual observation about something. This may generate an idea that spurs further, more

FIGURE I–10
The Unscientific Method.

directed observations. Eventually, some sort of tentative conclusion is reached, which generally takes the form of a **hypothesis**—a testable statement of what appears to be true. For example, suppose you come home to a darkened house and flip on a light switch only to find that the light doesn't work. Your mind instantly forms several hypotheses, the most readily testable of which is, "The power is out." To test this idea, you quickly try another light switch. Presto, that light works, so you dismiss the first hypothesis and shift to a second one: "The light bulb must be burned out." This is easily tested by replacing the bulb, and when this is done, you flip on the switch and the light comes on. You therefore conclude that the second hypothesis is correct. However, the fussy scientist deep inside your mind reminds you that the hypothesis is only *probably* correct. No hypothesis, no scientific theory or law, can be proven beyond any shadow of doubt. No matter how firm the scientific concept is, one exception can invalidate it. In the case of the light bulb hypothesis, it is quite possible (albeit unlikely) that in the time it took you to go from the first light switch (inoperative) to the second (operative), the power outage was corrected.

As we have seen in the light bulb example, hypotheses are testable. A statement such as "God exists" is not a hypothesis because we cannot test it through direct observation or appropriate experimentation. The truth of such statements must be supported by other means that lie outside the realm of science.

Testing a hypothesis involves the collection of data, or facts, that either substantiate it or make it necessary to modify or discard it. When the scientist is reasonably sure that the hypothesis is supported by the data, the next step is to tell the world by publishing the results. This is generally done by submitting a research paper to an appropriate scientific journal, which has editors and often peer reviewers. Their function is to ensure that the work is significant, of high quality, and that the author's conclusions are indeed supported by the data. If the paper fails in any of these regards, it will be rejected for publication.

An important inclusion in any scientific paper is a section on methodology. The scientist must include an accurate and detailed description of how the observations and/or experiments were conducted. Other scientists may then apply the published methods to verify, extend, or possibly reject, the published results. Thus, publishing the results of a scientific investigation is the formal solution to the type of problem illustrated in Figure I–10.

SCIENCE AND SOCIETY

In its purest form, science is amoral. It is shackled neither by value judgments nor moral restraints, for the fruit of scientific endeavor—knowledge—is neither good nor bad. **Technology,** on the other hand, deals with the application of ideas, often scientific ones,

FIGURE I–11
Protest. In some cases, technology and human values are in conflict.

There are several reasons for this. First, there is a creeping realization among scientists that most of the truly creative science has already been done, that the golden era of discovery is over. We have cracked the atom, we have laid bare the basic mysteries of life. What remains are filling in the details and, of course, applications of our current knowledge.

Second, answers to basic questions in science have always generated new questions, which invariably require greater ingenuity and resources. Often, new technical advancements must be made before science can move forward. As the questions become more sophisticated, so must the minds and equipment that answer them. This has necessitated greater specialization in training, teamwork, coordination of research efforts, and administrative superstructures—in short, a science bureaucracy that has made science more and more impersonal. Individual scientists are becoming further removed from their overall objective, and too often they are inundated with paper work and other details of "running the ship," leaving little time to reflect on their work.

Finally, the prevailing social and political environments specify the direction of research, largely by controlling the purse strings of science. Modern science is terribly expensive, and most of the funding comes from the government and, to an increasing extent, from industry. Thus, politicians and industrial magnates have a large say in what projects get funded. Scientists are painfully aware that if their research is not in currently favored areas, there is little chance to obtain financial support for it.

In modern practice, science has left the ivory towers and become intertwined in many complex moral issues. Test tube babies, genetic engineering, the search for extraterrestrial life, and behavior modification are just a few of the general issues that concern both science and society. One of the major questions has been, "Do scientists have a moral responsibility for their discoveries?" Most feel they do. The story goes that when Albert Einstein learned of the first nuclear explosion, he stated: "I wish I had become a blacksmith." And molecular biologists, at the height of advances being made in recombinant DNA technology in the mid-1970s, called a moratorium on their own research until appropriate restrictions and safety guidelines could be effected. Because the historical separation between theory and its application no longer exists, science has a direct impact on the real world, and thus, no longer enjoys moral immunity. Science and its effects are inextricably linked, and society has a stake in its accomplishments.

toward some practical end. Technology has given us computers to unburden our brains, medicines to cure and ease suffering, and high-yielding crops to feed our ever-burgeoning population. Technology has also brought us nuclear weapons, pollution, and many endangered species, including our own (Figure I–11). Unlike science, technology does have values, for its goal is to *improve* nature, to make the world a better place to live in. Unfortunately, people differ on what constitutes improvement, and in certain instances technology has generated more problems than it has solved. The goal of pure science, on the other hand, is simply to *know* nature—a pursuit which, in itself, is harmless.

Many scientists and nonscientists alike have emphasized this division between science and technology, between basic and applied research. But the division is an unrealistic one today. The vast majority of scientists are technologists of one stripe or another, not necessarily out of choice but rather circumstances.

UNIT I
Molecules and Cells

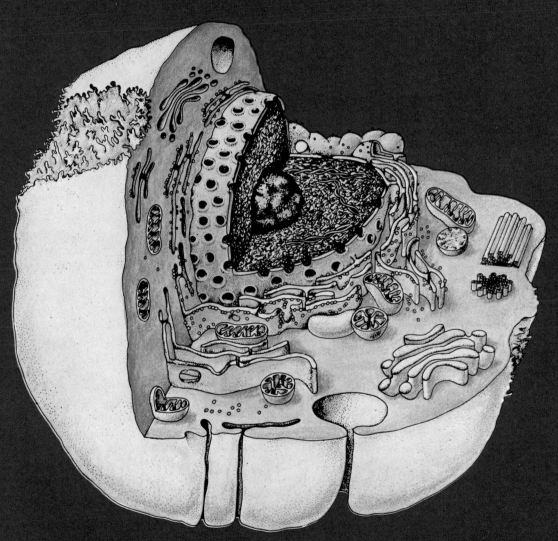

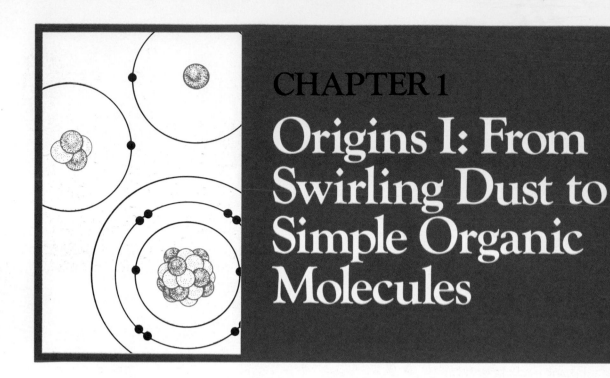

CHAPTER 1
Origins I: From Swirling Dust to Simple Organic Molecules

You may hear some people say that they understand life—they have it all figured out—but you probably will not find a biologist among them. Biologists are more inclined to believe that the questions about life outnumber the answers. Still, even the humblest biologist would agree that our knowledge of what life is and how it operates has increased dramatically in the past 100 years. In this single century more productive biological research has gone on than in all the previous years of human history combined. We have reached the point now where we can offer some sound ideas about how life began on this planet, and we can even test some of these ideas in the laboratory. In this chapter we will look at a few of the most important ideas about the origin of life on earth.

THEORIES ON THE ORIGIN OF LIFE

Some 4.6 billion years ago our earth formed from a swirling patch of cosmic dust and gases that gradually condensed into a solid sphere (see the Introduction). Eight hundred million years passed before the first known life forms emerged. What happened to change this once barren planet into the living earth? (Figure 1–1) How did life originate? We have no sure answers

to these questions, but thinkers over the ages have proposed a number of ideas.

We will limit our discussion of the origin of life to three possible explanations that demonstrate the range of ideas on the subject and their relationships to science. Two ideas, special creation and panspermia, propose that life appeared on earth instantaneously. A third idea, the theory of chemical evolution, holds that life originated by a slow, gradual, and indistinct transition of matter from nonliving to living states. Although the scientific evidence at present strongly favors chemical evolution, we are still a long way from being able to fill in all the gaps.

Special Creation: Outside the Realm of Science

Special creation, the notion that life was created by a supernatural being, has a long history, and it still enjoys substantial popularity. In particular, the Biblical idea that God created the earth and all of its inhabitants, especially human beings, has profoundly influenced our culture, our relationship to other species, and our relationship to our environment as a whole (Figure 1–2a).

The special creation doctrine has close ties with religion; its validity must be accepted on faith. The

FIGURE 1–1
The Transformation of a Planet. The earth existed for 800 million years as a barren, lifeless planet. How was it transformed into the "living earth"?

problem many scientists have with the idea of special creation is that no one could ever devise a set of scientific tests that would tell us whether special creation was even a possibility, much less an actual event. Science can deal only with natural phenomena—this is one of its most fundamental limitations. Because the idea of special creation cannot be tested through observation or experimentation, it lies outside the domain of science.

Panspermia

A more recent idea for "instant life" on earth was suggested originally by Arrhenius (1859–1927), a Swedish physicist and chemist. In a series of scientific papers and popular books published around the turn of the century, Arrhenius proposed that microscopic organisms from some distant planet were "blown" to earth. He called his theory **panspermia,** meaning roughly "seeds everywhere." Later, other supporters of panspermia suggested that meteorites probably brought these organisms to earth (Figure 1–2b).

Many biologists criticize the theory of panspermia by saying that it only transfers the problem of life's origin to another planet. One way to circumvent this criticism is to suppose that life per se had no beginning. In other words, life, like matter and energy, may have always existed, and through time it has spread to more and more planets.

A somewhat modified version of Arrhenius's idea—directed panspermia—has been proposed recently by Francis Crick and Leslie Orgel. Crick is best known for his collaboration with James Watson in the discovery of the structure of DNA, the universal genetic substance (see Chapter 11). Orgel has done much pioneering work in prebiotic ("before life") chemistry that ironically supports an alternative theory, chemical evolution. Crick and Orgel suggest that some distant civilization intentionally seeded our planet with a form of primitive microorganism (possibly a strain of bacteria) some 3.8 billion years ago. These advanced beings simply loaded their missionary rocket ships with dormant bacterial spores (a type of suspended animation for living cells) and guided at least one ship toward lifeless earth. Once in the earth's oceans, the bacterial spores presumably started to grow and reproduce, feeding on the various chemicals present. Without any natural obstacles to their growth, the transplanted bacteria proliferated throughout the seas, ultimately giving rise to all life on this planet.

Is this account too fantastic? For now, we can only say maybe. The theory of panspermia does have rather serious technological problems; nevertheless, it is somewhat attractive because it allows us to rationalize some puzzling facts about living organisms. For example, every organism, from bacterium to human being, has the same basic chemical code by which genetic information is transmitted into the activities of life (see Chapter 11). According to Crick and Orgel, the universality of this genetic code indicates that all present-day organisms have a single common ancestor, as would be the case with panspermia. If, however, the first living creatures arose from chemical evolution, there is no compelling reason to assume that this event occurred only once. On the contrary, chemical evolution would logically have occurred many times, resulting in many different genetic codes.

FIGURE 1–2
Theories on the Origin of Life. (a) The concept of special creation has permeated human thought and art throughout the ages. Shown here is the hand of God imparting life to man, a detail from Michelangelo's Sistine Chapel master-piece. (b) This meteor crater in Arizona is 1.2 kilometers ($\frac{3}{4}$ mile) across. Could simple life forms have been carried to earth by meteors, as some supporters of panspermia have suggested? (c) The theory of chemical evolution maintains that life originated as the culmination of purely chemical events: simple molecules in the prebiotic atmosphere and oceans reacted to form complex molecules, which eventually assembled into highly organized units. Ultimately, these units evolved the ability to maintain and reproduce them-selves, thereby earning the name, ''living things.''

(a)

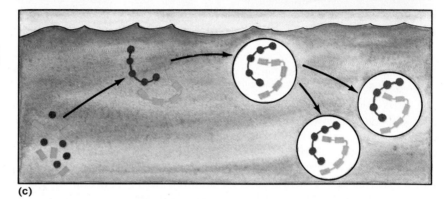

(c)

Was one genetic code so far superior to all others that it became the sole survivor?

Chemical Evolution

Despite the problem with the genetic code, most scientists today support the view that life originated on earth as a result of **chemical evolution.** This theory holds that over a long period of time, spontaneous chemical reactions occurring on the prebiotic earth gave rise to a great variety of substances necessary for life. These substances gradually increased in complex-ity and self-assembled into highly ordered configura-tions. When such assemblages were able to maintain and reproduce themselves, the line between nonliv-ing and living was bridged, and life began (Figure 1–2c).

The theory of chemical evolution has developed as an outgrowth of the enormous expansion in our understanding of how living systems function at the chemical level. This understanding in turn stems largely from earlier discoveries in a related discipline—chemistry. Today scientists recognize that all living things have a chemical basis. Among other things, this means that the basic laws governing the behavior of matter and energy apply just as much to trees, whales, bacteria, and mushrooms as they do to substances in test tubes. This was an important realization because

it opened the door for biologists to apply the chemist's knowledge and analytical tools to the study of life. Therefore, before continuing our discussion of the origin of life, you must first become familiar with some basic principles and vocabulary of chemistry.

ELEMENTS AND ATOMS

All matter in the universe is made up of distinct sub-stances called **elements,** such as gold, oxygen, and carbon. Each element has unique properties that set it apart from all the other elements: its form (solid, liquid, or gas) at normal temperatures, its melting and boiling points, its tendency to combine with other elements, and so on. Over the course of history, sci-entists have identified more than 100 elements, 92 of which occur in nature. About 30 of these natural ele-ments are found in living creatures (Table 1–1).

The distinctiveness of elements spurred the an-cient Greek philosophers to ask an important ques-tion: Can an element be divided indefinitely into smaller and smaller pieces without losing its basic properties, or is there some ultimate unit which, if further subdivided, would lose the qualities of the element? Aristotle and Plato contended that there were no such indivisible particles. To them—and to most other people for the next 2000 years—it seemed obvious that a piece of gold or iron must be continuous, like glass appears to be, not particulate, like sand. Not until the 1800s did this view change, when an Englishman named John Dalton published *Atomic Theory.* In his book, Dalton stated that every element is composed of tiny particles called **atoms,** and that all atoms of a particular element are identical. Dalton went on to say that each element has different properties because the atoms of each are different from those of other elements. For example, the element carbon is composed of many atoms, all of which differ physically from the atoms found in nitrogen, oxygen, or any other element.

Dalton's theory of the atom as an indivisible unit was an important step forward in our understanding of matter. We have since discovered, however, that atoms are composed of three even more basic particles: protons, neutrons, and electrons. Even so, this does not disprove Dalton's theory of atomic indivisibility, for an atom loses the unique properties of its element when it is split into subatomic particles. Knowing the structure of atoms helps us understand how atoms of different elements differ from one another and why each element behaves as it does.

In some ways, atoms resemble miniature solar systems (Figure 1–3). At the center of every atom is a dense core called the **nucleus,** which consists

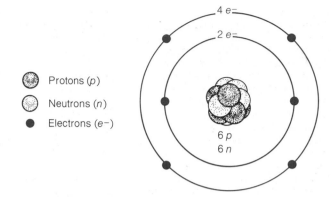

FIGURE 1–3
A Model of the Carbon Atom. The central nucleus of a carbon atom is composed of six protons and six neutrons, with six electrons orbiting the nucleus in different energy levels, or shells. Two electrons occupy the first shell and four electrons orbit in the second shell. This model is not drawn to scale: if we enlarged this carbon atom to the size of the earth, the nucleus would be only 200 feet in diameter, and could easily rest inside a small football stadium.

of one or more **protons** and a roughly equal number of **neutrons.** Each proton carries a positive electrical charge; neutrons carry no charge. Spinning about the nucleus are negatively charged particles called **electrons.** The electrons are held in orbit by their attraction to the positively charged protons. Atoms have no net electrical charge: for every positive proton in the nucleus, there is a negative electron orbiting it.

Protons and neutrons have essentially the same mass (weight): 0.00000000000000000000000016 gram. As small as this is, the mass of a proton or neutron is about 1835 times greater than the mass of an electron (a relative difference analogous to that of a horse and a mouse). Thus, for most purposes, an electron's contribution to the mass of an atom is negligible. For convenience, each neutron and proton is arbitrarily assigned a mass of one unit, and the **mass number** of an atom is simply the sum of its protons and neutrons (Table 1–2).

All protons are alike, no matter what kind of atom they appear in, and so are all neutrons and all electrons. What determines the particular properties of an element is the *number* of protons in its atoms (or, since there is one electron for every proton, the number of electrons). A carbon atom, for example, has six protons and six electrons; a nitrogen atom has seven of

TABLE 1–1
The Elements of Life. The elements in column I together comprise over 97% of the matter in all organisms. The elements in column II are also essential for life, but they make up less than 2% of the total mass of organisms. The trace elements in column III are occasionally found in living systems, and are found in very small amounts.

I The Main Four	II Also Essential	III Trace Elements
Carbon (C)	Calcium (Ca)	Arsenic (As)
Hydrogen (H)	Chlorine (Cl)	Barium (Ba)
Oxygen (O)	Cobalt (Co)	Boron (B)
Nitrogen (N)	Copper (Cu)	Bromine (Br)
	Iron (Fe)	Chromium (Cr)
	Magnesium (Mg)	Fluorine (F)
	Manganese (Mn)	Iodine (I)
	Nickel (Ni)	Lithium (Li)
	Phosphorus (P)	Molybdenum (Mo)
	Potassium (K)	Selenium (Se)
	Sodium (Na)	Silicon (Si)
	Sulfur (S)	Tin (Sn)
	Zinc (Zn)	Vanadium (V)

TABLE 1–2
Atomic Numbers and Masses of Some Common Elements.

Element and Symbol	Atomic Number (Number of Protons)	Number of Neutrons	Mass Number (Protons + Neutrons)
Hydrogen (H)	1	0	1
Helium (He)	2	2	4
Carbon (C)	6	6	12
Nitrogen (N)	7	7	14
Oxygen (O)	8	8	16
Sodium (Na)	11	12	23
Chlorine (Cl)	17	18	35

each; an oxygen atom has eight. These simple differences in number produce the vast chemical differences among the elements.

CHEMICAL BONDS

When two or more atoms exist in combined form, we call the combined unit a **molecule.** Some types of molecules are composed of multiple atoms of the same element. For instance, two hydrogen atoms can combine to form a molecule of hydrogen gas, symbolized H_2. Molecular hydrogen is a lighter-than-air gas that was used to inflate the early blimps. This was an extremely dangerous practice, however, because H_2 reacts explosively with another simple molecule, molecular oxygen (O_2), present in air. The explosion of the *Hindenburg* blimp in 1937 is a case in point.

Substances whose molecules are composed of different kinds of atoms are called **compounds.** For example, water is a compound whose molecules consist of two hydrogen atoms combined with one oxygen atom, and is symbolized H_2O. Carbon dioxide (CO_2) is another common compound consisting of molecules made up of one carbon atom and two oxygen atoms.

The forces that hold atoms together in a molecule are called **chemical bonds.** Chemical bonds, and the factors that control how they form and break apart, underlie the whole chemical basis of life. Let us now examine how the structure of an atom determines its ability to form chemical bonds with other atoms.

The electrons of an atom are distributed around the nucleus in regular intervals called **shells** (see Figure 1–3). Moving outward from the nucleus, each shell represents a successively higher energy level for the electrons in it. Electrons tend to occupy the *lowest* energy level available, so the first (or innermost) shell

will be completely filled before any electrons occupy the second shell. Similarly, the second shell will be filled to capacity before the third shell holds any electrons. The first shell of all atoms can accommodate only two electrons; the second shell can hold up to eight electrons. In the hydrogen atom, with its single proton and electron, the electron is always found in the first shell. In a sodium atom, which has eleven electrons, two electrons occupy the first shell, eight are in the second shell, and a single electron is in the third shell (Figure 1–4).

The chemical behavior of an element is closely tied to the number of electrons in the outermost shell of its atoms. For example, helium, a gas at normal temperatures and pressures, has two electrons that occupy the first shell and fill it to capacity. A full outermost shell is a stable condition for an atom, so helium does not tend to enter into chemical combinations with other atoms; that is, helium is an *inert* element. Other elements whose atoms' outermost shells are filled, such as neon and argon, are also chemically inert. In fact, these elements are so similar that they are grouped together as the **inert gases** (Figure 1–5a). Similarly, other groups of elements with the same number of electrons in their atoms' outermost shells have many physical and chemical properties in common. The **shiny metals** form another group of related elements: all of their atoms have single electrons in their outermost shells. This atomic structure renders these elements extremely *reactive*—they readily enter into chemical combinations with other elements (Figure 1–5b).

The difference in chemical reactivity between the inert gases, with their filled outer shells, and the shiny metals, with their unfilled outer shells, hints at a fundamental principle of atomic bonding: Atoms tend to combine with one another so as to fill their outermost shell with electrons. Atoms with partially filled

FIGURE 1–4
Electron Distribution in Atomic Shells of Four Common Elements.

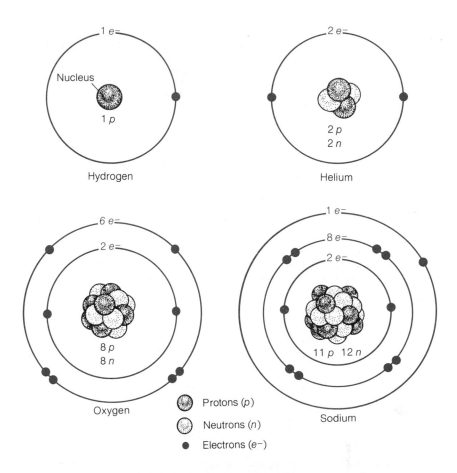

FIGURE 1–5
The Inert Gases and Shiny Metals. The two columns show how the distribution of electrons in atoms affects the properties of elements. For example, the shiny metals vary greatly in atomic number, but because they all have a single electron in their atoms' outermost shells, their properties are quite similar. Note that each shiny metal element has only one more proton and electron than a corresponding inert gas element, yet the physical and chemical properties of these two groups are vastly different.

(a) THE INERT GASES		(b) THE SHINY METALS	
Atomic number → (number of protons; also number of electrons) Chemical symbol for → element	2 Helium (He)	3 Lithium (Li)	
	10 Neon (Ne)	11 Sodium (Na)	
Properties 1. Outermost shell filled. 2. Gases. 3. Little or no tendency to undergo chemical reactions.	18 Argon (Ar)	19 Potassium (K)	*Properties* 1. One electron in outermost shell. 2. Shiny, soft, metallic solids. 3. Very reactive chemically; readily lose one electron; react vigorously with water.
	36 Krypton (Kr)	37 Rubidium (Rb)	
	54 Xenon (Xe)	55 Cesium (Cs)	

shells are chemically reactive; they become more stable by losing, gaining, or sharing their electrons with other atoms.

Covalent Bonds

Two atoms with unfilled outer shells may bond together by sharing one or more pairs of electrons, an arrangement called a **covalent bond.** For example, a hydrogen (H) atom has only one electron, so it can accommodate a second electron in its first (and only) shell. One way it can gain that electron is by joining with another hydrogen atom. Because both atoms are identical, neither can take over the other's electron. Instead, the two electrons orbit equally around both hydrogen nuclei, and a relatively stable hydrogen molecule (H_2) is formed (Figure 1–6).

Certain atoms can share more than one pair of electrons, producing double or even triple covalent bonds. The oxygen atom (O), for example, has six electrons in its second (and outermost) shell and needs two more to fill it. It can accomplish this by forming a double covalent bond with another oxygen atom so that one pair of electrons from each atom (a total of four electrons) is shared. Or, it can share electrons with one or more atoms of other elements (Figure 1–7).

Although all covalent bonds involve shared electrons, the sharing is more equal in some cases than in others. When the joined atoms have an equal attraction for the bonding electrons, the result is a **nonpolar covalent bond.** This equal sharing occurs in molecules such as H_2, O_2 and N_2, where the molecule consists of two identical atoms. It also happens in compounds where the bonded atoms are nearly equal in their electron-attracting abilities, such as in carbon dioxide (CO_2). In many other instances, however, one of the partner atoms has a much greater attraction than the other for the shared electrons. In water (H_2O), for example, the hydrogen atom is considerably weaker in attracting electrons than the oxygen atom is. The four shared electrons therefore spend more time orbiting the nucleus of the oxygen atom than that of either hydrogen atom. This gives the hydrogen atoms, on the average, slightly *less* than one negative electron each to counteract the positive charge of their single protons. As a result, each hydrogen atom in H_2O is slightly positive. On the other hand, the eight-proton oxygen atom has slightly *more* than eight electrons moving about its nucleus, so it is slightly negative. Because these electrical charges are distributed unevenly across the water molecule, it has **polarity**—two positive poles at the two hydrogen

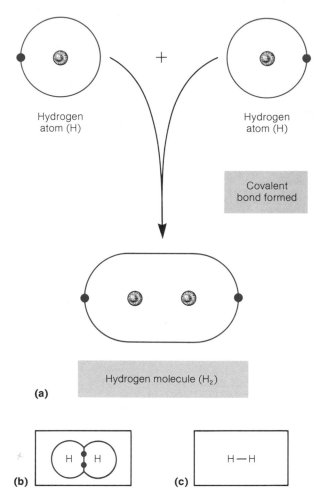

FIGURE 1–6
Molecular Hydrogen. (a) When two hydrogen atoms react, they form a molecule of H_2. Each atom shares its single electron with its partner atom, forming a covalent bond. Since both atoms have their first shells effectively filled, the molecule of H_2 is chemically stable. Other ways to diagram H_2 are shown in (b) and (c).

atoms and a negative pole at the single oxygen atom (Figure 1–8). Since the bond formed by such an unequal sharing of electrons between atoms results in a polar charge distribution, it is called a **polar covalent bond.**

Ionic Bonds

When the sharing of electrons between two atoms is so unequal that one atom completely gives up one or more of its electrons to its partner, the resulting

ATOMS

MOLECULES

(a) Water (H$_2$O)

(b) Molecular oxygen (O$_2$)

(c) Carbon dioxide (CO$_2$)

(d) Molecular nitrogen (N$_2$)

FIGURE 1–7
Covalent Bonding in Four Common Molecules. (a) Water has two single covalent bonds. Each bond is the result of one pair of electrons being shared between a hydrogen atom and the oxygen atom. (b) Molecular oxygen has one double covalent bond in which two pairs of electrons are shared. Each oxygen atom contributes one pair of electrons. (c) Carbon dioxide has two double bonds. (d) Molecular nitrogen has a triple bond that consists of three pairs of shared electrons. In each case, the formation of the covalent bond enables each atom to fill its outermost electron shell to capacity: two electrons for hydrogen and eight electrons for oxygen, carbon, and nitrogen. Only electrons in the outermost (bonding) shells are shown; the electrons of oxygen are shown in color throughout.

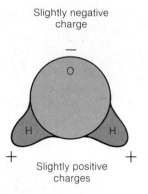

Slightly negative charge

Slightly positive charges

FIGURE 1–8
Polarity in a Water Molecule. The bonding electrons in a molecule of water orbit the oxygen atom more frequently than the hydrogen atoms. This unequal electron sharing generates two slightly positive poles at the hydrogen atoms, and a slightly negative pole at the oxygen atom.

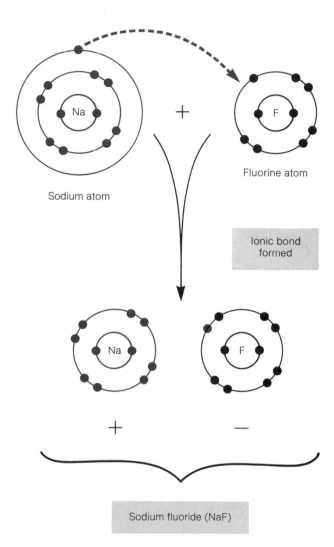

Sodium atom

Fluorine atom

Ionic bond formed

Sodium fluoride (NaF)

FIGURE 1–9
Ionic Bonding. The formation of an ionic bond involves the complete transfer of one or more electrons from one atom to another. Here the fluorine atom fills its outermost shell by accepting the single outermost electron from the sodium atom. This transfer yields a positively charged sodium ion (Na^+) and a negatively charged fluoride ion (F^-). Their opposite charges create an ionic bond joining Na^+ and F^- in a molecule of sodium fluoride (NaF), the cavity-fighting substance in toothpastes.

molecule is held together by an **ionic bond.** Consider what happens when a sodium atom (Na) reacts with a fluorine atom (F) to form sodium fluoride (NaF). Sodium atoms have eleven electrons—two, eight, and one in the first, second, and third shells respectively. According to our principle of atomic bonding, the sodium atom would tend to *lose* its third-shell electron to achieve a filled outer electron shell (the second shell). By the same principle, the fluorine atom, with its seven electrons in the second (and outermost) shell, would tend to combine with other atoms in order to *gain* one electron to fill its outer shell with eight electrons. Thus, fluorine has a strong attraction for electrons, whereas sodium is, in a sense, "looking for ways to get rid of" its outer-shell electron. When these two atoms collide, sodium's "extra" electron completely transfers to the fluorine atom (Figure 1–9). Now the sodium atom is electrically unbalanced: Its eleven protons and ten electrons add up to a net $+1$ charge. At the same time, the fluorine atom, with its extra electron, carries a -1 charge. Atoms that carry whole positive or negative charges are called **ions;** in our example, we have formed sodium ions (Na^+) and fluorine ions (F^-). Since ions of opposite charge attract one another (like the positive and negative poles of two magnets), the Na^+ is held to the F^- to form the NaF molecule. Thus, an ionic bond consists of an electrical attraction between oppositely charged ions; no electron sharing is involved.

When an ionically bound compound is dissolved in water, an interesting phenomenon takes place. The compound **ionizes,** or separates into its constituent positive and negative ions. The ions then become surrounded by layers of water molecules that tend to keep them in the ionized state. Thus, when NaF is dissolved in water, it ionizes into Na^+ and F^-.

Hydrogen Bonds

Covalent and ionic bonds are the most common means by which atoms join to form molecules. In addition to these, there are weaker bonds that also contribute to the structure and properties of compounds. One of the most important of these is the **hydrogen bond**—the sharing of a hydrogen atom between different regions of a single molecule or between two adjoining molecules.

In living systems, hydrogen bonds play important structural roles in two key types of biological compounds: proteins (see Chapter 2) and nucleic acids (see Chapter 11). At this time, however, let us examine the hydrogen bond in a simpler yet profoundly important substance to life: water. Recall that the two hydrogen atoms in water (H_2O) are bonded to the oxygen atom in a polar covalent fashion, resulting in partial positive charges associated with both the hydrogen atoms and a partial negative charge associated with the oxygen atom. As we have seen in ionic compounds, unlike charges attract one another. Thus, a slightly positive hydrogen atom in one H_2O molecule is attracted to the slightly negative oxygen atom of a neighboring H_2O molecule. The result is a hydrogen bond, or the sharing of that hydrogen atom between the two H_2O molecules. Because each H_2O molecule can form hydrogen bonds with up to four other H_2O molecules, the millions of individual molecules in a single water droplet tend to stick together, keeping the droplet intact (Figure 1–10). Although these hydrogen bonds are relatively weak and easily broken, they exert an enormous influence on the physical properties of liquid water, as we will see in Chapter 4.

CARBON AND ITS COMPOUNDS

Of the many different chemical bonds found in nature, none are more important than the covalent bonds involving carbon. With four electrons in its outer shell (see Figure 1–3), each carbon atom can bond with up to four other atoms. In living systems, carbon is typically combined with hydrogen, oxygen, nitrogen, sulfur, and other carbon atoms. In fact, nearly one out of every ten atoms in your body is a carbon atom. This bonding flexibility accounts for the enormous diversity of **organic compounds**—compounds that contain carbon. A good way to get a sense of the variety of organic compounds is to compare a few that are structurally similar. Just as the presence of one more proton and electron makes the difference be-

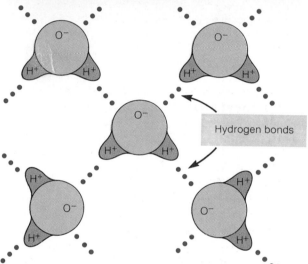

FIGURE 1–10
Hydrogen Bonding. The two slightly positive hydrogen atoms of a water molecule are attracted to the partially negative oxygen atoms of two other water molecules. Thus, each water molecule can join with up to four other water molecules by hydrogen bonding. This hydrogen bonding holds the individual molecules of a water droplet together, as shown in the photo.

tween carbon and nitrogen, or between nitrogen and oxygen, so too the properties of organic compounds can differ drastically because of the presence or absence of a single atom or a rearrangement of the same atoms. Let us start with a simple **hydrocarbon**—a type of compound that consists of hydrogen and carbon atoms only—and compare it to other molecules that have slightly different components.

Hydrocarbons are not usually found in living creatures, but they are often formed as breakdown products of dead organisms. Ethane, a 2-carbon hydrocarbon (C_2H_6), is a component of natural gas and can be burned as a fuel (Figure 1–11). If a **hydroxyl group** (symbolized by —OH) is substituted for one of the hydrogen atoms of ethane, however, the resulting compound is an **alcohol** known as ethyl alcohol (CH_3CH_2OH). At room temperatures, ethyl alcohol is a liquid. It may be familiar to you as the type of alcohol found in beer, wine, and liquor. If two hydrogen atoms in ethyl alcohol are replaced with an oxygen atom, a compound with considerably different properties is formed: acetic acid (CH_3COOH), the sour-tasting component of vinegar, which some of us enjoy on our spinach and tossed green salads. In the presence of oxygen (O_2), certain bacteria can convert ethyl alcohol to acetic acid, so recork your wine bottle if you don't want the wine to turn into vinegar! Acetic acid is an **organic acid,** a very common type of organic compound in living cells. Organic acids differ greatly in size and complexity, but all contain at least one **carboxyl group** (—COOH). (See Appendix B for a discussion of acids.)

Although we will discuss organic compounds in more detail in Chapter 2, two important points are worth noting here. First, the carbon atom's versatility in combining with other atoms means that organic compounds may contain dozens or even hundreds of atoms arranged in an almost infinite variety of patterns. Thus, organic compounds are extremely diverse. Second, although each organic compound is unique, those that have certain features in common, such as hydroxyl or carboxyl groups, also tend to have similar properties. For example, the 2-carbon acetic *acid* (CH_3COOH) is more similar in its physical and chemical properties to the 3-carbon propionic *acid* (CH_3CH_2COOH) than it is to the 2-carbon ethyl *alcohol* (CH_3CH_2OH). Because of these similarities in properties, organic chemists lump compounds with similar groups into classes such as hydrocarbons, alcohols, organic acids, and others.

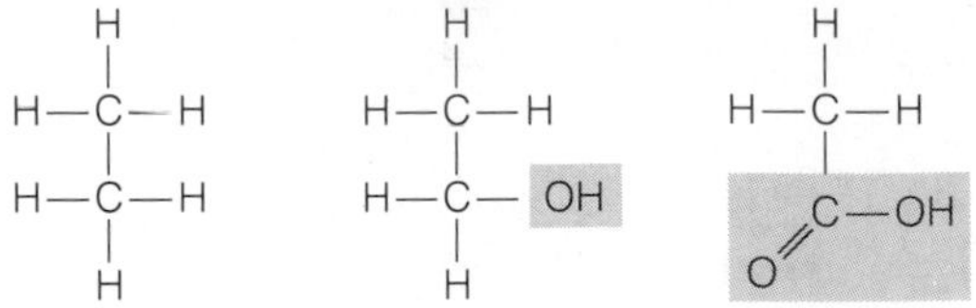

THE ORIGIN OF LIFE REVISITED

When we left our discussion on the origin of life, we had begun to consider the third and most scientifically popular theory concerning how life on earth began—chemical evolution. Recall that proponents of this theory envision a gradual, stepwise process in which

FIGURE 1–11
Simple Organic Compounds. A difference of only a few atoms results in compounds with very different chemical and physical properties.

small molecules reacted to form larger, more complex molecules, ultimately giving rise to supermolecular assemblages which evolved the characteristics we associate with living systems (see Figure 1–2c). But how did this happen? With your knowledge of basic chemistry, we can now explore this question.

Conditions on the Prebiotic Earth

Of the nine planets in our solar system, only earth has the physical conditions uniquely suited to support life (see Introduction). But the conditions that sustain life on earth today are not the same as those which apparently led to its chemical origin some 3.8 billion years ago. In the 1920s, a Russian, A. I. Oparin, and an Englishman, J. B. S. Haldane, independently proposed that the earth's atmosphere in prebiotic times contained an abundance of hydrogen-containing molecules, such as ammonia (NH_3), methane (CH_4), water vapor (H_2O), and molecular hydrogen (H_2) (see Introduction). This list has been modified somewhat by more recent geological data to include molecular nitrogen (N_2), carbon monoxide (CO), carbon dioxide (CO_2), and hydrogen cyanide (HCN). In addition, experts now believe that free hydrogen (H_2) was probably not as plentiful as Oparin and Haldane originally proposed. Most significantly, however, this prebiotic atmosphere lacked molecular oxygen (O_2); thus, it represents a sharp contrast to our current atmosphere of nearly 80% nitrogen, 20% oxygen, and trace amounts of other gases.

In addition to proposing a prebiotic atmosphere rich in hydrogen-containing gases, Oparin and Haldane went on to suggest that in the presence of energy, these simple gases spontaneously combined to form larger, more complex organic compounds. There was certainly no lack of energy on the early earth. The sun bathed the earth's surface with heat and ultraviolet radiation; volcanoes vented more heat; and violent thunderstorms sent lightning bolts slashing through the primitive atmosphere. While this scenario may seem quite hostile to life in its present forms, it was from just such a set of circumstances, Oparin and Haldane speculated, that life had its beginnings.

Chemical Evolution in the Laboratory

Was the Oparin–Haldane proposal merely armchair speculation, or was it a **hypothesis,** a testable scientific idea? Thirty years went by before anyone put their idea to the test. Then in 1953, Stanley Miller, a graduate student at the University of Chicago, built a closed system of glass tubes and flasks designed to simulate prebiotic conditions on earth. Miller positioned a pair of spark-generating electrodes (to simulate lightning) in one of the flasks, then injected water and a mixture of simple hydrogen-containing gases into his apparatus (Figure 1–12). After sparking this mixture periodically for a week, he analyzed a sample of the water and found, to his amazement, that the water contained a number of simple organic compounds, including organic acids and several kinds of *amino acids.* The appearance of amino acids in this mixture was a startling find, for amino acids are the building blocks of proteins, extremely important molecules for life (see Chapter 2).

Miller's elegant experiment not only supported the Oparin–Haldane hypothesis, but it opened the door to further laboratory studies on the theory of chemical evolution. Since 1953, many researchers have extended Miller's original findings. By exposing slightly different starting mixtures of gases to a variety of energy forms (such as heat, shock waves, ultraviolet radiation, and bombardment with high-energy particles), scientists have produced virtually all of the organic building blocks necessary for life. Because the conditions used in all of these studies simulated those we believe were present on earth before life, it is likely that these same organic substances formed by similar means on the prebiotic earth. Some of the simple organic compounds probably formed in the atmosphere, then washed into the oceans by driving rainstorms. Other reactions may have taken place directly in the water. In any event, we can imagine that after several million years of such chemical activity, the oceans became an "organic broth" containing a myriad of simple organic compounds. This was the first crucial step toward the origin of life.

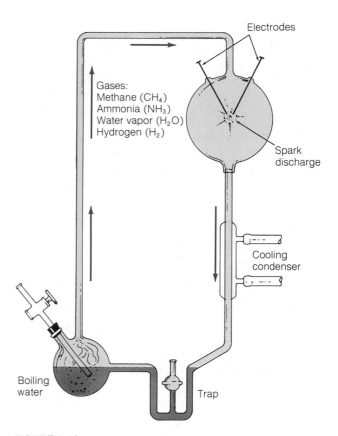

FIGURE 1–12
Miller's Experiment. Stanley Miller designed an experiment to test whether organic compounds would form spontaneously under the conditions presumed to exist on the prebiotic earth. He constructed an apparatus consisting of a gas-filled reaction vessel, a condenser, and a boiling water flask. At frequent intervals, a mixture of methane, ammonia, water vapor, and hydrogen in the reaction vessel was exposed to an electric spark. Any compounds that formed would dissolve in the water as it condensed back into liquid form. After a week, Miller analyzed the water and found several biologically important compounds, including amino acids.

One of the important stipulations of the Oparin–Haldane hypothesis is that the prebiotic atmosphere was devoid of free oxygen (O_2). If O_2 had been present then, it would have reacted with any spontaneously-formed organic molecules, degrading them back into simpler substances. Similarly, the organic compounds that make up this page are slowly decomposing by **oxidation** (a reaction with O_2), and over a period of several centuries the paper will turn from

ESSAY 1–1

IS CHEMICAL EVOLUTION OCCURRING TODAY?

Imagine a lifeless seashore—no kelp or driftwood lying about, no grasses growing in the cracks of rocks, no sounds of seagulls. Your senses are filled only with the surf pounding against the barren rocks, and the distant rumble of a thunderstorm. The land is perfectly still, quietly anticipating another drenching rainstorm. There are no seashores like this on earth, at least not anymore. But this is how we might imagine a coastline some 4.6 billion years ago when the earth was new. Only rock, water, and air stood in stark contrast to one another, but from these physical ingredients life was to emerge. According to the theory of chemical evolution, inorganic gases drifting in the atmosphere and dissolved in the seas reacted to form simple organic compounds, and these in turn reacted to form larger, more complex molecules and assemblages, until hundreds of millions of years later, the first living cells appeared.

Most scientists today believe that such a chemical origin of life could have occurred only under the rather special conditions that prevailed on prebiotic earth: an abundance of hydrogen-containing molecules and the absence of molecular oxygen. Therefore, all experiments designed to test the chemical origin theory have been carried out under strictly controlled laboratory conditions. Recent evidence indicates, however, that the earth may be conducting some of its own experiments. Scientists have discovered several underwater sites that have many of the characteristics believed necessary for a chemical evolution process.

The "Atlantis II deep," a research site in the middle of the Red Sea (see map) is a region geologists refer to as an axis of plate spreading. The earth's crust consists of many geological plates that butt up against each other like pieces of a jigsaw puzzle. These massive plates move relative to each other very slowly, causing major disturbances (such as earthquakes, volcanoes, and hot springs) in the regions where they join together. In the floor of the Red Sea, two plates are slowly

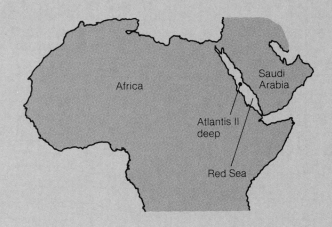

white to yellow, then to brown, and eventually crumble into ash. (You can speed up this oxidation process by lighting a match to the page.) Thus, our present atmosphere with its 20% O_2 content is not conducive to a chemical evolutionary process. (For a possible exception, see Essay 1–1.)

There is another reason why the chemical evolution toward new life forms cannot occur today, and it has to do with the presence of life itself. Micro-

organisms such as bacteria and fungi make their living by breaking down the organic matter of dead creatures into simpler inorganic components. Since these decomposers are found virtually everywhere on earth, they would surely degrade any new organic compounds that might arise spontaneously. But these biological "vacuum cleaners" were not present four billion years ago, nor was free oxygen. Thus, there were no impediments to a chemical origin of life.

spreading apart. As a result, molten rock and hot gases from below the earth's crust are gradually escaping to the seawater above. Temperatures over 56°C (133°F) have been recorded in water 2000 meters below the surface. Interestingly, the gases spewing out from the spreading seafloor are similar to those believed to have been present in the prebiotic atmosphere. They consist of simple hydrogen-containing compounds, such as methane, ethane, and hydrogen cyanide; no molecular oxygen has been detected. Furthermore, no living organisms inhabit the "deep." Thus, in the presence of an energy source (heat) and simple hydrogen-containing inorganic compounds, and in the absence of molecular oxygen and organisms, conditions seem favorable for the nonbiological synthesis of organic compounds. This is an ideal spot to monitor possible ongoing chemical evolution.

Scientists have been collecting water and mud samples from this area under conditions which ensure that the samples do not become contaminated with microorganisms as they are raised to the surface. Preliminary analyses of these sterile samples have revealed the presence

Sampling operation at the Atlantis II Deep site in the Red Sea.

of various cyanide-containing compounds thought to be crucial reactants in the early stages of organic compound formation. Most intriguingly, relatively high concentrations of glycine, a simple amino acid, are also present. Amino acids are very important to living systems because they are the building blocks of proteins. Current efforts are centered on finding other biologically significant compounds in this environment. In time, these studies could demonstrate that the chemical reactions necessary to form some of the compounds required by life are still occurring on earth by natural processes.

SUMMARY

About 100 different elements constitute all the matter in the universe. The unique properties of each element are determined by the number of protons ($+1$ charge), neutrons (no charge), and electrons (-1 charge), in its component atoms. The protons and neutrons make up the atom's nucleus; the electrons orbit the nucleus in various energy levels called shells.

How a given element reacts with other elements is determined primarily by the number of electrons in the outermost shell of its atoms. Compounds are made up of molecules consisting of two or more different atoms held together by chemical bonds.

A covalent bond is formed when two or more atoms share electrons, either equally (a nonpolar covalent bond) or unequally (a polar covalent bond). The covalent bond is the most common type of bond in organic molecules.

An ionic bond forms after one atom gives up one or more electrons to another atom. The atom that gains electrons becomes negatively charged, and its partner becomes postitively charged. Their opposite charges attract and ionically bond the two charged atoms, or ions, together. Ionic compounds ionize (separate back into ions) when they dissolve in water.

A hydrogen bond is the sharing of a hydrogen atom between different regions of the same molecule or between molecules. Such bonds arise when a hydrogen atom with a slight positive charge is attracted to a negatively charged atom. Although hydrogen bonds are relatively weak, they play an important role in the structure of certain key biological molecules.

The ability of the carbon atom to form four covalent bonds with other atoms, including other carbon atoms, accounts for the great diversity of organic (carbon-containing) compounds in nature. Compounds consisting of carbon and hydrogen atoms exclusively are called hydrocarbons. Compounds with one or more hydroxyl groups (—OH) are alcohols; those with one or more carboxyl groups (—COOH) are organic acids. Organic compounds with such groups in common share similar chemical properties.

Although we cannot be certain at present how life orginated on earth, several possibilities have been suggested. Special creation is the concept that life was created by a supernatural being. Panspermia is the theory that the earth was seeded with primitive life forms from another planet. Biologists currently favor the theory of chemical evolution, in which life is presumed to have arisen spontaneously by a gradual increase in the complexity of organic compounds, culminating in molecular assemblages with the capacity to maintain and reproduce themselves.

The atmosphere of the prebiotic earth probably consisted primarily of hydrogen-containing compounds such as water vapor, ammonia, methane, and hydrogen cyanide; also present were nitrogen, carbon monoxide, carbon dioxide, and hydrogen. Notably, oxygen was absent. Oparin and Haldane suggested that when this atmosphere was exposed to various forms of available energy (such as heat and ultraviolet rays), the organic compounds necessary for life would have formed spontaneously. Experiments designed to simulate these conditions indicate that simple organic substances do indeed form from inorganic starting materials and energy. As organic compounds accumulated in the earth's waters, they formed a dilute "organic broth," the first step toward life.

STUDY QUESTIONS

1. Why is the idea of special creation outside the realm of science?

2. How does the theory of panspermia help us rationalize the universality of the genetic code?

3. What are the three basic particles of which atoms are composed? Where are they located within the atom? How do these particles differ with respect to mass and charge?

4. Name three types of chemical bonds and indicate how they differ.

5. Which two of the following compounds would have similar properties?
(a) A 5-carbon hydrocarbon
(b) A 5-carbon alcohol
(c) A 6-carbon alcohol

6. What is the estimated age of the earth? When is it estimated that life began on the earth?

7. According to the theory of chemical evolution, how did the atmosphere of the prebiotic earth differ from our present one?

8. What was the first step toward the chemical origin of life, as proposed by Oparin and Haldane?

9. What is the significance of the first prebiotic synthesis experiment performed by Stanley Miller?

SUGGESTED READINGS

Crick, F. *Life Itself: Its Origins and Nature.* New York: Simon & Schuster, 1981. A popular account of the various theories on the origins of life, including an elaboration of directed panspermia. Enjoyable reading.

Dickerson, R. E. "Chemical Evolution and the Origin of Life." *Scientific American* 239 (1978): 70–86. This entire issue of *Scientific American* is devoted to the subject of evolution. Dickerson's article provides an in-depth treatment of the most probable chemical events occurring on the prebiotic earth, including a good discussion of the Oparin-Haldane proposal and Miller's first prebiotic synthesis experiment.

Dott, R., and R. Batten. *Evolution of the Earth.* 2nd ed. New York: McGraw-Hill, 1976. A well-written account of the principles of geology, including a concise introduction to the early physical and chemical evolution of earth and theories on its origin.

Masterton, W. L., and E. J. Slowinski. *Chemical Principles.* 4th ed. Philadelphia: W. B. Saunders, 1977. One of the best general chemistry texts of recent vintage.

Orgel, L. *The Origins of Life: Molecules and Natural Selection.* New York: Wiley, 1973. A clear description of chemical evolution.

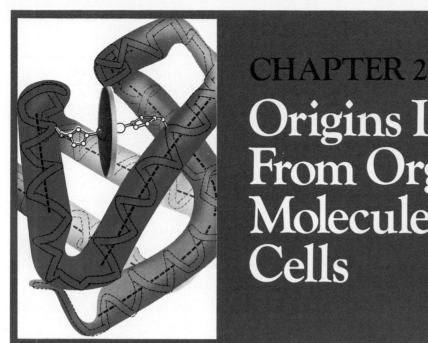

CHAPTER 2
Origins II: From Organic Molecules to Cells

Four billion years ago the earth was still a lifeless planet. According to the theory of chemical evolution, however, events were taking place then that would soon change this situation. Lightning, heat, and ultraviolet radiation were acting on an atmosphere rich in hydrogen-containing compounds, causing the formation of various simple organic compounds. As these compounds accumulated in the waters below, the oceans became an "organic broth" containing all the chemical units necessary for life. The next big step on the road to life must have been the linking together of these simple organic units into long chains, or **polymers**—a process called **polymerization.** One type of polymerization, the linking of amino acids into proteins, was an absolute prerequisite to life. Proteins are as central to life as carbon atoms are to organic compounds. Later in this chapter we will explore the characteristics of these vital substances in some detail. For now, let us examine the general mechanisms by which the first biopolymers presumably arose.

POLYMERIZATION: BUILDING LARGE MOLECULES

Although polymerization reactions were a necessary step in chemical evolution, they could not have taken place under the relatively dilute conditions that pre-vailed in the early oceans. We know from laboratory experiments that polymers are slow to form unless the concentrations of the reacting units are high. In other words, the reacting molecules must be very close together so that they can collide with sufficient frequency for the linking reactions to occur. On the prebiotic earth, then, there must have been ways for the organic broth to have become concentrated.

One of the simplest ways to concentrate a solution is to allow the **solvent,** or liquid (in this case, water) to evaporate. As the water from a dilute solution evaporates, the number of nonwater molecules per unit volume of water increases. You can test this quite easily. Dissolve a teaspoon of sugar in a cup of water and taste it. Now boil most of the water away and taste again. The sweet taste will be much stronger because the sugar is more concentrated. On the early earth, small tidepools rich in organic molecules may have become isolated by the fluctuating tides and been greatly concentrated by evaporation, perhaps even to dryness (Figure 2–1a and b).

Another concentrating mechanism, and perhaps the most likely candidate for prebiotic systems, is **adsorption**—the adhesion of a substance to a foreign surface. Clay particles, for example, are known to adsorb ions and certain organic compounds (Figure 2–1c). By adhering to such a surface, the adsorbed molecules are so close together that collisions among them are much more frequent.

17

FIGURE 2–1
Concentrating Mechanisms. On the prebiotic earth, the simple organic molecules present in the "organic broth" must have become concentrated in order for the polymerization reactions to have taken place. (a) An isolated tidepool could have evaporated, resulting in (b) a more concentrated solution of reacting molecules. (c) Alternatively, certain types of clay particles that bind organic molecules to their surfaces could have had a similar concentrating effect.

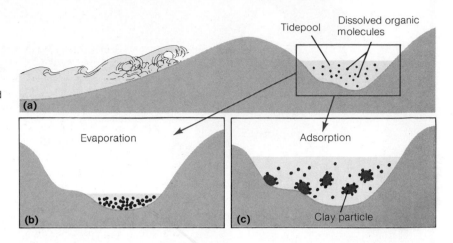

In addition to having the reacting molecules in a concentrated form, polymerization reactions require energy. In fact, any process of change from a simple state to a more complex one—whether linking organic molecules into polymers, or organizing words on a page to convey a thought—requires an input of energy from some outside source. On the prebiotic earth, energy for polymerization reactions was available in the forms of heat (from volcanoes, hot springs, and the sun), ultraviolet radiation, electrical energy (from lightning), and chemical energy inherent in organic compounds themselves. This energy acting on concentrated mixtures of organic building blocks apparently led to the formation of proteins from amino acids and the formation of other biologically important polymers from their corresponding units. Some of these polymers were certainly the major biological macromolecules that still make up the "stuff of life" today. Again, a detour into chemistry will help you understand how these complex molecules could have formed and evolved into the first living cells.

In Chapter 1 we saw that carbon atoms form stable combinations with other carbon atoms and with several other elements as well. Because of this versatility, carbon atoms serve as the backbone for an enormous number of different kinds of organic compounds. Yet organisms use relatively few types of organic compounds. Only four major types are found in abundance in all organisms: carbohydrates, lipids, proteins, and nucleic acids.

CARBOHYDRATES

Carbohydrates are the class of organic compounds that includes the sugars and starches. These compounds contain carbon, hydrogen, and oxygen atoms in the approximate ratio of 1 : 2 : 1. There are three major subgroups of carbohydrates which differ in molecular complexity. First and least complex are the **simple sugars,** or **monosaccharides** (*mono* = one, *saccharide* = sugar). The second level of complexity is represented by the **disaccharides** (*di* = two), each of which consists of two monosaccharide units joined together. Finally, when several to many sugar units are linked together in a chain, the resulting polymer is called a **polysaccharide** (*poly* = many), the highest level of carbohydrate complexity.

Simple Sugars, or Monosaccharides

The naturally occurring simple sugars, which we recognize by their sweet taste, have from three to seven carbon atoms. Two of the most common simple sugars are **glucose** and **fructose,** both 6-carbon molecules (Figure 2–2). Glucose is present in many ripe fruits and in "quick-energy" drinks often used by athletes; when it is dissolved in a saline (salt) solution, glucose is a life-sustaining food fed intravenously to hospital patients. Fructose is an intensely sweet sugar also found in many fruits. Both glucose and fructose give honey its sweet taste. Another important sugar, **deoxyribose** (see Figure 2–2), is a constituent of DNA, the hereditary material found in all living cells (to be discussed in Chapter 11).

Living creatures use these and scores of other simple sugars in two general ways. First, simple sugars form the building blocks from which many other biological substances are made. Second, the bonds holding the atoms of sugar molecules together are the major source of stored chemical energy for the work done within living organisms. When these bonds are broken, biologically useful energy is released (see Chapter 5).

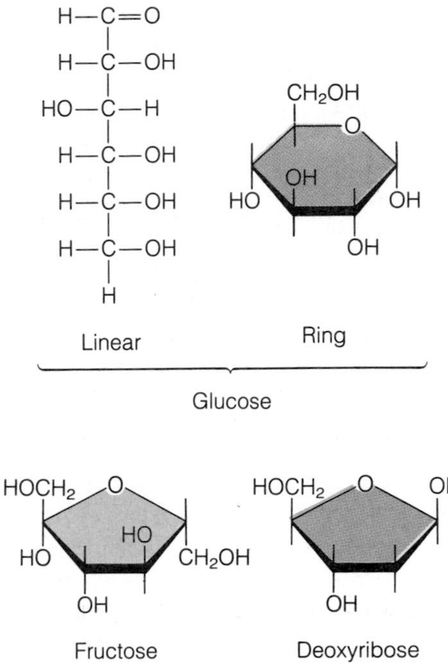

FIGURE 2–2

Three Simple Sugars. As a dry powder, sugar molecules have a linear shape, as shown here for glucose. When sugars are dissolved in water, each molecule forms a ring. In diagrams of the ring structure, it is conventional to represent carbon atoms in the ring simply as corners of the ring, and to omit any hydrogen atoms bonded directly to the ring carbons.

Both glucose and fructose have the same chemical formula ($C_6H_{12}O_6$), but their atoms are arranged differently, resulting in different chemical properties. Deoxyribose is a 5-carbon sugar ($C_5H_{10}O_4$).

Disaccharides

A disaccharide consists of two monosaccharides (either the same or different sugars) linked by a covalent bond. **Maltose** (malt sugar), a disaccharide composed of two glucose units, is produced in large quantities from the breakdown of starch in barley seeds during the making of beer. **Lactose** (the major sugar found in milk) consists of glucose and a closely related simple sugar, galactose. Common table sugar is the disaccharide **sucrose,** composed of glucose and fructose (Figure 2–3). Many plants transport carbohydrates in the form of sucrose; some plants store it. Sucrose is especially abundant in sugar beet roots and sugar cane stems, the two major sources of refined sugar.

The chemical reaction that joins two monosaccharides into a disaccharide is a type of polymerization called **dehydration synthesis.** The name comes from the fact that in the reaction one of the monosaccharides loses a hydrogen atom (H) and the other loses a hydroxyl group (—OH). The H and —OH form a molecule of water (H_2O) (*dehydration*) as the disaccharide is formed (*synthesis*) (Figure 2–3). The reversal of dehydration synthesis is called **hydrolysis** (water-splitting) because the degradation of a disaccharide into two monosaccharides requires the addition of a hydrogen atom and a hydroxyl group from water. Dehydration synthesis and hydrolysis are the most common means by which biological polymers are constructed and broken down.

Polysaccharides

When simple sugars link up into long chains, they form the third type of carbohydrate—polysaccharides. Depending on its structure, a particular polysaccharide

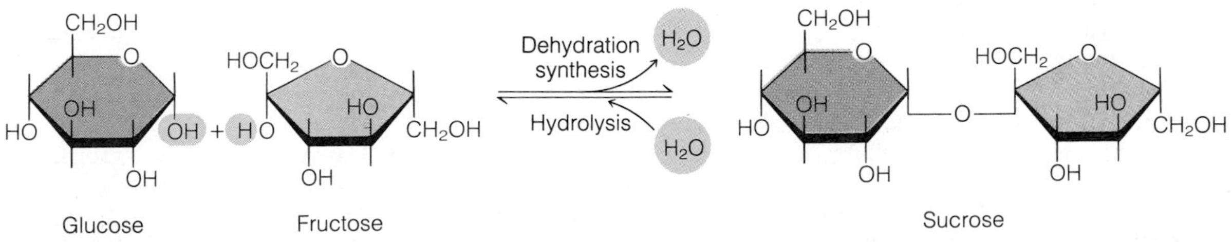

FIGURE 2–3

A Disaccharide. The disaccharide sucrose, common table sugar, is composed of glucose and fructose joined by dehydration synthesis. In dehydration synthesis, a hydrogen atom (H) and a hydroxyl group (—OH) (color) are removed from the reacting groups, forming water. The splitting of a disaccharide into its component simple sugars involves the addition of a water molecule, a reaction called hydrolysis.

may function as a storage form of food or as a structural component of a cell.

Plants typically store the carbohydrates they use for food in the form of **starch,** a polysaccharide composed entirely of glucose (Figure 2–4a). Starches make up roughly 80% of the human diet worldwide; the primary sources are the starch-storing cereal seeds such as rice, wheat, and corn. Animals and fungi store carbohydrates in the form of **glycogen,** a polysaccharide similar to starch except that it is considerably larger and more branched. In animals, glycogen is stored primarily as granules in liver and muscle cells (Figure 2–4b). When the level of glucose in the body

cells becomes low, the stored glycogen is broken down into its glucose units.

Plants, fungi, and many microorganisms derive structural support for their cells from polysaccharides present in their cell walls (outer cell coverings). The strongest, most rigid polysaccharide in cell walls is **cellulose,** a long straight chain of glucose units. The nature of the chemical links between the glucose units causes cellulose to be stiff and also allows individual cellulose chains to bind together by hydrogen bonds to form larger strands called microfibrils (Figure 2–5). Microfibrils in turn associate into larger fibrils, much like individual hemp fibers are woven into rope. Other

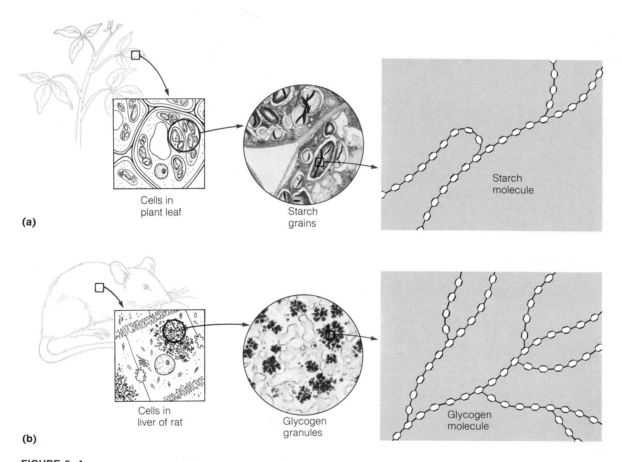

(a) Cells in plant leaf — Starch grains — Starch molecule

(b) Cells in liver of rat — Glycogen granules — Glycogen molecule

FIGURE 2–4
Two Storage Polysaccharides: Starch and Glycogen. Starch (found in plants) and glycogen (found in animals and fungi) are both storage polysaccharides composed entirely of glucose units. (a) Each grain of starch shown in the photo consists of thousands of starch molecules. (The black streaks are cracks in the starch grains.) The diagram shows part of an individual starch molecule. Each hexagon represents a single glucose unit. (b) This photo of part of a rat liver cell shows clusters of dark glycogen granules. The diagram of part of a glycogen molecule shows that it is more highly branched than a typical starch molecule.

FIGURE 2–5
A Structural Polysaccharide: Cellulose.
Cellulose, the most abundant organic substance on earth, is composed of very long, unbranched chains of glucose units. Because of their linear configuration, cellulose chains self-associate along their lengths to form rigid *microfibrils,* which in turn form larger units called *fibrils.* Individual fibrils can be seen in this photo of the surface of a plant cell wall.

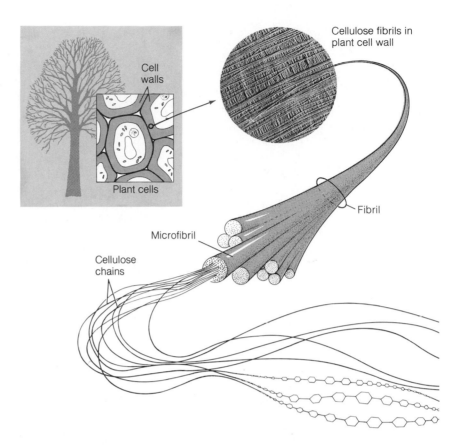

Cell walls

Plant cells

Cellulose fibrils in plant cell wall

Fibril

Microfibril

Cellulose chains

polysaccharides in the cell walls bind to the cellulose fibrils. This provides additional mechanical strength for the cell walls.

It is interesting to note that over half of the organic carbon on earth is found in only one compound: cellulose. The next most common carbon compound is starch. Actually, this is not so surprising when we consider that these two compounds are found in plants, and that plants make up the greatest bulk of living matter on our planet. The glucose connection has been very successful!

LIPIDS

Lipids, which include fats and oils, are compounds composed primarily of carbon and hydrogen with relatively little oxygen. Most lipids are insoluble in water. When mixed with water, they form droplets (like grease in cold water) or a film on the surface (like oil slicks on wet roads). Like the polysaccharides, some lipids serve as stored forms of food and others as important structural components of cells.

The main components of most lipids are fatty acids. A **fatty acid** is a hydrocarbon chain that ends with a carboxyl group ($-COOH$). Fatty acids vary in the length of the hydrocarbon chain (most have 12 to 22 carbon atoms) and in the number of carbon-to-carbon double bonds ($-C=C-$). A fatty acid with no carbon-to-carbon double bonds has the highest possible number of hydrogen atoms, so it is said to be **saturated** with hydrogen. **Unsaturated** fatty acids have one or more double bonds in the hydrocarbon chain. The double bonds create bends in the chain (Figure 2–6).

Triglycerides (Fats and Oils)

An important type of stored food in organisms is the group of lipids called triglycerides, better known as the fats and oils. A **triglyceride** consists of a glycerol molecule (a 3-carbon carbohydrate) attached to three fatty acids (Figure 2–6). One fatty acid is attached to each carbon atom of the glycerol unit. Triglycerides differ from one another by the nature of their fatty acids.

FIGURE 2–6
A Triglyceride. A triglyceride consists of three fatty acid units linked to a single glycerol unit (color).

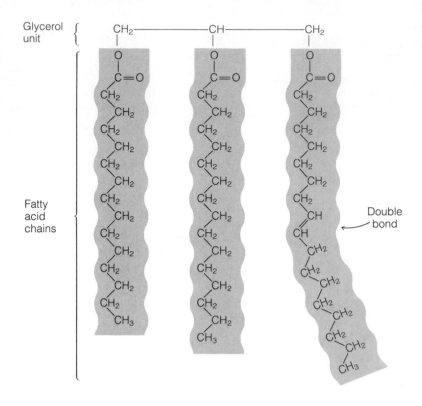

Generally speaking, animals produce and store triglycerides comprised of long, saturated fatty acid units; they are called **fats.** Since their straight fatty acid chains can pack together tightly, fats tend to be solids at room temperature. Lard and butter are two examples of animal fats. Plants, in contrast, usually produce unsaturated triglycerides, called **oils.** The bent fatty acid chains of oils cannot lie close together, so oils tend to be liquid at room temperature. The vegetable oils from olives, corn, safflowers, and peanuts often contain many double bonds and are referred to as **polyunsaturated oils.** Food manufacturers can artificially **hydrogenate** plant oils (that is, they can add hydrogen to saturate the double bonds), yielding solid fats such as those in vegetable shortening and margarine. The relationship between polyunsaturated oils and diet will be discussed in Chapter 21.

Phospholipids

Phospholipids, like triglycerides, consist of a glycerol backbone connected to fatty acids. Instead of three fatty acid chains, however, **phospholipids** have two fatty acids plus a **phosphate group** (Figure 2–7). In addition, most phospholipids have an organic group connected to the phosphate (see Figure 2–7). Since the phosphate and organic group usually have charges associated with them, this region of the phospholipid molecule is polar and is attracted to water. Because the fatty acid chains are very nonpolar, they repel water. As a result, phospholipids are two-faced when it comes to dissolving in water: the polar region readily associates with water, whereas the nonpolar fatty acid chains stick out above the water surface or associate with the fatty acids of neighboring phospholipid molecules (Figure 2–8). This characteristic makes phospholipids well suited to separating a living cell from its external environment, one of the most important functions of biological membranes. As we shall see in Chapter 3, phospholipids are structural components of all biological membranes.

Steroids

Steroids are a special class of lipids in which the ends of the hydrocarbon chains are joined together to form rings (usually four) (Figure 2–9). Like the fats and oils, steroids are quite insoluble in water.

Cholesterol is a very important steroid that is a structural component of animal cell membranes. It is

FIGURE 2–7
Phospholipid Structure. (a) A diagram of a phospholipid molecule, showing the structure of the charged phosphate and organic groups (polar head). Phospholipids vary by the nature of their fatty acids and organic group. (b) A simple diagram of a phospholipid.

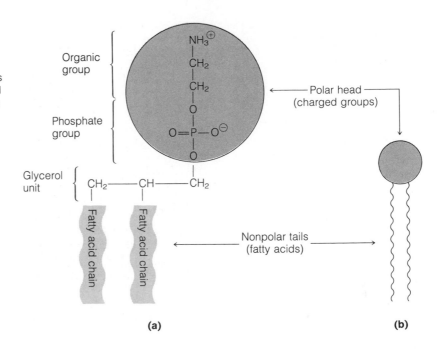

(a)

(b)

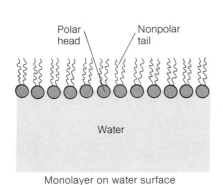

Monolayer on water surface

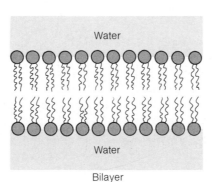

Bilayer

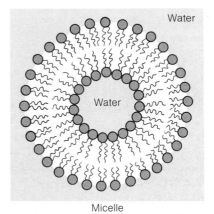

Micelle

FIGURE 2–8
Phospholipids in Water. When added to water, phospholipids may form a monolayer (single layer) on the water surface, where only the polar heads associate with the polar water molecules. When dispersed in water, phospholipids tend to form bilayers, with their nonpolar tails oriented toward each other and away from the water. Artificial bilayers usually exist as tiny droplets called micelles. In a similar fashion, phospholipids in cell membranes aid in separating the inside of a cell from its external environment. Note the similarity between the phospholipid bilayer and the cross section of a cell membrane shown in Figure 4–1.

FIGURE 2–9
Cholesterol. Like most steroids, cholesterol consists of four rings and a short hydrocarbon tail. In this diagram, each corner represents a carbon atom and its hydrogen atoms.

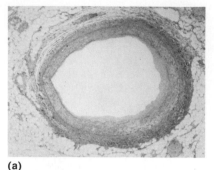

(a)

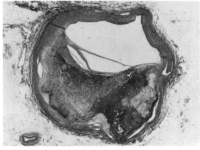

(b)

FIGURE 2–10
Atherosclerosis. This condition, in which the walls of the arteries become thickened and inelastic, is correlated with a high dietary intake of cholesterol. (a) Normal, healthy arteries become transformed into (b) hardened vessels as this disease progresses.

also the starting material for the synthesis of the steroid hormones in humans and other animals, such as the sex hormones (estrogen and testosterone) and the corticosteroids produced in the adrenal glands (see Chapter 24). On the negative side, high levels of cholesterol in the blood are correlated with atherosclerosis ("hardening of the arteries"), a condition that stems from excessive deposits of fats and cholesterol in the walls of the arteries (Figure 2–10). Hardening of the coronary arteries, vessels that deliver blood to the heart muscles, is the leading cause of heart attacks. The incidence of atherosclerosis is highest in Europe, Japan, and the United States, where the people eat substantial amounts of meat, cheese, and eggs, all rich sources of cholesterol.

PROTEINS

If you asked a biochemist to choose the most diverse class of biological compounds, he or she would undoubtedly name **proteins.** Every organism can pro-

duce thousands of different proteins, each with its own unique function in the life process. Proteins are so important to life that it would not be an exaggeration to say that what you are depends on what proteins you can make.

Amino Acids

Proteins are polymers made up of one or more chains of **amino acids.** All organisms use about 20 different kinds of amino acids to make proteins. As their name implies, amino acids contain an **amino group** ($-NH_2$) and a **carboxyl,** or organic acid, **group** ($-COOH$) (Figure 2–11a). Each amino acid differs by the nature of the additional group of atoms in the molecule (the R group in Figure 2–11a). The simplest amino acid is glycine, in which the R group is a single hydrogen atom; the largest is tryptophan, where the R group is a complex double-ring structure (Figure 2–11c).

With 20 different kinds of amino acids typically occurring in proteins, the potential for variety is enormous. Imagine a chain of five amino acids. With 20 amino acids to draw from, how many unique combinations of five are possible? The answer is 20^5, or over 3 million! When you consider that most biological proteins have 50 to several hundred amino acids, you can see that an organism with as many as 10,000 different proteins has still not exhausted the possibilities. However, just as the letters in a sentence are not chance associations of the 26 letters of the alphabet, the proteins in organisms are not random combinations of amino acids. Rather, amino acid sequences are constructed in specific patterns that determine the particular function of the completed protein.

Like simple sugars, amino acids can be broken down to yield energy. Their main function, however, is to serve as building blocks for proteins. During protein formation, amino acids are linked to each other by dehydration synthesis. The bond between two adjoining amino acids is called a **peptide bond** (Figure 2–11b). A long chain of amino acid units is called a **polypeptide** (Figure 2–11c). Some proteins consist of a single polypeptide chain, but many contain two or more chains connected by nonpeptide bonds. Insulin, for example, is an animal protein produced in the pancreas that helps regulate blood sugar levels. (This is the same insulin diabetics take because their pancreas does not produce enough.) It consists of two polypeptide chains, one with 30 amino acids and the other with 21, joined by disulfide bonds (Figure 2–12).

(a) General structure
of an amino acid

Amino
group

Carboxyl
group

R
group

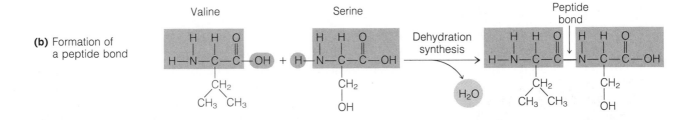

(b) Formation of
a peptide bond

Valine

Serine

Dehydration
synthesis

H_2O

Peptide
bond

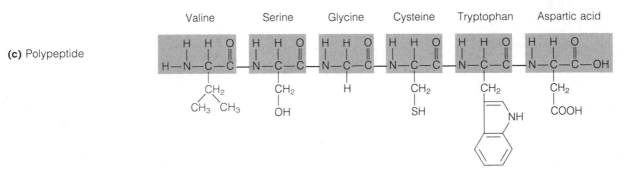

(c) Polypeptide

Valine   Serine   Glycine   Cysteine   Tryptophan   Aspartic acid

FIGURE 2–11
Amino Acids and Peptides. (a) All amino acids have an amino group ($-NH_2$) and a carboxyl group
($-COOH$) but differ by the nature of the R group. (b) The linking of two amino acids (here, valine
and serine) to form a dipeptide occurs by dehydration synthesis and results in a peptide bond (color).
(c) A polypeptide consists of many amino acids, each linked to the next by a peptide bond (color).
Note that each peptide bond extends between the carboxyl carbon of one amino acid and the amino
nitrogen (N) of the next amino acid on the chain.

FIGURE 2–12
The Amino Acid Sequence of Insulin.
Each circle represents an amino acid
with its three-letter abbreviation shown
(Gly = glycine, Ile = isoleucine, Val
= valine, etc.). Beef insulin is a protein
used in the treatment of human diabetes,
allowing many diabetics to lead normal
lives. Its two polypeptide chains (A and B)
are held together by disulfide bonds
($-S-S-$) between units of cysteine,
a sulfur-containing amino acid (color
circles).

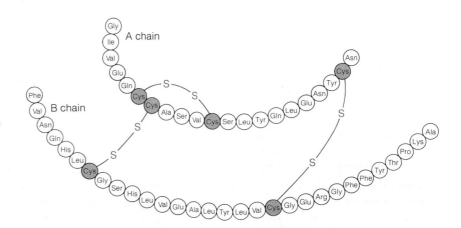

Protein Structure

Because of their large size, protein molecules can assume a number of configurations and shapes. It is useful, therefore, to describe them at three and sometimes four levels of structural organization (Table 2–1). The first level, or **primary structure,** refers to the sequence of amino acids in a protein. In the insulin molecule, for example, the amino acid sequence of the A chain is glycine-isoleucine-valine-glutamic acid-and so on (see Figure 2–12). The primary structure to a large extent dictates the higher levels of structure in a protein.

The **secondary structure** of a protein refers to the configuration of its polypeptide chain or chains. The bonding angles of most peptide bonds cause some regions along the chain to fold into a **helix,** or spiral (Figure 2–13a). These helices, held in place by hydrogen bonds, may be interspersed with straight regions. Human hair, for instance, is composed primarily of keratin, a protein whose secondary structure can have varying numbers of helical regions; it is this variation that makes hair either curly, wavy, or straight.

The **tertiary structure** of proteins can be of two general types. If the polypeptide chains, helical or otherwise, tend to stretch out into stiff, fiberlike structures, the proteins are said to be **fibrous** (Figure 2–13a). On the other hand, protein chains that bend and fold back upon themselves are termed **globular** (Figure 2–13c). Fibrous proteins, such as those that make up hair and fingernails, tend to be tough and resistant; globular proteins, such as those found in raw egg whites, are usually soft and mushy.

For proteins that consist of more than one polypeptide chain, the way the individual chains fit together

TABLE 2–1
Structural Levels of Proteins.

Structural Level	Definition	Types of Bonds Holding Structure Together	Example
Primary	Sequence of amino acids in a protein	Covalent (peptide) bonds	Gly — Ile — Val — Glu — Gln — ...
Secondary	Folding of the polypeptide chain into patterns, such as a helix	Hydrogen bonds	
Tertiary	Three-dimensional folding pattern of a single polypeptide chain	Hydrogen bonds, disulfide bonds, polar and nonpolar attractions	
Quaternary	Structural relationship between the polypeptide chains in a protein made up of two or more polypeptide chains	Same as tertiary structure	

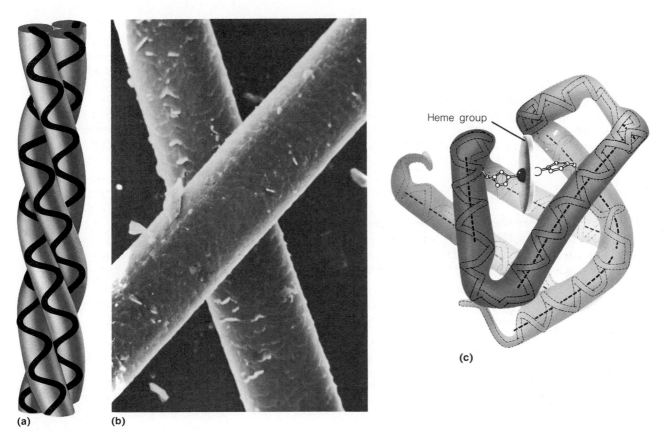

Heme group

(a) **(b)** **(c)**

FIGURE 2–13
Protein Structure. (a) Keratin, a fibrous protein, is the structural unit of hair, fur, and fingernails. It consists of three polypeptide chains, each with a helical secondary structure. Because it lacks tertiary folding (as in the molecule in part c), keratin has a tough, resilient character. (b) A close-up of two human hairs. Hair is strong and tough because it is made of fibrous proteins, not globular proteins. (Magnification: 260×.) (c) Myoglobin, an oxygen-storing molecule found in human muscle cells, is a globular protein that consists of a single polypeptide chain of 153 amino acids. The secondary structure of the polypeptide includes eight helical regions interspersed with straight or twisted regions. The polypeptide chain surrounds a heme group, a complex iron-containing unit that binds oxygen. The overall folding pattern of myoglobin gives it its globular tertiary structure.

determines that protein's **quaternary structure.** For example, hemoglobin, the oxygen-carrying protein in blood, consists of four chains, each with a tertiary, secondary, and primary structure (see Table 2–1).

Protein Function

Each different protein manufactured by an organism has a unique function directly related to its structure. First, some proteins perform a structural role by providing **protection or mechanical strength.** Keratin in hair, fingernails, and hooves, and the proteins of tendons and ligaments, serve such a function. Second, other proteins have a **storage** role. For instance, plant seeds contain packets of protein that become degraded once the seed germinates, thus providing the growing seedling with amino acids to make new proteins. The stored proteins in cereal grains, beans, and other crop seeds also can be eaten and digested by animals. Such seeds are a major source of protein in the human diet, particularly in developing countries where beef, poultry, fish, and animal products are difficult to obtain.

A third major function of proteins is **transport.** As mentioned earlier, hemoglobin transports oxygen in the circulatory system of higher animals; and as we shall see in Chapter 4, specific transport proteins in cell membranes regulate the flow of materials into and out of cells.

Finally, in their most important role, certain proteins **catalyze** chemical reactions. **Catalysis**—the speeding up of a chemical reaction by an agent that remains unchanged during the reaction—is carried out in organisms by a special class of proteins called **enzymes.** Without enzymes, cellular reactions would proceed too slowly to be of any use to the organism. We will discuss enzyme catalysis at more length in Chapter 5.

NUCLEOTIDES AND NUCLEIC ACIDS

Nucleotides are complex organic molecules that are probably best known as components of the **nucleic acids** DNA (deoxyribonucleic acid) and RNA (ribonucleic acid). DNA and RNA function in the storage, transfer, and inheritance of genetic information. The nucleic acids and their component nucleotides will be discussed in Chapter 11.

Two very important nucleotides are **ATP** (adenosine triphosphate) and **ADP** (adenosine diphosphate). These two nucleotides perform a vital role in the transfer of energy in all organisms. ATP can be split into ADP and a phosphate group, releasing biologically useful energy. That is, organisms can use this energy to drive the processes of life. A detailed discussion of ATP and energy appears in Chapter 5; for present purposes, you should keep in mind that ATP represents stored energy, and this energy is released to perform cellular work when ATP breaks down into ADP and phosphate. ATP or some similar compound undoubtedly provided energy for many of the energy-requiring chemical reactions on the prebiotic earth.

THE ORIGIN OF CELLS

As a prelude to life, the simple organic compounds, such as sugars and amino acids, first had to link together into their respective polymers, such as polysaccharides and proteins. The next step would have been for these polymers to self-associate into aggregates, or tiny droplets in the organic broth. Although the details of chemical evolution at this stage are sketchy, laboratory experiments have shown that such droplets do form under simulated prebiotic conditions. For example, biochemist Sidney Fox found that heating a concentrated mixture of amino acids yields polypeptides that, when wetted, aggregate into microscopic droplets Fox called **proteinoids** (Figure 2–14). Remarkably, such proteinoids exhibit some lifelike properties. First, the outer boundary of each droplet shows a degree of **selective permeability.** That is, certain substances can permeate (pass into) the droplet while others are excluded—a property crucial to the normal functioning of living cells. Second, proteinoids have been observed to fuse with one another and thereby to grow in size. Even their general appearance under the microscope has led some trained micro-

FIGURE 2–14
Proteinoids. (a) The droplets shown here are nonbiological aggregates of polypeptides. (Magnification: approximately 400 ×). Note the resemblance between these proteinoids and (b) cells of *Blastomyces dermatitidis*, a disease-causing fungus. (Magnification: approximately 3000 × .)

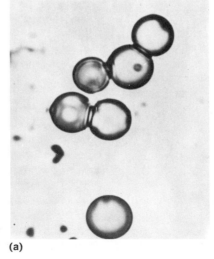

(a)

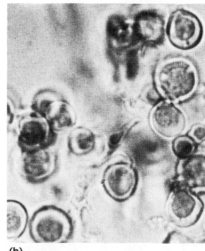

(b)

biologists to mistake them for a new species of bacterium.

These two characteristics—growth and selective permeability—were important to the origin of life, but other developments that have not been duplicated in the laboratory must also have occurred. For example, by accidents of structure some of the proteins in the prebiotic droplets must have been able to catalyze chemical reactions, a function that is now carried out by enzymes. However, the most crucial characteristic these droplets somehow developed was the ability to reproduce. One of the most challenging tasks facing those who study the origin of life is to determine how a self-replicating system might have come into being. No experiments have yet demonstrated the abiotic (without life) synthesis of a self-replicating molecule. Indeed, we have only become aware of the self-replicating mechanism in present-day cells within the last three decades. Nevertheless, some scientists have predicted that we will be able to create a self-replicating aggregate under abiotic conditions within the next 100 years, thereby proving that chemical evolution leading directly to the first living cells could have taken place on the prebiotic earth.

The Evolution of Metabolism

When the first self-replicating droplets emerged, simple organic compounds and polymers must have been still forming from the action of ultraviolet radiation, heat, and electrical energy in the atmosphere and on the surface of the oceans. But to the highly organized macromolecular droplets, such energy would have been more disruptive than formative, for it would have tended to degrade them (see Essay 2–1). We must imagine, then, that the droplets existed in an environment shielded from these disruptive forces, probably several feet below the surface of their watery habitat. However, the droplets still needed energy for the processes of growth and reproduction—energy to synthesize their nucleic acids, proteins, and the many other components of which they were constructed. This constant energy demand was undoubtedly met by high-energy compounds such as ATP dissolved in the organic broth. Having evolved the capacity to use the energy of ATP, the droplets presumably became entirely dependent on this energy source. After many generations of droplets, however, ATP was probably in short supply. Indeed, life itself

would have ended unless some chance variant among the droplets had an enzyme that catalyzed the regeneration of ATP from ADP and phosphate. This variant must have applied the energy of some other compound still in plentiful supply to regenerate its primary energy source, ATP. Thus, the new kind of droplet had an advantage over its predecessors, which were "starving" from the lack of ATP in the organic broth, and the new droplets replaced the old.

We can imagine that over time, even the secondary chemical energy sources were depleted, and new enzymes must have developed to regenerate these energy sources from still other compounds. Step by step, entire sequences of enzyme-catalyzed reactions, or **metabolic pathways** evolved, all in response to the dwindling stock of energy-rich organic compounds. There must have been much trial and error; success was measured simply by survival in the face of a diminishing nutrient supply. The surviving droplets became more and more sophisticated metabolically, until after a half-billion years or so, they could produce all of their own biological compounds from an external source of a few simple organic compounds and inorganic minerals. At that point, the primitive droplets crossed the threshold to become the first living cells. These early cells had the three key characteristics of life: growth, reproduction, and metabolism.

The Emergence of Photosynthesis

According to the theory of chemical evolution, the first living cells were **heterotrophic;** that is, they relied on outside sources (organic compounds already existing in the environment) to meet their needs for energy and raw materials. Over time, however, even the simplest organic compounds, such as sugars and amino acids, would have been used up as the rapidly proliferating heterotrophs metabolized them faster than the abiotic reactions could generate them (Figure 2–15). A dwindling food supply strongly favored the emergence of a new **autotrophic** cell, one that could synthesize its own organic nutrients (food) from inorganic compounds in vast supply.

Forming simple organic compounds from inorganic substances is not only chemically complicated, but energetically unfavorable. A substantial amount of energy is required to transform relatively low-energy inorganic compounds into relatively high-energy organic compounds; a mechanism for this transformation was essential for the continuation of life. What

FIGURE 2–15
Major Events in the Origin and Early Evolution of Life According to the Theory of Chemical Evolution. Chemical evolution not only gave rise to the first living cells on earth, but also provided a source of organic compounds to sustain these primitive life forms. As these early heterotrophs proliferated, they began to use up these compounds faster than they could be produced by the abiotic reactions, a situation that favored the development of photosynthesis. Oxygen (O_2), a by-product of photosynthesis, began to accumulate in the atmosphere and reached its present 20% level about a billion years ago. The presence of free oxygen permitted the development of respiration, the process by which all plants and animals extract biologically useful energy from organic compounds. Note that most of biological evolution has taken place at the level of single-celled organisms; the first multicellular life forms appeared only about 700 million years ago.

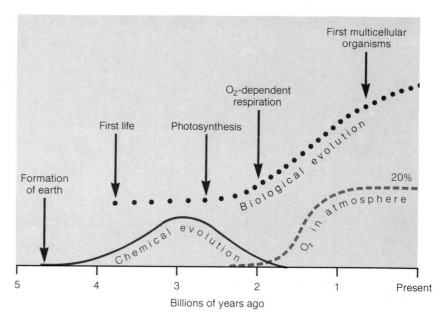

type of energy could these early cells use? Remember, the supply of chemical energy—in the form of organic compounds in the broth—was running out. Ultraviolet light would have caused more harm than good (see Essay 2–1), and heat and electrical energy were either harmful or unavailable to the cells in their watery environment. The only useful source, then, was visible light. When cells developed the ability to trap light energy and use it to synthesize organic compounds, they had the key to survival. They had evolved the process we call **photosynthesis.**

The most common form of photosynthesis involves the conversion of water and carbon dioxide (both inorganic substances) to sugar (an organic compound) and molecular oxygen:

$$6\,CO_2 + 6\,H_2O \xrightarrow{\text{Light energy}} C_6H_{12}O_6 + 6\,O_2$$

Carbon dioxide   Water   Sugar   Oxygen

Thus, photosynthesis enabled the autotrophs to manufacture food (sugar) from a very basic "diet" of carbon dioxide, water, and visible light, and ensured their continued survival. In addition, the autotrophs saved the day for the struggling heterotrophs, for the latter could now obtain their organic material from the photosynthesizers. This relationship has persisted throughout the history of life, as present-day animals (including you), fungi, and other heterotrophs owe their continued existence to the photosynthetic organisms, the green plants.

The Consequences of Oxygen Production

At the time photosynthesis evolved, there was little or no free oxygen (O_2) in the atmosphere. Oxygen atoms appeared only in combined forms like carbon dioxide (CO_2) and water (H_2O). However, since O_2 is a by-product of photosynthesis, it gradually accumulated in the air. The presence of free oxygen was new to life and required some adjustments. Recall that molecular oxygen tends to react with organic substances, degrading them into simpler molecules (see Chapter 1). Thus, organisms wage a constant battle against decomposition by this process, called oxidation. Presumably, many of the early life forms lost this battle; others survived until present times by living in habitats devoid of molecular oxygen. This latter group, all of which are microorganisms, can survive only under oxygen-free conditions. Other forms of primitive cells, including both autotrophs and heterotrophs, adapted to the presence of O_2 and even began using it to their advantage. They evolved **respiration** (Figure 2–15), a controlled metabolic process in which simple organic compounds are broken down in the presence of oxygen, giving off relatively large amounts of energy.

$$\text{Organic compounds} + O_2 \longrightarrow \overset{\text{Energy}}{CO_2 + H_2O}$$

Some of the energy released in respiration is used to combine ADP and phosphate to form ATP. The

ESSAY 2–1

THE EARTH'S OZONE LAYER: HERE TODAY! GONE TOMORROW?

In the past decade concern has developed over the future of the earth's ozone layer (see figure). We have learned that various fluorocarbons (such as trichlorofluoromethane) act as catalysts to speed the breakdown of atmospheric ozone. These fluorocarbons are released from aerosols, from certain solvents, and from the deterioration of some foam rubbers and plastics. As the ozone layer dissipates, the intensity of ultraviolet (UV) radiation reaching the earth's surface will increase, causing a host of problems for life on earth. These problems stem from the fact that UV light can break chemical bonds holding molecules together, thereby disrupting their structure and function. For living organisms the problem of UV irradiation is particularly acute, since UV chemically alters both DNA and protein, macromolecules crucial to life. In humans, overexposure to UV can lead to skin cancer, eye cataracts, and serious burns.

It is hard to believe that the fluorocarbons added to the atmosphere from hair sprays, deodorants, and other sources could affect something as vast and distant as the ozone layer. Part of the problem is that fluorocarbons are very stable compounds; that is, they persist and accumulate in the atmosphere. Further, since each fluorocarbon molecule acts as a catalyst, it can break down thousands of ozone molecules before it is destroyed itself. Even if the use of all aerosol sprays were to stop today, scientists have estimated that the fluorocarbon molecules already present in the atmosphere will destroy 2–3% of the ozone now shielding the earth.

Fortunately we may be able to avert the rather bleak outlook that seemed inevitable only a few years ago. The relatively early warnings provided by organizations such as the National Academy of Sciences in Washington, D.C., Britain's Department of the Environment, and the United Nations' Environment Program, have convinced many countries to reduce or ban the use of fluorocarbon propellants. Furthermore, the rate of ozone formation in the atmosphere by natural means and through human activities may have been underestimated. Indeed, one of the major ingredients of smog is ozone. Curiously, ozone in the air we breathe is a highly toxic pollutant, but in the upper reaches of the earth's atmosphere, it is a life saver.

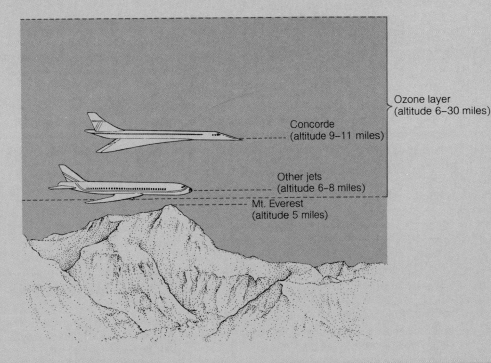

Ozone layer
(altitude 6–30 miles)

Concorde
(altitude 9–11 miles)

Other jets
(altitude 6–8 miles)

Mt. Everest
(altitude 5 miles)

rest is released as heat. Today most organisms depend on respiration to keep their cells continuously supplied with ATP. If respiration-linked ATP generation is blocked in any way—for example, by a lack of oxygen or by exposing the cells to poisons such as carbon monoxide or cyanide—the cells cannot perform work and soon die. It is interesting to note that although the earth's first cells could not have arisen had oxygen been present, most cells today cannot function without it.

A second important consequence of photosynthetic O_2 production was the gradual formation of an **ozone layer** in the earth's upper atmosphere. Ultraviolet radiation (UV) reacts with O_2 molecules to form ozone (O_3):

$$3\,O_2 \xrightarrow{\text{Ultraviolet radiation}} 2\,O_3$$

The ozone absorbs most of the sun's UV rays, screening them out before they reach the earth's surface (see Essay 2–1). Because UV radiation causes disruptive chemical changes, particularly in nucleic acids and proteins, organisms that existed before the ozone layer formed must have been restricted to a life zone several feet below the surface of the organic broth, where the upper layers of water would have screened out most of the UV radiation. Thus, the ozone layer permitted the early organisms to venture closer to the surface, and eventually to conquer the more exposed terrestrial habitat.

SUMMARY

Following the formation and accumulation of small organic compounds on the prebiotic earth, the next major step in chemical evolution was the polymerization of certain organic compounds into long chains. For the polymerization reactions to take place, however, the reacting units (such as amino acids) had to become highly concentrated, possibly by water evaporation or by adsorption of the organic building units to clay particles. In addition, energy was needed to join the units into chains.

Four important types of compounds that occur abundantly in all organisms are carbohydrates, lipids, proteins, and nucleic acids. Carbohydrates (such as sugars and starches) contain carbon, hydrogen, and oxygen atoms in the approximate ratio of $1:2:1$. They include the monosaccharides, with seven or fewer carbons in each molecule; the disaccharides, composed of two monosaccharide units; and the polysaccharides, composed of many sugar units. Carbohydrates function as structural units (such as cellulose in plant cell walls), as raw material for the synthesis of other compounds, and as sources of chemical energy.

Lipids are water-insoluble compounds comprised of a 3-carbon glycerol backbone attached to one, two, or three fatty acid units. They include the triglycerides (fats and oils), sources of stored chemical energy; the phospholipids, important structural components of biological membranes; and steroids such as cholesterol.

Proteins consist of one or more polypeptide chains, which in turn are polymers of amino acids. Proteins fulfill a wide variety of roles for organisms. These include structural roles (such as protein in hair), a storage form of amino acids, and various transport functions. Also, many proteins function as organic catalysts called enzymes, speeding up the rates of virtually all the chemical reactions that take place in living cells.

Nucleotides and nucleic acids make up the fourth major class of biological compounds. The nucleotides ADP and ATP play prominent roles in the transfer of chemical energy in all living cells, a function they undoubtedly fulfilled when life was emerging.

According to the theory of chemical evolution, as bioorganic compounds appeared on the prebiotic earth, they spontaneously aggregated into tiny droplets. As these droplets developed sophisticated metabolism and the abilities to grow and reproduce, they evolved into the first living cells. These primitive cells were heterotrophic, subsisting on organic compounds formed by the various abiotic reactions. As the supply of these preformed raw materials diminished, this environmental circumstance favored the evolution of an autotrophic means of nutrition—photosynthesis. Using visible light as the energy source, the photosynthesizers could manufacture organic compounds (such as sugar) from the plentiful inorganic substances carbon dioxide and water. Photosynthesis itself led to the formation and accumulation of free oxygen, and this set the stage for further changes. One change was the emergence of respiration, a process that uses oxygen to break down organic compounds and release energy. Another was the formation of an ozone layer in the upper atmosphere, which acts as a protective shield screening out most of the sun's ultraviolet radiation before it reaches the earth's surface.

STUDY QUESTIONS

1. Abiotic polymerization reactions require the reaction units to be relatively concentrated. Briefly describe two concentrating mechanisms that may have operated on the prebiotic earth.

2. What are the major classes of carbohydrates, and what are the distinguishing characteristics of each class?

3. What characteristics distinguish saturated fats from unsaturated fats?

4. How is the structure of a phospholipid related to its water-solubility characteristics?

5. Why do we find such an enormous potential for variety in protein structure?

6. What important role is played by ATP in living systems?

7. Why might the original heterotrophic life forms on earth have vanished abruptly if the autotrophic cell type had never evolved?

8. What were two important consequences of the production of oxygen by early photosynthesizers?

SUGGESTED READINGS

Goldsmith, D., and T. Owen. *The Search for Life in the Universe.* Menlo Park, CA: Benjamin/Cummings, 1981. A very readable, thought-provoking book. Chapters 8 and 9 deal with the origin of life.

Hanawalt, P. C., ed. *Molecules to Living Cells: Readings from Scientific American.* San Francisco: W. H. Freeman, 1980. A compilation of interesting articles from *Scientific American.*

Lehninger, A. L. *Principles of Biochemistry.* New York: Worth, 1982. A thorough introduction to biochemistry for advanced students.

Sharon, N. "Carbohydrates." *Scientific American* 243 (1980):90–116. A good account of the structure and function of these compounds in living organisms.

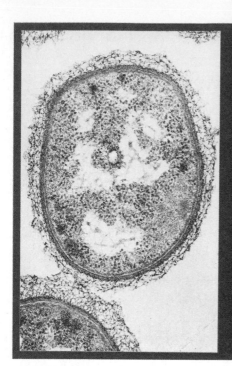

CHAPTER 3
The Organization of Cells

The first cells on earth crossed over the threshold of life when they gained the abilities to grow, metabolize, and reproduce. To carry out these processes, the early cells accumulated organic compounds from their liquid environment and used them for energy and raw materials. Things have not changed much at this level over the millenia. Modern cells, like their primitive ancestors, still grow, metabolize, and reproduce at the expense of a few select types of organic compounds they either manufacture or absorb from their surroundings. It makes no difference whether we talk about a lowly one-celled organism, a fern, or a seal; all of these creatures are composed of cells that carry out the same basic processes of life (Figure 3–1). It is not surprising to learn, then, that the cells of all organisms show more similarities than differences, both in terms of structure and function. The study of cells has reaffirmed a basic theme in biology: It is the unity of life, not its diversity, that is most striking.

THE DISCOVERY AND VISUALIZATION OF CELLS

Most early attempts to study life and its processes involved simple observation. This approach is still quite useful, for observation has always been the scientist's most powerful tool. But while the unaided human eye could discern leaves, skin, bones, and other large objects, it could not tell us how leaves, skin, and bones are constructed. This quest required the development of magnifying lenses to make objects appear larger.

Light Microscopy

In the seventeenth century the Dutch lens maker Anton van Leeuwenhoek invented the first single-lens light microscope (Figure 3–2a). Using this device with special back-lighting, he described many single-celled organisms, including some bacteria, which are among the smallest organisms on earth. Single-lens microscopes such as Leeuwenhoek's are called **simple microscopes. Compound microscopes,** the type found in laboratories today, have two lenses mounted in series (Figure 3–2c). The two lenses compound the magnification, so if one lens magnifies an object ten times (written 10×), and the other magnifies forty times (40×), then the final magnification will be 400× (10× times 40× = 400×).

At the same time that Leeuwenhoek was reporting his results with the simple microscope, Robert Hooke, an English physicist, was using a rudimentary compound microscope. In 1665 Hooke cut a thin

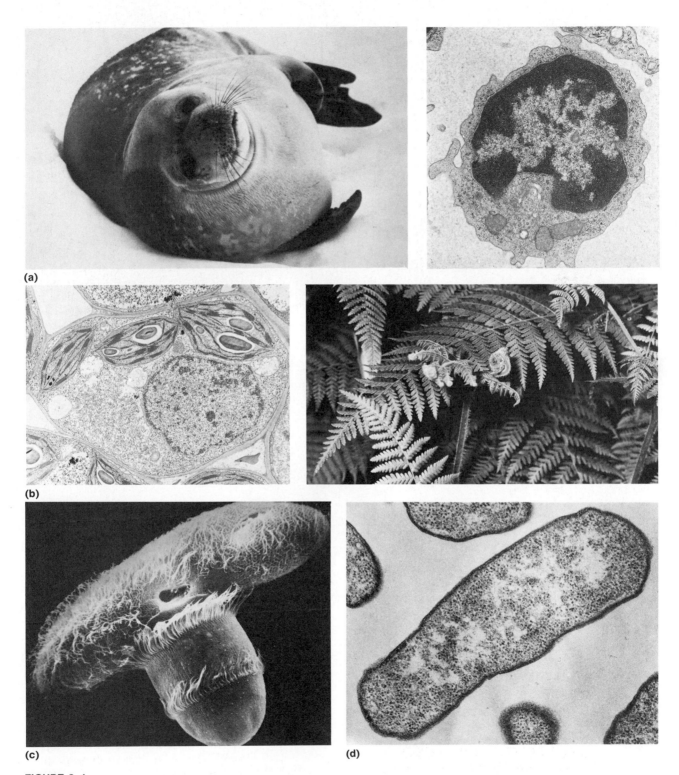

FIGURE 3–1
Cells. Despite the apparent diversity of living things, they are all built from the same fundamental unit—
the cell. (a) This Weddell seal, photographed in Antarctica, is made up of billions of individual cells.
(Magnification of cell: $13,100\times$.) (b) Plant cells fit together perfectly to form a delicate fern.
(Magnification of cell: $5500\times$.) (c) Two one-celled organisms locked in mortal combat. (Magnification:
$500\times$.) (d) A section through a bacterium, a simple one-celled creature. (Magnification: $38,000\times$.)

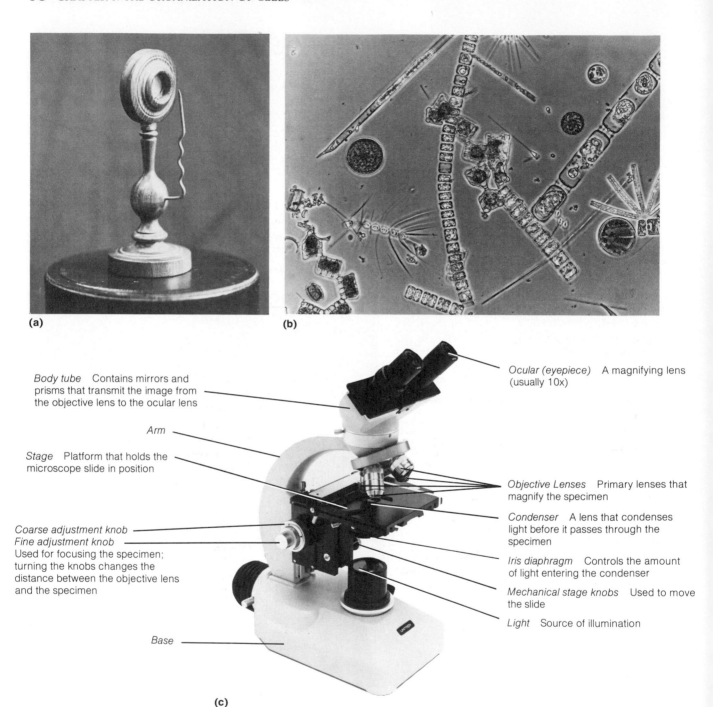

(a)

(b)

Body tube   Contains mirrors and prisms that transmit the image from the objective lens to the ocular lens

Arm

Stage   Platform that holds the microscope slide in position

Coarse adjustment knob
Fine adjustment knob
Used for focusing the specimen; turning the knobs changes the distance between the objective lens and the specimen

Base

Ocular (eyepiece)   A magnifying lens (usually 10x)

Objective Lenses   Primary lenses that magnify the specimen

Condenser   A lens that condenses light before it passes through the specimen

Iris diaphragm   Controls the amount of light entering the condenser

Mechanical stage knobs   Used to move the slide

Light   Source of illumination

(c)

FIGURE 3–2
Light Microscopy. (a) A simple microscope. This single-lens microscope is similar to the type used by Leeuwenhoek in the seventeenth century. The object to be examined is first attached to the tip of the wire in front of the lens. Then holding the microscope up to the observer's eye, the object is brought into focus by squeezing the wire. (b) Organisms in a drop of seawater, as seen through a compound light microscope. (Magnification: $150\times$.) (c) A compound microscope; note the two sets of lenses. The objective lens, near the object to be viewed, produces the primary enlargement of the object. This image is then further enlarged as it is transmitted through the ocular lens close to the observer's eye.

section of bark from a cork oak tree, placed it under his microscope, and reported that he could

> ... plainly perceive it to be all perforated and porous, much like a Honey-comb ... for the Interstitia, or walls (As I may so call them) or partitions of those pores were neer as thin in proportion to their pores, as those thin films of Wax in a Honey-comb (which enclose and constitute the Sexangular cells) are to theirs.

In his description of the cork bark, Hooke coined the terms "cell" and "wall" to refer to the basic unit of plant structure: the cell with its outer cell wall.

The Cell Theory

Over the next 170 years, many plant and animal structures were examined under microscopes and more and more observers noticed cells. Finally, in 1838 Matthias Schleiden, a German lawyer-turned-botanist, concluded that all plants were made of cells or derivatives of cells. A year later the German anatomist Theodor Schwann reached the same conclusion for animals. These two reports formed the foundation of the **cell theory:** *All plants and animals are composed of cells.*

The cell theory developed in an orderly, almost predictable fashion since it was based on a gradual accumulation of direct observations aided by improvements in microscope techniques. By contrast, the next great step forward in our understanding of cells was largely intuitive. In 1858 Rudolph Virchow, a German politician, anthropologist, and pathologist, claimed that *all cells are derived from preexisting cells,* an idea now known as the **theory of cell lineage.** Virchow's statement was clearly a display of great insight: the process by which cells give rise to new cells was not discovered until the 1880s, nearly 30 years later.

By the beginning of the twentieth century the cell theory was well entrenched in scientific thought. The fact that cells arise only from preexisting cells by the process of cell division (see Chapter 7) had been well-documented, and the larger components of cells had been described.

Electron Microscopy

Although the modern light microscope remains a valuable tool for the biologist, it has limitations. The best light microscopes have a **resolving power** of 0.2 micrometers (μm). (See Appendix A for metric units of length.) This means that any two points that are less than 0.2 μm apart will blur together and appear as a single point under a light microscope (0.2 μm is roughly 1/3000 the size of the period at the end of this sentence). By comparison, the human eye has a resolving power of 100 μm. This means that the light microscope is 500 times more effective at resolving detail than our eye. Unfortunately, the resolving power of light microscopes cannot be improved since the limiting factor is not the quality of the optical system, but rather the nature of light itself.

The 1930s marked the onset of a revolution in microscopy—the development of the **transmission electron microscope** (Figure 3–3). Using a beam of electrons instead of light, the electron microscope can resolve objects separated by a distance of only

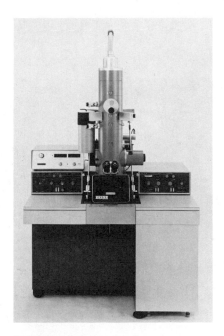

FIGURE 3–3
The Transmission Electron Microscope.
In an electron microscope, electrons are emitted from an electron gun located near the top of the instrument. Magnets are used to focus the electrons. The entire system operates under a high vacuum; otherwise air molecules would scatter the electrons, which would make precise focusing impossible. A modern electron microscope is typically about the size of a desk. The observer views the image on a fluorescent screen, and can make a permanent record of the image by exposing a photographic plate located beneath the fluorescent screen.

2 angstroms (Å) (0.0002 μm). This represents a resolving power 1000 times greater than that of the conventional light microscope, and 500,000 times greater than that of the naked eye.

One of the limitations of both light microscopes and transmission electron microscopes is that they reveal only a two-dimensional image of the object under observation. A relatively new instrument, the **scanning electron microscope,** overcomes these limitations by providing three-dimensional views of objects. A fine beam of electrons is passed back and forth over the surface of the specimen, resulting in an image with the quality of depth (Figure 3–4).

CELL SIZE

It should be clear from the preceding discussion that cells tend to be quite small—indeed, only a few unusu-

ally large cells are visible to the naked eye. The size of an organism thus depends mainly on the *number* of cells composing it. But why is a whale composed of billions of tiny microscopic cells instead of just one giant cell? Part of the answer has to do with the importance of surface area to a metabolically active cell. A cell that is busy building and breaking down organic compounds requires a relatively large surface area in relation to its volume. This is because the cell must exchange raw materials and wastes across its surface as rapidly as they are utilized or produced. By the laws of geometry, the smaller the cell is, the larger its surface area will be in relation to its volume, all other factors being equal (Figure 3–5). In a small cell, no part of its metabolically active volume is far removed from the cell's surface. But if our whale was just one large cell, nutrients entering across its surface would take days to move to the cell's center, which would literally starve the metabolic processes occurring there.

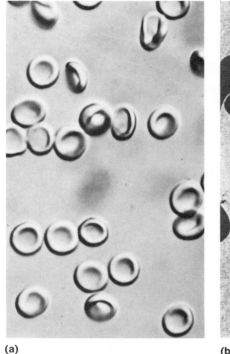
(a)

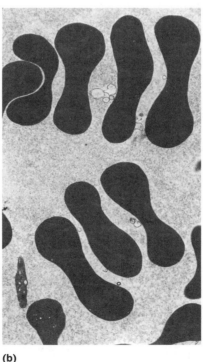
(b)

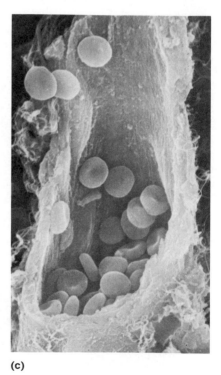
(c)

FIGURE 3–4

Comparison of Microscopic Images. (a) A light micrograph of human red blood cells (1875×).
(b) A transmission electron micrograph (TEM) showing a thin slice through several blood cells
(8200×). The cells in Figure 3–1a, b, and d are also TEMs. (c) A scanning electron micrograph
(SEM) of blood cells in a severed blood vessel, showing the surface of the cells (2000×.) Figure 3–1c
is also a SEM.

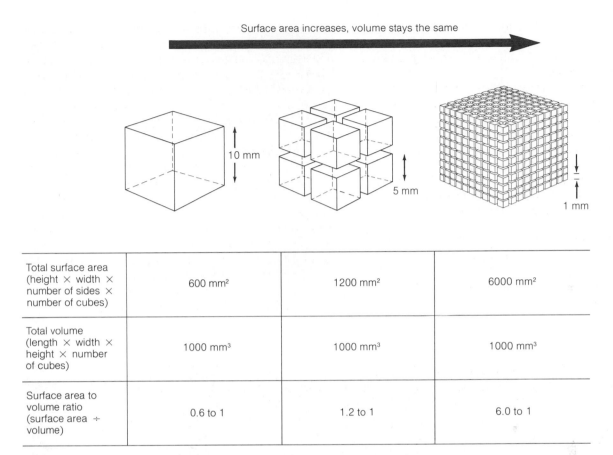

Surface area increases, volume stays the same

Total surface area (height × width × number of sides × number of cubes)	600 mm²	1200 mm²	6000 mm²
Total volume (length × width × height × number of cubes)	1000 mm³	1000 mm³	1000 mm³
Surface area to volume ratio (surface area ÷ volume)	0.6 to 1	1.2 to 1	6.0 to 1

FIGURE 3–5
Relationships of Surface Area to Volume. When a large cube is divided into smaller cubes, the total surface area increases, but the total volume stays the same. Thus, the surface-area-to-volume ratio increases as the linear dimensions of each cube decrease. The same general principle applies to cells, which is why large organisms are made up of many small cells, not one large cell.

Although most animals and plants are composed of comparatively small cells, there are some exceptions. These exceptions, however, do not invalidate our reasoning concerning ratios of surface area to volume. The only requirement is that the cell have an adequate surface area to meet the demands of its metabolically active volume. For example, the largest cells—that is, those with the smallest surface area in relation to volume—are the eggs of birds and reptiles. However, surface area is not critically important to these cells because most of an egg's volume is stored food, which is metabolically inactive. The metabolically active part of an egg is a small region adjacent to the cell's surface, which does have a high ratio of surface area to volume.

Probably the largest fully active cells are certain single-celled algae, such as *Valonia,* which can be up to 3 centimeters in diameter (Figure 3–6). In *Valonia* cells, most of the cell volume (up to 90%) is occupied by a central, chemically inactive sac called the vacuole (discussed later in this chapter). The metabolism of these cells takes place primarily in the thin layer of **protoplasm,** their living substance, sandwiched between the vacuole and the cell surface. Thus, these cells too have a relatively large surface area for their metabolically active volume.

The lower limit to cell size is set by a cell's need for space to house its metabolic machinery. Each cell must be large enough to contain all the internal membranes, enzymes, genetic material, and other structural components needed to carry out the processes of life. The smallest cells known are bacteria, which may be as tiny as 0.1 μm in diameter—about 1/100 the diameter

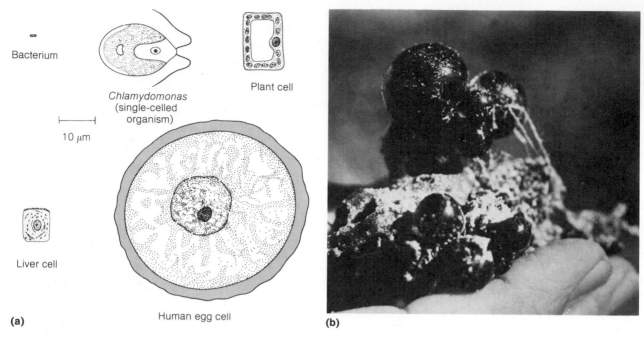

(a)

(b)

FIGURE 3–6
Cell Sizes. (a) Cells come in many different shapes and sizes, as illustrated by the various cells that
have been enlarged and drawn to scale here. (b) Cells of the green alga *Valonia* are among the largest
fully active cells. Because of their large size (up to 3 cm in diameter), *Valonia* cells are often used
by plant physiologists interested in the activities of a single cell. The diameter of a *Valonia* cell is over
600 times greater than that of the human egg cell shown in part (a).

of a typical animal cell. Over 3 million of these bacteria
could fit in the space of the period at the end of
this sentence.

CELL MEMBRANES

Despite differences in size and shape, all living cells
have certain structural features in common. One basic
structural component of all cells is their membranes.
The **plasma membrane,** also called the **cell membrane,** is the cell's outer envelope, separating it from
its external surroundings. Other internal membranes
divide the cell into tiny compartments. These compartments are extremely important for the normal functioning of a cell, for they allow different metabolic
activities to take place in different compartments so
that they do not interfere with each other.

Most biological membranes consist of a double
layer of phospholipids (see Figure 2–8) that is interspersed with proteins (Figures 3–7 and 4–1). Membranes are flexible sheets, like plastic wrap, but unlike
plastic, they are **selectively permeable** — their unique
structure allows certain substances to cross through

them rapidly, while other materials cross slowly or
not at all. Thus, the plasma membrane has the important function of regulating which substances enter or
leave a cell. We will explore the structure and function of membranes at greater length in Chapter 4.

Although all cells are bound by a plasma membrane, they may or may not have internal membrane-
bound compartments. On the basis of this primary
difference in internal organization, cells are classified
as either prokaryotic or eukaryotic. The prokaryotic
cell design is more primitive, so we will consider
it first.

PROKARYOTIC CELL STRUCTURE

According to the fossil record, the first living cells appeared on earth about 3.8 billion years ago. All fossils
from 1.4 to 3.8 billion years ago are similar in appearance to two major groups of organisms living today,
bacteria and cyanobacteria. Bacteria, cyanobacteria,
and their extinct relatives are grouped together because they share the prokaryotic cell design. **Prokary-**

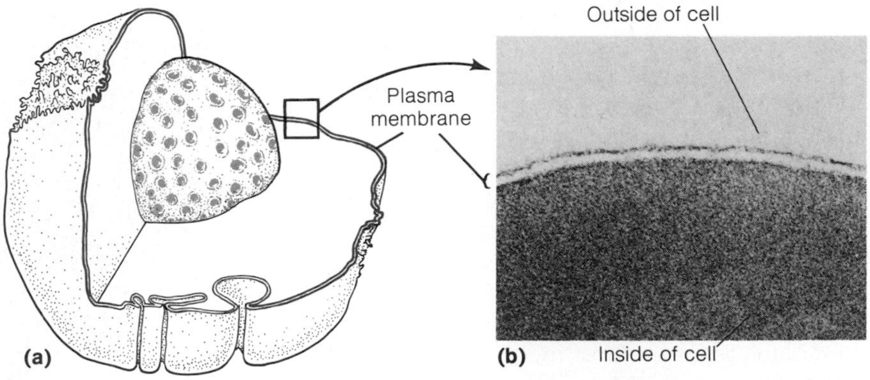

FIGURE 3–7
Plasma Membrane. (a) A cutaway view of a cell, showing where the photograph in part (b) came from. (b) This TEM of part of a red blood cell shows the plasma membrane as two dark bands, which together are about 90 Å in thickness; the bands are the double layer of phospholipids. The protein components of the membrane are not visible.

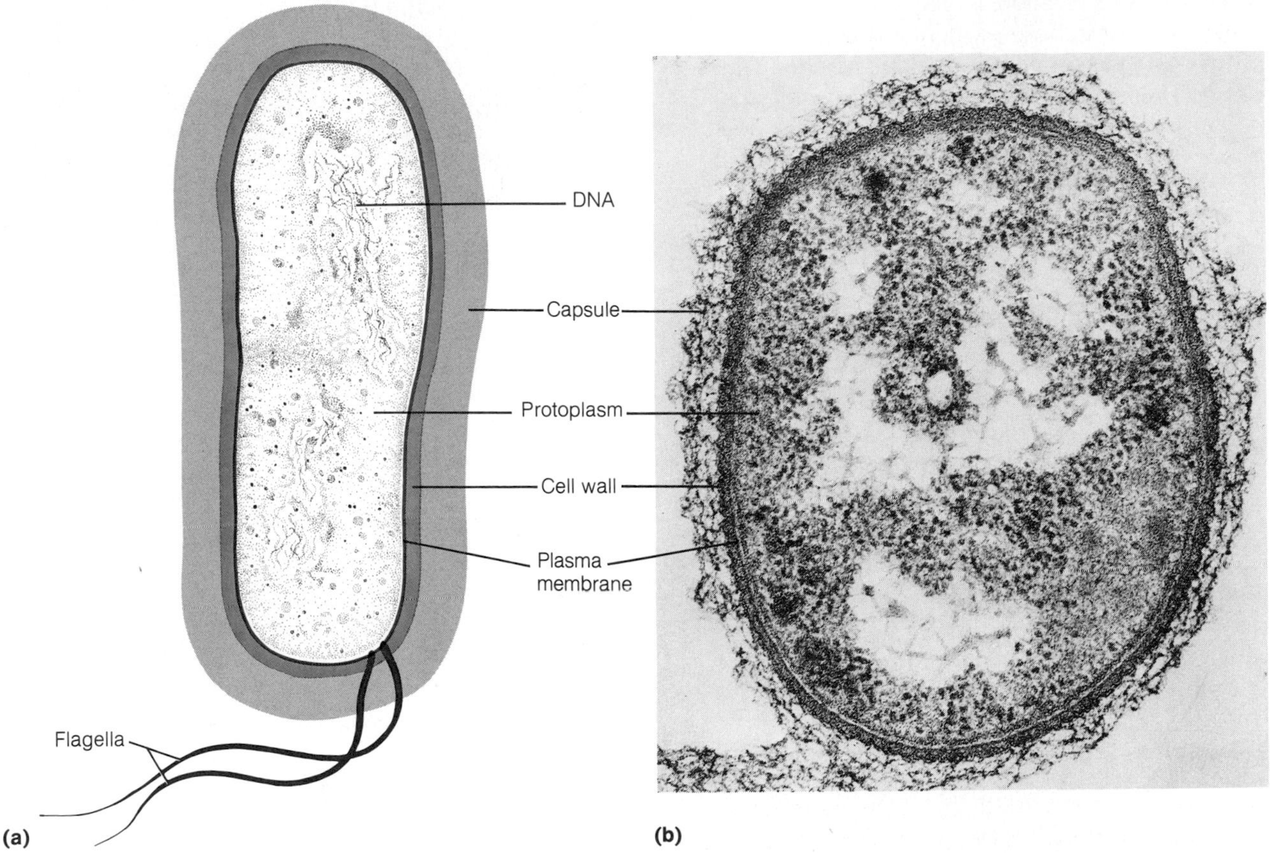

FIGURE 3–8
A Bacterial Cell. (a) Simplified diagram of a typical bacterial cell. (b) A TEM of a bacterial cell (74,400×).

otic, meaning "before nucleus," refers to a cell that lacks distinct membrane-bound compartments and, in particular, one that lacks a membrane-bound **nucleus.** The nucleus is the information storage center of the more advanced eukaryotic ("true nucleus") cells.

The characteristic features of modern prokaryotic cells include an outer capsule or sheath, a cell wall, and in some cases whiplike projections called flagella (Figure 3–8). If present, the **capsule** (as it is called in bacteria) or **sheath** (in cyanobacteria) is a slimy layer that covers the entire cell or, in some cases, covers a group of cells. The **cell wall** is a more rigid structure that lies just outside the plasma membrane. It provides the cell with some mechanical protection and gives it its characteristic shape. The **flagella,** if present, are shaped like long whips; their side-to-side motion propels the cell through its watery environment.

The inside of a prokaryotic cell is filled with undifferentiated protoplasm—a soupy liquid containing proteins, fats, and many other organic compounds. Within the protoplasm, the genetic material—DNA—appears as a thin and very convoluted fiber or fibers. No membrane separates the DNA from the rest of the protoplasm, as it does in eukaryotes. Some prokaryotes have extensive internal membrane systems where photosynthesis, respiration, or both take place (Figure 3–9). Even in these cells, however, many metabolic processes take place in the protoplasm, not within membrane-bound compartments. Such an "open" type of cellular organization is like a factory run in one large, open warehouse where management and assembly-line personnel work side by side. This may not seem like an efficient structural organization, but it has worked for the prokaryotes for at least 3.8 billion years.

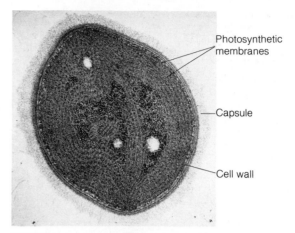

FIGURE 3–9

A Cyanobacterial Cell. This TEM of *Synechorystis,* a unicellular cyanobacterium, emphasizes the extensive array of photosynthetic membranes (20,000×).

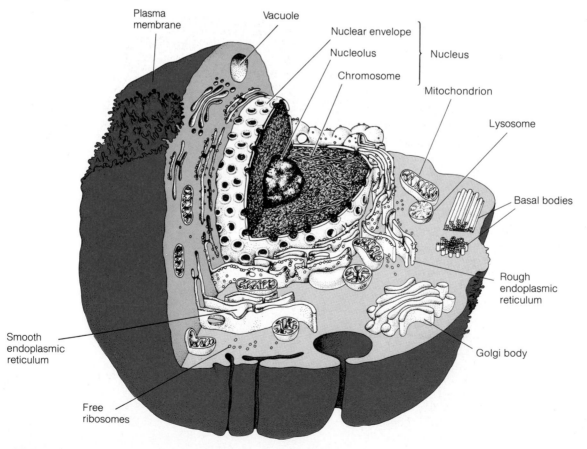

FIGURE 3–10

Generalized Animal Cell. This highly stylized diagram of an animal cell provides some perspective on the relative size and shape of its components.

EUKARYOTIC CELL STRUCTURE

The first eukaryotic cells appeared about 1.4 billion years ago, and today they are characteristic of all organisms except bacteria and cyanobacteria. **Eukaryotic cells** are distinguished from prokaryotic cells by the presence of membrane-bound bodies in their protoplasm: the nucleus and **organelles** such as mitochondria, lysosomes, and endoplasmic reticulum. These membrane-surrounded compartments permit a division of metabolic labors among specific regions of the cell, providing the cell with a means to control its chemical traffic. With all the chemical ingredients needed for a particular metabolic process concentrated

in one organelle, the process can proceed without interference from other activities in the protoplasm.

Generalized diagrams of an animal cell and a plant cell are shown in Figures 3–10 and 3–11, respectively. You should refer to these diagrams throughout the remainder of this chapter as we discuss the various parts of eukaryotic cells. In addition, Table 3–1 (located at the end of this chapter) summarizes the structure and function of these subcellular parts.

Nucleus: Control Center of the Cell

The protoplasm of eukaryotic cells can be subdivided into two regions: the nucleus and the cytoplasm. The nucleus is one of the most conspicuous structures within the cell. This large, usually spherical body is

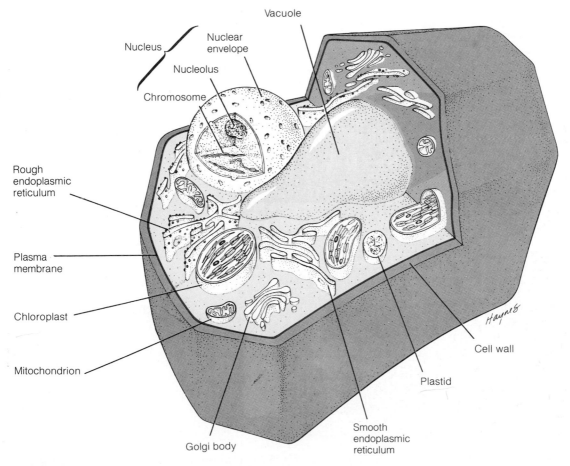

FIGURE 3–11
Generalized Plant Cell. In comparing this diagram with the one in Figure 3–10 of an animal cell, note the presence of plastids (a chloroplast is one type of plastid), a cell wall, and a large, central vacuole; also note the absence of lysosomes.

separated from the cytoplasm by a double membrane, the **nuclear envelope,** which has numerous pores extending through both membranes (Figure 3–12). It is through these pores that the nucleus and the cytoplasm "communicate," the pores serving as passageways for the movement of materials into and out of the nucleus. In some cells, the outer membrane of the nuclear envelope is joined to other intracellular membranes. Substances made in the nucleus can be channeled through these membranous corridors to other cellular compartments.

Inside the nuclear envelope are the **chromosomes,** structures composed of DNA and protein. The DNA component contains instructions for synthesizing the cell's proteins. These coded proteins in turn determine the activities that take place in cells (see Chapter 12 for details). This is why the nucleus is often called the control center of the cell.

Ovoid bodies known as **nucleoli** (singular **nucleolus**) are also found in the nuclei of most cells (Figure 3–12). Nucleoli are densely packed regions of certain chromosomes that are involved in the formation of ribosomes. (Ribosomes play an important role in the synthesis of proteins, as described in Chapter 12.)

Cytoplasm: Seat of Metabolic Activities

In eukaryotic cells, the protoplasm that lies outside the nucleus is called the **cytoplasm.** The cytoplasm consists of a watery liquid and the various organelles and other structures that are suspended in this liquid. Most of the cell's metabolic activities take place in the organelles. The major organelles in animal cells include mitochondria, endoplasmic reticulum, Golgi bodies, vacuoles, and lysosomes; other cytoplasmic structures include ribosomes, microfilaments, and microtubules. Plant cells have all of these organelles and structures except lysosomes; in addition, they have a cell wall (a rigid exterior covering that completely surrounds the plant cell) and plastids (see Table 3–1 and Figures 3–10 and 3–11). Plastids are organelles which include the all-important chloroplasts—the site of photosynthesis.

For the remainder of this chapter, we will examine the structure and functions of the various organelles and other cytoplasmic structures.

Mitochondria: Energy Factories

Mitochondria are called the energy factories of the cell because they are the sites of respiration. This important energy-converting process will be described fully in Chapter 5; briefly, it involves the breakdown of small organic compounds into carbon dioxide and water. During this process, chemical energy is released, which is used to form ATP. The importance of this energy transformation cannot be overemphasized, for ATP is the major form of chemical energy used by the cell to drive its energy-requiring processes. Con-

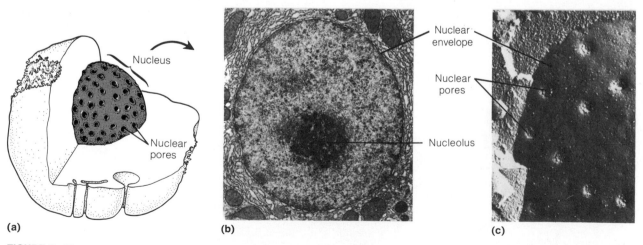

(a)   **(b)**   **(c)**

FIGURE 3–12

Nucleus. (a) Cutaway diagram of a cell showing its nucleus. (b) TEM of a thin section of a nucleus. The granular material in the nucleus are chromosomes in their fibrious, dispersed state (see Chapter 7) (8200×). (c) Freeze-fracture TEM showing a surface view of the nuclear envelope with its prominent pores (39,000×). Freeze-fracture entails first freezing a tissue, then mechanically fracturing the frozen tissue to expose a hidden surface.

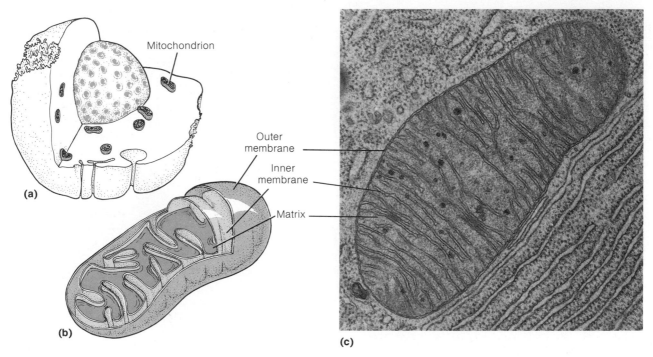

FIGURE 3–13
Mitochondrion. (a) Diagram showing relative size of mitochondria within a cell. (b) Cutaway diagram of a mitochondrion. (c) This TEM of a section through a mitochondrion from a guinea pig pancreas cell clearly shows the double membrane, the inner membrane of which is folded inwardly (38,400×).

sequently, mitochondria are most numerous in cells that expend a great deal of energy, such as muscle cells, where they may number in the thousands.

Mitochondria are small, rod-shaped organelles that range in length from 0.2 to 3.0 μm (about the size of many bacterial cells). They have two membranes: a smooth outer membrane and an inner membrane which folds inwardly (Figure 3–13). The space enclosed by the inner membrane, the **matrix,** contains most of the enzymes that catalyze the various steps of respiration.

Plastids: Storage and Photosynthesis

Plastid is a general term for a class of double-membraned oval bodies found in most living plant cells. They are considerably larger than mitochondria and are easily seen in the light microscope. The various types of plastids are distinguished by their contents and function. **Chromoplasts** are brightly colored organelles that store pigments. Carrot roots are orange, tomatoes are red, and banana skins are yellow because their cells contain an abundance of these plastids. Another type of plastid, the colorless **leucoplasts,**

function as storage depots for food molecules, such as starches or oils.

The most important plastid, not only for plants but for all life on earth, is the **chloroplast** (Figure 3–14). Chloroplasts give leaves their green color. They are the site of photosynthesis, the process by which green plants convert light energy to chemical energy. This important organelle and its functions are discussed in greater detail in Chapter 6.

Both chloroplasts and mitochondria contain small amounts of DNA, the same type of information-storing molecule that is found in the nucleus. This and other observations about mitochondria and chloroplasts provide the basis for some interesting theories on the evolutionary origin of these two organelles in eukaryotic cells (Essay 3–1).

Endoplasmic Reticulum: Protein Synthesis and Transport

Distributed throughout much of the cytoplasm is a system of membranes called the **endoplasmic reticulum (ER).** The ER appears in stacks or as a network of branched tubules (Figure 3–15). The form and

FIGURE 3–14
Leaf Cell with Chloroplasts. This TEM of a plant leaf cell shows five chloroplasts in cross section. The extensive internal membrane system (appearing as dark bands) in each chloroplast is where light is trapped and converted to chemical energy. The large central cavity of this cell is the vacuole (6,000×).

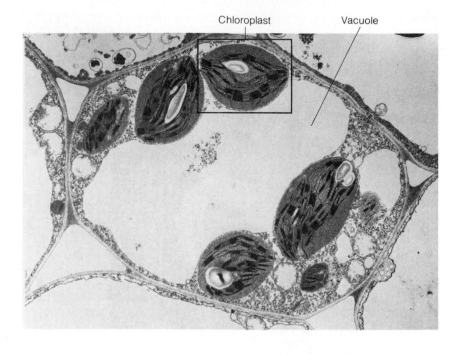

Chloroplast   Vacuole

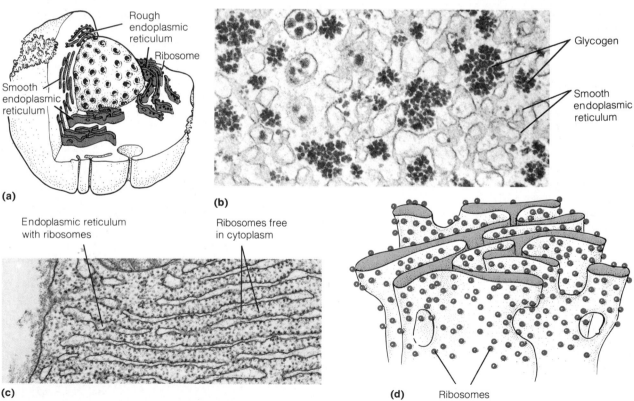

(a)

Rough endoplasmic reticulum

Ribosome

Smooth endoplasmic reticulum

(b)

Glycogen

Smooth endoplasmic reticulum

Endoplasmic reticulum with ribosomes

Ribosomes free in cytoplasm

(c)

(d)   Ribosomes

FIGURE 3–15
Endoplasmic Reticulum. (a) Diagram showing relative size of the endoplasmic reticulum within a cell. (b) Smooth ER shown in a TEM of a rat liver cell (50,000×). Note the glycogen, which is stored and broken down in smooth ER. (c) A TEM of rough ER in a cell of rat pancreas (24,000×). (d) Diagram of rough ER showing that the ribosomes are attached only to the outer surface of the ER membrane.

ESSAY 3–1

THE ORIGIN OF ORGANELLES

The fossil record suggests that the first life forms on earth were simple prokaryotic cells similar to present-day bacteria. Appearing some 3.8 billion years ago, these ancient prokaryotes preceded the eukaryotic cell type by over 2 billion years. The evolutionary progression from a cell without membrane-bound organelles (prokaryotic) to one whose cytoplasm is divided and subdivided into many compartments by enclosed membranes (eukaryotic) is in keeping with the general evolutionary trend from the simple to the complex. However, scientists are still unsure how this major advance in cellular architecture came about.

One theory holds that the first organelles evolved as a result of the **invagination** (folding in) of the plasma membrane of an ancestral prokaryote. In this model, part of the cell's outer membrane curved inward, forming a pocket within the cell. When the pocket pinched off, it formed a small compartment (see figure). Initially these cytoplasmic compartments contained the same components as the original plasma membrane; over time, however, other components became incorporated into these rudimentary organelles, giving them the ability to carry out a specific metabolic function. For example, we can imagine that enzymes and other molecules necessary for respiration became concentrated in certain compartments, giving rise to mitochondria. Once the new compartment could be replicated within the cytoplasm, it could be called an organelle.

Another theory popular among biologists suggests that some organelles—mitochondria and plastids in particular—arose by **endosymbiosis.** That is, large

prokaryotic cells were invaded by, and became hosts for, other smaller prokaryotes (see figure). For example, a small, photosynthetic, single-celled cyanobacterium (also known as a blue-green alga) might have been ingested by a larger, non-photosynthetic cell, which by some physiological malfunction failed to digest it. Instead the photosynthetic cell was retained within the cytoplasm of its host cell to their mutual benefit. The photosynthetic products of the alga could be used by the host cell, which in turn kept the alga alive by providing it with other nutrients and a place to live. In theory, the algal cell then divided and proliferated in step with the host cell divisions such that later generations maintained this endosymbiotic relationship. The cyanobacterial invaders, then, can be thought of as the forerunners of the chloroplasts in modern photosynthetic eukaryotes. In a similar fashion, mitochondria may have

originated from heterotrophic prokaryotes.

Evidence consistent with the endosymbiosis theory includes the following: (1) mitochondria and plastids are semi-independent, arising only from preexisting mitochondria and plastids, and controlled only partly by the nucleus; (2) protein-free DNA occurs in plastids and mitochondria, as it does in bacteria (chromosomal DNA in the nucleus is bound tightly to proteins); (3) the sizes of plastids and mitochondria are similar to those of bacterial and cyanobacterial cells.

Most biologists today favor the endosymbiosis model for the origin of mitochondria and plastids, but general agreement among biologists is certainly not in itself strong evidence. More studies are needed before we can eliminate one model in favor of the other, or possibly, replace both with an entirely new idea.

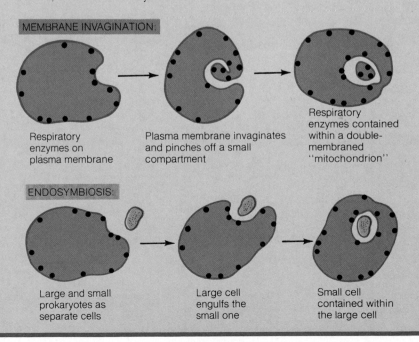

MEMBRANE INVAGINATION:

Respiratory enzymes on plasma membrane

Plasma membrane invaginates and pinches off a small compartment

Respiratory enzymes contained within a double-membraned "mitochondrion"

ENDOSYMBIOSIS:

Large and small prokaryotes as separate cells

Large cell engulfs the small one

Small cell contained within the large cell

function of the ER vary from one type of cell to another. In the cells of gonads (sexual organs of animals), the ER is where sex hormones are synthesized. In muscle cells, the ER is a system of tubules that plays a crucial role in muscle contraction. In liver cells, the ER has a number of functions, including the storage and breakdown of glycogen. The liver ER also plays a role in the detoxification (chemical breakdown) of harmful substances, such as alcohol and heroin. In alcoholics and drug addicts, the ER in liver cells proliferates into massive networks of membranes that break down these poisons.

The examples of endoplasmic reticulum we have described so far are all forms of **smooth ER** (Figure 3–15). Many cells also have **rough ER,** so called because tiny granules called **ribosomes** are found adhering to the outer surface of the ER membrane (Figure 3–15c,d). Ribosomes may also occur freely in the cytoplasm. In either case, they play an important role in protein synthesis (see Chapter 12). The proteins manufactured on the rough ER are generally destined for release from the cell (see next section); those produced on the free ribosomes function within the cell.

Golgi Bodies: Packaging and Secretion

When proteins are synthesized on the rough ER, they are often carried in small membranous spheres called ER vesicles to another system of membranes, the **Golgi bodies** (Figure 3–16). In electron micrographs, each Golgi body appears as a stack of flattened sacs and associated vesicles. Proteins arriving in the ER vesicles become incorporated into the Golgi sacs where they may be chemically modified. The processed proteins are packaged into tiny Golgi vesicles that pinch off from the sacs. The vesicles are then transported toward the plasma membrane. Upon reaching the plasma membrane, the Golgi vesicles fuse with it, causing the protein cargo to be released from the cell. The transport and release of proteins or other large molecules by means of vesicles is called **secretion.** In

FIGURE 3–16

Secretion. In this highly diagrammatic representation of the events leading to secretion, the process begins when proteins synthesized on the rough ER are packaged into ER vesicles. These vesicles then fuse with the Golgi body where the proteins are processed and packaged into Golgi vesicles. These vesicles fuse with the plasma membrane, thereby discharging their cargo outside the cell. In some instances, materials manufactured in the Golgi body (such as cell wall polysaccharides) may be exported without involvement of the ER.

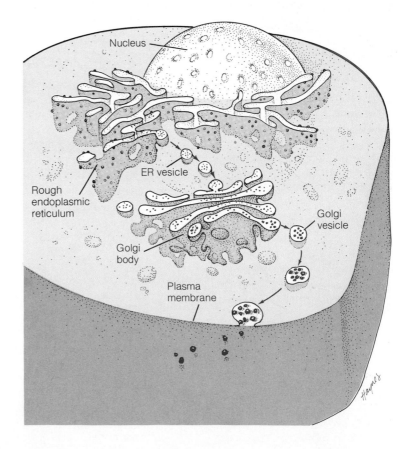

plants, the Golgi bodies are also involved in the synthesis and secretion of polysaccharides during cell wall formation.

Vacuoles: Storage and Water Balance

Vacuoles are spherical, single-membraned organelles that may originate from the Golgi bodies, or in some cases from the inward pinching of the plasma membrane. Their roles vary depending on the type of cell and how they were formed. In an amoeba, a single-celled pond organism that "eats" by engulfing small bits of food, a region of the plasma membrane closest to a food morsel envelops the food, then pinches off toward the cytoplasm to form a **food vacuole** (see Figure 4–11). The paramecium, another unicellular pond dweller, has a **contractile vacuole** which periodically expels excess water that has flowed into the cell (see Figure 13–13). Without this water-pumping vacuole, the tiny paramecium would swell and eventually burst.

When a typical plant cell grows, tiny vacuoles dispersed throughout its cytoplasm come together to form a single vacuole. As the cell continues to expand by taking in water, the vacuole also expands and ultimately may occupy up to 90% of the mature cell's volume (see Figure 3–11). This large, central vacuole serves as a storage site for salts, various waste products (plants have no means for excreting wastes into their surroundings), and oftentimes crystals and pigments. Beets and the outer layers of certain kinds of onions are purple because their cells store large amounts of pigment in their vacuoles. Many poisonous plants store toxic substances in their vacuoles, safely sequestered from their own metabolic machinery but available to cause problems for any animal that munches on their tissues. The vacuoles of plant cells also play a structural role. When filled with water, vacuoles provide hydrostatic pressure that helps support and give shape to cells. When the cells of a leaf, for example, lose water, their vacuoles shrink and the leaf collapses, or wilts (see Chapter 4).

Lysosomes: Intracellular Digestion

Lysosomes are a special kind of vacuole that contain digestive enzymes—agents that catalyze the breakdown of proteins, fats, polysaccharides, and other large molecules. Although absent in plant cells (they store digestive enzymes in their vacuoles), some animal cells may have up to several hundred lysosomes. Their function, as their contents imply, is digestion. For example, the food particles engulfed by an amoeba are broken down into sugars, amino acids, and other digestion products when lysosomes fuse with the food vacuoles. But lysosomal enzymes do not discriminate between foreign and domestic macromolecules. If the enzymes escape into the cell's cytoplasm, they will degrade proteins, membranes, and other components of the cell. Injured and old cells are often destroyed by their own lysosomes; in some cases whole tissues are digested by their lysosomes. Such is the fate of the tadpole's tail as this swimming larva metamorphoses into a frog. Similarly, human fingers and toes are formed during embryonic development by lysosomal digestion of the tissues that lie between these digits. Because of this potential to digest their own cellular environment, lysosomes have been nicknamed "suicide bags."

The Cytoskeleton: Intracellular Movement

When you study the photomicrographs of cells shown in this and other chapters, you might get the impression that cells are static entities with their nuclei and organelles seemingly fixed in place. This would be a false impression, for the contents of most living cells are in perpetual motion. When viewed carefully in the microscope, many organelles and smaller structures can be seen moving along in a steady stream. This movement, called **cytoplasmic streaming,** is powered by tiny contractile fibers called microfilaments. Microfilaments, together with the larger microtubules, compose the cell's **cytoskeleton** (Figure 3–17). The cytoskeleton gives a cell its shape. Changes in the position of the microfilaments and microtubules lead to changes in shape and various forms of movement, both inside and outside the cell body.

MICROFILAMENTS. Microfilaments are extremely thin fibers made of the proteins *actin* and *myosin* or closely related proteins. When supplied with energy from ATP, actin and myosin filaments can slide by one another, causing a localized contraction. Millions of these filaments occur in muscle cells, and their simultaneous contractions cause the shortening of individual muscle cells that result in the movement of an arm, leg, eyebrow, or whatever (see Chapter 26). In other types

FIGURE 3–17
The Cytoskeleton. The cell's cytoskeleton consists of microfilaments, microtubules, and the strands that connect them. Organelles such as endoplasmic reticulum and mitochondria are supported and moved by the cytoskeleton.

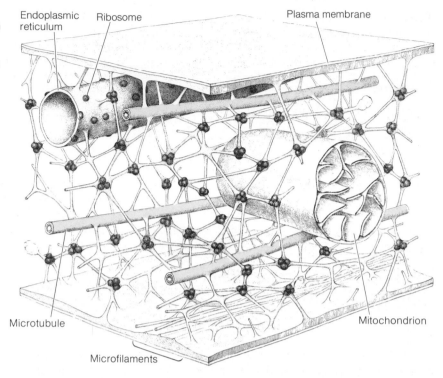

Endoplasmic reticulum

Ribosome

Plasma membrane

Microtubule

Microfilaments

Mitochondrion

FIGURE 3–18
Microfilaments. Microfilaments in this cell are apparent because of a special visualization technique used in combination with a compound microscope. Note how they radiate throughout the cytoplasm (1175×).

of cells, however, the microfilaments are much fewer in number, and often appear in sheets of parallel fibers that stretch in various directions across the cell (Figure 3–18). Through a sequence of contractions, the network of microfilaments power cytoplasmic streaming.

Microfilament action also appears to be involved in localized changes in cell shape, particularly the kind associated with amoeboid movement. Amoeboid movement involves cellular shape changes that enable an amoeba, for instance, to surround and eventually engulf a bit of food (see Figure 4–11). In a similar process, white blood cells in your body move about and engulf foreign material such as bacteria.

MICROTUBULES. As their name suggests, **microtubules** are tiny hollow cylinders, or tubes (Figure 3–19). They consist of protein subunits called *tubulins,* which have the remarkable capacity to self-assemble. When tubulin molecules attach at one end of the tube, the microtubule grows in length; when tubulin units dissociate from the other end, the tube shortens (Figure 3–19). The interaction of many microtubules in close array provides a structural framework, or scaffolding, for the cell that gives it shape and rigidity. Through assembly and disassembly of individual microtubules in the scaffolding network, the shape of the cell can change.

Microtubules also play a key role in the movements of chromosomes and various organelles. During cell division, microtubules assemble into a football-shaped structure, which is called the **spindle apparatus,** and guide the movement of chromosomes (see Chapter 7).

Cilia and Flagella: Cell Locomotion

So far we have seen that microfilaments and microtubules are involved in movements that occur within cells. But many types of cells can also move through their environments. Among eukaryotes, these motile cells propel themselves by means of threadlike extensions from their surfaces: **cilia** (singular, cilium) or **flagella** (singular, flagellum). Cilia are short hairlike projections that generally occur in large numbers on the surface of some cells (see Figure 3–1c). Ciliated cells move by the wavelike beating of their cilia, much like wheat stalks waving in a breeze. Some ciliated cells remain stationary and use their beating cilia to move materials past them. In humans, for example, cilia lining the respiratory tract sweep mucus-entrapped foreign particles (such as dust, pollen, and tobacco tar from cigarettes) upward into the throat, where they can be swallowed or spat out (see Chapter 22).

FIGURE 3–19
Microtubules. (a) TEM of microtubules shown in longitudinal section (67,000×). (b) Cross section of a single microtubule in which the individual subunits can be seen (830,000×). (c) This diagram of a microtubule shows how the dimeric units of tubulin, the protein subunit, fit together to form a hollow cylinder. The units assemble at one end of the microtubule and disassemble at the other end.

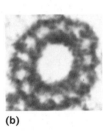

(b)

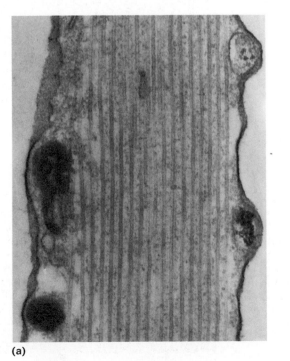

(a)

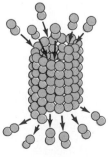

Assembly end

Disassembly end

(c)

Flagella are much longer than cilia and are usually present singly or as two flagella per cell. Their whiplike motion propels many kinds of unicellular organisms and the sperm cells of many plants and animals (including humans) through their liquid environment.

In terms of internal architecture, cilia and flagella are very similar. Both have nine pairs of microtubules that are arranged in a circular pattern, and these surround two central microtubules (Figure 3–20). This 9 + 2 arrangement is universal, occurring in the cilia of the lowly paramecium as well as the flagellum of a human sperm. Located on one side of each outer pair of microtubules are armlike extensions, or cross-bridges. It is believed that these tiny arms can creep along the neighboring microtubule, causing a localized bending in the cilium or flagellum. As this wave of bending migrates rapidly down its length, the cilium or flagellum moves in its characteristic beating or whip-like manner.

At the base of each cilium and flagellum is a tiny cylindrical structure called a **basal body** (see Figure 3–20). The basal body anchors the cilium or flagellum to the cell. A basal body, like a cilium and flagellum, has nine sets of microtubules arranged circularly near

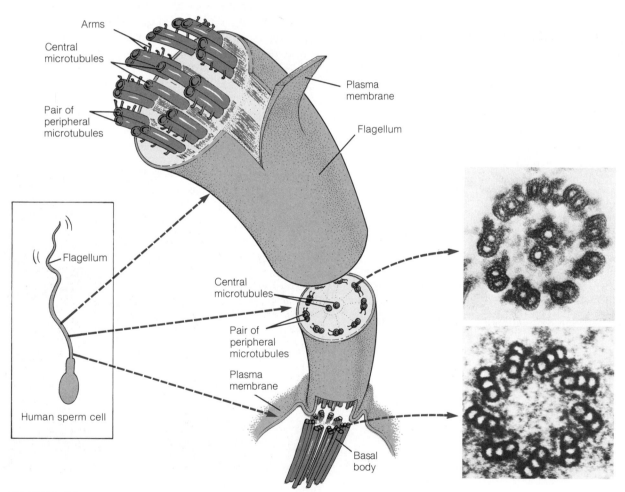

FIGURE 3–20
Flagellum with Basal Body. The flagellum of a sperm cell has an ordered arrangement of 20 micro-tubules: two central microtubules and nine pairs located near the periphery. Each peripheral pair has a series of cross-bridges, or arms, which can creep along the neighboring microtubule to cause bending of the flagellum. The flagellum is anchored to the cell by a basal body, which consists of nine triplets of microtubules arranged in a circle. All flagella and cilia, and their associated basal bodies, show these same microtubular arrangements.

its outer edge. Scientists think that when a cilium or flagellum is formed, the basal body microtubules serve as a starting point for the assembly of their peripheral microtubules.

Basal bodies also occur freely in the cytoplasm of animal cells. During cell division, two pairs of basal bodies known as **centrioles** apparently organize the assembly of microtubules into the spindle apparatus (see Chapter 7).

Cell Wall: Protection and Support

Cell walls are complexes of polymers that lie just outside their cell's plasma membrane. These rigid external coverings lend support and shape to cells, and protect them against infection and mechanical damage. Cell walls are found in all prokaryotes, fungi, and plants, but are absent in animals.

In higher plants the cell wall is deposited in distinct layers (Figure 3–21). The first and outermost layer, the **middle lamella,** is composed primarily of *pectin,* a gummy polysaccharide that cements the walls of adjoining cells. Pectins are extracted from fruits and used as a thickening agent in jams and jellies. Just interior to the middle lamella is the second layer, called the **primary cell wall.** This layer is composed of rigid cellulose fibers (see Figure 2–5) that are woven together with various complex polysaccharides (including pectin) and a small amount of protein. The arrangement of these polymers is such that the primary cell wall can be loosened and stretched to accommodate cell expansion. Finally, in certain plant cells specialized for mechanical support, a third layer called the **secondary cell wall** is laid down inside the primary wall. With its greater cellulose content (the secondary walls of cotton fibers are nearly pure cellulose) and thicker dimensions, secondary walls are very tough and rigid. Wood and bark are composed primarily of dead cells that have thick secondary walls, a feature that permits the stems of tall trees to support their tremendous weight. It was the thick secondary walls of cork that led to Hooke's discovery of cells, for all that this pioneering biologist could see in his single-lens microscope was a honeycomb network of secondary walls enclosing empty spaces.

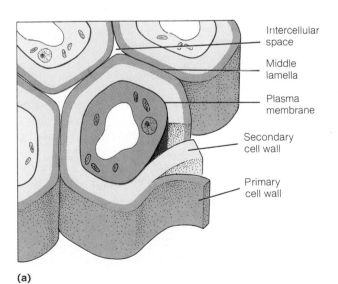

Intercellular space

Middle lamella

Plasma membrane

Secondary cell wall

Primary cell wall

(a)

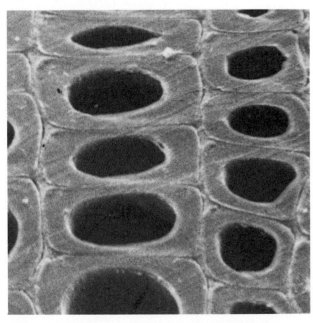

(b)

FIGURE 3–21
The Plant Cell Wall. (a) This diagram of several plant cells illustrates their multi-layered walls. The middle lamella is a very thin layer which occurs between adjacent cells and holds them together. The primary cell wall is characteristic of all plant cells, and a few cell types also produce a thick secondary wall. (b) SEM of water-conducting cells found within wood. At maturity these cells lack protoplasm; all that remains are cell walls.

TABLE 3–1
The Components of Eukaryotic Cells.

Component	Structure	Function
Plasma membrane	Phospholipid bilayer interspersed with proteins	Regulates flow of substances into and out of cell
Nucleus	Double-membraned sac containing chromosomes	Stores genetic information
Organelles		
Mitochondrion	Double-membraned sac	Site of respiration
Chloroplast*	Double-membraned sac with extensive inner membrane differentiation	Site of photosynthesis
Endoplasmic reticulum	Membranous tubules and flattened sacs	Various metabolic activities; when associated with ribosomes, is a site of protein synthesis; secretion
Golgi body	Membranous flattened sacs and vesicles	Packaging and secretion of macromolecules
Vacuole	Sac containing various metabolites, salts, or pigments	Storage; water balance
Lysosome†	Sac containing digestive enzymes	Intracellular digestion
Other structures		
Ribosome	Particle composed of ribonucleic acid (RNA) and protein	Protein synthesis
Microfilament	Thin fiber composed of actin, myosin, or other similar proteins	Contractions power cytoplasmic streaming and shape changes of cells
Microtubule	Hollow cylinder of protein subunits called tubulins	Skeletal framework of spindle apparatus, cilia, and flagella; involved in cell locomotion
Basal body	Tiny cylinder with circular array of microtubules	Anchors cilia and flagella to cell body; assembly of microtubular arrays in cilia, flagella, and the spindle fibers in animal cells
Cell wall*	Complex of various polysaccharides (pectin, cellulose, etc.) and protein	Mechanical support and protection

*Not present in animal cells
†Not present in plant cells

SUMMARY

With improvements in microscopes, scientists had discovered by the mid-nineteenth century that all organisms are composed of cells (the cell theory), and that all cells are derived from preexisting cells (the theory of cell lineage). The advent of the electron microscope in the 1930s permitted biologists to probe even deeper into the internal anatomy of cells, a technological breakthrough that has greatly aided our understanding of cell structure and function. Although electron microscopy may give the false impression that cells are static, we know they are actually active, streaming masses of protoplasm that constantly exchange energy and matter with their surroundings. This exchange takes place at the cell surface, so each living cell must have a large enough surface area in relation to its volume to meet the demands of its met-abolic processes. For this reason, most cells are quite small.

The two basic cell types found in nature are the prokaryotic cell (found in bacteria and cyanobacteria) and the eukaryotic cell (found in all other organisms, including plants and animals). Both prokaryotic and eukaryotic cells have a protoplasm bound externally by a plasma membrane, but the protoplasm of eukaryotes is organized into various kinds of membrane-bound compartments: the nucleus, which houses the information-storing chromosomes, and several different types of organelles. The various organelles are the structural sites of different cellular activities.

Cellular energy metabolism takes place in two types of double-membraned organelles: the mitochondria, where respiration occurs, and chloroplasts, a type of plastid found in plants where photosynthesis takes place. The chloroplasts of photosynthetic eukaryotes construct organic compounds at the expense of light energy, and the mitochondria of all eukaryotes break down these compounds to keep the cell supplied with ATP.

Another organelle, the endoplasmic reticulum, consists of flattened membranous sacs that conduct various types of biosynthetic activities. When ribosomes are associated with its outer surface, the endoplasmic reticulum carries out protein synthesis. In general, the proteins synthesized here are transported via tiny vesicles to Golgi bodies, where they are packaged into Golgi vesicles for export from the cell (a process called secretion).

Vacuoles in their most common form are storage sacs containing salts, pigments, or waste products. Some have specialized functions, such as the food vacuoles formed in an amoeba, and the water-expelling contractile vacuole of the paramecium. A very specialized type of vacuole, the lysosome, stores a cache of digestive enzymes that degrade large molecules. Found only in animal cells, the lysosomes can digest bits of food taken into the cell, or even the cell itself if its enzymes are released into the cytoplasm.

In addition to organelles, eukaryotic cells have proteinaceous filaments and tubules that function in various movement phenomena. Microfilaments are thin contractile fibers that power cytoplasmic streaming and cause changes in cell shape associated with amoeboid movement. Microtubules are hollow cylinders composed of protein subunits that have the ability to self-assemble. They make up the spindle apparatus that moves chromosomes during cell division, and are the motile units in cilia and flagella. Microtubules are also found in basal bodies, structures that anchor cilia and flagella and are involved in the assembly of the microtubular skeletons of cilia and flagella.

Cell walls are found in most major groups of organisms except animals. Lying outside the cell's protoplasm, cell walls provide mechanical support and protection.

STUDY QUESTIONS

1. What concepts of fundamental importance are stated by the cell theory and the theory of cell lineage?

2. What are the main differences between the workings of a light microscope and an electron microscope? What are the two types of electron microscopes, and how do they differ?

3. What factors influence the size of a cell?

4. What are some of the structural features common to all cells?

5. How are prokaryotic cells different from eukaryotic cells?

6. What are the two major regions of an animal cell? What are six of the major organelles in an animal cell?

7. What features of a typical plant cell make it different from a typical animal cell?

8. How do microfilaments and microtubules differ in structure? What types of movement are effected by each?

SUGGESTED READINGS

Albersheim, P. "The Walls of Growing Plant Cells." *Scientific American* 232(1975):80–95. This article is one of the first attempts to describe the detailed architecture of cell walls and the mechanism by which walls enlarge during plant cell growth.

Dustin, P. "Microtubules." *Scientific American* 243(1980): 66 76. The view that microtubules form an internal framework within cells—a cytoskeleton—is relatively new. This recent article nicely summarizes this viewpoint.

Hanawalt, P. C., ed. *Molecules to Living Cells: Readings from Scientific American.* San Francisco: W. H. Freeman, 1980. This collection of up-to-date articles with outstanding illustrations was written by leading scientists for the informed layperson.

Kessel, R. G., and R. H. Kardon. *Tissues and Organs: A Text-Atlas of Scanning Electron Microscopy.* San Francisco: W. H. Freeman, 1979. A beautiful book well worth browsing through in the library.

Ledbetter, M. C., and K. Porter. *Introduction to the Fine Structure of Plant Cells.* New York: Springer, 1970. An atlas of outstanding electron micrographs.

Porter, K. R., and J. B. Tucker. "The Ground Substance of the Living Cell." *Scientific American* 244(1981):56–67. An up-to-date description of the cell cytoskeleton with micrographs from a high-voltage electron microscope.

Raven, P. H. "A Multiple Origin for Plastids and Mitochondria." *Science* 169(1970):641–646. This article summarizes some of the evidence related to the origin of organelles. The speculative nature of the subject makes for interesting reading.

CHAPTER 4

How Materials Move Across Membranes

Understanding the structure of cells helps us see how some of the basic processes of life are carried out. In the preceding chapter we focused on the parts of the cell, emphasizing their structure and general function. Equally important, though, is how the cell as a whole interacts with its surroundings.

To sustain its living state, the organism must continuously exchange water, minerals, gases, and various organic molecules with its environment. It must take in nutrients and expel metabolic waste products. In multicellular organisms, these functions are carried out by each and every living cell in their bodies. Yet how do cells discriminate between nutrients and waste products, so that the former gain entry while the latter are excluded or expelled into the surroundings? This ability to discriminate is a major feature of the membranes of cells. Thus, it is appropriate to begin our study of how substances get into and out of cells by investigating the nature of this selective boundary, the membrane.

MEMBRANE STRUCTURE

Chemical analyses tell us that plasma membranes and membranes surrounding organelles are composed of approximately equal parts of phospholipid and protein. We are still not sure how these components are arranged, however. Over the past 50 years or so, many structural models of membrane architecture have been proposed, all supported to varying degrees by experimental data. Although none of these models is entirely satisfactory, biologists currently favor the **fluid mosaic model** of membrane structure (Figure 4–1). The fluid mosaic model suggests that the phospholipids are arranged in a bilayer, with their nonpolar fatty acid chains facing in toward each other (see Figure 2–8). Thus, the bilayer is like a "butter sandwich"—the fatty acid chains form the central "butter" layer, and the polar heads of the two phospholipid layers make up the "slices of bread." Proteins are scattered throughout this phospholipid bilayer, held in place by weak chemical forces. These attractive forces do not completely restrict protein movement, however, so the proteins tend to move around within the bilayer. The fluid mosaic model derives its name from the *mosaic* pattern of protein arrangement and the *fluidity* of both the protein and phospholipid components.

The functions of a particular membrane are determined largely by the cast of proteins in the membrane. Some of the proteins are enzymes that catalyze reactions; others are carriers that transport specific substances across the membrane; still others extend beyond the outer surface of the membrane and serve as "flags" for the recognition of other cells (see Figure 4–1). The specific proteins in a given membrane set

FIGURE 4–1
The Fluid Mosaic Model of Membrane Structure. In this model for the plasma membrane, the membrane consists of a phospholipid bilayer interspersed with various kinds of proteins. The cell surface recognition molecules are proteins from which short chains of sugars extend; they serve as "flags" that identify the cell.

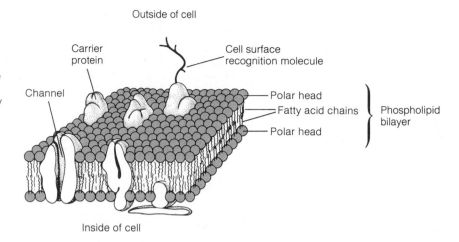

it apart functionally from other types of membranes in the same cell, and also from membranes in other types of cells.

As a test of its validity, the fluid mosaic model of membrane architecture had to account for **differential permeability,** a general property of all cellular membranes. Differential permeability means that different substances pass through membranes at different rates. How fast a given molecule passes through a membrane depends largely on the molecule's size and on the presence of charged groups. In general, small uncharged substances move through membranes easily; large and/or charged molecules move through sluggishly, if at all. There are some notable exceptions to this rule, however, because

there are different mechanisms by which substances get across membranes. We will examine four general ways that materials move into and out of cells: diffusion, facilitated diffusion, active transport, and vesicular transport.

DIFFUSION

According to a law of nature, all substances tend to move from an area in which they are concentrated to any adjoining region in which they are less concentrated, as long as there is no barrier preventing such movement. When they do so, we call this net movement **diffusion** (Figure 4–2). For instance, perfume

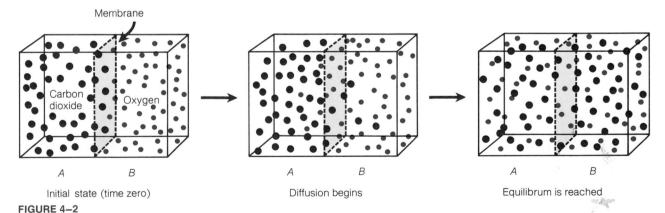

FIGURE 4–2
Diffusion. At time zero, chambers *A* and *B* are filled with carbon dioxide and oxygen, respectively. The chambers are separated by a membrane that permits the passage of both gases. As the gases move about by random motion, they tend to spread evenly throughout all the space available to them. This means that carbon dioxide molecules will diffuse toward chamber *B*, and oxygen molecules will diffuse toward chamber *A*. These diffusions continue until the concentrations of each substance are equal in both chambers.

on a cheek diffuses into the surrounding air, often to the attention of nearby noses. Sugar crystals at the bottom of a glass of iced tea spread out as they dissolve, a rather slow diffusion process that can be greatly increased by stirring. These diffusion processes occur because the random kinetic motions of molecules tend to disperse them evenly throughout the space available, particularly when they are gaseous or dissolved in a liquid such as water. If the molecules are not evenly dispersed, such that there is a difference in the concentrations of a substance between two adjacent areas, we say that a **concentration gradient** exists. Under these circumstances, there will occur a net movement of the substance toward the less concentrated area until the concentration gradient is eliminated. The greater the concentration gradient, the faster diffusion will take place. Once the gradient no longer exists, the system will be in **diffusion equilibrium** and diffusion stops. This is not to say that the molecules stop moving, however. All molecules are in constant motion, driven by their inherent kinetic energy. But at diffusion equilibrium, there is no *net* movement of molecules in one direction or another.

Let us illustrate the concept of diffusion with the amoeba, a single-celled organism that lives in lakes and ponds. As the amoeba respires, carbon dioxide, a by-product of respiration, builds up inside its cell. This creates a concentration gradient favoring the diffusion of carbon dioxide out of the amoeba toward the lesser concentration of carbon dioxide in the pond water (Figure 4–3). Since the pond is large relative to this tiny organism, the carbon dioxide concentration in the pond will always be lower than that in the amoeba. Hence, the amoeba can rid itself of this metabolic waste product by a rather simple process that requires neither energy nor special cellular structures. In fact, all organisms rely on simple diffusion to rid their cells of excess carbon dioxide.

For diffusion of a given substance to occur between a cell and its environment, not only is a concentration gradient necessary, but the plasma membrane must be permeable to that substance. Small, nonpolar molecules such as carbon dioxide and oxygen can move through the plasma membrane rather easily. As the molecular size of nonpolar substances increases, however, the rate of diffusion through the membrane decreases. Thus, for such substances, the plasma membrane acts as a sieve, much as a flour sifter discriminates between fine grains and clumps.

The rates at which charged substances (polar or ionic) (see Chapter 1) move into or out of cells is highly variable, depending on the cell type and the

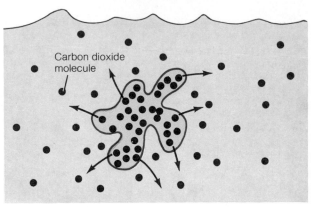

FIGURE 4–3
Diffusion from an Amoeba. As an amoeba respires, carbon dioxide builds up inside it. Since the concentration of carbon dioxide inside the amoeba is greater than it is in the pond water, this gas diffuses out of the amoeba into the pond.

substance. Most must be shuttled across the membrane by a special carrier protein. A notable exception, however, is water, a polar substance that moves through membranes quite readily, unaided by any special mechanism.

Osmosis

The net movement of water across a membrane is a special type of diffusion called **osmosis.** Exactly how water molecules negotiate a membrane is still debated among biologists. Because they are polar, water molecules are repelled by the nonpolar fatty acid layer of the membrane, just as they are repelled by household wax (a lipid). Yet water penetrates membranes with surprising ease. To account for this behavior, one popular theory holds that water moves through small channels, or pores, created by the protein components of the membrane (see Figure 4–1). In any case, the important point is that the unrestricted movement of water into and out of cells occurs and is crucial to organisms, for water is the medium of cellular processes (see Essay 4–1).

In order to understand why osmosis occurs, it will be helpful to introduce the concept of water potential. **Water potential** is a measure of the tendency for water to move from one region to another. If two regions have different water potentials, and there is no barrier to prevent water flow between them, then water will move from the region of high water potential to the region of low water potential. For example, water flows downstream because it has a higher water potential on the mountain than it does in the valley; water evaporates from your skin (high water potential) to the surrounding air (low water potential), also because of a water potential gradient that exists between these two regions. Similarly, the net flow of water through a membrane (osmosis) always occurs down a water potential gradient. In essence, then, water "runs downhill," or down the water potential gradient. But what are the factors that determine water potential?

Many forces influence water potential, including gravity, the concentration of water, and the presence of materials that bind or absorb water. The two most important factors that affect the water potential of *cells* are **pressure** and **solute concentration** (the number of dissolved substances per unit volume of

solution). Let us examine these latter two in more detail.

Pressure affects water potential directly: increases in pressure raise the water potential, and decreases in pressure lower the water potential. If all other factors are equal, water will flow from a cell having a higher water pressure to a cell with a lower water pressure (Figure 4–4.)

The relationship between solute concentration and water potential is a little more confusing because it is an inverse function: a solution with a high concentration of solutes will have a relatively low water potential. For example, consider what would happen if a cell with a relatively high solute concentration was placed into pure water (Figure 4–5). Assuming that the water pressures of the cell and pure water were the same, and that only water (not solutes) could pass

FIGURE 4–4
Effect of Pressure on Water Potential.
Pressure affects water potential directly. The cell at the left has a higher water potential than the cell on the right because of the outside pressure. Thus, water flows from left to right (see arrows).

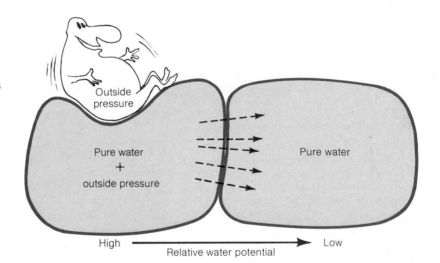

Outside pressure

Pure water
+
outside pressure

Pure water

High → Low
Relative water potential

FIGURE 4–5
Effect of Solutes on Water Potential.
The presence of solutes decreases water potential by an amount that depends on the concentration of solutes. (a) When a cell containing solutes is placed into pure water (no solutes), osmosis occurs toward the cell (arrows) down the water potential gradient. (b) If the same cell is dropped into a sugar solution that has a higher solute concentration (hence, lower water potential), water will move out of the cell (arrows).

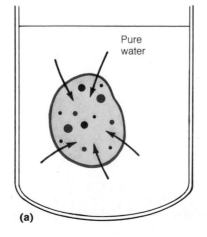

Pure water

(a)

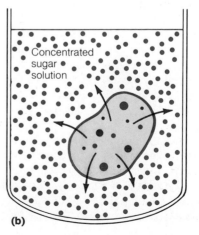

Concentrated sugar solution

(b)

ESSAY 4-1

WATER: THE INDISPENSABLE LIQUID

Water is the most abundant liquid on earth, covering nearly three-fourths of our planet's surface. It comprises 50–95% of the weight of every living organism. Many creatures live in it, others only drink it, but none can survive without water.

Because all life forms on earth are so dependent on water, many scientists believe that this relationship is universal. Thus, in our ongoing search for life on other planets, one of the first orders of business is to find evidence of water. The discovery of polar ice caps on Mars several years ago lent exciting support to the possibility that Martian creatures might exist. To date, however, no life has been found on Mars. In any event, it is difficult to imagine life without this amazing liquid—a liquid whose physical and chemical properties are so well suited to the processes of life.

Recall that a water molecule consists of two hydrogen atoms joined to an oxygen atom by polar covalent bonds (see Chapter 1). The unequal sharing of electrons gives the water molecule polarity—a negatively charged region (oxygen atom) and two slightly positive "corners" (hydrogen atoms). Because opposite charges attract each other, neighboring water molecules bind to one another by hydrogen bonds (the slightly positive hydrogen atom of one water molecule "sticks" to the slightly negative oxygen atom of another). Compared to other types of intermolecular attractive forces, the hydrogen bond is fairly strong. Because each water molecule can form up to four hydrogen bonds with other water molecules, liquid water is exceptionally cohesive. Cohesion is the force which holds molecules of the same substance together. You have probably noticed that when two drops of water touch, they immediately merge into a single drop. Water's cohesion also makes it possible to fill a glass with water such that the water goes slightly over the rim but does not spill. This strong cohesive nature of water underlies many of water's unique physical features that make it a very convenient liquid for life. Let us now explore some of these features.

The cohesive hydrogen bonds in water give this substance some special attributes with regard to heat. For one, water has an unusually high *specific heat:* it takes one calorie of heat to raise one gram of liquid water one degree Celsius. This means that water can absorb a relatively large amount of heat with little increase in temperature, because much of the added heat goes into disrupting hydrogen bonds rather than increasing molecular motion. Thus, water protects against large temperature changes in the bodies of creatures (remember, organisms are mostly water). Also, aquatic organisms are not subjected to large temperature changes in their watery environment.

Water also has a high *latent heat of vaporization:* to convert one gram of liquid water at room temperature to water vapor (steam) requires about 580 calories of heat. Again, much of this required heat goes into breaking hydrogen bonds. Many land-dwelling animals take advantage of this property of water to dissipate excess body heat. The evaporation of water from an animal's surface (for example, by perspiring or panting) takes with it a lot of heat, which cools the body.

Water not only sticks to other water molecules quite well, but it also *adheres* strongly to other polar or ionic substances. The spreading of water across a paper towel and its spontaneous movement up a thin glass tube are both examples of *capillary action.* Capillary action

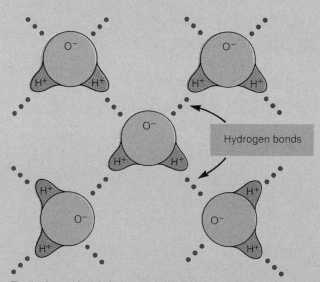

The hydrogen bonds between water molecules make liquid water very cohesive.

Hydrogen bonds

The evaporation of perspiration has a cooling effect.

When dry seeds (*left*) imbibe water, they greatly expand (*right*).

Ice floats because it is lighter than liquid water. This property has important consequences for life on earth.

is the result of water's high degree of adhesion. Water becomes available to the roots of plants when it moves by capillary action through tiny spaces between soil particles. Adhesion also accounts for the way wood and other "water-loving" materials soak up water (as wooden doors do during a rainstorm, making them difficult to open or close). This phenomenon, called *imbibition,* is of special importance to many plants. Dry seeds that imbibe water enlarge to several times their original size. Enough internal pressure is built up to break open the seed coat, thereby permitting germination.

Because of its attraction to other polar or ionic substances, water has the ability to surround and separate the molecules or ions of many other compounds; that is, it dissolves them. Water does this so well that it is commonly known as the *universal solvent.* When dissolved substances, or *solutes,* are dispersed in *solution* (solutes plus solvent), they are free

to move about, collide, and possibly react with one another. Thus, water's solvent properties are extremely important for life because most of the chemical reactions carried out by living cells take place in water. Furthermore, when compared to other common solvents, water is relatively unreactive. This permits a whole host of other reactions to occur in water without any chemical interference from the water.

A final unusual property of water is that it *expands when it freezes.* Most substances become more dense when they go from liquid to solid form, but water is densest at 4°C. Because the solid form of water is lighter (less dense) than the liquid form, ice floats. If water was like most other substances, the ice formed during the winter in temperate regions would sink to the bottom of lakes and ponds. After several years these freshwater bodies would be solid masses of ice during the winter, and during the summer only

the top foot or so would melt. Because ice floats, however, this scenario cannot occur. Only small ponds freeze over entirely during the winter. Ice formed on the surface of lakes during the cold season insulates the liquid water below from the cold air; and when spring arrives, it is the surface ice that is exposed directly to the warm air and the sun's radiation, causing it to melt quickly. As a result of this amazing characteristic of water, freshwater organisms are not frozen out of their habitat during winter.

After considering the many special properties of water, you can understand why it is indispensable for life on earth. And when you further consider the widespread occurrence of hydrogen and oxygen in the universe, it is probably safe to assume that other life systems scattered throughout the universe also depend on this exceptional liquid.

through the cell membrane, then water would move from the region of pure water (high water potential) to the solution in the cell (low water potential). Of course, the cell would expand as the water entered by osmosis. Now consider the consequences of placing a similar cell into a sugar solution that has a higher solute concentration (lower water potential) than the cell (Figure 4–5b). Osmosis would take place toward the outside sugar solution, and the cell would shrink as it lost water. In both cases, osmosis would continue until the water potential gradients were eliminated (until the water potential of the cell was equal to the water potential of the external fluid). At that point, the cell would be in **osmotic equilibrium** with its surroundings.

Most living cells are at or very close to osmotic equilibrium with their external environment, but this equilibrium state may be reached in different ways. Let us examine some typical cases for plant and animal cells.

HYPERTONIC PLANT CELLS. Plant cells are generally **hypertonic** (*hyper* = higher, *tonic* = tension) relative to their immediate environment; that is, the solute concentration inside the cell is higher than it is outside. (For most plants, "outside" refers to the cell walls and spaces between cells.) If this difference in solute concentrations was the only factor governing osmosis, then water would flow into the hypertonic plant cell, causing it to expand until the protoplasm became diluted to the same solute concentration that existed on the outside. For example, if the cell's internal solute concentration was twice that of its surroundings, osmosis would occur until the cell's volume doubled. In actuality, however, plant cells can expand only a small amount because they are bound by relatively rigid cell walls. Water "tries" to enter the cell but the cell wall resists expansion; the result is a buildup of internal hydrostatic pressure called **turgor** (Figure 4–6). Therefore, although the higher internal solute concentration tends to draw water in, the turgor pres-

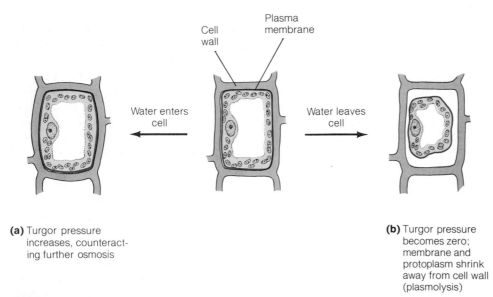

(a) Turgor pressure increases, counteracting further osmosis

(b) Turgor pressure becomes zero; membrane and protoplasm shrink away from cell wall (plasmolysis)

FIGURE 4–6
Turgor Pressure and Plasmolysis. (a) Because plant cells are usually hypertonic relative to the fluid surrounding them, water tends to flow into the cell. This tendency is countered by the rigid cell wall, which only permits a small amount of cell expansion. The result is the development of turgor, an internal hydrostatic pressure that tends to force water back out of the cell. (b) If plant cells are exposed to solutions that have a higher solute concentration than the cells (such as when you cover lettuce with a very salty salad dressing), they will lose water by osmosis. Loss of cellular water may also occur by simple evaporation when cells are exposed to dry air. When water moves out of a plant cell, the vacuole and protoplasm shrink in volume, causing the plasma membrane to pull away from the rigid cell wall. This shrinkage is called *plasmolysis.* At the tissue or organ level, plasmolysis leads to severe wilting.

sure acts against this force by pushing water out. It is as if you were to try to blow up a balloon inside a wooden box. As you blew up the balloon, you would soon reach a point where the rigid walls of the box exerted so much back pressure that you could not get any more air into the balloon: your cheeks would get bigger and your face would turn red, but the balloon would not expand. All your effort would only increase the pressure inside the balloon. In plant cells, the greater the solute concentration difference is across the plasma membrane, the higher the turgor pressure inside the cell will be.

Turgor pressure in plant cells can be considerable, often exceeding 80 pounds per square inch (compare that to a typical automobile tire pressure of 30 pounds per square inch). Such high internal pressure is crucial to a plant cell's normal functioning, since losses in turgor (called wilting) interfere with growth, photosynthesis, and the absorption of nutrients. Turgor also gives plants much of their strength and shape, which you can appreciate if you have ever forgotten to water your houseplants.

HYPERTONIC AND ISOTONIC ANIMAL CELLS. Although they do not have cell walls, hypertonic animal cells have other ways of stopping excessive water gains during inward osmosis. The problem is perhaps most acute for single-celled pond dwellers such as the amoeba and paramecium, which are both hypertonic relative to their freshwater environment. Water continually enters these organisms by osmosis. Without cell walls to prevent them from expanding, one might imagine that these cells would eventually burst. But these organisms do not burst because they have contractile vacuoles that continually pump water out of the cell, keeping pace with the inward osmotic flow (see Figure 13–13b).

The cells of higher animals are usually **isotonic** (*iso* = same) with their surrounding blood and lymph fluid; that is, the solute concentration within their cells is the same as the solute concentration of the fluid surrounding the cells. However, since the body fluids of animals are subject to changes in solute concentration (for example, when the animal eats or drinks), animals have ways to keep the concentration of solutes in their blood and lymph fluids nearly constant. This role is performed by organs such as the kidneys, which remove excess water or solutes from the blood.

Under experimental conditions, animal cells can be removed from their isotonic surroundings and transferred to solutions that have either a lower or higher solute concentration; the results show why relatively constant isotonic environments are so important. When human red blood cells are placed in distilled water where they are hypertonic, water quickly enters the cells, causing them to swell and eventually burst. On the other hand, when red blood cells are transferred to a solution that has a significantly higher solute concentration than the cells, the cells are **hypotonic** (*hypo* = lower). The cells lose water by osmosis and shrink in size.

Within certain limits, animal cells can adapt to a range of external solute concentrations, adjusting their own internal concentrations to reach osmotic equilibrium and function normally. However, without a cell wall to restrict its expansion, the animal plasma membrane is too fragile to expand indefinitely as water flows in. Neither plant nor animal cells can survive significant contractions in volume brought about by external fluids with very low water potentials (high solute concentrations).

Facilitated Diffusion

Water is a polar substance whose movement across cellular membranes is not typical of charged substances. Other polar (or ionic) substances used by cells, all of which are larger than the water molecule, also cannot penetrate the "butter" layer of membranes, and they would tend to stick to the charged regions of exposed membrane proteins. Thus, we would predict that the movement of such substances across membranes would be sluggish at best. For many biologically important molecules and ions, however, their passage is not only rapid, but often many times faster than would be predicted by a simple diffusion process. The uptake of glucose into the cells lining the intestines of animals is a case in point. When cells isolated from the small intestine of an animal are exposed to glucose, they take up this sugar much faster than would be expected by a simple diffusion process. As the investigator increases the external glucose concentration, the rate of uptake increases steadily. Ultimately, however, there is a point beyond which increasing the sugar concentration does not affect how fast it enters the cells (Figure 4–7). In other words, the glucose absorption mechanism displays *saturability*. This fact taken together with the greatly enhanced rate of glucose uptake strongly suggests the involvement of membrane **carrier proteins** in glucose transport. Such carriers facilitate the transport of glucose across the intestinal cell's plasma membrane, and being finite in number, become saturated with glucose at

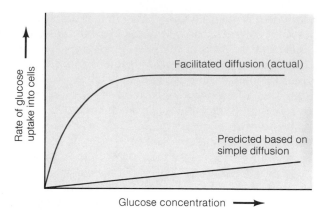

FIGURE 4–7
Comparison of Facilitated and Simple Diffusion Processes. The uptake of glucose into the intestinal cells of many animals occurs by facilitated diffusion, a carrier-mediated process that occurs much faster than simple diffusion.

in such cases merely facilitate the movement of their substances down their concentration gradients, facilitated diffusion does not require an input of energy.

Facilitated diffusion is advantageous to organisms because it greatly increases the rates at which certain molecules are absorbed or expelled by cells. In our example, glucose is absorbed rapidly by the intestinal cells, from where it enters the bloodstream to provide nourishment for the rest of the animal's body cells. If intestinal cells had to rely solely on simple diffusion, the slow uptake of glucose would not only slow down the metabolism of the entire organism, but most of the glucose would pass through the digestive system unabsorbed.

As we said, biologists believe that facilitated diffusion is operated by carrier proteins in the membrane. The current model to explain how carrier proteins work suggests that part of the carrier molecule extends from the surface of the membrane and binds to the specific type of molecule it transports (Figure 4–8). As a result of this binding, the shape of the carrier changes in such a way that the substance is shuttled across the membrane and released on the other side. The carrier then returns to its original shape, ready to transport another molecule. Carrier proteins involved in facilitated diffusion are something like revolv-

high external concentrations of this sugar. The process whereby glucose or some other substance is carried across a membrane down its concentration gradient, and at a rate faster than predicted by simple diffusion, is called **facilitated diffusion.** As the carrier proteins

FIGURE 4–8
Model of Facilitated Diffusion. The carrier protein involved in the facilitated diffusion of a substance binds to the substance on one side of the membrane, which causes a change in the shape of the carrier. The shape change creates a channel through which the substance can move to the other side of the membrane.

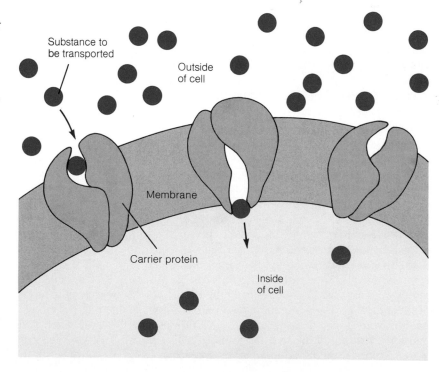

ing doors. Although they are capable of transport in either direction across the membrane, the actual net transport always occurs down the concentration gradient for the substance being transported. This is so because the frequency of binding that initiates transport will be greater on the side of the membrane where the substance is higher in concentration.

ACTIVE TRANSPORT

Thus far, we have considered only how substances move across a membrane down their concentration gradients, and how water moves by osmosis down its water potential gradient. These are passive processes that follow the laws of diffusion; no energy is needed to make them occur. In many instances, however, cells accumulate substances *against* their concentration gradients. In particular, a number of mineral ions (such as potassium and chloride ions) and small organic compounds (such as amino acids), which are essential to normal cell functioning, are at higher concentrations inside most cells than outside. Furthermore, many cells actively transport certain substances out of the cell. For example, plant root cells actively transport sodium ions outward, thereby maintaining a lower concentration of this ion inside the root than exists in the soil (Figure 4–9).

The energy-dependent movement of a substance across a membrane against its concentration gradient is called **active transport.** Because the direction of transport is opposite to the direction of diffusion, it's an uphill battle, so active transport must be driven by metabolic energy, usually provided by the splitting of ATP into ADP and inorganic phosphate (P_i). This process is so effective that many cells can establish and maintain very large concentration gradients—the concentration of a substance inside a cell may be 100 times greater or less than the concentration outside the cell.

FIGURE 4–9
Simple Diffusion versus Active Transport. (a) Since the cell membranes of plant root cells are permeable to sodium ions, we would expect the concentration of sodium ions to be the same in the soil solution and in the cells if simple diffusion was operating. (b) Since the concentration of sodium ions in the cells is many times lower than the concentration in the soil, the cells must be pumping sodium ions out against a concentration gradient. This process is called active transport.

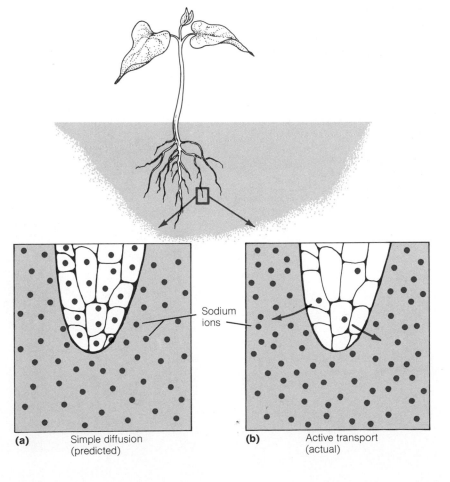

Sodium ions

(a) Simple diffusion (predicted)

(b) Active transport (actual)

As in facilitated diffusion, active transport involves carrier proteins which bind to specific substances and transport them across the membrane. The active transport carriers, however, are not like revolving doors. They transport their molecules in one direction only, and each shuttle requires the energy of ATP. Associated with each carrier is an ATP-splitting enzyme (called ATPase), which transfers the energy of ATP splitting to the carrier (Figure 4–10).

VESICULAR TRANSPORT

Some cells produce molecules that are transported out of them en masse (that is, in quantities much larger than one molecule at a time). Often these molecules are much too large to move through the plasma membrane directly, such as the digestive enzymes produced in animals' pancreatic cells that are secreted into the

small intestine. In such cases, the proteins are synthesized on the rough endoplasmic reticulum and are encapsulated in tiny membranous vesicles. Then they move through the Golgi bodies for further processing. Golgi vesicles containing the proteins then fuse with the plasma membrane to release their cargo outside the cell (see Figure 3–16). This fusion of the vesicle with the plasma membrane is called **exocytosis,** or secretion.

Endocytosis, a similar but reverse process, allows large amounts of substances or large particles that are outside the cell to get into the cell. In this case, a region of the plasma membrane extends around the material to be taken in (Figure 4–11). When the two extended arms of the plasma membrane meet, they fuse and pinch off a vesicle inwardly, leaving the membrane-encapsulated cargo in the cytoplasm. The membrane around the vesicle is digested (usually by enzymes in lysosomes) and the transported material

FIGURE 4–10
Mechanical Model of Active Transport.
This imaginative model shows ATPase, an enzyme that catalyzes the splitting of ATP, transferring the energy generated by that reaction to a carrier protein in the membrane. The energized carrier then transports a particular substance across the membrane against its concentration gradient.

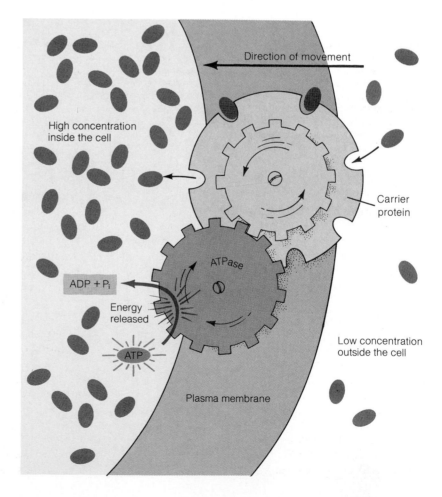

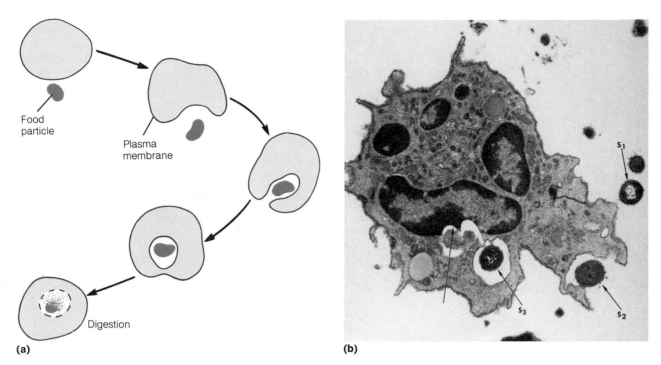

(a)

Food
particle

Plasma
membrane

Digestion

(b)

FIGURE 4–11
Endocytosis. (a) In a special form of endocytosis called phagocytosis, a food particle bumps into
the plasma membrane, which triggers the invagination of the membrane at that point. As the membrane
completely encircles the particle, a vesicle containing the particle pinches off, and the plasma
membrane reseals itself. Eventually, digestive enzymes in the cytoplasm break down the vesicle
membrane and the food particle. (b) The sequential steps of phagocytosis are clearly represented
in this electron micrograph of a white blood cell engulfing bacteria ($23,500\times$). S_1 is a free bacterium,
S_2 is a bacterium that has been partially engulfed, and S_3 is a bacterium completely engulfed. (The
unlabeled arrow points to an area of further digestion.)

is thus liberated inside the cell. When the material
being transported inward is an undissolved particle
(such as a food granule), the cell is performing a special
type of endocytosis called **phagocytosis.** Amoebas
obtain food by phagocytosis, and white blood cells
engulf and destroy foreign matter, such as bacteria,
and damaged body cells by the same process.

Recently, biologists have been taking advantage of
the plasma membrane's ability to fuse with itself or
other membranes. They are using membrane fusion
techniques to get materials into cells that normally
would not pass through the plasma membrane. Artificial membrane vesicles are prepared from a mixture
of phospholipids, other membrane components, and
the substance to be delivered to the cells. During
vigorous agitation, the lipids self-assemble into tiny
vesicles called **liposomes,** with the substance to be
delivered trapped inside them. The liposomes can
then be added to a suspension of cells (animals cells,

or plant cells with their cell walls removed), where
they fuse with the cells' plasma membranes, releasing
their cargo into the cells.

With the liposome technique at hand, scientists
are now searching for possible medical applications.
One possibility often discussed is to construct liposomes that would fuse only with a selected cell type,
such as cancerous cells. The liposomes would be
loaded with a cell poison that could neither leak out
of the liposome nor enter a cell in its free form. These
liposomes would be injected into a patient with cancer. The liposomes would recognize and fuse specifically with cancer cells, releasing the cell poison into
the protoplasm where it would exert its deadly action
(Figure 4–12).

This procedure is actually a little trickier than it
may sound. Presently, the major obstacle standing in
the way of developing anti-cancer liposomes is the
matter of selective fusion. Obviously, any liposomal

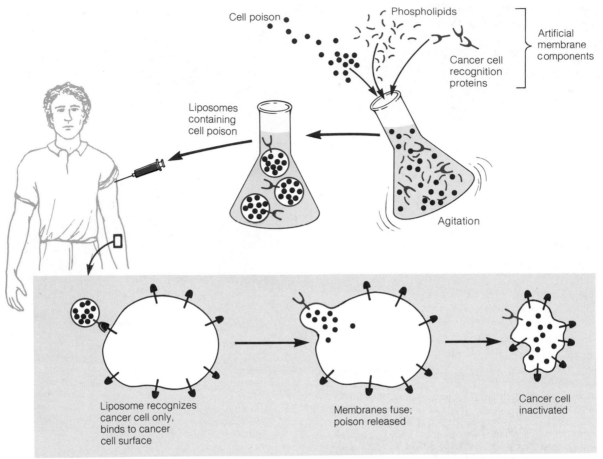

FIGURE 4–12
Cancer-Hunting Liposomes. In the future, liposomes containing a cargo of poison may be created
that bind specifically to cancer cells. As the liposomes fuse with the cancer cell membrane the poison
enters the cancer cell and kills it.

"cancer bullet" would be of value only if it fused specifically with cancer cells and not to any normal body cells. In that regard, scientists are encouraged by the recent discovery that the plasma membranes of certain cancer cells contain some unique cell surface recognition molecules (see Figure 4–1). Cell biologists think of these special molecules as "flags" that essentially communicate to other cells: "Hey, I'm a cancer cell." The liposome researchers would like to take advantage of this by producing liposomes which contain molecules on their outer surface that recognize and bind to the cancer cell surface molecules. Hopefully, this binding would be necessary to trigger membrane fusion so that the liposomes would fuse only with cancer cells, not normal cells. We would

then have cancer-hunting liposomes that could be sent out on "search and destroy missions." Cancer treatment would be revolutionized!

SUMMARY

The movement of water, nutrients, waste products, and other substances between the cell and its environment is regulated largely by its differentially permeable plasma membrane. This quality of selective permeability, crucial to maintaining a proper chemical balance in cells, is a result of the chemical composition and architecture of membranes. The currently favored view of membrane structure—the fluid mosaic model—holds that membranes consist of a fluid phospholipid bilayer in which various types of proteins are embedded. Some of these proteins are enzymes; others perform a transport function; still others serve as recognition sites for cell-to-cell communication.

The movement of many substances across a membrane takes place by simple diffusion. The rate at which a particular substance diffuses depends on its concentration gradient (the difference in its concentrations between two areas). When diffusion is occurring across a membrane, membrane permeability is also a factor that affects rate. In general, membranes are most permeable to small, nonpolar molecules such as oxygen and carbon dioxide. Larger molecules and electrically charged substances (polar compounds and ions) diffuse across membranes slowly, if at all.

Although water is a slightly polar compound, it moves across membranes quite readily, probably through special channels in the membrane proteins. The net movement of water across a membrane is a special type of diffusion called osmosis. The direction of osmosis is determined by the water potential gradient: water always moves from an area of high water potential to an area of low water potential. Water potential is affected directly by pressure. The presence of solutes decreases water potential (the higher the solute concentration is, the lower the water potential will be).

Plant cells are usually hypertonic (higher in solute concentration) relative to their surroundings. This difference in solute concentration tends to cause water to flow into the plant cell, but the rigid cell wall resists increases in cell volume, and hence resists water entry. The resulting water pressure inside the cell is called turgor.

Most animal cells are isotonic (equal in solute concentration) with the fluid that surrounds them. This balance prevents excessive cell swelling or shrinking that might otherwise occur by osmosis in hypertonic or hypotonic animal cells, respectively. Those animal cells that are hypertonic relative to their environment (such as the single-celled paramecium) have contractile vacuoles to expel the excess water that perpetually enters them.

In certain cases, the diffusion of a substance across a membrane is assisted by carrier proteins. This carrier-mediated transport of a substance down its concentration gradient is called facilitated diffusion.

All living cells have the capacity to transport certain substances (such as mineral ions, sugars, and amino acids) across their membranes against a concentration gradient. This process, also carrier-mediated, is called active transport, and it requires energy, which is usually provided by the splitting of ATP.

Another type of transport mechanism involves the movement of materials en masse into a cell (endocytosis) or out of a cell (exocytosis). Phagocytosis is a special type of endocytosis in which whole cells or large undissolved particles are engulfed by a cell.

STUDY QUESTIONS

1. What are two functions of membrane proteins?

2. What kinds of substances move across the plasma membrane by diffusion?

3. What do you think would happen if sea urchin eggs, normally in equilibrium with their salt-water environment, were placed in fresh water? Explain.

4. How is water potential affected by pressure? By solute concentration?

5. How does having a hypertonic protoplasm and the presence of a cell wall contribute to the development of turgor in plant cells? What is the role of turgor in plant cells?

6. What distinguishes active transport from facilitated diffusion?

7. Vanadate is an inhibitor of the enzyme ATPase, preventing the breakdown of ATP. Knowing this, design an experiment using vanadate to determine if glucose diffuses or is actively transported into bean roots.

SUGGESTED READINGS

Christensen, H. N. *Biological Transport.* 2nd ed. Reading, MA: Benjamin/Cummings, 1975. Written for the serious student, this text describes the molecular details of the various membrane transport processes.

Finean, J. B., R. Coleman, and R. H. Michell. *Membranes and Their Cellular Functions.* 2nd ed. Oxford: Blackwell, 1978. This nicely illustrated paperback provides the interested student with an advanced treatment of membrane structure and function.

Karp, G. *Cell Biology.* New York: McGraw-Hill, 1979. A comprehensive text covering all aspects of cell structure and function.

Satir, B. "The Final Steps in Secretion." *Scientific American* 233(1975):28–37. A well-illustrated article describing how large substances are exported from cells.

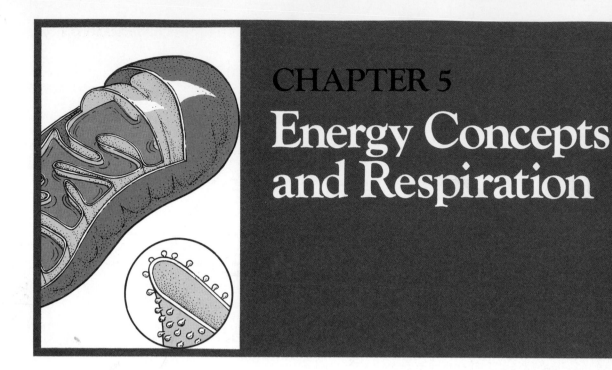

CHAPTER 5
Energy Concepts and Respiration

According to the theory of chemical evolution, the first organisms on earth were single-celled heterotrophs—creatures incapable of manufacturing their own food. They survived by absorbing organic compounds from their aquatic environment. By breaking down these substances to simpler organic compounds, a process called **fermentation,** these early cells obtained energy to regenerate ATP from ADP and phosphate. (ATP, you recall, is a molecule that stores chemical energy in a form cells can readily use to perform biological work, such as growth, movement, and active transport.) These primitive heterotrophs were obviously successful with this way of life, for they ultimately gave rise to all later forms of life. But their success created some problems for them as well. As they multiplied in the primordial oceans, these simple organisms began to use up their food supply. Life would have ended by starvation had it not been for the evolution of a new type of cell—one that could produce its own food from abundant inorganic substances and the energy of sunlight. With this new type of cell came the nutritional process we call photosynthesis. In its most familiar form, photosynthesis can be outlined as follows:

$$CO_2 + H_2O + \text{Light energy} \longrightarrow \text{Carbohydrates} + O_2$$

Carbon dioxide Water Oxygen

As the early photosynthetic organisms proliferated in the seas, the oxygen they produced began to accumulate in the previously **anaerobic** (oxygen-free) atmosphere. This gradual buildup of oxygen in turn set the stage for another major evolutionary event: the advent of **respiration,** a metabolic process whereby organic substances (such as carbohydrates) are completely broken down in the presence of oxygen to simpler, inorganic compounds (such as CO_2 and H_2O). Respiration releases much more energy, and thus yields much more ATP, than the fermentation process used by the first heterotrophs. Therefore, respiration opened the door for the evolution of more complex forms of life, just as the invention of the gasoline-powered engine permitted the development of more complex transportation systems.

The respiration of carbohydrates in the presence of oxygen can be summarized as follows:

$$\text{Carbohydrates} + O_2 \longrightarrow CO_2 + H_2O + \text{Chemical energy}$$

If we compare this summary reaction for respiration with that for photosynthesis, we discover an interesting relationship: with one important exception, the respiration of carbohydrates is photosynthesis in reverse! In other words, the final products of photosynthesis (carbohydrates and O_2) are the initial reactants of respiration, and vice versa. The exception

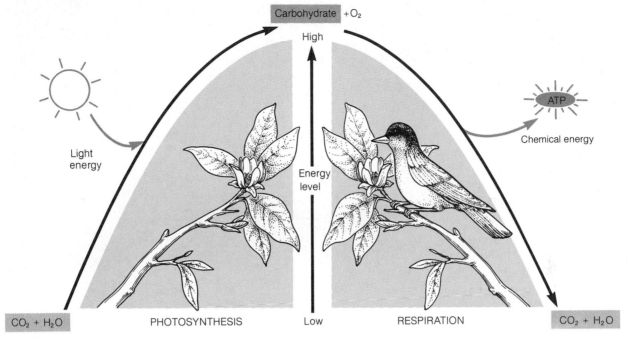

FIGURE 5–1
The Relationship between Photosynthesis and Respiration. In photosynthesis, light energy is used by plants to convert compounds relatively low in energy (CO_2 and H_2O) into comparatively high-energy carbohydrates. Molecular oxygen (O_2) is a by-product. When carbohydrates are respired in both plants and animals, their stored chemical energy is in turn transferred to ATP, the major source of energy for cellular work.

is that while light energy is consumed in photosynthesis, it is not produced during respiration. Respiration generates chemical energy instead, namely ATP (Figure 5–1).

Since all photosynthetic organisms manufacture their own carbohydrates, then break them down either partially or completely, the net result of their energy metabolism is the conversion of light energy into the chemical energy of ATP. Heterotrophs, on the other hand, cannot manufacture their own food. They rely on the photosynthesizers for carbohydrates (or other food substances that can be broken down, such as fats and proteins) to meet their energy needs. Thus, humans, as well as all the other animals, fungi, and heterotrophic bacteria, ultimately depend on the green plants (or other photosynthetic organisms) for food. If the photosynthesizers stopped harvesting light, life on earth would end.

HAVE YOU THANKED A
GREEN PLANT TODAY?

ESSENTIAL CONCEPTS
OF ENERGY METABOLISM

Before we examine the processes of respiration later in this chapter and photosynthesis in Chapter 6, we need to discuss two concepts that are essential for understanding these major forms of energy metabolism. First, how is ATP formed, and how is it used in the cell? Second, how do enzymes work, and why are they critical to metabolic processes?

ATP and Energy Coupling

We have mentioned that the energy released from the partial or complete breakdown of carbohydrates is used to form ATP. But why do cells bother to convert the chemical energy stored in carbohydrates to chemical energy in ATP? Why is ATP so special? A mechanical model might be helpful in explaining the special role of ATP in cells.

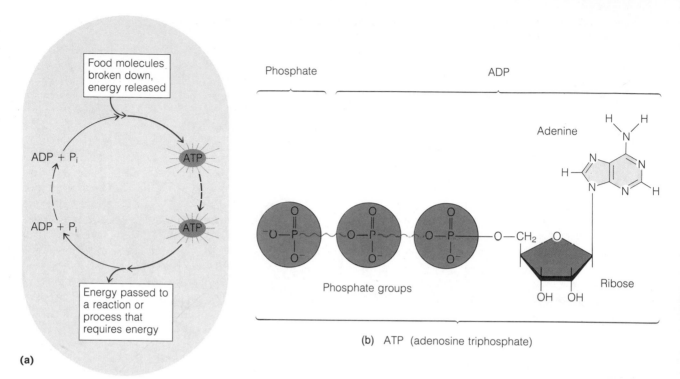

FIGURE 5–2
ATP: The Link between Respiration and Cellular Work. ATP serves as an energy shuttle in cells.
(a) Energy released during the respiration of carbohydrates (or other food sources) is coupled to
the formation of ATP from its breakdown products, ADP and P_i. The ATP can diffuse to various parts
of the cell; when it is split, the energy released can be applied to perform cellular work. When ATP
splits, ADP and P_i are regenerated. (b) In this diagram of ATP structure, the colored wavy lines
indicate unstable, high-energy bonds.

The water that rumbles down a waterfall releases a tremendous amount of energy, but the energy in this form would not help you start the engine of your car. However, if we used the falling water to turn a hydroelectric generator, the electric energy produced could be used to charge many batteries. Now a battery would be quite useful in starting your car's engine, or supplying energy for the car's radio and lights. In applying this model to the living cell, you might think of the water poised at the top of the falls as the stored energy of carbohydrates, and the falling water as the carbohydrates being broken down. The energy given off by the falling water is ultimately transformed into battery power, just as respiration transforms the energy of carbohydrates into ATP energy. And like the battery that helps us conduct work, ATP is a convenient form of energy that cells use to carry out their biological work. Because it is used for virtually all energy-consuming processes in the cell, ATP is often referred to as the "energy currency" of living cells.

It may also be helpful to think of ATP as a small, unstable energy carrier that shuttles back and forth within a cell, picking up energy in one place and releasing it in another. At the sites in the cell where food molecules are broken down, the energy released is used to link adenosine diphosphate (ADP) and inorganic phosphate (P_i) to form ATP. At other sites where energy is needed, ATP is split into ADP and P_i, giving up its stored energy to some other reaction or process that requires energy (Figure 5–2a).

To understand why ATP is especially suited for its role as a cellular energy shuttle, we must examine its structure. ATP is composed of a nitrogenous base, *adenine;* a 5-carbon sugar, *ribose;* and three *phosphate groups* (Figure 5–2b). All of the atoms in a molecule of ATP are held together by covalent bonds, but the bonds linking the last two phosphate groups are particularly unstable. The negative charges on the oxygen atoms in these groups repel each other, which tends to push the phosphate groups apart. If the outermost phosphate bond is broken (which happens very frequently in cells), some of the energy of this repulsive force is liberated (Figure 5–3). When the terminal phosphate group splits off, adenosine *tri*phosphate (ATP) becomes adenosine *di*phosphate (ADP) and P_i. The splitting of ATP is called an **exergonic** reaction because energy is released in the process.

If ATP was allowed to react in a test tube to yield ADP and P_i, all the energy given off would be released as heat. Heat is not a usable form of energy in living systems because organisms cannot transform heat energy into other forms of energy. Therefore, to make use of the stored energy in ATP, cells have **energy-coupling mechanisms.** These coupling mechanisms transfer the energy released during the splitting of ATP

directly to an **endergonic,** or energy-requiring, process or reaction.

One of the most common ATP energy-coupling mechanisms involves the transfer of the terminal phosphate group of ATP to one of the reacting molecules of an endergonic reaction. This mechanism is used in the first steps of the respiration of glucose (see Figure 5–9); it is also used by plant cells to synthesize starch and by animal cells to synthesize glycogen, two polysaccharides made from glucose units. In the latter cases, the terminal phosphate group of ATP is transferred to a glucose molecule, yielding glucose-1-phosphate and ADP (Figure 5–4). This energized form of glucose can then link up to a chain of glucose units in a molecule of starch or glycogen, releasing the phosphate group in the process. Thus, the endergonic process of producing these polysaccharides is made feasible with help from the energy of ATP. The net result is the transfer of energy from ATP to the stored chemical

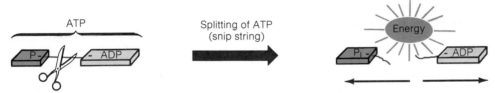

FIGURE 5–3
Model of ATP Energy. We can compare ATP to two bar magnets attached with their negative poles facing each other. The repulsive force between the magnets, like that between the negatively-charged phosphate groups in ATP, creates tension on the string holding the magnets together. When the string is cut, that tension is converted into mechanical energy, which pushes the magnets apart. In the case of ATP, the energy liberated when the terminal phosphate bond is "snipped off" provides energy for work processes in the cell.

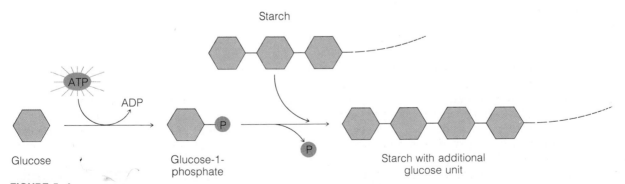

FIGURE 5–4
The Role of ATP in Starch Biosynthesis. The incorporation of glucose into a growing starch molecule is an endergonic process that requires energy from ATP. It begins when ATP reacts with a glucose molecule, yielding the more reactive glucose-1-phosphate and ADP. In this chemically activated form, the glucose unit can be added to the end of a starch chain as the phosphate group is released. The synthesis of glycogen occurs by an analogous mechanism.

energy of the glucose-to-glucose bonds in starch or glycogen. These storage polysaccharides represent reserve energy for the cell, and this stored energy can be harvested when starch or glycogen is disassembled into glucose molecules, which are then respired.

Another form of cellular work driven by the energy of ATP is active transport, the movement of substances across a membrane against their concentration gradients (see Chapter 4). In this case, the splitting of ATP activates a carrier protein in the membrane, perhaps by changing its shape so that it can transport a particular molecule or ion across the membrane. Once the substance has been released on the other side, the carrier protein returns to its nonactivated shape, ready to become energized by another ATP molecule and shuttle another molecule or ion across the membrane.

As you have seen, ADP, P_i, and ATP play a pivotal role in the energy metabolism of cells. By accepting energy from the exergonic breakdown of food molecules in respiration, ADP and P_i become joined into ATP. ATP can then transfer its captured energy to any one of many endergonic metabolic reactions associated with cellular work—work that must be performed to keep the cell (and the organism) alive. But the metabolic reactions crucial to life have other hurdles to overcome, as you will see in the next section.

Activation Energy and Enzymes

Hydrogen gas reacts with molecular oxygen to form water:

$$2 H_2 + O_2 \longrightarrow 2 H_2O$$

This reaction is very exergonic—so much so that it is explosive! Yet if you were sitting in a room filled with H_2 and O_2, you wouldn't notice a thing—at least until you lit a match. Then your last sensations might be a feeling of dampness in the air, since water is a product of this reaction, and openness in a room now without walls or a ceiling. What happened? The heat from the match provided the **activation energy** needed to get a few million molecules of H_2 and O_2 to react. The heat energy given off by these reacting molecules then activated a billion more, which in turn activated a trillion more, and so on, until a split second later all the free H_2 and O_2 had reacted. In other words, the lit match sparked a chain reaction explosion.

All reactions, whether exergonic or endergonic, require a certain amount of activation energy to proceed. This energy is needed to disrupt specific chemical bonds in the reactants so that new bonds and new products can form. Heat can supply the necessary activation energy by raising the kinetic energy (energy of motion) of the reactants. This increases both the frequency and the violence of collisions between the reactants.

There is another way to encourage the reaction between H_2 and O_2: simply add a piece of platinum to the gas mixture. The platinum will act as a **catalyst**, triggering the same explosive result. Catalysts facilitate chemical reactions by lowering the amount of activation energy needed to start the reaction (Figure 5–5). Most catalysts operate as surface agents; that is, the reactants bind to the surface of the catalyst so that they are positioned correctly to react. In addition, the catalyst may place stress on the bonds that must be broken for the reaction to proceed. Although the catalyst interacts directly with the reactants, it remains unchanged during the course of the reaction and therefore is available to act again and again.

In living cells, chemical reactions are catalyzed by a class of proteins called enzymes. Each enzyme catalyzes a specific reaction (or type of reaction), and each has a unique **active site** region that recognizes and binds to one specific set of reactants (Figure 5–6). After one set of bound reactants have undergone the enzyme-catalyzed reaction, the products are released and the active site becomes available to bind a second set of reactants. A single enzyme molecule can repeat this sequence millions of times per minute.

All enzymes are globular proteins with very intricate folding patterns. These folds are held in place by relatively weak chemical forces occurring across different regions of the protein chain or chains, such as hydrogen bonding and attractions between opposite charges. Because of the weakness of these forces, enzymes are quite fragile. Their three-dimensional shapes are very sensitive to changes in temperature, acidity, and the concentrations of ions in their vicinities. And minor changes in the shape of an enzyme molecule can have a considerable effect on its ability to catalyze. For example, let us consider the effect of temperature on enzyme activity. In general, enzymes are most effective in the temperature range of 20–45°C (68–113°F) (Figure 5–7). When the temperature is above 45°C, the weak bonds that give the enzyme its particular folding pattern begin to break, including those in the region of the active site. Changes in the shape of the active site interfere with its ability to bind the reactants; hence, the enzyme suffers a reduced catalytic activity. When the temperature exceeds 60°C (140°F), most enzymes undergo large, irreversible shape changes and lose all catalytic function.

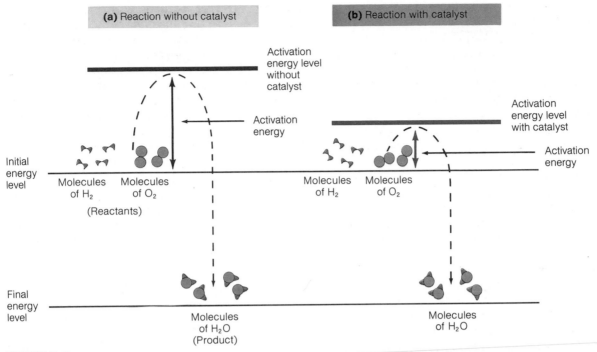

FIGURE 5–5

Activation Energy and Catalysis. The exergonic reaction between H_2 and O_2 to form H_2O releases a large amount of energy. It cannot proceed, however, unless the H_2 and O_2 molecules collide into each other with sufficient force to react. This force is called the activation energy; it is the energy necessary to initiate the reaction. Catalysts lower the activation energy needed for a given reaction by increasing the frequency of successful collisions between reactants, thereby increasing the rate of reaction.

FIGURE 5–6

Induced Fit Model of Enzyme Catalysis.
In enzyme-catalyzed reactions, the active site of the enzyme binds the specific reactants in a way that ensures their collision. This binding causes a change in the shape of the enzyme (induced fit) which facilitates the reaction. After the reaction has taken place, the products are released and the enzyme returns to its original configuration, ready to bind the next set of reactants.

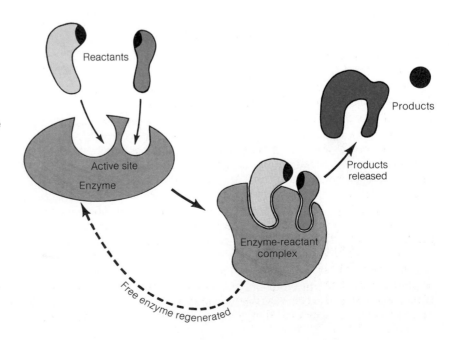

FIGURE 5–7
The Effect of Temperature on Enzyme Activity. In general, increases in temperature between 0°C and 40°C lead to small changes in the shapes of enzymes that increase their effectiveness as catalysts. Enzymes are generally most effective between 20°C and 40°C. When exposed to temperatures of 45°C or higher, most enzymes undergo shape changes that result in the loss of enzyme activity.

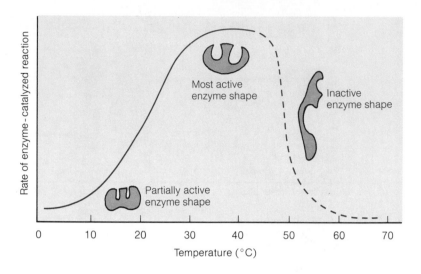

The chemical reactions that constitute respiration, photosynthesis, and all other metabolic processes are catalyzed by specific enzymes. If enzymes were not present to catalyze each step in such processes, the reactions would proceed too slowly to be of any use to the organism. This fact provides a basis for the effective control of cellular chemistry. By regulating the production or activities of specific enzymes, the cell's metabolism can change, often as a response to changing conditions. We will have a closer look at mechanisms of enzyme regulation in Chapter 12. For now, let us examine one of the many metabolic processes that depends heavily on enzymes—respiration.

THE RESPIRATION OF GLUCOSE

Three major types of organic compounds can be respired in cells: sugars, amino acids, and fatty acids. Each gives up stored energy as it is completely broken down in the presence of oxygen to simpler inorganic substances. Glucose, a 6-carbon sugar, is the most common fuel for respiration, so we will limit our detailed coverage of respiration to the breakdown of glucose.

Overview of Glucose Respiration

For our purposes, we will consider that the respiration of glucose occurs in three distinct stages. In the first stage, called **glycolysis*,** the 6-carbon glucose mole-

cule is split into two molecules of the 3-carbon pyruvic acid (Figure 5–8). Along the way a small amount of the original energy in the glucose molecule ends up in two molecules of ATP and four electrons that are transferred to electron carriers. Most of the glucose energy, however, is now in the two pyruvic acid molecules.

The second and third stages of respiration occur only in aerobic organisms (those normally exposed to oxygen), and only when oxygen is available. They are called the **Krebs cycle** (second stage) and **respiratory electron transport** (third stage); in eukaryotic cells, both processes take place in the mitochondria. In the Krebs cycle pathway, the pyruvic acid molecules produced by glycolysis are completely dismantled, forming carbon dioxide as a by-product. Chemical energy, in the form of electrons, is transferred from pyruvic acid to specific electron carriers.

In the third and final stage of respiration, the electrons captured by the electron carriers in glycolysis and the Krebs cycle are transferred to the respiratory electron transport chain. Here the electrons are passed

---

*Biochemists define respiration as the complete breakdown of an organic fuel in the presence of O_2, a process that takes place exclusively in the mitochondria of eukaryotes. Because glycolysis occurs outside the mitochondria, and because it can proceed in the absence of O_2 (see "The Fermentation of Glucose" section), many biochemists do not consider glycolysis to be part of respiration. In aerobic organisms, however, the cells are normally exposed to oxygen, and so electrons removed during glycolysis are ultimately transferred to oxygen in the mitochondria. Under these conditions, then, glucose is an organic fuel that is completely broken down with the participation of oxygen. For this reason, we shall refer to glycolysis as the first stage of carbohydrate respiration.

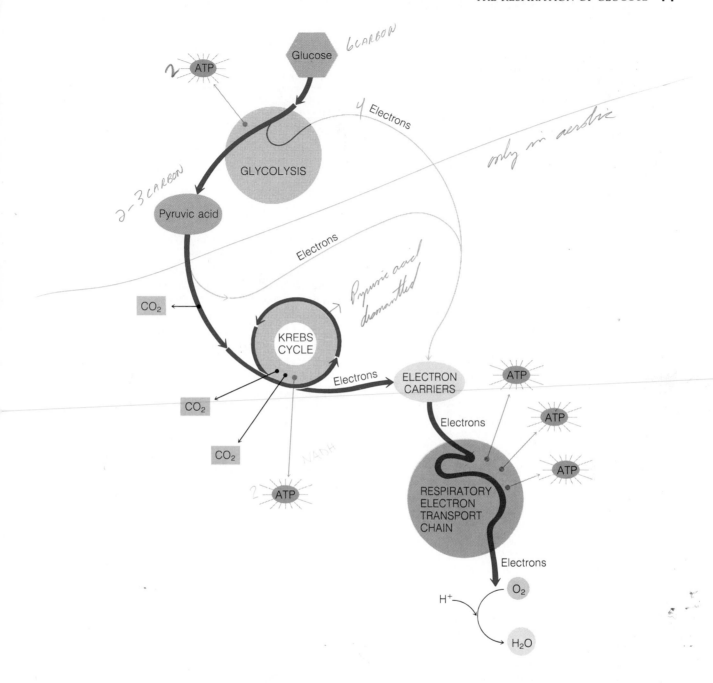

FIGURE 5–8
Overview of Glucose Respiration. The three stages of glucose respiration are shown here in abbreviated form. In glycolysis, glucose is broken down to two molecules of pyruvic acid. In the process two pairs of electrons are transferred to electron carriers, and two ATPs are formed. In the next stage, called the Krebs cycle, the pyruvic acid molecules are completely degraded to CO_2 as more electrons are transferred to electron carriers and two more ATPs are formed. Finally, the electrons now in the pool of electron carriers are donated to the respiratory electron transport chain, where they are ultimately passed to O_2. As the electrons move down the chain, a considerable amount of chemical energy is released, some of which is captured in the formation of many more ATP molecules (34 per glucose).

from one component in the chain to another, ultimately joining with oxygen (O_2) and hydrogen ions (H^+) to form water (H_2O). During this electron passage, a large amount of energy is released, which is used to form many ATP molecules. In fact, about 90% of the energy transfer from glucose to ATP during respiration occurs in the electron transport phase. And because electron transport can take place only when oxygen is present to accept the electrons, we must once again thank the green plants. Not only do plants provide respirable fuels for all heterotrophs, but they also give off oxygen during photosynthesis. We and other aerobic organisms utilize this oxygen to get the maximum amount of energy out of the breakdown of food molecules. Let us now look at the stages of respiration in more detail.

Glycolysis

Glycolysis is a metabolic pathway consisting of 10 sequential reactions. Each reaction is catalyzed by a different enzyme, and all the enzymes exist freely in the cytoplasm of a cell (they are not associated with any organelle). Virtually all living organisms, even the anaerobic ones that ferment glucose, have this set of enzymes.

Recall that for every glucose molecule that enters glycolysis, two molecules of pyruvic acid are formed. If you compare the chemical formula for glucose ($C_6H_{12}O_6$) to that of pyruvic acid ($C_3H_4O_3$), you can get a general idea of what glycolysis accomplishes. The 6-carbon glucose molecule is split in half, yielding two 3-carbon molecules that collectively have only eight hydrogen atoms, four less than the original glucose. The removal of hydrogen atoms (actually, electrons and hydrogen ions) occurs in one reaction in glycolysis (and in several steps of the Krebs cycle); the transfer of electrons to electron carriers plays a key role in the energy retrieval aspect of respiration, as we shall see.

The glycolytic pathway is shown in detail in Figure 5–9. We have presented this pathway in its entirety not for you to memorize, but to emphasize that glycolysis, like all metabolic pathways, is comprised of a steplike series of reactions. In all such pathways the product of one reaction becomes the reactant in the next reaction, and so on. Each individual reaction contributes a minor chemical change, but over an entire pathway these changes can add up to major revisions of the original reactant.

As you examine this pathway, note especially reaction 6. Here, two electrons and a hydrogen ion are

TABLE 5–1
Balance Sheet for Glycolysis.

Net Input		Net Output
1 Glucose ($C_6H_{12}O_6$)	$\longrightarrow$	2 Pyruvic acids ($C_3H_4O_3$)
2 NAD^+	$\longrightarrow$	2 NADH
2 ADP + 2 P_i	$\longrightarrow$	2 ATP

transferred from PGAL to the electron carrier **NAD^+** (nicotinamide adenine dinucleotide), forming **NADH.** Note also that a phosphate group is added to PGAL, yielding DiPGA, a very high-energy compound. The remaining steps of glycolysis, culminating in the formation of pyruvic acid, are concerned essentially with the transfer of high-energy phosphates from DiPGA to ADP, forming ATP. Four ATP molecules per glucose are produced in these steps, but note that two molecules of ATP are consumed at the beginning of the pathway. Thus, glycolysis yields a net gain of two ATPs per glucose.

Table 5–1 presents a balance sheet of the net input and output for the entire glycolytic pathway. Overall, glycolysis releases about 20% of the total chemical energy available in the glucose molecule. Some of this released energy is lost as heat, but most of it is conserved as chemical energy, now in molecules of NADH and ATP. The remaining 80% of the energy is in the form of the two pyruvic acid molecules. As we shall now see, much of this energy will be converted to other chemical forms when pyruvic acid is completely dismantled in the Krebs cycle, the second phase of respiration.

The Krebs Cycle

The Krebs cycle, named after Sir Hans Krebs who played a major role in the elucidation of this pathway, takes place in the matrix (liquid core) of the mitochondrion in eukaryotic cells (Figure 5–10). As in glycolysis, each of the steps of this cycle is catalyzed by a specific enzyme or multienzyme complex. The overall result of the progressive actions of these enzymes is the transfer of hydrogen atoms from pyruvic acid to two types of electron carriers: NAD^+ (as in glycolysis) and **FAD** (flavin adenine dinucleotide). When FAD accepts two hydrogen atoms, it becomes **$FADH_2$** which, like NADH, represents stored chemical energy. As the hydrogen atoms are transferred to these electron carriers, the carbon and oxygen atoms originally part of pyruvic acid are released as carbon dioxide

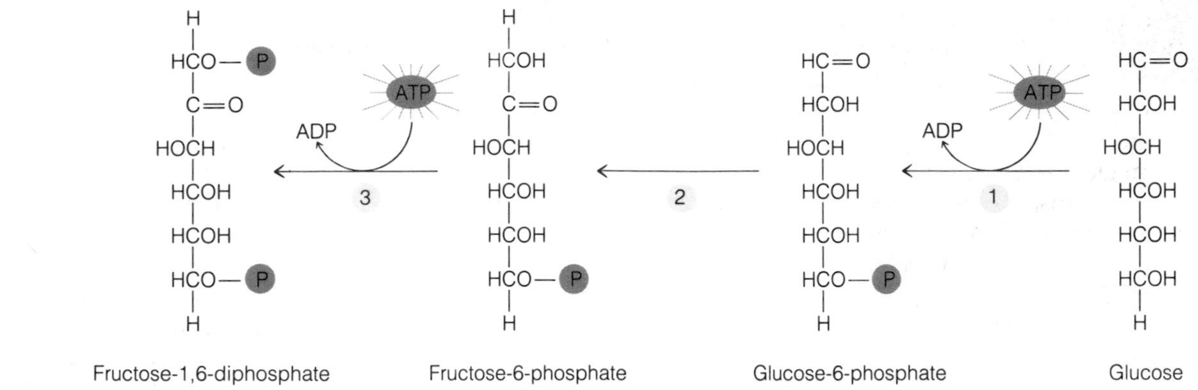

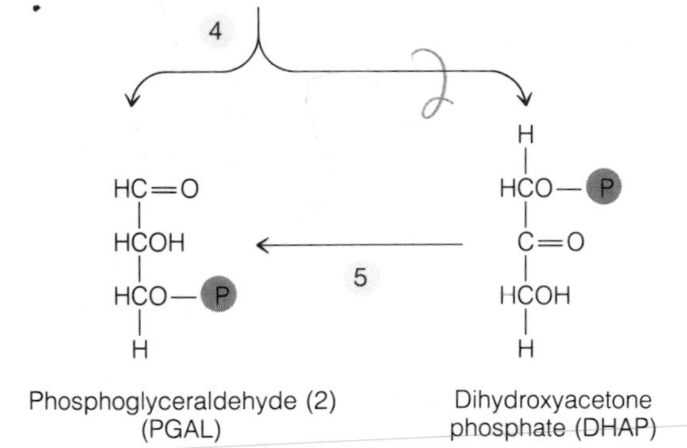

FIGURE 5–9
Glycolysis.

Reactions 1–3: Glucose is Activated. Sugar phosphates are chemically more reactive than their corresponding sugars, so the first three reactions in effect "activate" the glucose in preparation for subsequent energy-releasing steps. ATP provides the phosphates for these steps.
Reaction 1: The terminal phosphate (P) of ATP is transferred to glucose.
Reaction 2: The activated glucose is rearranged.
Reaction 3: A second ATP donates its terminal phosphate.

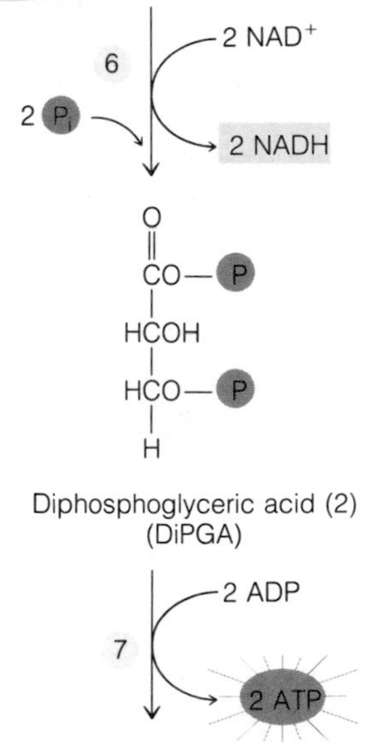

Reactions 4 and 5: A Six-Carbon Sugar-Diphosphate is Split in Half.
Reaction 4: A six-carbon sugar is split into two 3-carbon compounds, PGAL and DHAP.
Reaction 5: DHAP is rearranged to become PGAL. In effect, two PGAL molecules are obtained for each glucose molecule that enters glycolysis.

Reaction 6: NADH and DiPGA are Formed. In a complicated reaction, two electrons and a H^+ are transferred from PGAL to NAD^+, producing the electron carrier NADH. Next, inorganic phosphate is added to the PGAL, yielding DiPGA, a very high-energy compound. Note that from now on, there are two molecules of everything.

Reactions 7–10: The Energy of DiPGA is Cashed in to Make ATP. Two molecules of the high-energy compound DiPGA transfer their high-energy phosphate groups to four ADP molecules, producing four ATP molecules.
Reaction 7: The first P is transferred, yielding ATP and 3-PGA.
Reaction 8: The P in 3-PGA is rearranged.
Reaction 9: A water molecule is removed from 2-PGA to yield the high-energy PEP molecule.
Reaction 10: In the final step of glycolysis, PEP donates its P group to ADP, producing pyruvic acid and ATP.

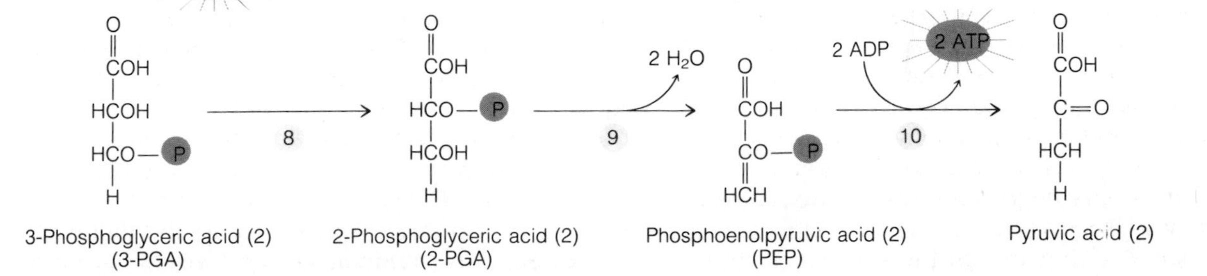

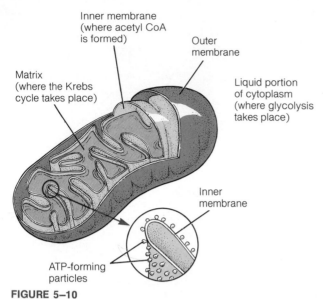

FIGURE 5–10
The Compartments of a Mitochondrion.
The mitochondrion consists of an outer membrane and a greatly infolded inner membrane. The inner membrane encloses a liquid core called the matrix, where most of the enzymes that function in the Krebs cycle are located. The enzyme complex that acts on pyruvic acid, and the components of the electron transport chain, are embedded in the inner membrane. The particles protruding from the surface of the inner membrane (see detail) are associated with the electron transport components; they function in ATP formation that is coupled to electron transport.

molecules (CO_2). Carbon dioxide gas escapes from the mitochondria and cells by diffusion, and in higher animals, enters the bloodstream where it is transported to the lungs. This is the same carbon dioxide that we exhale during breathing.

When pyruvic acid enters the mitochondrion, it is acted upon by a multienzyme complex located in the mitochondrion's inner membrane. This complex catalyzes a rather complicated series of reactions. First, the 3-carbon pyruvic acid is split into CO_2 and a 2-carbon **acetyl group** (Figure 5–11). As this splitting takes place, two electrons and a hydrogen ion are transferred to NAD^+, forming NADH. Next, the acetyl group is joined to a molecule of **coenzyme A** (abbreviated CoA), forming **acetyl CoA.** (The function of CoA is like that of a phosphate group—it activates

the organic substance to which it is attached.) Finally, the acetyl CoA is released from the enzyme complex into the mitochondrial matrix where it enters the Krebs cycle.

The Krebs cycle begins when the 2-carbon acetyl group borne by CoA is transferred to the 4-carbon **oxaloacetic acid** to yield a 6-carbon molecule, **citric acid,** and free CoA (see Figure 5–11). Then, through a cyclic series of reactions, citric acid is progressively transformed chemically to oxaloacetic acid, the starting point of the Krebs cycle. Along the way, two more CO_2 molecules are produced as the hydrogen atoms end up in three molecules of NADH and one molecule of $FADH_2$. In addition, one ATP molecule is formed for each acetyl CoA that enters the cycle. After one complete turn of the cycle, the regenerated oxaloacetic acid can become joined to another acetyl group, and thus keep the cycle turning.

Note in Figure 5–11 that there are four steps at which two electrons and a hydrogen ion are transferred to NAD^+ to form NADH: one when pyruvic acid is converted to acetyl CoA, and three in the Krebs cycle. Each turn of the cycle also produces one molecule each of $FADH_2$ and ATP. Up to this point, then, the mitochondrion has "squeezed" the chemical energy out of pyruvic acid and transferred it to four molecules of NADH, one $FADH_2$ and one ATP for every pyruvic acid molecule broken down. In the final stage of respiration—respiratory electron transport—the energy in NADH and $FADH_2$ is "cashed in" for ATPs.

The Respiratory Electron Transport Chain

As we emphasized earlier, the primary function of glucose respiration is the transfer of the stored energy in glucose to a more usable chemical form, namely ATP. So far, however, we have seen only four ATPs produced for each glucose broken down to carbon dioxide: two in glycolysis and two more in the Krebs cycle (one ATP is generated in the Krebs cycle for each pyruvic acid, and there are two pyruvic acid molecules formed per glucose during glycolysis). There is a simple explanation for this: most of the original chemical energy in glucose now rests in the molecules of NADH and $FADH_2$. This energy can be harvested for ATP formation when NADH and $FADH_2$ pass their electrons (and hydrogen ions) to oxygen by way of the electron transport chain.

If NADH or $FADH_2$ passed its electrons directly to oxygen, a large amount of heat would be given

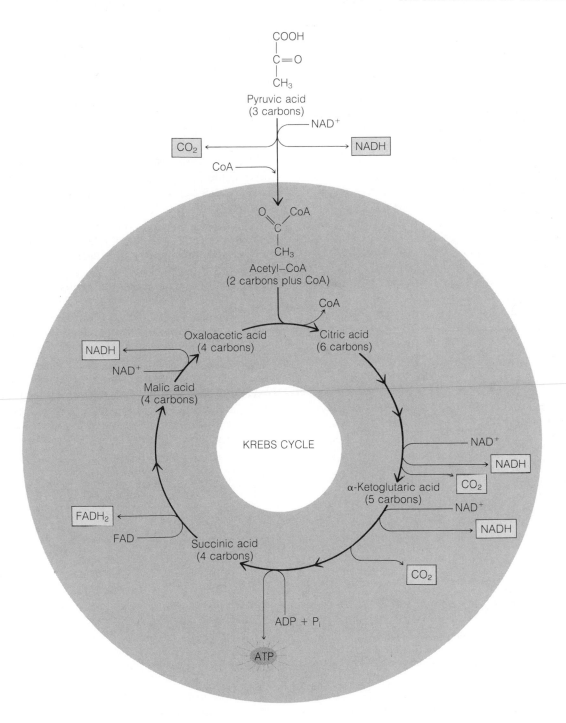

FIGURE 5–11

The Krebs Cycle. In this abbreviated diagram, pyruvic acid is converted to acetyl CoA and CO_2 as a pair of electrons are transferred to NAD^+, forming NADH. Acetyl CoA then reacts with a 4-carbon compound (oxaloacetic acid) to form a 6-carbon compound (citric acid). In a cyclic series of reactions that follows, 3 NADH, an $FADH_2$, an ATP, and 2 CO_2 are formed. Oxaloacetic acid is regenerated to complete the cycle.

off. Such a reaction would be too exergonic for cells to capitalize on the large chunk of energy released. Indeed, they would burn up from the heat. Living systems have sidestepped this potential problem by transferring electrons from these high-energy electron carriers to a system of electron-transporting molecules situated in the inner membrane of the mitochondrion. The components of this electron transport chain represent energy levels intermediate between NADH (high energy) and oxygen (low energy). As the electrons are passed "downhill" in energy along this chain, the energy is released in small, discrete packets, and the mitochondrion couples some of these small energy releases to the formation of ATP. The energy transfer to ATP is more efficient when the electrons are moved through a series of small energy-releasing steps, just as the energy imparted to a paddlewheel by falling water is more efficiently transferred when the water cascades down many small paddles rather than splashing onto one large paddle.

Figure 5–12 shows the various components of the electron transport chain, which are arranged according to their relative energy levels. Note that there are three sizable energy drops that occur when electrons are passed from NADH to oxygen, each of which releases enough energy to form one ATP. Thus, for each pair of electrons donated by NADH, three ATPs

can be formed (see Essay 5–1 for an exception to this rule). When the pair of electrons comes from $FADH_2$, however, the electrons enter the transport chain at a slightly lower energy level. Consequently, only two ATPs can be formed for each electron pair passed down the chain from $FADH_2$.

The importance of the electron transport stage becomes clear when you consider that 90% of the energy transferred to ATP during respiration occurs here. Table 5–2 summarizes the energy transfers that take place during glycolysis, the Krebs cycle, and electron transport. Note that the respiration of one glucose molecule can yield 38 ATPs, 34 (90%) of which are formed during mitochondrial electron transport. It is small wonder, then, that the mitochondrion is called the energy factory of the cell.

Maintaining a positive ATP balance is critical to the normal functioning of cells. If the ATP supply remains low, cells cannot perform biological work and they soon die. Since respiration is the means by which nearly all the cellular ATP is reformed from ADP and P_i, you can appreciate why the continuous flow of electrons from glucose (or other respirable fuel) to oxygen is so important to aerobic organisms, particularly active ones like us. If this flow is retarded or interrupted for some reason, the consequences can be serious. For example, cyanide (the executioner's

TABLE 5–2
Energy Recovered During the Respiration of Glucose.

Process	Electron Carriers Produced	ATP Produced
Glycolysis:		
Glucose $\longrightarrow$ 2 pyruvic acid	2 NADH	2 ATP
Krebs Cycle:		
2 Pyruvic acid $\longrightarrow$ 2 acetyl CoA + 2 CO_2	2 NADH	
2 Acetyl CoA $\longrightarrow$ 4 CO_2	6 NADH	2 ATP
	2 $FADH_2$	
Respiratory Electron Transport Chain:		
10 NADH + 5 O_2 $\longrightarrow$ 10 NAD$^+$ + 10 H_2O		30* ATP
2 $FADH_2$ + O_2 $\longrightarrow$ 2 FAD + 2 H_2O		4 ATP
	Total:	38 ATP

*The NADH molecules generated in glycolysis cannot get into the mitochondria to transfer their electrons directly to the electron transport chain. Instead, there are electron shuttle reactions that transfer the electrons across the mitochondrial membrane. In some cells (particularly liver and heart), the shuttle of each pair of electrons from glycolytic NADH results in the formation of 3 ATPs, as we have assumed in our calculations above. In other cells (such as muscle), the NADH electrons are shuttled into the electron transport chain at the level of coenzyme Q (see Figure 5–12), resulting in the formation of only 2 ATPs per NADH. In these cells the complete breakdown of glucose yields a total of 36 ATPs.

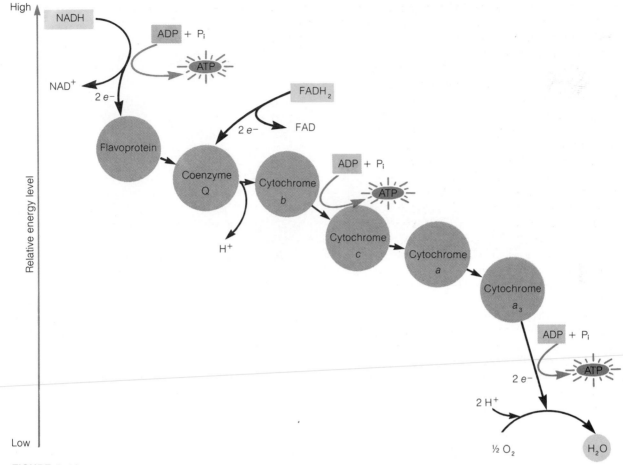

FIGURE 5–12

The Electron Transport Chain. This highly stylized diagram of the respiratory electron transport chain of mitochondria emphasizes the energy relationships among the individual components of the chain. At the top of the energy staircase, NADH donates its pair of electrons and a H^+ to a flavoprotein. Enough energy is given off during this transfer to join an ADP and P_i to form ATP. The flavoprotein then donates its pair of electrons and H^+s to coenzyme Q. At the next step in the chain, the H^+s are released but the electrons continue to be passed through a series of cytochromes, ultimately joining with an oxygen atom and two H^+s to form H_2O. The transport of two electrons from cytochrome *b* to oxygen results in the formation of two ATPs. Thus, for each pair of electrons donated by NADH, a total of three ATPs are formed.

When $FADH_2$ passes its two electrons and H^+s to the chain, coenzyme Q is the initial acceptor. From here, the sequence of events is the same as described for NADH, starting at the level of coenzyme Q. Note that for each pair of electrons coming from $FADH_2$, only two ATPs can be formed.

gas) and carbon monoxide (an equally lethal gas found in the exhaust of automobiles) can kill aerobic organisms by blocking respiratory electron transport. Withholding oxygen for extended periods (five minutes for humans) can have the same fatal effect, again by stopping the ATP-generating electron transport system. Thus, asphyxiation causes death because cells run out of ATP.

THE FERMENTATION OF GLUCOSE

Most organisms, including all plants and animals, carry out the aerobic form of respiration and therefore require a continuous supply of oxygen. There are circumstances, however, that can lead to temporary

ESSAY 5–1

HOT PLANTS: A RESPIRATORY ODDITY

Heat is a universal product of metabolism, and respiration produces more heat than any other metabolic process. More than half the chemical energy stored in respirable fuels is released as heat during cellular respiration, all of which eventually escapes into the organism's surroundings. In mammals, birds, and other "warm-blooded" animals, respiration is the major source of body-warming heat. But have you ever felt a warm plant? Probably not, because in plants heat production (respiration) is too slow and the dissipation of heat from their surfaces is too rapid to permit them to "store" heat. For every general observation on nature, however, there are always exceptions. In this case, the exceptions are a few unusual plants that get hot.

Although plants cannot maintain constant body temperatures like the higher animals, some of the arum lilies, a group which includes the familiar calla lily and philodendron, have specialized floral structures that can generate surprisingly large amounts of heat. One of these plants, the voodoo lily, has a spadix (see diagram) that can reach temperatures 10–15°C (18–27°F) warmer than the surrounding air. Another member of this family, the skunk cabbage, melts the early spring snow when it flowers. Why do these spadix tissues emit so much heat? Well, if you ever smelled a flowering voodoo lily, skunk cabbage, or other arum, you would have a major clue to answering this question. Put simply, these plants stink! The heat generated in the spadix causes the volatilization of various nitrogen-containing compounds (including ammonia), which have an odor reminiscent of dung or decaying

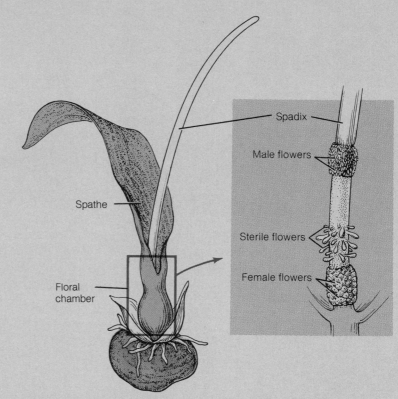

Reproductive organ of the voodoo lily.

flesh. But the same stench that moves us upwind from these plants is quite effective in attracting flies and various kinds of beetles that pollinate the flowers. As these insects visit one foul-smelling plant after the next, they inadvertently cross-pollinate the flowers, just as bees, hummingbirds, and other types of insects cause the transfer of pollen among sweet-smelling flowers of other species (see Chapter 14). And cross-pollination has an extremely important role in the continued survival and evolution of plant species.

To ensure adequate levels of cross-pollination, the timing of events in the floral apparatus of most

arum lilies is closely regulated. The heat production and odor emanation from the spadix begins just as the female flowers reach maturity, so the flies and beetles are attracted at the very moment these flowers are receptive to pollen. When the insects land on the spadix, they encounter tiny oil droplets on the surface that make gaining a foothold difficult at best. Sooner or later the insects tumble down to the base of the floral chamber where the female flowers are located (see diagram). These flowers produce a sticky, sweet fluid that the insects feed on. Once inside the floral chamber, however, the insects soon discover that they are prisoners. The only way out is to climb up the oily spadix, an escape that is made even more difficult by a group of sterile flowers protruding from the spadix. Later that night, with the insects still captive, the male flowers reach maturity and shed pollen downward into the floral chamber, showering the insects. When the insects resume their escape attempts the following morning, they find the going has been made a little easier by the partial wilting of the sterile flowers. Laden with their pollen cargo, they fly away, only to be allured once again to another odoriferous arum plant. In such a manner, the flies and beetles transfer pollen from one individual plant to the next, tolerating the temporary imprisonment in exchange for food.

Obviously, one of the key ingredients in this rather exotic adaptation for cross-pollination is the generation of a pungent odor, which in turn hinges on a hot spadix. How is so much heat generated? In studying the respiratory pathway of the spadix when its temperature is rising, scientists have discovered a rather unusual electron transport system in the mitochondria. Instead of electrons moving through the regular chain of cytochrome molecules on their way to molecular oxygen, there is a bypass that diverts the electrons directly to oxygen (see figure). Since two of the ATP-forming steps associated with the cytochromes are bypassed, this means that more of the respiratory energy ends up as heat rather than

ATP. Furthermore, the operation of this alternate pathway does not depend on a supply of ADP and phosphate, the availability of which normally limits the rate of respiration. Thus, the respiration of stored foods occurs much faster, and this of course contributes to even more heat production.

Thus, at the appropriate moment the spadices of arum lilies switch to an energy-wasting, heat-generating form of respiration to help volatilize ammonia and other smelly substances. The odor in turn allures flies and beetles to its flowers. The cost to these plants is the rapid depletion of stored foods in the spadix, but this is a small price to pay for the benefits of cross-pollination.

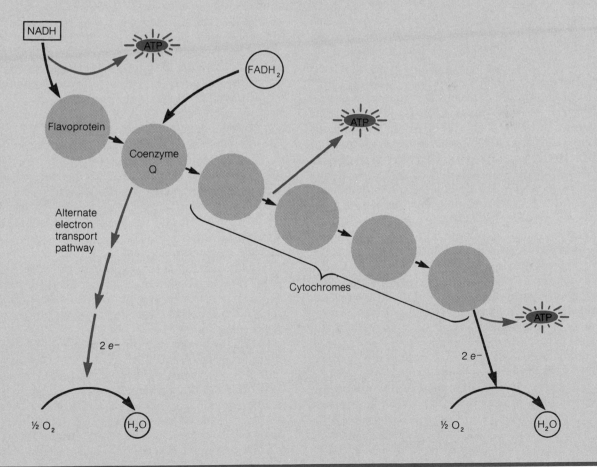

shortages of oxygen in the tissues of these organisms. For example, when the soil surrounding the root system of a land plant becomes waterlogged, the rate of oxygen diffusion to the roots is greatly retarded, resulting in an oxygen deficit (anaerobic conditions) in the root cells. Similarly, an athlete exercising strenuously may consume oxygen in respiration faster than the bloodstream can deliver it to the muscles, producing an anaerobic environment for these hard-working tissues. If oxygen deprivation in either case is prolonged, the cells, tissues, or even the organism itself may die. But cellular damage under anaerobic conditions is temporarily forestalled by a metabolic escape valve—fermentation. Let us first see how oxygen deficits affect the normal respiratory pathways, and then look into how fermentation provides some relief from this dangerous situation.

In the absence of free oxygen, the respiration of pyruvic acid comes to a rapid halt. Oxygen is the terminal electron acceptor on the electron transport chain, so without it the individual electron transport components have no place to unload their electrons. By analogy, a water bucket brigade would come to an immediate halt if no one was at the end to dump the water. When the components of the electron transport chain are full of electrons, they cannot accept electrons from NADH and $FADH_2$ produced in the Krebs cycle. Thus, NAD^+ and FAD molecules needed in the Krebs cycle are not regenerated, so the Krebs cycle shuts down too. In the absence of oxygen, a monumental electron traffic jam develops in the mitochondrion.

Glycolysis would also come to a halt from the lack of NAD^+ (needed in reaction 6 of this pathway; see Figure 5–9) were it not for the fermentation detour. In the absence of oxygen, the NAD^+ outside the mitochondria can be regenerated from NADH through a reaction that converts pyruvic acid to lactic acid, a process called **lactic acid fermentation** (Figure 5–13a). This type of fermentation is used by animals under oxygen stress and by various microorganisms. Since lactic acid is toxic to living cells in high concentrations, however, its accumulation during extended periods of anaerobic conditions can have undesirable side effects, such as muscle cramping in animals.

Other microorganisms and most plants respond to oxygen deficits by engaging a two-step process to regenerate extramitochondrial NAD^+: **alcoholic fermentation.** First, pyruvic acid is split into carbon dioxide and acetaldehyde, a 2-carbon compound (Figure 5–13b). Then acetaldehyde is converted to ethyl alcohol, a reaction in which NADH is converted to NAD^+.

The purpose of both types of fermentation is to regenerate extramitochondrial NAD^+ from NADH so that glycolysis can go on in the absence of oxygen. Fermentation yields much less usable energy than respiration—only two molecules of ATP for each molecule of glucose fermented, compared to 38 during respiration. The energy demands of higher plant and animal cells are typically much higher than what fermentation can provide, so extended anaerobic conditions will lead to ATP starvation and ultimately to cell death. In contrast, some microorganisms manage to survive quite well on the limited amounts of ATP provided by fermentation. Let's look at a few examples.

Yeasts respire sugars aerobically when oxygen is available, but they can live for long periods under anaerobic conditions by fermenting sugars to ethyl alcohol. Their growth rate in the absence of oxygen is slowed dramatically, but the mere fact that they can survive when their ATP regeneration rate is so low distinguishes them from strictly aerobic creatures.

Yeasts, as you know, play a key role in the production of alcoholic beverages. When yeast cells are added to a vat of crushed grapes (for wine production) or barley seeds (for beer), they absorb sugars from these plant sources and respire them aerobically. Under such conditions the yeast grows and multiplies rapidly. Eventually the runaway yeast population depletes the supply of oxygen in the vat, forcing trillions of yeast cells to switch metabolic "gears" and begin fermentation. The ethyl alcohol so produced diffuses into the surrounding medium and is allowed to accumulate to an appropriate level (4–5% by volume for beer; 10–12% for wine), at which point the brewer stops the process. Some yeasts can withstand alcohol concentrations of up to 14%, but above that the alcohol becomes too toxic and the yeast cells die. Thus, any alcoholic beverage produced by yeast in a natural fermentation process alone will not exceed 14% alcohol by volume. So-called hard liquors (vodka, whiskey, rum, and so forth) with much higher alcohol contents are produced from a natural ferment by distillation (separation of the alcohol from the water to make it more concentrated).

Certain bacteria do not use oxygen at all. These **obligate anaerobes** obtain energy to make ATP by transferring electrons and hydrogen ions from organic compounds to other substances (such as nitrates) instead of oxygen. As a matter of fact, exposure to oxygen kills them, so these bacteria are found only in anaerobic environments, such as beneath the surface of soils or buried in decaying vegetation. Some of these anaerobes are especially dangerous. For exam-

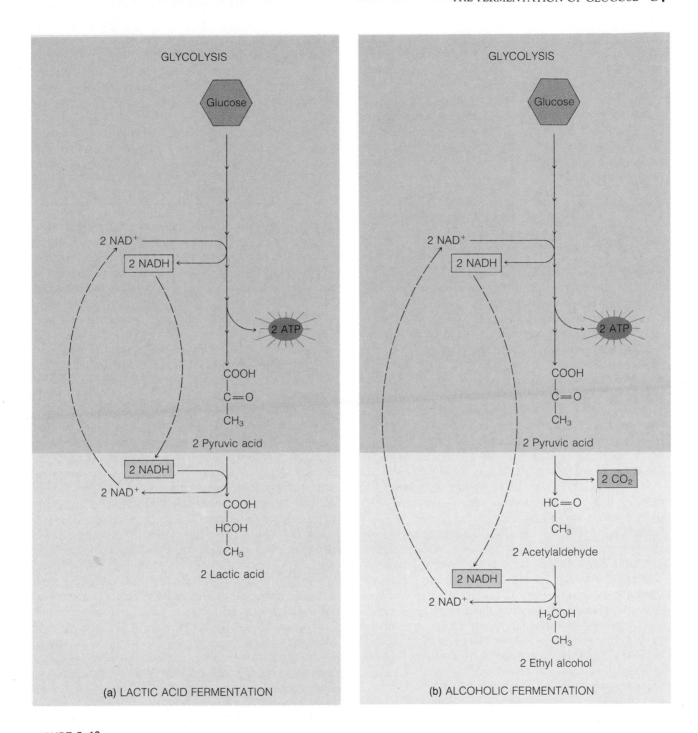

FIGURE 5–13
The Fermentation of Glucose. The two major types of glucose fermentation both involve the glycolytic production of pyruvic acid, but differ thereafter. (a) In lactic acid fermentation, pyruvic acid is converted to lactic acid when it accepts two electrons and a H+ from NADH. (b) In alcoholic fermentation, pyruvic acid is split into CO_2 and the 2-carbon acetaldehyde. Acetaldehyde then accepts the electrons and H+ from NADH to become ethyl alcohol. Note that in both fermentation pathways, NAD+ is regenerated. This permits glycolysis to continue even in the absence of oxygen, and two ATPs per glucose fermented can be formed.

ple, two species of *Clostridium* produce two of the deadliest poisons known: the tetanus and botulism toxins. Tetanus infections can cause uncontrolled muscle spasms and rigidity of muscles, particularly those of the neck and lower jaw—hence the common name, lockjaw. Of the 100 or so human cases of tetanus that occur each year in the United States, about 30% end in death. Botulism is the most severe form of food poisoning known; the fatality rate in humans approaches 100%.

THE MOBILIZATION OF POLYSACCHARIDES AND FATS FOR RESPIRATION

We would feel justified in limiting our discussion of respiration to glucose breakdown if organisms truly lived by glucose alone. The fact is, however, that food energy comes in many forms, and organisms have the ability to convert a wide array of organic substances into respirable compounds. The two most common forms of "energy foods" are polysaccharides (such as starch and glycogen) and fats. Before these large substances can be used for respiration, however, they must first be dismantled into their constituent units, a process called **digestion.**

The digestion of starch or glycogen is normally a hydrolysis process (see Figure 2–3) in which the glucose-to-glucose bonds are broken, releasing free glucose units. The glucose products can then enter glycolysis (Figure 5–14).

In the digestion of fats, also a hydrolysis process, the fatty acid units are cleaved from the 3-carbon glycerol unit. Glycerol can enter glycolysis by being converted to one of the 3-carbon intermediates (DHAP) on this pathway. The fatty acids take a different route. They enter the mitochondria where they are progressively degraded into acetyl CoA molecules. Acetyl CoA, of course, enters the Krebs cycle directly (see Figure 5–14). Thus, digesting starch and fats is like chopping wood into pieces small enough to enter the respiratory stove.

Although individual cells carry out all phases of digestion on a small scale, many organisms conduct large-scale digestion at a site removed from the cells that ultimately utilize the digestion products. For example, animals digest foods in their guts and the digestion products are absorbed into the bloodstream where they are transported to the respiring cells. Plants often have specialized structures that store energy foods, such as the tuber of a potato, the bulb of an onion, and the enlarged root of a carrot plant. When the plant enters an active growth period (usually in spring), the stored foods are digested within the storage organ, and the products are then whisked off to the growing regions of the plant. Here they can be respired by the growing cells, or used as chemical building blocks for the synthesis of new cellular structures.

SUMMARY

With the emergence of photosynthesis on earth over 2 billion years ago, molecular oxygen began to accumulate slowly in the atmosphere. The presence of free oxygen made possible the evolution of respiration, a process in which simple organic compounds are completely broken down to inorganic substances (such as CO_2 and H_2O) in the presence of oxygen. Respiration releases a great deal of energy, and couples some of this energy to the formation of ATP molecules.

The splitting of ATP into ADP and P_i is an exergonic reaction, releasing energy to perform cellular work. For instance, the transfer of ATP's terminal phosphate to a glucose molecule increases the latter's chemical energy. Without the additional energy provided by the attached phosphate, glucose could not participate in many biosynthetic reactions, such as the formation of starch or glycogen. The energy of ATP is also applied to the endergonic transport of ions or molecules across membranes against their concentration gradients (active transport). Without a constant supply of ATP to drive these various forms of cellular work, the cell would die. For aerobic creatures, the ATP supply is maintained by respiration.

All reactions, whether endergonic or exergonic, have an activation energy barrier; that is, they need energy to get the reactants reacting. In the laboratory the activation energy is often supplied in the form of heat; in living cells enzymes perform this role through a process called catalysis. In enzyme catalysis, the activation energy requirement of a given reaction is lowered, which greatly increases the rate of that reaction. Virtually all chemical reactions that occur in organisms, including the many reactions of respiration, are catalyzed by enzymes.

The respiration of glucose can be divided into three sequential stages: glycolysis, which occurs in the soluble portion of the cytoplasm; the Krebs cycle, which takes place in the mitochondrial matrix; and respiratory electron transport, the components of which are located in the inner mitochondrial mem-

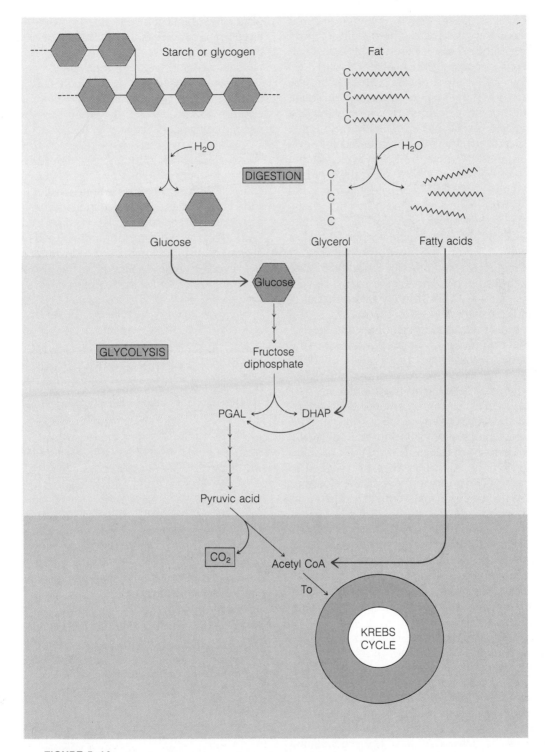

FIGURE 5–14
The Relationship between Digestion and Respiration. When starch or glycogen is digested, the resulting glucose units may enter glycolysis directly. The digestion of fats yields glycerol and fatty acids. Glycerol can be converted to DHAP, an intermediate in glycolysis. The fatty acids are broken down in the mitochondria to acetyl CoA units, which can enter the Krebs cycle.

brane. In glycolysis, a series of 10 enzyme-catalyzed steps, glucose is converted to two molecules of pyruvic acid. Some of the energy released in this pathway is transferred to two molecules each of NADH and ATP.

In aerobic organisms supplied with oxygen, the pyruvic acid enters the mitochondria, where it is completely broken down to carbon dioxide. In the first step, pyruvic acid is split into a molecule of CO_2 and a 2-carbon acetyl group that becomes acetyl CoA. The acetyl group then enters the Krebs cycle when it joins with a 4-carbon compound to yield the 6-carbon citric acid. As the Krebs cycle proceeds, two of the carbon atoms from citric acid come off as CO_2 molecules, and the other four are recycled to regenerate oxaloacetic acid. In the process, electrons and hydrogen ions are transferred to the electron carriers NAD^+ and FAD, forming NADH and $FADH_2$. In addition, one ATP is formed for every pyruvic acid that is broken down in this pathway. Thus, the chemical energy of pyruvic acid is transferred to ATP (1), NADH (4), and $FADH_2$ (1) during this second stage of respiration.

In the electron transport phase of respiration, the chemical energy of NADH and $FADH_2$ is "cashed in" for ATPs when these electron carriers pass their electrons to oxygen via the electron transport chain. The components of this chain can accept and donate electrons, each at a particular energy level. As the electrons move from one component to the next on the chain, a little bit of energy is released with each transfer. Some of these energy-releasing steps are coupled to the formation of ATP. As oxygen accepts the electrons at the end of the chain, it joins with hydrogen ions to become water.

For each glucose molecule that is completely broken down to CO_2 and H_2O molecules by the sequential actions of glycolysis, the Krebs cycle, and respiratory electron transport, 38 ATP molecules can be formed. The vast majority of these ATPs are produced in the mitochondria during the electron transport stage of respiration.

Under anaerobic conditions (no O_2), the Krebs cycle and electron transport cannot proceed because there is no terminal electron acceptor (the role of O_2 under aerobic conditions) to collect the electrons ultimately coming from glucose. Under these conditions, aerobic organisms are forced to ferment glucose, forming lactic acid (lactic acid fermentation) or ethyl alcohol (alcoholic fermentation). By regenerating NAD^+ needed in the glycolytic pathway, fermentation allows cells to produce 2 ATP molecules for each glucose that is converted to one of these fermentation products. While this is not enough ATP to sustain plants and animals for extended time periods, yeasts can limp by for weeks in the absence of oxygen, using the smaller amounts of ATP generated through alcoholic fermentation to keep them alive.

Obligate anaerobes, certain types of bacteria that cannot tolerate oxygen, make use of other terminal electron acceptors (not O_2) when they break down organic compounds for energy to make ATP.

The two most common types of stored energy fuels are polysaccharides and fats. Before entering one of the respiratory pathways, however, these large compounds must be digested into smaller units. Starch and other storage polysaccharides are digested into their constitutent sugar units. Fats are hydrolyzed to yield glycerol and fatty acids; fatty acids are further degraded into 2-carbon acetyl CoA units. The sugars and glycerol enter respiration at the glycolysis stage; acetyl CoA units enter the Krebs cycle.

STUDY QUESTIONS

1. Describe the relationship between respiration and photosynthesis in terms of what these processes consume and produce.

2. Why is ATP referred to as a cellular energy shuttle? Cite one example of how ATP provides energy for an endergonic reaction or process.

3. What structural features of enzyme molecules make them so susceptible to temperature changes?

4. How does an enzyme speed up a specific chemical reaction?

5. What are the net products of glycolysis? In which of these products lies most of the original chemical energy of the glucose molecule?

6. Where in a eukaryotic cell do glycolysis, the Krebs cycle, and respiratory electron transport take place?

7. What is the fate of the hydrogen atoms in pyruvic acid when it is broken down in the Krebs cycle? In what form do the carbon atoms appear?

8. What is the role of oxygen in the electron transport chain?

9. What are the metabolic consequences of oxygen deprivation in an animal cell? a yeast cell?

10. Why can't you live without oxygen?

11. At what step in respiration do the digestion products of starch enter? What are the points of entry for digested fats?

SUGGESTED READINGS

Becker, W. M. *Energy and the Living Cell: An Introduction to Bioenergetics.* New York: Harper & Row, 1977. This short text, available in paperback, nicely integrates information and concepts to provide an outstanding overview of how cells handle energy.

Hinkle, P. C., and R. E. McCarty. "How Cells Make ATP." *Scientific American* 238(1978):104–123. This article examines the details of ATP production in bacteria, mitochondria, and chloroplasts. Its outstanding illustrations are especially helpful in providing an understanding of this extremely complicated and important phenomenon.

Krebs, H. A. "The History of the Tricarboxylic Acid Cycle." *Perspectives in Biology and Medicine* 14(1970):154ff. H. A. Krebs was the discoverer of the tricarboxylic acid cycle (also called the Krebs Cycle). This article therefore provides a unique perspective on how metabolic pathways are unraveled.

Lehninger, A. L. *Principles of Biochemistry.* New York: Worth, 1982. This is an advanced text written by a leading biochemist. Nevertheless, the clarity of its narrative makes many chapters understandable for the motivated beginning biology student.

Stryer, L. *Biochemistry.* 2nd ed. San Francisco: W. H. Freeman, 1980. This introductory biochemistry text has an especially lucid description of cellular respiration with clear illustrations.

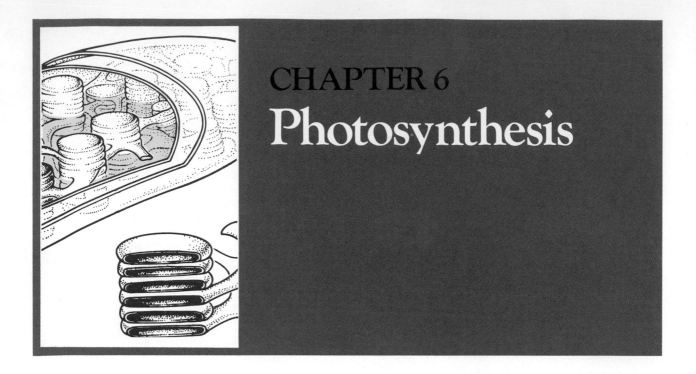

CHAPTER 6
Photosynthesis

It would be difficult to overemphasize the importance of photosynthesis to life on earth, for this process provides the biological energy to sustain, directly or indirectly, all but a few types of organisms. **Photosynthesis** is the capture of light energy and its conversion into the chemical energy of organic compounds such as carbohydrates. These carbohydrates are then stored, used as fuel for respiration, or used as chemical building blocks for growth. The photosynthesizers themselves—green plants and various one-celled organisms—benefit directly from photosynthesis, but all animals and fungi benefit indirectly from photosynthesis by consuming plants or other organisms that eat plants. Thus, without the constant flow of light and its conversion into the chemical energy of organic compounds, virtually all life forms would starve to death.

The question of how plants obtain their food has intrigued observers for thousands of years. The early Greeks believed that plants received their nourishment from the soil. They thought that because plants grow in the ground, the soil is their "placenta of life"; that is, they absorb food into their roots. Actually this suggestion was not unreasonable considering the state of scientific knowledge at the time, and the so-called soil-eater concept persisted until the early seventeenth century. Perhaps it was fostered by the common observation that fertilizers added to the soil improved the growth and overall healthy appearance

of plants. The soil-eater notion was finally laid to rest, however, when the Belgian physician Jean Baptiste van Helmont (1577–1644) reported the results of a simple but elegant experiment:

> That all vegetable matter immediately and materially arises from the element of water alone I learned from this experiment. I took an earthenware pot, placed in it 200 lb. of earth dried in an oven, soaked this with water, and placed in it a willow shoot weighing 5 lb. After five years had passed the tree growth therefrom weighed 169 lb. and about 3 oz. But the earthenware pot was constantly wet only with rain or (when necessary) distilled water; and it was ample in size and imbedded in the ground; and, to prevent dust flying around from mixing with the earth, the rim of the pot was kept covered with an iron plate coated with tin and pierced with many holes. I did not compute the weight of the deciduous leaves of the four autumns. Finally, I again dried the earth of the pot, and it was found to be the same 200 lb. minus about 2 oz. Therefore, 164 lb. of wood, bark, and root had arisen from the water alone.*

Though van Helmont was correct in concluding that plants are not soil eaters, he was mistaken in

*From Jean Baptiste van Helmont, "By Experiment, that All Vegetable Matter Is Totally and Materially of Water Alone," translated by Naphtali Lewis in Mordecai L. Gabriel and Seymour Fogel, eds., *Great Experiments in Biology,* © 1955, p. 155. Englewood Cliffs, NJ: Prentice-Hall.

assuming that water alone was the source of his willow plant's increased mass. What he overlooked was the air around the plant. We now know that atmospheric carbon dioxide is of prime importance to photosynthesis. In sunlight, plants absorb carbon dioxide (CO_2) and expel molecular oxygen (O_2). Both the carbon atoms and oxygen atoms of CO_2 are incorporated into organic compounds during photosynthesis. Water, on the other hand, contributes only its hydrogen atoms. Thus, the bulk of the dry weight of a plant is derived from carbon dioxide. We can summarize photosynthesis as follows:

$$6\,CO_2 + 6\,H_2O \xrightarrow{\text{Light}} C_6H_{12}O_6 + 6\,O_2$$

Carbon   Water   Carbohydrate   Oxygen
dioxide

One way to appreciate the magnitude of this process is to consider how much carbon dioxide plants incorporate (fix) into organic compounds: in a single year, plants take in an estimated 170 billion tons of carbon dioxide in photosynthesis. Our present atmosphere contains only about 700 billion tons of carbon dioxide, so over 20% of the earth's supply of this gas is converted into living matter by photosynthetic organisms each year. This fixed carbon dioxide is not lost forever from the atmosphere, however; it is constantly replenished by processes that release carbon dioxide—primarily respiration (see Chapter 5), and to a lesser but increasing extent, the combustion of fossil fuels by one organism, *Homo sapiens* (Figure 6–1).

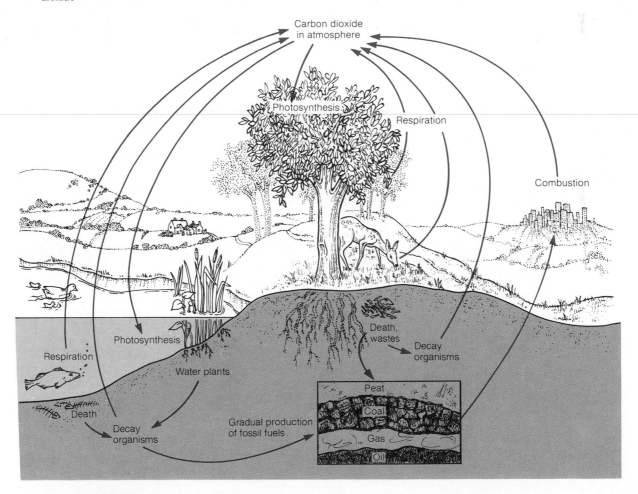

FIGURE 6–1

The Carbon Cycle. Since the amount of carbon dioxide in our atmosphere remains relatively constant, carbon dioxide incorporation by photosynthesis must be balanced by processes that release carbon dioxide, some of which are shown here. As in all recycling systems, the balance can be disturbed if any part of the cycle is eliminated, partly destroyed, or enhanced.

The energy that drives photosynthesis is visible light. Thus, in order to understand the process of photosynthesis, it will be helpful to have a brief look at the nature of light and the photosynthetic pigments that absorb it.

LIGHT AND THE PHOTOSYNTHETIC PIGMENTS

The natural source of light for photosynthesis is the sun. Sunlight is not uniform; rather, it is a mixture of various types of radiation, each with different energy values and wavelengths (Figure 6–2). We humans are not ordinarily aware of the complete range of sunlight because our eyes are sensitive only to the relatively narrow band of radiation called visible light. The longest wavelengths we perceive are red; the shortest are violet. We also recognize colors of intermediate wavelengths—orange, yellow, green, and blue. But what portion of the sun's total spectrum is used in photosynthesis? We can answer this question experimentally by exposing different plants to various parts of the spectrum (ultraviolet light, visible light, and so on) and then seeing whether any of the plants carry out photosynthesis. The results of this experiment show that only the visible wavelengths of light are effective in photosynthesis.

Since we know that visible light contains a mix of wavelengths, or colors, we can now ask which wavelengths of visible light are most effective in causing photosynthesis. This question was answered by T. W. Englemann, a German physiologist, in 1882. Englemann placed a green alga (a tiny aquatic plant) on a glass slide and used a specially constructed microscope to view the alga from above while illuminating it from below. Using a prism to separate white light into its component colors, he irradiated the alga with a rainbow of colors. To determine which colors most effectively stimulated the alga to give off oxygen, a measure of photosynthesis, Englemann introduced oxygen-loving bacteria onto the slide. These bacteria were most stongly attracted to the areas where red and blue light struck the algal filament (Figure 6–3), so Englemann concluded that these colors were most effective in stimulating photosynthesis. Englemann's conclusion has since been verified and refined for many kinds of plants with the aid of more sophisticated instruments.

For visible light to affect any biological process, including photosynthesis, it must first be absorbed by compounds called **pigments.** Different pigments absorb different wavelengths of light. The pigments involved in absorbing light energy for photosynthesis in higher plants are the **chlorophylls** (*a* and *b*) (Figure 6–4) and the **carotenoids.** The chlorophylls give plants their characteristic green color; the yellow, orange, and red carotenoids are generally masked by the abundant chlorophylls, but show their colors in autumn leaves, in the skins of bananas, tomatoes, and other fruits, and in the roots of carrots.

One clue that the chlorophylls and carotenoids are active in photosynthesis comes from the parallel between the absorption spectra of these pigments and the action spectrum of photosynthesis. An **action spectrum** is a graph that shows the relationship between some *action* affected by light (such as photosynthesis) and the different wavelengths of light (different parts of the radiation *spectrum*). The results of Englemann's experiment can be plotted in the form of an action spectrum for photosynthesis, which shows that the rate of photosynthesis is highest for violet-blue and red wavelengths (see Figure 6–3b). An **absorption spectrum,** on the other hand, is a graph of a pigment's ability to *absorb* different parts of the spectrum. Each type of pigment absorbs different

FIGURE 6–2

The Solar Radiation Spectrum. Radiation moves in waves, and the distance between waves (wavelength) depends on the type of radiation. The solar radiation spectrum extends from the shortest wavelength (highest energy) gamma rays to the longest wavelength (lowest energy) radio waves. The human eye can detect only a small band of solar radiation called the visible spectrum, which extends from the violet rays (wavelength 380 nm) to the far-red (wavelength 750 nm).

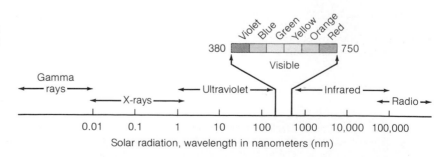

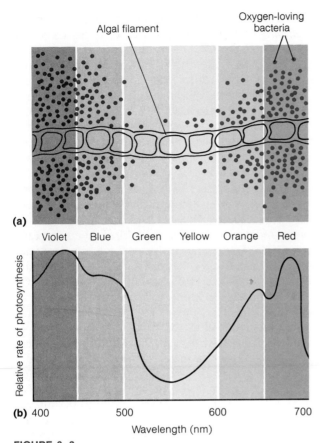

Algal filament

Oxygen-loving bacteria

(a)

Violet   Blue   Green   Yellow   Orange   Red

Relative rate of photosynthesis

(b) 400   500   600   700

Wavelength (nm)

FIGURE 6–3
Action Spectrum of Photosynthesis.
(a) In 1882, T. W. Englemann placed a microscope slide containing a green algal filament and oxygen-loving bacteria over a light source that had been split into its component wavelengths by a prism. Since most of the bacteria clustered around parts of the alga illuminated by violet-blue and red light, Englemann reasoned correctly that these wavelengths were most effective in causing photosynthesis (oxygen production) in the alga. (b) With the aid of more sophisticated instruments, scientists have been able to measure more accurately the effect of different wavelengths on the rate of photosynthesis. Note how this action spectrum of the same green alga used by Englemann is a good reflection of his results obtained with the bacteria.

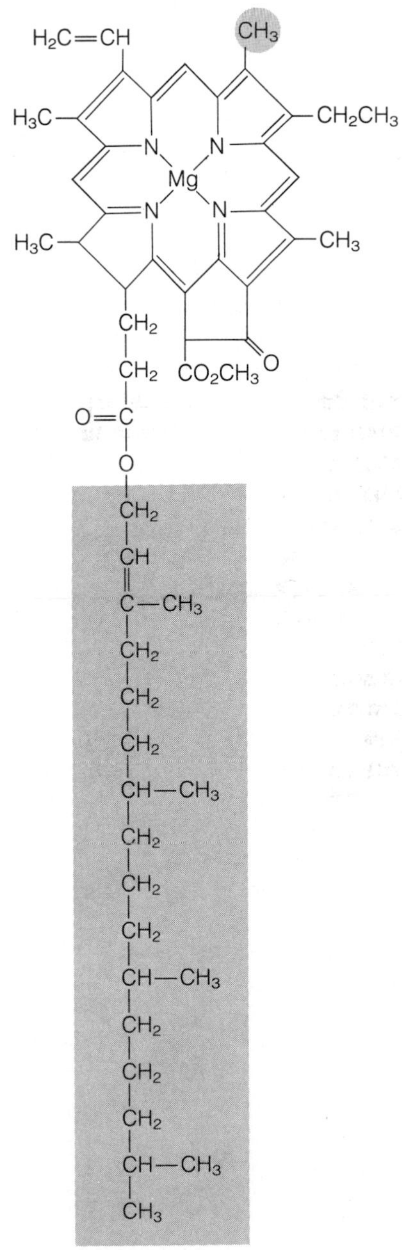

$H_2C=CH$

CH_3

H_3C

CH_2CH_3

N   N

Mg

N   N

H_3C

CH_3

CH_2

CH_2 CO_2CH_3 O

$O=C$

O

CH_2

CH

$C-CH_3$

CH_2

CH_2

CH_2

$CH-CH_3$

CH_2

CH_2

CH_2

$CH-CH_3$

CH_2

CH_2

CH_2

$CH-CH_3$

CH_3

FIGURE 6–4
Chlorophyll *a*. The chlorophyll *a* molecule consists of a long hydrocarbon "tail" (gray) joined to a group of rings containing a magnesium (Mg) at its center. It is the group of rings in chlorophyll that absorbs light. Chlorophyll *b* differs from chlorophyll *a* in having a —CHO group in place of the —CH_3 (color).

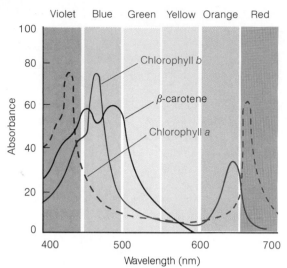

FIGURE 6–5
Absorption Spectra of Three Photosynthetic Pigments. Chlorophylls *a* and *b* both exhibit peaks of light absorption in the violet-blue and red regions of the visible spectrum, while absorbing green light poorly. Because the chlorophylls reflect more green light than other parts of the spectrum, these pigments (and the leaves that contain them) appear green to our eyes. *β*-carotene, a carotenoid, absorbs only in the blue-green region, and reflects yellow light the best.

wavelengths with a different degree of effectiveness, absorbing certain wavelengths well and others poorly or not at all. The absorption spectra for the chlorophylls and the carotenoids taken together matches the action spectrum for photosynthesis very closely (compare Figure 6–3b and Figure 6–5). Hence, we may tentatively conclude that the chlorophylls and carotenoids are the light-absorbing pigments active in photosynthesis—a conclusion that has been substantiated by many other kinds of evidence.

In plants, the photosynthetic pigments are located within chloroplasts. You may recall from Chapter 3 that chloroplasts, organelles found in photosynthetically active plant cells (see Figure 3–14), are bound by two membranes (Figure 6–6). The inner membrane is connected to a series of internal membranous sacs called **thylakoids.** Most of the thylakoids are arranged into interconnected stacks called **grana** that are suspended throughout the soupy chloroplast matrix, or **stroma.** The chlorophyll and carotenoid pigments are embedded within the thylakoid membranes in close association with various electron-transporting compounds. The intimate association of the light-harvesting pigments with these electron carriers is critical to the conversion of light energy into stable chemical energy. Let us now see how this life-sustaining conversion works.

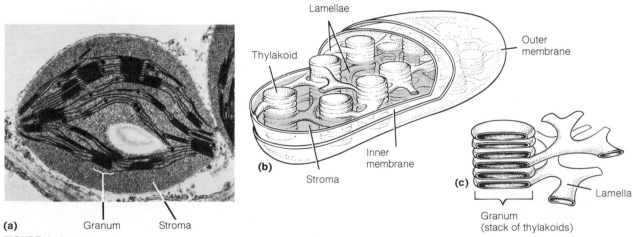

FIGURE 6–6
The Chloroplast. (a) This electron micrograph of a chloroplast from a leaf cell shows the boundary double membrane and the complex series of internal membrane sacs, the thylakoids. The thylakoid membranes contain the chlorophylls and carotenoids that capture light in photosynthesis. Many of the thylakoids are arranged in stacks called grana, which are suspended in a liquid matrix termed the stroma. (Magnification: 14,300×.) (b) This artist's conception of a chloroplast provides a three-dimensional view of how the grana are interconnected to one another by lamellae, a feature emphasized in (c), a cutaway diagram of a granum and its associated lamellae.

THE LIGHT AND DARK REACTIONS

Overview of Photosynthesis

The individual steps of photosynthesis fall into two general sets of reactions: the light and dark reactions. In the **light reactions,** light energy is captured by pigments and transformed into chemical energy in the forms of ATP and NADPH (nicotinamide adenine dinucleotide phosphate; similar to NADH of respiration). During this process water molecules are split, releasing molecular oxygen (O_2) as a by-product (Figure 6–7). In the **dark reactions,** carbon dioxide (CO_2) is converted to carbohydrate. By producing carbohydrates, the dark reactions provide the food that supports virtually all organisms, including you.

The light and dark reactions of photosynthesis take place within different compartments of the chloroplast. The light reactions occur on or within the thylakoid membranes, where the pigments and electron carriers are located; the dark reactions are catalyzed by enzymes, which occur in the stroma. These two sets of processes are linked chemically by the diffusion

of the products of the light reactions, ATP and NADPH, from the thylakoid surfaces to the stroma, where they are used to drive the conversion of CO_2 into carbohydrate. The use of ATP and NADPH in the dark reactions also generates products that are necessary for the light reactions: ADP, P_i, and $NADP^+$ (see Figure 6–7). The light reactions not only provide the chemical energy to drive the dark reactions of photosynthesis, but also generate practically all the oxygen in the atmosphere needed by aerobic organisms for respiration. Thus, photosynthesis provides all respiring organisms with fuel (carbohydrates) and the oxygen needed to "burn" that fuel. Let us now examine the light and dark reactions in more detail.

The Light Reactions

As we said, the light reactions begin with the absorption of light by pigments and end with the formation of ATP and NADPH. The light-harvesting chlorophylls and carotenoids are organized into units called photosystems located in the thylakoid membranes. Each **photosystem** is made up of several hundred antenna pigment molecules, including chlorophyll *a*, chloro-

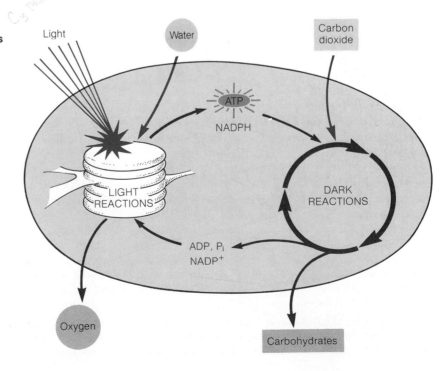

FIGURE 6–7
Overview of the Light and Dark Reactions of Photosynthesis. Photosynthesis takes place in the chloroplasts of plant cells. The light reactions, which occur in the thylakoids, convert light energy into the chemical energy of ATP and NADPH. During this process, water is split and oxygen is given off. The dark reactions, which take place in the stroma, utilize ATP and NADPH to convert carbon dioxide to carbohydrate, the basic food source for all organisms. ADP, P_i, and $NADP^+$ are recycled from the dark reactions to the light reactions.

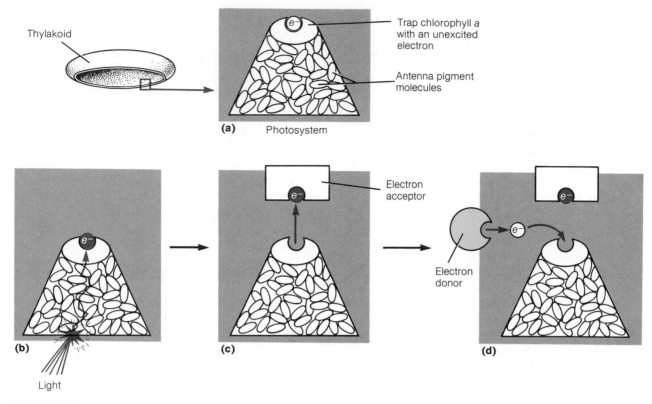

FIGURE 6–8
How a Photosystem Works. (a) A photosystem consists of several hundred antenna pigment molecules and a single molecule of trap chlorophyll *a*. Photosystems are located in the thylakoid membranes. (b) When light strikes a photosystem, light absorbed by one of the antenna pigment molecules is passed from pigment to pigment until it excites an electron in the trap chlorophyll *a*. (c) The excited trap chlorophyll *a* electron is lost to an electron acceptor substance. (d) The missing electron in trap chlorophyll *a* is replaced by an electron given up by an electron donor.

phyll *b,* and various types of carotenoids. In addition, each photosystem has a single molecule of a special type of chlorophyll called **trap chlorophyll *a*** (Figure 6–8a). When a unit of light strikes a photosystem, an electron in one of the antenna pigment molecules becomes excited to a higher energy level, and the energy of this excited electron is passed from one antenna molecule to another until it reaches and excites an electron in the trap chlorophyll *a* (Figure 6–8b). The excited trap chlorophyll *a* electron is then transferred to a nearby electron acceptor (Figure 6–8c). This leaves the trap chlorophyll *a* with an electron hole (a missing electron), which is quickly filled by an electron coming from an electron donor (Figure 6–8d). The trap chlorophyll *a* has now come full circle and is ready to accept another unit of energy from its associated antenna pigments. These events are repeated over and over, and each time a unit of light

energy is captured by a photosystem, an electron is boosted from an electron donor to a relatively higher energy level in an electron acceptor.

In the currently favored model of the light reactions, there are two different kinds of photosystems that work together. The two photosystems, designated **photosystem I** (**PS I**) and **photosystem II** (**PS II**), are linked by a chain of electron transporting molecules; there is also a short electron transport chain associated with the PS I electron acceptor. This arrangement of the photosystems with their associated electron transport chains is depicted in a model called the **Z-scheme** (so called because it looks like a Z on its side) (Figure 6–9).

When light shines on a plant leaf, both photosystems operate simultaneously to pump excited electrons to their corresponding electron acceptors. Each high-energy electron is then passed down an electron

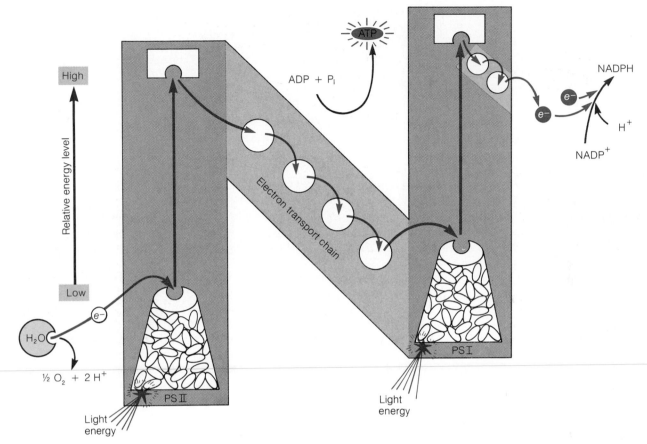

FIGURE 6–9
The Z-scheme. The Z-scheme shows how the two photosystems (I and II) are joined by an electron transport chain. There is also a short electron transport chain associated with the electron acceptor of photosystem I. The arrows indicate the direction of electron flow. Note that electrons begin at a low energy level (the electron donor of photosystem II) and end up at a high energy level (in NADPH). The input of light energy makes this ''uphill'' electron flow possible. The electron transport between photosystem II and I is ''downhill,'' and some of the energy released during this process is used to join ADP and P$_i$ into ATP.

transport chain. When two electrons coming from PS I have traveled down the short electron transport chain, they join with NADP$^+$ and a hydrogen ion (H$^+$) to form NADPH. Meanwhile, the electrons lost from PS I are replaced by ones coming from PS II by way of the long electron transport chain linking PS II to PS I. As electrons move down in energy along this chain, some of the released energy is captured when ADP and P$_i$ join into ATP.

We now have achieved the purpose of the light reactions: ATP and NADPH have been formed. There is one item of unfinished business, however: the electron hole in the trap chlorophyll *a* of PS II. Water is the electron donor for PS II, and in an enzyme-catalyzed reaction, a water molecule is split into two

electrons, two hydrogen ions, and an oxygen atom. These electrons (one at a time) fill the electron hole of PS II trap chlorophyll *a*, and the hydrogen ions are used in the formation of NADPH from NADP$^+$. For every two water molecules that are so split, two oxygen atoms are generated, which then unite to form a molecule of oxygen (O$_2$).

We have presented the light reactions as if they occur in stages, but in actuality it would be more accurate to envision this process as a continuous flow of electrons from water to NADP$^+$, powered by the energy of sunlight. The participation of two cooperating photosystems is necessary to boost these electrons from their relatively low energy level in a water molecule to the comparatively high energy state of NADPH, and as the electrons move from PS II to PS I, a little extra energy is released for ATP formation. Thus, at the expense of water molecules and visible light, the thylakoid components generate chemical energy in the forms of ATP and NADPH, which are then put to work in the second phase of photosynthesis—the dark reactions.

The Dark Reactions

The dark reactions of photosynthesis are concerned with the conversion of CO_2 to carbohydrate. The dark reactions do not utilize light energy directly, but they will come to a rapid halt in the absence of light because they depend on a supply of ATP and NADPH generated by the light reactions. Thus, the term *dark reactions* is somewhat misleading.

The individual steps of the dark reactions were worked out by a group of scientists headed by Melvin Calvin at the University of California, Berkeley in the early 1950s. Calvin and his coworkers took advantage of the newly discovered radioactive isotope of the carbon atom—^{14}C (reads "carbon fourteen"). Starting with $^{14}CO_2$, these workers successfully traced the path of carbon through the individual steps of the dark reactions (Essay 6–1). In honor of his contribution in describing this photosynthetic pathway—now known as the **Calvin cycle**—Calvin received the Nobel Prize in Chemistry in 1961.

THE CALVIN CYCLE.   In order to understand what happens in the Calvin cycle, it is helpful to divide this process into three major steps: CO_2 fixation, carbon reduction, and the rearrangement of sugars. In the **CO_2 fixation** reaction, CO_2 becomes *fixed* into organic form. In the chloroplast stroma, CO_2 reacts with the 5-carbon sugar RuBP (ribulose bisphosphate), forming a 6-carbon compound (Figure 6–10). This compound is very unstable and immediately splits in half to form two molecules of a 3-carbon acid, PGA (phosphoglyceric acid). One of the PGA molecules contains the newly fixed CO_2 group.

The next reaction of the pathway is the **carbon reduction** step. The two molecules of PGA are chemically reduced to the 3-carbon sugar PGAL (phosphoglyceraldehyde) by the addition of electrons and hydrogen ions from NADPH. This step requires two molecules each of ATP and NADPH, products generated by the light reactions of photosynthesis (in the absence of light, the Calvin cycle becomes blocked at this step due to the lack of ATP and NADPH).

The remaining reactions of the Calvin cycle involve the rearrangement of the carbon atoms of PGAL, a 3-carbon sugar, into other types of sugars, including the 6-carbon sugar glucose. Most of the sugars are converted back to RuBP, thereby recycling the 5-carbon sugar that attaches to CO_2. Figure 6–11 summarizes these major steps of the Calvin cycle.

To gain some perspective on the Calvin cycle, you should realize that each chloroplast in a plant cell contains hundreds of copies of the enzymes that participate in these reactions, and literally millions of molecules of PGA, PGAL, RuBP, and the other sugars on the pathway. The fixation of each CO_2 molecule in a chloroplast represents a net addition of one carbon atom (and two oxygen atoms) to the pool of sugars in the stroma. If an illuminated chloroplast incorporates 6000 CO_2 molecules every second, it would show a profit of 1000 glucose molecules ($C_6H_{12}O_6$) per second. Thus, while most of the carbon atoms are recycled back to RuBP (necessary for CO_2 fixation to continue), one net glucose molecule forms and leaves

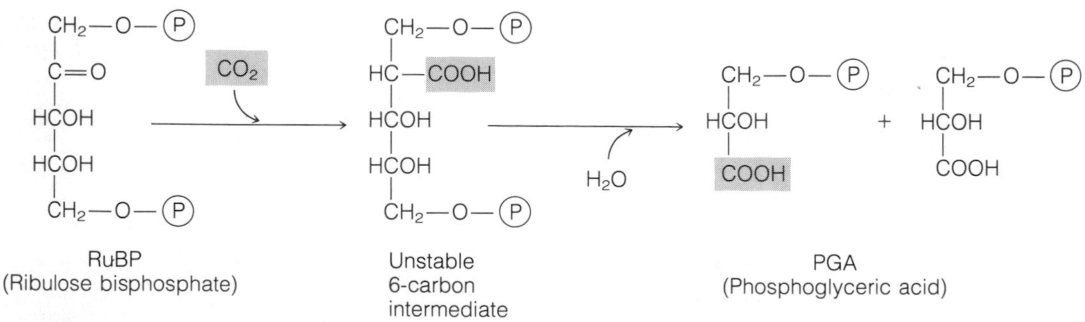

FIGURE 6–10
CO_2 Fixation Step of the Calvin Cycle. This two-step reaction results in the formation of two molecules of PGA (phosphoglyceric acid), one of which has the newly-fixed CO_2 group (gray).

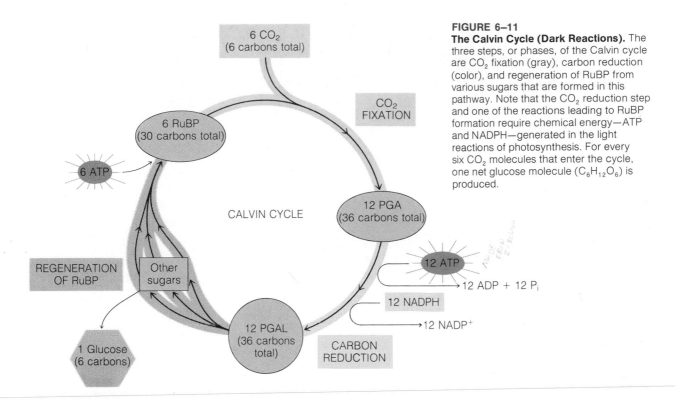

FIGURE 6–11
The Calvin Cycle (Dark Reactions). The three steps, or phases, of the Calvin cycle are CO_2 fixation (gray), carbon reduction (color), and regeneration of RuBP from various sugars that are formed in this pathway. Note that the CO_2 reduction step and one of the reactions leading to RuBP formation require chemical energy—ATP and NADPH—generated in the light reactions of photosynthesis. For every six CO_2 molecules that enter the cycle, one net glucose molecule ($C_6H_{12}O_6$) is produced.

the cycle for each six CO_2 molecules that enter it (see Figure 6–11).

C_4 PHOTOSYNTHESIS. All green plants carry out the Calvin cycle. Many plants, however, have an additional component in their dark reactions that helps in the capture of CO_2. These plants fix CO_2 initially into a 4-carbon acid (oxaloacetic acid), which is subsequently converted to other 4-carbon compounds (Figure 6–12). Because they initially fix CO_2 into a 4-carbon compound, these plants are called **C_4 plants.** Many of the tropical grasses, such as sugarcane, bermuda, corn, sorghum, and some species of bamboo, are C_4 plants. (Plants that lack this C_4 component fix CO_2 directly into the 3-carbon PGA molecule via the Calvin cycle; they are called C_3 plants.)

The 4-carbon compounds produced by C_4 plants are not converted directly to sugars. Instead, they are transported from the leaf cells in which they are formed (cells which do not have Calvin cycle activity)

FIGURE 6–12
Initial CO_2 Fixation Step in C_4 Plants.
In C_4 plants, CO_2 is initially fixed into a 4-carbon compound (oxaloacetic acid). The reaction involves the addition of CO_2 to the 3-carbon PEP(phosphoenolpyruvic acid) molecule. The oxaloacetic acid is then converted to other 4-carbon compounds (malic acid or aspartic acid).

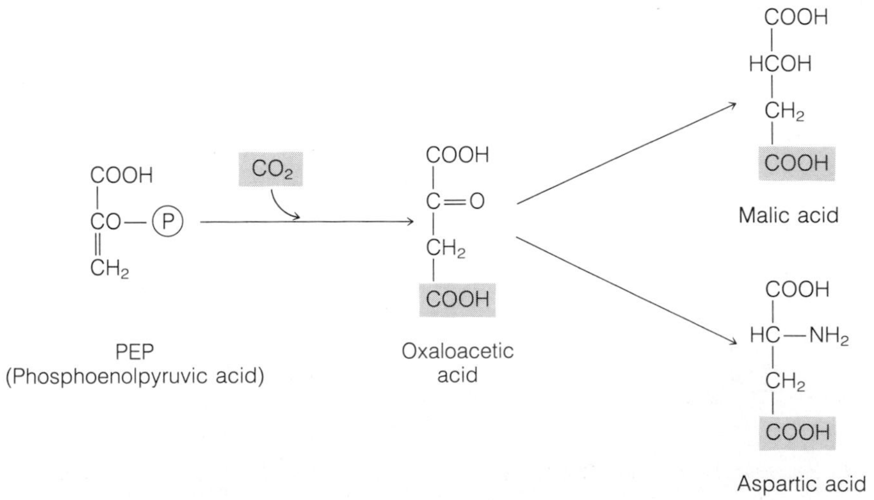

ESSAY 6-1

THE UNRAVELING OF C_3 PHOTOSYNTHESIS

Being a good biochemist, Melvin Calvin knew that the photosynthetic conversion of carbon dioxide to sugar was not a simple, one-step reaction. He reasoned that CO_2 must react initially with some organic compound in the chloroplast, and then the product of this CO_2 fixation reaction must be converted into sugar by some yet-undiscovered metabolic pathway. When the ^{14}C isotope of carbon (see figure) became available shortly after World War II, Calvin and his coworkers at the University of California, Berkeley, had the "tag" they needed to trace the pathway of the carbon atom from CO_2 to sugar. The rationale for the experiments they conducted goes as follows: When a photosynthesizing plant is exposed to $^{14}CO_2$, this radioactive molecule will react with compound X (an unknown) to form compound Y. Compound Y will then be converted to a second compound on the pathway, and this to a third compound, and so on. One by one the entire sequence of chemical intermediates on the pathway will become labeled with ^{14}C, eventually leading to radioactive sugars and even radioactive starch. Thus, the longer the exposure to $^{14}CO_2$, the more compounds containing ^{14}C should appear. On the other hand, shorter exposure times should yield fewer labeled compounds, so by decreasing the time that the plants are exposed to $^{14}CO_2$, one should be able to determine the order of intermediates through which carbon atoms flow from CO_2 to sugars. Now let's have a look at how Calvin's group put this reasoning to the test.

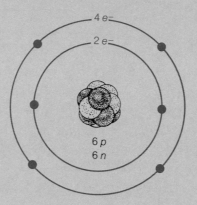

^{12}C
Stable isotope of carbon

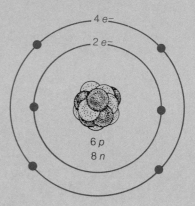

^{14}C
Radioactive isotope of carbon

Decays to a nitrogen atom, releasing a detectable energy particle

^{14}C and ^{12}C are known as different **isotopes** of carbon because they differ only by the number of neutrons in their atoms. Their chemical properties, which hinge on the number of protons and electrons, are identical.

The Berkeley scientists grew a population of *Chlorella,* a single-celled green alga common to many ponds and lakes, in a glass container. Then they injected $^{14}CO_2$ into the algal suspension. At various times thereafter (after short, medium, or long exposure times), samples of the algae (now containing ^{14}C-labeled compounds by virtue of photosynthesis) were released into a flask of boiling alcohol (see figure). The hot alcohol not only stopped all metabolic reactions but also dissolved all the small organic compounds present in these now-dead cells.

Next, they separated the scores of organic compounds dissolved in the alcohol by a technique known as paper chromatography. To do this, the researchers concentrated

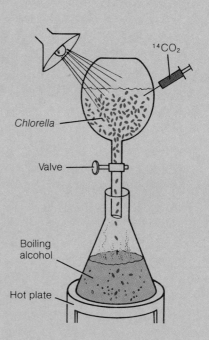

$^{14}CO_2$

Chlorella

Valve

Boiling alcohol

Hot plate

the alcohol extract by evaporation, then applied a small drop of the concentrate near one of the corners of a sheet of fibrous paper. When the spot dried, they placed the paper into a sealed glass chamber containing a solvent so that the bottom edge of the paper was immersed in the solvent, and the spot (alcohol extract) was just above the liquid (see figure). Slowly, the solvent crept up the paper by capillary action, dissolving the various compounds in the spot and carrying them upward.

Paper chromatography is a powerful tool used by biochemists to separate different compounds in a mixture from one another. Since different compounds have different abilities to stick to the paper, and also have different solubilities in the solvent traveling up the paper, they will migrate up the paper at different rates. For example, a compound that has little tendency to adhere to the paper and is very soluble in the moving solvent will be carried up the paper rapidly. It will thus separate from other compounds that have greater adhesions to the paper and/or less solubilities in the solvent. The Berkeley group used this separation technique to great advantage. After chromatographing their extract in one direction to achieve an initial separation of the *Chlorella* compounds, they subjected the resulting line of partially separated compounds to a second chromatography at right angles to the first, using a different solvent.

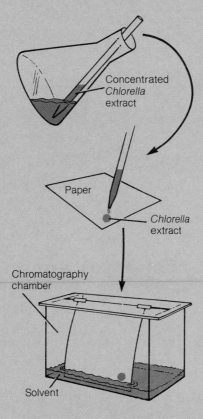

Concentrated
Chlorella
extract

Paper

Chlorella
extract

Chromatography
chamber

Solvent

Next, Calvin's group sprayed the chromatography paper with chemical agents that react with specific types of compounds to produce colored products. They could then identify the colored spots by comparing their position on the paper with those of known compounds that had been similarly chromatographed. For example, glucose, whether from a bottle on the shelf or a chloroplast, always ends up at the same spot on a chromatogram when the conditions of chromatography are the same.

Finally, Calvin's group determined which of the compounds in the *Chlorella* extract were radioactive— that is, which compounds became labeled with ^{14}C atoms coming from $^{14}CO_2$. They did this by placing a sheet of radiation-sensitive film over the chromatogram, a technique called autoradiography. Radioactivity from any ^{14}C-labeled compound caused the film to become developed over the spot containing that compound, just as light causes camera film to become developed. As predicted, longer exposures of *Chlorella* to $^{14}CO_2$ resulted in greater numbers of radioactive compounds. When the exposure of the algae to $^{14}CO_2$ was limited to a mere three seconds, only one compound was significantly radioactive—phosphoglyceric acid (PGA), Calvin's compound *Y* and the first stable product of CO_2 fixation (see Figure 6–10). Slightly longer exposures resulted not only in ^{14}C-PGA, but also other intermediates further along the pathway to sugar formation.

By taking advantage of the discovery of ^{14}C, and the advents of paper chromatography and autoradiography, Calvin and his coworkers were able to elucidate the entire path of carbon in what is aptly called the Calvin cycle.

to another type of leaf cell nearby, where they are broken down to CO_2 and a 3-carbon compound (Figure 6–13). This CO_2 is then affixed to RuBP in the Calvin cycle taking place in the chloroplasts of these cells. The 3-carbon unit returns to the first cell type. Thus, C_4 plants fix CO_2 twice in two different types of leaf cells.

You are probably wondering why some plants have this "extra" mechanism for fixing CO_2 in photosynthesis. Why not fix CO_2 directly via the Calvin cycle and be done with it? The answer has to do with efficiency. In the tropical habitats where most C_4 plants grow, the vegetation is often subject to very high temperatures and periodic droughts. Under these conditions, C_4 plants are more efficient at fixing CO_2 than C_3 plants. With their higher rates of food manufacture, C_4 plants can grow and reproduce more effectively than C_3 plants living under the same stressful conditions.

An example of the C_4 competitive edge occurs in lawns across the United States during the summer. Most lawns consist primarily of C_3 grasses, such as fescue, rye grass, and Kentucky bluegrass. In areas where the summers are hot and dry, however, the lawns often become overrun by a yellowish, thick-stemmed pest called crabgrass. As you may have suspected, crabgrass is a C_4 plant. The crabgrass, which most people regard as an eyesore, outcompetes the prettier C_3 grasses.

THE USES OF SUGAR

The glucose molecules produced in the Calvin cycle represent the profit a plant reaps from an initial investment of H_2O, CO_2, and a little light energy. These sugar molecules have several possible fates in the plant

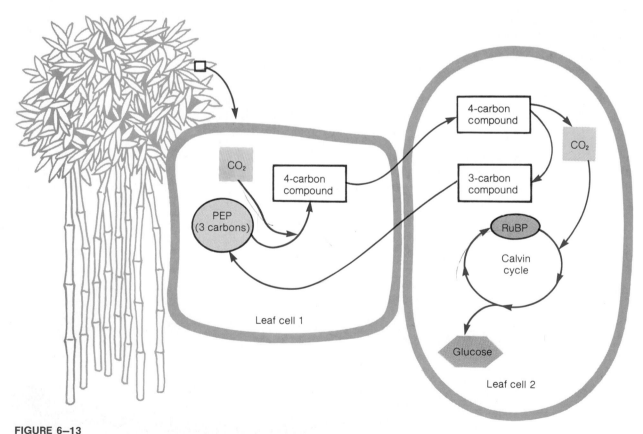

FIGURE 6–13
The Dark Reactions of C_4 Plants. In the leaves of C_4 plants, CO_2 is first fixed into 4-carbon compounds in one type of cell (leaf cell 1), which are then transported to another type of cell (leaf cell 2). The 4-carbon compounds here are split into a 3-carbon compound (which is transported back to leaf cell 1) and CO_2. The CO_2 then enters the Calvin cycle occurring in leaf cell type 2.

(Figure 6–14). When photosynthesis rates are high, some of the glucose is (1) polymerized into starch, which accumulates as visible starch grains inside the chloroplast (see Figure 2–4a). Most of the glucose, however, moves out of the chloroplast into the leaf cell's cytoplasm. Here it can be used as: (2) a chemical energy source for the production of ATP in respiration; (3) a chemical building block for the synthesis of other cellular constituents, such as amino acids, lipids, nucleotides, or even the pigments necessary for photosynthesis itself; or (4) a starting material for the synthesis of sucrose (glucose-fructose disaccharide). Sucrose is the major type of sugar that is transported from the leaves to the nonphotosynthetic regions of the plant—roots, stems, flowers, and fruits. The cells on the receiving end utilize sucrose for the same activities that occur in leaf cells: storage, respiration, and biosynthesis.

When sugars are used as chemical building blocks, they enter into various combinations with mineral elements (nitrogen, phosphorus, sulfur, and others) obtained from the soil to form all the organic compounds of which plants are composed. Animals, fungi, and decaying bacteria depend on these plant compounds, directly or indirectly, for their nutrition. By eating or absorbing plant matter (or animal matter from

FIGURE 6–14
The Fates of Glucose. Under optimal conditions for photosynthesis, a single chloroplast produces thousands of glucose molecules each second. Some of this sugar is stored in the chloroplast in the form of starch (1), but most of it enters the leaf cell's cytoplasm where it may be respired to regenerate ATP (2), or used as a chemical building block for the synthesis of other cellular compounds (3), including sucrose (4). Sucrose is transported out of the leaves to nonphotosynthetic regions of the plant.

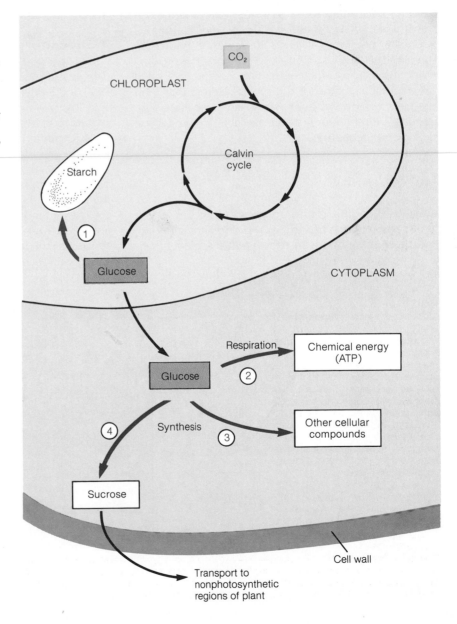

plant eaters), these heterotrophs obtain chemical energy and raw material to support respiration and biosynthetic activities in their cells. Thus, photosynthesis is important for all organisms, not just plants. As sophisticated as human technology is, we have yet to come up with a better way to manufacture food.

THE LIMITING FACTORS OF PHOTOSYNTHESIS

Photosynthesis is the means by which plants make the basic organic compounds needed for maintenance and growth. Thus, the overall productivity of a plant, a garden, or an entire forest to a large extent hinges on the rate of photosynthesis. For this reason, agricultural scientists have sought ways to optimize photosynthesis in the field in an effort to enhance the productivity of our crops. Much of this research has depended on an understanding of the natural factors that limit the rate of photosynthesis. These factors include light intensity, temperature, water availability, and carbon dioxide concentration.

Light Intensity

Obviously, plants cannot photosynthesize without light. Light makes ATP and NADPH formation possible, and these compounds in turn provide the energy to convert CO_2 to carbohydrate. Predictably, therefore, as the light intensity applied to a plant increases, the rate of photosynthesis increases proportionally. Most plants, however, reach a point above which further increases in light intensity do not increase photosynthesis. We call this the **light saturation point.** Plants vary greatly with respect to their light saturation points. For example, corn, a sun-loving plant, does not become light saturated even at full sunlight intensity; *Oxalis rubra,* a shade-loving plant, becomes saturated at relatively low light intensities (Figure 6–15). Interestingly, not only does the level of light required to saturate photosynthesis vary with the plant species, but individual plants can adjust somewhat to changes in light intensity with time.

Temperature

Temperature can also affect the rate of photosynthesis, often indirectly. Under high temperatures, plant leaves may lose more water by evaporation than they receive by transport from the roots. This net water loss usually causes wilting of the leaves and the closure of their stomates. **Stomates** are specialized gas exchange units on the leaf surface consisting of two guard cells surrounding a pore (see Figure 15–3). When the stomates close, the pores become sealed, which greatly retards the entry of CO_2 into the leaf. With the availability of CO_2 to the chloroplasts lowered, the photosynthetic rate drops correspondingly.

FIGURE 6–15
Light Intensity and Photosynthesis. The effect of increasing light intensity on photosynthetic rate varies with plant species. In *Oxalis,* a shade-loving plant, maximum rates of photosynthesis are achieved at relatively low light intensities (shaded conditions). Corn is a sun-loving plant whose photosynthetic rate climbs continuously as the light intensity is increased. Note that at low intensities, *Oxalis* is a more efficient photosynthesizer than corn. At high intensities, however, corn's photosynthetic rate is much higher, as is its growth rate.

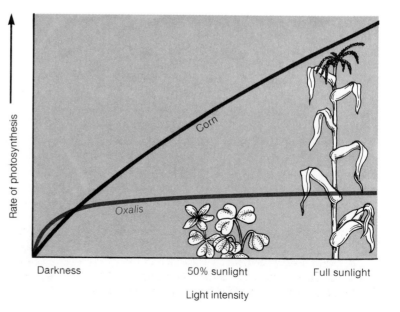

At the other extreme, very cold temperatures also reduce the rate of photosynthesis, but the effect of cold is more direct. The enzymes of the Calvin cycle, like virtually all enzymes, exhibit much reduced catalytic activities at low temperatures, and virtually no activity at freezing temperatures (see Figure 5–7). Thus, low temperatures interfere directly with the enzyme-catalyzed conversion of CO_2 to carbohydrate.

There are some interesting exceptions to these general effects of temperature. For example, some bacteria and cyanobacteria can carry out photosynthesis at temperatures as high as 75°C (167°F). At the other extreme, antarctic lichens reach their highest photosynthetic rate at 0°C (32°F) and can fix CO_2 at temperatures as low as −18°C. The mechanisms that make such adaptations possible are unknown.

Water Availability

Although water is a reactant in photosynthesis, it never reaches a low enough concentration in a leaf cell to limit the rate of photosynthesis directly. Nevertheless, a low supply of water in the soil often limits photosynthesis. When there is too little water in the soil for the plant to replace what it loses by evaporation, the stomates of the leaves close. As we have seen, stomatal closure restricts the entry of atmospheric CO_2 into the leaf. Thus, low water content limits photosynthesis by reducing the CO_2 concentration in the leaf.

Carbon Dioxide Concentration

In conditions where plants have adequate water and light and a suitable temperature, the availability of CO_2 is often the limiting factor of photosynthesis. Plant scientists have discovered that by increasing the level of CO_2 above its normal atmospheric concentration (a paltry 0.04%), the rates of photosynthesis in many plants can be enhanced. Hence "CO_2 fertilization" is sometimes used by horticulturalists in greenhouses. For example, by enriching the greenhouse air with CO_2, many plants, such as carnations, grow more rapidly. Not all plants respond favorably to higher CO_2 concentrations over the long run, however. Some species respond by producing small leaves, which results in an overall reduction in photosynthesis. Other plants respond by producing more carbohydrates and less protein.

It seems paradoxical that plants have evolved such that their photosynthetic apparatus performs best at CO_2 levels several times higher than the normal 0.04%.

However, there appears to be a good physiological basis for this. Although higher levels of CO_2 can increase photosynthesis over the short term in a greenhouse or laboratory, many plants eventually react adversely to the resulting changes in their metabolic pathways and growth patterns. Thus, although the atmospheric level of CO_2 may not maximize a plant's rate of photosynthesis, it seems to be better for its overall growth and development.

SUMMARY

Virtually all living organisms depend directly or indirectly on photosynthesis for food. Many years and the work of many scientists have gone into developing our understanding of photosynthesis; contrast the simple "soil-eater" concept espoused by the ancient Greeks to our current knowledge of the light and dark reactions.

Visible radiation, particularly the red and blue wavelengths, is required for plant photosynthesis. By comparing the action spectrum of photosynthesis with the absorption spectra of various plant pigments, scientists have determined that the chlorophylls and carotenoids in higher plants and many algae harvest light for photosynthesis. These pigments are organized into photosystems within the thylakoid membranes of the chloroplast, the site of the light reactions. Two photosystems (I and II) work together, each with a set of antenna pigment molecules and a trap chlorophyll *a* molecule. As light of the appropriate wavelength strikes a pigment molecule, an electron is raised to a higher energy level. This absorbed energy is transferred from pigment to pigment until it ends up in a trap chlorophyll *a* molecule. The trap chlorophyll *a* loses its excited electron to an electron acceptor and retrieves an electron from an electron donor. The net result of these light-initiated events is that electrons from water molecules are transferred to NADP+ to form NADPH, and molecular oxygen is produced as a by-product. In addition, as electrons move energetically downhill from photosystem II to photosystem I, the energy released is used to form ATP from ADP and P_i. ATP and NADPH are stable forms of chemical energy that are used in the second phase of photosynthesis, the dark reactions.

The dark reactions take place in the stroma of the chloroplast, are enzyme-catalyzed, and are responsible for the fixation and chemical reduction of CO_2 to carbohydrate. In C_3 photosynthesis, CO_2 becomes affixed to a 5-carbon sugar (RuBP) to form two molecules of PGA, a 3-carbon acid. This process is called CO_2 fixation. Subsequently, PGA is chemically reduced to PGAL, with electrons donated by NADPH and energy supplied by ATP (carbon reduction). Most of the PGAL so formed is chemically rearranged to form more RuBP, but some PGAL ends up as glucose.

Some plants, including most of the tropical grasses, exhibit the C_4 mode of fixing CO_2. In such plants, CO_2 is initially incorporated into a 4-carbon compound, then transported to other cells where the 4-carbon compound is dismantled to form a 3-carbon compound and CO_2. The 3-carbon compound is recycled, and the CO_2 is then fixed again in the C_3 pathway (Calvin cycle). C_4 plants are more efficient than C_3 plants at trapping CO_2, particularly at high temperatures or low water levels, both of which tend to reduce the availability of CO_2 to the photosynthetic cells.

The net glucose molecules generated in photosynthesis represent the major starting material for the synthesis of all other organic compounds in plants. Indirectly, the glucose from photosynthesis also serves as the starting point for the organic compounds in animals and other heterotrophs. Plants can polymerize glucose into starch for storage, convert it to sucrose for long-distance transport to nonphotosynthetic areas of the plant, and, in combination with the various mineral elements, transform glucose into amino acids, lipids, nucleotides, and other cellular components. All of these biochemical processes require energy in the form of ATP, most of which is provided by the respiration of glucose.

Environmental factors such as light intensity, temperature, water availability, and CO_2 concentration may limit the rate of photosynthesis and, consequently, plant productivity.

STUDY QUESTIONS

1. Which processes contribute to the release of carbon dioxide into the atmosphere? Which of these has exhibited the greatest increase in recent times?

2. What conclusions did van Helmont reach in his willow tree experiment? In what ways was the experiment carefully designed? What factor(s) did he fail to take into account?

3. What is an action spectrum? What is an absorption spectrum? How have scientists used these to determine which pigments are involved in photosynthesis?

4. What are the reactants and products of the light reactions? Of the dark reactions?

5. How did the ^{14}C isotope of carbon help the Calvin group elucidate the dark reactions of photosynthesis?

6. What is meant by C_3 and C_4 photosynthesis? How are they similar? How do they differ?

7. List the possible metabolic fates of a glucose molecule just manufactured in a plant leaf chloroplast.

8. How do environmental factors such as light intensity, temperature, water availability, and CO_2 concentration influence photosynthesis?

SUGGESTED READINGS

Anderson, L., A. Ashton, A. Mohamed, and R. Scheibe. "Light/Dark Modulation of Enzyme Activity in Photosynthesis." *BioScience* 32(1982):103–107. Explains how light activates some enzymes and inactivates others in chloroplasts, and how this modulation is important in photosynthesis.

Bjorkman, O., and J. Berry. "High-Efficiency Photosynthesis." *Scientific American* 229(1973):80–93. Interesting article explaining how some plants manage to survive under very harsh conditions.

Mooney, H., and S. Gulmon. "Constraints on Leaf Structure and Function in Reference to Herbivory." *BioScience* 32(1982):198–206. Shows how leaves differ in photosynthetic capacity and how this is linked to plant-herbivore interactions.

Rabinowitch, E., and Govindjee. *Photosynthesis.* New York: Wiley, 1969. Chapters 1 and 2 are good accounts of the history and overall biological significance of photosynthesis. Remaining chapters are lucid but relatively advanced in level.

Zelitch, I. "The Close Relationship between Photosynthesis and Crop Yield." *BioScience* 32(1982):796–802. Shows how selective breeding may bring about large increases in crop yield, especially in C_3 species.

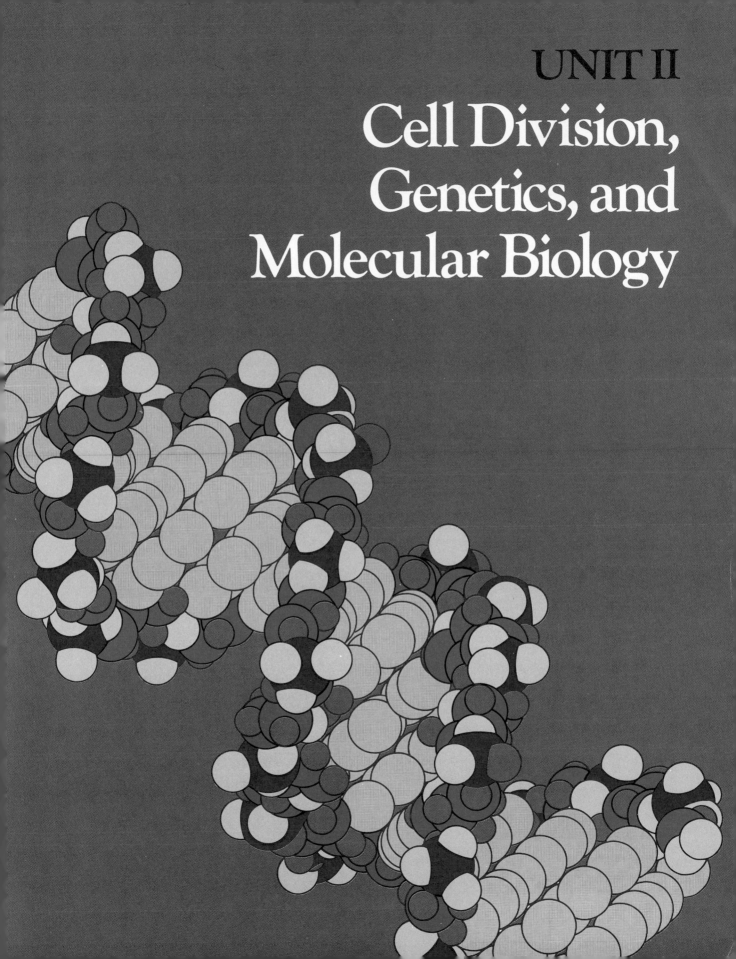

UNIT II
Cell Division,
Genetics, and
Molecular Biology

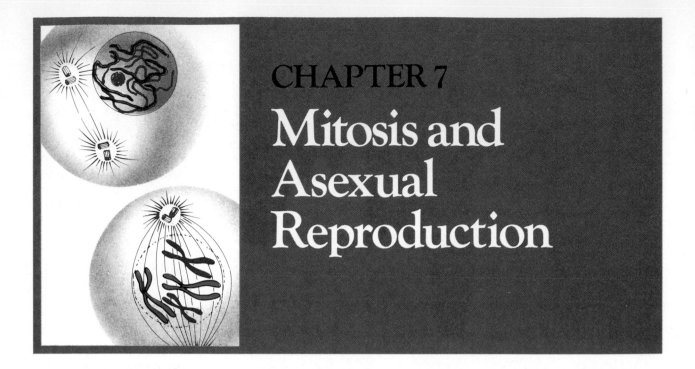

CHAPTER 7
Mitosis and Asexual Reproduction

Most biologists would agree that no other quality more clearly distinguishes the living from the nonliving than the capacity for reproduction. And as we shall see in Chapter 28, reproduction is necessary for biological evolution; it is the thread that links present-day life forms to their past. A continuum of life runs from the first living creatures to all organisms alive today. All of us are continuations of the life process of our parents, their parents, and so on. The lineage has and will be maintained by the process of reproduction. In this chapter we will begin to explore this unique phenomenon by examining how reproduction works at the cellular level.

The process by which cells reproduce (that is, give rise to new cells) is called **cell division.** Three key events proceed in sequence as a cell divides. First, the DNA, which encodes the genetic instructions of a cell, is replicated, a process discussed in Chapter 11. Second, the replicated copies of DNA separate as they are moved to opposite ends of the cell. In this manner the DNA copies will be partitioned to the two new daughter cells when, in the third stage of cell division, the protoplasm divides in half. Each daughter cell thus receives a complete copy of DNA and about half of the parent cell's cytoplasmic components, including organelles. The result is two cells, each with a full set of genetic instructions and biochemical machinery to carry out an independent existence.

Apart from some minor differences, the replication of DNA in all organisms is very similar. Variation does exist, however, in the mechanism for separating the replicated copies of DNA and in the manner in which the protoplasm splits in half. In this and the following chapter, we will examine the major types of cell division, both in prokaryotes and eukaryotes. Since prokaryotic bacteria have the simplest form of cell division, let us begin our study with these primitive organisms.

CELL DIVISION IN BACTERIA

The genetic material in modern bacteria exists in the form of a single, circular molecule of DNA that is attached at one point of the circle to the plasma membrane (Figure 7–1a). At the onset of cell division in these cells, the DNA molecule begins to replicate, ultimately forming two identical DNA circles. As replication proceeds, the new loop of DNA also becomes attached to the plasma membrane (Figure 7–1b). Now we have the two copies of DNA (still forming) attached at different points along the midregion of the plasma membrane. The plasma membrane between these two points then grows by the addition of new membrane material, and this causes the DNA copies

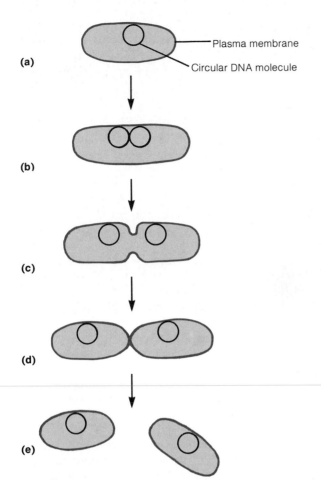

(a) Plasma membrane

Circular DNA molecule

(b)

(c)

(d)

(e)

FIGURE 7–1
Cell Division in Bacteria. (a) A bacterial cell containing one circular DNA molecule attached to the plasma membrane. (b) As the DNA is replicated, the new DNA copy attaches to a different point on the plasma membrane. (c) The membrane between the two attachment points grows, separating the DNA molecules, and the membrane begins to furrow. (d) The membrane fuses, dividing the parent cell into two daughter cells, each with an identical DNA molecule. New cell wall material is laid down between them. (e) The new cells separate.

to separate from one another. When separation is complete, the plasma membrane between the two DNA circles furrows inwardly from opposite sides (Figure 7–1c). The membrane furrows meet near the center of the cell and fuse, thereby dividing the parent protoplasm in half (Figure 7–1d). Each half is now a

daughter cell, and after each synthesizes a new cell wall between them, they separate and become independent cells (Figure 7–1e).

The type of cell division just described for bacteria is called **fission.** When fission is complete, the daughter cells enter a period of growth that usually culminates in another fission. When these two cells divide there is a total of four cells; when these divide there are eight cells, and so on. Under favorable conditions, some bacteria can divide every 20 minutes. Thus, in just 8 hours, one bacterial cell could give rise to 16,777,216 cells! And because the human body is a favorable environment for many types of bacteria, it's no wonder that we can feel perfectly healthy one day and rotten the next, suffering from a population explosion of bacteria in our system.

CELL DIVISION IN EUKARYOTES

When the first eukaryotes appeared about 1.4 billion years ago, the earth already had about 2.4 billion years of biological history behind it. Cells were increasing in complexity, and their genetic information content was expanding. Eventually, this information could no longer fit neatly into a single DNA molecule. Thus, the genetic information in the comparatively complex eukaryotic cells of today is stored in multiple units of DNA called chromosomes, which are located in the nucleus (see Chapter 3). Each chromosome (literally "colored body") consists of hundreds to several thousands of **genes,** the units of genetic information. Human cells, for example, have 46 chromosomes carrying a total of about 100,000 genes. The number of chromosomes varies for different types of organisms. For example, the cells of goldfish contain 94 chromosomes; fruit flies, 8; tobacco and chimpanzees, 48; and onions, 16.

While the advent of multiple chromosomes may have solved the problem of storing large amounts of genetic information, it created another potential problem. How could a cell with multiple chromosomes ensure that each daughter cell produced during cell division received one copy of each replicated chromosome? It seems that one daughter cell might end up with both copies of a given chromosome, leaving the other daughter cell with a missing segment of genetic information. Evolution provided the answer in the form of an orderly chromosome segregation process called **mitosis.** Before investigating how mitotic cell division works, let us first examine the structure of the chromosome.

Chromosome Structure

Chromosomes are present at all stages in the life of a eukaryotic cell. Their physical appearance differs, however, depending on whether or not the cell is undergoing cell division. When the cell is not undergoing mitosis, each chromosome exists as an ex-

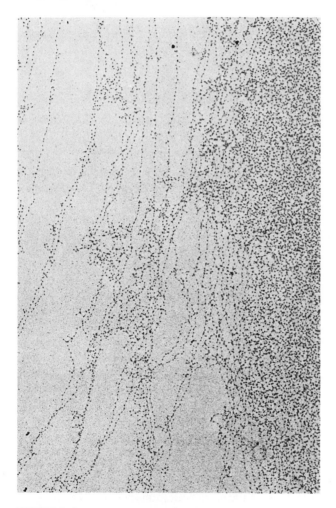

FIGURE 7–2

Chromosome Fibers. This transmission electron micrograph shows part of a single chromosome that has been folded over many times. The bead-like structures are composed largely of proteins, whereas the slender thread connecting them is DNA. (Magnification: 39,460×.)

tremely long, threadlike structure within the nucleus. In highly magnified electron micrographs, these chromosome fibers look like beads on a string (Figure 7–2). With the aid of specific enzymes that digest DNA or protein, scientists have determined that the string is made of a single molecule of DNA, whereas the beads are partly DNA but mostly protein. That is, when isolated chromosomes are treated with a protease (protein-digesting enzyme), then examined under the electron microscope, the beads originally present appear as a very faint outline (due to the small amount of DNA remaining), while the string remains visibly intact. On the other hand, if the chromosomes are treated with DNAase (a DNA-digesting enzyme) before microscopic examination, the string disappears altogether, but the proteinaceous beads are still quite visible.

When a cell is about to divide, the appearance of its chromosomes changes. After the DNA is replicated and the process of mitosis begins, the chromosomes reorganize into very compact, thick structures—so thick that they are easily visible under conventional light microscopes. In their compact form, the chromosomes are more easily segregated during mitosis. (Imagine the tangled mess if chromosomal replicates were pulled to opposite ends of a dividing cell in their fibrous state.)

Exactly how chromosomes shorten and thicken is not yet understood. Some observations suggest that the individual fibers coil tightly, like a spring; others suggest that the beaded string folds back on itself, like a water hose on a fire truck. Perhaps both (or neither) of these mechanisms are correct.

Mitosis

Several hours after DNA replication is complete, the process of mitosis begins. The time spent in mitosis varies greatly by cell type and environmental conditions (such as temperature). In general, animal cells complete the mitotic events in about one hour and plant cells take two to three hours. These events occur in a continuous sequence without pause, but biologists generally describe them in terms of four sequential stages: prophase, metaphase, anaphase, and telophase. Keep in mind that these mitotic stages are only arbitrary designations that serve as reference points for discussion—the end of prophase and the beginning of metaphase are in the eye of the observer only; they are not distinct chromosomal events.

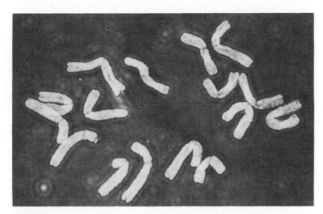

FIGURE 7–3
Prophase Chromosomes. This light micrograph of a dividing onion cell shows its 16 chromosomes during prophase of mitosis. Each chromosome consists of two identical chromatids.

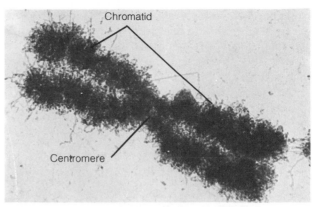

FIGURE 7–4
Chromosome Close-up. Electron micrograph of a human chromosome showing the two identical chromatids joined at the centromere (Magnification: 28,000×.)

PROPHASE. The first stage of mitosis—prophase—starts when the chromosome fibers begin to coil and fold back upon themselves, becoming visible chromosomes when viewed under the light microscope (Figure 7–3). As they continue to shorten and widen, each chromosome can be seen to consist of two sausage-shaped structures, called **chromatids,** held together at a constricted region of the chromosome known as the **centromere** (Figure 7–4). Each chromosome during prophase is actually two identical chromatids, the products of DNA replication.

While the chromosomes become compacted in the nucleus, several other events take place in the cytoplasm. Two centrioles (two pairs of basal bodies; see Chapter 3) lie just outside the nuclear envelope; microtubules, the elongate tubes of protein that function in various types of movement (see Figure 3–18), begin to form from these centrioles. While this is occurring, the centrioles migrate toward opposite sides of the nucleus, still serving as focal points for the radiating microtubules, now called **spindle fibers.** When the centrioles become situated, the spindle fibers can be seen to stretch across the nucleus, forming the **spindle apparatus.** Dividing plant cells also form a spindle apparatus, but they do so without benefit of centrioles, structures that are absent in these cells.

During the latter part of prophase, the nucleolus (see Chapter 3) disappears from view (apparently becoming incorporated into one or more of the compacting chromosomes), and the nuclear envelope disintegrates. The spindle fibers then attach to each chromosome at the centromere, marking the end of prophase.

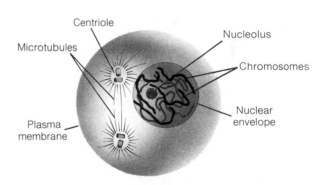

Early prophase

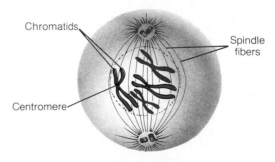

Late prophase

METAPHASE. As prophase comes to a close, metaphase begins. During this rather short stage of mitosis (usually lasting 5–10 minutes), the chromosomes are drawn to the midplane of the cell by the spindle fibers. The mechanism behind this movement is still a mystery. By the end of metaphase, the centromere of each chromosome lies in the midplane of the cell.

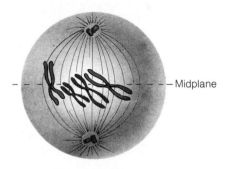

Metaphase

ANAPHASE. The third stage of mitosis, called anaphase, begins as all the identical chromatids are pulled apart from one another. Each chromatid is now referred to as a **daughter chromosome,** and each newly-forming daughter cell will receive one complete set of daughter chromosomes. The actual mechanisms involved in the initial separation of chromatid sets and the subsequent movement of the daughter chromosomes to opposite ends of the cell are not known. We do know that both the spindle fibers and the centromeres participate, but whether the daughter chromosomes are actually *pulled* by the spindle fibers or merely *guided* by them is still an open question. At the close of anaphase, the two sets of daughter chromosomes are clustered at opposite ends of the cell, still recognizable as compact units.

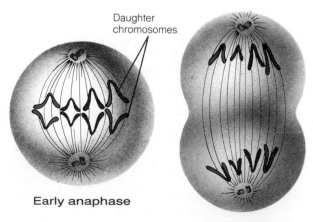

Daughter chromosomes

Early anaphase

Late anaphase

TELOPHASE. In the final stage of mitosis—telophase—the separated daughter chromosomes unravel back into their threadlike, fibrous appearance. A new nuclear envelope forms around each set of chromosomes. At the end of telophase, one or more nucleoli become visible in each daughter nucleus, but distinct chromosomal structures are no longer discernible in the light microscope. This signals the end of mitosis.

The photomicrographs in Figure 7–5 show the sequence of mitotic stages as they actually appear during plant cell division.

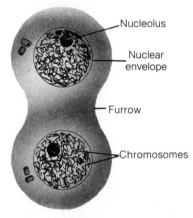

Nucleolus

Nuclear envelope

Furrow

Chromosomes

Telophase

Cytokinesis

At the end of mitosis, the two daughter nuclei are situated at opposite ends of the cell. Cell division is not complete, however, until the parent cell cytoplasm is split in half, producing two separate daughter cells. This process of cytoplasmic division is called **cytokinesis.**

In the cells of all organisms except plants, cytokinesis involves an inward furrowing of the plasma membrane (Figure 7–6). The plasma membrane pinches inward at the equator until the cytoplasm is completely divided in half. As a result, each half contains a daughter nucleus and about half of the parent cell's cytoplasm. It is as if you tightened a string around the middle of an inflated balloon until you effectively ended up with two balloons. The mechanism that causes this membrane furrowing is not precisely known, but it apparently involves the contraction of microfilaments (see Figure 3–17), which pulls the membrane inward.

In higher plants, cytokinesis is quite different. Instead of the plasma membrane furrowing, a **cell plate** forms along the midplane of the cell (Figure 7–7). The cell plate consists of small membranous

FIGURE 7–5
Mitosis. These photomicrographs show mitosis occurring in a living cell of an African blood lily.

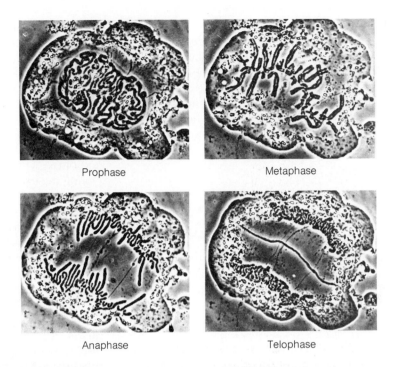

Prophase

Metaphase

Anaphase

Telophase

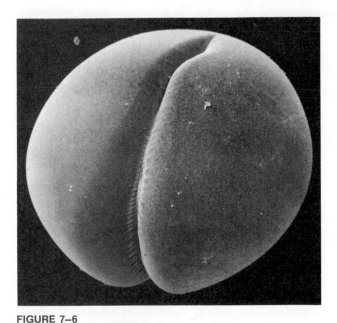

FIGURE 7–6
Cytokinesis in a Frog Zygote.
When an egg and sperm unite, the resulting cell is called a zygote. The scanning electron microscope caught this dividing frog zygote midway through the inward furrowing of its plasma membrane. Under normal circumstances this single cell would soon give rise to two cells; those two would divide to become four cells, and so on, until an adult frog was formed. (Magnification: 60×.)

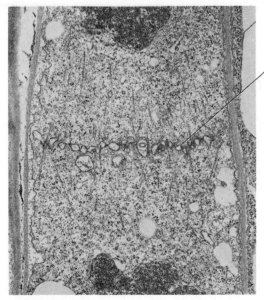

Cell Plate

FIGURE 7–7
Cell Plate Formation. This transmission electron micrograph shows the midregion of a dividing cell from silver maple caught at late telophase of mitosis. The line of vesicles across the middle of the photo is the cell plate. Note also the microtubules that are oriented perpendicularly to the cell plate, and portions of the daughter nuclei (dark material) at the top center and bottom center. (Magnification: 11,400×.)

vesicles that originate from Golgi bodies. Their alignment at the cell's midplane seems to be under the influence of microtubules. The vesicles ultimately fuse together to form a large flattened sac. The sac's membranous sides become new portions of plasma membrane for both daughter cells. The enclosed contents of the sac, a mixture of polysaccharides, becomes the beginning of the cell walls between the new cells. When the vesicles are completely fused, the cell division process is complete.

THE CELL CYCLE

We have seen that mitotic cell division involves an orderly sequence of events beginning with DNA replication, followed by mitosis, and ending with cytokinesis. But the chromosomal DNA is not the only component of the cell that must be replicated when new cells form. All the organelles and other structures within the cell must double in amount between each cell division. If this did not occur, succeeding generations of cells would have fewer and fewer organelles, and the cells would not be able to function. Therefore, the synthesis of mitochondria, ribosomes, endoplasmic reticulum, and other cytoplasmic components must also take place at some point in the cell's life cycle.

The sequence of events that takes place from the time a cell is first produced until it divides is referred to as the **cell cycle.** (It might be more appropriate to call it the cell division cycle, since *individual cells* go through the entire sequence only once, as they give rise to new cells.) For the cells of most organisms, the cell cycle takes anywhere from 10 to 30 hours to complete, but this varies considerably; as we saw, some fast-growing bacteria can divide every 20 minutes. During the usual 10 to 30 hours, mitosis lasts from one to three hours and the rest of the cycle is spent in **interphase** (*inter* = in between). During interphase, the events associated with cytoplasmic growth (such as the duplication of organelles) and DNA replication occur.

The first stage of interphase is called **Gap 1 (G_1)** (Figure 7–8). During G_1 most of the processes associated with cytoplasmic growth take place, including the uptake of water for cell expansion and the biosynthesis of RNA, lipids, proteins, and carbohydrates. Many of these organic molecules are used for building membranes (such as endoplasmic reticulum and Golgi bodies) and for duplicating organelles (such as mitochondria). This is a busy time for the cell and, like the other stages of interphase, should not be regarded as a "resting stage" before mitosis.

G_1 is followed by the **S** phase; S stands for synthesis, specifically nuclear DNA synthesis. During this period, which typically lasts six to nine hours in mammalian cells, the cell's chromosomal DNA is replicated (see Chapter 11). Cytoplasmic growth processes also continue during the S phase.

The completion of DNA replication marks the end of the S phase and the beginning of **Gap 2 (G_2)**. During the two to six hours of G_2, the cell speeds up its synthesis of certain proteins involved in mitosis. For example, tubulin, the protein subunit of microtubules, is produced in large quantities and assembled into microtubules, which will form the spindle fibers during mitosis.

Scientists have spent a great deal of effort studying the cell cycle and how it is regulated. The primary question is: What event or events commit certain cells to continue dividing over and over, whereas others stop dividing? For example, human skin cells just below the surface divide continuously to replace the outer skin cells as they are worn away. Liver cells typically do not divide in an adult, yet if part of the liver is injured, the surrounding cells will divide to replace the damaged part. In contrast, brain cells are at best sluggish dividers; if part of the brain is destroyed, its associated functions might be lost indefinitely. As yet, we have only a sketchy understanding of cell cycle control.

Detailed knowledge of the mechanisms by which the cell cycle is regulated may help us understand cell division in its uncontrolled state, such as in cancer. Cancerous cells divide rapidly, often crowding and impairing the functions of healthy tissue. Although this disease takes many forms, there is good reason to believe that some basic cause is behind them all— possibly some malfunction at the level of cell cycle regulation.

ASEXUAL REPRODUCTION

For multicellular organisms, mitotic cell division generates new cells for growth and for replacement of dead or injured cells. All multicellular organisms that reproduce by means of **sexual reproduction** begin life as a single cell, the product of a union between a sperm and an egg cell. This first cell, the fertilized egg, divides to form two cells; these two divide to form four cells, and so on. Eventually the multicellular adult stage is reached as the result of countless mitotic

FIGURE 7–8
The Cell Cycle.

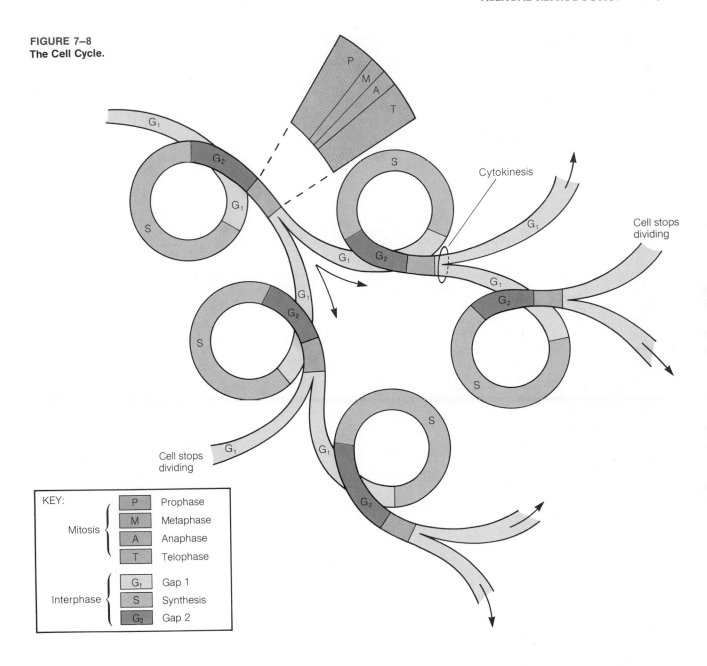

KEY:

Mitosis
P	Prophase
M	Metaphase
A	Anaphase
T	Telophase

Interphase
G₁	Gap 1
S	Synthesis
G₂	Gap 2

cell divisions. Every cell generated during growth has the same genetic information originally present in the fertilized egg because DNA replication followed by mitosis yields genetically identical daughter cells. Growth occurs when cells produced by mitosis remain together as an integrated unit in the parent organism.

In many eukaryotic organisms, mitotic cell division also serves as the basis of **asexual reproduction,** the general process by which new individuals are produced that are identical to the parent organism. In asexual reproduction, the cell or cells produced by mitosis become separated from the parent structure and establish an independent existence. Many examples of asexual reproduction can be found among the lower organisms. Single-celled yeast cells propagate asexually by pinching off a cytoplasmic "bubble" containing a daughter nucleus, a process called **budding** (Figure 7–9a); filamentous forms of algae and fungi can **fragment,** forming two or more smaller filaments that enlarge by mitotic cell divisions. Algae and fungi also produce **asexual spores,** single cells that can germinate and grow into new individuals (Figure 7–9b).

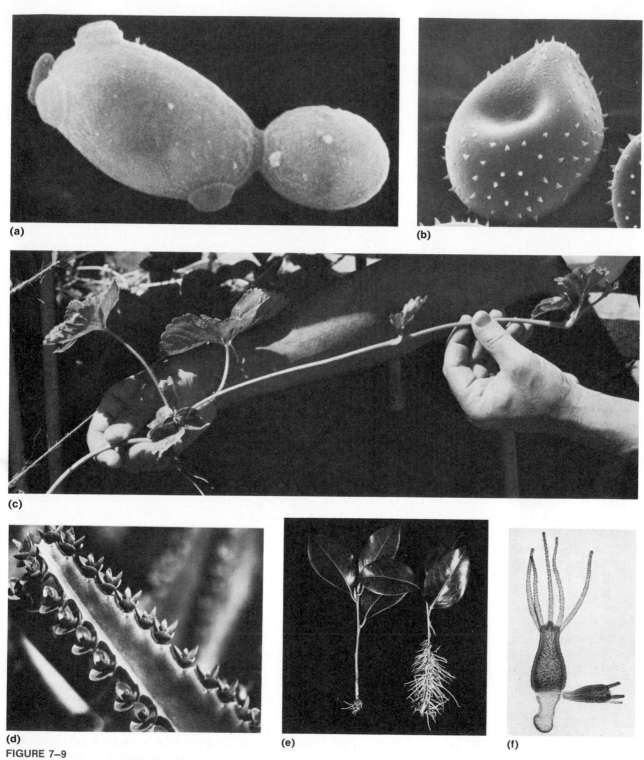

FIGURE 7–9

Modes of Asexual Reproduction. (a) Budding of a yeast cell. (b) An asexual spore of a rust, a type of fungus. (The spikes are extensions of the cell wall.) (c) Runners from a strawberry plant. (d) *Kalanchoe,* the maternity plant, with plantlets. (e) Roots on stem cuttings of camellia. (f) A new offspring forming on a *Hydra.*

Some fungi reproduce by asexual means exclusively.

Quite a number of higher plants also reproduce asexually, particularly when environmental conditions are favorable for growth. The strawberry plant sends out **runners,** horizontal stems that grow away from the parent plant (Figure 7–9c). The tip of the runner forms its own shoot and root system, and when the runner dies back, a new plant is born. *Kalanchoe,* commonly known as the maternity plant, produces small plantlets on the margins of its mature leaves (Figure 7–9d). Each plantlet can drop to the ground and grow into a new plant. Humans have taken advantage of the ease with which many plants propagate asexually, for both agricultural and horticultural purposes. For example, for centuries farmers have been propagating potatoes by slicing up the potato into pieces that each contain an "eye." The eye of the potato is a bud capable of growing into a new shoot, and ultimately into a new potato plant. A potato is actually a *tuber,* an underground stem modified for food storage. Other plants that produce underground storage organs (such as rhizomes, bulbs, and corms) can be propagated asexually by similar means, and many plants can be propagated from stem cuttings (Figure 7–9e).

Animals rely primarily on sexual reproduction for increasing their numbers, but a few also reproduce asexually. For example, *Hydra,* a small freshwater animal, can produce offspring asexually (Figure 7–9f). New individuals start out as small outgrowths on the parent's body, enlarge by mitotic cell divisions, and eventually dislodge from the parent to lead an independent existence. Higher animals, including humans, do not reproduce asexually. In the years ahead, however, biological research may provide the means to overcome this natural restriction. We already have the technology to keep human cells alive for decades in the laboratory (see Essay 7–1). But generating a new, independent human from a collection of cells in a test tube shall remain within the province of science fiction, at least until we have a better understanding of how human development works.

SUMMARY

Reproduction at the cellular level is known as cell division, and it entails three major events that occur in the following sequence: (1) DNA replication, (2) segregation of the DNA copies to opposite ends of the cell, and (3) cytokinesis, the division of the cytoplasm.

The simplest form of cell division, called fission,

occurs in bacteria. As the bacteria's circular DNA replicates, the two DNA copies become attached to different points on the plasma membrane near the center of the cell. As the plasma membrane between these two points grows and eventually furrows inwardly, the DNA copies become segregated into the newly forming daughter cells.

In eukaryotes, the nuclear DNA is associated with proteins in structures called chromosomes. When the nucleus is not dividing, each chromosome appears as a long "string" of DNA with "beads" that are made up largely of protein. The DNA is replicated in this fibrous state prior to the onset of mitosis.

Mitosis in the eukaryotic cell is the means by which the two strands of replicated DNA (identical chromatids) are segregated from one another. The mitotic events occur in recognizable stages. Prophase, the first stage, begins as the chromosomal fibers coil back on themselves to form shortened, thick chromosomes. Each chromosome consists of two sausage-shaped chromatids held together at a constricted region called the centromere. Later in prophase, microtubules assemble into spindle fibers that ultimately stretch from one pole of the cell to the other. After the nuclear envelope breaks apart, spindle fibers emanating from both poles attach to the centromeres and, in some unknown way, orient the chromosomes toward the cell's equator.

Prophase is followed by metaphase, the stage at which the chromosomes come to lie at the equator, still attached to the spindle fibers. The sister chromatids then separate from one another; they are now called daughter chromosomes. In anaphase, the daughter chromosomes move toward opposite poles of the cell, being guided (perhaps pulled) by the spindle fibers. Shortly thereafter, the separated chromosomes uncoil, returning to their fibrous condition, and a nuclear envelope forms around each new daughter nucleus. This final stage of mitosis is called telophase.

Cytokinesis, or the division of the parent cytoplasm, begins during telophase. In all organisms except plants, this involves the inward furrowing of the plasma membrane at the cell's equator. When the membrane fuses in the center, two daughter cells result. Plant cell cytokinesis is characterized by the formation of an equatorial cell plate, which consists of Golgi vesicles containing cell wall substances. The vesicles fuse to form the beginning of new cell walls and new plasma membrane, which separate the two daughter cells.

All dividing cells go through a sequence of cytoplasmic and nuclear events collectively referred to

ESSAY 7-1

HeLa CELLS IN MEDICAL RESEARCH: A USEFUL TOOL OR LABORATORY PEST?

Henrietta Lacks died in 1952, a victim of cancer. Ironically, the cancer that caused her death lives on. How is this possible? Let's go back to 1951 when the 31-year-old Mrs. Lacks entered the medical clinic at Johns Hopkins University in Baltimore. During an examination the doctor discovered a small lesion of purple-colored tissue within Henrietta Lacks' cervix. Suspecting cancer, the doctor removed a small piece of the abnormal tissue and sent it to the pathology lab. The results of the biopsy confirmed his suspicions. Moreover, the cancer was malignant, prompting a prescription of radiation therapy. A few days later (but before radiation treatment was begun), the doctor removed another small sample of the cancerous tissue and submitted it to a group of medical researchers at the university. These scientists were attempting to grow tumor cells in culture as a means of studying cancer in the laboratory, an effort that up until then had met with little success. The cancerous cells from Henrietta Lacks changed all this.

Now known as HeLa cells (named after Henrietta Lacks), these tumor cells grew splendidly in culture, doubling in number every 22 hours. Unlike other human cells in culture at that time (both normal and cancerous), the HeLa cells did not die after several months in isolation. Indeed, from the original few thousand or so cells cultured in 1951, untold trillions of HeLa cells now exist scattered among petri dishes and culture tubes throughout the world. These remarkably durable cells have been the subject of count-

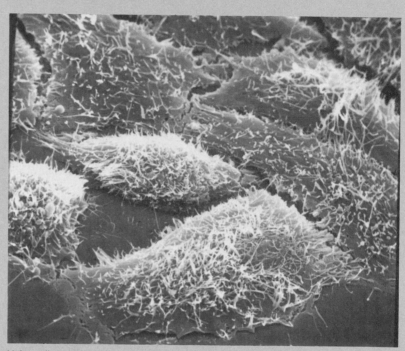

HeLa cells growing in culture. The projections from the plasma membrane shown here are characteristic of cultured cells.

less studies into the mechanisms of cancer and cell aging, mainly because they are so easy to grow, are cancerous, and simply do not age like most other human cells. Their durability not only explains why the radiation treatment given Henrietta Lacks was ineffective in arresting the spread of this cancer throughout her body, but also why these cells have become a serious problem in cell culture laboratories worldwide. We will look into the nature of this problem shortly, but first let us provide some background on the cell culture technique itself.

Cell cultures are started when a

small population of cells from a plant or animal are transplanted onto an artificial, sterile culture medium. The medium typically contains an energy source (usually glucose), vitamins, lipids, salts, amino acids, and various growth-stimulating factors. In such a nutrient-rich medium, the transplanted cells usually undergo many mitotic cell divisions and thereby grow in number. New cultures can be started by transferring small portions of the daughter cells onto fresh media. This process can be repeated ad infinitum as long as the transferred cells continue to divide.

The technique of cell culture is an extremely valuable tool in biological research. It permits the investigator to study the activities of cells under controlled environmental conditions, away from the unknown and unpredictable influences of the organism's body. In addition, cultured cells can be exposed to various treatments (such as poisons, harmful X-rays, and other nasty agents) that cannot be performed on whole organisms, particularly humans. The only possible drawback is in relating results obtained from cultured cells to the living organism. Because of their special environment, the activities of cultured cells may not always reflect what goes on in the organism's body.

Through the use of cell culture, we have learned a great deal about life at the cellular level. For example, biologists have used cell cultures to study the interactions of cells in populations. When several different types of cells are grown together in culture, cells of the same type often display the ability to "recognize" and associate with one another. In so doing, the various cells become organized into a mosaic of rudimentary "tissues." From such studies has come a wealth of information concerning the cellular recognition process. Other cell cultures have played key roles in the research and production of vaccines, and in the diagnosis of genetic defects in unborn fetuses (see Chapter 25).

One of the most interesting uses of cell culture in biological research has been in the area of aging and death. Cells cultured from non-cancerous human tissues typically go through a finite number of cell divisions (perhaps 50) before "pooping out." The reason for this limited life span is still unclear. Some scientists favor the notion that cell death is caused by the gradual accumulation of toxic metabolic substances over time; others cite the buildup of harmful changes in the genetic substance (DNA) in such cells. Many types of cancerous cells, however, have apparently escaped this "programmed death" so characteristic of normal cells. The HeLa cancer is a prime example of a cell culture line that will not die. The tenacity of these cells has made them a popular research tool—perhaps too popular. HeLa cells have been showing up in places where they do not belong.

In 1974 a cancer researcher at one of the major universities in California had cultured what seemed to be a unique human breast cancer cell. One of its unusual characteristics was a chromosome shaped like Mickey Mouse ears. When a colleague, Dr. Walter Nelson-Rees, got wind of this, he remembered seeing one of these odd-shaped chromosomes in a culture line of human embryo kidney cells. Were these new breast cancer cells actually kidney cells? This seemed highly unlikely, so Nelson-Rees had a closer look at all the chromosomes of both cell cultures. You can imagine his surprise when he discovered that neither cell line was what it was supposed to be—they were both HeLa cultures! This revelation sent a shock wave through the cancer research establishment. How many human cancer cell lines had been overrun with HeLa cells? How many published research papers dealing with various cancer lines were in reality describing work on HeLa cells and not knowing it? This prompted Nelson-Rees to do some detective work on other standard human cancer lines. As of 1981, Nelson-Rees had screened several hundred lines of human cancer in culture and found 90 imposters—all variants of the original HeLa cell line. It had become clear that Henrietta Lacks' legacy to cancer research had become a laboratory pest perpetuated by sloppy techniques of careless laboratory workers.

Perhaps some good has come out of the HeLa cell mismanagement. Although the reputations of a few cancer researchers have been tarnished by this episode, an important lesson has been learned. More cell culture scientists are taking extra precautions in their work, including greater care in the transfer of cells from flask to flask, and periodically checking their cultures for authenticity. Many labs have decided to work on only one cell culture line at a time, making sure to thoroughly sterilize all laboratory equipment and glassware before switching to a different type of cultured cell. While such measures may be inconvenient and time-consuming, they must be strictly adhered to, particularly in this era when new life forms are being engineered daily. To err is human, but there are some errors we just can't afford to make.

as the cell cycle. Mitosis alternates with interphase, and the substages of interphase—Gap 1, S, and Gap 2—are characterized by various cellular growth processes, particularly biosynthetic events. It is during interphase that the cytoplasmic organelles are duplicated (G_1), DNA is replicated (S), and the microtubules are synthesized (G_2).

Mitotic cell division is the cellular basis both of growth in multicellular organisms and asexual reproduction in all organisms that can propagate asexually. Essentially, growth occurs when mitotically generated daughter cells remain associated with the parent organism. Asexual reproduction occurs when a cell or cells produced by mitotic cell division become separated from the parent structure and establish an independent existence. Various types of asexual reproduction occur in nature, particularly among the lower organisms and plants. These include budding, fragmenting, producing asexual spores, sending out runners, splitting underground storage organs, and rooting stem cuttings.

STUDY QUESTIONS

1. What are the three major cellular events that take place during cell division?

2. How are identical DNA copies segregated into daughter cells during bacterial cell division?

3. Describe the structure of a eukaryotic chromosomal fiber.

4. Indicate the stage of mitosis during which the following events occur:
   (a) Assembly of the spindle apparatus
   (b) The movement of daughter chromosomes toward opposite ends of the cell
   (c) Nuclear envelope disintegration
   (d) The return of chromosomes to their fibrous condition
   (e) The alignment of chromosomes at the cell's midplane

5. How does cytokinesis differ between plant and animal cells?

6. During which phases of the cell cycle do the following events take place?
   (a) DNA synthesis
   (b) Tubulin synthesis
   (c) The segregation of daughter chromosomes
   (d) The beginning of organelle duplication

7. Name two types of asexual reproduction that occur among algae and fungi (lower organisms).

8. How are strawberries propagated asexually? Potatoes?

9. In general, do plants or animals utilize asexual reproduction more?

SUGGESTED READINGS

Giese, A. C. *Cell Physiology.* 5th ed. Philadelphia: W. B. Saunders, 1979. An advanced treatment for the motivated undergraduate, Giese's text provides detailed information on all aspects of cellular physiology, including the cell division process.

Mazia, D. "The Cell Cycle." *Scientific American* 230(1974): 53–64. Well-written account of the events occurring during interphase and mitosis, including experimental evidence for the processes and their control.

Richards, V. *Cancer: The Wayward Cell.* 2nd ed. Berkeley and Los Angeles: University of California Press, 1978. A general introduction to the kinds, causes, and treatment of cancer.

Wolfe, S. *Biology of the Cell.* 2nd ed. Belmont, CA: Wadsworth, 1981. Current and detailed treatment of mitotic cell division.

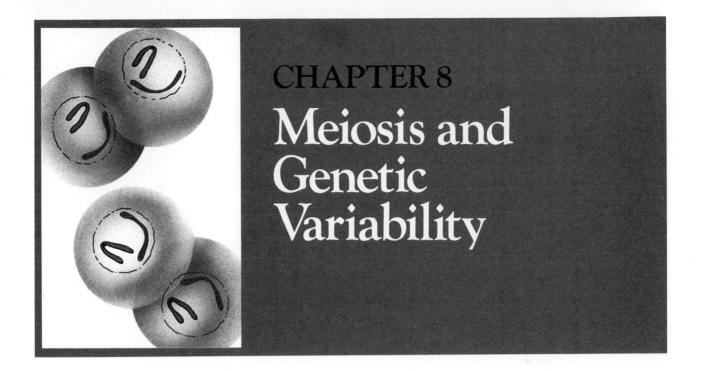

CHAPTER 8
Meiosis and Genetic Variability

As we have seen in Chapter 7, mitotic cell division leads to the proliferation of cells for growth and asexual reproduction in organisms. Since mitosis involves the segregation of identical chromatids, all the daughter cells will be genetically identical—they will all have the same set of genes in their nuclei. But what about sexual reproduction, the process by which parents who are genetically different produce a new individual? The offspring produced by sexual reproduction are not identical to either parent, nor are they exact blends of the two parents. If the latter were true, then all siblings (brothers and sisters) would look alike and have characteristics exactly intermediate between those of their parents. Since this is not the case, there must be another type of cell division process underlying sexual reproduction—a cell division process that does not produce genetically identical daughter cells.

There is another aspect of sexual reproduction that implies the operation of a fundamentally different type of cell division process, and that is the problem of chromosome number. Sexual reproduction in all plants and animals involves two major processes: (1) the formation of **gametes** (usually eggs and sperm), and (2) the union of gametes, a process called **fertilization,** to form a single-celled **zygote** (fertilized egg). When a sperm cell unites with an egg cell, their nuclei fuse to form the single nucleus of the zygote.

Thus, the chromosomes of the sperm and egg come together, forming a nucleus that has twice as many chromosomes as either gamete. But if the zygote (the first cell of the new individual) is to have the same number of chromosomes as its parents, there must be a mechanism to *halve* the number of chromosomes appearing in the gametes. The reason for this should become clear in the following example.

Human body cells each contain 46 chromosomes in their nuclei. If human eggs and sperm were produced by *mitotic* cell division, they would also contain 46 chromosomes. The union of these hypothetical gametes would result in a zygote having 92 chromosomes. In other words, the new individual would start out life with twice as many chromosomes as either parent! If such a pattern continued, the chromosome number in each succeeding generation would double, so that within ten generations, the original 46 chromosomes per cell would mushroom to 23,552. Since this clearly does not occur, we can deduce (as scientists did a century ago) the intervention of a special form of cell division which yields gametes having half the normal number of chromosomes. **Meiosis** is this special type of cell division; in humans, meiosis generates gametes that have 23 chromosomes each (Figure 8–1). We will discuss how meiosis accomplishes this reduction in chromosome number shortly, but

FIGURE 8-1
Meiosis in Humans. The number of chromosomes in human cells is 46; this number remains constant from generation to generation. Since new individuals are generated by the union of egg and sperm, each of these gametes must contain half of 46, or 23 chromosomes. The type of cell division that halves the number of chromosomes per cell is called meiosis. In mitosis, the chromosome number remains constant with each cell division. Mitotic cell division is responsible for the transformation of a single-celled zygote into a human being with approximately 10 trillion cells.

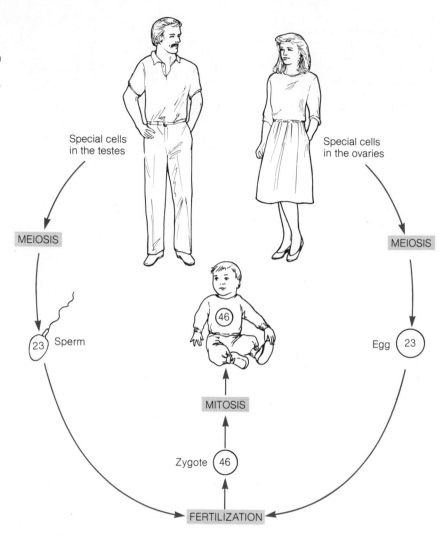

Special cells in the testes — MEIOSIS — 23 Sperm

Special cells in the ovaries — MEIOSIS — Egg 23

46

MITOSIS

Zygote 46

FERTILIZATION

first let's take a closer look at the problem of chromosome number itself.

HAPLOID AND DIPLOID CHROMOSOME NUMBER

Each species of organism has a characteristic number of chromosomes in the nuclei of their **somatic cells.** In animals and higher plants, somatic cells include all the cells of the body except gametes and gamete-forming cells. The somatic cells of a corn plant have 20 chromosomes, goat cells have 60, onions have 16, and, as you know, humans have 46. When the chromosomes in somatic cells are examined carefully

under the microscope, an interesting pattern emerges. With one possible exception, the chromosomes can be matched up in pairs—each member of the pair is very similar in size and shape to the other member (Figure 8–2). The one possible exception is the **sex chromosomes,** which determine whether an individual will be male or female. The sex chromosomes are either similar (two X chromosomes) or dissimilar (one X and one Y chromosome). In humans and many other animals, males have one X and one Y chromosome, and females have two Xs.

The two chromosomes of each matched pair are known as **homologous chromosomes** or **homologues** (*homo* = same), and they contain genes that affect the same traits. For instance, if a given chromo-

FIGURE 8–2
Chromosomes from a Human Male.
An arrangement of chromosomes such as this is called a karyotype. To obtain a karyotype, a blood sample is drawn and a chemical is added that causes blood cells to stop dividing at metaphase (see Chapter 7). The metaphase chromosomes are photographed and enlarged. The chromosomes are then cut out of the photograph and arranged into their 23 pairs. Note that each chromosome consists of two chromatids and that the two paired chromosomes are strikingly similar in 22 of the pairs. In human males, the chromosomes of the 23rd pair do not look alike. These two chromosomes are called the sex chromosomes because they determine whether the individual will be male or female. Human males have one X and one Y chromosome (as shown here); females have two X chromosomes.

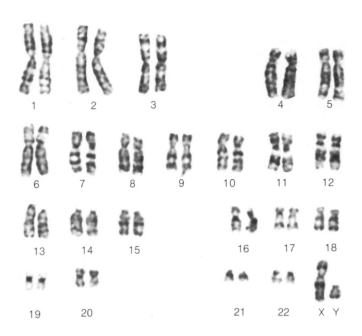

some has a gene that affects eye color, then its homologue will also have an eye color gene. When present together, these two homologous genes interact to determine the individual's eye color. (Exactly how genes determine traits will be discussed in Chapters 9 and 10.) Therefore, your somatic cells do not contain 46 unrelated chromosomes, but 23 *pairs* of chromosomes. One chromosome of each pair was originally contributed by your father's sperm and the other by your mother's egg. Thus, your somatic cells have 23 paternal chromosomes and 23 maternal chromosomes.

Cells that contain a double set of chromosomes (both chromosomes of each homologous pair) are called **diploid** (meaning double), or **2n.** Cells with only one chromosome from each pair (such as gametes) are called **haploid** (meaning single), or **n.** When plants and animals reach sexual maturity, certain diploid cells located in their reproductive organs undergo meiosis, yielding haploid cells that have only one chromosome of each homologous pair. These haploid cells give rise directly or indirectly to the haploid gametes. Let us now see how this chromosome-halving process of meiosis works.

MEIOSIS

Recall from Chapter 7 that sometime before mitosis begins, the chromosomal DNA is replicated. Similarly, the DNA is replicated before meiosis begins. Thus, by the onset of meiosis, each chromosome consists of two identical chromatids.

Overall, meiosis consists of two nuclear divisions that result in four haploid nuclei (Figure 8–3). During the first meiotic division, *homologous chromosomes* are segregated from one another. Each of the chromosomes at this stage still consists of two chromatids. During the second division, the *chromatids* are segregated from one another. The stages of meiosis associated with the first division are denoted by the Roman numeral I (prophase I, metaphase I, and so on); the second division stages are denoted by II (prophase II, etc.). The entire sequence of meiotic stages is diagrammed in Figure 8–4; you should refer to this figure as you read about each stage.

The First Meiotic Division

PROPHASE I. As in mitosis, the beginning of meiosis—prophase I—is marked by the organization of the chromosomal fibers into visible chromosomes. Unlike mitosis, however, meiotic prophase I is marked also by the pairing of homologous chromosomes; that is, the paternal and maternal chromosomes of each homologous set become physically attached to each other, forming a structural unit called a **bivalent.** Remember that each chromosome consists of two

FIGURE 8–3

Overview of Meiosis. In this example, there are two chromosomes in the diploid cell: one maternal (color) and one paternal (black) chromosome. After the DNA is replicated, each chromosome consists of two chromatids. During the first meiotic division, the homologous chromosomes are segregated, but the sister chromatids remain attached to each other at the centromere. During the second meiotic division, the chromatids of each cell segregate, resulting in four haploid cells with one chromosome each.

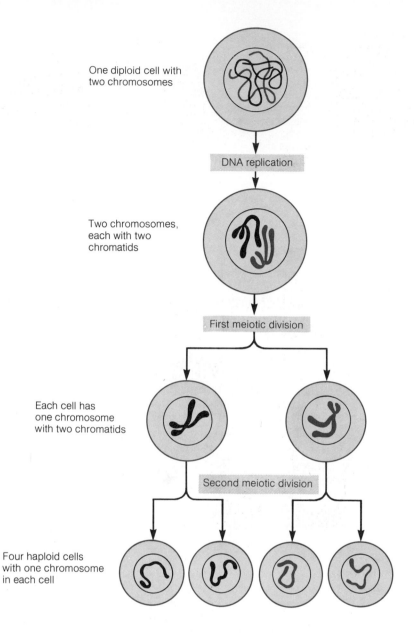

One diploid cell with two chromosomes

DNA replication

Two chromosomes, each with two chromatids

First meiotic division

Each cell has one chromosome with two chromatids

Second meiotic division

Four haploid cells with one chromosome in each cell

chromatids; so each bivalent consists of four chromatids; in human meiosis there would be 23 bivalents.

As the chromosomes continue to shorten and thicken, a spindle forms outside the nuclear envelope just as it does during mitosis. The nuclear envelope then disintegrates, and the spindle fibers attach to the centromeres of each chromosome.

METAPHASE I. As in metaphase of mitosis, metaphase I of meiosis is characterized by the alignment of the chromosomes at the midplane of the cell. However, this is where the similarity ends, because each bivalent behaves as a unit during the alignment process. Thus, at metaphase I of a human cell, there will be 23 bivalents occupying the midplane, not 46 independent chromosomes, as would be the case for metaphase of mitosis. Furthermore, the centromere of each chromosome is connected to spindle fibers that stretch toward only one pole of the cell, not toward both poles.

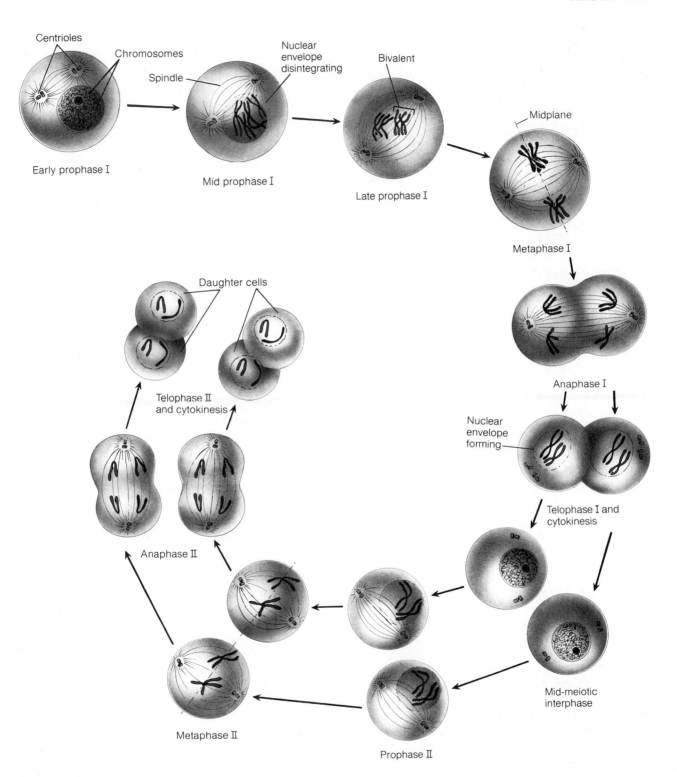

FIGURE 8–4
The Stages of Meiosis. In this diagram, there are four chromosomes in the diploid cell (two pairs of homologous chromosomes).

ANAPHASE I. Because the chromosomes of each bivalent are connected to opposite poles of the cell by the spindle fibers, it is the homologous chromosomes, not the chromatids, that are segregated during anaphase I. That is, the paternal chromosome (still consisting of two chromatids) of each bivalent is pulled in one direction, while its homologous maternal chromosome moves in the opposite direction. Although this is true for any given homologous pair, it does not mean that all maternal chromosomes move toward one pole and all paternal chromosomes move toward the other pole. The direction of movement of one maternal chromosome is independent of the direction taken by all the other maternal chromosomes. The same is true for the paternal chromosomes. Thus, the maternal and paternal chromosomes are distributed randomly between the two daughter cells resulting from the first meiotic division. Later in this chapter we will see that this independent assortment of chromosomes during meiosis is a primary source of genetic variability among members of a species.

TELOPHASE I. Telophase I marks the end of the first meiotic division. The homologous chromosomes now situated at opposite poles of the dividing cell begin to unwind, returning to their fibrous state, and a nuclear envelope forms around each new nucleus. Cytokinesis then splits the parent cytoplasm in half, forming two daughter cells. Each daughter cell contains only half the original number of chromosomes (one chromosome from each homologous pair), but all the chromosomes still consist of two chromatids. The chromatids are then separated in the second meiotic division.

The Second Meiotic Division

The second division of a meiosis takes place rather rapidly and is mechanically similar to mitosis. In most organisms, the daughter cells from the first division go through a short interphase period before they begin the second meiotic division; the chromosomes then coil and fold to become visible chromosomes during prophase II. In other organisms, however, the telophase I chromosomes remain in their contracted state, skip interphase, and proceed directly to prophase II.

PROPHASE II. In prophase II, spindles form and the nuclear membranes disappear, permitting the spindle fibers to become attached to the centromeres. In this case, spindle fibers radiating from both poles attach to each centromere, as in mitosis.

METAPHASE II. In metaphase II, the chromosomes move toward the midplanes of their cells. When the chromosomes are aligned along the midplanes, the individual chromatids begin to move away from each other.

ANAPHASE II. During anaphase II, each chromosome splits at its centromere, and the chromatids move toward opposite ends of the cells. Each chromatid is now referred to as a daughter chromosome.

TELOPHASE II. Meiosis ends with telophase II, the stage in which the daughter chromosomes unravel to become long, thin fibers. A nuclear envelope forms around each new daughter nucleus (there are now four nuclei), and cytokinesis begins to divide the cytoplasm of the two cells, forming a total of four daughter cells.

A comparison of meiosis and mitosis is presented in Figure 8–5.

MEIOSIS AND GENETIC VARIABILITY

We mentioned at the beginning of this chapter that sexual reproduction produces individuals that are different genetically from their parents and siblings. This genetic variation plays a very important role for the long-term survival of a species. The greater the genetic variation within a species, the more likely it will be that some of the individuals can survive in a new or changing environment. The same principle is applied by stock brokers when they advise their investors to diversify ("don't put your eggs all in one basket"). In this way, the roof may fall in on one investment, but the others will keep you afloat financially. Through sexual reproduction, the species distributes its pool of genes into many different combinations (each individual is a unique combination of genes), and in so doing, gains an edge in the struggle against extinction. Let us illustrate this important concept with an example.

Imagine that there exist two varieties of strawberry plants, one able to grow well in rocky soils, the other able to survive with minimal amounts of water. Neither may fare well in dry, rocky soil, but if they are partners in sexual reproduction, some of their offspring may inherit the right combination of genes that permits them to survive in and/or colonize a rocky soil with little water. It is also likely that some of the new offspring will inherit neither trait. Thus,

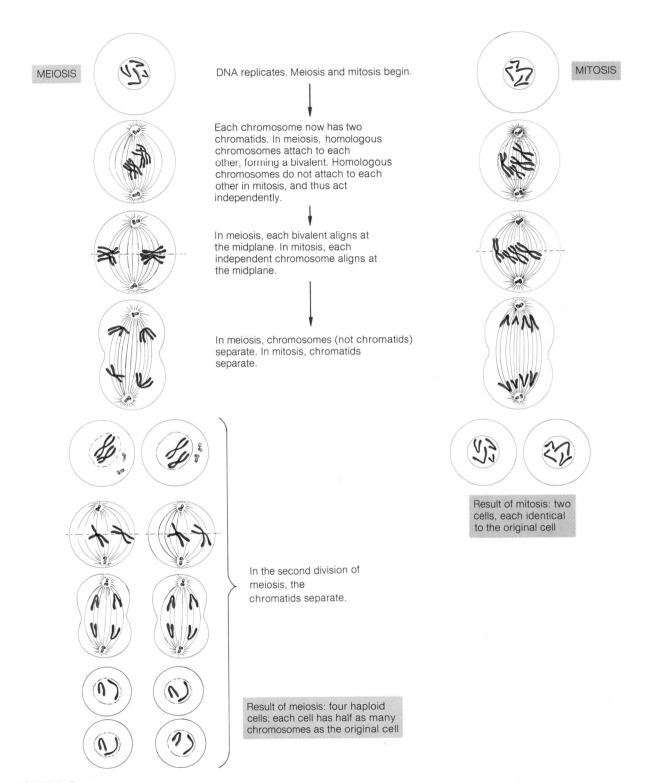

MEIOSIS

DNA replicates. Meiosis and mitosis begin.

MITOSIS

Each chromosome now has two chromatids. In meiosis, homologous chromosomes attach to each other, forming a bivalent. Homologous chromosomes do not attach to each other in mitosis, and thus act independently.

In meiosis, each bivalent aligns at the midplane. In mitosis, each independent chromosome aligns at the midplane.

In meiosis, chromosomes (not chromatids) separate. In mitosis, chromatids separate.

Result of mitosis: two cells, each identical to the original cell

In the second division of meiosis, the chromatids separate.

Result of meiosis: four haploid cells; each cell has half as many chromosomes as the original cell

FIGURE 8–5
Comparison of Meiosis and Mitosis. Both types of nuclear division are preceded by DNA replication, resulting in two chromatids for every chromosome. Meiosis entails two divisions, reducing the diploid number of chromosomes by half to the haploid number. Mitosis, on the other hand, involves only one division, producing daughter cells with the same chromosomal content as the parent cell.

although sexual reproduction generates a variety of different gene combinations among the offspring, it does not guarantee that *favorable* gene combinations will always appear. Let us now see how meiosis contributes a major share of the genetic variability found among organisms.

When two gametes from different individuals fuse, the resulting zygote receives half of its genes from one parent and the other half from the other parent. Even if the same two parents produce thousands of offspring, however, the chances that any two of them will be genetically identical are extremely slim. Let's see why this is true. For any particular gene that affects a given trait, a diploid cell has two copies of that gene — a paternal and maternal form — located at similar sites on the homologous chromosomes. These homologous genes may be identical or nonidentical, but they always affect the same trait (such as eye color in humans, height in corn plants, or whatever). The haploid gametes generated through meiosis will have only one copy of this gene because each receives only one chromosome of each homologous pair, and hence only one gene of each homologous pair of genes. The same is true for the thousands of other genes located on all the chromosomes — each gamete has either the maternal or paternal copy of each gene, not both.

It is important to realize that the haploid gametes all have one copy of *every* gene; that is, they have a single but complete set of genetic instructions. The exact complement of that complete set of genes will vary, however, because events occurring during meiosis scramble the various maternal and paternal gene copies into unique combinations. The two events in meiosis contributing to this reassortment of genes are (1) crossing over, and (2) the random alignment of the bivalents at metaphase I, which results in the independent assortment of the chromosomes in the haploid cell products.

Crossing Over

Recall that during prophase I of meiosis, the homologous maternal and paternal chromosomes come together to form bivalents. The points of attachment between homologous chromatids, called **chiasmata** (singular, chiasma), can vary both in number and position along the lengths of the joined chromatids. Some bivalents may be held together by five or six chiasmata, others by only one. Geneticists refer to chiasmata as crossover points, because it is here that the joined

chromatids may break and then reunite with the "wrong" chromatid segment. As a result, homologous segments of two chromatids are exchanged, a process called **crossing over** (Figure 8–6a). The maternal chromatid now has a segment containing paternal genes, and conversely, the paternal chromatid has the corresponding maternal segment. Consequently, the four haploid cells produced in this case will all be different with respect to the set of genes located on this chromosome (Figure 8–6b).

In the example shown in Figure 8–6, we show the results of a single crossover event occurring in one bivalent in one cell undergoing meiosis. Other cells may have crossovers occurring at one or more different sites within this bivalent, leading to different mixtures of paternal and maternal genes on this chromosome in the haploid cells produced. And when you consider that there are other bivalents also undergoing random crossover patterns during meiosis, the number of genetically distinct gametes possible becomes enormous. In humans, with their nearly 100,000 genes distributed over 23 pairs of chromosomes, the number of different gene combinations generated by crossovers is virtually infinite.

Independent Assortment of Chromosomes

While crossing over can lead to new *gene* combinations within a single chromosome, the independent alignment of the bivalents during metaphase I of meiosis can yield a variety of *chromosome* combinations appearing in the gametes. As the bivalents line up in the midplane of the meiotic cell during metaphase I, their orientations relative to one another will be completely random. Because of this random alignment, the two daughter cells resulting from the first meiotic division will each receive a random assortment of paternal and maternal chromosomes. This concept is perhaps best illustrated with a hypothetical example (Figure 8–7). Imagine a diploid organism that has four chromosomes (two maternal and two paternal chromosomes), and thus will form two bivalents in meiosis. During metaphase I of meiosis, when the bivalents align along the cell's midplane, the orientation of one of the maternal chromosomes will be independent of the orientation of the other maternal chromosome. If both maternal chromosomes happen to become attached to spindle fibers emanating from one pole, then one of the daughter cells produced in the first meiotic division will have both maternal chromosomes, and

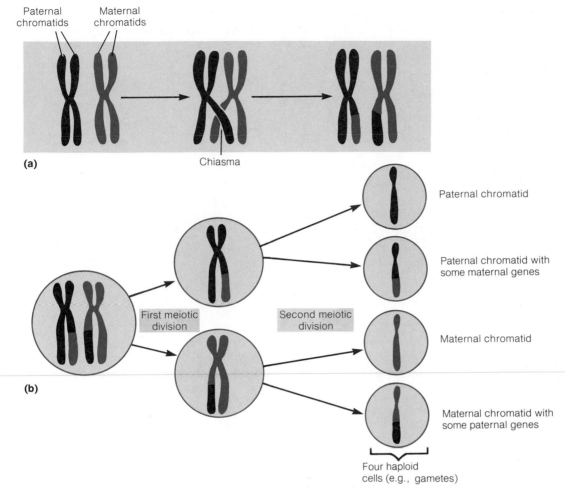

(a)

Paternal chromatids

Maternal chromatids

Chiasma

(b)

First meiotic division

Second meiotic division

Paternal chromatid

Paternal chromatid with some maternal genes

Maternal chromatid

Maternal chromatid with some paternal genes

Four haploid cells (e.g., gametes)

FIGURE 8–6
Crossing Over. (a) During prophase I, homologous chromatids cross over each other and exchange chromosomal segments. In this way new combinations of genes are created. (b) As a result of crossing over, all four haploid cells at the end of meiosis are genetically different from each other.

the other will have both paternal chromosomes (Figure 8–7a). Alternatively, if during metaphase I these same bivalents align so that one maternal chromosome is oriented toward one pole and the other toward the opposite pole, then the first meiotic daughter cells— and the gametes that ultimately form—will contain one maternal and one paternal chromosome (Figure 8–7b). Since there are only two bivalents in our hypothetical cell and two possible orientations for each bivalent during metaphase I of meiosis, then the possible number of different chromosomal combinations in the gametes is 2^2, or four. How many chromosomal combinations are possible in human gametes? An

enormous 2^{23}, or 8,388,608! When you add to this all the crossover patterns possible, you can see that no two haploid gametes are likely to be genetically identical. Moreover, when you consider that fertilization occurs between gametes produced by genetically different parents, it is easy to see why all the individuals of a species produced through sexual reproduction are genetically different. Incidentally, identical twins are not an exception to this rule. Identical twins originate from the same zygote by the splitting in two of a young, developing embryo (see Chapter 25). Thus, the generation of identical twins from a single embryo is actually a form of asexual reproduction.

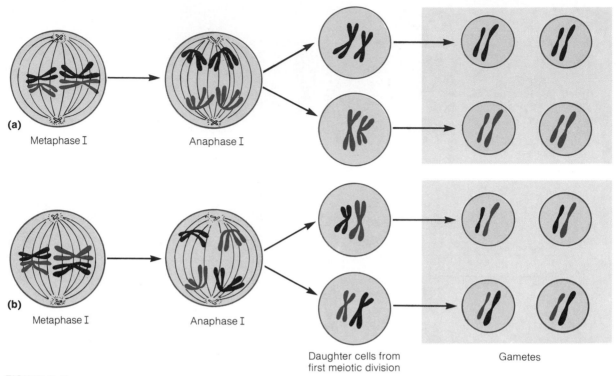

FIGURE 8–7

Independent Assortment of Chromosomes. During metaphase I each maternal chromosome (color) is paired with its homologous paternal chromosome (black) (chiasmata not shown), and all the bivalents align at the equator. However, the orientation of one bivalent relative to another is random. Thus, the haploid cells may contain both maternal and paternal chromosomes, only maternal chromosomes, or only paternal chromosomes. This independent assortment of chromosomes contributes to the genetic variety among the gametes produced.

ERRORS IN MEIOSIS

As we have seen, meiosis is a very complex and intricate process. The separation of homologous chromosomes must occur without a hitch if completely normal gametes are to be formed. Thus, it is not surprising that mistakes do occur. In humans, the rate of spontaneous abortions (miscarriages) is estimated to be approximately 69%, most of which go unnoticed because they occur shortly after conception. Examinations of naturally aborted fetuses reveal a high frequency of chromosomal abnormalities, most of which result from mistakes in meiosis.

One of the most common errors in meiosis (and mitosis) is a phenomenon called **nondisjunction**—the failure of a pair of homologous chromosomes (or chromatids) to detach and segregate from one another (Figure 8–8). This results in gametes that are either

missing a chromosome, or have an extra chromosome in their nuclei. If such a gamete unites with a normal gamete, the resulting zygote will have the same chromosome abnormality (see Figure 8–8), and usually will abort during early fetal development. However, certain types of chromosome deletions or additions are not fatal. In humans, for example, there are individuals who either lack or have extra copies of one of the sex chromosomes (see Chapter 10). Many of these individuals survive to adulthood, but they are generally sterile and display other characteristic features. An even more dramatic case of chromosomal nondisjunction results in a disorder called Down syndrome.

Down syndrome is characterized by mental retardation (IQ rarely exceeds 60) and various physical abnormalities (Figure 8–9). All parts of the body (head, neck, torso, arms, fingers, legs, and toes) are shortened, the face is broad and flat, and the eyes may be slanted.

FIGURE 8–8
Meiotic Nondisjunction. If the two chromosomes making up a given homologous pair do not segregate during anaphase I of meiosis, then two of the resulting gametes will be missing that chromosome altogether, and the other two gametes will have two copies of it. Fertilization of either type with a normal gamete will result in a zygote having either one or three copies of that chromosome, a situation that is usually lethal.

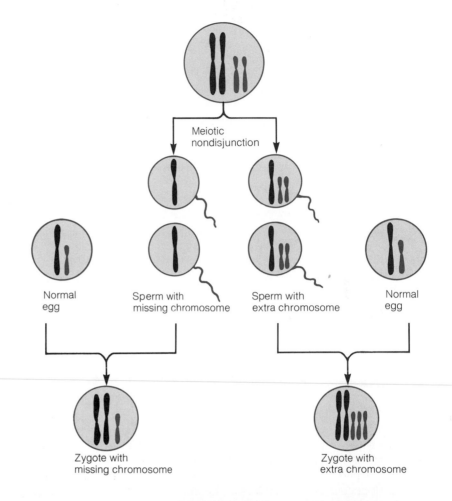

FIGURE 8–9
Down Syndrome Children. These children are in a specially designed preschool.

(In the past, Down syndrome was also called *mongolism* because of the superficially Oriental appearance of the eyes.) Down syndrome children also tend to have inborn defects of various internal organs, and have a higher than normal mortality rate in the first few years of life.

About 1 in every 700 live births is a Down syndrome child. The overwhelming majority of these are caused by the presence of an extra chromosome number 21, a situation known to geneticists as **trisomy-21.** In other words, the Down syndrome child has three chromosome 21s instead of the normal two. In about 25% of the cases, the extra chromosome derives from the father's sperm; in the other cases it is the egg that has the extra chromosome 21. The frequency of trisomy-21 increases only slightly with paternal age, but increases dramatically with maternal age (Figure 8–10). For this reason, pregnant women over the age of 35 are encouraged to undergo **amniocentesis** (see Essay 25–1), a relatively minor procedure carried out at about the 16th week of pregnancy. During amniocentesis fetal cells are withdrawn and examined for chromosomal abnormalities. If trisomy-21 shows up in the fetal cells, a therapeutic abortion can be performed.

At one time, individuals with Down syndrome were usually institutionalized. It is now known, however, that Down syndrome children make much more progress with home care and special infant stimulation programs. As we learn more about the processes involved in child development, we can provide even more assistance to Down syndrome children and other developmentally disadvantaged children. And as we learn more about the causes of errors in meiosis and mitosis, we may someday be able to prevent some of these errors from occurring.

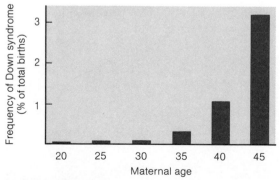

FIGURE 8–10
Estimated Rate of Down Syndrome Occurrence for Different Maternal Ages.

SUMMARY

The somatic cells of higher plants and animals each contain a diploid number of chromosomes. That is, they contain two complete sets of chromosomes—one maternal set and one paternal set. For every maternal chromosome, there is a homologous paternal chromosome. These homologous pairs have corresponding gene pairs that affect the same traits.

Since sexual reproduction entails the union of gametes (egg and sperm), there must be a cell division process that halves the diploid number of chromosomes during gamete formation. Meiosis does this, giving rise to four haploid cells. The haploid cell products then develop (directly or indirectly) into gametes, each of which contains a single set of chromosomes.

Before meiosis begins, all the chromosomal DNA is replicated. This is followed by two meiotic divisions; the first division entails the segregation of homologous chromosomes, and the second division effects the segregation of the chromatids. Like mitosis, each of the meiotic divisions has the recognizable stages of prophase, metaphase, anaphase, and telophase (each followed by I or II for the first or second division, respectively).

During prophase I, the homologous chromosomes become physically linked to form bivalents while they are condensing into visible chromosomes. After the nuclear membrane ruptures, spindle fibers attach to the centromeres (two centromeres per bivalent) and move the bivalents toward the midplane of the cell (metaphase I). The homologous chromosomes are then segregated to opposite ends of the cell (anaphase I) where they disperse back to their fibrous state (telophase I). Nuclear envelopes form around both daughter nuclei while cytokinesis divides the parent cell in half. At this stage, each chromosomal fiber still consists of two chromatids.

The second meiotic division is very similar to mitosis. The chromosome fibers in both daughter nuclei aggregate into visible chromosomes once again (prophase II) and align at the cells' midplanes (metaphase II). Next, the chromatids move to opposite poles along the spindle (anaphase II). Finally, the chromatids (now recognized as chromosomes) unravel into chromosomal fibers, now in four daughter nuclei (telophase II). Meiosis is complete when cytokinesis splits these two cells into four separate compartments—the four haploid daughter cells.

The haploid cells generated by meiosis tend to be genetically different because of the recombination of genes that occurs during crossing over and the independent assortment of chromosomes. Crossing over takes place during chiasma formation in prophase I of meiosis when regions of homologous chromatids are exchanged. The independent assortment of paternal and maternal chromosomes into the haploid products occurs because of the random alignment of the bivalents during metaphase I. Fertilization also adds to the genetic variability among members of a species because unique gametes from genetically different parents fuse to yield unique combinations of chromosomes (and genes) in the resulting zygote.

Sometimes errors occur in meiosis (and mitosis). One of the most common errors is nondisjunction—the failure of a pair of homologous chromosomes (or chromatids) to detach and segregate from one another. If the resulting gamete unites with a normal gamete to form a zygote, the zygote will be chromosomally abnormal. One type of nondisjunction results in Down syndrome. Down syndrome is the result of three chromosome 21s, a situation known as trisomy-21.

STUDY QUESTIONS

1. If the cells of an organism contain 17 pairs of chromosomes, what is the diploid ($2n$) number of chromosomes? What is the haploid (n) number?

2. During the first meiotic division, _____ chromosomes are segregated; the segregation of _____ occurs in the second meiotic division.

3. Make a diagram of metaphase I of meiosis for a diploid cell that has 6 chromosomes. How would this diagram differ from one showing metaphase of mitosis for this cell?

4. Haploid cells cannot divide meiotically. Why not?

5. At which stage of meiosis do formerly identical chromatids cease to be identical?

6. Describe how crossing over contributes to genetic variability in the haploid cells generated by meiosis.

7. If an organism had a diploid number of chromosomes equal to six, how many different combinations of maternal and paternal chromosomes are possible in the haploid gametes? (Assume no crossing over occurs.)

8. Discuss why genetic variability is important to the long-term survival of a species.

SUGGESTED READINGS

Ayala, F. J. and J. A. Kiger. *Modern Genetics.* 2nd ed. Menlo Park, CA: Benjamin/Cummings, 1984. The first chapter in this well-written genetics text has a good concise description of meiosis and its significance.

Giese, A. C. *Cell Physiology.* 5th ed. Philadelphia: W. B. Saunders, 1979. The introductory chapters of this well-known text provide good coverage of cells and cell division.

Karp, G. *Cell Biology.* New York: McGraw-Hill, 1979. This advanced text has several good chapters on cell division.

Smith, D. W., and A. A. Wilson. *The Child with Down's Syndrome.* Philadelphia: W. B. Saunders, 1973. A book written for the lay public.

Strickberger, M. W. *Genetics.* 2nd ed. New York: MacMillan, 1976. Good discussion of meiosis and its importance.

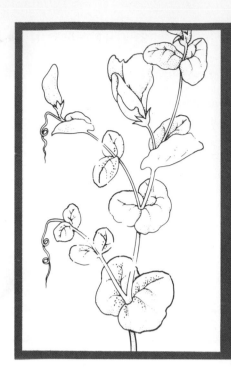

CHAPTER 9
Mendelian Genetics

Have you ever wondered why a child looks like a particular parent or grandparent, or how it is possible for two brown-eyed parents to have a blue-eyed son or daughter? For hundreds (perhaps thousands) of years such questions have intrigued curious minds. Farmers in particular have been interested in these matters, for our high-yielding crops and improved breeds of animals have arisen largely through **selective breeding**—individuals with desirable traits are mated to produce offspring with those traits. Before the twentieth century, however, plant and animal breeders did not know *why* selective breeding worked. They knew then (as we know today) that selective breeding is the key to improving milk production in cows or increasing the size of a wheat grain, but beyond recognizing that "like begets like," they knew little else. When it came to explaining patterns of inheritance, the early breeders spoke of the mixing of blood, a terminology that lingers on in colloquialisms such as "blood relatives" and "blood lines."

The fundamental mechanisms underlying heredity were first discovered by Gregor Mendel. In the years 1856–1864 Mendel, an Austrian monk who lived and worked in a small monastery in Brno, Czechoslovakia, performed his now-famous experiments on the garden pea (Figure 9–1). Mendel was certainly not the first to conduct genetic experiments, but unlike his predecessors, he was fortunate (perhaps wise is more apt) to have selected a relatively simple genetic system for his studies. By examining the inheritance patterns of relatively well-defined traits, Mendel obtained clear-cut results that led him to the discovery of several principles of inheritance. These principles are now the cornerstone of modern genetics, and Gregor Mendel is generally regarded as the "Father of Genetics."

MENDEL'S LAWS OF INHERITANCE

At the time Mendel began his experiments on inheritance in 1856, it was widely known that in sexual reproduction both parents play a part in determining the characteristics of their offspring and that the parental "influences" are transmitted by the gametes. Mendel's greatest contribution to the field of genetics was his discovery that the hereditary influences carried by the gametes exist in discrete units, or "factors" as Mendel called them. Furthermore, he hypothesized

FIGURE 9–1
Gregor Mendel (1822–1884). Gregor Mendel was the son of a peasant farmer in a small village north of Vienna. Young Gregor showed promise in his early school years, and at considerable expense and personal sacrifice by members of his family, Gregor was sent away to high school. After graduation, he decided to become a priest, at least partly to continue his education without causing his family further financial hardship. In 1843, he entered a monastery in Brno, now in Czechoslovakia, and in 1847, he was ordained. At his monastery's expense, he studied mathematics and natural science at the University of Vienna for two years. Mendel then returned to the monastery and taught natural science at the technical high school until 1868, although he never passed the exam for a teacher's certificate.

Mendel began his genetics experiments in 1856. His laboratory was the monastery garden, where he followed patterns of inheritance in garden peas. In his work, Mendel showed that traits were inherited in logical and predictable patterns. Although his findings were published in 1865, their significance eluded the notable scholars of his day.

In 1868, Mendel was elected abbot of the Brno monastery. His administrative duties and his long battle over whether monasteries should be taxed by the government kept him away from his research garden. He died in 1884, 16 years before his research on the mechanisms of inheritance was discovered by the scientific community.

correctly that these genetic factors become segregated and reassorted during gamete formation, and then are recombined in fertilization. We now know that Mendel's "factors" are genes, the units of genetic information carried on chromosomes.

Mendel's most famous experiments were performed with the common garden pea. Peas were a good choice, for they are easy to grow and are fairly resistant to pests. Peas also have a reasonably short generation time—they complete their life cycle (from seed to next-generation seed) in three months. Thus, results from breeding experiments could be obtained fairly rapidly. Most importantly, the peas Mendel chose for his studies were **purebreeding** strains—ones that "bred true" for easily discernible variations of specific traits. For example, Mendel had a purebreeding strain of red-flowered peas that produced red-flowered offspring exclusively when allowed to self-pollinate. (In peas, pollination and fertilization usually occur *within* each flower, not *between* flowers of separate plants.) Similarly, a purebreeding white-flowered strain yielded only white-flowered progeny (offspring) when it self-pollinated.

The Monohybrid Cross

One of the first questions Mendel asked himself was: "What would happen if a purebreeding red-flowered pea was pollinated by (crossed with) a white-flowered purebred?" To perform this cross, Mendel first removed all the anthers (pollen-forming organs) from

FIGURE 9–2

Mendel's Peas. (a) Cutaway diagram of a pea flower emphasizing the reproductive structures. The anthers are part of the male reproductive structures where sperm-producing pollen are formed. The pistil, the female reproductive structure, consists of the pollen-receptive stigma, a neck-like style, and the ovary. The pea ovary contains several ovules, each of which produces one egg. (b) After pollination and fertilization occur, the fertilized ovules develop into seeds. As the seeds grow, the surrounding ovary develops into a fruit, the pea pod.

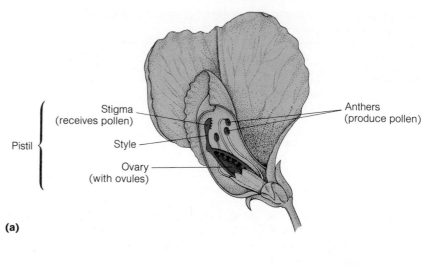

(a)

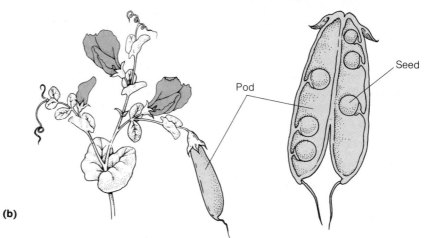

(b)

the red flowers of one individual before they produced pollen (Figure 9–2). This eliminated any chance of self-pollination. He then collected pollen from the mature anthers of a white-flowered individual and transferred this pollen to the altered red flowers. Finally, he covered the cross-pollinated red flowers with small bags to keep out any insect-dispersed or wind-blown pollen from other pea plants. This ensured that all the offspring were the result of "red-flowered" eggs fertilized by "white-flowered" sperm (from the pollen). About four weeks later, Mendel collected the seeds that, in breeder's terminology, were the **F₁ generation** (F₁ = first filial; from Latin *filius,* meaning son). The parents belonged to the **P generation** (P = parental).

Mendel's F₁ seeds were **hybrids;** that is, they resulted from a cross between genetically different purebreds. Because Mendel was following only one

trait (flower color), the F₁ offspring were specifically **monohybrids. (**As you undoubtedly are beginning to see, geneticists have a vocabulary of their own. Table 9–1 lists the terms we have introduced already and several that we will discuss shortly.) Mendel was not particularly interested in what to call his F₁ plants; rather, he wanted to know the color of their flowers. To find out, he planted the seeds.

In about eight weeks, Mendel had his answer: all the F₁ hybrids blossomed red (Figure 9–3a). Does this mean that flower color in peas is controlled by the eggs, not the pollen? To test this idea, Mendel performed the reciprocal cross—he transferred pollen from a red-flowered purebred to the flowers of a white purebred from which he had removed the immature anthers. The results turned out the same—all the F₁ hybrids had red flowers. Thus, it is the red-flowered parent that has the dominating influence over flower

TABLE 9–1
Some Common Terms Used by Geneticists.

Term	Definition
Purebred	An individual or genetic strain that, if self-pollinated or inbred, produces only one type of offspring with respect to a given trait or traits.
Hybrid	Offspring of two parents (or strains) that differ in one or more traits.
Monohybrid	Hybrid for a single trait being examined.
P	Parental generation.
F₁	First-filial generation; offspring of the P generation.
F₂	Second-filial generation; offspring of the F₁ generation.
Gene	A hereditary unit that governs one or more traits; occurs in a single copy in haploid cells (such as gametes), and in two copies in diploid cells.
Allele	One of two or more alternative forms of a gene.
Genotype	The composition of alleles present for a given gene or genes.
Phenotype	Observable expression of a genotype for a given trait or traits.
Homozygous	Refers to a diploid genotype in which a single allele is present in double dose (such as *AA, bb, OO*).
Heterozygous	Refers to a diploid genotype in which two different alleles are present for a given gene (such as *Aa, Bb, Cc*).
Dominance–recessiveness	An interaction between two alleles in which the expression of the recessive allele is masked by the presence of the dominant allele; the phenotype of a hybrid will be the same as the parent having the dominant phenotype.
Incomplete dominance	An interaction of two alleles in which neither allele is dominant or recessive.

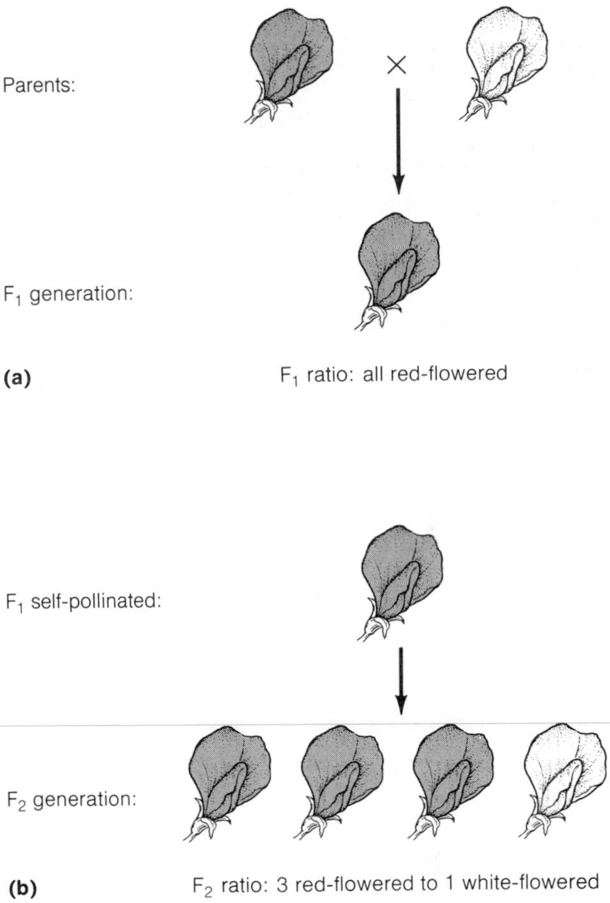

(a) F₁ ratio: all red-flowered

(b) F₂ ratio: 3 red-flowered to 1 white-flowered

FIGURE 9–3
A Monohybrid Cross. (a) When Mendel crossed a purebreeding red-flowered plant with a purebreeding white-flowered plant, all of the offspring (the F₁ generation) had red flowers. Mendel concluded that red-flower color is dominant over the recessive white-flower color. (b) When the red-flowered F₁ plants self-pollinated, they produced both red-flowered plants and white-flowered plants in a 3:1 ratio.

color in the F₁ generation, regardless of the type of gamete (egg or sperm) it contributes. Mendel concluded that red flower color is **dominant** over white flower color in peas. We can also say that white flower color is **recessive** to red, since it does not show up in the hybrid.

If Mendel had ended the experiment at this point, he would have been no closer to understanding the basis of heredity than his contemporaries. But Mendel forged on. He was curious to know whether the red-flowered hybrids would breed true for red flowers, so he put bags over their flowers to ensure self-pollination. This self-pollination of the F₁ hybrids is equivalent to a **monohybrid cross,** a cross between two monohybrid individuals. The results of this experiment gave Mendel the information he needed to understand inheritance of flower color in peas. The F₁ generation did *not* breed true. Instead, the **F₂ generation** (second filial) was composed of both red-flowered and white-flowered individuals, which appeared in the ratio of three red-flowered plants for every one white-flowered plant (3:1) (Figure 9–3b). (This result can also be expressed as $\frac{3}{4}$ red-flowered plants and $\frac{1}{4}$ white-flowered plants, or 75% and 25%.)

When Mendel carried out analogous crosses with peas that bred true for variations of other traits (such as yellow versus green seed color, tall versus dwarf stems, etc.), he always observed the same general pattern of inheritance. In each case, the recessive character was masked in the F_1 generation, but reappeared in 25% of the F_2 generation (Figure 9–4). These results led Mendel to propose an explanation for the inheritance of such traits.

THE LAW OF SEGREGATION. To account for the results of his monohybrid crosses, Mendel proposed that the hereditary factors controlling specific traits occur in pairs. For the trait of flower color in peas, each plant has two factors, which become segregated during the formation of the gametes. Thus, each egg or sperm carries only one flower color factor. When an

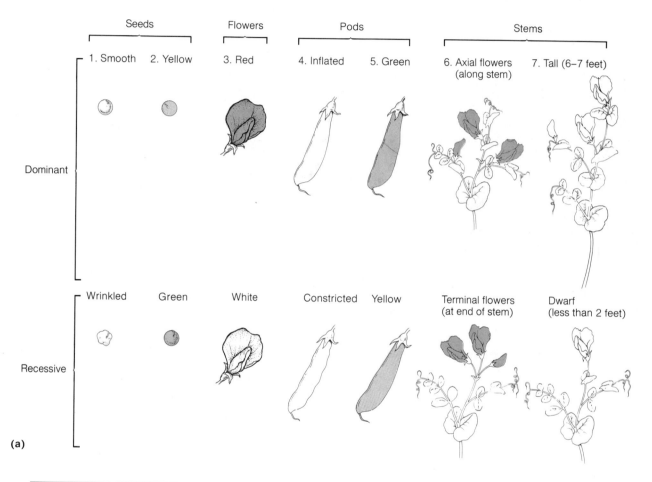

(a)

(b)

Trait	Dominant Form	Recessive Form	F₂ Generation Dominant	F₂ Generation Recessive	Ratio of Dominant to Recessive
1. Seed texture	Smooth	Wrinkled	5474	1850	2.96:1
2. Seed color	Yellow	Green	6022	2001	3.01:1
3. Flower color	Red	White	705	224	3.15:1
4. Pod form	Inflated	Constricted	882	299	2.95:1
5. Pod color	Green	Yellow	428	152	2.82:1
6. Flower position	Axial	Terminal	651	207	3.14:1
7. Stem length	Tall	Dwarf	787	277	2.84:1

FIGURE 9–4
Pea Traits. (a) Some of the traits of garden peas Mendel studied. (b) Mendel's data for the F_2 generation, obtained by examining thousands of seeds and hundreds of plants. Note that in all cases the ratio of the dominant form to the recessive form in the F_2 generation approximates 3:1.

egg and sperm unite, each contributes its single factor to the fertilized egg. Thus, the new individual will contain two flower color factors just like its parents. The idea that a given trait is determined by the interaction of two factors, and that these factors become separated from one another during gamete formation, is Mendel's first principle of inheritance: the **law of segregation.**

To put the law of segregation into the form of a simple model, let us assume (as did Mendel) that the red-flowered purebred has two identical flower color factors. If we let R stand for the red flower color factor, then this plant has a gene composition, or **genotype,** of RR. We can also assume that white flowers must be governed by a different flower color factor, which we will designate r. Thus, the genotype of the white-flowered purebred is rr. (It is common practice to use the uppercase letter for the dominant form and the lowercase letter for the recessive form.) In modern terminology, the R and r factors are called **alleles**—alternative forms of a particular gene (in this case, the flower color gene) that affect a specific trait in different ways. Each parental pea strain is pure-breeding because it has two identical alleles for flower color. An individual with identical alleles for a trait is called **homozygous** (*homo* = same). Consider the red-flowered purebred with its RR genotype. All the gametes produced by this plant bear a single R allele, so self-pollination or the crossing of two RR peas always results in an R-containing egg fertilized by an R-containing sperm. Thus, the offspring all have the RR genotype and a **phenotype,** or visible characteristic, of red flowers.

In Mendel's experiments, when a purebreeding red-flowered parent was crossed with a purebreeding white-flowered parent ($RR \times rr$), the F_1 progeny received one R flower color allele from the red-flowered parent and one r flower color allele from the white-flowered parent, yielding an F_1 genotype of Rr. Since all the F_1 peas were phenotypically red-flowered, Mendel concluded that the R allele is dominant (r is recessive).

Let us now see if Mendel's model predicts the outcome of the monohybrid cross ($Rr \times Rr$). Since the F_1 plants are **heterozygous** (*hetero* = different) for flower color, each plant can produce two types of gametes—half the gametes have the R allele and the other half have the r allele. Assuming that gamete union is totally random, Mendel's model predicts that 25% of the F_2 generation will be RR, 50% will be Rr, and 25% will be rr (Figure 9–5). Because R is dominant, both the RR and Rr offspring (75% of all

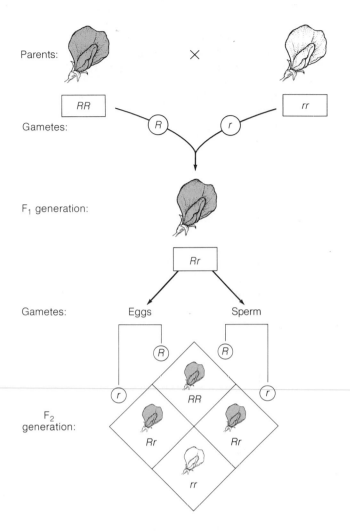

F$_2$ phenotype ratio:
3 red-flowered to 1 white-flowered

FIGURE 9–5
Mendel's Explanation for the Monohybrid Cross. (Top) In the cross between the two purebreeding parents, the F$_1$ offspring received a different flower color allele from each parent, giving them a genotype of *Rr*. (Bottom) Having both alleles present, the F$_1$ hybrids could produce eggs and sperm that have either the *R* or *r* allele. When these gametes united randomly, the F$_2$ peas were produced in the proportion of three red-flowered plants for every one white-flowered plant.

the F$_2$ plants) are red-flowered. Only the rr plants (25%) show white flowers. Thus, the F$_2$ phenotype ratio should be 75% red and 25% white—a 3:1 ratio—just as Mendel observed.

Mendel's model of gene segregation not only accounted for the observed phenotype distribution in his peas, but it also predicted meiosis, the cell division process that precedes the formation of gametes (see Chapter 8). We have since learned that genes in eukaryotic organisms (such as the pea) are parts of larger structures called chromosomes, which exist in homologous pairs in the diploid state (see Chapter 8). Let us now bring together your knowledge of meiosis with that of the genetic law of segregation to see why Mendel got the results he reported.

Peas have fourteen chromosomes in the nuclei of their cells—seven paternal and seven maternal chromosomes, or seven homologous pairs. The flower color alleles are located on one of the chromosome pairs. In the case of Mendel's F_1 hybrid, the maternal chromosome of this pair had the R allele (inherited from its red-flowered parent), and the paternal homologue had the r allele (inherited from its white-flowered parent). When the hybrid produced gametes by meiosis, the R- and r-bearing homologues (as well as the other six homologues) segregated and distributed into separate gametes. For each diploid cell in the hybrid that divided by meiosis, four haploid cells were produced, two carrying the R allele and two carrying the r allele (Figure 9–6). Thus, equal proportions of R and r gametes were formed. When they combined randomly in fertilization, the results were a 3:1 ratio of flower color phenotypes in the F_2.

A large part of Mendel's success stemmed from the fact that he worked with large sample sizes. He crossed many plants and scored hundreds of offspring in every experiment. Mendel knew that his model could only predict the *probability* of a particular phenotype ratio—just as the "law of averages" predicts that there is an equal chance of a tossed coin coming up "heads" or "tails." We all know, however, that probability and actuality may differ. If you flip a coin four times, for example, there is an outside chance it will come up tails each time. Based on this experience alone, you might conclude that both sides of the coin are tails. There is much less chance of reaching such an erroneous conclusion if you increase the sample size to 100 flips. Thus, Mendel avoided "errors of chance" by examining large sample sizes of his experimental peas.

The Testcross

If you happen upon a red-flowered pea growing in a vacant lot, how can you tell if the plant is heterozygous (Rr) or homozygous (RR) for red flowers? A geneticist would approach this problem by performing a testcross. In a **testcross,** the individual in question is crossed with a homozygous recessive individual, in this case a white-flowered pea which can have only one possible genotype—rr. If the red-flowered plant is a hybrid (Rr), then approximately half of the testcross offspring will have red flowers

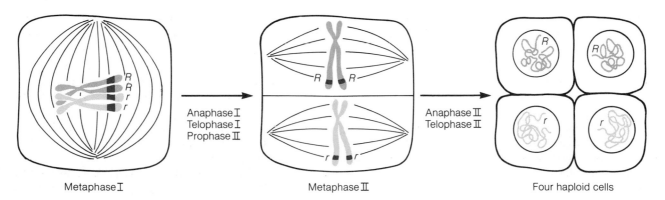

Metaphase I   Anaphase I / Telophase I / Prophase II   Metaphase II   Anaphase II / Telophase II   Four haploid cells

FIGURE 9–6
Segregation of Alleles during Meiosis. Peas have seven pairs of chromosomes, one pair of which (shown here) bears the flower-color alleles. During meiosis in the F_1 hybrid, the homologous chromosomes (each consisting of two chromatids) pair up, including the maternal and paternal homologues bearing the R and r alleles. During the first division of meiosis, the homologues segregate into separate cells, as do the alleles they carry. When meiosis is complete, each haploid cell formed has one copy of each of the seven chromosomes, and hence, a single flower-color allele (either R or r). The haploid cells develop into gametes.

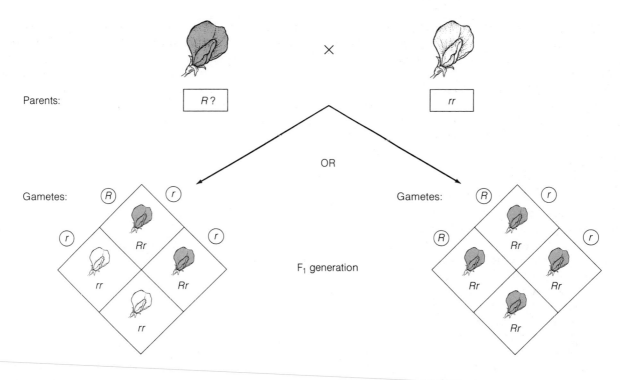

(a) F$_1$ phenotype ratio:
1 red-flowered to 1 white-flowered

(b) F$_1$ phenotype ratio:
all red-flowered

FIGURE 9–7
The Testcross. A testcross between an individual with the dominant phenotype (such as red flowers) and a recessive homozygote (such as *rr*) will reveal whether the individual in question is heterozygous or homozygous dominant. (a) If the red-flowered parent is heterozygous (*Rr*), half of the offspring will have red flowers and half will have white flowers. (b) If the red-flowered parent is homozygous (*RR*), all of the offspring will have red flowers.

(*Rr*) and half will have white flowers (*rr*). On the other hand, if your unknown plant is homozygous (*RR*) for red flowers, then all the offspring will have red flowers (*Rr*) (Figure 9–7). Of course, a testcross only works if the trait in question is governed by a single gene for which there are two possible alleles. Also, the sample size must be large enough to ensure confidence in the results. (If your testcross results in five offspring, all of which produce red flowers, can you be sure that the original red-flowered pea you found is homozygous?)

The Dihybrid Cross

After successfully describing the inheritance pattern for single traits in peas, Mendel went on to tackle how two unrelated traits are inherited simultaneously. In one of his most well-known experiments, Mendel crossed a pea that bred true for both yellow seed color (genotype *YY*) and smooth seed texture (*SS*) with one that bred true for seeds that are green (*yy*) and wrinkled (*ss*) (Figure 9–8). All the F$_1$ progeny had yellow and smooth seeds, indicating to Mendel that these phenotypes are dominant and that green seed color and wrinkled seed texture are obviously recessive. The F$_1$s are **dihybrids** (genotype *YySs*)—they are heterozygous for both traits.

Mendel then carried out the dihybrid cross (*YySs* × *YySs*). The most striking result was the appearance of two new phenotypes in the F$_2$ (actually, new phenotype combinations): yellow-wrinkled seeds and green-smooth seeds (see Figure 9–8). A total of four phenotypes appeared in the approximate ratio of 9:3:3:1; this ratio was characteristic of all dihybrid crosses performed by Mendel in which a simple dominance–recessive pattern existed for both traits in question.

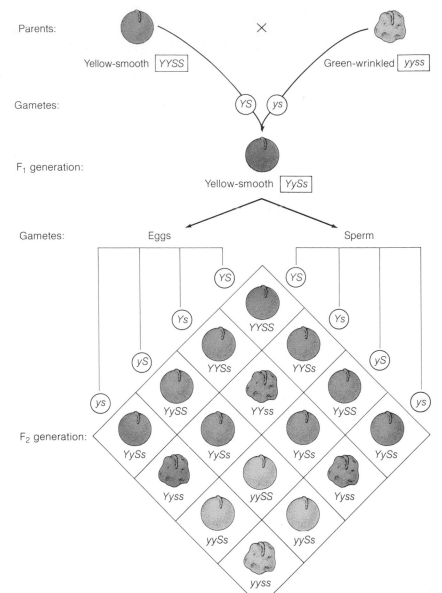

FIGURE 9-8
A Dihybrid Cross. Mendel crossed a pure-breeding strain of peas having yellow-smooth seeds (*YYSS*) with a purebreeding strain having green-wrinkled seeds (*yyss*). The resulting F₁ plants were heterozygous for both traits (*YySs*). Mendel then allowed the F₁ to self-pollinate. The genotypes and phenotypes of the predicted F₂ offspring are indicated in the boxes.

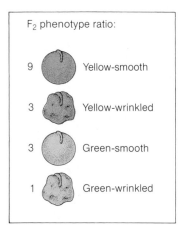

F₂ phenotype ratio:

9 Yellow-smooth

3 Yellow-wrinkled

3 Green-smooth

1 Green-wrinkled

THE LAW OF INDEPENDENT ASSORTMENT. The results of Mendel's dihybrid cross can be rationalized by making two assumptions: (1) the two alleles of each gene segregate during gamete formation, and (2) the inheritance of seed color occurs independently of seed texture. This second assumption became Mendel's second principle, the **law of independent assortment:** the inheritance of either allele for one gene has no influence on which allele is inherited for a second gene. To find out why this is true, let us go back to meiosis.

Recall that during meiosis, the chromosomes of one bivalent (paired homologous chromosomes) assort independently of the chromosomes making up the other bivalents (see Figure 8–7). Among the seven pairs of chromosomes in peas, one pair carries the seed color gene and another the seed texture gene. During the first division of meiosis in the dihybrid (*YySs*), the orientation of the bivalents at metaphase I may be such that the chromosome bearing the *Y* allele moves to the same pole as the chromosome having the *S* allele; the *y* and *s* chromosomes would thus move together toward the opposite pole (Figure 9–9). In this case, the genotypes of half the gametes formed will be *YS*; the other half will be *ys*. There is another chromosomal arrangement that is just as likely, however. The metaphase alignment of the seven bivalents could also result in the *Y* and *s* chromosomes moving together, while the *y* and *S* chromosomes move in the opposite direction. This pattern will yield equal numbers of gametes having the *Ys* and *yS* genotypes. And because the relative orientations of the bivalents at meiotic metaphase I is entirely random, the four genetically distinct gametes—*YS, ys, Ys,* and *yS*—

FIGURE 9–9
Meiosis and Independent Assortment.
In a pea plant heterozygous for both seed color and texture (*YySs*), the independent alignment of bivalents during metaphase I of meiosis accounts for Mendel's law of independent assortment. This diagram shows only two of the seven pairs of pea chromosomes—those carrying the alleles for seed color (*Y* or *y*) and seed texture (*S* or *s*).

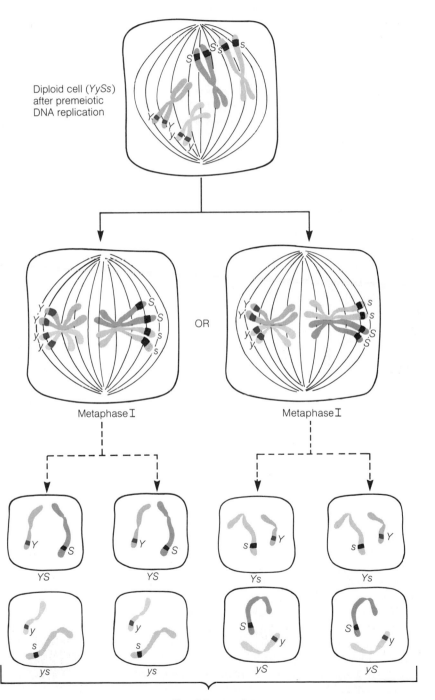

Diploid cell (*YySs*) after premeiotic DNA replication

Metaphase I   OR   Metaphase I

YS   YS   Ys   Ys

ys   ys   yS   yS

Possible gametes

should be produced in equal amounts. Thus, because the alleles are integral parts of the independently assorting chromosomes, they too assort independently of one another.

If we assume that the four different types of gametes unite randomly (there is no reason to assume otherwise), then the predicted F_2 phenotypes should be $\frac{9}{16}$ yellow-smooth seeds, $\frac{3}{16}$ yellow-wrinkled, $\frac{3}{16}$ green-smooth, and $\frac{1}{16}$ green-wrinkled (see Figure 9–8).

MENDEL'S LAWS EXTENDED

As mentioned earlier, the well-defined varieties of peas Mendel studied were fortunate choices for basic experiments on inheritance. For one, all the traits he examined showed an "all-or-none" character—flowers were red or white, never pink; seeds were yellow or green, never yellowish-green; etc. That is, there was never a problem in distinguishing phenotypes. Secondly, each trait in question was governed by a single gene for which only two alleles existed (R or r for flower color, Y or y for seed color, etc.). Finally, when Mendel followed the inheritance of two traits simultaneously, the alleles controlling those traits always assorted independently. We say Mendel was fortunate because he could have chosen any number of other traits in peas (or other organisms) and not observed such distinctive and regular patterns. As we shall see, not all traits are governed by single genes, nor are all genes represented by only two alleles. Furthermore, the presence of two different alleles in a hybrid does not always result in simple dominance. These exceptions to the patterns Mendel observed do not invalidate his laws of inheritance. On the contrary, they strengthen and extend them.

Incomplete Dominance

Recall that in Mendel's F_1 peas, the presence of the dominant allele masked the expression of its recessive partner. The interactions of two alleles in a hybrid are not always so clear-cut, however. For example, when purebreeding red-flowered and white-flowered snapdragons are crossed, the F_1 hybrids all produce pink flowers. In this case the phenotype of the heterozygote is intermediate to those of the homozygotes, an allele interaction called **incomplete dominance.** Neither allele dominates over the expression of the

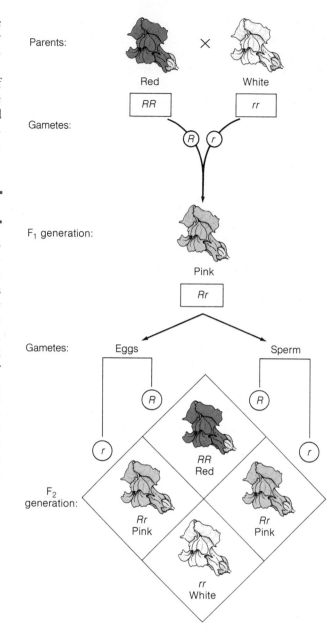

F_2 genotype ratio: 1 RR to 2 Rr to 1 rr
F_2 phenotype ratio: 1 red to 2 pink to 1 white

FIGURE 9–10
Incomplete Dominance in Snapdragons.
When purebreeding red- and white-flowered snapdragons are crossed, the F_1 hybrids are pink, an intermediate color. This indicates that in snapdragons neither allele for flower color is dominant. When the F_1 hybrids are allowed to self-pollinate, they produce F_2 plants that exhibit a Mendelian ratio of phenotypes.

other. This does not mean that the mode of flower color inheritance in snapdragons is any different than that in peas. When two pink-flowered snapdragons are crossed, the phenotype pattern in the F_2 generation is exactly what Mendel's law of segregation predicts: one red-flowered (homozygous) plant to two pink-flowered (heterozygous) plants to one white-flowered (homozygous) plant (Figure 9–10).

Cases of incomplete dominance are quite rare, at least when phenotypes are monitored as visible characteristics. When we examine cases of apparent dominance at the molecular level, however, we often find that heterozygotes for a given gene are quantitatively different than either type of homozygote. Tay-Sachs disease in humans is a case in point. Tay-Sachs is a genetic disorder caused by having two harmful recessive alleles (genotype *aa*). Tay-Sachs children appear normal at birth; near the end of their first year, however, they become listless and begin to show other signs of brain deterioration. With each passing month, their health worsens; they gradually lose the abilities to move, see, breathe, even eat. Death occurs before the age of five. The progressive deterioration of the brain stems from the accumulation of certain lipids, called gangliosides, in the nerve cells. Individuals with genotypes *AA* or *Aa* are not affected because they have an enzyme, ganglioside GM$_2$-hexosaminidase (abbreviated Hex A), that breaks down these lipids before they can build up in the nerve cells. In this respect, the *A* (normal) allele is clearly dominant over the *a* (Tay-Sachs) allele. On the other hand, blood tests reveal that heterozygotes (*Aa*) have much less Hex A than do the homozygous normal (*AA*) individuals. Thus, in terms of Hex A levels, the *A* allele is incompletely dominant. Even with a reduced level of this enzyme, however, *Aa* individuals are able to keep excess amounts of the gangliosides in check, and hence, display a normal phenotype.

Parenthetically, the measurement of blood serum Hex A levels is used as a diagnostic test to identify carriers of the recessive Tay-Sachs allele (individuals with genotype *Aa*). Since the incidence of the Tay-Sachs allele is highest among Jews (1 out of every 30 American Jews is a carrier), genetic counselors urge that all Jewish couples have a simple Hex A blood test taken before having children. As long as one member of each couple has an *AA* genotype, there is no chance of producing a Tay-Sachs child because all of their children will receive at least one *A* allele. What are the chances of producing a Tay-Sachs child if both parents are carriers (*Aa* × *Aa*)? The answer is 25%.

Multiple Alleles

Thus far we have considered only single-gene traits governed by two possible alleles. There are, however, quite a number of traits controlled by single genes for which three or more alleles exist. The *ABO* blood group in humans is an example of a **multiple allele** system consisting of three alleles: the *A*, *B*, and *O* alleles. An individual may be homozygous for any one of these alleles (*AA, BB,* or *OO*), or may have any combination of two alleles (*AB, AO,* or *BO*). The six genotypes result in four phenotypes, or **blood types,** because the *O* allele is recessive to both the *A* and *B* alleles, whereas if the *A* and *B* alleles are together, both are expressed (Table 9–2).

Blood types are important when it comes to matching blood donors with recipients for a blood transfusion. People with blood type A (genotype *AA* or *AO*) have a specific type of molecule, called the **A antigen,** that extends from the surface of their red blood cells. Type B people have instead the **B antigen,** and type AB individuals produce both types of antigens (see Table 9–2). These blood cell molecules are called antigens because when they are injected into the bloodstream of an animal (such as a rabbit, mouse, human, or other higher animal) that does not produce them, the recipient's immune system is alerted to manufacture specific **antibodies** against them (see Chapter 20). These antibodies bind to the surface antigens in a way that causes the blood cells to agglutinate (clump), and eventually be destroyed. (The same principle is involved in the immune defense against bacteria, viruses, and other infectious agents.) Thus, a type B person will form antibodies that agglutinate and destroy type A blood cells, and vice versa—a potentially dangerous situation that could easily lead to fatal blood clots. Since type O blood has neither A nor B antigens, it will not evoke antibody formation

TABLE 9–2
The ABO Blood Group in Humans.

Genotypes	Phenotypes (Blood Types)	Red Blood Cell Antigen(s) Present
OO	O	None
AA, AO	A	A
BB, BO	B	B
AB	AB	A and B

in any recipient. Type O people, accordingly, are called "universal blood donors." However, we now know that other antigens, apart from those coded by the ABO gene, appear on red blood cells. Thus, blood typing today involves more than a simple screening for the A and B antigens.

Figure 9–11 shows two examples of how the ABO blood group is inherited.

Pleiotropy

Phenylketonuria (PKU), like Tay-Sachs, is a genetic disorder caused by a double dose of a harmful recessive allele. PKU infants lack an enzyme needed to convert the amino acid phenylalanine to tyrosine, also an amino acid. As a result, abnormally high levels of phenylalanine accumulate in the blood, which leads to other biochemical consequences. For one, the excess phenylalanine interferes with the synthesis of

melanin, the dark pigment found in skin, hair, and eyes. PKU children thus tend to have very fair skin and light blond hair. More serious, however, is the conversion of the excess phenylalanine into related substances that retard brain development. PKU individuals have reduced head sizes and are severely retarded (a mean IQ of 20). This is an example of a single gene that obviously affects more than one trait, a situation geneticists refer to as **pleiotropy.** Fortunately, the harmful characteristics associated with PKU can be avoided if this disorder is diagnosed early in infancy. The child is kept on a low-phenylalanine diet until at least age three, by which time brain development is complete. This so-called "phenotypic cure" alleviates the problems created by high phenylalanine levels in the blood but, of course, does not change the affected person's genes. At this point in time, "genotypic cures" are merely a hope for the future.

Many human genetic disorders result from single pleiotropic genes. The few listed in Table 9–3 are each caused by a double dose of defective allele.

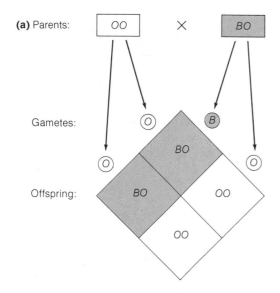

(a) Parents: OO × BO

Gametes: O B O

Offspring: BO BO OO OO

F_1 genotype ratio: 1 *BO* to 1 *OO*
F_1 phenotype ratio: 1 type B to 1 type O

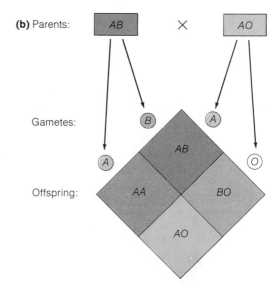

(b) Parents: AB × AO

Gametes: B A A O

Offspring: AB AA BO AO

F_1 genotype ratio: 1 *AB* to 1 *AA* to 1 *AO* to 1 *BO*
F_1 phenotype ratio: 1 type AB to 2 type A to 1 type B

FIGURE 9–11
Inheritance of the ABO Blood Group. The human ABO blood group is an example of a multiple allele system. In the human population there are three possible alleles, *A, B,* and *O*. Of course, any one individual has only two alleles. The inheritance of blood type follows Mendel's laws, as demonstrated by the two examples given here.

TABLE 9–3
Single Gene Disorders That Have Pleiotropic Effects in Humans.

Disorder	Molecular Block	Effect	Symptoms	Treatment
Phenylketonuria	Lack enzyme to convert phenylalanine to tyrosine.	Reduced levels of tyrosine; excess phenylalanine converted to phenylpyruvic acid.	Reduced melanin synthesis results in light-colored skin, hair, and eyes; phenylpyruvic acid retards brain development, resulting in mental retardation.	Keep infant on a low phenylalanine diet until brain development complete (at least three years of age).
Tay-Sachs	Lack enzyme to break down certain gangliosides, a type of lipid.	Accumulation of gangliosides in brain cells interferes with brain functions.	Paralysis, blindness, severe feeding difficulties; total immobility by two years of age, death before five years of age.	None
Classical Albinism	Lack enzyme to convert tyrosine to precursors of melanin.	Absence of melanin pigments.	White skin and hair, pink eyes; serious visual problems; sunburn easily.	Protect skin and eyes from sunlight.
Galactosemia	Lack enzyme needed to metabolize lactose (milk sugar).	Breakdown product of lactose—galactose—accumulates in body.	In infants, severe vomiting, diarrhea, eye lens cataracts, liver damage, mental retardation, and death if not treated.	Replace milk with other foods that lack lactose and galactose.
Diabetes mellitus (certain forms)	Inability to produce enough insulin.	Poor assimilation of glucose; glucose excreted in urine.	Fatigue, weight loss, brain damage, blindness, dehydration.	In moderate cases, controlled by diet; severe cases require insulin injections.

Polygenic Inheritance

Just as pleiotropy refers to cases in which one gene exerts an influence over more than one trait, there are many examples of the reverse situation, called **polygenic inheritance,** in which two or more genes affect a single trait. For example, height in humans is influenced by at least ten different genes. Because of the number of genes involved, each of which may have two or more allelic forms, height is inherited in an additive manner. In other words, tall people tend to have a preponderance of "tall" alleles for the height-controlling genes, whereas individuals having a balance of "tall" and "short" alleles are correspondingly medium in height. With so many allelic combinations possible, humans display a wide range of variability in height (Figure 9–12).

Nature versus Nurture

Height also has an environmental component. Nutrition, disease, injury, and other factors can have significant effects on human growth. For example, the average height of Japanese-Americans is greater than their relatives raised in Japan, a difference that apparently is related to differences in diet.

There are many other human traits that have a final outcome very much affected by environmental factors. Recall that the symptoms of PKU can be averted by keeping the affected child on a low phenylalanine diet during the early years of brain development. Human skin color, a polygenic trait, is also subject to modification by exposure to the sun's ultraviolet rays. The most interesting traits with respect to genetic versus environmental control of phenotype, or nature versus nurture, are those relating to human behavior (intelligence, emotions, personality, and so forth). How much variation for a given trait is determined genetically, and how much is determined by environmental influences? These are very tricky questions for biologists because so little is known about the genetics of behavior. Furthermore, we simply cannot subject people to controlled experiments designed to examine the influence of environment on their behavioral development. Our ignorance on this subject is as vast as the brain is complex. Lacking reliable scientific data, virtually anyone can express an opinion on these matters, and they usually do.

Number of individuals:	1	0	0	1	5	7	7	22	25	26	27	17	11	17	4	4	1
Height in inches:	58	59	60	61	62	63	64	65	66	67	68	69	70	71	72	73	74

FIGURE 9–12
Polygenic Inheritance. Single traits that are controlled by two or more genes (polygenic inheritance) typically show a wide range of phenotypic variation within a population. When such variations are plotted, they form a bell-shaped curve. Height in humans is a polygenic trait, shown here by the approximate bell-shaped curve formed by 175 men recruited for the United States Army near the turn of the century.

The more we examine patterns of inheritance, the more we find that the principles of genetics are not quite as simple as Mendel's original experiments suggested. Nevertheless, Mendel's studies were an important breakthrough in the history of genetics, and they laid the foundation for many further discoveries. We will describe some of these discoveries in Chapter 10.

SUMMARY

Our current understanding of heredity owes a great deal to the mid-nineteenth century experiments of Gregor Mendel. Mendel crossed pea plants that were purebreeding for contrasting phenotypes, such as red versus white flower color, or yellow versus green seed color. The first generation offspring all resembled one parent, rather than some resembling each parent or all showing an intermediate phenotype. From this, Mendel concluded that the parental phenotype that appeared in the F_1 generation was dominant, and that the other parental phenotype was recessive. The F_1 hybrids were then allowed to self-pollinate (a

monohybrid cross). The resulting F_2 plants appeared in the approximate ratio of three dominant to one recessive phenotype. This led Mendel to propose the law of segregation: Each (diploid) organism carries two copies of a given gene, which segregate into single copies in the gametes. Accordingly, monohybrids can produce two types of gametes, each containing a different allele for the gene in question. By random union of such gametes, one-fourth of the progeny are homozygous dominant, one-half are heterozygous dominant, and one-fourth are homozygous recessive. Thus, three-fourths will show the dominant phenotype and one-fourth will show the recessive phenotype, thereby accounting for the observed three-to-one ratio in the F_2 generation.

When Mendel performed dihybrid crosses between peas that differed in two traits, each of which showed a simple dominance-recessive pattern, he obtained a 9:3:3:1 ratio of phenotypes in his F_2 plants. This result suggested to Mendel that the alleles of one gene assort independently of the alleles for a different gene, a principle called the law of independent assortment. We know today that independent assortment has its basis in the random alignment of bivalents during meiosis preceding gamete formation.

Twentieth century geneticists have extended Mendel's work to include other types of inheritance patterns. Certain traits, such as flower color in snapdragons, are inherited in an incompletely dominant fashion. The F_1 hybrids produced in a cross between purebreeding strains of red-flowered and white-flowered snapdragons display pink flowers, a phenotype intermediate between their parents. In many other traits, such as Tay-Sachs, the heterozygotes show the dominant phenotype but have intermediate levels of the molecular determinant (for instance, an enzyme) of that phenotype.

Other variations from Mendel's described patterns of heredity include genes for which there exist multiple alleles; pleiotropy; and polygenic inheritance. The ABO blood group in humans is a multiple allelic trait regulated by a single gene for which there are three possible alleles. An individual's blood type (phenotype) is determined by which two of these alleles are present. Pleiotropy refers to a situation in which one gene affects two or more apparently unrelated traits. For example, phenylketonuria is a single-gene disorder that, if untreated, causes mental retardation and light coloration of the skin, hair, and eyes. The reverse situation, in which two or more genes affect a single trait, is called polygenic inheritance. Height and skin color in humans are examples of polygenic traits.

A few traits, especially polygenic ones, are influenced by the external environment. The extent of environmental determination of human behavior (such as intelligence and emotions) is an important question that has far-reaching sociological implications.

pital, Mrs. Brown suspected that there had been a mixup in the nursery and that they had given her the wrong baby. This baby's blood type was O and Mr. and Mrs. Brown had type A and B blood, respectively, Mr. Paige had blood type O and Mrs. Paige was type AB. Did the hospital goof and switch the Brown and Paige babies?

5. In cocker spaniels, the allele for black coat color (*B*) is dominant over the allele for red coat color (*b*). Suppose a black-coated male mates with a red-coated female. The female later gives birth to three black-coated pups. Could you conclude that the male parent was homozygous for black coat color? Explain.

6. Inheritance of flower color in snapdragons follows an incomplete dominance pattern. Homozygotes have red or white flowers, and the heterozygote has pink flowers. A pink-flowered snapdragon is crossed with a white-flowered individual. What phenotypes and what ratio of occurrence would you expect in the offspring?

7. In tomatoes, tallness is governed by a single gene for which the *T* allele specifies tall, and the *t* allele specifies a dwarfed condition. *T* is dominant over *t*. Also, the hairy-stem characteristic (*H*) is dominant over the hairless condition (*h*). Suppose you crossed a tall, hairy tomato (*TTHb*) with a tall, hairless plant (*TThh*). What would be the expected ratio of phenotypes in the offspring?

8. In pea plants, the allele for smooth seed texture (*S*) is dominant over the allele for wrinkled seed texture (*s*), and the yellow seed color allele (*Y*) is dominant over the green seed color allele (*y*). Imagine that you cross a plant with smooth-yellow seeds with one having wrinkled-green seeds. The offspring include 2008 plants with smooth-yellow seeds and 2015 with smooth-green seeds. What are the most likely genotypes of the two parent peas?

STUDY QUESTIONS

1. Imagine that you have a red-flowered pea plant and wish to find out whether this plant is homozygous (*RR*) or heterozygous (*Rr*) for flower color. What results would you predict for each possibility from (a) a self-pollination, and (b) a testcross with a white-flowered pea?

2. Predict the ratio of genotypes in offspring produced by a cross between a homozygous red-flowered pea and a heterozygous red-flowered pea. What is the phenotypic ratio?

3. If a man with blood type A and a woman with blood type O have a child with blood type O, what is the genotype of the father? Explain.

4. Mrs. Paige and Mrs. Brown had baby girls on the same day at Mercy Hospital. A few days after she left the hos-

SUGGESTED READINGS

Ayala, F. J., and J. A. Kiger, Jr. *Modern Genetics*. 2nd ed. Menlo Park, CA: Benjamin/Cummings, 1984. A good general genetics text written for the life science major. Chapter 2 deals with Mendelian genetics.

Jenkins, J. B. *Human Genetics*. Menlo Park, CA: Benjamin/Cummings, 1983. Besides having introductory chapters covering the Mendelian rules of inheritance, this well-written book has excellent coverage of nondisjunction, blood groups, and the more common human genetic disorders.

Peters, J. A., ed. *Classic Papers in Genetics*. Englewood Cliffs, NJ: Prentice-Hall, 1959. Included in this collection of original articles is Mendel's first paper (1865) describing his experiments with peas.

ANSWERS TO STUDY QUESTIONS

1. (a) If parent is *RR*, all offspring would produce red flowers. If parent is *Rr*, would expect $\frac{3}{4}$ red and $\frac{1}{4}$ white flowers.
   (b) If parent is *RR*, all offspring would produce red flowers. If parent is *Rr*, would expect $\frac{1}{2}$ red and $\frac{1}{2}$ white flowers.

2. Cross: *RR* × *Rr*. Genotype ratio: 1 *RR* to 1 *Rr*. Phenotype: all red-flowered.

3. Father's genotype: *AO*. Since the child receives one of its blood group alleles from its father and the child is *OO*, the father must be heterozygous for blood type A.

4. The hospital did *not* make a mistake. Mr. and Mrs. Paige could have type A (*AO*) or type B (*BO*) children only. Mr. and Mrs. Brown could have a type O baby since their genotypes could be *AO* and *BO*, respectively.

5. No. The male parent may have been heterozygous for black coat color (*Bb*), but by an error of chance, all three pups were conceived by fertilizations involving B-carrying sperm.

6. Cross: pink-flowered (*Rr*) × white-flowered (*rr*). Expected phenotypes of the offspring: $\frac{1}{2}$ pink (*Rr*) and $\frac{1}{2}$ white (*rr*).

7. Phenotypes of offspring: $\frac{1}{2}$ tall-hairy and $\frac{1}{2}$ tall-hairless.

8. Genotype of smooth yellow-seeded parent: *SSYy*; wrinkled green-seeded parent: *ssyy*.

Meiosis
Mitosis > process by which somatic cells divide

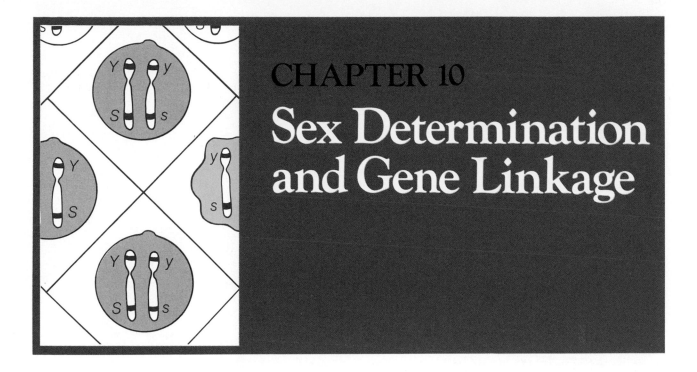

CHAPTER 10
Sex Determination and Gene Linkage

In the previous chapter we took advantage of your knowledge of chromosomes and meiosis to help explain Mendel's principles of gene segregation and independent assortment. Mendel, of course, did not have the same advantage. In fact, it was not until 1902 that Walter S. Sutton, in the United States, and Theodor Boveri, in Germany, independently proposed that genes are located on chromosomes. This suggestion, which became known as the **chromosome theory of heredity,** was based on the parallel behavior of genes (as inferred from inheritance patterns) and chromosomes (as seen in the microscope) during meiosis and fertilization. In other words, Sutton and Boveri noted that like genes, chromosomes exist in pairs (homologues) and segregate during meiosis, then recombine at fertilization.

Further support for the chromosome theory came from studies on the inheritance of sex and the discovery that the inheritance of some traits is not independent of certain others. These studies indicated that specific genes are always associated with specific chromosomes—a phenomenon known as **gene linkage.** In this chapter, we will examine the effect of gene linkage on heredity and consider the special case of linkage involving the sex-determining chromosomes.

THE INHERITANCE OF SEX

In the cells of most animals and quite a few plants, the male and female of the species differ by one pair of chromosomes. These chromosomes carry the major sex-determining genes and are thus called the **sex chromosomes.** In all mammals including humans, females have a matched pair of sex chromosomes called the X chromosomes. Females are therefore designated **XX.** Males, on the other hand, have a nonhomologous pair of sex chromosomes called X and Y; males are designated **XY.** The X chromosome in males is indistinguishable from the X chromosomes in females, but the Y chromosome is quite different in appearance (Figure 10–1). In humans, then, females have 23 matched pairs of chromosomes per cell, including the XX pair; males have 22 matched pairs plus the nonhomologous XY pair.

The inheritance of sex in humans (and other mammals) is quite straightforward. During meiosis the sex chromosomes behave like all the other homologous pairs and segregate into separate daughter cells. Thus, each gamete carries a single sex chromosome. Since females are XX, the eggs they produce bear a single X

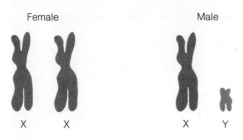

FIGURE 10–1
Sex Chromosomes in Humans.

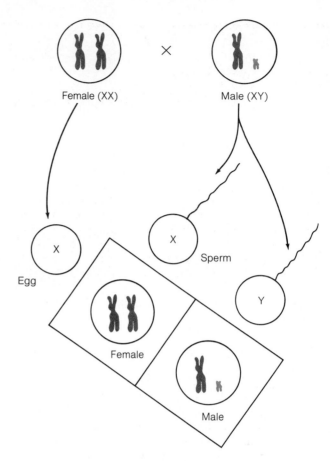

FIGURE 10–2
Inheritance of Sex. In humans and other mammals, the sex of an offspring is determined by the presence of either the X or Y chromosome in the sperm that fertilizes the egg.

chromosome plus a single copy of the 22 other chromosomes. The XY males, however, can produce two different types of sperm with respect to sex chromosome content: half the sperm will have the X chromosome and the other half will have the Y chromosome. If an egg happens to be fertilized by a sperm bearing an X chromosome, the zygote will be XX and female. Fertilization by a Y-bearing sperm will yield an XY zygote, which will be male (Figure 10–2). As you can see, it is the nature of the sperm (X or Y) that determines the sex of the offspring.

Occasionally, meiotic nondisjunctions of the sex chromosomes occur, yielding gametes that lack or have extra copies of the X or Y chromosomes. (Recall that a nondisjunction involving human chromosome 21 can result in a child with three copies of this chromosome—a disorder called Down syndrome; see Figure 8–8.) Even with their abnormal complements of sex chromosomes, such gametes are usually viable. Their union with normal gametes produces individuals with some unusual characteristics (Essay 10–1).

THE INHERITANCE OF SEX-LINKED TRAITS

Since the determination of sex in most animals is governed by the complement of sex chromosomes, it follows that these chromosomes bear genes that affect the type of gonads (ovaries and testes) that develop and the appearance of the secondary sex characteristics in humans (such as breast and hip enlargement in women and the growth of facial hair in men). The sex chromosomes in humans and other animals also carry genes that control other traits not related to sex. The entire complement of genes carried on the sex chromosomes forms a single **linkage group,** and all the traits they affect are called **sex-linked traits.**

The first demonstration of a sex-linked trait was made by Thomas Hunt Morgan and his students at Columbia University in 1910. Working with *Drosophila melanogaster,* the common fruit fly (Figure 10–3), Morgan provided a clear exception to the Mendelian law of independent assortment.

Demonstration of a Sex-Linked Trait in Fruit Flies

While examining his cultures of fruit flies one day, Morgan noticed a male fly that had white eyes instead of the usual red eyes. He realized that this mutant fly would be useful in determining how eye color in

ESSAY 10–1

HUMAN SEX CHROMOSOME ABNORMALITIES

In humans, XX females and XY males are the norm. Occasionally, however, meiotic nondisjunctions occur, resulting in gametes that either lack a sex chromosome or have an extra one or more. The fusion of such gametes yields individuals with abnormal sex chromosome complements and, often, distinctive phenotypes. For example, the XXY chromosome combination, or karyotype, produces a male with Klinefelter syndrome. This occurs in about 1 out of every 1000 male births. The adult XXY male is taller than average, is sterile, and has small testes, long limbs, sparse body hair, and often noticeable breasts. Males with Klinefelter syndrome can be treated with testosterone, the male steroid hormone, to promote the development of secondary sex characteristics and inhibit breast enlargement. Even without hormone treatment, however, young XXY males often experience normal sexual behavior, and many of them marry and raise adopted children. In later years, however, impotence can be a problem.

Another abnormal male karyotype, XYY, occurs at a higher frequency—about 1 in 700 males. Aside from being somewhat taller and less intelligent than average, the vast majority of XYY males appear to lead fairly normal lives. This picture is far more tempered than the one that emerged in the late 1960s. In a case study published in 1965, the karyotypes of 197 institutionalized males with "dangerous, violent or criminal propensities" were examined; seven were found to be XYY (a startling frequency of 3.5%). This finding precipitated further studies of men confined to mental and penal institutions; soon after, articles linking the extra Y chromosome to criminal and violent behavior began appearing in the popular press. In several widely publicized murder trials in 1968, defense attorneys argued that their XYY defendants were not legally responsible for their actions because of a genetic predisposition toward aggressive and violent behavior.

Since the original flood of reports in the 1960s, more studies on the XYY syndrome have been carried out. These later studies have largely deflated the earlier claims that the extra Y chromosome drove their bearers to violent, aggressive behavior. Although there is clearly a higher-than-random incidence of XYY males in criminal institutions, there is no hard data suggesting that these inmates are any more aggressive or violent than the XY prisoners. It is now generally believed that the higher crime rate of XYY males is due to their lower-than-average IQs.

Of the abnormal female karyotypes, the most common is XXX (about 1 in 950 female births). In most cases, XXX females have no distinctive traits. They are generally fertile and develop normal secondary sex characteristics, although some have difficulties associated with menstruation. Another female karyotype, XO (a missing X chromosome), does produce a pattern of unusual characteristics collectively called Turner syndrome. Females with Turner syndrome are typically short (average adult height is under 5 feet), have short, broad necks, and are invariably sterile. Their undeveloped ovaries produce neither eggs nor the female sex hormones called estrogens. Lacking these hormones, the XO female requires treatment with synthetic estrogens to encourage development of the secondary sex characteristics. Treatment with synthetic estrogens does not restore fertility but does make normal sexual relations possible.

The frequency of Turner syndrome is estimated to be approximately 1 in 5000 female births. This low frequency is not surprising in light of the fact that the XO karyotype is the most common chromosome abnormality found in spontaneously aborted fetuses. It is estimated that about 98% of all XO embryos are aborted early in pregnancy. If true, this would put the rate of XO conceptions at 1 per 100 potential females.

The discovery of abnormal sex karyotypes in humans has led to at least one undisputed conclusion: the presence of a Y chromosome renders a male. Females are females because they lack a Y chromosome, regardless of the number of X chromosomes they have. Exactly how fewer or extra sex chromosomes produce physiological malfunctions is only beginning to be understood.

FIGURE 10–3
Morgan and His Fruit Flies. In studying the inheritance patterns of many traits in the common fruit fly (*Drosophila melanogaster*), Thomas Hunt Morgan (1866–1945) provided supporting evidence for the chromosome theory of heredity. Under optimal conditions, fruit flies have a generation time of only two weeks, and each impregnated female may lay hundreds of eggs. Because each fly is only about 3 mm in length, Morgan could keep an entire population of flies in a half-pint bottle containing nutrient medium. As an experimental organism, the fruit fly had many advantages over Mendel's peas.

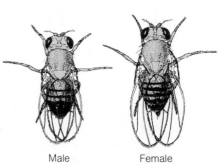

Male   Female

Drosophila is inherited, so he crossed this white-eyed male with a purebreeding red-eyed female. All their offspring (F_1) had red eyes, indicating that red eye color was dominant over white. Morgan then crossed his F_1 flies and obtained the following results:

F_2 generation: Red-eyed females   2459
                   White-eyed females   0
                   Red-eyed males   1011
                   White-eyed males   782

The most striking feature of these results was that all the white-eyed flies were males. For Morgan, the conclusion was clear: The traits of eye color and sex were somehow linked, and the genes that determined these traits were inherited as a unit.

At the time Morgan began his experiments, the existence of sex chromosomes in *Drosophila* was known. Sex in fruit flies, like humans, is determined by the presence of two X chromosomes for females and an X and Y chromosome for males (Figure 10–4). Combining this fact with the inheritance pattern for eye color, Morgan proposed that the eye color gene in *Drosophila* is carried on the X chromosome but is not present on the Y chromosome. Accordingly, male flies would have only one allele for eye color, because they have only one X chromosome. Whether that allele is for the dominant red-eye phenotype or the white-eye phenotype, it will be expressed in males; there is no second allele to mask it. Females, on the other hand, have two X chromosomes and, therefore, two eye color alleles. Any female with at least one red eye color allele will have red eyes.

Let us examine Morgan's hypothesis using the geneticist's notation for the sex chromosomes and the eye color alleles. If we let *R* stand for the dominant red

eye color allele and *r* equal the white eye color allele, then Morgan's original cross can be represented as follows:

White-eyed male × Purebreeding red-eyed female
$\quad$ $X^r Y$ $\qquad\qquad\qquad\qquad$ $X^R X^R$

In this notation, the eye color alleles are shown as superscripts of the X chromosomes; the Y chromosome is included even though it lacks the eye color gene. Figure 10–5a shows the predicted outcome of this cross. All the males have the genotype $X^R Y$ and all the females are heterozygous $X^R X^r$. Since *R* is domi-

Female   Male

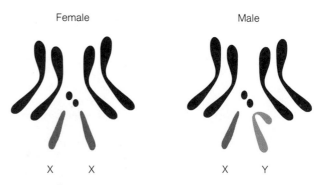

X   X   X   Y

FIGURE 10–4
Chromosomes in Fruit Flies. Fruit flies have eight chromosomes ($2n = 8$). These chromosomes are arranged into four homologous pairs in the diagrams. The pair of sex chromosomes consists of two X chromosomes in the female and an X and Y chromosome in the male. In contrast to the case in humans, the *Drosophilia* Y chromosome is larger than the X chromosome.

nant, both sexes of these F₁ flies would have red eyes, which is consistent with Morgan's actual results. This result does not prove a link between sex determination and eye color, however. The same F₁ phenotype is predicted by simple Mendelian rules of inheritance.

When we examine the phenotypes of the F₂ flies, however, it becomes clear that the genes for sex determination and eye color do *not* assort independently. The only way to account for the observed F₂ phenotype ratio is to assume (as did Morgan) that the eye color gene is linked to the X chromosome in *Drosophila.* By so doing, one would predict a 3 : 1 ratio of red to white eye color, with all the females having red eyes, and the males split evenly between red and white eyes (Figure 10–5b). (The less-than-expected number of white-eyed males appearing in the F₂—782 white-eyed *versus* 1011 red-eyed—was due to the fact that white-eyed flies are more likely to die before they

hatch.) From this and other crosses performed with fruit flies, Morgan established a definite relationship between chromosomes and Mendel's hereditary factors.

Sex-Linked Traits in Humans

The inheritance of sex-linked traits in humans occurs by the same mechanism described for *Drosophila.* This means that a recessive allele for any sex-linked gene is expressed as the recessive phenotype in males but is masked in heterozygous females.

Red-green color blindness is a recessive sex-linked trait that renders individuals unable to distinguish shades of red or green—both appear gray. About 8% of American males are color-blind, in contrast to only 0.6% of females. This statistic alone suggests that the color vision gene is sex-linked; other evidence

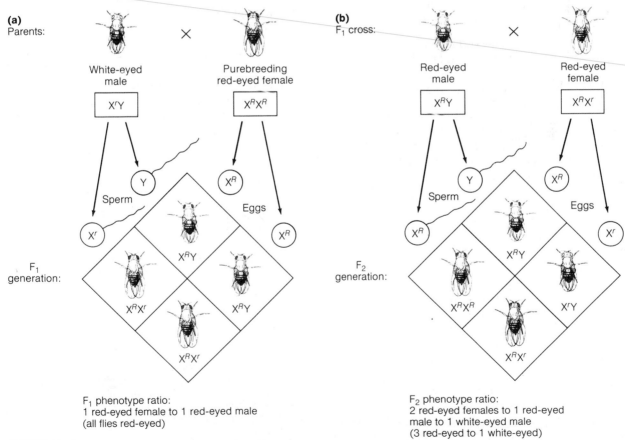

(a) Parents:

White-eyed male — X^rY

Purebreeding red-eyed female — X^RX^R

Sperm: Y, X^r
Eggs: X^R, X^R

F₁ generation: X^RY, X^RX^r, X^RY, X^RX^r

F₁ phenotype ratio:
1 red-eyed female to 1 red-eyed male
(all flies red-eyed)

(b) F₁ cross:

Red-eyed male — X^RY

Red-eyed female — X^RX^r

Sperm: Y, X^R
Eggs: X^R, X^r

F₂ generation: X^RY, X^RX^R, X^rY, X^RX^r

F₂ phenotype ratio:
2 red-eyed females to 1 red-eyed male to 1 white-eyed male
(3 red-eyed to 1 white-eyed)

FIGURE 10–5
Morgan's Model for Eye Color Inheritance in *Drosophila.* (a) A cross between a white-eyed male and a purebreeding red-eyed female. The F₁ offspring of both sexes have red eyes. (b) The F₁ cross. Note that the predicted 25% occurrence of white-eyed flies in the F₂ are exclusively male.

comes from the analysis of **pedigrees,** or family trees, such as that shown in Figure 10–6. (The analysis of the inheritance of human traits relies heavily on pedigree constructions because human geneticists do not have the option of conducting selective matings.)

To understand why color blindness occurs much more frequently in males, let us examine the types of parental combinations that can produce color blindness in sons and daughters. A son inherits the color-blind allele from his mother, who may be either color-blind herself (remember, 0.6% of American females are color blind; genotype X^cX^c) or, more likely, a normal-sighted carrier (about 16% of American females: X^CX^c). There is a 50% chance that a son will inherit the X^c chromosome from a carrier mother. Whether or not the father is color-blind (X^cY or X^CY) has no bearing because a son receives only the Y chromosome from his father. For a daughter to be color-blind, however, not only must her mother have at least one allele for color blindness, but her father must be color-blind. As this combination of parents occurs rather infre-

quently, there are few color-blind females in the human population.

Another well-studied sex-linked trait in humans is **hemophilia,** a disorder that renders the individual less able to form blood clots. This is a serious medical problem because in hemophiliacs, or "bleeders," even a minor injury can result in a major loss of blood. Because the hemophilia allele is recessive and carried on the X chromosome, hemophilia is predominantly a male disorder. In fact, hemophilia is extremely rare in females because there are so few hemophiliac males that father children. The faulty allele is maintained in human populations largely by the phenotypically-normal female carriers. Queen Victoria of nineteenth century England was such a carrier, and this had fateful consequences for many of Europe's royal families, most notably the Russian monarchy (Essay 10–2).

Well over 100 traits in humans are now known with certainty to be sex-linked, and most of these have nothing to do with sex determination. Besides blood clotting and color vision, this cluster of genes also affect

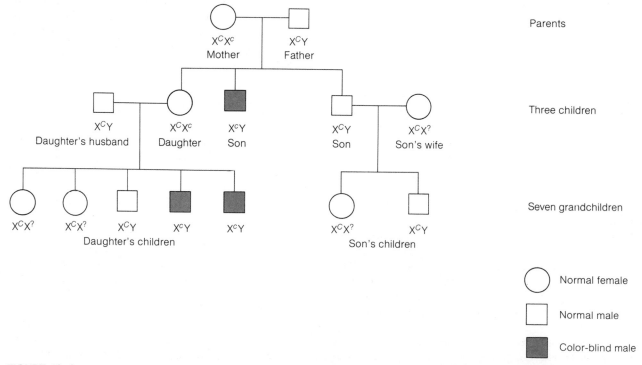

FIGURE 10–6

Hypothetical Pedigree for the Inheritance of Color Blindness. In this three-generation pedigree for color blindness, the phenotypes and known genotypes are indicated. Assigning genotypes to the males is easy: Color-blind males have the color-blind allele (X^c) and normal-sighted males have the normal allele (X^C). Normal-sighted females may be either homozygous (X^CX^C) or heterozygous (X^CX^c). The latter are carriers of the color-blind allele, whose identity as such usually hinges on their giving birth to color-blind sons or to daughters who have color-blind sons. A female whose genotype is uncertain is indicated by $X^CX^?$.

traits as diverse as glucose metabolism, eye pigmentation, hearing, skin texture, teeth, the nervous system, the immune system, and the muscular system. Harmful recessive forms of these genes are responsible for various disorders, including ocular albinism (the lack of eye pigments), ichthyosis (a severe scaly skin condition), and the Duchenne form of muscular dystrophy. Predictably, all of these sex-linked disorders occur predominantly in males.

AUTOSOMAL LINKAGE

All of the chromosomes of an organism *except* the sex chromosomes are termed **autosomes.** Both males and females of a species have the same autosomes; there are 44 autosomes (22 pairs) in humans and six (three pairs) in fruit flies. All the genes located on a given homologous pair of autosomes form an autosomal linkage group, and like the sex-linked genes, they are inherited as a unit; that is, they cannot assort independently of one another.

The Effect of Autosomal Linkage on Inheritance Patterns

Recall that Mendel's discovery of the law of independent assortment resulted from his crosses of garden peas that differed in two traits, such as seed color and seed texture. In his dihybrid cross, the alleles for yellow and green seeds assorted independently of those for smooth and wrinkled seeds, resulting in a F_2 phenotype ratio of $9:3:3:1$ (see Figure 9–8). To account for this ratio, we assumed that the seed color and texture genes were situated on different chromosomes, and thus could assort into four different combinations in the gametes: *YS, Ys, yS,* and *ys.* But for the sake of discussion, suppose that the seed color and texture genes in peas are linked. If so, would you predict the same phenotypic outcome in the F_1 and F_2 generations observed by Mendel? The following discussion addresses this question.

Starting with the two purebreeding strains of peas, we can designate the genotype of the yellow smooth-seeded parent to be *YS/YS,* and the green wrinkled-seed parent to be *ys/ys.* (The slash shows that the *Y* and *S* alleles are linked and the *y* and *s* alleles are linked.) Each of these parents can produce only one type of gamete with respect to the traits in question: *YS* from the yellow-smooth purebred and *ys* from the green-wrinkled purebred. Therefore, all the F_1 offspring will

be *YS/ys* (dihybrid) and have yellow-smooth seeds (Figure 10–7). As you can see, the genotype and phenotype of the F_1 are the same for both linkage and nonlinkage cases.

If we continue our hypothetical cross to the F_2 generation, however, we would not expect a $9:3:3:1$ ratio. Instead, only two phenotypes—yellow-smooth seeds and green-wrinkled seeds—should occur in a $3:1$ ratio, respectively (see Figure 10–7). Why? We must go back to meiosis in the dihybrid (*YS/ys*). Because the *Y* and *S* alleles are physically joined to one chromosome, they cannot segregate from each other during meiosis (see next section for exceptions). Similarly, the *y* and *s* alleles are joined together on the homologous chromosome. Consequently, when the first division of meiosis segregates the homologous chromosomes, the *Y* and *S* alleles move as a unit toward one end of the dividing cell, and the *y* and *s* alleles move with their chromosome toward the opposite end. Since there is no recombination of the dominant and recessive alleles for these two genes, only two gamete genotypes are possible: *YS* and *ys* (see Figure 10–7). With these two types of gametes uniting randomly, the predicted genotypes of the offspring are: 25% *YS/YS,* 50% *YS/ys,* and 25% *ys/ys.* This corresponds to a $3:1$ ratio of yellow-smooth seeds to green-wrinkled seeds.

Recombination of Linked Genes

The hypothetical inheritance pattern just described for the two pea seed genes is just that—purely hypothetical. In actual cases involving a pair of linked genes, a certain amount of gene recombination does occur during meiosis. Let us illustrate with a real example. Corn plants have two linked genes that affect the color and size of their kernels. Colored kernels (*C*) is dominant over white kernels (*c*); nonshrunken kernels (*N*) is dominant over shrunken kernels (*n*). In one experiment, a dihybrid strain (*CN/cn*) was crossed with a homozygous recessive strain (*cn/cn*). The results shown in Table 10–1 indicate that the vast majority of the offspring were either colored and nonshrunken (48.9%) or white and shrunken (48.1%), the two phenotypes one would predict if the kernel color and size genes were linked. In addition, however, a small number of two recombinant phenotypes (1.5% each) also appeared. Whereas this is not enough to suggest that an independent assortment of the alleles occurred (in which case there would have been a $1:1:1:1$ phenotype ratio), it is enough to conclude that a low frequency of gene recombination did occur.

FIGURE 10–7
Hypothetical Dihybrid Cross Involving Gene Linkage. If we assume that the genes for seed color and seed texture in peas are linked on the same chromosome, then Mendel's dihybrid cross would have yielded a 3:1 phenotypic ratio in the F_2 generation instead of the 9:3:3:1 ratio he actually observed. This is because alleles that are linked on the same chromosome cannot assort independently during gamete formation.

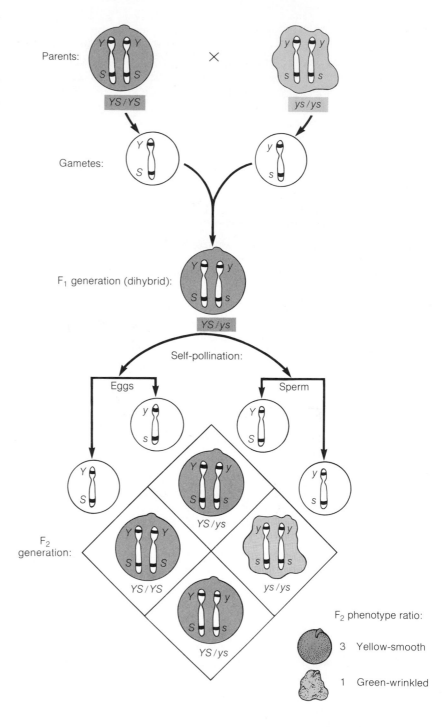

What possible mechanism could account for this recombination of genes? The answer is crossover. Recall that crossover is the exchange of chromosomal sections between homologous chromatids, an event that takes place during prophase I of meiosis (see Figure 8–6). If a crossover occurs between two linked genes in a doubly heterozygous individual, then two of the four resulting gametes will have new combinations of the alleles in question. For example, if a single crossover takes place between the kernel color and size genes during meiosis in the dihybrid (*CN/cn*), then four different gamete genotypes will be produced: *CN, cn, Cn,* and *cN* (Figure 10–8). In the absence of crossover, only the *CN* and *cn* gametes are

TABLE 10–1
Results of a Cross Suggesting Recombination Between Two Linked Genes. Parents: *CN/cn* (colored-nonshrunken kernels) × *cn/cn* (white-shrunken kernels). The *C* and *N* alleles are dominant.

	Colored-nonshrunken kernels	White-shrunken kernels	White-nonshrunken kernels	Colored-shrunken kernels
			Recombinants	
Number of kernels	21,408	21,060	638	662
Percent of total	48.9	48.1	1.5	1.5
Presumed genotype	*CN/cn*	*cn/cn*	*cN/cn*	*Cn/cn*

FIGURE 10–8
Crossover Is the Source of Recombinant Phenotypes for Two Linked Traits. In the cross *CN/cn* × *cn/cn* (corn plants), a crossover occurring between the kernel color and size genes in the dihybrid yields two recombinant gametes (*cN* and *Cn*). Union of the recombinant gametes with *cn* gametes from the other parent generates the recombinant phenotypes in the offspring. Presumably, crossovers also occurred during meiosis in the *cn/cn* parent, but they would have had no effect on the genotypes of the gametes—all gametes would be *cn* with or without crossover.

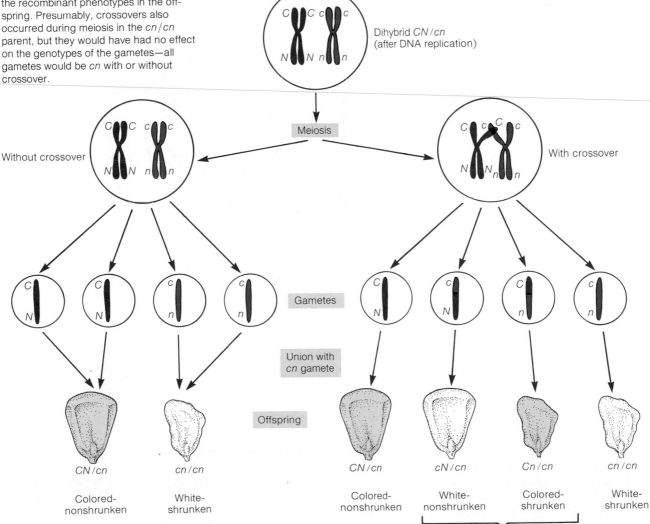

ESSAY 10-2

HEMOPHILIA AND THE RUSSIAN REVOLUTION

On November 1, 1894, at the age of 26, Nicholas II inherited the throne of Russia. It was not a particularly good time to be Czar, for there were rumblings of revolution among the Russian commoners. Nicholas's father, Czar Alexander III, had been a strong and repressive ruler who crushed any whisper of dissent and imprisoned many political activists. Although Nicholas also believed in a strong autocracy, his temperament and training did not prepare him to handle the growing resentment and restlessness of the masses.

A few weeks after his coronation, Nicholas II married Alexandra, a granddaughter of England's Queen Victoria and the daughter of Alice of Hesse and Louis, Grand Duke of Hesse (Germany). Nicholas loved Alexandra deeply and frequently bowed to her wishes. Hence, Alexandra wielded great influence over Nicholas, and actually made many of the important policy decisions of the time.

In the late 1890s and early 1900s, dissent and revolutionary fervor increased in Russia, including student demonstrations and acts of terrorism. Elected legislatures were formed, but they were consistently too liberal for Nicholas, who refused to accede to their wishes on matters such as dividing large estates of land among the peasantry, providing equal rights for Jews, dissenters, and minorities, and granting amnesty to political prisoners.

Nicholas II and Alexandra had four daughters in succession, but it was deemed crucial for the survival of the Russian monarchy that a son inherit the throne. Finally, a son, Alexis, was born in 1904. For a brief time great happiness reigned in the imperial court. It ended rather dramatically, however, when they discovered that Alexis was an invalid, a hemophiliac "bleeder."

Among the royal families of Europe, this dreaded blood disease had first appeared in Leopold, Duke of Albany, a son of Queen Victoria (see the royal pedigree). Caused by a sex-linked recessive allele, hemophilia occurs almost exclusively in males, although it is transmitted through heterozygous females. Queen Victoria was the first known carrier among European royalty. Of particular relevance to Russian history, the hemophilia allele was passed from Queen Victoria to her daughter, Alice of Hesse, then to Alexandra of Russia, and finally to little Alexis.

Alexandra was distraught; her concern for Alexis was neurotic. She sought help from physicians; but when that failed, she turned to soothsayers, quacks, and mystics in a desperate effort to cure the crown prince. In this state of mind she came under the influence of the "evil monk" Rasputin (1871–1916), a self-proclaimed faith healer of peasant stock who had great aspirations. His hypnotic personality captivated and controlled the thinking of many followers. He convinced Alexandra that he could keep Alexis's bleeding under control. Rasputin's influence at court became a public scandal,

Czar Nicholas II and his wife Alexandra with their children—
Olga, Tatiana, Maria, Anastasia, and Alexis.

ROYAL PEDIGREE

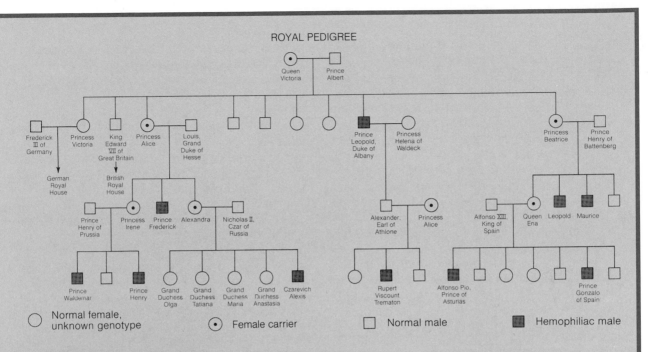

○ Normal female, unknown genotype   ◉ Female carrier   □ Normal male   ■ Hemophiliac male

Grigori Efimovich Rasputin.

but he was protected by Alexandra, who viewed him as a saint sent by God to save the troubled dynasty.

Nicholas did not approve of Rasputin, but his devotion to Alexandra was overpowering. Consequently the mystic gained increased influence in official matters. Many competent officials were dismissed and replaced with scalawags recommended by Rasputin. This weakened the government and paved the way for its overthrow. Rasputin created an atmosphere of intrigue and distrust in court, alienating millions of Russians. Hence, the internal workings of the monarchy were further weakened during a time of increasing tension within the country and all of Europe.

When World War I broke out in 1914, Alexandra persuaded Nicholas to take personal charge of military operations in the field and to leave her in control of domestic affairs. With Nicholas away from the palace, Rasputin had little to hinder his nearly complete control. Young Alexis, whose hemophilia was indirectly responsible for Rasputin's rise, became a secondary problem. Resentment against the monarchy was building among the Russian people. Their czar was away fighting the Germans and had left their country in the hands of a czarina of German descent who, besides being suspect herself, was under the influence of a scheming monk believed by some to be a German agent. Extreme measures were considered and finally taken by loyalists and dissidents alike. Finally, on the night of December 29, 1916, Rasputin was shot to death.

As the months passed, the number of strikes and demonstrations increased, each more intense than the previous one. Finally, on March 8, 1917, the palace was overrun. On March 14, the provisional Kerensky government was set up, and the next day Czar Nicholas II abdicated the throne to his brother, who wisely declined the honor. Just over a year later Bolshevik revolutionaries executed the entire royal family—Nicholas, Alexandra, their daughters, and young Alexis.

Students of Russian history point out that although the czar's abdication was not a direct result of Alexis's hemophilia, the boy's illness was in some measure responsible for Rasputin's rise to disruptive influence. Indeed, the mere fact that the heir apparent was an invalid undoubtedly played a role.

Eight months after Nicholas's abdication, the Bolshevik revolution—engineered by V. I. Lenin and Leon Trotsky—began.

possible. Thus, the *Cn* and *cN* gametes are recombinants, and their unions with *cn* gametes (from the *cn/cn* parent) leads to the recombinant phenotypes in the progeny: colored-shrunken kernels (*Cn/cn*) and white-nonshrunken kernels (*cN/cn*). Apparently, the crossover frequency between the two kernel genes is low, as the offspring show a total of only 3% recombinant phenotypes.

The inheritance patterns for autosomal- and sex-linked traits described by Morgan and others did much to establish the chromosome theory of heredity. By 1925, geneticists had a good understanding of how traits are transmitted from one generation to the next. Mendel's garden pea and Morgan's fruit fly served well in these early experiments, but in the years to follow, they were to be replaced by even simpler "guinea pigs." Bacteria and viruses were used when geneticists began asking: What is a gene and how does it work? These questions are the subject of Chapters 11 and 12.

SUMMARY

In the early 1900s, the principles of heredity originally put forth by Mendel were rediscovered and extended. Sutton and Boveri, in 1902, proposed that Mendel's hereditary factors are associated with the chromosomes of cells, an idea that became known as the chromosome theory of heredity. Evidence in support of this theory was soon forthcoming when scientists discovered the chromosomal basis of sex determination. Furthermore, the demonstration that certain traits were inherited together suggested that the genes controlling them were linked to the same chromosome.

The sex of most species of animals (and some plants) is determined by the individual's complement of sex chromosomes. In mammals (including humans) and fruit flies, females have two X chromosomes (XX) and males have one X and one Y chromosome (XY). The sex of offspring is determined by which of the father's sex chromosomes, X or Y, is transmitted by the sperm.

The first demonstration of a sex-linked trait was made by T. H. Morgan in 1910. Through his experiments on *Drosophila,* the common fruit fly, Morgan showed that the inheritance of eye color and sex occurs in a coordinate fashion. He reasoned correctly that the eye color gene is located on the X chromosome but is not present on the Y chromosome. This meant that the recessive allele specifying white eyes is always expressed in males but could be masked by the presence of a dominant red-eye color allele in heterozygous females.

Since Morgan's original demonstration of a sex-linked trait, many other sex-linked traits have been discovered. Humans are known to have well over 100 sex-linked genes, most of which govern traits that are not related to sex. Included among these are the well-known disorders of red-green color blindness and hemophilia. Since these disorders are caused by a recessive sex-linked allele, their frequencies of occurrence are much higher in males. Males inherit the harmful allele from their mothers, who in most cases are phenotypically-normal carriers.

The genes (traits) associated with the sex chromosomes form one linkage group—they are inherited as a unit. Similarly, the genes clustered on any other given pair of homologous chromosomes, called autosomes, also belong to a linkage group. Because any two such genes cannot assort independently during meiosis, the inheritance of the two traits in question does not follow simple Mendelian rules. With the possibility of crossover, however, the physical linkage of two genes on a single chromosome is not absolute. The crossover exchange of sections between homologous chromatids can result in a recombination of alleles for two different genes in a linkage group, leading to a certain frequency of recombinant phenotypes.

STUDY QUESTIONS

1. In humans, is it the sperm or egg that determines the sex of the offspring? Explain your answer by diagramming the sex chromosome complements of two parents, their gametes, and their potential offspring.

2. In fruit flies, red eye color is dominant over white eye color, and the eye color gene is located on the X chromosome. List the phenotypes and their ratios expected in the following crosses:
   (a) Red-eyed male × white-eyed female.
   (b) White-eyed male × heterozygous red-eyed female.

3. A color-blind man marries a woman known to be homozygous dominant (normal vision) for this sex-linked trait. They have one daughter. If the daughter also marries a color-blind man:
   (a) What possible phenotypes would you expect in their children?
   (b) Would the expected frequencies be different if this were not a sex-linked trait?

4. In the hypothetical pedigree for sex-linked hemophilia shown in Figure 10–9, the phenotype of each individual is indicated. Fill in the genotype of each individual.

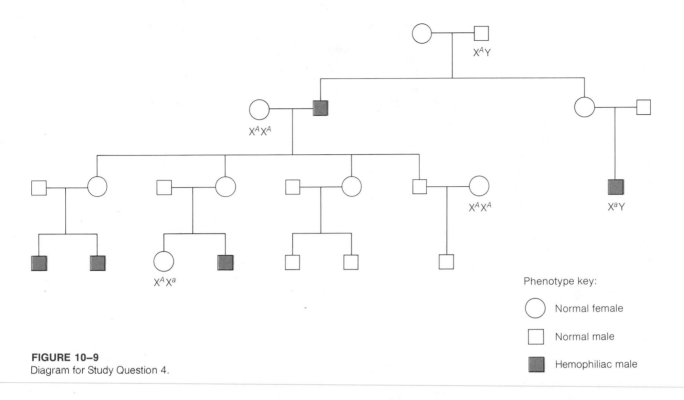

FIGURE 10–9
Diagram for Study Question 4.

Phenotype key:

○ Normal female

□ Normal male

■ Hemophiliac male

5. In sweet peas, purple petals (P) is dominant over red petals (p), and elongate pollen (R) is dominant over round pollen (r). A dihybrid ($PpRr$) is crossed with a double recessive pea having red petals and round pollen.
   (a) If the petal color and pollen shape genes were not linked, what would be the expected genotypes and phenotypes (and their ratios) in the offspring?
   (b) Predict the genotypes and phenotypes of the offspring if the dihybrid had a PR/pr linkage pattern, and no crossover occurred.
   (c) Name the recombinant phenotypes appearing in the offspring assuming linkage *and* crossover.

SUGGESTED READINGS

Jenkins, J. B. *Human Genetics.* Menlo Park, CA: Benjamin/Cummings, 1983. Chapter 4 provides excellent coverage of autosomal and sex-linked inheritance, and Chapter 13 gives an interesting discussion of the genetics of human behavior.

Lerner, I. M., and W. J. Libby. *Heredity, Evolution, and Society.* 3rd ed. San Francisco: W. H. Freeman, 1976. A very readable account of current genetic issues. Intended for the non-biology major.

Peters, J. A., ed. *Classic Papers in Genetics.* Englewood Cliffs, N.J.: Prentice-Hall, 1959. In addition to Mendel's classic paper, this collection has Sutton's and Morgan's original works.

ANSWERS TO STUDY QUESTIONS

1. Refer to Figure 10–2.
2. (a) All white-eyed males; all red-eyed females.
   (b) $\frac{1}{2}$ red-eyed, $\frac{1}{2}$ white-eyed males; $\frac{1}{2}$ red-eyed, $\frac{1}{2}$ white-eyed females.
3. (a) $\frac{1}{2}$ normal, $\frac{1}{2}$ color-blind males; $\frac{1}{2}$ normal, $\frac{1}{2}$ color-blind females.
   (b) No.
4. *See* Figure 10–10 (next page).
5. (a) 25% *PpRr*: purple petals, elongate pollen.
   25% *Ppr*: purple petals, round pollen.
   25% *ppRr*: red petals, elongate pollen.
   25% *pprr*: red petals, round pollen.
   (b) 50% *PR/pr*: purple petals, elongate pollen.
   50% *pr/pr*: red petals, round pollen.
   (c) Purple petals, round pollen.
   Red petals, elongate pollen.

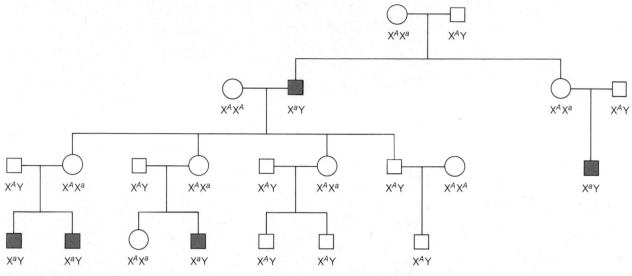

FIGURE 10–10
Answer to Study Question 4.

CHAPTER 11
DNA Structure and Function

At the time T. H. Morgan and his students were demonstrating that genes were situated on chromosomes, major advances were also being made in biochemistry. Biochemists were developing techniques to take apart cells and analyze their chemical components, particularly the macromolecules. Soon the spotlight fell on the proteins. Proteins were not only abundant in cells, but they showed an enormous diversity in structure and function. By the 1920s, it became apparent that proteins played a central role in the determination of traits and that many genetic disorders could be traced to the absence of certain enzymes. The knowledge that proteins were integral components of chromosomes strongly recommended this macromolecule for the role of hereditary agent. In fact, many prominent protein chemists suggested that the proteins of chromosomes were master copies of the enzymes in the cytoplasm. This hypothesis, though quite popular for many years, was proved to be wrong.

ON THE TRACK OF DNA

We now know that the inheritable genetic substance in cells is deoxyribonucleic acid (DNA). A wealth of evidence has come from some rather elegant and ingenious experiments begun more than 50 years ago.

These experiments still stand as monuments to the resolving power of science and the cleverness of the human mind.

Evidence from Bacteria

In 1928, Frederick Griffith, a British bacteriologist, reported the results of some interesting experiments he had conducted on *Streptococcus pneumoniae,* a bacterium that causes the most common form of pneumonia. In his search for a pneumonia vaccine, Griffith worked with several different strains of this bacterium, two of which are important for our discussion. One was called the "S" strain, because its cells have a gelatinlike coating that imparts a smooth appearance to the colonies growing on culture plates. The other strain lacks this smooth coat and its colonies have a rough texture; it was named the "R" strain. These two strains also differ with respect to pathogenicity (ability to cause disease). When Griffith injected S cells into laboratory mice, these animals quickly developed pneumonia and died (Figure 11–1). On the other hand, mice injected with R cells showed no ill effects. Griffith also demonstrated that the pathogenic S cells could be rendered harmless when exposed to heat. Heat-killed S cells do not cause pneumonia.

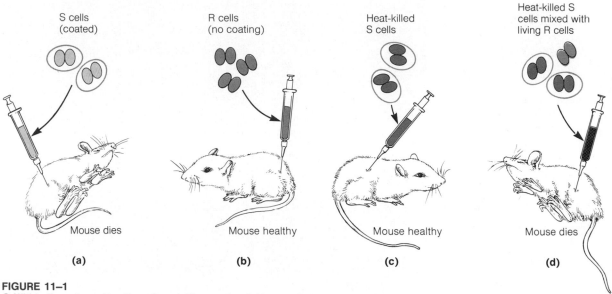

FIGURE 11–1
Griffith's Transformation Experiment. The results of this experiment suggested that some chemical factor from the heat-killed S cells entered the normally harmless R cells and transformed them into disease-causing cells. The chemical factor was later identified as DNA.

These results were not in any way surprising. They merely served as **controls** (standards of comparison) for the next treatment. Griffith mixed some of the live R cells with some heat-killed S cells, then injected this mixture into mice. Incredibly, many of the mice died of pneumonia. When samples of blood were taken from the diseased mice, hordes of live S cells were found. What happened? Recall that injections with either R cells or heat-killed S cells alone do not cause pneumonia. Yet, when these cells were injected together, living S cells appeared and the mice contracted pneumonia. Griffith was left with only one reasonable interpretation: Something must have passed from the heat-killed S cells to the nonpathogenic R cells, transforming the latter into mouse-killers. Furthermore, this transformation of R cells must have entailed a genetic change, for the "new" S cells produced more S cells in succeeding generations. The next question was obvious: What is the nature of the **transforming factor?** The answer to this question would yield the nature of the genetic substance, at least for bacteria.

The transforming factor was identified in 1944 by Oswald Avery and his co-workers at Rockefeller University. Avery's lab isolated and purified DNA from heat-killed S cells, which, when incubated with live R cells, transformed them into S cells. The transforming ability of the purified DNA was lost when it was pretreated with a DNA-digesting enzyme, but not by pretreatment with a protein-digesting enzyme. Thus, it was shown that it is the S-cell DNA, not S-cell proteins, that had the genetic information specifying pathogenicity.

Evidence from Viruses

An additional line of evidence equating DNA with the genetic substance stems from work on viruses. Viruses are extremely tiny particles consisting of a nucleic acid core (usually DNA) and a protein coat (Figure 11–2). They are not independent organisms, but require the metabolic machinery of their particular host cells (bacteria, human, etc.) to reproduce (to make copies of the nucleic acid and proteins that self-assemble into new virus particles). But, which part of the virus, nucleic acid or protein, stores the instructions for replication? In 1952, working with a DNA virus that infects bacterial cells, Alfred Hershey and Martha Chase answered this question. They showed in a cleverly designed experiment that when a virus infects the bacterial cell, only the viral DNA enters the cell; the viral proteins are shed and remain outside of the bacterium. Shortly after in-

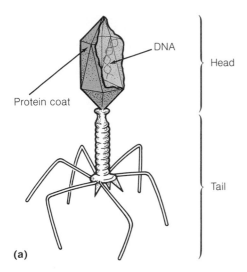

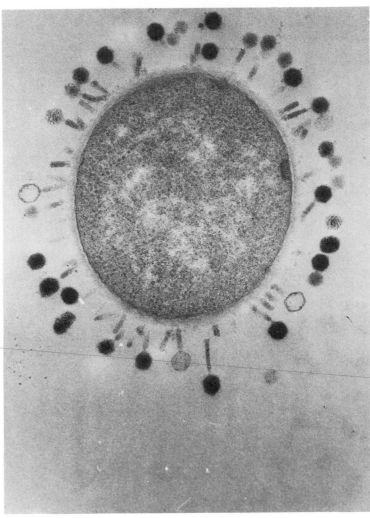

FIGURE 11–2
The Structure of a Virus. (a) In this diagram of a virus that infects bacterial cells, note that the head region (cutaway view) contains the DNA, which is surrounded by a protein coat. The tail region also is made of protein. (b) This electron micrograph shows many viruses attached to the surface of a bacterium. The viruses are injecting their single DNA molecules into the bacterium. (Magnification: 47,000×.)

fection, new viruses are manufactured within the bacterial cell, apparently from information encoded in the viral DNA that entered the host (Figure 11–3).

Although the experiments with *Streptococcus pneumoniae* and the bacteria-infecting virus elegantly demonstrated that DNA is the genetic substance, they did not help us to understand how DNA stores genetic information or how this substance is replicated. Perhaps knowing the chemical structure of DNA would help in this regard. So thought a few investigators in the early 1950s, but the task seemed awesome. Scientists feared that a chemical substance responsible for storing vast amounts of genetic information must be so complex that any attempts to unravel its structure would surely end in frustration. Fortunately, this was not the case. In fact, the structure of DNA turned out to be surprisingly simple.

THE STRUCTURE OF DNA

In the early 1950s, James Watson, an American biologist, and Francis Crick, an English biophysicist, met at Cambridge University in England, where they held research internships. Each had a keen interest in DNA, Watson from the viewpoint of genetics and Crick from

FIGURE 11–3

The Hershey–Chase Experiment. This experiment used viruses in which the DNA molecule was labeled with radioactive phosphorus (^{32}P) and the protein contained radioactive sulfur (^{35}S). (No cross-labeling was possible because DNA lacks sulfur and proteins lack phosphorus.) The labeled viruses were then allowed to infect bacteria. The new viruses formed in the bacterial cells contained ^{32}P radioactivity only, indicating that the viral DNA—not the protein—coded for the construction of new viruses. (In this diagram, the tails have been deleted for clarity.)

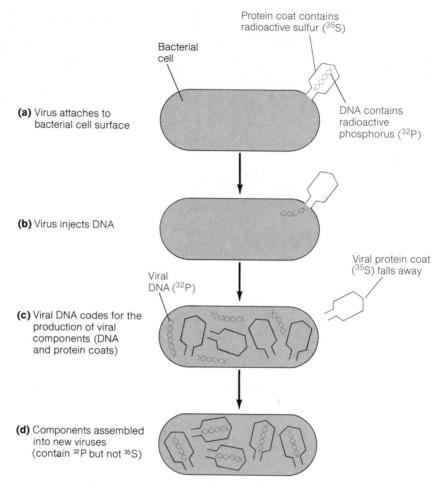

(a) Virus attaches to bacterial cell surface

(b) Virus injects DNA

(c) Viral DNA codes for the production of viral components (DNA and protein coats)

(d) Components assembled into new viruses (contain ^{32}P but not ^{35}S)

Protein coat contains radioactive sulfur (^{35}S)

Bacterial cell

DNA contains radioactive phosphorus (^{32}P)

Viral protein coat (^{35}S) falls away

Viral DNA (^{32}P)

---

his interest in the structure and physical properties of macromolecules. They believed that knowing the structure of DNA was the key to unlocking the mysteries of heredity and many aspects of cell function. In what began as a side project, Watson and Crick tried to construct a wire model of DNA that would account for the few facts known about this substance. What were these facts?

To start with, it was known that DNA was a large polymer composed of four different kinds of nucleotides. Each nucleotide in DNA consists of a **phosphate group,** a 5-carbon sugar called **deoxyribose,** and a nitrogen-containing **base** (Figure 11–4). The nucleotides of DNA differ by the nature of their bases—**cytosine (C), guanine (G), adenine (A),** and **thymine (T).**

Another bit of information was provided by Erwin Chargaff, a biochemist who had studied the base composition of DNA. Chargaff noted that in all the DNA samples he analyzed, the amounts of adenine and thymine were equal (A = T), as were the amounts of cytosine and guanine (C = G). These base equivalencies played a significant role in the final model-building efforts of Watson and Crick.

Watson and Crick had a molecular jigsaw puzzle on their hands. They had all the pieces; now they had to figure out how the nucleotides fit together to form the DNA molecule. Their efforts were guided to a large extent by X-ray diffraction patterns of crystallized DNA provided by Rosalind Franklin and Maurice Wilkins, also of Cambridge. The X-ray patterns showed that DNA is helical (spiral) in shape, and gave our model builders crucial data on the molecule's dimensions and the spacing of the nucleotides.

After much trial and error, Watson and Crick, in 1953, assembled their wire nucleotides into a model of DNA that made sense. Their model not only accounted for the base equivalencies (A = T and G = C), but it also formed a perfect helix in accordance with the X-ray diffraction data (see Figure 11–5a and the computer-generated model of DNA in the Introduction). For their roles in discovering the structure of DNA—a momentous event in the history of biology—Watson, Crick, and Wilkins were awarded the Nobel Prize in Chemistry in 1962.

The major features of the Watson–Crick model of DNA are listed on the next page. As you study them, it will be helpful to refer to Figure 11–5.

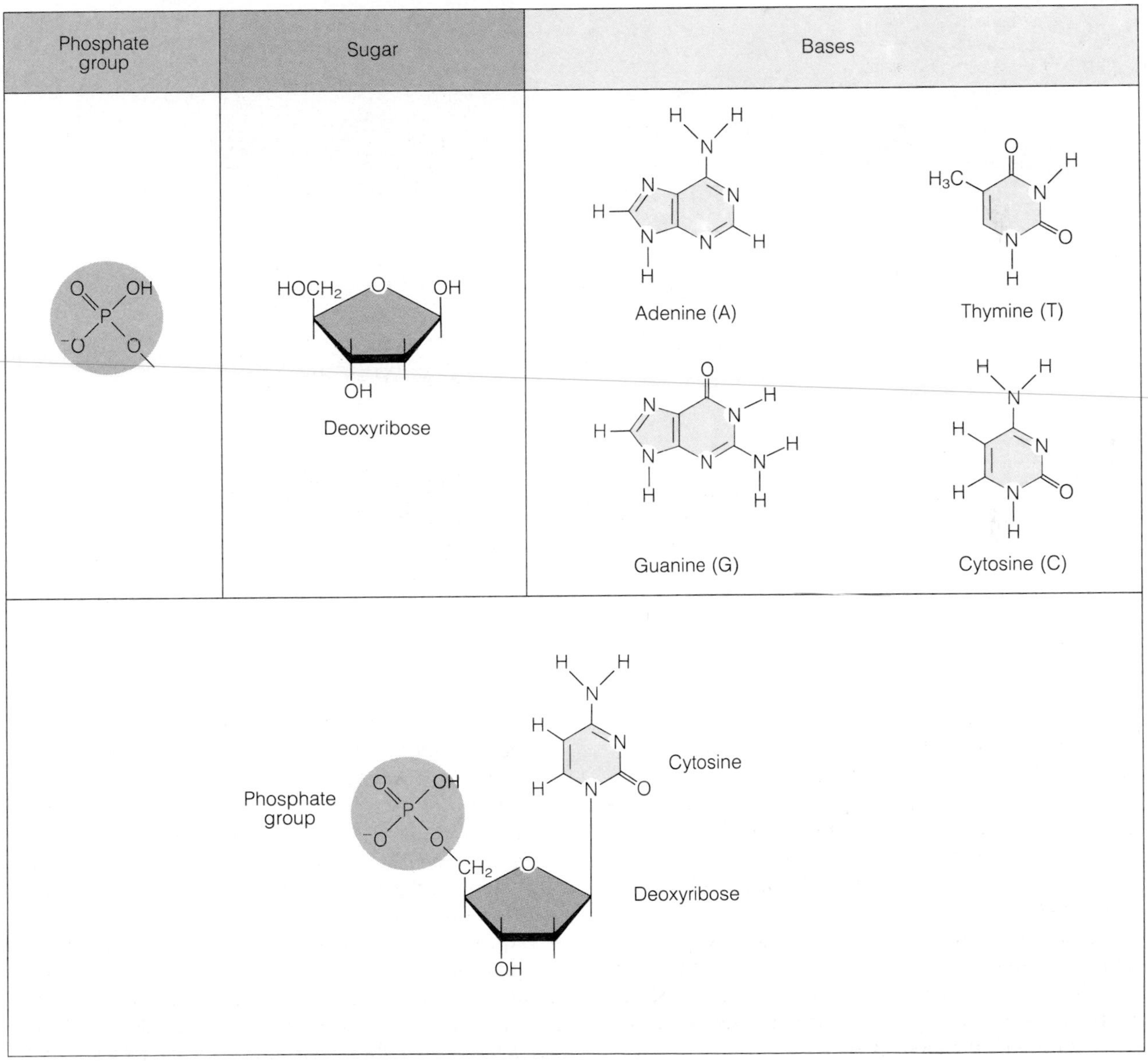

FIGURE 11–4

Nucleotide Structure. (*Top*) The components of a DNA nucleotide. Each nucleotide contains one of the four possible bases. (*Bottom*) A cytosine-containing nucleotide.

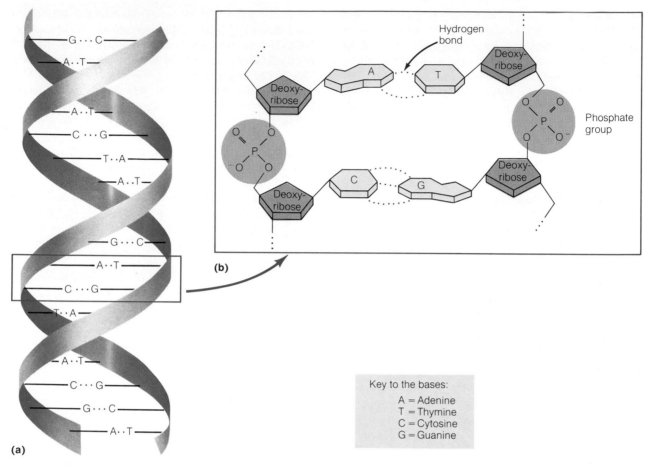

Key to the bases:

A = Adenine
T = Thymine
C = Cytosine
G = Guanine

FIGURE 11–5
DNA Structure. (a) In this diagram of DNA, the ribbons represent the sugar–phosphate backbones, and the bases (A, T, C, and G) are oriented toward the core. Note how the complementary strands twist around one another to form a helix. (b) Detail of two base pairs. A is bonded to T by two hydrogen bonds, and C and G share three hydrogen bonds.

1. DNA is composed of two polynucleotide chains (long strands of many nucleotides linked together). These two chains twist around one another to form a helix. Hence, DNA is often called the "double helix" (Figure 11–5a).
2. The individual nucleotides in each strand are linked covalently between the phosphate group of one nucleotide and the deoxyribose (sugar) of the next nucleotide (Figure 11–5b).
3. These linked sugars and phosphates form the outer backbones of the double-stranded DNA. The bases are oriented inward, forming the molecule's core.
4. One polynucleotide strand is held to the other strand by hydrogen bonds between their bases, forming something analogous to a twisted ladder—

the railings of the ladder are the sugar–phosphate backbones, and the rungs are the paired bases.
5. The bases on opposing strands form specific base pairs—adenine is always opposite thymine and guanine is always opposite cytosine. Thus, each adenine and thymine form a base pair unit, and each guanine and cytosine form a base pair unit. This strict base pairing rule underlies Chargaff's base equivalency data.

The base-pairing feature of the Watson–Crick model means that the two polynucleotide strands are complementary to each other. Knowing the sequence of bases on one strand allows you to deduce the sequence on the other strand. For instance, if the partial base sequence on one DNA strand is –ACTTG–, then the base sequence on the complementary strand must be –TGAAC–. This characteristic gives the molecule an appealing orderliness—a quality one might expect of the master genetic substance of life. Moreover, DNA's "pretty structure," as Watson called it, provided the first clue as to how DNA is replicated, as we shall now see.

DNA REPLICATION

Long before the discovery of DNA's structure, biologists knew that the genetic information stored in chromosomes had to be faithfully duplicated prior to cell division. If this did not occur, the resulting daugh-

ter cells would end up with something other than a "full deck" of instructions. Thus, whatever the genetic substance turned out to be, it had to be capable of self-replication. One of the most aesthetically pleasing features of the Watson–Crick model of DNA was that a mechanism for its replication was implied in its structure.

The key to self-replication of DNA, as Watson and Crick pointed out, lies in the base pairing. Because the two DNA strands are held to each other by relatively weak hydrogen bonds between complementary bases, they imagined that the two strands could unwind and separate rather easily, thereby exposing the bases. Each strand could then act as a **template** to guide the formation of a new complementary strand (Figure 11–6). In other words, each exposed adenine on both DNA

FIGURE 11–6
DNA Replication as Envisioned by Watson and Crick. When the complementary strands of a parent DNA molecule separate, the bases on each parental strand are available to form complementary base pairs with free nucleotides dissolved in the nucleus. Once aligned, the free nucleotides are linked together into the daughter strands.

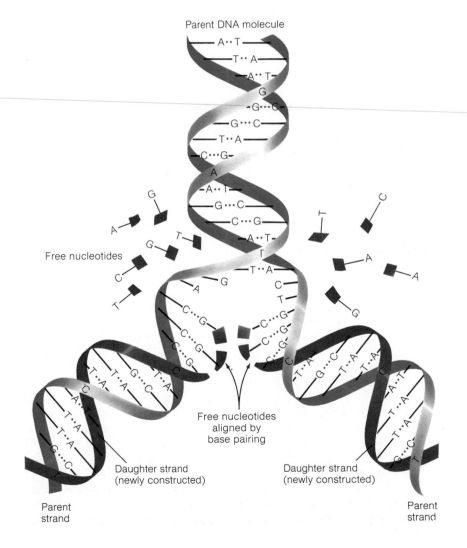

strands could pair (form hydrogen bonds) with a complementary thymine-containing nucleotide dissolved in the nucleus, and conversely, exposed thymines could pair with free adenine-containing nucleotides. Guanine- and cytosine-containing nucleotides could align in a similar fashion with exposed cytosine and guanine bases, respectively. Once the free nucleotides are in place, their juxtaposed sugar and phosphate groups could be linked together (dehydration synthesis) to form the new polynucleotide chains. After the entire lengths of the parental DNA strands have been replicated, the final result would be two double-stranded DNA *molecules,* each consisting of one new and one parental strand.

With the exception of a few details, this general description for DNA replication turned out to be correct. As we presently understand the process in eukaryotes, the parental DNA molecule unwinds at many sites along its length simultaneously, forming what are termed **replication bubbles** (Figure 11–7). Molecules of the enzyme, DNA polymerase, bind to these bubble regions where they first promote the base-pairing alignment of the free nucleotides, then their polymerization into short stretches of polynucleotides. These polynucleotide fragments are then joined into the completed daughter strands by another enzyme, DNA ligase (*ligate* = to bind). With all of these events occurring simultaneously at many sites along the parent DNA, the replication of an entire DNA molecule consisting of perhaps several million nucleotides takes only a few hours.

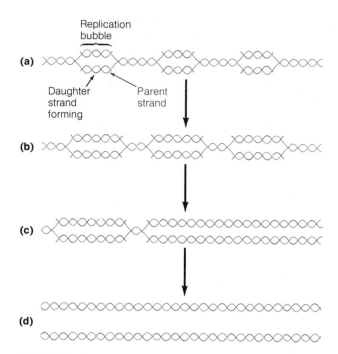

FIGURE 11–7
Current Model of Eukaryotic DNA Replication. (a) The replication of DNA begins at many specific points on the parent DNA molecule, forming many replication bubbles. (b, c) DNA replication then proceeds in both directions along each bubble. The bubbles enlarge as replication continues. (d) When replication is complete, each DNA molecule contains one parental strand and one newly-synthesized daughter strand.

INFORMATION STORAGE

Although knowledge of the structure of DNA quickly led scientists to its mechanism of self-replication, how DNA functioned as an information-storing molecule was not immediately apparent. In fact, many scientists felt that the structure of DNA was *too* simple. How could a substance composed of only four different kinds of subunits (nucleotides) have the structural diversity to encode literally thousands of different genetic messages in each organism?

The Nucleotide Alphabet

The solution to the information-storing enigma unfolded when scientists realized that the genetic code is something like a written language: The diversity of its messages lies not in the number of different letters in its alphabet, but in how the letters are arranged. Thus, it is the *sequence* of nucleotides along a strand of DNA that determines the nature of a given genetic message.

To demonstrate the potential for information storage in DNA, let us begin by considering a part of a DNA strand only two nucleotides long. Choosing from the pool of four different nucleotides, we can construct 4^2 (4×4), or 16 different two-nucleotide sequences (Figure 11–8). If we increase the length of this strand to three nucleotides, the number of unique sequence possibilities jumps to 4^3 ($4 \times 4 \times 4$), or 64. Actually, a unit of genetic information—a single gene—is typically

	Second nucleotide			
	A	T	G	C
A	AA	AT	AG	AC
T	TA	TT	TG	TC
G	GA	GT	GG	GC
C	CA	CT	CG	CC

(First nucleotide)

FIGURE 11–8
Unique Two-Nucleotide Sequences.

about 900 nucleotides long (that is, it is made up of 900 base pairs in the double-stranded DNA molecule). In this case, the unique nucleotide sequences possible is 4^{900} (read "4 to the 900th power"), an astronomical number! Clearly then, there is more than enough variety built into the structure of DNA to account for all the different genes in all the different organisms on earth. The great diversity of life forms would not be possible without this large number of different possible genes.

The Nature of Genetic Information

A gene is a single message unit of a DNA molecule, and genes differ by their particular sequences of nucleotides. But this still leaves the question of the message itself—what is the nature of genetic information? Put simply, *the information in DNA determines the structure, and hence the function, of proteins.* Proteins in turn regulate the activities of cells. In fact, proteins are involved in just about every facet of cell function. Transport proteins in the plasma membrane determine to a large extent which substances enter or leave the cell. Regulatory proteins stimulate or inhibit key metabolic processes; the contractile proteins of microfilaments, microtubules, and muscle fibers power various types of movement. And finally, virtually all chemical reactions that take place in cells are catalyzed by special proteins called enzymes. Thus, it would not be an exaggeration to say that what a cell is and what it can do is dictated by the complement of proteins present. Almost all of the differences that distinguish one cell type from another, you from your neighbor,

or a cat from a dog, can be traced back to differences in proteins produced in each.

Now, let us be a little more specific. Recall from Chapter 2 that each protein consists of one or more polypeptides; each polypeptide, in turn, is a chain of amino acids. Since there are 20 different kinds of amino acids found in proteins, the structure and function of any given protein hinges on the specific sequence of amino acids making up its polypeptide chain or chains. The amino acid sequence of a polypeptide is dictated by the specific sequence of nucleotides in a gene. What a simple yet elegant relationship between DNA and proteins! We can now refine our definition of a gene: *A gene is a segment of a DNA molecule that determines the structure (amino acid sequence) of a particular polypeptide.* This is often abbreviated as, "one gene—one polypeptide." For a protein that consists of two or more different polypeptides, there are two or more genes that determine its structure.

Proteins and Traits

In 1865, Gregor Mendel published the first report demonstrating that traits are determined by discrete factors we now call genes. It took nearly a century of further research before biologists discovered *how* genes govern traits. The final key that unlocked this mystery of life was provided in the early 1960s when, through the combined efforts of many scientists, the mechanism by which DNA coded for protein structure was revealed (see Chapter 12). But the relationship between proteins and traits has a longer history.

In the early 1900s, Archibald Garrod, an English physician, was investigating the causes of several hereditary diseases, including albinism (see Table 9–3). In these disorders, which Garrod called **inborn errors of metabolism,** he reasoned that each was caused by a block in one of the steps of a metabolic pathway. Such a metabolic block would prevent the formation of any substance on the pathway that appeared *after* the blocked step. In classical albinism, for example, Garrod suggested that the metabolic block occurred on the pathway that formed melanins, the pigments normally found in skin, hair, and eyes. Unable to synthesize melanins, albinos have very pale skin, white hair, and pink eyes.

How were these metabolic blocks manifested? Garrod knew that individual reactions of metabolic pathways are catalyzed by specific enzymes. Considering this and other factors, Garrod proposed that inborn

(a)

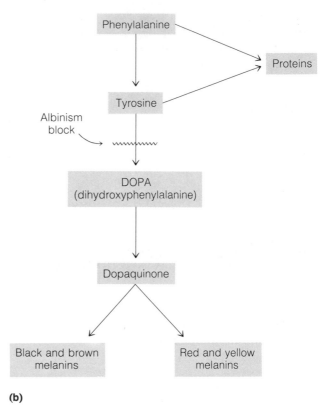

(b)

FIGURE 11–9
Albinism. (a) The rock star Johnny Winter is an albino. (b) Classical albinism is caused by the lack of tyrosinase, an enzyme that converts tyrosine to DOPA. This metabolic block precludes the synthesis of melanins.

errors are caused by missing or defective enzymes. Many years later, this proposal was proved correct. In classical albinism, for example, the absence of melanin pigments is due to the lack of tyrosinase, an enzyme required for the second step on the pathway from phenylalanine to the melanins (Figure 11–9).

Garrod, like Mendel, was far ahead of his time. It was not until the 1940s that other investigators began thinking seriously about the connections among genes, enzymes, and phenotypic traits. By the time Watson and Crick began assembling their wire models, geneticists had established that biochemical processes consist of individual, enzyme-catalyzed reactions that are under genetic control and that each enzyme is produced by a single gene. (This "one gene—one enzyme" idea was later broadened to include all proteins (not just enzymes), then narrowed to the current "one gene—one polypeptide" relationship.)

An understanding of the biochemical basis of a genetic disorder often aids in the treatment of such inborn defects. The successful treatment of phenylketonuria, galactosemia, and diabetes mellitus (see Table 9–3) bears witness to this fact. Another approach to combating genetic defects, and certainly the preferred one, is prevention through genetic counseling. At present, this is our only weapon against such diseases as Tay-Sachs, muscular dystrophy, albinism, and hemophilia. For these there are no phenotypic cures, although the dangers of hemophilia can be lessened by frequent blood transfusions. Perhaps the ultimate hope for treating these and many other genetic disorders lies in recombinant DNA research, a new and fast-growing field that may some day provide us with the tools to replace defective genes. The theory and applications of recombinant DNA technology is our next topic of discussion.

RECOMBINANT DNA

You have probably read in newspapers or magazines about some of the recent breakthroughs in the area of molecular biotechnology—articles that report the synthesis of genes in test tubes, the transfer of genes from one type of organism to another, and the "creation" of bacteria that produce human hormones. All of this is quite true. Over the past 15 years or so, molecular biologists have learned how to "recombine" fragments of DNA from various sources into the native DNA molecules of bacteria (or other organisms), creating new bacterial strains that produce some rather unlikely proteins, at least from the bacterium's viewpoint. In effect, these biologists are changing "Mother Nature." This in itself is not unnatural—one of the maxims of living systems is change. These scientists are merely accelerating the rate of change and directing the changes toward specific ends.

The purposeful manipulation of genes, or **genetic engineering,** clearly involves some moral issues, particularly when the organism being manipulated is human. There is also the issue of safety, for certain types of recombinant DNA studies could pose health risks. As citizens, we must weigh those risks against the potential benefits (and there are many) to be derived. On the positive side, recombinant studies will provide us with the means to produce large quantities of biological substances, such as insulin for diabetics. We can also look toward the improvement of the genetic quality of crop plants for food, creating microorganisms that "eat" oil slicks or attack pests, and many medical applications. On the other side of the scale, however, some scientists have warned that harmful genes capable of causing widespread disease could be created inadvertently if stringent safety measures are not exercised. Before you make a judgment, first examine how most recombinant DNA studies are conducted.

The organism most frequently used in recombinant DNA research is the bacterium *Escherichia coli* (abbreviated *E. coli*). This is the same species of bacterium that inhabits the human intestine. Most of the *E. coli* genes are located on its large, circular chromosome. However, some cells have smaller, extra pieces of circular DNA called **plasmids** (Figure 11–10). Plasmids contain genes that are not essential to the bacterium's growth, so plasmids can be modified, lost, or gained without disrupting the bacterial life cycle. Plasmids are thus ideal for recombinant DNA experiments because they can serve as vehicles to carry foreign DNA into bacterial cells without upsetting the recipient's growth and reproduction.

In a typical recombinant DNA experiment (Figure 11–11), bacterial cells are broken open and the plasmids are separated from the larger chromosomes. Next, a particular type of enzyme is added to the plasmid suspension, which splits each circular plasmid into

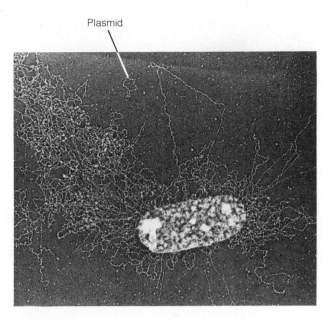

Plasmid

FIGURE 11–10
E. coli **DNA.** This *E. coli* cell was specially treated, causing it to expel its DNA. The tangled white threads are part of its single circular chromosome. The small circular piece of DNA at the top of the photo is a plasmid.

one or more linear pieces of DNA. The plasmid pieces are then mixed with the desired pieces of foreign DNA, which have been snipped from the DNA of another organism or, in some cases, synthesized in the laboratory. Another type of enzyme is added to seal the mixture of pieces back into circular plasmids, some of which now include the foreign DNA. Finally, the reconstituted plasmids are added to a colony of living bacterial cells. By a mechanism not yet understood, some of the plasmids are taken up into some of the bacterial cells. (The mechanism is presumably analogous to the transformation of R cells by S-cell DNA in the Avery experiment.) When these transformed cells replicate their DNA prior to cell division, they also replicate the new plasmid containing the foreign piece of DNA. Thus, succeeding generations of this new bacterial strain will have the foreign DNA and, with it, the ability to manufacture the foreign protein coded by it. Because bacteria multiply so rapidly, scientists can grow huge vats of these modified bacteria, all busily producing a desired protein (Figure 11–12).

Several research groups have already created new variants of bacteria that, by virtue of their recombined plasmids, synthesize insulin, growth hormone, and

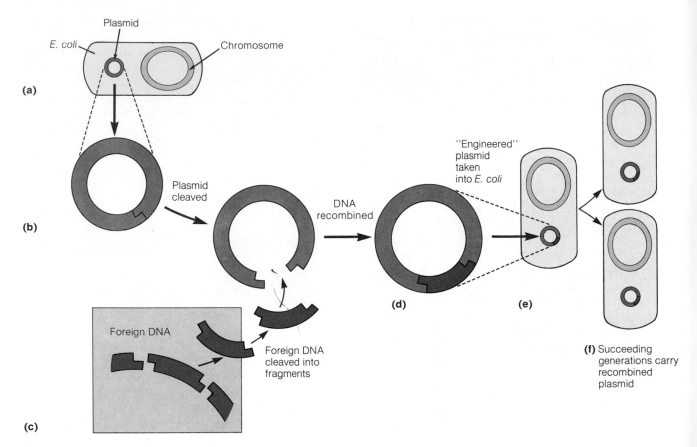

FIGURE 11–11
A Typical Recombinant DNA Experiment. Plasmids from *E. coli* cells are isolated and then cleaved by special enzymes into one or more linear pieces. The same types of enzymes are used to prepare DNA fragments from another (foreign) source. The opened plasmids are mixed with the foreign DNA fragments and sealed together by DNA ligase, the nucleotide-linking enzyme. The reconstituted plasmids are then introduced to fresh *E. coli* cells, some of which absorb the plasmids. These transformed bacteria multiply, producing millions of cells that contain the same segment of foreign DNA. If the alien DNA codes for a protein not normally synthesized in the bacterium, a new strain of bacteria has been created.

interferon. Insulin, a hormone that regulates blood sugar levels, is extremely important in the treatment of diabetes. Growth hormone, as its name implies, helps children grow normally; a synthetic growth hormone is being used to treat children who have a certain kind of dwarfism. Interferon is a substance that is produced by the human body in extremely minute amounts. It is being tested as a treatment (not a cure) for various types of cancer, multiple sclerosis, and hepatitis B; early reports suggest that interferon also has potential in the treatment of certain virus-related diseases. A vaccine that protects cattle against foot-and-mouth disease has also been developed by recombinant DNA techniques—the first vaccine to be so produced. The chief advantages that recombinant DNA methods offer for the production of such chemicals are that the hormones and other products so generated are much easier to purify and their yield is so much greater than when similar chemicals are extracted from natural sources. This translates into greater availability and cheaper prices, a welcome addition to the ever-burgeoning expense of medical care.

The "gene-splicers" have a seemingly inexhaustible list of medical, industrial, and agricultural applications of the new technology, and many private research companies have virtually sprung up overnight and sold millions of dollars worth of stock on the promise of developing new solutions to old problems.

carry out this research must be licensed. Such measures were called for by many leading scientists in the field. who pointed out some of the dangers inherent in gene-splicing experiments. For example, as mentioned earlier, the favorite "guinea pig" of recombinant DNA studies is *E. coli,* the intestinal bacterium of humans and other mammals. If a disease-causing strain of *E. coli* were accidentally created, which then escaped the confines of the laboratory, the result could be an epidemic of gravest magnitude. For this and other reasons, recombinant DNA laboratories are equipped with special isolation areas, and researchers must use variants of bacteria that cannot possibly survive outside their protected laboratory environment. Most scientists believe that the current safeguards are sufficient.

SUMMARY

Tracking down the identity of the genetic substance in organisms began in 1928 with the demonstration by Griffith that a substance could pass from a pathogenic strain (S strain) of *Streptococcus pneumoniae* to a non-pathogenic strain (R strain) of this bacterium, genetically transforming the latter into pneumonia-causing S cells. Sixteen years later, Avery and his co-workers showed that the transforming factor was DNA. Further evidence for DNA as the genetic substance was provided by Hershey and Chase in 1952, when they showed that it was the DNA component of their bacterium-infecting virus that contained the information to construct new viral particles in its host cells.

The structure of DNA was deduced in 1953 when Watson and Crick, using wire models of DNA's four constituent nucleotides, constructed two polynucleotide chains that twisted around each other to form a double helix. Each chain of the DNA molecule consists of nucleotides linked via their phosphate and deoxyribose (sugar) groups. The two chains are held together by hydrogen bonds between complementary bases that extend toward the center of the molecule. The base adenine on one chain bonds to the base thymine on the complementary strand; similarly, guanine forms a base pair with cytosine.

During DNA replication in eukaryotes, the double helix unwinds at many sites along its length. The exposed bases act as a template to guide the alignment of free nucleotides along each untwisted strand. DNA polymerase facilitates the alignment process, and catalyzes the linking of the free nucleotides into new polynucleotide chain fragments. Replication is com-

FIGURE 11–12
A Vat of Interferon. Interferon, a chemical produced in very small amounts in the human body, has been used in test studies to treat various forms of cancer and virus-related diseases. Interferon obtained from natural sources could cost up to $30,000 per patient, but the same substance produced en masse by the recombined bacteria growing in this vat is much cheaper (about $300 per patient) and plentiful.

Economic growth in this field was spurred in 1982 when the U.S. Supreme Court ruled that at least certain types of organisms altered by genetic manipulation are patentable. Thus, the controversy over whether an individual or group can claim exclusive rights to a living organism has been temporarily defused.

In light of the potential hazards associated with recombinant DNA work, however, government agencies have established stringent guidelines concerning the types of organisms and foreign DNA that can be used in these experiments. Also, investigators that

pleted after DNA ligase links the new chain fragments into single long chains. Each new strand remains hydrogen bonded to its template strand, resulting in two DNA molecules. Each DNA molecule thus consists of one new and one parental strand.

The information stored in DNA is in the form of specific nucleotide sequences on one of the DNA strands. Each unit of information—a gene—specifies the amino acid sequence of a polypeptide. For proteins that consist of two or more different polypeptides, there are two or more genes that code for their structures. This is the thread that links genes to traits, for it is proteins with their enzymatic, transport, regulatory, and contractile roles that determine the appearance and functions of cells and, ultimately, of the whole organism.

Over the past 15 years, molecular biologists have developed the means to insert foreign DNA into the native DNA of living cells, particularly into the plasmids of bacteria. The recombined plasmids are replicated during the bacterial cell cycle, so succeeding generations of bacteria retain the genetic instructions to synthesize the foreign protein or proteins. Recombinant DNA researchers have already created artificial strains of bacteria that synthesize a variety of medically important substances, including insulin, growth hormone, and interferon. In the future, we can look toward using recombinant DNA techniques to improve the genetic quality of crop plants, and, possibly, to correct certain human genetic disorders by gene transplantation. The potential risks of creating disease-causing strains of microorganisms through recombinant DNA research are minimized through restrictions and regulations governing this type of research.

STUDY QUESTIONS

1. Cite evidence from bacterial studies that supported DNA as the genetic substance.

2. How did Hershey and Chase demonstrate that it is the DNA of their virus, not the protein coat, that encodes the information to construct new viral particles?

3. What was known about the structure of DNA when Watson and Crick began building their models?

4. Draw a simple model of a double-stranded DNA molecule that has three nucleotides per strand. Use Ⓟ for phosphate, Ⓢ for deoxyribose, and A, T, G, and C for the bases.

5. What structural feature of the DNA molecule suggests how DNA serves as a template for its own replication?

6. What is the nature of information stored in DNA?

7. Hemoglobin, the oxygen-carrying protein in blood, consists of four polypeptide chains—two identical α chains and two identical β chains. How many genes code for the hemoglobin protein?

8. Use the example of albinism to explain the relationships among DNA, proteins, and traits. Do the same using the Tay-Sachs example (see Chapter 9 for a discussion of Tay-Sachs).

9. What is a plasmid? Why is it useful in recombinant DNA research?

10. Give some examples of how recombinant DNA research has provided some practical benefits. What are some of the potential applications of this research?

11. What kinds of precautions must be taken when scientists use *E. coli* as a recipient of recombinant DNA?

SUGGESTED READINGS

Olby, R. *The Path to the Double Helix.* Seattle, WA: University of Washington Press, 1975. Olby, a science historian, traces the development of ideas, experiments, and prejudices concerning the genetic substance. The human side of science is a keynote in this fascinating book.

Scientific American 245(1981). This entire issue is devoted to industrial microbiology and genetic engineering.

Watson, J. D. *The Double Helix.* New York: Atheneum Publishers, 1968. A brash and engaging account of the trials, frustrations, and ultimate success of the Watson and Crick adventure.

Watson, J. D. *Molecular Biology of the Gene.* 4th ed. Menlo Park, CA: Benjamin/Cummings, 1984. A classic in its field for the advanced student. This is an in-depth account of molecular biology.

Watson, J. D., and J. Tooze. *The DNA Story: A Documentary History of Gene Cloning.* San Francisco: W. H. Freeman, 1981. Outlines the techniques used in recombinant DNA research and what has been accomplished using these techniques. Most of the book is devoted to a discussion of the debate in the last decade over the potential hazards of recombinant DNA research, including many articles published during this time.

CHAPTER 12
DNA→RNA→Protein: Information Flow and Its Regulation

With the discovery of the structure of DNA in 1953, a wave of excitement rushed through the scientific community. Not since the rediscovery of Mendel's work in 1900 had geneticists so many questions to ponder, ideas to formulate, and experiments to run. It was the beginning of a golden age for biology.

One of the more obvious questions sparked by the double helix was: How does the structure of DNA determine the structure of proteins? Recall that at the time of Watson and Crick's model-building days, the relationship between DNA and proteins was firmly established, and there was little doubt as to the general nature of that relationship. Scientists quickly adopted the working hypothesis that the base sequence in DNA in some manner determined the amino acid sequences of proteins. This implied the existence of a **genetic code,** whereby the "language" of DNA (base sequence) is translated into the "language" of proteins (amino acid sequence). But what were the DNA "codewords," and how did the translation process work?

THE TRIPLET CODE

Based on purely theoretical considerations, Crick had proposed early on that the four-letter alphabet in DNA (A, T, G, and C) was "read" in groups of three, or **triplets.** The rationale was quite simple. Because there are 20 different amino acids in proteins, a one-to-one coding system is clearly unworkable—the four bases could specify only four different amino acids. Similar reasoning discounts a two-letter code, for there are only 16 possible two-letter combinations (see Figure 11–8), still not enough to code for the 20 amino acids. A triplet code, however, would generate 4^3 ($4 \times 4 \times 4$), or 64 possible DNA codewords. Although this is more than enough, a three-letter combination of bases permits at least one unique codeword for each amino acid.

Direct evidence for the triplet code hypothesis did not surface until the early 1960s, several years after it was first proposed. There was good reason for the delay. First, scientists had to sort out the actual mechanism by which information stored in DNA is used by cells to construct proteins. This was no easy task, for the process of information flow turned out to be quite complex. One of the keys to understanding this process lay in recognizing the roles played by another kind of nucleic acid, **ribonucleic acid (RNA).**

RNA AND TRANSCRIPTION

Like DNA, RNA is a polymer of nucleotides linked together by their sugar and phosphate groups. Unlike

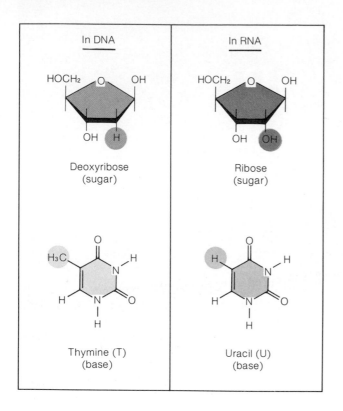

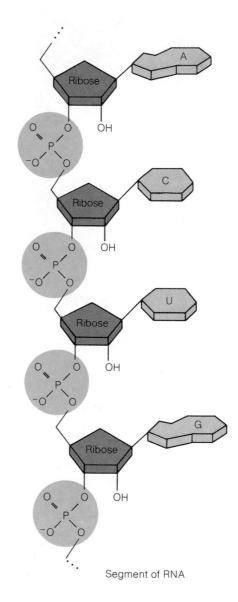

Key to the bases:
A = Adenine
C = Cytosine
U = Uracil
G = Guanine

Segment of RNA

FIGURE 12–1
RNA Structure.   In contrast to DNA, RNA is single-stranded, ribose replaces deoxyribose as the sugar component, and uracil replaces thymine as one of the bases. The nucleotides of RNA are linked into a chain by the same sugar–phosphate connections as in DNA.

DNA, however, RNA molecules are **single-stranded** (rather than double-stranded), and their nucleotides contain the sugar **ribose** (instead of deoxyribose). Another difference is that RNA contains the base **uracil (U)**, but not thymine; DNA and RNA both have the bases adenine, cytosine, and guanine (Figure 12–1).

Types of RNA

Most of the cellular RNA is found in the cytoplasm, where it exists in three general forms: messenger RNA (mRNA), ribosomal RNA (rRNA), and transfer RNA (tRNA). Each performs a different role in protein synthesis.

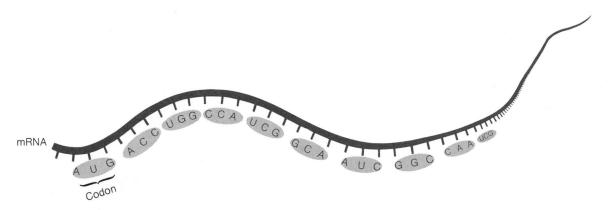

FIGURE 12–2
Messenger RNA. Messenger RNA molecules are long, linear chains of nucleotides. Each linear triplet of bases is a codon (gray) that codes for one of the 20 amino acids.

MESSENGER RNA. As its name implies, **messenger RNA** carries the "messages" of genes. In eukaryotic cells, each mRNA molecule is *transcribed* from a specific segment of DNA in the nucleus (as discussed in the "Transcription" section that follows). Each mRNA molecule then moves into the cytoplasm to serve as a template for the ordering of amino acids into a polypeptide. Thus, information flow is from DNA to mRNA to polypeptide. Different genes code for different kinds of mRNA, each of which carries a unique sequence of triplet base combinations, called **codons,** that codes for a unique polypeptide (Figure 12–2). Each codon specifies a particular amino acid. Messenger RNA is the "middle man" in the information transfer process between DNA and protein.

RIBOSOMAL RNA. Ribosomes (see Figure 3–15) are tiny particles that assist in the translation of an mRNA message into a polypeptide. Each ribosome consists of two subunits, one large and one small, which in turn are composed of proteins and **ribosomal RNA** (Figure 12–3). It is believed that the rRNA components aid in the binding of ribosomes to mRNA, forming an **mRNA–ribosome complex.**

TRANSFER RNA. Transfer RNA molecules are the *adapters* in protein synthesis—they transfer the individual amino acids to the mRNA–ribosome complex for assembly into polypeptides. Although single-stranded, a tRNA molecule is generally folded into a cloverleaf pattern, which is held in place by hydrogen bonds between complementary base pairs (Figure 12–4a). At one end of each tRNA molecule is an adenine base, the attachment site for a particular amino acid. At another region of the tRNA molecule is a loop of un-

paired bases that includes the **anticodon**—a group of three bases that varies with each type of tRNA. With their amino acids attached, and with the assistance of ribosomes, the anticodons of the tRNA molecules pair with complementary codon bases on mRNA (Figure 12–4b). In this way, the tRNA molecules "deposit" their amino acids to a growing polypeptide in an order that reflects exactly the sequence of codons in the mRNA transcript.

Much of the accuracy of protein synthesis is built in at the level of tRNA. For each of the 20 different amino acids, there is at least one unique type of tRNA and one specific enzyme that catalyzes the attachment of the amino acid to its type of tRNA. For example, the amino acid serine can be attached to seryl-tRNA because a specific enzyme that catalyzes this attachment recognizes only serine and seryl-tRNA, no others. In an analogous manner, all the other amino acids become attached to their corresponding tRNA molecules, all with unique anticodons that pair with specific mRNA codons. Thus, the fidelity of the genetic code resides in the matching of amino acids to specific tRNAs.

FIGURE 12–3
Ribosome.

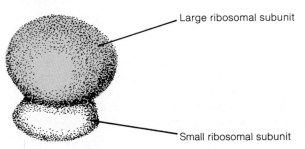

Large ribosomal subunit

Small ribosomal subunit

FIGURE 12–4
Transfer RNA for the Amino Acid Serine.
(a) The three-dimensional shape of a tRNA molecule resembles a cloverleaf. Hairpin loops are formed by hydrogen bonds between complementary bases. Note the attachment site for the amino acid (serine) and the anticodon (AGA).
(b) This diagram shows the base pairing between the anticodon and the mRNA codon, which specifies the insertion of serine into a polypeptide.

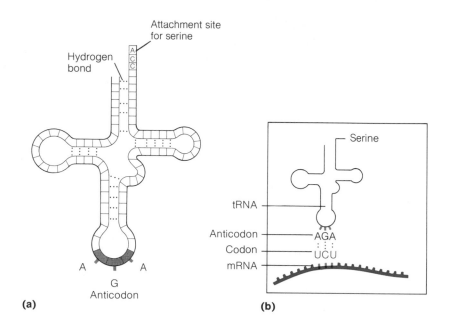

Transcription

Transcription refers to the synthesis of any type of RNA from a DNA template. In many ways the process is similar to DNA replication. It begins when a localized region of a DNA molecule at the beginning of a gene unwinds, exposing the bases on both DNA strands (Figure 12–5). One of these strands—the **sense strand**—is used as a template for constructing the RNA molecule. The sense strand is the gene. The other so-called **antisense strand** is not transcribed. Nevertheless, it is important because it serves as the template upon which another sense strand is made during DNA replication.

After part of the DNA molecule unwinds, an enzyme called **RNA polymerase** binds to the sense strand. RNA polymerase then moves along the DNA chain, helping to align the free RNA nucleotides with their complementary DNA bases. RNA polymerase also catalyzes the linkage of the ribose and phosphate groups of the RNA nucleotides, forming an RNA chain. The base pairing rules are the same as in DNA replication, with one exception: in transcription, the DNA base adenine pairs with the RNA nucleotide that contains uracil (see Figure 12–5). Once the entire gene has been transcribed, the completed RNA molecule is released from the RNA polymerase–DNA complex.

All the types of RNA so transcribed then move into the cytoplasm where they function in protein synthesis. Ribosomal RNA transcripts associate with ribosomal proteins to form ribosomes. Transfer RNA

molecules exist freely in the cytoplasm, where they can be linked to their corresponding amino acids. Both ribosomes and the various tRNAs must be continuously available for protein synthesis to take place, and both participate in the synthesis of *all* proteins. With few exceptions, a given mRNA transcript is involved in the synthesis of only *one* protein (or more accurately, one polypeptide). Thus, the specific proteins produced in cells generally depend on which genes have been transcribed into mRNA molecules.

PROTEIN SYNTHESIS

The transfer of information from mRNA to protein is a complicated process involving not only mRNA, the tRNAs, and ribosomes, but also energy, many different enzymes, and other cytoplasmic factors. Understandably, putting together all the pieces into a neat picture took many years and the dedicated efforts of many researchers. By the mid-1960s, the protein synthesis puzzle was complete enough for scientists to break the genetic code.

The Genetic Code

As Crick predicted, the genetic code is based on triplet combinations of bases. Each mRNA message consists of a linear series of three-letter words (codons), and each word codes for a particular amino acid. For ex-

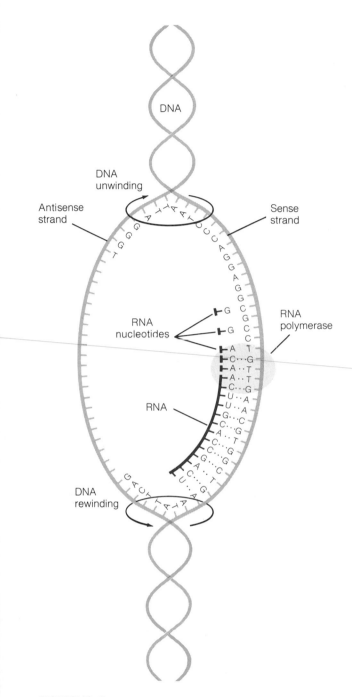

FIGURE 12–5
Transcription. During transcription, a segment of double-stranded DNA unwinds. The exposed bases on the sense strand are thus available for base pairing with complementary RNA nucleotides. RNA polymerase facilitates the alignment and linking together of the nucleotides to form the single-stranded RNA molecule. As the completed part of the RNA chain releases from the DNA template, the two strands of DNA rewind. Note that the DNA base adenine pairs with the RNA nucleotide containing uracil.

ample, the codon UCU (uracil–cytosine–uracil) codes for the amino acid serine; GGU (guanine–guanine–uracil) codes for glycine; and so on for all 20 amino acids (Table 12–1).

Three of the 64 codons—UAG, UAA, and UGA—do not specify any amino acid. These are **stop codons** on the message, and whenever one of these appears as a codon on the mRNA chain, amino acid assembly stops, signaling completion of the polypeptide. Each of the 61 other codons specifies one (and only one) of the 20 amino acids. Since there are more codons than amino acids, we say that the code is **degenerate.** All of the amino acids except methionine and tryptophan have more than one codon. Arginine, leucine, and serine each have six different codons, and all the rest have between two and four.

Elucidation of the genetic code came from some rather ingenious biochemical experiments. In these experiments, researchers broke open bacterial cells and isolated various parts of their protein synthesis machinery—ribosomes, various enzymes, all the tRNAs, and other factors. A suspension of these components was placed in a test tube containing the 20 different amino acids and an energy source (ATP). Then, synthetic mRNA of a known base sequence was added, which directed the synthesis of a particular polypeptide. Now, since the amino acid sequence of any polypeptide produced under these conditions will be determined by the base sequence (known) of the synthetic mRNA, analyzing the amino acid composition of the polypeptide product should reveal the mRNA codons for those amino acids. For example, in the first successful experiment of this kind, the synthetic mRNA used was **poly U,** a long chain of uracil-containing nucleotides (Figure 12–6). The polypeptide formed in this case consisted of many units of a single type of amino acid, phenylalanine. Thus, UUU must be the codon for phenylalanine. By using this technique and others, all the mRNA codons were eventually identified. A major mystery of life had been solved.

The original studies on breaking the genetic code were carried out with bacteria, but shortly thereafter, researchers began to investigate the genetic code in various plants, animals, and fungi. It soon became evident from these studies that the genetic code is the same in all organisms. The code is universal! This discovery was hailed as one of the great unifying principles of biology, since it implies that all present-day creatures share a common ancestry.

It seems that for every rule of nature, there are exceptions. In 1979, several reports appeared indicat-

TABLE 12–1

Amino Acids and Their mRNA Codons. The information transcribed onto mRNA is translated into a specific amino acid sequence of a polypeptide by reading triplet codons. The triplet code has three major features:

1. The code is degenerate; that is, with the exception of methionine and tryptophan, two or more triplet codons specify the same amino acid. The code is not ambiguous, however, because no single codon specifies more than one amino acid.
2. There are 64 possible three-base sequences (codons). The three codons UAG, UAA, and UGA serve as ''stop'' signals in the message to terminate polypeptide synthesis. The codon AUG (for methionine) signals the start of polypeptide synthesis.
3. The code is directional. Messenger RNA is read in one direction only, so that UUC, which codes for phenylalanine, will not be confused with CUU, which codes for leucine.

Amino Acid and Abbreviation	mRNA Codon	Amino Acid and Abbreviation	mRNA Codon
Alanine (ala)	GCU GCC GCA GCG	Lysine (lys)	AAA AAG
Arginine (arg)	AGA AGG CGU CGC CGA CGG	Methionine (met)	AUG
		Phenylalanine (phe)	UUU UUC
		Proline (pro)	CCU CCC CCA CCG
Asparagine (asn)	AAU AAC	Serine (ser)	UCU UCC UCA UCG AGU AGC
Aspartic acid (asp)	GAU GAC		
Cysteine (cys)	UGU UGC		
Glutamic acid (glu)	GAA GAG	Threonine (thr)	ACU ACC ACA ACG
Glutamine (gln)	CAA CAG		
Glycine (gly)	GGU GGC GGA GGG	Tryptophan (trp)	UGG
		Tyrosine (tyr)	UAU UAC
Histidine (his)	CAU CAC	Valine (val)	GUU GUC GUA GUG
Isoleucine (ile)	AUU AUC AUA		
		Stop	UAG UAA UGA
Leucine (leu)	UUA UUG CUU CUC CUA CUG		

ing that the genetic code for human and yeast mitochondria differs in certain respects from the universal code. (Mitochondria have small amounts of DNA and their own protein-synthesizing machinery that is structurally and functionally separate from the protein synthesis system in the cytoplasm; see Essay 3–1.) The differences are small but significant: the mitochondrial UGA codon (a stop codon in the universal code) specifies tryptophan, and AUA (universally an isoleucine codon) codes for methionine. Does this throw a wrench into the principle that all modern organisms have a common ancestry? Not at all. The differences

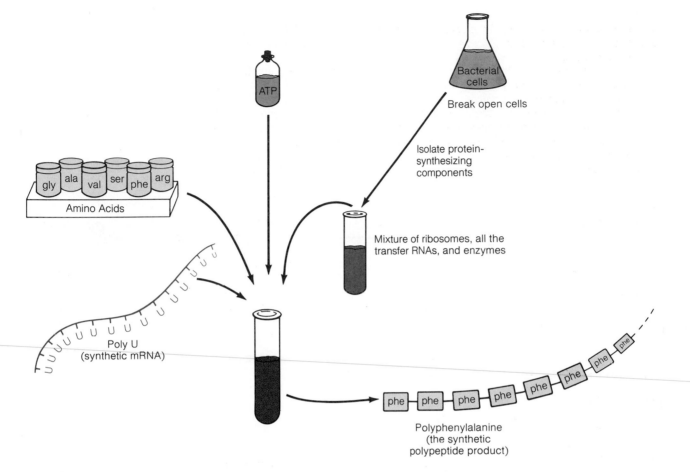

FIGURE 12–6
Breaking the Genetic Code: Cell-Free Protein Synthesis. This diagram summarizes the first suc-
cessful experiment aimed at elucidating the genetic code. In the presence of all 20 amino acids,
ATP, and the protein-synthesizing machinery obtained from broken bacterial cells, the synthetic poly
U transcript directed the assembly of phenylalanine (phe) molecules into a polyphenylalanine chain.
Therefore, UUU is the codon for phenylalanine.

between the cytoplasmic and mitochondrial codes
simply suggest that the mitochondrial genetic code
may have evolved along a separate line, one which
presumably originated very early in the history of life
on earth.

Translating the Message: The Mechanics of Protein Synthesis

In keeping with the metaphor of "language" and
"messages," biologists refer to the transfer of information from mRNA to a polypeptide as **translation.**
We have already introduced the principal players in
this drama. It is now time to put them into action.

The **initiation** of translation begins when a ribosome binds to an mRNA molecule near an AUG codon,
forming an mRNA–ribosome complex (Figure 12–7a).
In eukaryotes, AUG indicates the beginning of a gene,
so it is known as the **start codon.** Translation always
begins at an AUG codon. Since AUG specifies the
amino acid methionine (met), the first amino acid
in a polypeptide is always methionine. Methionine is
brought to the mRNA–ribosome complex by its tRNA,
abbreviated met-tRNA. The anticodon of met-tRNA
is UAC, which pairs perfectly with the codon AUG
(Figure 12–7b).

Next, a second tRNA, with its amino acid in tow,
binds to the second codon on the mRNA molecule.
For example, if the second codon is UUU, a tRNA with
the anticodon AAA will pair there. In this case, the
tRNA carries the amino acid phenylalanine (phe). With

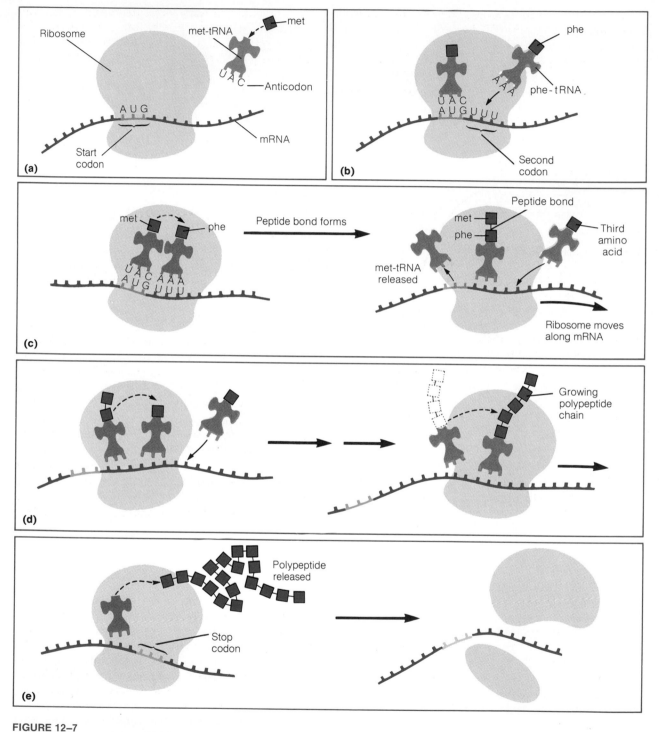

FIGURE 12–7

Protein Synthesis. This diagram summarizes the important steps by which amino acids are assembled into a polypeptide. (a) and (b) Stages in chain initiation. (c) The first peptide bond is formed. (d) Chain elongation. (e) Chain termination. (Met = methionine; Phe = phenylalanine. See text for details.)

these two tRNAs now in place, their attached amino acids (methionine and phenylalanine) lie close enough together to be joined by a peptide bond. The formation of the peptide bond is catalyzed by an enzyme in the ribosome (Figure 12–7c).

As the peptide bond is formed, the bond linking methionine to its tRNA is broken. As a result, the dipeptide (met-phe) is attached only to the second tRNA. The first tRNA then detaches from the ribosome. The ribosome now shifts to the next codon, ready to bind a third tRNA with its specific amino acid. Again, the mRNA codon determines which tRNA will be next in line, and thus, which amino acid will be next in the polypeptide chain. When the third tRNA is in place, its amino acid is joined to the second amino acid (phe), forming a tripeptide. This process is repeated over and over again as the ribosome moves along the mRNA chain, adding amino acids to the growing polypeptide chain (Figure 12–7d). This phase of translation is called **chain elongation.**

Chain elongation continues until the ribosome encounters one of the three possible stop codons. There are no tRNAs that bind in response to a stop codon, so the polypeptide chain is terminated (Figure 12–7e). The ribosome dissociates from the mRNA, and the completed polypeptide is released. This step is appropriately called **chain termination.**

In many instances, a newly-synthesized polypeptide may undergo **post-translational modifications.** One of the more common modifications is an enzyme-mediated excision of a short segment of amino acids from the end of the completed polypeptide. If the excision occurs at the start of the polypeptide chain, the initiation amino acid, methionine, will be lost with the excised peptide.

A single mRNA molecule may be translated into hundreds of copies of a particular polypeptide, each with an identical amino acid sequence. The accuracy of this process hinges on the specific attachment of amino acids to their corresponding tRNAs, and the specific base pairing of the tRNA anticodons to complementary mRNA codons. The flow of information from DNA to protein also depends on the accurate transcription of DNA genes into mRNA molecules. Although there are many places for mistakes to occur in the information transfer process, they seldom do.

VARIATIONS ON INFORMATION FLOW

The picture of information flow (transcription and translation) presented thus far is based primarily on studies conducted during the 1950s and 1960s with bacteria and viruses. Although our description holds quite well for these simple life forms, studies on higher organisms have shown that information flow in eukaryotes, although having the same basic characteristics already described, can exhibit additional complexities. In particular, two recently discovered phenomena, mRNA editing and gene rearrangement, represent variations from this general theme.

Messenger RNA Editing

In 1977, researchers began reporting that some eukaryotic genes were veritable "patch works" of stored information. In these cases, only certain stretches of bases along the gene were used to construct a polypeptide; between these message-carrying units were intervening sequences of nucleotides representing islands of noninformation (Figure 12–8). How could such genes give rise to a continuous, translatable mRNA transcript? Investigators discovered the answer when they compared a given gene's mRNA transcript isolated from the nucleus to the same gene's mRNA transcript isolated from the cytoplasm—the nuclear mRNA transcript was considerably longer! Hence, they concluded that the original mRNA transcript must be cut and spliced within the nucleus to remove the noninformational nucleotide sequences, much as movie film is edited. This process is called **messenger RNA editing.** The edited mRNA, now composed of spliced segments of the original transcript, is then transported to the cytoplasm, where it directs the synthesis of a polypeptide.

Gene Rearrangements

Another recent discovery indicates that genes, or parts of genes, can become rearranged within a chromosome. This phenomenon has been documented for a group of genes that determine the structures of *antibodies,* proteins that play a crucial role in the body's defense against foreign cells or substances (see Chapter 20). Mammals, including humans, can produce up to ten million different kinds of antibodies. Each kind of antibody recognizes and initiates the destruction of a particular type of alien cell or substance. Recent evidence indicates that this enormous diversity in antibody structure is due in part to the rearrangement of the genes that code for different regions of the antibody molecule.

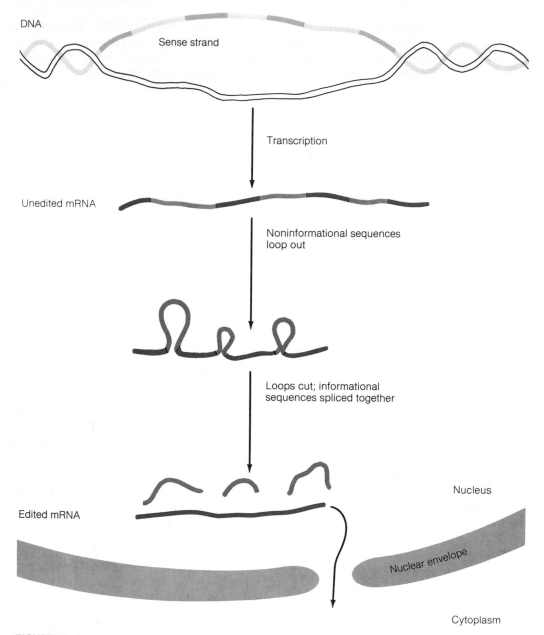

FIGURE 12–8
Messenger RNA Editing. Some eukaryotic genes contain informational stretches of DNA bases (light color) that are interposed with noninformational stretches (light gray). The original transcript (unedited mRNA) produced from such a gene undergoes processing in the nucleus. The noninformational mRNA sequences form loops, bringing the informational sequences (color) close together. Enzymes then cut the loops and splice the informational sequences together into a continuous mRNA molecule. This edited mRNA then moves into the cytoplasm to direct the synthesis of its coded polypeptide.

Many different kinds of antibodies share a common amino acid sequence in part of their protein structure, called the **constant region.** This constant region of the molecule is linked to a **variable region,** the amino acid sequence of which varies according to the type of antibody (Figure 12–9). It is this variable region, then, that accounts for the millions of different kinds of antibodies possible. Researchers now believe that early in the development of an antibody-producing cell, a particular variable-region segment of DNA becomes rearranged with other DNA segments that code for the constant region. This variable–constant DNA unit can then be transcribed onto a single mRNA molecule that directs the synthesis of an antibody polypeptide with a specific variable region. Thus, this cell produces a particular kind of antibody. Presumably, different cells produce different antibodies because during their development, different variable-region DNA segments became inserted next to the constant-region genes.

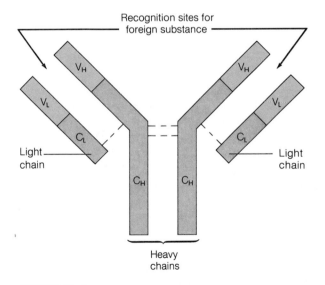

FIGURE 12–9
Antibody Structure. A typical antibody molecule is a protein composed of four polypeptides: two light chains and two heavy chains. Each chain consists of a *constant region* (denoted C_L or C_H) and a *variable region* (V_L or V_H; shown in color). The variable regions include the recognition sites for a specific foreign substance. Different kinds of antibodies differ primarily in the amino acid sequences of their variable regions.

The significance of mRNA editing and the control mechanisms leading to gene rearrangements are not yet understood. Perhaps in the future these pieces of the puzzle will fit neatly into our picture of life at the molecular level, but by then, we will surely have other stray pieces.

MUTATIONS

The ability of an organism (or cell) to manufacture a particular protein depends not only on the faithful transcription and translation of the appropriate gene or genes, but also on maintaining the base by base integrity of the gene. After all, the specific message encoded by a gene depends entirely on its specific and unchanging base sequence. Occasionally, however, chemical changes do occur in DNA that result in altered base sequences; we call such changes **mutations.**

Mutations generally show up as altered phenotypes. For example, if a purebreeding line of red-flowered peas produces 600 red-flowered progeny and one white-flowered individual, the latter is regarded as a genetic **mutant.** In searching for the cause of the white flower phenotype, we would undoubtedly find that a protein (presumably an enzyme) involved in the formation of red flower pigment has been so altered in its amino acid sequence that it no longer functions in pigment synthesis. This change must be rooted in an altered DNA base sequence of the flower color gene.

In terms of their effects on organisms, mutations can be either beneficial, neutral, or harmful. Mutations that benefit organisms include those that in some way increase their survivability and reproductive success. Changes in genes that lead to the production of a more efficient enzyme, or possibly a new enzyme that renders the organism better adapted to its environment, are examples. Such mutations are the wellspring of evolutionary progress. Neutral mutations are changes in DNA that have little or no bearing on the fitness of an organism. They include base sequence modifications that yield such minor changes in the structure of proteins that protein function, and hence phenotype, is not altered. Alternatively, neutral mutations may cause phenotypic changes, but they will be ones that do not confer any advantage or disadvantage to the organism. For example, the mutation leading to white flower color in peas might be classified as neutral if white-flowered peas are just as fit to survive and reproduce as are red-flowered peas. Many mutations,

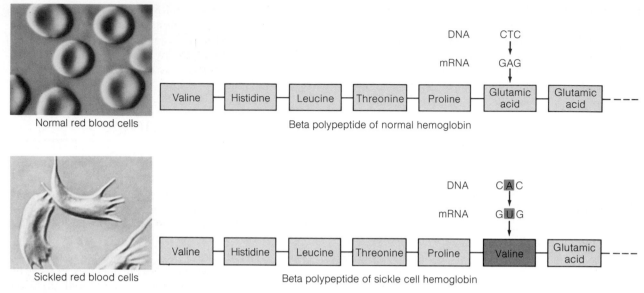

FIGURE 12–10
Sickle Cell Anemia.   The substitution of valine for glutamic acid in the two beta polypeptide chains of hemoglobin is the cause of sickle cell anemia. A portion of the altered beta polypeptide is shown here. When the oxygen content of the blood in an individual with sickle cell anemia reaches a critically low level, the red blood cells collapse into a sickled shape. These deformed cells can clog small capillaries, thus depriving tissues of vital oxygen.

however, are harmful because changes in the amino acid sequence of a protein, even minor ones, can affect its function in a negative way. If the affected protein is vital to some biological process, the well-being of the organism is in jeopardy. Sickle cell anemia is a case in point.

Sickle cell anemia is a genetic disorder in humans that affects the ability of blood to transport oxygen. Any form of physical exertion in affected individuals can lead to restricted blood flow and oxygen deprivation in the body's tissues; severe attacks can be fatal. The disorder is caused by an abnormality in hemoglobin, the oxygen-carrying protein of red blood cells. The hemoglobin protein consists of four polypeptide chains: two identical alpha chains and two identical beta chains. The abnormality in sickle cell hemoglobin exists in the two beta polypeptides. Instead of glutamic acid as the sixth amino acid in these chains (the normal situation), the sickle cell beta polypeptides have valine (Figure 12–10). This single amino acid substitution in each beta chain consisting of 146 amino acids is sufficient to alter the chemical properties of hemoglobin. Under low oxygen concentrations in the blood, a condition often caused by physical exertion, the abnormal hemoglobin molecules in each red blood cell clump together, causing the normally disk-shaped cells to collapse into a sickle

shape (see Figure 12–10). These sickled cells get lodged in small capillaries and block the flow of blood to the tissues, causing minor, and sometimes major, strokes wherever the blockage occurs.

The single amino acid error in the beta polypeptide chains of hemoglobin can be traced to a single base mutation in the beta polypeptide gene. In normal individuals, the sixth codon of the beta chain mRNA is GAG (codon for glutamic acid), but it is GUG (codon for valine) in sickle cell anemia patients. Thus, we can infer that sickle cell anemia is the result of one DNA base being replaced by another. Apparently, a thymine (which codes for the normal GAG codon) was replaced by an adenine (which codes for the sickle cell GUG codon) in the original mutation that caused sickle cell anemia.

Types and Causes of Mutations

The type of mutation that produced the sickle cell anemia gene—one base replaced by another—is called **base substitution** (Figure 12–11). Single base substitutions are generally the least consequential of all the various types of mutations because they usually result in a single amino acid substitution in the protein product. In some cases, these mutations have no effect

(a) No mutation:

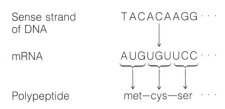

Sense strand
of DNA   T A C A C A A G G · · ·

mRNA   A U G U G U U C C · · ·

Polypeptide   met—cys—ser · · ·

(b) Substitute G for A:

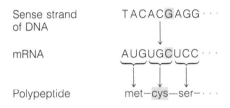

Sense strand
of DNA   T A C A C G A G G · · ·

mRNA   A U G U G C U C C · · ·

Polypeptide   met—cys—ser— · · ·

Result: No effect because same amino acid is coded

(c) Substitute C for A:

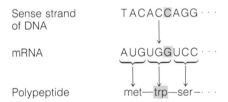

Sense strand
of DNA   T A C A C C A G G · · ·

mRNA   A U G U G G U C C · · ·

Polypeptide   met—trp—ser— · · ·

Result: Single amino acid substitution

(d) Substitute T for A:

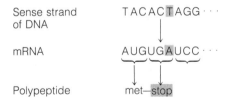

Sense strand
of DNA   T A C A C T A G G · · ·

mRNA   A U G U G A U C C · · ·

Polypeptide   met—stop

Result: Premature termination of the polypeptide
        because a "stop" codon is produced

FIGURE 12–11

Base Substitution Mutations. This diagram illustrates the possible effects of a single base substitution on the sense strand of DNA. (See Table 12–1 for amino acid abbreviations.)

whatsoever on the function of a protein; in other cases, however, the protein's function is altered significantly, as in sickle cell anemia.

As a rule, other types of mutations alter the order of amino acids so dramatically that the affected protein's function is totally lost. These mutations are usually harmful to the organism. **Base deletion** and **base insertion,** which involve the elimination or addition of one or more bases in a gene, respectively, are examples of the more drastic types of mutations. Their effects are particularly dramatic when they occur near the beginning or middle of a gene, because all of the codons of the gene's mRNA transcript that fall after the point of mutation will be out of phase. This results in an entirely new amino acid sequence for the latter part of the polypeptide chain (Figure 12–12). An interesting exception is the addition or deletion of three consecutive or closely clustered bases. In such cases, only one to several amino acids would be added to or deleted from the polypeptide. The rest of the mRNA codons would remain in phase. In fact, Crick and his colleagues noted in 1961 that the addition or deletion of three bases in a localized region of a T4 bacteriophage (virus) gene caused much less of a phenotype change than did one- or two-base insertions or deletions. Crick correctly interpreted this as genetic evidence for a triplet code.

The substitution, insertion, and deletion of nucleotides in a gene arise most frequently during DNA replication. For instance, a base substitution can occur if a noncomplementary nucleotide aligns with an exposed DNA base and is then incorporated into the new DNA strand. Consider what will happen if a guanine-containing nucleotide is inserted opposite a thymine base on a replicating DNA strand (Figure 12–13). When the new DNA strand (with its incorrect G) undergoes replication, its complementary strand will have a C inserted opposite the G, and all the cells that derive from this original mutant cell will have the G-C base pair instead of the normal A-T pair. Because of their heritability, mutations occurring early in the development of an organism have more potential impact than those occurring in adults, for they are passed on to a larger population of future cells by mitotic cell division.

We do not know why mutations occur, but we do know that they occur at very low but measurable rates and that some genes have higher mutation rates than others. Also, there are certain environmental conditions that increase the rate of mutations. For example, high-energy radiation (such as X rays and ultraviolet light) and harsh chemicals (such as nitrous oxides, certain hydrocarbons, and other chemicals

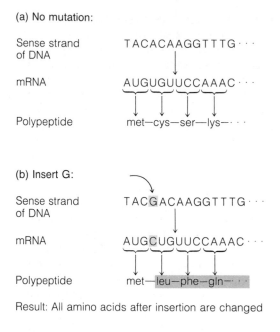

(a) No mutation:

Sense strand
of DNA   T A C A C A A G G T T T G · · ·

mRNA   A U G U G U U C C A A A C · · ·

Polypeptide   met—cys—ser—lys— · · ·

(b) Insert G:

Sense strand
of DNA   T A C G A C A A G G T T T G · · ·

mRNA   A U G C U G U U C C A A A C · · ·

Polypeptide   met—leu—phe—gln— · · ·

Result: All amino acids after insertion are changed

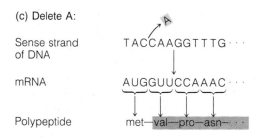

(c) Delete A:

Sense strand
of DNA   T A C C A A G G T T T G · · ·

mRNA   A U G G U U C C A A A C · · ·

Polypeptide   met—val—pro—asn— · · ·

Result: All amino acids after deletion are changed

FIGURE 12–12
Base Insertion and Deletion Mutations.
Note how the addition or deletion of a
single base in DNA shifts the reading
phase of the mRNA codons which appear
after the point of mutation. As a result,
the amino acid sequence of the polypeptide is drastically changed. (See Table
12–1 for amino acid abbreviations.)

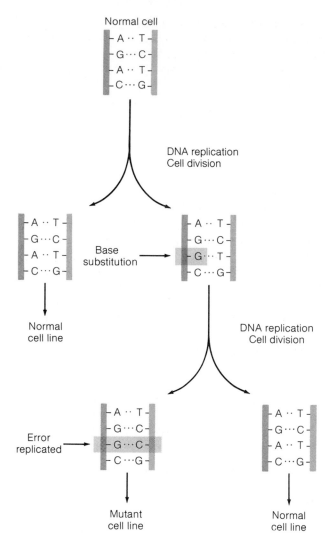

FIGURE 12–13
Development of a Mutant Cell Line.
This diagram illustrates how a single base
substitution occurring in a single cell during
DNA replication will be propagated to later
generations of cells.

found in smog, pesticides, cigarette smoke, polluted
water, and even in food) increase mutation rates
(Essay 12–1). Even though these agents, called
mutagens, increase the risk of mutation, most mutations occur spontaneously and randomly.

Mutations, Alleles, and Variation

Throughout the history of life on earth, mutations
have played an important role in the evolutionary
process, for mutations are the ultimate source of new
alleles. Recall from Chapter 9 that alleles are alternative
forms of genes (such as the *R* and *r* flower color alleles
in Mendel's peas), and that the specific complement
of alleles present for a given gene (the genotype)
determines the organism's phenotype (physical appearance) for the corresponding trait. If a particular
gene mutates to a form that codes for a functionally
altered protein, then the mutation has yielded a new
allele.

For a new allele to be passed on to future generations of sexually-reproducing organisms, it must appear in the gametes. A mutation occurring in your big toe cannot have any bearing on the genetic composition of your children, because toe cells do not give rise to eggs or sperm. Only mutations that occur in gametes or gamete-producing cells in the gonads can be transmitted to the offspring. Even if a mutant allele is transmitted via a gamete to a new individual, however, its effect will usually be masked by the presence of a "normal" allele for that gene contributed by the other parent. In other words, mutated genes are generally recessive.

As the ultimate cause of genetic change in organisms, mutations underlie the diversity of life forms that exist and have existed. Each tiny amoeba, a rose, an elephant, in short, every creature on earth, has its own unique set of alleles that determines its unique characteristics. The range of biological variation is enormous, and it all has its roots in chemical "mistakes" in DNA that have occurred infrequently but constantly throughout time.

THE REGULATION OF GENE EXPRESSION

We have seen that individual organisms vary by their specific complements of alleles, but what about the variation that exists *within* an individual? For example, the human body is composed of hundreds of different kinds of cells, each specialized to perform one or more specific functions. Red blood cells transport oxygen, kidney cells filter toxic wastes out of the blood, muscle cells contract to cause movement—the list could go on and on. As different as these cells are, all of them are derived ultimately from one tiny cell, the zygote. And since all of these cells were generated through mitotic cell divisions, which (barring mutations) produced genetically identical daughter cells, then the blood, kidney, and muscle cells must all have identical genetic instructions—the same as those originally present in the zygote. Thus we have a paradox: although all the cells of an individual have the same genetic information, they are not identical in form or function.

Although the resolution of this apparent paradox remains a major goal of biological research, some facts are now clear. To begin with, we know that individual cells store a complete set of genetic instructions, but only a very small percent of these instructions are actually *expressed* (decoded into proteins) at any point in time. Almost all living cells express those genes that code for proteins involved in general cellular activities, such as respiration, the transport of substances across membranes, and other "housekeeping" functions; however, such genes represent only a small fraction of the cell's total genetic information. The remaining genes code for proteins that perform more specialized enzymatic, structural, and regulatory functions. Various portions of this information may or may not be expressed, depending on the location (or environment) of a particular cell in the organism. For example, both an eye cell and liver cell respire, absorb nutrients, and have the genetic plans to produce the pigments involved in vision. However, since the eye cell is surrounded and influenced by other eye cells, light, and a host of other environmental factors not present in the liver, only the eye cell will express the genes related to visual pigments. At the same time, the liver cell produces specialized proteins not found in eyes. Hence, cells with the same genetic plans can differ because they use different parts of those plans to produce different proteins.

But, how is the expression of genetic information regulated so that each cell produces the appropriate kinds and numbers of proteins to meet its needs? This is a major unsolved question, but over the past 25 years or so, many types of regulatory mechanisms have been uncovered. We now know of at least two major levels at which cells regulate the formation of proteins: the **transcriptional level** (regulating mRNA production) and the **translational level** (regulating the assembly of amino acids into proteins). Of the two, transcriptional control is more widespread in nature. This is what we would expect from the standpoint of biological economy, for it makes sense to restrict the production of mRNA in the first place rather than permit transcription of mRNA that may not be translated. As an example of transcriptional control, let us examine a specific case discovered in bacteria.

Transcriptional Control in Prokaryotes: The Lac Operon

Like most questions in molecular biology, the regulation of transcription has been investigated most intensively in viruses and bacteria, largely because these creatures are biochemically simpler than eukaryotes. The first and most thoroughly studied of the transcriptional control mechanisms was discovered in *E. coli.*

You may recall that *E. coli* is a bacterium commonly found in the intestines of mammals (including humans), where it lives on food passing through the digestive system. Since newborn mammals subsist solely on milk, these bacteria must be able to utilize lactose (milk sugar) as their main food source. When nonhuman mammals are weaned, however, they no longer drink milk. Hence, later generations of *E. coli* in their guts have no further need for the various enzymes that assimilate lactose. As you will see, these bacteria are capable of responding to their changing nutrient environment by "turning on and off" the genes coding for the lactose-assimilating enzymes.

In the early 1960s, two French biologists, François Jacob and Jacques Monod, grew *E. coli* cells on culture dishes containing glucose as the only energy source. When they replaced the glucose with lactose, they noted that within minutes the cells began producing three enzymes involved with lactose utilization (lac enzymes). Apparently the presence of lactose had somehow triggered the bacteria to synthesize these enzymes.

In subsequent experiments, Jacob and Monod provided evidence for what was to become the classic example of **enzyme induction,** a type of feedback system regulating transcription. In their model for the control of lac enzyme synthesis, they proposed the existence of three structural genes situated end to end on the *E. coli* DNA molecule (Figure 12–14a). These genes code for the three lac enzymes, and they are transcribed as a unit onto a single mRNA transcript.

Based on genetic analyses of many mutant strains of *E. coli,* Jacob and Monod discovered that expression of the lac structural genes involves three other regions of the DNA molecule: the promoter, the operator, and the regulator gene (see Figure 12–14a). The **promoter** is the "start-up" site for transcription of the lac genes; that is, it is the binding site for RNA polymerase. The **operator** lies between the promoter and the first structural gene. Its role is described in the following discussion. Jacob and Monod called this linear sequence of units—the promoter, operator, and three structural genes—the **lac operon.**

The lac **regulator gene** is located at a different region of the chromosome. This gene codes for the synthesis of lac repressor, a protein that is produced continuously in the cell. In the absence of lactose, the repressor binds to the lac operator, physically blocking RNA polymerase from reaching the lac structural genes (Figure 12–14b). Under these circumstances, the structural genes cannot be transcribed;

hence, no lac enzymes are produced. This situation changes rapidly, however, when the cells are transferred to a medium containing lactose.

Within 10 minutes after exposure to lactose, the lac enzymes begin appearing in the *E. coli* cells. To account for their synthesis, Jacob and Monod proposed that lactose binds to the repressor, changing its shape such that it can no longer attach to the operator (Figure 12–14c). With the operator unencumbered, RNA polymerase can initiate transcription of the lac structural genes, yielding an mRNA transcript that directs the assembly of the lac enzymes. This bold hypothesis was eventually proven correct when the repressor was isolated and shown to have the appropriate binding properties.

As long as the bacterial cells are growing in an environment containing lactose, the lac enzymes continue to be produced. When the lactose supply runs out, the repressor protein is free to bind to the lac operator and thereby shut down the transcription of the lac structural genes. In this way, the biochemical resources used in mRNA and protein synthesis are not "wasted" in the production of enzymes the cell does not need.

The discovery of the lac operon system, for which Jacob and Monod won the Nobel Prize in 1966, opened the door to many other investigations into the regulation of information flow in cells. Although other bacterial (and viral) operons have been found, scientists have yet to find anything analogous to an operon system in eukaryotic cells. This is not to say that eukaryotes lack transcriptional controls involving gene repression. Rather, it means that eukaryotic gene regulation is more complicated, and certainly more difficult to study, than the simpler systems operating in prokaryotes. We will conclude our discussion of gene regulatory mechanisms by briefly considering what we know (and do not know) about the regulation of gene activity in eukaryotes.

Transcriptional Control in Eukaryotes

Because we know that proteins are directly involved in regulating gene expression in prokaryotes, much of the early work on eukaryotic gene regulation centered on the roles played by the protein components of chromosomes. One group of chromosomal proteins, the **histones,** were found to inhibit transcription markedly. Could they be the repressor proteins in eukaryotic cells? Unfortunately, the initial excitement over histones as possible transcription regulators was short-lived. Researchers soon learned that all the differ-

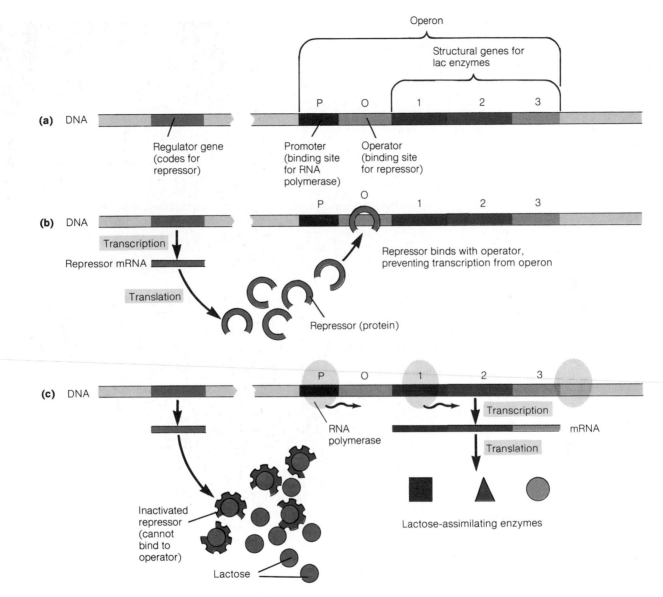

FIGURE 12–14
The Lac Operon.   (a) Structural arrangement of the regulator gene and the components of the lac operon on the *E. coli* chromosome. (b) In the absence of lactose, the regulator gene product—the repressor—binds to the operator and prevents the transcription of the lac structural genes. (c) Lactose, when present, binds to and inactivates the repressor so it can no longer bind to the operator. This permits RNA polymerase to bind to the promoter and transcribe the structural genes. The mRNA is then translated into the three lac enzymes.

entiated types of cells in a multicellular organism had strikingly similar histone proteins, even though these cells produced vastly different types of cytoplasmic proteins. Thus, there was no variation in histones that could account for the variation in gene expression underlying cellular differentiation.

On the other hand, there are many other types of **nonhistone proteins** also associated with eukaryotic chromosomes. Some of these types of proteins may be present in one cell type but absent in others. In general, the variation in patterns of the nonhistone proteins correlates better with the variation in cell

ESSAY 12–1

WILL IT CAUSE CANCER?

Remember Agent Orange (dioxin), Love Canal, and Three-Mile Island? How about the ban on saccharin and the Surgeon General's warning on cigarettes? And when are "they" going to do something about smog? Every day we are exposed to harsh chemicals in the air we breathe, coffee we drink, and foods we eat. How are these "bad" chemicals identified and what health risks do they pose?

The major risk of environmental pollutants is cancer. Thousands of chemicals are screened every year to determine if they are **carcinogenic** (cancer-causing), and if so, at what levels. Until recently, such tests were always performed on laboratory animals such as mice or rats. The knowledge gained from these tests, although beneficial in cancer research, is costly and time-consuming to obtain. Chemicals must be introduced to the animals in large amounts, and weeks or even months pass before the results are in. Now there is a new test to screen for potential carcinogens that utilizes bacteria rather than animals. Called the **Ames test,** this screening procedure takes advantage of the fact that most carcinogens are also **mutagens,** substances that increase the rate of mutations.

The Ames test uses a mutant strain of bacteria (*Salmonella*) that cannot synthesize the amino acid histidine. This strain has suffered a mutation in one of the genes for the histidine pathway, resulting in the absence of one of the enzymes necessary for histidine synthesis. Consequently, these bacteria can grow only on a culture medium that includes histidine. In the Ames test, the suspected chemical mutagen is first mixed with an extract prepared from rat liver tissue. This step is important because many mutagens (carcinogens) are active in mammals only after they have been converted to other substances by the cells exposed to them. The liver extract is rich in enzymes that can catalyze such activations, and thus its inclusion makes the Ames test more sensitive and reliable.

Mutant bacteria are then incubated with the liver extract and the suspect chemical. If the chemical is a mutagen, then the number of random mutations occurring in the bacterial cells will be higher than background levels. Occasionally, a mutation will occur in the faulty histidine gene that restores this gene to its original, nonmutant form. These genetic revertants are easily identified by placing the bacterial mixture onto a culture medium that lacks histidine. Only those cells that have suffered a histidine reversion mutation will grow into new colonies, or populations, of cells. The number of colonies that form is directly proportional to the number of histidine revertant mutations, which in turn depends on the potency of the mutagen. Controls in which the mutagen is withheld are run simultaneously to give a background rate of mutation, which is always very low.

Although the Ames test is considered a preliminary screening procedure, about 90% of the substances found to be mutagenic in the bacteria have also been shown to be carcinogenic in animals.

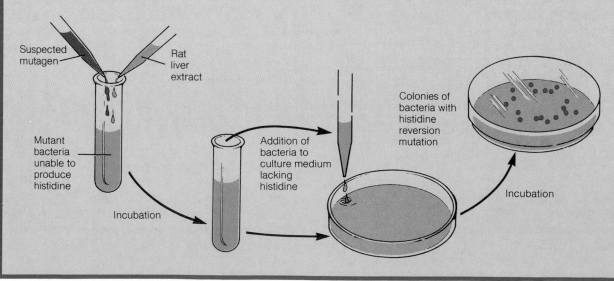

Suspected mutagen

Rat liver extract

Mutant bacteria unable to produce histidine

Incubation

Addition of bacteria to culture medium lacking histidine

Colonies of bacteria with histidine reversion mutation

Incubation

types and functions, but exactly how this group of proteins functions is not yet understood. Moreover, since the nonhistone proteins are themselves products of genes, we are also left with the question of how their synthesis is regulated.

Adding another complication, the eukaryotic chromosome, unlike its prokaryotic counterpart, contains many repetitive nucleotide sequences in its DNA. In other words, the same gene may appear in several or even hundreds of copies scattered throughout the chromosomes. Some researchers think these repetitive sequences may play an important role in eukaryotic gene regulation, but what that role might be remains unclear.

Translational Control in Prokaryotes and Eukaryotes

Both prokaryotic and eukaryotic cells exhibit control mechanisms that operate at the level of amino acid assembly on the mRNA–ribosome complex. For example, a specific mRNA transcript may have a high incidence of a particular codon for which the appropriate tRNA is in low supply. Whenever the ribosome reaches one of these codons, translation comes to a halt until that tRNA (with its attached amino acid) becomes available. Thus, the rate at which the cell can synthesize a polypeptide is limited by the availability of the required type of tRNA. Translation would also slow down for other mRNA transcripts requiring that particular tRNA.

We have touched on only a few types of control mechanisms operating at the various levels of information transfer in a cell, and there are many others we are just beginning to understand. Conducting basic research in this field, called regulatory biology, is not only important to our understanding of life itself, but will also generate many practical benefits. Before we can directly attack problems such as cancer, genetic diseases, and other forms of abnormal development—all of which involve a breakdown in the normal regulation of cellular activities—we must first understand how the normal control mechanisms work. We believe it is realistic to anticipate some major breakthroughs before the end of the twentieth century (Essay 12–2).

SUMMARY

The flow of information from DNA to protein has two parts: transcription and translation. In transcription, the genetic message carried by one or more genes is transcribed from an unwound strand of DNA onto messenger RNA (mRNA). The sequence of bases on the DNA strand determines the order of the mRNA nucleotides. The mRNA in turn serves as a template for translation, the ordering of amino acids into one or more polypeptides.

To translate the base sequence of mRNA into a polypeptide, the mRNA message is "read" in groups of three bases called codons, each different codon specifying a particular amino acid. Transfer RNA molecules link with their "partner" amino acids and move to the mRNA–ribosome complex. There they bind to the ribosome where the three-base anticodon region of the tRNA pairs with the mRNA codon. Since each of the many types of tRNA can have only one type of amino acid linked to it, a given mRNA molecule is always read into a specific amino acid sequence.

For some eukaryotic genes, an additional step in the information flow process is evident: messenger RNA editing. In these cases, the original mRNA transcript is cut up by enzymes to remove intervening sequences of nucleotides, and then the information-bearing pieces are spliced together to yield an edited transcript that directs the synthesis of a polypeptide.

Another complication for some eukaryotes is the phenomenon of gene rearrangements, a mechanism that underlies the millions of different kinds of antibodies that mammals can potentially synthesize. It involves the rearrangement of DNA segments on a chromosome into many possible combinations, each combination yielding a specific mRNA transcript that codes for an antibody polypeptide.

Chemical changes in the base sequence of genes, called mutations, occur infrequently but constantly. They generally involve substitutions, deletions, or insertions of nucleotides that can have minor or major effects on the proteins coded by the altered genes. In terms of their effects on organisms, mutations can be either beneficial, neutral, or harmful. Since they are the source of new alleles, mutations play an important role in the evolutionary process by providing variations among organisms.

Regulatory mechanisms operating at various levels of the information transfer process permit cells to respond to changing environments and guide cellular specialization during the development of multicellular organisms. Transcriptional controls are the most widespread type of regulation of gene expression in bacteria, as exhibited by the lac operon. Translational controls regulate the expression of mRNA into protein. Although the structure and operation of genes is much more complicated in eukaryotic cells, similar mechanisms probably play an important role in the differentiation of cells.

ESSAY 12-2

ONCOGENES AND CANCER: AN EXCITING NEW RESEARCH DEVELOPMENT

Over the past few decades, progress on the front of cancer research has been slow and steady, at least until recently. In the last few years the pace has quickened tremendously, spurred on by recent breakthroughs in molecular biology. As of this month (February 1983), some very prominent cancer researchers have shed their usual cloak of cautious optimism and boldly predicted that we will understand the molecular basis of cancer within three years, maybe less. Armed with such knowledge, researchers will be in an excellent position to devise strategies for the treatment and cure of this deadly disorder.

The excitement in cancer research has been generated along many lines, including a better understanding of the role of viruses in causing some forms of cancer, and recent developments in producing monoclonal antibodies targeted to destroy cancer cells (see Essay 20-1). But clearly the most exciting development is the recent identification of **oncogenes,** segments of DNA whose unregulated expression causes cancer. So far, 15 oncogenes have been found in animals, and 11 of these have counterparts in human chromosomes. According to Robert Weinberg, a leading cancer researcher at MIT, each one of us "carries the seeds of our own cancer in our genes."

In normal cells the oncogenes are either "turned off" (not transcribed), or active only under special controlled circumstances. In a cancer cell, however, an oncogene is "switched on" at the wrong time, or its expression somehow eludes the cell's normal control mechanisms and far too much oncogene-coded protein is produced.

One of the most attractive features of the oncogene theory of cancer is that it serves as a general mechanism through which many cancer-causing agents may operate. Spontaneous mutations, viruses, radiation, and chemical carcinogens can all cause damage to DNA. If the damage leads to the uncontrolled activation of an oncogene, or an altered gene product, the normal cell becomes cancerous. A single base substitution in one oncogene in bladder cells has already been documented as the cause of bladder cancer. And since any change in the DNA of a cell would be passed on through cell division, we can understand why a single cancer cell gives rise to more of its kind, eventually crowding out normal cells.

The two most important questions facing cancer researchers right now are: What activates the oncogenes, and once activated, how can we turn them off? The second question has obvious implications for the treatment and cure of cancer, but it may not be so easy to answer. It is one thing to understand what cancer is and how it operates; it is quite another to find ways to short-circuit its mechanism without simultaneously destroying normal cells. But with the enormous progress made in just a few years, and every expectation that the progress will continue, some of the more optimistic researchers look forward to a cancer-free twenty-first century.

STUDY QUESTIONS

1. Name three structural differences between DNA and RNA.

2. Which type of RNA encodes the information used to order amino acids into a polypeptide?

3. How is transcription similar to DNA replication? How do they differ?

4. Given the following base sequence on the sense strand of DNA, determine the amino acid sequence for which it would code. (Refer to Table 12–1 for the amino acids and their corresponding mRNA codons.)

   G G C T A G C A T A G C C G A T T T

   What would be the amino acid sequence if all the thymine nucleotides were replaced by adenine? What would be the sequence if the second guanine was deleted?

5. Explain the role of transfer RNA in protein synthesis. What roles does the ribosome play?

6. The sense strand of a given DNA gene in a mouse is 3600 nucleotides long, but its coded polypeptide is only 400 amino acids long. Give a reasonable explanation for this apparent discrepancy. What percent of the sense strand is informational?

7. What criteria are applied in determining whether mutations are beneficial, neutral, or harmful?

8. Why do mutations involving a single base insertion (or deletion) often cause such drastic changes in the coded protein?

9. How do you rationalize the apparent paradox that a human liver cell and a human muscle cell, although vastly different in form, function, and types of proteins produced, both possess the same genetic information?

10. How is the lac operon system in *E. coli* like a thermostat-regulated heater (one that automatically switches on when it's cold and turns off when it's warm)?

SUGGESTED READINGS

Abelson, J. "A Revolution in Biology." *Science* 209(1980): 1319. A brief account of how techniques for the sequencing of DNA have led to a variety of important discoveries.

Chambon, P. "Split Genes." *Scientific American* 244(1981): 60–71. A clearly written summary of the phenomenon of messenger RNA editing in eukaryotic nuclei.

Davern, C. I., Introduction. *Genetics: Readings from Scientific American.* San Francisco: W. H. Freeman, 1981. This collection of articles covers DNA structure and the genetic code, and includes discussions of their implications for applied genetics.

Langone, J. "Robert Weinberg: Scientist of the Year." *Discover* January 1983: 37–44. In a tribute to cancer researcher Robert Weinberg, this article summarizes the recent evidence for oncogenes as built-in cancer "switches," and Weinberg's contribution to this theory. Written for the lay person.

Stent, G. S. *Molecular Genetics.* San Francisco: W. H. Freeman, 1978. The author does a good job of weaving the facts into an historical context, and helping the reader interpret experimental results.

Watson, J. D. *Molecular Biology of the Gene.* 4th ed. Menlo Park, CA: Benjamin/Cummings, 1984. This is an excellent advanced text written by the codiscoverer of DNA structure.

UNIT III
Microorganisms and Plants

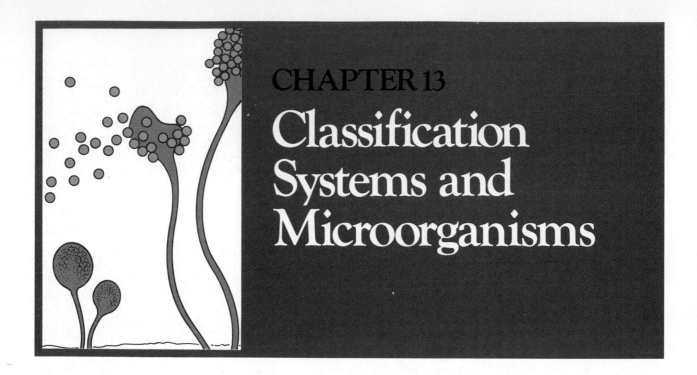

CHAPTER 13

Classification Systems and Microorganisms

Imagine yourself for a moment thrust back in time to an early agricultural society. You must survive on the plants you grow, the animals you have domesticated, and the success of your hunting or fishing. Your tools and other implements come from the rocks, plants, and animals around you. In such a setting you would quickly become familiar with the particular kinds of materials that can be fashioned into arrowheads or cutting tools. You would learn which kinds of plants and animals are edible and which ones are foul-tasting or dangerous. You would soon identify those plants that are useful in constructing shelters. In a relatively short time you would be able to recognize many kinds of organisms, particularly those that were important to your survival and comfort.

The situation just described, though imaginary to you, is a real one for many people. In a study conducted several years ago that compared the species of plants recognized by an aboriginal tribe with those identified by scientists, the natives were as discriminating as the scientists for those species that are important to their daily lives. Their discrimination was less precise, however, for plants that serve them no particular purpose. To the pure scientist, however, all creatures are of interest, whether useful or not. The branch of biology called **taxonomy** or **systematics** presently recognizes about 350,000 species of plants and over 1,250,000 species of animals.

THE EARLY CLASSIFICATION SYSTEMS

The earliest forms of classifying organisms, particularly with respect to plants, were utilitarian, emphasizing medicinal and other practical properties (Figure 13–1). Beginning in the Renaissance, the emphasis of botanical and zoological classification gradually changed as utilitarianism gave way to increased interest in the intrinsic properties of organisms.

The culmination of this development was the work of the Swedish scientist Carolus Linnaeus (1707–1778), a great pioneering taxonomist. Although primarily a botanist, Linnaeus also classified animals, rocks, and even diseases. His most important work was *Species Plantarum* (1753), a book in which he named and described 7300 plant species. Linnaeus's system of classifying plants is sometimes called a sexual system because he grouped the species according to the number of male and female parts within their flowers. Using this system and Linnaeus's books, anyone who knew the parts of a flower could identify an unknown plant, so the scheme stimulated much interest in the botanical sciences. It was strictly artificial, however, for it did not necessarily group related species in the same general category. This did not bother the scientists of the time, however, because

FIGURE 13-1
A Medicinal Plant? Many early botanists were also physicians who used plant-derived drugs to treat various ailments. Note the suggested medicinal value of this plant growing in Kew Gardens near London.

MIRABILIS JALAPA
"MARVEL OF PERU"
"PERUVIAN LILY"
TWO DRAMS OF THE ROOTE TAKEN
INWARDLY DOTH VERY NOTABLY
PURGE WATERISH HUMOURS"
PARKINSON 1629.

most of them believed that species arose by special creation. Thus, they were not concerned with classifying organisms on the basis of their ancestry.

THE MODERN CLASSIFICATION SYSTEM

The foundation of modern classification systems derives from the pioneering work of Charles Darwin. In his famous book *The Origin of Species* published in 1859, Darwin presented his **theory of evolution,** which states that all present-day species are derived from earlier species and all organisms, past and present, share a common ancestry. Darwin's theory of evolution (see Chapter 28) has become the most unifying theme in biology, and it is the organizing principle of modern taxonomy. Thus, taxonomists today classify organisms in a way that reflects their biological ancestry. The ancestral relationships are complex, so classification schemes are often complex. Nevertheless, a general understanding of how organisms are classified can provide considerable insight into the unity and diversity of life.

All organisms presently are classified and named according to an international system of criteria first established in the early part of this century. Since then, the system has been revised and updated regularly at scientific meetings. The rules make no attempt to pass judgment on the acceptability of a name proposed for a newly discovered organism. Rather, they establish the procedure that must be followed and the categories to be used when an organism is classified and named. These rules apply only to formal scientific names, not to common names.

You may remember that the scientific name of the fruit fly studied by Thomas Hunt Morgan is *Drosophila melanogaster* (see Chapter 10). The first element of the name is the **genus** (*Drosophila*); the second element (*melanogaster*) is the **specific epithet.** Thus, the genus and the specific epithet together form a binomial, which is the name of the **species.** The genus *Drosophila* includes several different types of fruit flies besides the *D. melanogaster* species commonly used for research. The species is usually the finest category into which scientists classify most organisms, although subspecies are occasionally recognized. The generally-accepted criterion used in defining a species is that organisms of the same species can interbreed under natural conditions to yield fertile offspring. Individuals of different species normally do not mate, and if they do, the mating is either unsuccessful or the offspring is sterile. For example, a horse (*Equus caballus*) can be mated to a donkey (*Equus assinus*) to produce a mule, but mules are sterile and cannot reproduce. Therefore, the horse and donkey are classified as different

Alligator	Category	Wheat
Animalia	Kingdom	Plantae
Chordata (chordates)	Phylum Division	Tracheophyta (vascular plants)
Vertebrata (vertebrates)	Subphylum Subdivision	Spermatophytina (seed plants)
Reptilia (reptiles)	Class	Angiospermae (flowering plants)
Archosauria (archosaurs)	Subclass	Monocotyledoneae (monocots)
Crocodilia (crocodiles, alligators, and gavials)	Order	Commelinales (grasses, sedges, rushes, bromeliads, spiderworts)
Crocodylidae (crocodiles, alligators)	Family	Poaceae (grasses)
Alligator	Genus	*Triticum*
A. mississipiensis	Species	*T. aestivum*

FIGURE 13–2
Classification of an Alligator (*Alligator mississipiensis*) and a Wheat (*Triticum aestivum*).

species. A quarterhorse and a thoroughbred, on the other hand, can mate and produce fertile offspring, and so are classified as the same species—*E. caballus.*

Although the horse and donkey are recognized as separate species, they share many features in common that indicate they are closely related ancestrally. Thus, we recognize this close relationship by classifying both in the same genus (plural genera). As we move up the taxonomic scale, similar genera are grouped within a single **family,** similar families within an **order,** and similar orders within a **class.** Ancestrally-related classes of plants and fungi are grouped into **divisions,** and similar classes of animals and animallike organisms are grouped into **phyla** (singular **phylum**). The largest and most general category is the **kingdom.** Thus, the hierarchy of classification categories is as follows:

Kingdom
  Phylum or Division
    Class
      Order
        Family
          Genus
            Species

Subgroups are occasionally designated for each of these categories, such as subkingdom, subphylum, and so forth. Examples of how this system works are given in Figure 13–2.

Although scientific classification may seem rather rigid, in fact it is continually revised and reevaluated as new information becomes available. For instance, prior to 1969 all organisms were recognized as either animal or plant (Figure 13–3). This system functioned quite well to distinguish elephants from oak trees, but a number of organisms caused concern among the taxonomists. *Euglena,* for example, is a genus of single-celled freshwater organisms that swim about by moving their whiplike flagella (Figure 13–4). Some species of *Euglena* are photosynthetic and have chlorophyll pigments chemically identical to those in higher plants. Other species lack chlorophyll and feed by ingesting organic matter from their aquatic environment. And species of *Euglena* do not synthesize starch (like plants) or glycogen (like animals), but rather store carbohydrate in an entirely different form. In short, these organisms exhibit some plantlike properties and some animallike properties, in addition to some properties that characterize neither group.

FIGURE 13–3
The Two-Kingdom System. Until 1969, most biologists assigned all organisms to either the animal or the plant kingdom on the basis of the characteristics listed here.

Animals	Plants
No chlorophyll	Chlorophyll
Heterotrophic	Autotrophic
Motile	Nonmotile
No cell wall	Cell wall
Limited growth in higher forms	Continuous growth in many species
Glycogen as storage carbohydrate	Starch as storage carbohydrate
Centrioles	No centrioles
Most have nervous system	No nervous system

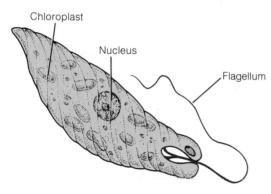

FIGURE 13–4
Euglena. This organism has both plantlike characteristics (such as chloroplasts) and animallike characteristics (such as motility).

Recognizing the problems that creatures like *Euglena* create, most biologists now divide all organisms into five kingdoms: **Animalia, Plantae, Fungi, Protista,** and **Monera** (Table 13–1 and Figure 13–5). According to the five-kingdom system, kingdom Animalia now includes only the heterotrophic, multicellular animals, most of which obtain their food by ingestion (eating). Higher forms have sophisticated nervous and muscle systems for sensing and responding to their environment. The animals will be discussed in detail in Chapters 18–27.

Kingdom Plantae includes the algae, mosses, liverworts, ferns, and all seed-producing plants. These organisms manufacture their own food by photosynthesis and they lack special sensory, nervous, or muscle systems. The plants will be discussed in Chapters 14–17. Organisms that are neither fully plantlike nor fully animallike fall into the kingdoms Monera, Protista, and Fungi. We will dedicate the balance of this chapter to a discussion of the organisms comprising these three less-familiar kingdoms.

TABLE 13–1
Major Characteristics of the Five Kingdoms.

Kingdom	Includes	Cell Type	Type of Nutrition	Type of Motility	Cell Wall Composition	Body Form
Monera	Bacteria and cyanobacteria	Prokaryotic	Autotrophic or heterotrophic	Mostly nonmotile, some with flagella	Peptidoglycan (repeating units of a peptide and sugars)	Unicellular (solitary, colonial, or filamentous)
Protista	Slime molds, amoebas, sporozoans, ciliates, and diatoms	Eukaryotic	Autotrophic or heterotrophic	Some with flagella or cilia; some amoeboid	Absent in most; silica in diatoms	Unicellular (solitary or colonial)
Fungi	Yeasts, molds, mildews, mushrooms, and others	Eukaryotic	Heterotrophic by food absorption	Mostly nonmotile	Chitin or cellulose	Mostly filamentous, some solitary unicellular
Plantae	Algae, mosses, liverworts, ferns, and seed-producing plants	Eukaryotic	Autotrophic by photosynthesis	Mostly nonmotile	Cellulose	Multicellular
Animalia	Invertebrates and vertebrates	Eukaryotic	Heterotrophic by food ingestion	Mostly motile	Absent	Multicellular

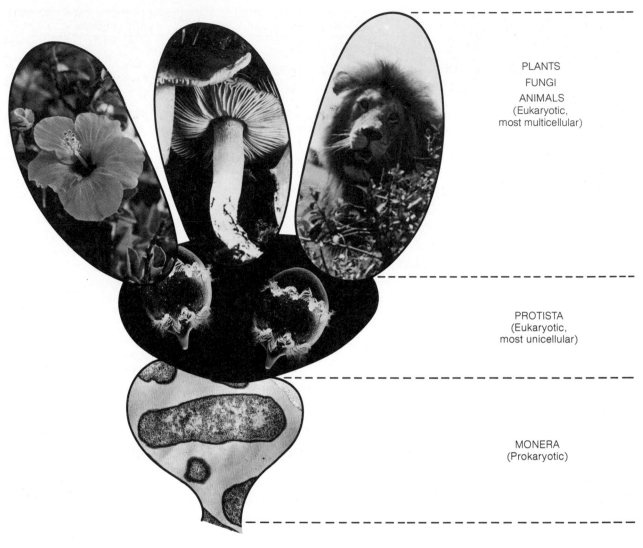

PLANTS
FUNGI
ANIMALS
(Eukaryotic,
most multicellular)

PROTISTA
(Eukaryotic,
most unicellular)

MONERA
(Prokaryotic)

FIGURE 13–5
The Evolutionary Relationships among the Five Kingdoms.

KINGDOM MONERA

Kingdom Monera is composed exclusively of prokaryotic organisms. (Recall that the cells of prokaryotes lack a nucleus and organelles.) In fact, because the architecture of their cells is so different from that of all other organisms (eukaryotes), this feature is the sole criterion for separating monerans from the other kingdoms. The two major groups of monerans are the bacteria and the cyanobacteria.

Bacteria

To study bacteria one must think small, for most of the nearly 2000 known species are less than 2 micrometers in diameter. All bacteria are unicellular, but in certain species the cells tend to stick together, giving the impression of a multicellular mass or filament even though each cell actually functions independently. Figure 13–6 illustrates the basic shapes of bacterial cells: spherical, rod-shaped, and spiral-shaped. Refer to Figure 3–8 for the basic structure of a bacterial cell.

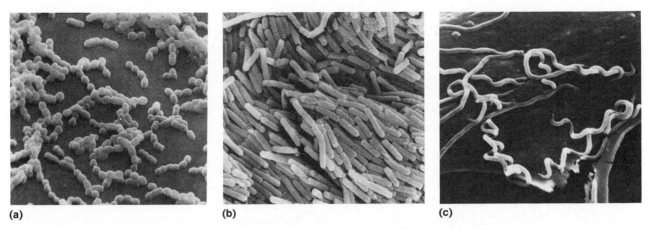

(a) (b) (c)

FIGURE 13–6
The Shapes of Bacteria. (a) Spherical, (b) rod-shaped, and (c) spiral-shaped bacteria.
(Magnifications: (a) 2500×; (b) 1250×; (c) 4000×.)

Although most bacteria are heterotrophic and absorb food from their environment, some are photosynthetic. Bacterial photosynthesis, however, is quite different from the process we described for plants (see Chapter 6). Almost all photosynthetic bacteria are anaerobic and do not produce oxygen as a by-product. In fact, exposure to oxygen kills them.

Bacteria usually reproduce by asexual fission (see Figure 7–1): once the bacterial DNA is replicated, the two identical DNA molecules are partitioned to opposite sides of the inward-furrowing plasma membrane, creating two genetically identical daughter cells. Under ideal conditions, some bacteria can undergo this process every 20 minutes. Usually reproduction

is not nearly this rapid because the cells are limited by food supply and by the accumulation of waste products in their surroundings. Nevertheless, the ability of bacteria to divide rapidly plays a major role in food spoilage and the spread of disease.

Under rare circumstances, DNA from one bacterial cell can be transferred to another cell (Figure 13–7). Since such a transfer can alter the genetic makeup of the recipient cell, it can be viewed as a special kind of sexual reproduction. However, fertilization and meiosis, as we know them in eukaryotes, do not occur among the Monera.

As you know, many types of bacteria cause diseases in humans and other organisms, including

FIGURE 13–7
Transfer of Genetic Information between Bacterial Cells. When this electron micrograph was prepared, these two different strains of bacteria were about to transfer DNA through the cytoplasmic bridge connecting them.

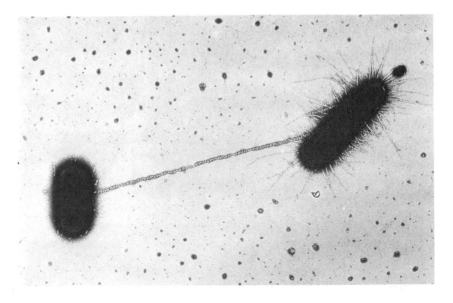

tuberculosis, diphtheria, pneumonia, typhoid fever, tetanus, syphilis, gonorrhea, and various types of food poisoning. Despite the bad reputation that the disease-causing species have given this group, many bacteria are beneficial. One of their most important roles is the recycling of organic matter on earth. Bacteria in the soil, oceans, lakes, and ponds are agents of decay, breaking down organic compounds in dead organisms into inorganic substances—carbon dioxide, water, ammonia, and mineral ions—that can be reused (see Chapter 31). If this recycling did not occur, essential nutrients would remain locked up in dead bodies and be unavailable to plants. Thus, death and decay are important to the continuation of life on earth.

Several types of bacteria have commercial applications. Some are used in the production of vinegar, yogurt, and some types of cheeses, and a group of bacteria called the actinomycetes are a major source of antibiotics used in medicine. Other species of bacteria serve as "guinea pigs" in recombinant DNA studies (see Chapter 11). Through the genetic manipulation of bacterial plasmids, important biological products (such as insulin and interferon) are manufactured in large quantities at a relatively small cost. This new technology will almost certainly bring about an explosion of more commercial uses of bacteria in the near future.

Finally, several bacterial species, most notably *Escherichia coli,* can be found in the intestines of mammals, including humans. They survive on the unabsorbed nutrients that pass through the digestive system. In return the animal obtains certain vitamins synthesized by the bacteria (see Chapter 21).

Cyanobacteria

The cyanobacteria (*cyano* = blue) were formerly known as the blue-green algae because of their color and plantlike form of photosynthesis. The new term recognizes that these organisms are more closely related to the bacteria than to other algae, particularly with respect to their prokaryotic cell structure.

Cyanobacteria of various kinds can be found in almost any damp environment—in salt water and fresh water, in moist soil, on damp rocks and tree trunks, and even in hot springs with temperatures up to 85°C (185°F). Many species are unicellular and solitary; others form round colonies consisting of several to many cells; still others are filamentous (composed of cells attached end to end like beads on a string) (Figure 13–8a).

As prokaryotes, the cyanobacteria lack an organized nucleus. Their genetic information is stored in

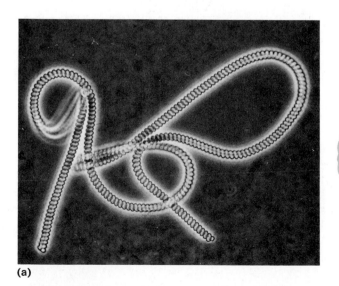

(a)

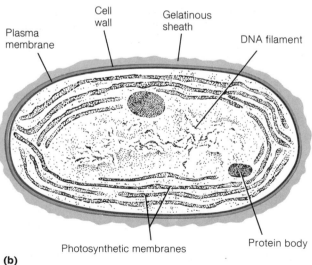

(b)

FIGURE 13–8
Cyanobacterial Cells. (a) These cyanobacterial cells (*Spirulina*) form long filaments. (b) This diagram of a typical cyanobacterial cell shows the outer covering, which consists of a gelatinous sheath, surrounding a more rigid cell wall. Note the photosynthetic membranes and centrally-dispersed DNA filaments.

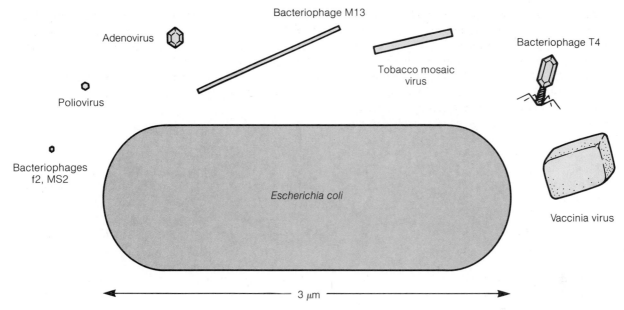

Adenovirus

Bacteriophage M13

Poliovirus

Tobacco mosaic
virus

Bacteriophage T4

Bacteriophages
f2, MS2

Escherichia coli

Vaccinia virus

3 μm

FIGURE 13–9
Comparative Sizes of Several Viruses and the Bacterium *Escherichia coli*.

many relatively short filaments of DNA. As in plants, cyanobacterial photosynthesis produces oxygen as a by-product and it takes place within an extensive system of membranes located near the outer edge of the cell's protoplasm (see Figure 13–8b and 3–9). Biologists believe that cyanobacteria living more than 2 billion years ago were the main source of molecular oxygen that so dramatically changed the earth's atmosphere, ultimately permitting the evolution of eukaryotic organisms (see Chapter 2).

Reproduction in cyanobacteria is asexual, often occurring by a fission process much like that described for bacteria. In filamentous forms, new individuals can arise by simple fragmentation of the parent filament.

Many cyanobacteria play important ecological roles. They are a food source for many other small, aquatic organisms. Also, certain species convert atmospheric nitrogen (N_2) into ammonia (NH_3), an activity particularly important in rice fields where they provide usable nitrogen for the rice plants. Cyanobacteria can be pests as well. In aquatic environments having a high mineral content, often caused by heavy sewage contamination or fertilizer run-off, cyanobacteria multiply rapidly into a large population. When these cells die, a second explosion of decay bacteria follows, which feed by decomposing the dead cyanobacteria. In high concentrations, the bacteria produce toxins and deplete the dissolved oxygen, rendering the hab-

itat unfit for fish and other life forms. This poisoning of the aquatic environment is a common symptom of polluted ponds and lakes.

VIRUSES

Viruses are tiny, noncellular particles consisting of nucleic acid (DNA or RNA) and a protein coat. They come in a variety of shapes and sizes, but all are considerably smaller than a typical cell (Figure 13–9). All viruses are obligatory parasites—they cannot replicate outside of a living host cell because they have no biosynthetic machinery of their own. Because of their absolute dependence on their host cell's protoplasm, some biologists do not consider viruses to be living organisms. We describe them here near the monerans because of their extremely simple design, and for lack of a better place to put them.

Unfortunately, many viruses cause disease in their hosts. In humans, viral diseases include influenza, polio, chicken pox, the common cold, mumps, measles, infectious hepatitis, some types of cancer, and of particular concern to sexually-active individuals, herpes (Essay 13–1). Treatment of viral infections is made difficult by the fact that viruses are so simple in design. Any medication that breaks their reproductive cycle must do so by interfering with the metabolism

ESSAY 13-1

HERPES: A VIRUS ON THE RAMPAGE

A pregnant woman enters labor and learns that her birth canal harbors an infectious disease. A caesarean section must be performed to spare her infant from an often fatal infection in newborns. A teenager stares into the mirror at the painful and unsightly cold sore on his lip. A young man who suffers from recurrent outbreaks of genital lesions learns that he has an incurable disease. These people with diverse symptoms have something in common: they all have herpes. Herpes infections are not only incurable at the moment, but they stay in the body forever, bouncing back episodically in the form of skin eruptions.

Herpes is a virus called *Herpes simplex*. It lives and reproduces in various regions of the body, primarily skin tissues, where it destroys cells. At least five different types of herpes viruses can infect humans, of which two are quite common. Type 1, or oral herpes, usually shows up in cold sores on the lips. In rare instances, oral herpes can cause a deadly form of meningitis, and type 1 infections of the eye are a leading cause of infectious blindness. Type 2 herpes tends to "hit below the belt." Also known as genital herpes, this variant causes painful blisters on the thighs, buttocks, and genitals. Estimates put the number of Americans having genital herpes somewhere between 10 and 20 million, making it the number one sexually transmitted disease of today. And because it is incurable, the numbers keep growing—about 500,000 new cases appear every year in the United States.

The sores caused by both oral and genital herpes are similar. Their arrival is first signaled by an itching or tingling sensation in the skin. Within a day or two, red bumps appear and these develop into painful blisters filled with a virus-laden fluid. As the body's immune system responds to the intruder, the blisters accumulate pus and eventually rupture, leaving open sores that are painful and highly contagious. The sore then crusts over and begins to heal. For initial infections of genital herpes, the first bout lasts about three weeks. Subsequent outbreaks last only about five days.

Herpes is most commonly transmitted by direct contact between a blistered area and noninfected skin regions. The virus can be transferred by a kiss, touch, or sexual intercourse. Even though oral and genital herpes are recognized as different strains of the virus, they seem to be interchangeable. For example, type 1 herpes in a cold sore on the lip can be transmitted as genital herpes to another individual during oral sex. Thus, during the blister and open sore stages of infection, any type of physical contact between the infected area and other areas of the skin, particularly the eyes and genital areas, should be avoided. After treating a cold sore or other lesion with an ointment or drug, the hands should be washed with soap and water.

Herpes is devious and tenacious. Once the sores have healed, the virus particles that have escaped attack by the immune system migrate through nerve cells and "hibernate" within nerve ganglia deep within the body. Oral herpes usually retreat to ganglia located at the base of the brain. Genital herpes hide in the sacral ganglia at the base of the spine. Here they are safe from the body's immune attack. It has been estimated that up to 50% of the American population has latent oral herpes, and 10–15% have genital herpes sequestered in the sacral ganglia. What brings them out of hiding to cause the recurrent skin eruptions is still a mystery, but contributing factors may include overexposure to the sun, emotional stress, poor nutrition, and a generally run-down condition.

At the present time there is no cure available for herpes. On the bright side, however, researchers in Britain have recently reported (June 1983) a very effective treatment which, if it withstands rigorous government testing, could become available in the United States within five years. Frustrated by the lack of a medical cure, many herpes sufferers turn to home remedies that include buttermilk, herbs, seaweed, earwax, peanut butter, and baking soda. The best a home remedy can offer is to ease the discomfort of the painful sores. The pain of the lesions can be lessened by keeping the sores clean and dry, and the discomfort of genital herpes can be reduced by wearing loose, comfortable clothing. Counseling sessions to ease the anxieties and emotional stresses of herpes sufferers are available throughout the country.*

*The Herpes Resource Center (260 Sheridan Avenue, Palo Alto, CA: 415-328-7710) provides information about herpes, support groups, and a telephone information, counseling, and referral service. The VD (Venereal Disease) National Hotline can be dialed toll-free 8:00 A.M. to 8:00 P.M. weekdays and 10:00 A.M. to 6:00 P.M. weekends. The number is 1-800-227-8922 (in California, 1-800-982-5883).

of their host cells. Furthermore, they are insensitive to the bacteria-killing antibiotics. For these reasons, treatment of viral infections is generally limited to relieving the symptoms or bolstering the immune system.

Viruses are host-specific; that is, each type of virus has a limited range of cell types that it can infect. Human viruses specialize in humans, bacteriophages (or simply *phages*) specifically invade bacterial cells, and so on. The degree of host-specificity for viruses varies tremendously, however. For example, the T4 phage infects only *E. coli* bacteria. In contrast, the rabiesvirus can attack and be passed among many different mammals, including humans.

The formation of new viruses begins when the viral nucleic acid is injected into a host cell (see Figure 11–3). In the case of DNA viruses, the DNA serves as a template both for its replication and its transcription into viral messenger RNAs. The mRNA transcripts are then translated into viral proteins, among which are the structural components of the protein coat and enzymes that function in the assembly of these coat proteins and viral DNA into new viruses. Once released, each new virus can repeat the entire sequence by infecting another host cell. All of the metabolic events associated with viral replication require the metabolic components of the host cell.

KINGDOM PROTISTA

Because it includes a wide variety of organisms, some of which exhibit both plantlike and animallike characteristics, the kingdom Protista has caused the most trouble for taxonomists. The current consensus favors classifying most eukaryotic unicellular and colonial organisms as protistans. (Recall that the cells of eukaryotic organisms have a distinct nucleus and membrane-bound organelles.) Thus, the problem of classifying creatures such as *Euglena* is solved rather neatly: it is neither a plant nor an animal, but a protistan. Other problems do arise, however. For example, slime molds are unicellular and mobile at one stage of their life cycle, gathering up food from their environment by absorption (Figure 13–10). When they exhaust their food supply, however, the individual cells aggregate into a multicellular blob that eventually produces a plantlike stalk. Consequently, the slime molds do not fit perfectly into any kingdom. Should they have their own separate kingdom? If so, how many other kingdoms are necessary to accommodate other very specialized life forms?

No doubt friendly debates will continue for some time over what are the most appropriate criteria for classifying the protistans. Given that viewpoints differ,

FIGURE 13–10
The Life Cycle of a Slime Mold. The slime mold *Dictyostelium discoideum* inhabits moist forest soils where there is decaying plant material. Its life cycle begins when tiny spores germinate and grow into large, amoeboid cells. When the food supply becomes limiting, these cells aggregate into organized clumps that occasionally take the form of a slug. Eventually a fruiting body grows upward from the mass, terminating in a spore-producing cap. Spores are released and the life cycle begins anew.

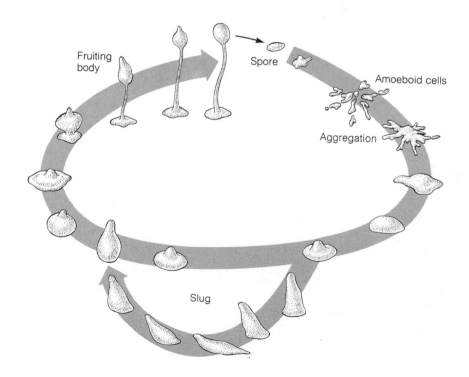

Fruiting body

Spore

Amoeboid cells

Aggregation

Slug

our brief discussion here will focus on just four of the least ambiguous groups of protistans. Three of these are animallike (amoebas, sporozoans, and ciliates), and the fourth is plantlike (diatoms).

Amoebas

Amoebas are unicellular organisms best known for their irregular and constantly changing shape (Figure 13–11). Amoebas and their relatives, collectively called the **Sarcodina,** move through the pond or lake bottoms in which they live by extending and retracting their cytoplasm in temporary "arms" called **pseudopods** (*pseudo* = false; *pods* = feet). Many amoebas live mainly on a diet of smaller protistans, bacteria, and other small cells, which they surround with a pseudopod and then ingest by phagocytosis (see Figure 4–12).

Most amoebas are bound by a membrane only. However, one group of amoebalike organisms—the **radiolarians**—secrete a hard shell that is often quite intricate (Figure 13–12). Pseudopods protrude through holes in the shell and trap food on their sticky surface. When radiolarians die, their shells remain intact and, in certain regions of the earth, contribute to limestone or chalk deposits.

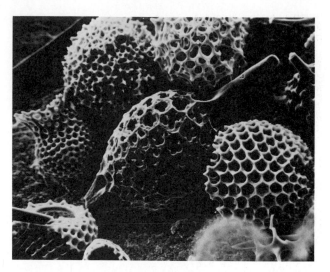

FIGURE 13–12
Radiolarians. Radiolarians produce hard, delicately structured shells through which pseudopods can extend. (Only the shells can be seen in this micrograph.)

Sporozoans

The sporozoans are unicellular internal parasites that lack a means of locomotion. They have quite complex life cycles which ensure that they survive, reproduce, and are passed from host to host. One example of a sporozoan is *Plasmodium,* the organism responsible for malaria. *Plasmodium* is transferred to humans by *Anopheles* mosquitos, which live mostly in the tropics. When an infected mosquito bites a human, it disperses *Plasmodium* cells into the individual's bloodstream In the human host, the *Plasmodium* cells produce a toxin that causes the characteristic chills and fever of malaria. Eventually some *Plasmodium* cells form gametes by meiosis, which may enter another mosquito that bites the human carrier. In the newly-infected mosquito, the gametes fuse into a zygote that eventually reproduces asexually, forming new cells ready to infect another human being.

Ciliates

Whereas amoebas move very slowly by creeping pseudopods, and sporozoans cannot move on their own at all, the ciliates swim relatively rapidly by the coordinated beating of many cilia distributed over at

Pseudopods

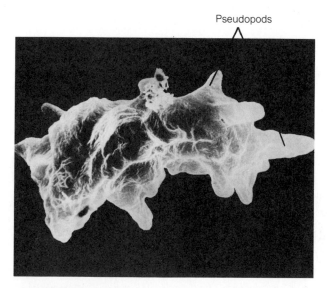

FIGURE 13–11
An Amoeba. The cytoplasm of an amoeba streams into temporary pseudopods for locomotion and food capture.

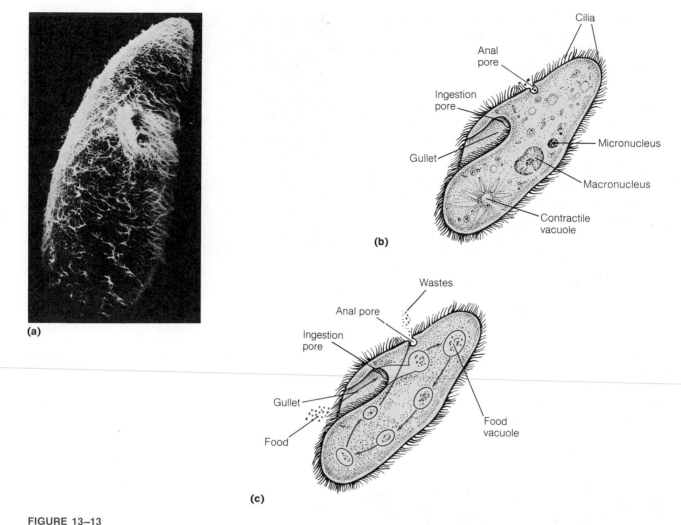

FIGURE 13–13
Paramecium. (a) This scanning electron micrograph emphasizes the ciliated surface of a *Paramecium*. (Magnification: 490×.) (b) Although it is only a single cell, *Paramecium* has specialized structures for food intake (ingestion pore), release of undigested materials (anal pore), and the regulation of water balance (contractile vacuole). Protein synthesis is associated with the macronucleus; the micronucleus is involved in sexual reproduction. (c) This diagram illustrates the intake and processing of food particles in *Paramecium*.

least part of their bodies (see Figure 3–1c). *Paramecium,* probably the most familiar of the ciliates, lives in ponds (Figure 13–13a). Cilia not only propel this slipper-shaped creature through the water, but also serve to sweep food particles into its gullet (Figure 13–13b, c). *Paramecium* and other ciliates also have contractile vacuoles to expel water that constantly enters the cell by osmosis.

Many of the ciliates have a complex internal orga-

nization; a few (such as *Paramecium*) even have two different types of nuclei. Some biologists therefore prefer to regard them as acellular rather than single-celled creatures—that is, as organisms that have evolved subcellular units within a "cell-less" body. The details of this argument are beyond our scope here, but there is no doubt that the ciliates, if we choose to call them single-celled, exhibit the most complex organization of any living cells.

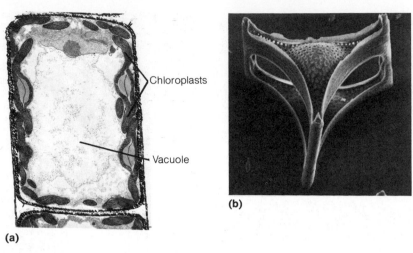

FIGURE 13–14
Diatoms. (a) Electron micrograph of a diatom showing many chloroplasts. (b), (c) These scanning electron micrographs of two diatom species show the elaborate patterns of their siliceous cell walls.

Diatoms

Diatoms are plantlike protistans that are found in both fresh and salt water. All diatoms are photosynthetic and contain chlorophyll in well-organized chloroplasts (Figure 13–14a). The most unique feature of diatoms is their rigid cell wall made of silica. This wall is often quite ornate, consisting of two almost identical halves, one of which is slightly larger than the other. The two halves fit together almost like the shell of a clam or the upper and lower halves of a petri dish. Each half-wall (or valve) has numerous pores that allow for direct contact between the cell membrane and the outside environment (Figure 13–14b, c).

During the millions of years that diatoms have inhabited the earth, their virtually indestructible walls have collected in great abundance in certain regions. One diatom deposit in Lompoc, California, is over 3000 feet thick. These collections of diatomaceous earth are mined and used in water filters, as abrasives for polishing, and as strengthening agents in paper, insulation, paint, and many other products.

KINGDOM FUNGI

About 100,000 species of fungi are currently recognized, ranging from tiny yeasts and water molds to large mushrooms, morels, and bracket fungi (Figure 13–15). Probably thousands more species are still undiscovered. Kingdom Fungi is so diverse that it is difficult to pick out characteristics common to all of its members. Nevertheless, we can cite several general distinctions of this group. First, most fungi are filamentous (the unicellular yeasts are an exception). Although their filaments, called **hyphae,** do not form complex tissues like those found in higher plants and animals, some species display rather large and complex networks of tangled hyphae. The conspicuous structures of mushrooms, morels, and bracket fungi are examples (see Figure 13–15). Second, many fungi at some point in their life cycle are **heterokaryotic,** a condition in which genetically different nuclei coexist within the cells of a single individual. The heterokaryotic state results from the fusion of hyphae from two different individuals, after which the nuclei from one flow into the cytoplasm of the other. Heterokaryon formation is part of sexual reproduction in these fungi. Third, the cell walls of many (but not all) fungi are composed of **chitin,** the same tough polysaccharide that is found in the hard shells of lobsters, crabs, insects, and other arthropods. Finally, all fungi are heterotrophic and absorb nutrients from the organic material on which they grow. To obtain nutrients, the fungus first secretes digestive enzymes that degrade the large organic polymers. The fungal cells then absorb the digestion products (sugars, amino

FIGURE 13–15
Fungi. (a) The thorax of this mosquito larva is heavily infected with sporangia of an aquatic fungus. (b) *Amanita,* an extremely poisonous mushroom. (c) Morels, considered a delicacy by many, are often found on rotting tree logs. (d) Bracket fungi derive their nutrition from tree bark.

acids, and so on) and other small compounds to support their metabolism.

Some fungi are **parasitic,** living on or within another organism to the latter's detriment. The fungi that cause athlete's foot, ringworm, late blight of potatoes, and the mildews that attack grapes and roses are examples. But like many of the heterotrophic bacteria, the majority of fungi are decomposers. They are particularly evident in the warm, moist areas of the world (such as the tropics), where they will grow on virtually anything organic: books, crops, wood, leather, clothes, even crude oil.

In addition to their important role as decomposers, fungi are also important to humans in the production of many beverages and foods. *Saccharomyces cerevisiae,* or brewer's yeast, is the fermenting agent in the production of wine and beer, and the leavening agent that causes bread dough to rise. Certain species of *Penicillium* produce the characteristic flavors of blue cheese, Roquefort, Camembert, and Gorgonzola cheeses; *Penicillium chrysogenum* is a source of penicillin, the bacteria-killing antibiotic. Many fungi are edible, such as the common field mushroom, *Agaricus*

campestris, some of the morels, and the much sought-after truffle, which grows underground in association with roots of forest trees. But with the good there is always some bad. About 70 species of mushrooms are poisonous, the most dangerous of which belong to the genus *Amanita* (see Figure 13–15b). Some of the *Amanita* species look very much like edible mushrooms, and making the distinction requires an experienced eye, careful attention to detail, and a good mushroom guidebook. The only rule of thumb is simply, "When in doubt, leave it alone." If you make a mistake, it could be your last.

The fungi's success as a biological group is due in large part to their enormous powers of propagation, particularly by asexual means. Yeasts and a few other unicellular forms reproduce asexually by **budding** (see Figure 7–9a). Many filamentous fungi propagate by **fragmentation** of their hyphae, whereby each fragment becomes a new individual. But the most widespread form of asexual reproduction in fungi is the production of **asexual spores.** A spore is a reproductive unit (usually a single cell) that is capable of giving rise to a new individual. The common bread mold, *Rhizopus,* produces spores within a special structure called a sporangium. Tens of thousands of spores may be produced per square inch of hyphae (Figure 13–16a). Because of their tiny size, spores become air-borne and are easily blown around by the slightest of air currents. It is no wonder that any slice of exposed bread can turn into a hairy mat of *Rhizopus* in just a few days, producing millions more of the tiny black spores.

Some form of sexual reproduction occurs in most

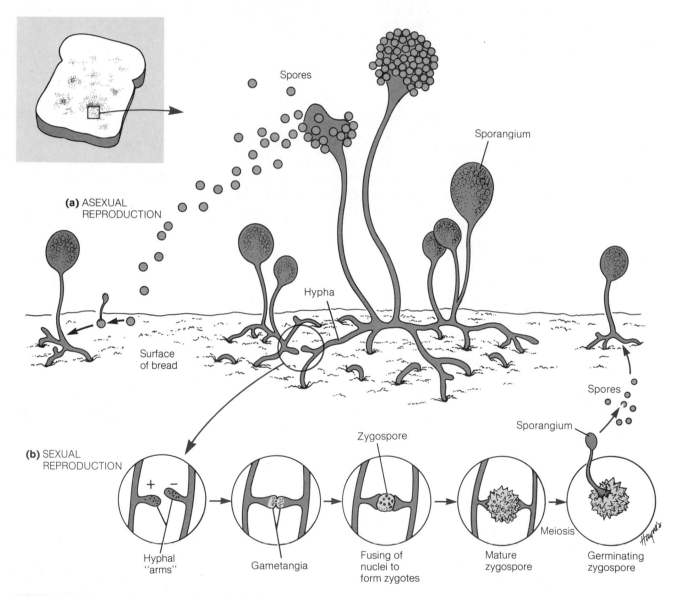

FIGURE 13–16
The Life Cycle of the Common Bread Mold (*Rhizopus*). (a) The spread of *Rhizopus* over the surface of bread occurs by growth of the hyphae and the production of asexual spores. The spores are borne on the surface of sporangia (''spore-bearing sacs'') that extend upward from the hyphae. Each spore is capable of growing into a new system of hyphae and sporangia. (b) Sexual reproduction occurs when the hyphae from opposite mating types (+ and −) meet and form gametangia (gamete-bearing sacs). The gametangia fuse and + and − nuclei unite to form several zygotes enclosed within a single cell called the zygospore. The zygospore develops a thick, ornate cell wall and may remain inactive for an extended period of time. Eventually the zygotes undergo meiosis and, in the simplest of cases, a single haploid cell germinates to form a sporangium.

fungi, though there are quite a few species for which no sexual life cycle has been observed. The details of sexual reproduction displayed in this kingdom are almost as diverse as the fungi themselves, so we will restrict our coverage to the pattern for *Rhizopus* and related molds. The conspicuous structures of the bread mold—hyphae, sporangia, and spores—all contain haploid nuclei. The nuclei in the hyphae are potential gametes under the appropriate conditions—that is, when the hyphae from two individuals of opposite sexes, or **mating types** (designated + and − types), grow near one another. The proximity of + and − hyphae causes a chemical signaling event that triggers the outgrowth of cytoplasmic "arms" from the two filaments toward each other (Figure 13–16b). When these arms meet, a crosswall forms at the tip of each, creating a **gametangium** (gamete-bearing sac) that contains several haploid nuclei (now called gametes). The two gametangia then fuse and the + and − nuclei unite to form diploid nuclei (zygotes). The fused gametangia, now called a **zygospore,** enlarges and develops a thick, spiked cell wall. The zygospore is resistant to harsh environmental conditions and can remain inactive for long periods of time. If conditions are favorable for growth, the zygotes contained within undergo meiosis, yielding many haploid nuclei. In the simplest of cases, one of these nuclei divides mitotically many times and directs the formation of a sporangium, which grows out of the zygospore. Many of the nuclei migrate into the sporangium, where each becomes encapsulated in a small amount of cytoplasm and a cell wall to become a spore. When released, each spore can grow into a new filamentous mold.

The scientific names of organisms are composed of their genus and specific epithet names.

Until 1969, biologists recognized only two kingdoms of organisms: plants and animals. Most now believe that the evolutionary relationships among organisms is best represented by a five-kingdom system: Monera, Protista, Fungi, Plantae, and Animalia.

Kingdom Monera is composed exclusively of prokaryotic organisms, and includes the bacteria and cyanobacteria. Most are unicellular but some species form simple filaments or colonies. All the cyanobacteria and some of the bacteria are photosynthetic; the remaining bacteria are heterotrophic absorbers that either parasitize living organisms or decompose dead ones. Viruses, considered by some to be nonliving, consist only of nucleic acid and a protein coat. All viruses require the biosynthetic machinery of a living host cell to replicate.

Members of the kingdom Protista are all eukaryotic and mostly unicellular organisms. Some of the protistan groups have plantlike features (such as the diatoms); others resemble the animals (such as the amoebas, sporozoans, and ciliates). Still others, like *Euglena,* have both plantlike and animallike characteristics.

The fungi are also eukaryotic and, with few exceptions (such as the yeasts), are multicellular and filamentous in form. All fungi are heterotrophic and absorb nutrients from the organic substrate they live on. Some are parasitic but most fungi are decaying agents. The latter, together with the decay-causing bacteria, play an important role in the recycling of organic material in dead organisms. The fungi are prolific asexual reproducers, and most have a sexual stage in their life cycles.

SUMMARY

Long before they kept written records, people named and classified organisms in ways that reflected their impacts on human activities. Over time, the rationale and details of classification schemes have changed dramatically. With the general acceptance of evolutionary theory in the mid-nineteenth century, taxonomists began to group organisms in ways that reflected their ancestral relatedness. Modern classification systems include this approach, and also make use of Linnaeus' proposed hierarchy of categories. From the broadest to the most specific, these categories are kingdom, phylum (in animals) or division (in plants and fungi), class, order, family, genus, and species.

STUDY QUESTIONS

1. What is the underlying principle of the modern classification system for organisms?

2. *Homo sapiens* is the scientific name for humans. Which two taxonomic categories are represented by this name?

3. Name the five kingdoms of organisms, and give an example of one representative organism in each.

4. If you had a single-celled organism under a high-powered light microscope, what feature would you look for to prove that the cell was a protistan rather than a moneran?

5. Name two major groups of organisms belonging to kingdom Monera. What are some of the problems they cause and benefits they provide?

6. Which group of protistans has the most complex internal organization of its cell?

7. Which protistan group is characterized by pseudopods?

8. Name one example of a plantlike protistan.

9. List the general characteristics of fungi.

10. In terms of their roles in the biological world, what do heterotrophic bacteria and fungi have in common?

SUGGESTED READINGS

Barnes, R. D. *Invertebrate Zoology*. 4th ed. Philadelphia: Saunders, 1980. This is an advanced textbook but it contains a very readable account of the protistans.

Christenson, C. M. *Molds, Mushrooms, and Mycotoxins*. Minneapolis: University of Minnesota Press, 1975. A very informative book about the activities of fungi. Entertaining reading.

Cohen, S. S. "Are/Were Mitochondria and Chloroplasts Microorganisms?" *American Scientist* 58(1970):281–289. This article contains some interesting speculation on the evolution of microorganisms.

Hopwood, D. A. "The Genetic Programming of Industrial Microorganisms." *Scientific American* 245(1981):90–102. This is but one article in an entire volume on industrial microbiology.

McNeil, W. H. *Plagues and People*. Garden City, NY: Anchor Press, 1976. This book presents fascinating historical accounts of the impact of plagues.

Strobel, G. A., and G. N. Lanier. "Dutch Elm Disease." *Scientific American* 245(1981)56–83. Dutch elm disease threatens to eliminate our native elms. This interesting article describes the disease and discusses possible ways to combat it.

CHAPTER 14

The Plant Kingdom

In the hierarchy of life forms on earth, one major group of organisms is truly indispensable: the autotrophs, organisms that can use geochemical or light energy to convert simple inorganic substances into organic compounds. Because of this ability, the autotrophs do not depend directly on other creatures to meet their nutritional needs. All the heterotrophs—the animals, fungi, and most of the bacteria and protistans—owe their continued existence to autotrophic organisms. It is inconceivable that life on any planet could exist for very long without autotrophs.

The overwhelming majority of **photoautotrophs,** that is, organisms which use light as an energy source, are plants. As you recall from Chapter 13, certain monerans (cyanobacteria and photosynthetic bacteria) and protistans (diatoms, some euglenoids, and others) manufacture their own food, but all told their contribution to global photosynthesis is small. Thus, it is primarily the plants that support the extraordinary number of animals and fungi.

In addition to their photosynthetic mode of nutrition, plants have several other characteristics in common. With the exception of a few algae, all plants are multicellular. Their cells are eukaryotic and are surrounded by a cell wall containing cellulose. Finally, in the sexual life cycles of plants, distinct haploid (n), and diploid ($2n$) generations alternate.

The plant kingdom includes many taxonomic groups that are generally distinguished from one another by differences in structures and details of their life cycles. In this chapter we will introduce the major groups of plants in roughly the order of their evolutionary appearance, emphasizing the major trends of change. Since any discussion of plant group comparisons involves an understanding of plant life cycles, let us first consider the concept of alternation of generations.

ALTERNATION OF GENERATIONS

All plants show a clear alternation of haploid and diploid phases of their sexual life cycles (Figure 14–1). This alternation of generations is related to the fact that in plants, meiosis yields haploid spores, not gametes. The spores develop into multicellular structures called **gametophytes** (literally, "gamete-producing plants"), all the cells of which are also haploid. In some plants, the gametophytes are independent organisms. When fully developed, the gametophyte produces gametes mitotically. The union of two haploid gametes (usually an egg fertilized by a sperm) results in a diploid zygote, which marks the beginning of the diploid generation.

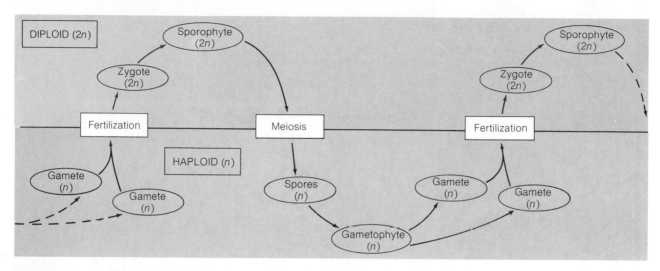

FIGURE 14–1
Generalized Sexual Life Cycle of Plants.

The zygote then grows into the **sporophyte** ("spore-producing plant"). When it reaches maturity, the sporophyte produces haploid spores by meiosis, thus completing the reproductive cycle. Note that the transitions between the gametophyte and sporophyte generations always involve changes in chromosome number (see Figure 14–1).

The relationship between the gametophyte and sporophyte generations is quite variable among different plant groups, and these variations have been useful in classifying them. As we examine the various types of plants, keep an eye out for a major trend in plant evolution: in the primitive plant groups the gametophyte generation dominates the life cycle, but in the more advanced groups the sporophyte generation is predominant. This evolutionary reduction of the gametophyte has reached its climax in the flowering plants, where the gametophyte consists of only a few cells that are totally dependent on the sporophyte for survival.

THE NONVASCULAR PLANTS

Botanists today recognize two major groups of plants: **nonvascular** and **vascular plants** (Figure 14–2). The nonvascular plants are so named because they lack **vascular tissues**—tissues specialized for the long-distance transport of water and nutrients. The nonvascular plants include the algae and the land-dwelling bryophytes (mosses and liverworts). Generally speaking, nonvascular plants have a flattened, relatively thin body form called a **thallus,** which assures that no cell is very far removed from the plant's aquatic or semi-aquatic environment. Without vascular tissues, nonvascular plants have no true roots, stems, or leaves.

Algae

Of all plant groups, algae are the oldest, flourishing in the oceans and freshwater environments for well over a billion years. A few types live on land, but only in areas where there is plenty of moisture. Algae come in many sizes and shapes, ranging from unicellular forms to the massive marine kelps that grow as long as 60 meters (roughly 200 feet). Sexual reproduction among the algae requires water, for in most cases at least one of the gametes (usually the sperm) is motile and must swim to the site of fertilization.

The major types of algae can be distinguished by the kinds of photosynthetic pigments they contain and their mode of reproduction. On the basis of these and other criteria, taxonomists divide the algae into three major divisions: Chlorophyta (green algae), Phaeophyta (brown algae), and Rhodophyta (red algae).

CHLOROPHYTA: GREEN ALGAE.   The green algae are the largest algal group, including about 7000 species.

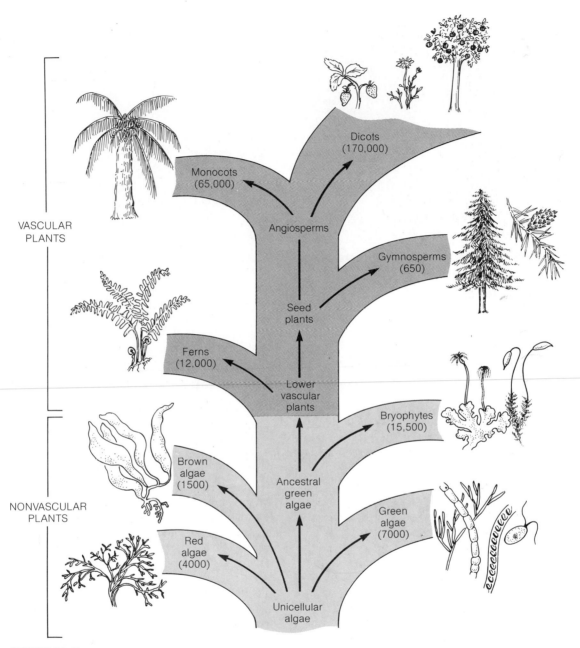

FIGURE 14–2
The Major Plant Groups and Their Evolutionary Relationships. The approximate number of living species in each group is given in parentheses.

An abundance of chlorophyll gives them the green color that is their most distinctive trait. Most green algae live in freshwater habitats, but a few species can be found in moist terrestrial spots, and even fewer species live in salt water. They vary greatly in size and shape. Some are unicellular, whereas others grow in colonies, filaments, or sheetlike forms (Figure 14–3). The smaller forms tend to be free-floating, but the larger filamentous and sheetlike green algae are often found attached to rocks, other plants, or even animals such as snails and turtles. One of the most common free-floating, filamentous forms is *Spirogyra,* the "pond

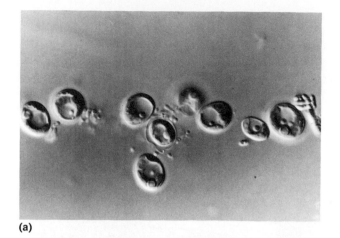

(a)

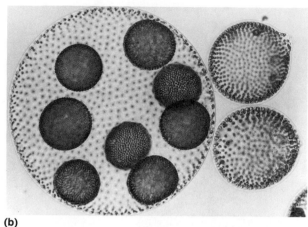

(b)

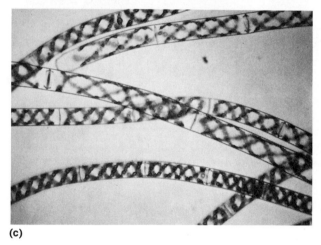

(c)

(d)

FIGURE 14–3
Some Common Green Algae.
(a) *Chlorella* is a unicellular form having a single chloroplast. This is the alga used by Calvin and his coworkers to trace the path of carbon in photosynthesis (see Essay 6–1). (b) *Volvox* is a colonial green alga consisting of many individual cells held togéther by a gelatinous matrix. The colonial spheres may become as large as 1 mm in diameter. There are several daughter colonies inside the larger sphere. This is a form of asexual reproduction in *Volvox*. (c) *Spirogyra*, or ''pond scum,'' is a multicellular filament. Note the spiral-shaped chloroplast in each cell.
(d) Commonly known as sea lettuce, *Ulva* is representative of the sheetlike green algae.

scum'' found on the surface of ponds, aquariums, and occasionally neglected swimming pools. *Chlorella,* the single-celled green alga used by Calvin's group in tracing the path of carbon in photosynthesis (see Essay 6–1), may soon be used in the manufacture of oil (Essay 14–1).

On the basis of similarities at the cellular level, many biologists believe that the higher plants arose from an ancestral line similar to the present-day green algae. Like the higher plants, green algae use chlorophylls *a* and *b* as well as carotenoid pigments in photosynthesis, and their primary storage carbohydrate is starch.

Green algae reproduce sexually by a number of different though related methods. (*Oedogonium,* an unbranched filamentous alga found in fresh water, is one example (Figure 14–4). The filament is haploid and therefore belongs to the gametophyte generation. The *Oedogonium* gametophyte is hermaphroditic, meaning that each individual filament can produce

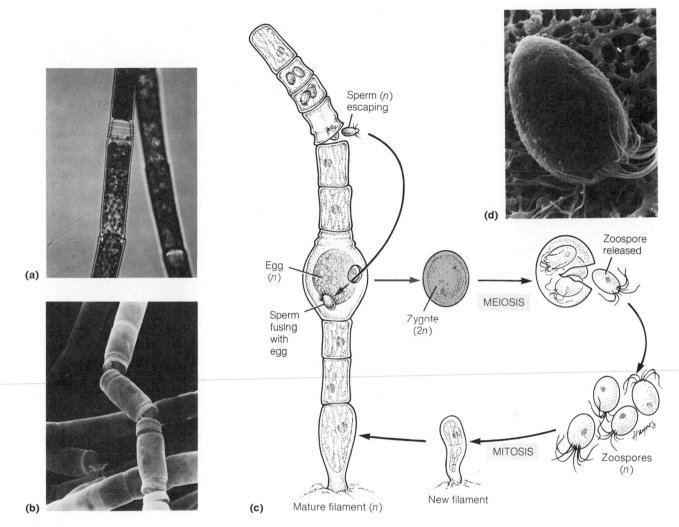

FIGURE 14–4

***Oedogonium,* a Filamentous Green Alga.** The photomicrograph (a) and scanning electron micrograph (b) show the haploid filaments of *Oedogonium*. (c) During its sexual life cycle, a motile sperm fuses with a stationary egg cell to form a zygote, the only sporophytic structure. After a period of rest, the zygote divides meiotically to yield four haploid zoospores, each of which can grow into a new filament. (d) This scanning electron micrograph shows a zoospore with its many flagella.

both types of gametes: small, motile sperm and a larger, nonmotile egg. The sperm are released directly into the water, where they swim toward an egg cell still contained within the filament, apparently guided by chemical attractants released by the egg. After fertilization, the zygote is shed from the filament and can remain dormant (inactive) for some time. Eventually, the zygote divides by meiosis to produce four haploid **zoospores.** After a period of swimming about, the zoospores settle onto a suitable surface, where they

give rise to new filaments by mitotic cell division. The zygote of *Oedogonium* is thus the only structure of the sporophyte generation.

The production of dissimilar gametes, especially when one is relatively large and nonmotile, is considered to be an evolutionary advancement. Because the egg is large, it can store food to support early development of the zygote after fertilization. For this reason *Oedogonium* is considered to be a relatively advanced member of the green algae.

ESSAY 14-1

CHLORELLA: ALGAE TO FUEL THE WORLD?

Chlorella may turn out to be the biggest name in gasoline since Standard Oil. This tiny, one-celled green alga has a chance to become a major competitor in the alternative fuels race. Isolated from fresh water lakes in the spring and autumn, *Chlorella* grows quickly in culture and requires little attention. In Japan, *Chlorella* has been used as protein and vitamin supplements in foods like ice cream, yogurt, and bread, but it may soon be filling gas tanks instead of stomachs.

Recently, two Canadian scientists from the University of Toronto reported on a microbial fuel system consisting of *Chlorella* and a modified bacterium. *Chlorella,* as the photosynthetic connection, plays the principal role of energy supplier. The scientists use this tiny alga to fix carbon dioxide into carbohydrates, which then become the food source for *Arthrobacter* AK 19, an oil producing bacterium.

AK 19 stores oil like a sponge holds water. It can accumulate over 85% of its body weight as fat. Under the electron microscope, AK 19 looks like a tiny plastic bag almost bursting with round oil droplets. Remarkably, the AK 19 oil is similar to crude oil, and it can be refined to yield an energy-packed substitute for gasoline.

How much oil can the *Chlorella* and AK 19 tandem produce? The scientists calculate that although they have to use a lot of light (200 watts per square meter), for each hectare of growing microbes it is possible to produce the equivalent of 50,000 barrels of crude oil a year.

How does this microbial oil compare to gasohol made from the ethanol of maize or sugar cane? Microbial oil production does not require fertile land, so it is not produced at the sacrifice of agricultural soil that could be used to grow food for people or livestock. And although both gasohol and microbial oil need carbon dioxide and sunshine, some algal species are so photosynthetically efficient that they can produce oil crops every two days, whereas maize and sugar cane require months of cultivation.

It is possible that microbial oil will not only compete with petroleum fuels, but it may also supply the food industry with a substitute for animal fats and vegetable oils. However, the Toronto team thinks that the microbial fuel may best be put to use as the liquid energy required for space travel.

PHAEOPHYTA: BROWN ALGAE. The brown algae are generally found attached to rocks and reefs along the coastlines of the oceans. All of the approximately 1500 species are multicellular, and nearly all are marine. Their brown color comes from an abundance of one particular carotenoid pigment.

The marine kelps are the most familiar of the brown algae, growing in large beds along temperate coastlines. Superficially the kelps resemble higher land plants: they are attached to rocks with a **holdfast** that looks much like a root system, and their **stipe** and **blades** outwardly resemble a stem and leaves (Figure 14–5). These similarities, however, should not be taken as evidence of a close ancestral link between the higher plants and the brown algae. Distantly related organisms often evolve similar structural features as a result of adapting to similar environmental circumstances. In this case, the marine kelps adapted to stay anchored in one place despite wave and tidal motion, and to position photosynthetic tissue near the water's surface to maximize sunlight exposure. These factors combined to give rise over time to the higher-plant-like structure of many brown algae.

Some of the kelps are economically useful as a food supplement for livestock, as fertilizer, and as a source of various natural products. *Macrocystis,* a giant kelp, is harvested in large quantities off the coast of California and extracted for algin, a slimy, viscous compound deposited in the cell walls. Algin has commercial value as an emulsifier and stabilizer in paint, ice cream, cheeses, and various cosmetics; it imparts a creamy texture to these products.

RHODOPHYTA: RED ALGAE. Just as the brown algae take their color from a brown carotenoid, most of the red algae have an abundance of a red-colored accessory pigment. A few red algae live in fresh water, but most are marine. Some can be found 150 meters (500 feet) or more below the ocean's surface. They are able to live at these great depths because of the capacity of their accessory pigments to absorb green light, the only part of the visible spectrum that can ·penetrate water so deeply.

With a few exceptions, the red algae are multicellular and filamentous. Often, the filamentous red algae do not *look* filamentous because the individual filaments may be packed together very tightly, yielding highly complex patterns (Figure 14–6). This is certainly true of the coralline red algae, whose structures are further complicated by deposits of calcium carbonate between their filaments. The calcium carbonate gives these marine plants a hard, shell-like appearance, and they become good surfaces for other plants and animals to grow on. In particular, corals (which are animals) often become attached to coralline algae. Coralline algae form the structural framework of coral reefs and atolls in many tropical regions of the world (see Figure 33–4).

Some types of red algae (such as *Porphyra*) are eaten by humans, especially in the Orient; others are harvested for the natural products they contain—agar, for example, is a cell wall polysaccharide commercially extracted from *Gelidium*. When mixed with water, agar forms a gel that biologists use as a culture medium

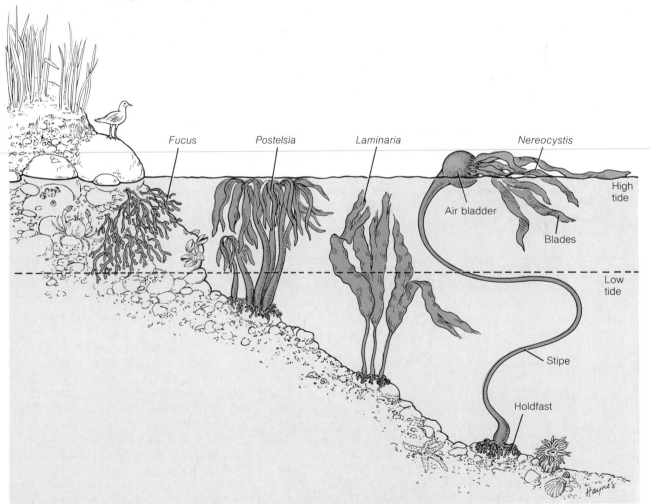

FIGURE 14–5

Kelps. Kelps are brown algae that vary greatly in size and shape, from such giants as *Nereocystis* to the relatively small *Laminaria* and the even smaller *Fucus* and *Postelsia*. Myriad other algae—green, brown, and red—and a host of attached and swimming animals would live on and around the four brown algae shown here.

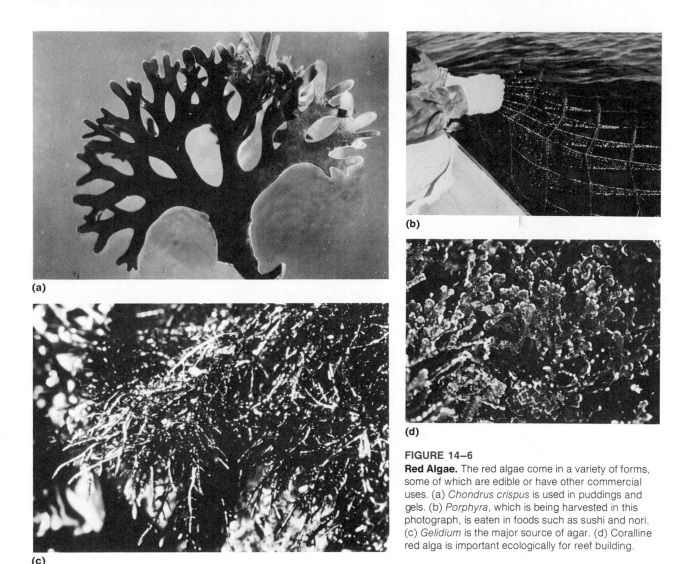

FIGURE 14–6
Red Algae. The red algae come in a variety of forms, some of which are edible or have other commercial uses. (a) *Chondrus crispus* is used in puddings and gels. (b) *Porphyra*, which is being harvested in this photograph, is eaten in foods such as sushi and nori. (c) *Gelidium* is the major source of agar. (d) Coralline red alga is important ecologically for reef building.

for growing fungi and bacteria. It is also used for industrial purposes.

Bryophytes: Mosses and Liverworts

About 425 million years ago, an important evolutionary event took place: plants invaded the land. The invasion was gradual, probably beginning near the water's edge with plants that were relatively small. These first terrestrial plants must have been similar to the **bryophytes,** which today are the simplest land plants. The most widespread and familiar bryophytes are the mosses and liverworts (Figure 14–7). These plants are small and generally restricted to damp, shady places where they can retain moisture. (Remember, they have no vascular tissue to transport water.) Although they have a leafy appearance, bryophytes lack true leaves, stems, and roots. Their body design is more akin to the thallus of algae.

The sexual life cycle of a moss is diagrammed in Figure 14–8. A moss gametophyte begins as a single-celled haploid spore, which germinates and grows into a branching filament (see right side of Figure 14–8). Some of the branches penetrate the ground and become **rhizoids,** rootlike projections that absorb minerals and water from the soil. On the upper surface of the filament, buds are formed that grow into the erect, "leafy" branches we generally recognize as the moss plant.

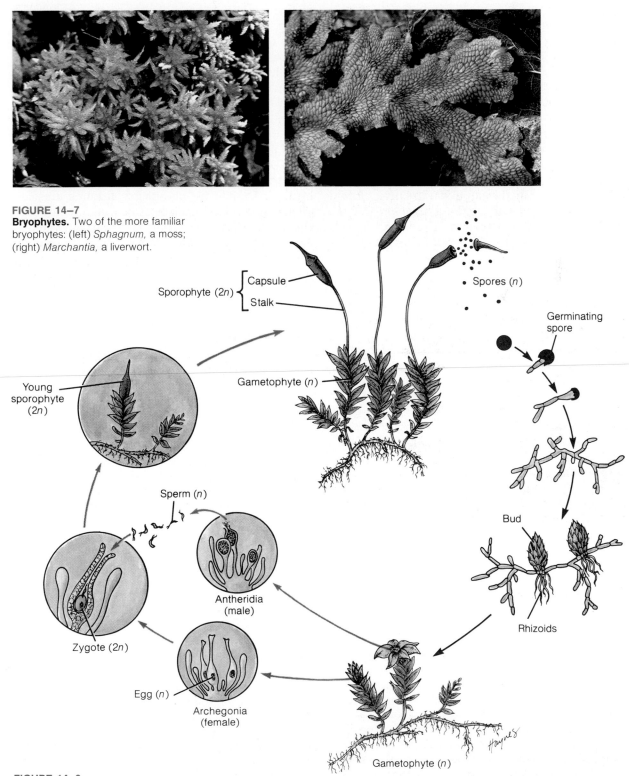

FIGURE 14–7
Bryophytes. Two of the more familiar bryophytes: (left) *Sphagnum,* a moss; (right) *Marchantia,* a liverwort.

Sporophyte (2n) {
Capsule
Stalk

Spores (n)

Germinating spore

Gametophyte (n)

Young sporophyte (2n)

Bud

Sperm (n)

Antheridia (male)

Rhizoids

Zygote (2n)

Egg (n)

Archegonia (female)

Gametophyte (n)

FIGURE 14–8
The Life Cycle of a Moss. The gametophyte generation predominates in the moss life cycle. It is photosynthetic, lives longer than the sporophyte, and has rhizoids that absorb water and minerals and anchor it to the soil or rock. The sporophyte is dependent on the gametophyte for its nutrients.

When these shoots mature, they produce multi-cellular sex organs at their tips. In most moss species, sperm-producing **antheridia** develop on certain branches, and flask-shaped **archegonia,** each of which harbors a single egg cell, appear on others (see Figure 14–8). In the presence of water each antheridium ruptures, exposing the flagellated sperm within. Each sperm must then somehow get over to an archegonial branch to rendezvous with an egg. Raindrops provide the means for overcoming this obstacle of distance. A drop of rain may fall on the tip of an antheridial branch and splash some sperm onto an archegonial branch. The sperm then swim down the neck of the archegonium to reach the egg. Fertilization results in a diploid zygote, marking the beginning of the sporophyte generation.

By normal processes of cell division, cell enlargement, and differentiation, the zygote develops into a mature sporophyte. The moss sporophyte, however, is not an independent plant. It starts life within the gametophyte archegonium, eventually forming a long stalk that extends several centimeters above the leafy gametophyte (see Figure 14–8). The stalk terminates in a rather elaborate sac called a **capsule,** resembling an old-fashioned street light atop a lamppost. The capsule contains cells that undergo meiosis to yield haploid spores, which are then released. If the spores land on moist soil, they germinate and give rise to more branched filaments, completing the moss's life cycle.

Economically, the bryophytes are of little importance to humans. The only significant exception is *Sphagnum,* or peat moss. Dried peat is used as a fuel in some countries, and occasionally as packing material for nursery plants and as a mulch to improve soils.

Now that you have a picture of what a moss is, we should mention a few so-called mosses that are not mosses at all. Reindeer moss, commonly found growing on rocks in the northern latitudes, is actually a **lichen,** an organism composed of both a fungus and an alga (see Chapter 32). Spanish moss, which droops from tree branches and telephone wires in the southeastern United States and elsewhere, is actually a flowering plant.

THE ORIGINS OF VASCULAR PLANTS

According to the fossil record, the bryophytes originated some 425 million years ago, probably from an ancestral stock of multicellular green algae. At about the same time, another line from the green algae began, culminating in the vascular plants. The vascular plants quickly spread over a wide range of terrestrial habitats, transforming the once barren rock and sand into a lush green landscape.

In the conquest of land by plants (and later by animals), two major obstacles had to be overcome: dryness and gravity. Aquatic organisms are bathed in water, which both prevents them from dehydrating (both plants and animals are 60–95% water) and provides buoyancy. Given the advantages of an aquatic environment, you might wonder why land plants ever evolved. Despite its inhospitable features, the land offered some distinct advantages of its own to plant life. The fossil record tells us that plants and animals were abundant in the lakes, streams, and marine coastlines around the time plants first established themselves on shore. Competition for light and minerals and predation by aquatic animals must have been intense. In contrast, the nearby land was wide open. The first land plants could take advantage of direct sunlight and an abundance of soil minerals without the threat of being eaten. Once established on land, they quickly adapted and diversified into the many terrestrial habitats available. A major part of this adaptation process must have included structural changes to compensate for the lack of water and the ever-present force of gravity; most of these structural changes are embodied in the vascular plant design.

Land plants exist simultaneously in two quite different environments—the soil and the air. The major source of water and minerals for these plants is the soil. For any land plant to attain a size greater than that of bryophytes, it must have a system of roots specialized to absorb water and dissolved minerals from the soil. The roots also serve to anchor the plant to the ground. And since their source of energy is sunlight, plants also need exposure above the ground. As a result, land plants have evolved a separation of functions: the roots have become specialized for absorption, and the aerial parts, most notably the leaves, have become specialized for photosynthesis. Because these two major plant activities are separated spatially, there must be a transport channel through which materials can flow as needed. Roots must receive organic nutrients produced by photosynthesis in the leaves, and leaves must be supplied with water and mineral nutrients absorbed by the roots. For this reason, the evolution of vascular tissues designed for the long-distance transport of water and nutrients was an early and very important event in the transition from water plants to land plants. Furthermore, because the vascular tissues are composed of relatively thick-walled cells (wood, for example, is a vascular tissue), they added mechanical strength to roots, stems, and leaves to counteract gravity.

Another important adaptation to life on land is the presence of a noncellular waxy **cuticle** that covers the surface of all air-exposed plant parts. The cuticle drastically reduces the amount of water a plant loses by evaporation. Finally, successive advances in modes of reproduction liberated the more advanced vascular plants from requiring surface water for fertilization.

LOWER VASCULAR PLANTS

The first and hence the most primitive vascular plants are collectively referred to as the **lower vascular plants.** The fossil record abounds with these land pioneers, but today only a comparatively small number of species remain. One of the surviving groups, the ferns, will serve as our example of the lower vascular plants.

Most ferns are found in the tropics, but many also grow in moist, temperate regions along streambeds or roadsides and in meadows. Ferns range in size from some no bigger than a thumbnail to the giant tree ferns of tropical forests, which can grow to be 18 meters (about 60 feet) high, with leaves about 4 meters (12 feet) long (Figure 14–9).

(b)

(a)

(c)

FIGURE 14–9
Ferns. Living representatives of the most primitive vascular plants include (a) sword ferns; (b) tree ferns; and (c) a staghorn fern.

The conspicuous generation of ferns is the sporophyte, which consists of roots, stems, and leaves. Most temperate-region ferns have an underground, horizontal stem called a **rhizome,** from which the roots and leaves originate. If you look at the undersurface of the older leaves, you will usually find brown or black patches. Each patch consists of clusters of **sporangia,** or spore-bearing sacs. Within each sporangium are specialized diploid cells that undergo meiosis to yield haploid spores (Figure 14–10, upper right).

The spores are released to the ground, often catapulted by a springlike action of the sporangium. Under suitably moist conditions, they germinate and grow into a multicellular gametophyte plant, which is heart-shaped and about the size of a fingernail. The mature gametophyte gives rise to both egg-producing archegonia and sperm-producing antheridia on its undersurface (see Figure 14–10). The sperm released from the antheridia must swim through a film of water adhering to the lower surface of the gametophyte to

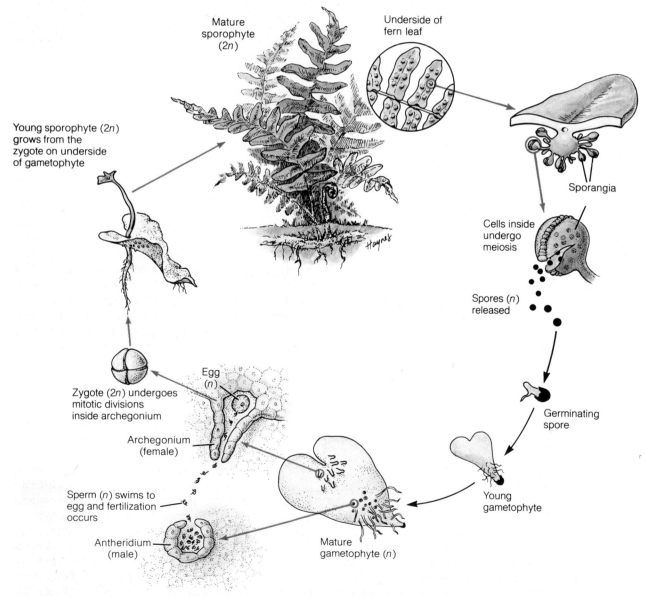

Mature sporophyte (2n)

Underside of fern leaf

Young sporophyte (2n) grows from the zygote on underside of gametophyte

Sporangia

Cells inside undergo meiosis

Spores (n) released

Germinating spore

Young gametophyte

Mature gametophyte (n)

Antheridium (male)

Sperm (n) swims to egg and fertilization occurs

Archegonium (female)

Zygote (2n) undergoes mitotic divisions inside archegonium

Egg (n)

FIGURE 14–10
The Life Cycle of a Fern. The gametophyte of ferns is only a few millimeters in diameter. It is photosynthetic and independent from the sporophyte. The embryonic sporophyte, however, lives parasitically off the gametophyte until its first leaf emerges and begins to photosynthesize.

(a) (b) (c) (d)

FIGURE 14–11
Some Representative Gymnosperms. (a) Cycads such as this sago palm resemble palm trees in their overall form, although true palms are flowering plants. (b) *Ginkgo biloba,* the maidenhair tree, is often used in urban landscaping because it can withstand smog and other atmospheric pollutants better than most other trees. (c) Pines, such as these lodge pole pines growing in Yellowstone National Park, are probably the most familiar of the gymnosperms. (d) Giant Sequoia are among the oldest creatures on the earth. Both pines and sequoias are conifers.

reach the egg-containing archegonia. Although several archegonia may be present on a single gametophyte, and several eggs (one per archegonium) may become fertilized, usually only one zygote develops completely into an embryo, the young sporophyte. The embryo grows at the expense of the withering gametophyte, gradually developing roots, a stem (rhizome), and leaves to become an adult sporophyte.

Although the ferns are generally found in moist environments, they do have vascular tissues and a cuticle; thus they are better adapted to dry land than are the bryophytes. In the evolutionary conquest of land by plants, however, ferns are only a transitional group. The most advanced land dwellers are the seed plants.

SEED-PRODUCING VASCULAR PLANTS

One of the truly remarkable "inventions" of nature is the seed. In life cycle terminology, a **seed** is an enclosed embryonic sporophyte. It usually contains almost no water at maturity, and exists in a physiological state of suspended animation, undergoing no growth and carrying on only a minimal level of metabolism. Thus, seeds are convenient vehicles of dispersal for their species: they can be scattered by wind, rain, animals, and other agents to facilitate the spreading of a species over a wide area. Seeds are also clearly an adaptation to a changeable terrestrial environment. In their dry, inactive state, seeds can withstand droughts or very cold temperatures. If a seed encounters unfavorable growing conditions, it simply remains in suspended animation until better conditions (such as rain or warmer weather) prevail. At that time, the embryo grows at the expense of stored food within the seed to become a seedling.

Gymnosperms

Plants that produce seeds on the surfaces of scales or on short stalks are called **gymnosperms** (literally, "naked seeds"). They generally have **cones** rather than flowers, and their seeds are not enclosed in fruits. The three major groups of living gymnosperms are the cycads, the ginkgo (only one species), and the conifers (Figure 14–11). Although the gymnosperms are a fairly small group of plants today, what they lack in diversity they make up for in total numbers. The northern temperate forests (primarily in Canada, the United States, and the Soviet Union) are populated chiefly by the conifers, which include the familiar pines, firs, cedars, spruces, junipers, hemlocks, larches, and yews as well as the majestic redwoods. They have no value as a human food source, but the

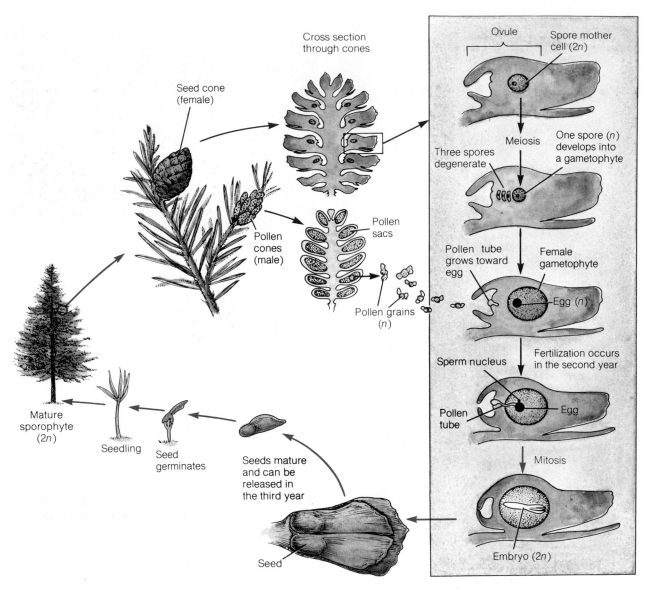

FIGURE 14–12
The Life Cycle of a Pine. The formation of seed cones to the release of seeds normally requires three growing seasons in most pines.

conifers contribute more than 80% of the lumber to the wood-products industry in the United States.

The gymnosperms differ from the ferns, bryophytes, and algae in three major ways:

1. They reproduce by means of seeds.
2. Their reproductive cycle eliminates an independent gametophyte.
3. They do not require surface water for fertilization to take place.

All these characteristics are apparent in the life cycle of a typical gymnosperm, the pine (Figure 14–12).

The sporophyte of pines is the familiar large plant consisting of roots, stems, needle-shaped leaves, and cones. **Pollen cones** contain male sporangia, and **seed cones** contain female sporangia. Meiosis within the sporangia yields haploid spores—the beginning of the gametophyte generation. The spores develop directly into male and female gametophytes within their respective sporangia.

(a)

(b)

FIGURE 14–13
Pine Cones. (a) Pollen cones generally form in clusters near the tips of branches. (b) Seed cones, usually borne singly, are shown here in their third year of growth. Note that by this stage the scales have separated to permit the dissemination of seeds.

The microscopic male gametophytes, or **pollen grains,** are released during the spring and dispersed by air currents. Some of the air-borne pollen grains adhere to a very sticky resinlike material produced by the seed cones, and are drawn into the cone between the tightly packed scales as the resin dries. This transfer of pollen from the male sporangia to the female sporangia is called **pollination,** which represents a major advance in the adaptation of plants to land. Pollination circumvents the requirement of liquid water for sperm transfer, as is the case for bryophytes and ferns.

Seed cones require three years to reach seed-bearing maturity (Figure 14–13). During the first season, the seed cones are small ovoid structures about 1 to 3 cm (0.4 to 1.2 inches) long. Pollination takes place late in the first year of development. About a month later, meiosis occurs in each of the female sporangia, or **ovules,** producing spores that develop into female gametophytes. During the next year, two or three archegonia develop inside each female gametophyte. Each archegonium contains one egg cell. In the meantime, pollen grains that were drawn into the seed cone have been growing **pollen tubes** through the ovular tissues toward the archegonia (see Figure 14–12). When a pollen tube reaches an egg, it releases a sperm nucleus to fertilize the egg. The resulting zygote then develops into an embryo, a process that takes another year and culminates in a seed. Even though more than one egg may be fertilized, only one embryo per ovule develops fully. During the last stages of the seed cone's development, the central region of the cone lengthens, separating the scales and exposing the mature seeds for dispersal. Each seed represents a potential new pine tree.

Angiosperms

Angiosperms (literally, "encased seeds") are the flowering plants, represented by approximately 235,000 known species. They are clearly the most diverse and abundant land plants now in existence.

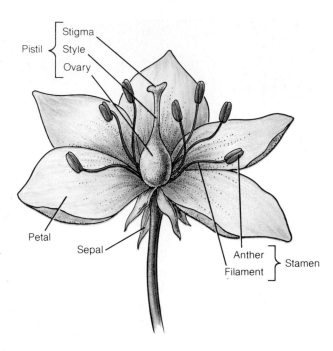

FIGURE 14–14
Flower Structure. This highly diagrammatic flower exhibits the four organs found
in a complete flower: sepals, petals,
stamens, and a pistil.

Angiosperms are found in virtually all land habitats, including such hostile areas as the arctic tundra and the subtropical deserts. A few species occupy marine and freshwater habitats, and at least one flowering plant, Spanish moss, lives virtually in air, obtaining water from the atmosphere and minerals from airborne dust while suspended on tree limbs or telephone wires.

The two most distinctive features of the angiosperms are flowers and a seed encased by a fruit. Both of these unique structures play important roles in the life cycle of angiosperms, which is dominated by the sporophyte generation. The male and female gametophytes are microscopic and completely enclosed by the sporophyte tissues from which they derive nutrition. Because the gametophytes are produced within specialized floral organs, we will turn our attention first to the flower itself.

FLOWERS AND POLLINATION. When a flowering plant enters its reproductive phase, some or all of its vegetative buds transform into floral buds. Vegetative buds produce new stems and leaves; floral buds produce the various floral organs. The first of the floral organs to develop are the leaflike **sepals,** which envelop the new flower bud. Eventually the sepals fold back to expose the **petals,** generally the most attractive part of the flower. Their showy colors and patterns help attract pollinating creatures such as birds and insects. The actual reproductive function of the flower, however, is carried out by two types of floral organs called the stamens and the pistil (Figures 14–14 and 14–16).

Stamens are the male reproductive organs. Usually each stamen consists of a slender filament terminating in a four-lobed sac called the **anther.** Each lobe is initially an immature sporangium containing many cells that undergo meiosis to yield haploid **microspores.** Each microspore then develops by mitosis into a male gametophyte, or pollen grain (Figure 14–15, left side). The pollen grain is actually two nuclei within a single, wall-bound cell.

At this point the anther splits open, exposing the pollen. In **self-pollinating** species, the pollen usually remains in the flower that produced it. In **cross-pollinating** species, the pollen may be blown or carried to other flowers on nearby plants. The pollen does not normally develop any further until the grain has landed on the receptive surface of a pistil.

The **pistil** is the last floral organ produced in a developing flower. It is centrally located and surrounded by the stamens. Pistils are generally flask-shaped: the **ovary** makes up the enlarged base, the **style** is the slender stalk, and the **stigma** forms the top (see Figure 14–15, right side). The upper surface of the stigma is often quite hairy and sticky, and is thus well designed to trap pollen grains. Once a pollen grain lands on the stigma, it grows a pollen tube through the pollen wall into the tissue of the stigma (see Figure 14–15, bottom left). The pollen tube continues to grow down through the style, digesting tissue of the style for nutrients as it grows. As the tube approaches its goal—an ovule in the ovary—one of the nuclei in the pollen grain divides mitotically to yield two sperm nuclei. At this stage the male gametophyte has reached full maturity.

In the meantime, events taking place in the ovary are leading to the development of the female gametophyte. Within each immature ovary, one or more ovules differentiate. Each ovule in turn houses a single large cell which divides meiotically to produce four haploid **megaspores** (see Figure 14–15, right side). One of the megaspores develops into the female gametophyte; the other three degenerate. The nucleus of the surviving megaspore then divides mitotically, ultimately generating eight haploid nuclei enclosed in

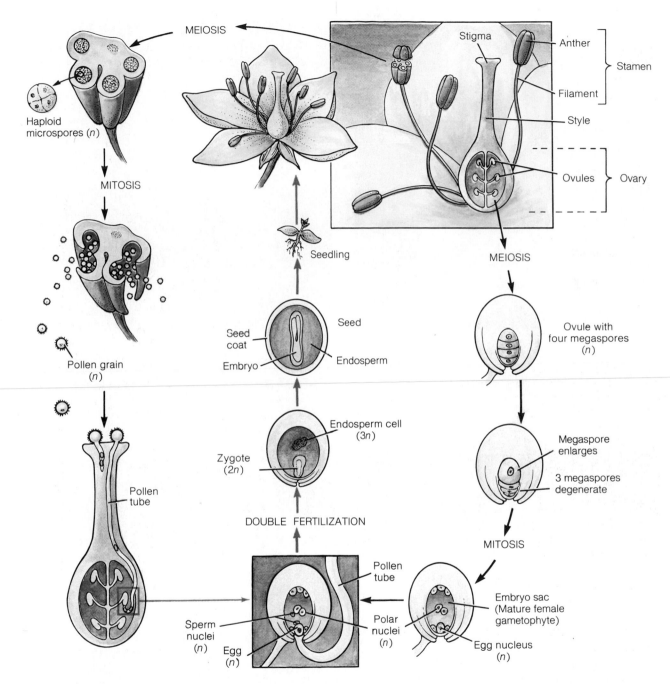

FIGURE 14–15
The Life Cycle of a Flowering Plant. This figure summarizes the life cycle of a typical angiosperm.
The structures and events shown here are described in the text.

the original megaspore cell wall. Of these eight nuclei, one becomes the egg nucleus, and two others become the polar nuclei; the other five eventually degenerate. The fully developed female gametophyte, known as the **embryo sac,** is now ready for fertilization.

FERTILIZATION AND SEED DEVELOPMENT. The union of gametes in angiosperms is unique. When the pollen tube finally reaches the ovule, the two sperm nuclei are released into the embryo sac, and **double nuclear fusion** takes place. One sperm fertilizes the egg to

FIGURE 14–16
Variations on a Theme. Flowers display themselves in many fashions as shown by these photographs of (a) a poppy native to California; (b) a water lily; (c) a rose; (d) a passionflower; and (e) a day-lily.

(a) (b) (c) (d) (e)

yield the zygote; the other sperm fuses with the two polar nuclei, resulting in the triploid (3*n*) **endosperm cell** (Figure 14–15, middle). The zygote then develops into the multicellular embryo (the young sporophyte), and the endosperm cell grows into the nutrient-storing endosperm tissue. Both are surrounded by a thin layer of ovular cells that becomes the **seed coat,** the protective outer covering of the seed. During the development of most angiosperm seeds, the embryo digests the nearby endosperm as it expands, obtaining nutrients in the process. In these cases, the mature seed consists of merely an embryo enclosed by the seed coat.

The two major classes of flowering plants—the **dicotyledons** (or dicots) and **monocotyledons** (or monocots)—derive their names from differences in seed structure. The dicots have two embryonic leaves called **cotyledons;** the monocots have only one cotyledon (Figure 14–17). During seed development in most dicots, the cotyledons expand greatly at the expense of the withering endosperm, accumulating a storehouse of nutrients to support the growing seedling after germination. In contrast, monocot seeds retain their endosperm at maturity, and their single cotyledon has a nutrient-absorbing function during seedling growth.

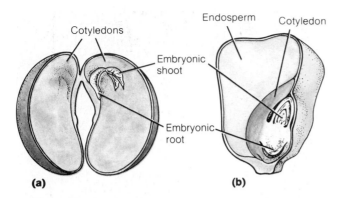

FIGURE 14–17
Seed Structure. (a) A typical dicot seed (bean) with its two cotyledons. (b) A typical monocot seed (corn) with its single cotyledon and endosperm.

FRUITS. Unlike gymnosperm seeds, which develop on the surface of scales, angiosperm seeds are completely enclosed in ovarian tissues as they mature. As the seed or seeds grow, the ovary also expands, often growing thousands of times larger than it was at fertilization. This enlarged ovary is called the **fruit.**

As you are well aware, fruits come in all sizes, shapes, and colors. Many are fleshy and edible, such as apples, oranges and avocados; others are hard and fibrous, such as coconuts and peanut shells. There are a few fruits that develop without benefit of fertilization and therefore lack seeds. Bananas, navel oranges, Thompson seedless grapes, and tangelos are examples of seedless fruits. There are even some fruits that we mistakenly call vegetables, such as string beans, pea pods, tomatoes, corn, cucumbers, and squash.

Fruits do not exist to nourish the sprouting seed—that is the role of the endosperm or cotyledons. Why, then, do angiosperms produce fruits? In many cases, the fruit appears to play a role in seed dispersal. Edible fruits attract animals, which eat the fruit and pass its seeds undamaged through their digestive tracts. As animals move around, the seeds may be deposited at a site some distance from the parent plant. Some fruits have feathery appendages that enable them to be carried by the wind; others have barbs that catch on the fur of animals (Figure 14–18). The coconut fruit floats on water and can be carried thousands of miles by ocean currents to settle eventually on some distant shore.

FIGURE 14–18
Fruits as Seed Dispersers. Many fruits are adapted to use animals as agents of dispersal. For example, (a) seeds can pass unharmed through the digestive tract of a fruit-eating animal; (b) seeds may travel great distances by hitching a ride in an animal's fur. Other fruits are adapted for wind dispersal, such as (c) the fruits of the maple and (d) the fruits of the dandelion.

(a)

(b)

FIGURE 14–19
Pollination Mechanisms. (a) Many wind-pollinated flowers, such as grasses, have small, rather plain flowers that produce abundant amounts of pollen. The stigma is often feathery or hairy to help snare wind-blown pollen grains. (b) Likely candidates for visits from hummingbirds are flowers that are tubular, pendant (drooping), and red. (Bees generally avoid red-flowered species.) c) The flowers of certain orchids, such as the *Caladenia lobata* pictured here, bear a striking resemblance, in terms of both appearance and odor, to the females of the wasp species that pollinates them. The flowers mature at the same time the male wasps reach sexual maturity and slightly before the females are receptive. The male wasp tries to mate with the orchid and in the process deposits pollen from other orchid flowers. It also picks up pollen to be deposited on other flowers during further mating attempts. Thus, by "deceit" on the orchid's part, cross-pollination is achieved.

(c)

Pollinating Mechanisms

Cross-pollination—the transfer of pollen from one individual plant to another of the same species—is widespread among the flowering plants. It offers the advantage of genetic recombination, leading to greater genetic variety among the offspring and therefore to an increased capacity to adapt to a changing environment (see Chapter 8). The difficulty of cross-pollination, of course, is that it requires the physical transfer of pollen from one plant to another. A variety of pollen transfer "strategies" have evolved to overcome this problem.

All gymnosperms and many angiosperms rely on air currents to effect cross-pollination. Among the angiosperms, wind-pollinated flowers are typically small and inconspicuous, and they generally have neither nectar, fragrance, nor brightly colored petals (Figure 14–19a). Their strategy is to produce copious amounts of pollen, often to the discomfort of hay-fever sufferers.

The vast majority of flowers are pollinated by insects. Bees, butterflies, wasps, moths, flies, and beetles fill this important role, as do a few species of birds and mammals, such as hummingbirds, bats, and some rodents (Figure 14–19b). To attract pollinators, flowers usually produce a sweet, edible nectar in specialized glands at the base of the petals. The pollinator must brush up against the pollen-laden anthers to reach the nectar, and when it leaves it carries away some of the sticky pollen grains to the next flower it visits.

Because the simultaneous transfer of pollen and gathering of food are extremely important activities for the survival of the plant and the pollinator, respectively, evolution has tended to cospecialize their roles. That is, many flowers have evolved specific forms, color patterns, and scents to attract a particular type of pollinator (Figure 14–19c). At the same time, pollinators have undergone adaptive changes suited to the types of flowers they visit. Some species of flowering plants and insects have coadapted to such a point that the elimination of one species would surely eliminate the other.

SUMMARY

In their sexual life cycle, the plants' sporophyte ($2n$) generation alternates with a gametophyte (n) generation. The sporophyte produces spores by meiosis, and these develop into gametophytes. At maturity the gametophyte produces the gametes (usually egg and sperm), which fuse to form the zygote, the beginning of the next sporophyte generation. Many of the criteria used in classifying plants are based on structures associated with their specific sexual life cycles.

The first plants that appeared on earth, the unicellular algae, were the ancestors of the major groups of algae existing today. Today's algae occupy both freshwater and marine habitats; some are even found in moist terrestrial surroundings. They have a thallus body design (no roots, stems, or leaves) and they lack vascular tissue. The various divisions of algae can be distinguished by the type of photosynthetic pigments they contain and by their mode of reproduction.

Green algae, the most diverse of the algal groups, resemble the higher plants by having an abundance of chlorophylls and carotenoids. They are found primarily in fresh water, although a few species inhabit terrestrial or marine environments. Unicellular, colonial, filamentous, and sheetlike forms are known. The brown algae are virtually all marine and include the familiar kelps. Some forms resemble the higher plants by having a rootlike holdfast for anchorage and a stemlike stipe from which leaflike blades grow. Some of the kelps are harvested commercially for use in livestock feeds and in various natural products. Most of the red algae are marine, some living at great depths on the ocean floor. Their accessory pigments capture the deepest-penetrating light wavelengths for photosynthesis. Some sheetlike forms exist, but most red algae are filamentous.

About 425 million years ago, two major groups of plants began to appear on land: the nonvascular bryophytes and the first vascular plants. The early bryophytes were ancestors of the modern mosses and liverworts, all of which are quite small and generally restricted to damp, shady habitats. Like their green algae ancestors, the bryophytes have a thallus body structure, and their sperm transfer requires water.

Also evolving from ancestral green algae, the vascular plants went far beyond the bryophytes in adapting to the environmental conditions on land. Their most significant development was the evolution of vascular tissues to transport water, minerals, and organic nutrients. The vascular tissues also provide mechanical support for the stems and leaves, and a waxy cuticle retards water loss from their surfaces. Other higher plant adaptations to land are seeds and pollen (Table 14–1).

Botanists generally recognize two major groups of vascular plants: the lower vascular plants and the seed plants. The ferns are the most diverse and well known of the lower vascular plants. They exhibit a distinct separation of sporophyte and gametophyte generations, the former being the most conspicuous. Since the sperm produced by the gametophyte must swim to the eggs to fertilize them, ferns generally grow in the moist soils typical of the tropics and temperate forests.

Gymnosperms comprise a relatively small group of plants dominated by the conifers, which reproduce by means of pollen cones and seed cones. Pollen is carried by air currents to the seed cones, is drawn in toward the ovules, and eventually produces sperm. Fertilization of an egg in an ovule marks the beginning of seed development. The mature seeds are borne on the surfaces of seed cone scales.

Angiosperms, the most diverse of all plant groups, are distinguished by their flowers and by their seeds enclosed within a fruit. The flowers generally consist of leaflike sepals, showy petals, stamens, and one or more pistils. After pollen is transferred from an anther to the stigma of the same (self-pollinating) or a different (cross-pollinating) individual, a pollen tube grows toward an ovule within the flower's ovary. Double fertilization marks the beginning seed development. As the seed develops, the surrounding ovarian tissue enlarges into a fruit. Fruits often aid in seed dispersal.

Many flowers depend on animals for pollination, and have evolved specialized features to cause them to be recognized and visited by pollinators.

TABLE 14–1
Evolutionary Trends in the Plant Kingdom. In this summary of the key features of the major plant groups, the characteristics listed at the left are considered to be structural adaptations to the terrestrial environment. Most of these are absent from the green algae, but all are features of the most advanced group, the angiosperms. In general, evolutionary trends toward advancement appear on this table from left to right (plant groups) and from top to bottom (characteristics).

Characteristics	Green Algae	Bryophytes	Ferns	Gymnosperms	Angiosperms
Internal fertilization	In some	X	X	X	X
Degree of tissue specialization	Very little	Little	Moderate	Very high	Very high
Vascular tissues			X	X	X
Cuticle			X	X	X
Sporophyte predominant	Rarely		X	X	X
Gametophyte dependent on sporophyte				X	X
Pollen instead of swimming sperm				X	X
Seeds				X	X
Fruits					X

STUDY QUESTIONS

1. What two key events in a plant's life cycle separate the sporophyte and gametophyte generations? Explain by using the life cycle of *Oedogonium* as an example.

2. What characteristics distinguish the algae from other plant groups? What characteristics distinguish the green, brown, and red algae from each other?

3. Compare the mosses and ferns on the basis of their adaptations to existence on land.

4. What characteristics of the gymnosperms make them better adapted to land than the ferns?

5. The seed cones of mature pine trees are often as much as three years old. What events associated with reproduction occur during the first year of seed cone development? the second year? the third year?

6. Diagram and label a typical flower. Indicate which structures are responsible for producing the male and female gametophytes.

7. Why are seeds considered an evolutionary advancement?

8. Many fruits have specialized structures to aid in seed dispersal. Describe some of these structures and explain how they work.

SUGGESTED READINGS

Banks, H. P. "Early Land Plants: Proof and Conjecture." *Bioscience* 25: (1975) 730–737. Discusses the fossil evidence which relates to the origin of land plants.

Dawes, C. J. *Marine Botany.* New York: Wiley, 1981. Presents an in-depth discussion of marine algae and their environment.

Heywood, V. H. *Flowering Plants of the World.* New York: Mayflower Books, 1978. Encyclopedic coverage of the plant families with beautiful color photographs.

Perl, P. "Ferns." *Time-Life Encyclopedia of Gardening.* Alexandria, VA: Time-Life Books, 1977. This informative book has an attractive section on the use of ferns as ornamentals.

Proctor, M., and P. Yeo. *The Pollination of Flowers.* New York: Taplinger, 1973. Discusses all aspects of pollination biology.

Raven, P. H., R. F. Evert, and H. Curtis. *Biology of Plants.* 3rd edition. New York: Worth, 1981. Chapter 19 of this well-illustrated text has a good description of angiosperm evolution and pollination ecology.

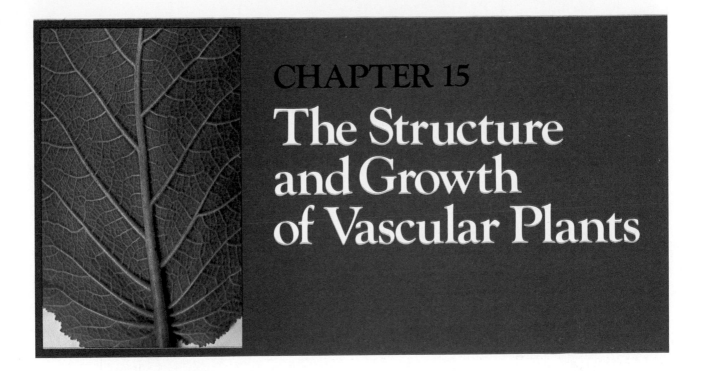

CHAPTER 15

The Structure and Growth of Vascular Plants

Out of the evening sky a shining silver disk slowly descends onto a grassy clearing. As a hatch door swings open, astonished onlookers witness the appearance of two green, humanlike creatures. These visitors, it turns out, are green because they have chlorophyll in their skins. The chlorophyll traps light energy, enabling these creatures to make their own food—they are entirely autotrophic!

Is this stretching the imagination too far, or is it theoretically possible for a creature of human proportions to be a self-sufficient food producer? Could there be a life-bearing planet in some distant galaxy that is inhabited by photosynthetic animals and no plants? Let's have a look.

Under the most favorable natural conditions on earth, the most efficient plants carry on photosynthesis at a rate that yields 0.2 milligrams of sugar per hour for each square centimeter of leaf surface area. Let's be generous and assume that our green visitors from outer space can photosynthesize at a similar rate. A 160-pound human being can expose about 17,000 square centimeters of skin to sunlight at any one time, so a green humanlike creature of similar size could generate 3400 milligrams (0.2 × 17,000) of sugar per hour, or approximately 41 grams (about 3 tablespoons) in a 12-hour day. Is this enough for survival?

Forty-one grams of sugar represents about 157 kilocalories of food energy. (Kilocalories are equivalent to the calories familiar to dieters.) However, a 160-pound human requires about 1800 kilocalories a day just to maintain body temperature, breathe, and keep the heart pumping and the bowels moving. If the person carries out any physical activities—walking, talking, and so on—the daily requirement increases to 2000–4000 kilocalories. Obviously our photosynthetic aliens could not produce enough food energy to meet normal human demands. But possibly with a few alterations here and there, we can bring our green visitors closer to a balance between food productivity and energy demand.

First, the green aliens would not have to be very active because they would not need to hunt, gather food, or even go to the supermarket. If they lived in a warm, tropical environment, they could expose their photosynthetic skin to the sun year-round, eliminating the need for clothing, shelter, and heating fuel. Their basic requirements would include only carbon dioxide and oxygen from the air, and water containing a healthy supply of dissolved minerals. But even if drinking was their only physical activity, they would still need to meet the basic requirement of 1800 kilocalories per day. We must look for ways to either increase their photosynthetic productivity or to decrease their energy consumption.

One possible way to enhance productivity would be to increase their photosynthetic surface area. This cannot be done simply by increasing body size because body volume, and hence energy demand, would increase at a greater rate than the food-producing surface area (see Figure 3–5). Decreasing the visitors' size would be a step in the right direction, but before they reached a suitable ratio of surface area to volume, they would become microscopic. Thus, the only way to increase surface area without sacrificing size would be to provide our green humanoids with thin, flattened tissue extensions radiating out from their torsos. Such fragile solar collectors would greatly hamper movement, so to avoid injury our green aliens would do well to anchor themselves to the ground. This they could accomplish by another set of appendages extending into the soil. Such underground structures could double as water-absorbing and mineral-absorbing organs, eliminating the need for a mouth.

A second approach to redesigning our green creatures toward self-sufficiency would be to reduce their bodily demands for energy. We could begin by eliminating the energy-demanding structures and processes that are not necessary for a sedentary lifestyle. For example, they would not need muscles or a system of nerves to coordinate their movements. For that matter, the entire nervous system, including the brain, could be scrapped. Their digestive systems would have nothing to digest because they don't eat other organisms. Gas exchange could be handled by a simple diffusion process because the high ratio of surface area to volume would mean that none of their cells would be very far removed from atmospheric gases. And since these modified green aliens would not be very active, they would have no need for an energy-consuming heart to keep their body fluids racing through their vessels. The active circulatory system could be replaced by a simpler passive system.

Now, what kind of animals have our modified green space creatures become? They are no longer animals at all—in fact, they more closely resemble the vascular plants on earth. In other words, the most efficient design for a multicellular photosynthetic organism has already been achieved in the evolution of the vascular plants. Let us now look more closely at this successful design for self-sufficient living—the modern vascular plant.

THE STRUCTURE OF VASCULAR PLANTS

All vascular plants have several structural features in common. As you might expect, these are the very same features that we imposed on our green visitors to make them more autotrophically self-sufficient. Recall that the aliens needed a large photosynthetic sur-

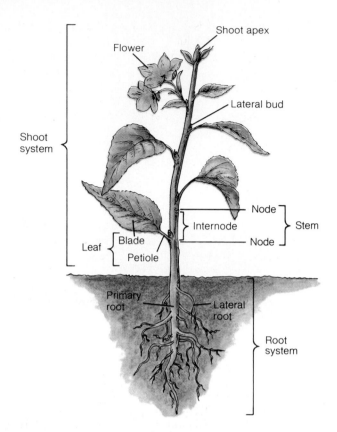

FIGURE 15–1
General Structure of an Angiosperm (Dicot).

face area, so we provided them with "thin, flattened tissue extensions"—leaves. The extensions radiated out from their torsos, which correspond to the stems of plants. And, of course, the appendages that anchored our friends in the ground and absorbed water and minerals serve the same function as roots of plants. The aliens' passive circulatory system correlates with the vascular system of higher plants.

Leaves, stems, and roots are the major organs found in all vascular plants (Figure 15–1). The leaves, stems, and any reproductive structures present (flowers, fruits, or cones) make up the **shoot system.** The **root system** consists of the **primary root** and **lateral roots** that branch off from the primary root. Let us now examine these structures in more detail, starting with the leaf.

Leaves

As we discussed, the ability of plants to be self-sustaining by means of photosynthesis hinges on exposing a large amount of surface area to sunlight. In the vast majority of plants this functional requirement is accomplished by the production of leaves. A typical leaf consists of a thin, flattened **blade,** which is joined to the stem by a stalklike extension called the **petiole** (see Figure 15–1). Vascular tissues in the stem run through the petiole into the blade, where they form a highly branched system of veins. In monocots, such as grasses, lilies, and palms, the veins run parallel to one another in the typically long and narrow blades (Figure 15–2a). Dicot leaf blades are usually broad and have a netted vein pattern (Figure 15–2b). These general leaf forms, while typical for angiosperms, are supplemented by many variations, particularly among the ferns and gymnosperms. Many ferns have blades that are divided and subdivided into exquisitely symmetrical patterns (Figure 15–2c). Gymnosperms also exhibit a wide variety in leaf shapes, including the palmlike leaves of cycads (see Figure 14–11a), the slender needles of cedars, and the tiny scale leaves of junipers (Figure 15–2d–e).

(a)

(b)

(c)

(d)

(e)

FIGURE 15–2
The Shapes of Leaves. (a) Narrow, elongate blades with parallel veins are typical of monocots.
(b) Dicot leaf blades tend to be broad and have a netted venation pattern. (c) Fern leaves are
subdivided into many "leaflets." (d) These short needle-shaped leaves are from the Deodar cedar.
(e) The tiny scale leaves of juniper completely cover the stem.

LEAF ANATOMY.   Since the main function of leaves is to carry out photosynthesis, let us see how the anatomy of a leaf relates to this important activity.

The leaf is covered on its top and bottom surfaces by a protective cuticle and epidermis (Figure 15–3a). The **cuticle** is a layer of waxes that helps retard water loss from the leaf. The **epidermis** is the outermost layer of cells, which is responsible for the deposition of the cuticle. Dotting the lower epidermis (in some species also the upper epidermis) are numerous pores called **stomates** (Figure 15–3b), which permit the diffusion of gases (carbon dioxide, oxygen, and water vapor) into and out of the leaf. We will have more to say about stomates shortly.

Sandwiched between the upper and lower epidermis are the chloroplast-laden **mesophyll** cells (see Figure 15–3a). These cells are packed with chloroplasts, the organelle where photosynthesis takes place.

In many leaf blades, particularly those of dicots, the mesophyll is divided into a **palisade layer,** consisting of one to three tiers of columnar cells situated just below the upper epidermis, and a **spongy layer** of irregular-shaped cells located in the lower half of the blade. As the name implies, the spongy layer of cells is suspended in a vast system of interconnected air spaces, so all the mesophyll cells in the leaf are in direct contact with air. This "open" arrangement of the mesophyll tissue aids in the rapid exchange of carbon dioxide and oxygen during photosynthesis.

To complete our picture of leaf anatomy, we need only add the "plumbing." Running through the middle of the mesophyll are veins, each consisting of two types of vascular tissue—xylem and phloem (see Figure 15–3a). The **xylem** is the "water conduit," supplying the leaf with water and dissolved minerals originally absorbed from the soil. The **phloem** trans-

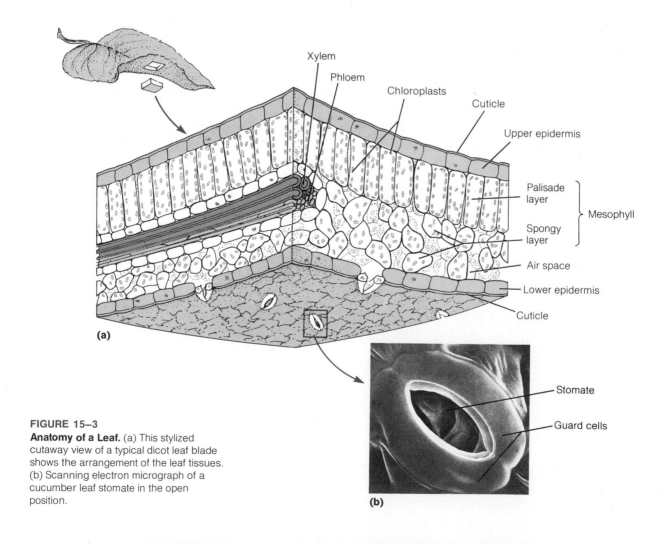

FIGURE 15–3
Anatomy of a Leaf. (a) This stylized cutaway view of a typical dicot leaf blade shows the arrangement of the leaf tissues. (b) Scanning electron micrograph of a cucumber leaf stomate in the open position.

ports sugars produced during photosynthesis in the mesophyll to other nonphotosynthetic regions of the plant.

THE OPENING AND CLOSING OF STOMATES. For photosynthesis to proceed efficiently, the leaf mesophyll must receive a continuous supply of carbon dioxide from the atmosphere. The epidermis with its cuticle presents a formidable barrier to carbon dioxide diffusion, however. Consequently, vascular plants have evolved stomatal pores on the surface of their leaves (and photosynthetic stems) to permit rapid carbon dioxide entry and oxygen release. One disadvantage of these openings, however, is that they allow water vapor in the air spaces of a leaf (or stem) to escape into the atmosphere, rendering the leaf susceptible to dehydration, particularly during hot, dry conditions. For all vascular plants, this problem has been partly solved by the evolution of a mechanism to control the opening and closing of the stomates, but other "solutions" have evolved as well (Essay 15–1). Under moderate conditions of soil moisture, humidity, and temperature, the stomates are open during the day and closed at night. Thus, carbon dioxide can enter the leaf during periods of photosynthesis, and the loss of water vapor is greatly curtailed at night when there is no need for carbon dioxide. Let us now investigate how this remarkable mechanism works.

Each stomate on the leaf surface is surrounded by a pair of specialized epidermal cells called **guard cells** (Figure 15–3b). The guard cells separate from one another when they swell up with water. This separation creates a pore—the stomate—between them. When the guard cells lose water and deflate, they collapse on each other, closing the pore.

How is the uptake and loss of water in the guard cells regulated? According to current theory, there are ion pumps located in the plasma membranes of the guard cells that, with the expenditure of ATP energy, transport potassium ions inward. At daybreak (with stomates initially closed), light activates these pumps and potassium ions begin to accumulate in the guard cells (Figure 15–4). Recall from Chapter 4 that solutes decrease water potential; as the potassium ion concentration increases in the guard cells, their water potential drops to below that of the surrounding epidermal cells. This water potential gradient favors the entry of water into the guard cells, causing them to swell and separate, creating a pore in the process (see Figure 15–4). At nightfall, the potassium pump is no longer activated. The potassium ions diffuse out of the guard cells, water follows, and the deflating cells collapse over the pore.

The question remaining is: How does light activate the ion pump? Although we have no clear answer to this question, there are some clues. The guard cells, unlike all other epidermal cells, have chloroplasts. Perhaps these chloroplasts provide the ATP energy

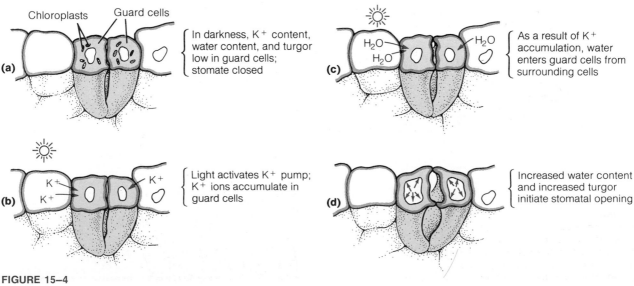

FIGURE 15–4
The Mechanism of Stomatal Opening.

ESSAY 15-1

STRATEGIES OF DESERT PLANTS

To the uninitiated, vast expanses of desert often appear bleak and lifeless. Rocky rubble or barren dunes may extend for miles, broken only by a distant mountain range. Temperatures typically climb high during the day and plunge at night. The air is dry; the annual rainfall is usually less than three inches and unpredictable as well. You might question how any plant could survive under such conditions, especially with the lack of water. Yet much of the desert harbors a unique and specially adapted assemblage of plants and animals. In this essay we will look at some of the special adaptations of plants to this harsh environment.

In the desert, competition for water is intense, as the wide spacing between shrubs reveals. Most of these shrubs, such as mesquite, have deep root systems that penetrate to the permanent ground water lying deep below the surface of most deserts. Other drought-tolerant species, such as cacti, have shallow but very extensive surface root systems. These plants are able to absorb available moisture quickly over a wide area, even from the lightest rain, and then store the water in succulent (juicy) stems or leaves for use during dry periods.

Some desert plants are actually drought evaders, spending the dry periods of the year as dormant bulbs or root stocks. When conditions become favorable, they sprout and grow rapidly, taking full advantage of the brief time before drought conditions return (figure a). Another group of desert plants are drought deciduous, losing their leaves during unfavorable times of the year (figures

The Mojave Desert in Southern California.

b and c). Still others produce two kinds of leaves: larger, more typical leaves during the early spring, when water is generally available, and small, relatively thick leaves when summer begins. The cells of these summer leaves have a high concentration of solutes, which slows the evaporation of water. In addition, the leaves have a low surface-area-to-volume ratio and thick, waxy cuticles which help to reduce water loss. Some plants, like the agave, have thick, leathery leaves year-round that also operate as water-storage organs (figure d). Cacti exhibit an additional type of leaf adaptation: the leaves are modified into spines (figure e). Photosynthesis takes place in the thickened, water-storing stems, and the spines discourage predators.

The structural adaptations of leaves just discussed are often accompanied by alterations of internal structure. For example, the cell walls of drought-adapted leaves are usually thick, and the stomates are recessed in cavities where air currents cannot sweep away the water vapor as quickly. The mesophyll cells are usually tightly packed, resulting in fewer intercellular spaces from which water vapor can escape. Finally, abundant sclerenchyma fibers impart mechanical strength and help to prevent injury due to wilting.

Even this brief examination of desert strategies testifies to the varied and well-adapted plant forms native to the desert. The hostile desert environment has indeed failed to exclude life. On the contrary, a diverse group of tenacious plants have met the challenge of survival.

Adaptations of Desert Plants. (a) Desert dandelions sprout and grow rapidly during the short spring season on the Mojave Desert. (b) The straggly ocotillo can grow and lose several sets of leaves in response to alternating wet and dry conditions. (c) Close-up of an ocotillo branch. (d) Agave plants have true leaves but reduce water loss by means of a thick cuticle. (e) The spines of cacti are modified leaves. The thick stems store water.

needed to drive the potassium pump, a process that could take place only when light is available. There is also evidence suggesting that low CO_2 concentrations in the guard cells trigger stomatal opening. Light-activated photosynthesis in the mesophyll tissue and guard cells would certainly contribute to lowering the leaf CO_2 concentration by simply fixing it into carbohydrate.

Roots

While the leaves are busy producing carbohydrates for the plant, the roots are performing other vital functions. The two primary roles of the roots are: (1) to provide anchorage for the shoot system, and (2) to absorb water and minerals from the soil. Many plant roots are also modified to store water and food, particularly in crop plants such as turnips, beets, and carrots.

When a plant seed germinates (begins to grow), the first embryonic part to emerge is usually the primary root. In plants that have a **taproot system,** the primary root continues to dominate as the plant grows, while smaller lateral roots branch from it (Figure 15–5a). In other plants, notably the grasses, the primary root soon becomes inconspicuous in a tangled mat of new roots that originate from the base of the stem. The result is a **fibrous root system** (Figure 15–5b).

The extent of a plant's root system can be enormous, as you may have discovered if you have ever tried to dig up a large shrub or tree. In measurements taken on a fairly small sedge plant with a total leaf surface area of 50 square centimeters, the total root surface area was 350 square centimeters, and the combined length of all the roots was over 10 meters (about 40 feet).

THE ROOT SURFACE. As roots grow through the soil, their root tips are subject to abrasion from the stationary soil particles. As a form of protection, the root tip is covered by a mound of cells called the **root cap,** which continually produces new cells to replace those that have worn away (Figure 15–6a). Several millimeters above the root cap are **root hairs** extending out from the root surface. Each root hair is the outgrowth of a single epidermal cell. Root hairs present a large surface area in direct contact with the soil, and thereby facilitate the absorption of water and dissolved minerals by the root's outermost layer of cells, the **epidermis.** The epidermis also serves as a barrier to the entry of infectious microorganisms.

ROOT ANATOMY. Just inside the root epidermis is a region called the **cortex,** which is composed primarily of parenchyma cells (see Figure 15–6a, top). **Parenchyma** are thin-walled, relatively unspecialized cells that can have a variety of functions depending on their location in the plant. The parenchyma cells of the root cortex function primarily in the active uptake of mineral ions that have moved across the epidermis, but they may also be modified to store water or starch.

The central core of the root is occupied by a complex array of tissues collectively known as the **vascular cylinder** (Figure 15–6b). The outermost tissue of this cylinder is the **endodermis,** a one-cell-layer sheath that completely encloses the inner tissues. The endodermal cells are unique in that part of their cell

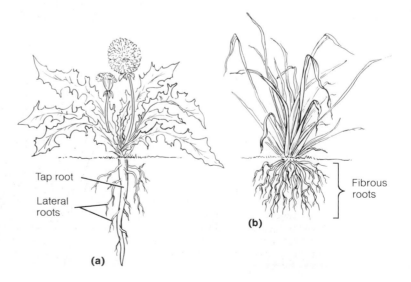

FIGURE 15–5
Root Systems. (a) Taproot systems are characteristic of many dicot species. (b) Fibrous root systems are common among monocots.

Tap root
Lateral roots

Fibrous roots

(b)

(a)

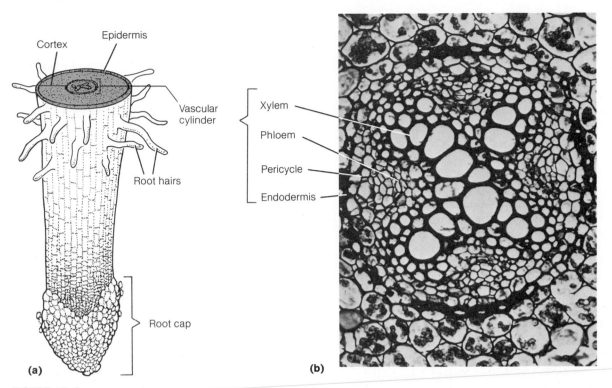

(a)

Cortex

Epidermis

Vascular cylinder

Root hairs

Root cap

(b)

Xylem

Phloem

Pericycle

Endodermis

FIGURE 15–6
Anatomy of the Root. (a) Three-dimensional view of a root tip. (b) Cross-sectional view of the root's vascular cylinder. Note the large water-conducting cells of the xylem.

walls are impregnated with a waxy material impervious to water and dissolved minerals. This collar of wax known as the **Casparian strip** regulates the flow of water and minerals from the cortex to the inner tissues of the vascular cylinder (see Chapter 16 for more details).

Interior to the endodermal sheath is the **pericycle,** a tissue made up of one to several layers of parenchyma cells. Pericycle cells retain the potential to undergo cell divisions, and under the appropriate circumstances, they divide to produce lateral roots (Figure 15–7). In woody plants, the pericycle also initiates cell divisions that ultimately result in the secondary growth of roots—a type of growth that is responsi-

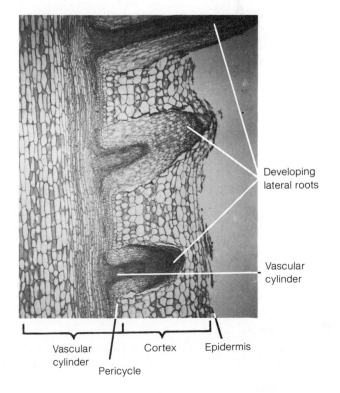

Developing lateral roots

Vascular cylinder

Vascular cylinder

Pericycle

Cortex

Epidermis

FIGURE 15–7
Formation of Lateral Roots. Lateral roots originate from cell divisions begun in the pericycle, and grow through the cortex and epidermis before emerging in the soil. Note that each lateral root develops its own vascular cylinder, which remains connected with the vascular cylinder of the main root.

ble for increases in root diameter (see "Secondary Growth" section at the end of this chapter).

Inside the pericycle are the two vascular tissues, xylem and phloem. In most plants, the xylem forms the central, often star-shaped core of the root, and the phloem occurs in strands positioned just outside the xylem core (see Figure 15–6b). Let us now discuss these two important tissues in more detail.

Xylem and Phloem

The xylem is a complex tissue that runs continuously from the roots through the stem to the leaves (or other shoot structures, such as flowers and fruits). Its primary function is the long-distance transport of water and dissolved minerals. Located in the xylem are two cell types specialized for water conduction: tracheids and vessel elements. **Tracheids** are long, thin, tapering cells with thick cell walls (Figure 15–8a). Characteristic pits, or thin areas on adjoining walls, permit the passage of water and minerals from one tracheid to the next. Tracheids overlap one another along their tapered end walls, forming columns of cells running lengthwise through the roots and stems. **Vessel elements** (see

Figure 15–8a) are generally shorter and broader than tracheids, and during their maturation their end walls become largely if not completely digested away. Stacked end to end like barrels, the vessel elements form a continuous tube, or vessel, for the rapid transit of water and minerals. At maturity both tracheids and vessel elements are dead: they lack protoplasm and consist entirely of cell walls.

Like the xylem, the root phloem is continuous with phloem in the stems, leaves, and other shoot structures. The phloem sap contains sugars and minor amounts of other organic compounds, and these substances are transported in conducting cells called **sieve tube members** (Figure 15–9). These are elongated cells whose tapered end walls are dotted with tiny pores; the pores of one member align with those of the next. Long columns of sieve tube members aligned end to end form a continuous transport channel called a **sieve tube.** At maturity, sieve tube members lack nuclei and many of their original cytoplasmic organelles—structures that, if present, would impede the flow of sugar sap. Sieve tube members are often associated with companion cells (see Figure 15–9), the function of which remains unclear.

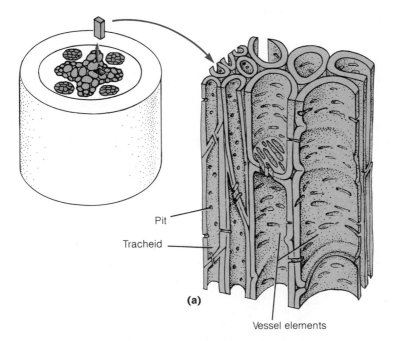

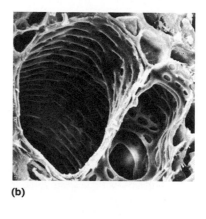

FIGURE 15–8

Water-Conducting Cells of the Xylem. (a) This cutaway view of the root xylem shows the long, narrow tracheids with pits in their thick cell walls. Water and dissolved minerals move between adjoining tracheids through the pits. Also shown are two types of vessel elements. The slightly tapered end walls of the vessel elements on the left have large, oblong pores for the passage of water. The shorter, broader elements on the right have no end walls and offer less resistance to the passage of water. (b) Scanning electron micrograph of two large vessel elements of a cucumber root. Note the thick, highly sculptured cell walls.

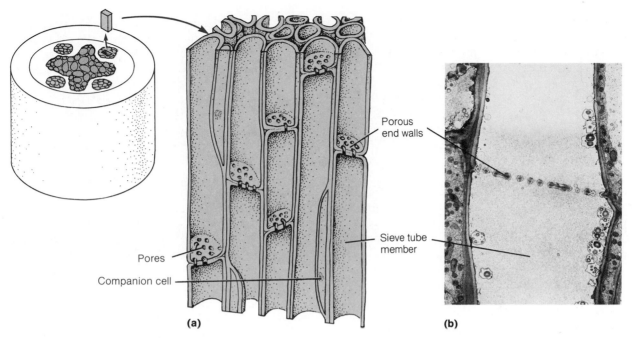

FIGURE 15–9
Food-Conducting Cell of the Phloem. (a) This diagram shows part of the root phloem consisting of sieve tube members stacked end to end. Adjoining end walls have common pores. Each sieve tube member is associated with a companion cell. (b) Electron micrograph of adjoining sieve tube members showing the pores in their end walls. Note the scarcity of cytoplasmic components in these sugar-conducting cells.

Stems

In their most usual forms, stems are the trunks and branches of trees and shrubs, the climbing vines of grapes and ivy, and the upright, leaf-bearing stalks of herbs and grasses. Their primary purpose is to support and distribute the leaves in such a way that the plant can take maximum advantage of available sunlight. In other words, the leaves are generally spread out such that they do not shade one another.

Structurally, all stems are divided into **nodes,** where one or more leaves are attached, and **internodes,** the stem region between nodes (see Figure 15–1). At each node just above the attachment point of a leaf is one or more **lateral buds.** These buds are young, inactive shoots that can grow into new leaf-bearing branches. Each stem terminates in a **shoot apex.**

Some not-so-common stems include the horizontal runners of strawberries and bermuda grass, the underground rhizomes of iris and most ferns, the tubers of potatoes and Jerusalem artichoke, and the corms of gladiolus, crocus, and taro. Rhizomes, tubers, and corms are underground stems, many of which are modified for storage of food.

STEM ANATOMY. Like all plant organs, the surface of young stems is covered by an epidermis and its waxy secretion, the cuticle. Just inside the epidermis is the cortex (Figure 15–10). The stem cortex is composed primarily of parenchyma cells, which generally have a storage function. In many plants, the outermost tiers of cortical parenchyma contain chloroplasts, which impart a green color to the stem. These cells supplement the leaves' production of food by carrying out photosynthesis. Also found frequently in the stem cortex are strengthening tissues called **sclerenchyma** (Figure 15–11). With their thick walls, these cells provide mechanical support for stems.

In stems, as in roots, the xylem is located interior to the phloem. Instead of forming a central vascular cylinder, however, the stem xylem and phloem are arranged into **vascular bundles.** In dicots these bundles form a ring between the cortex and the centrally located **pith;** in monocots the vascular bundles are scattered throughout the stem (see Figure 15–10).

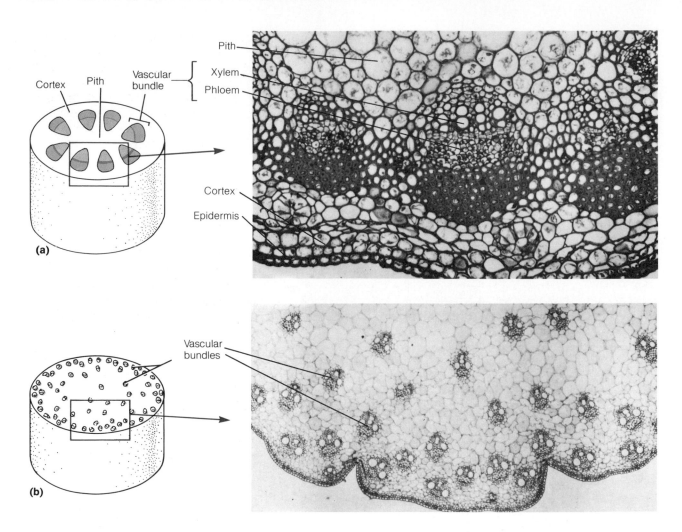

FIGURE 15–10

Anatomy of Dicot and Monocot Stems. (a) Cross section of the stem from a dicot. The vascular tissues are arranged in bundles, which in turn are arranged in a cylinder between the cortex and the pith. The photomicrograph shows part of a cross section from a dicot stem. (b) This diagram and photomicrograph of a corn stem in cross section demonstrate the scattered arrangement of vascular bundles typical of monocots.

FIGURE 15–11

Sclerenchyma. The major types of sclerenchyma cells are sclereids and fibers. They are noted for their very thick cell walls. (a) Sclereids are especially common in the tough fruit walls of nuts and in the seed coats of many species. (b) Fibers are very long and narrow and have tapered end walls. In some plants, fiber cells may attain lengths of up to 20 cm (nearly 8 inches).

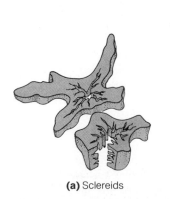

(a) Sclereids

(b) Fibers

THE GROWTH OF VASCULAR PLANTS

The arrangements of tissues in leaves, roots, and stems discussed previously do not just happen. Rather, these tissues with their specialized structures and functions result from an orderly, genetically-programmed sequence of growth and developmental events. In this last section of the chapter we will consider the various growth processes that shape a typical vascular plant.

Plants exhibit a type of growth pattern called **indeterminant.** This means that unlike animals, which have a distinct growth period that culminates in some genetically-determined mature size, plants grow throughout their lifetimes. In plants, new cells of all types are produced continuously in special localized regions called **meristems.** There are no such regions of persistent growth in animals.

Vascular plants have two major types of meristems: apical meristems and lateral meristems. The activities of **apical meristems** result in the increase in *length* of stems and roots and the production of leaves and floral organs. These processes are called **primary growth.** Increases in the *diameter* of stems and roots occur through the activities of **lateral meristems,** and are called **secondary growth.**

Primary Growth

Apical meristems are located at the tips, or apices (singular apex), of shoots and roots (Figure 15–12). Each apical meristem is composed of several thousand meristematic cells. Divisions of the meristematic cells yield new cells that subsequently enlarge and differentiate to become specialized cells. If a daughter cell is destined to become a stem or root cell (produced by a

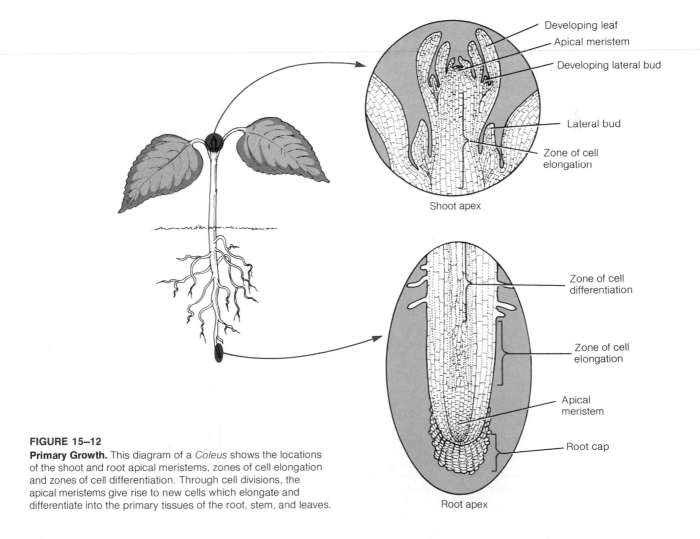

FIGURE 15–12
Primary Growth. This diagram of a *Coleus* shows the locations of the shoot and root apical meristems, zones of cell elongation and zones of cell differentiation. Through cell divisions, the apical meristems give rise to new cells which elongate and differentiate into the primary tissues of the root, stem, and leaves.

shoot or root apical meristem, respectively), it will enlarge predominantly in the longitudinal direction. This elongate growth occurs in a region just behind the apical meristem known as the **zone of cell elongation** (see Figure 15–12). These cells may attain a final length of up to 200 times their original size. During the final stages of cell elongation, the cell begins to take on its specialized shape and function. By the time this maturation process is in full swing, the cell will be located still farther away from the apex in what is called the **zone of cell differentiation.** At the end of this developmental series of events, the cell will be fully enlarged and differentiated to carry out a specific role in the plant. For example, a maturing cell located on the external surface near the root tip will develop into an epidermal cell, and may grow a root hair; another cell maturing just below the mature zone of stem

xylem may develop a thick cell wall, lose its protoplasm, and become a water-conducting tracheid. Thus, what a cell develops into depends on its specific location in the zone of differentiation in either a stem or root tip.

Secondary Growth

In virtually all monocots and in most herbaceous dicots (such as peas and beans) only primary growth takes place. All the cells result from the activities of apical meristems. In gymnosperms (such as pines, firs, and cedars) and woody dicots (such as oaks, elms, and other hardwoods), however, lateral meristems develop soon after the primary tissues are formed. One of these lateral meristems, the **vascular cambium,** produces the major share of secondary growth, which accounts

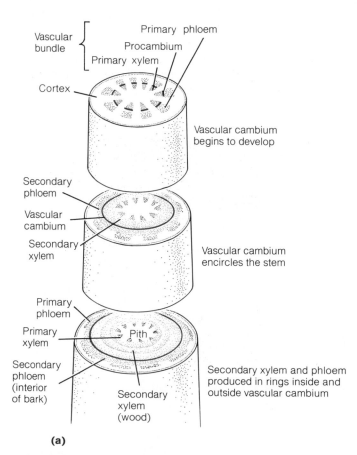

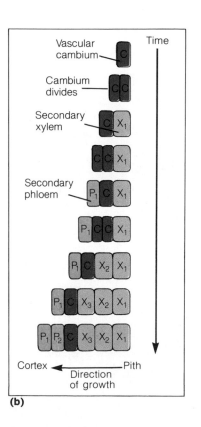

FIGURE 15–13

Secondary Growth. (a) The vascular cambium arises initially from the procambium that lies between the primary xylem and phloem of the vascular bundles, but it soon extends between the bundles to form a continuous ring. (b) In this time-sequence diagram, one of the daughter cells produced by the division of a vascular cambium cell (C) differentiates into either a secondary xylem (X) or secondary phloem (P) cell; the other daughter cell remains a cambium cell.

for most of the increases in root and stem diameters. In roots, the vascular cambium originates from the pericycle (see Figure 15–7). In stems, the vascular cambium arises initially from the **procambium,** a layer of cells located between the xylem and phloem in the vascular bundles (Figure 15–13a, top). However, the vascular cambium soon extends between the vascular bundles to form a continuous ring of dividing cells (Figure 15–13a, middle). For the sake of brevity, we will restrict our comments regarding secondary growth to what happens in stems; keep in mind, though, that similar events take place in roots.

In stems the vascular cambium gives rise by cell divisions to an inner layer of **secondary xylem** and an outer layer of **secondary phloem** (see Figure 15–13a). That is, when a cambium cell divides, one of the daughter cells remains a cambium cell, whereas the other differentiates into a secondary xylem cell (if it lies toward the interior of the stem) or a secondary phloem cell (if it lies toward the exterior) (Figure 15–13b). The secondary xylem is more commonly known as **wood;** the secondary phloem becomes the innermost layers of **bark.** Recently produced secondary xylem and phloem serve in the conduction of

water and sugars, respectively. As concentric layers of secondary xylem and phloem cells accumulate inside and outside the vascular cambium, the stem increases in diameter.

As a result of the pressure created by the expanding secondary xylem, the tissues external to the vascular cambium eventually get crushed and crack. If this situation proceeded unchecked, the stem tissues would dehydrate and be subject to infection. Protection is afforded, however, by the development of another lateral meristem called the **cork cambium.** The cork cambium originates in the cortex outside the vascular cambium and produces **cork** cells by cell division. The cell walls of cork cells are impregnated with the same waxy substance found in root endodermal cell walls. The cork thus replaces the crushed and cracked epidermis in its function of preventing excessive water loss and infection.

With yearly additions of secondary xylem, the cork and secondary phloem tissues produced earlier also become crushed. A new cork cambium forms every several years inside the older, crushed tissues. The outer, cracked tissues eventually slough off in the form of peeled bark. Technically, all the tissues lying

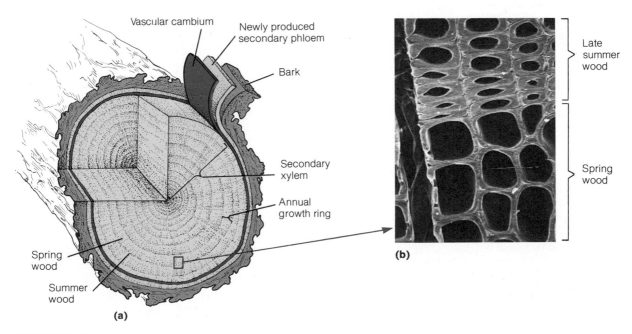

(a)

(b)

Vascular cambium
Newly produced secondary phloem
Bark
Secondary xylem
Annual growth ring
Spring wood
Summer wood
Late summer wood
Spring wood

FIGURE 15–14
Annual Growth Rings. (a) Annual rings result from a size difference in the secondary xylem cells that develop during the growing season. (b) As can be seen in this scanning electron micrograph of a small portion of one ring, relatively large cells are produced in the spring and early summer months, and smaller, thicker-walled cells form in the late summer. The latter cells account for the dark rings separating each annual growth increment.

outside the vascular cambium—secondary phloem, cork, and cork cambium—constitute the bark.

When a tree trunk or branch is cut and viewed in cross section, the wood often shows a concentric pattern of alternative light and dark bands. Each pair of light and dark bands is an **annual growth ring,** the secondary xylem produced during one entire growing season (Figure 15–14a). The light-colored bands, called **spring wood,** are composed of relatively large cells with medium-thick cell walls; smaller, thicker-walled cells comprise the dark bands, or **summer wood** (Figure 15–14b). During the late autumn and winter in temperate regions, cell divisions in the vascular cambium cease. Because of this seasonal variation in secondary growth, you can estimate the age of a section of stem (or root) by counting the number of annual rings. The widths of the individual rings can vary yearly in response to environmental conditions such as the availability of water or minerals, and temperature. Thus, scientists can get a general idea of recent past climate patterns in a given area by analyzing the size of growth rings in woody plants.

The oldest, innermost part of a woody stem often appears dark where various tars and resins have accumulated; this region is called **heartwood.** The secondary xylem produced during the most recent two or three years of growth is called **sapwood,** for it is the functional xylem that conducts sap (water and minerals) from the plant's roots to its shoot system. As a tree increases in diameter, its heartwood increases proportionally, whereas the thickness of the sapwood remains relatively constant.

SUMMARY

The structure of higher plants reveals many of their adaptations to an autotrophic existence. An extensive root system, with its multitude of tiny root hairs, is well suited to its role of absorption and anchorage. The internal structure of stems is clearly designed for support and transport. The leaf, with its large surface area, chloroplast-rich cells, and numerous intercellular spaces and stomates, is a highly efficient photosynthetic organ. Over 400 million years have gone into establishing the adaptive design evident in living land plants.

In the leaf, one or more vascular bundles from the stem carry over through the petiole, branching into tiny veins in the leaf blade. The veins are arranged parallel to one another in monocot leaves, but form a netted pattern in dicots. Each vein, composed of xylem and phloem, is embedded in the photosynthetic mesophyll tissue. The proximity of the mesophyll cells to both air spaces and the vascular tissues facilitates the efficient exchange of gases and the transport of materials into and out of the leaf. The mesophyll is in turn sandwiched between the upper and lower epidermis.

In vascular plants, stomates are present in the leaf epidermis to permit the exchange of carbon dioxide and oxygen between the leaf and the atmosphere. Each of these pores is surrounded by two guard cells, which when fully turgid creates a wide pore. Loss of water from the guard cells causes them to collapse and close the pore. The stomates are generally open during the day (which is convenient for photosynthesis) and closed during the night.

Both roots and stems have a protective epidermis, which overlies a cortex. In roots, the vascular tissues—xylem and phloem—occupy the central region. They are surrounded by the pericycle and endodermis, forming a vascular cylinder. In stems, the vascular tissues occur in bundles. These vascular bundles form a cylinder in dicot stems and have a scattered arrangement in monocots.

In higher plants, cell divisions occur primarily in meristems. Apical meristems located at the tips of roots and shoots generate new cells, which subsequently elongate and differentiate into the primary tissues of the plant.

In gymnosperms and woody dicots, secondary growth in stems and roots occurs as a result of cell division in two lateral meristems called the vascular cambium and the cork cambium. In stems, the vascular cambium arises from the procambium (in the vascular bundles) and the adjoining cortex cells. Forming a continuous ring of cells, the vascular cambium produces secondary xylem (wood) inwardly and secondary phloem toward the stem's periphery. The cork cambium originates from a ring of cortex cells and generates cork cells by cell division. The cork seals up cracks and fissures created by the internal expansion of the secondary xylem. The secondary xylem is produced seasonally, each growing season accounting for one annual ring. The number of rings corresponds to the age in years of the stem.

STUDY QUESTIONS

1. Describe how the anatomy of a leaf is suited to the leaf's function of photosynthesis.

2. Why are stomates important to the survival of land plants?

3. Name two primary functions of root systems. How do tap-root systems differ from fibrous root systems?

4. How are the conducting cells of the xylem and phloem different in structure and function? How are they similar?

5. Which tissues in roots are not found in stems? How do the arrangements of vascular tissues in these organs differ?

6. Distinguish between a monocot and a dicot in terms of their stem cross sections and their leaf structures.

7. What is the relationship between meristems and the indeterminate growth pattern of plants?

8. What are the distinctions between primary and secondary growth? Which kinds of plants exhibit secondary growth?

9. Explain how the light and dark bands of annual growth rings are formed.

SUGGESTED READINGS

Esau, K. *Anatomy of Seed Plants.* 2nd ed. New York: Wiley, 1977. This advanced text is nicely illustrated and has a very complete discussion of secondary growth (Chapters 8–12).

Fritts, H.C. *Tree Rings and Climate.* New York: Academic Press, 1976. This book discusses in detail how tree rings reflect the environmental conditions a plant has grown under during its lifetime.

Raven, P.H., R.F. Evert, and H.A. Curtis. *Biology of Plants.* 3rd ed. New York: Worth, 1981. Chapters 22–24 of this beautifully illustrated textbook discuss plant structure and growth.

Ray, P.M., T.A. Steeves, and S.A. Fultz. *Botany.* Philadelphia: Saunders, 1983. Chapters 8–13 of this text provide an in-depth look at plant structure and function.

Zimmermann, M.H., and C.L. Brown. *How Trees Grow.* Studies in Biology, No. 39. Baltimore: University Park Press, 1973. This small book is a good introduction to the details of secondary growth.

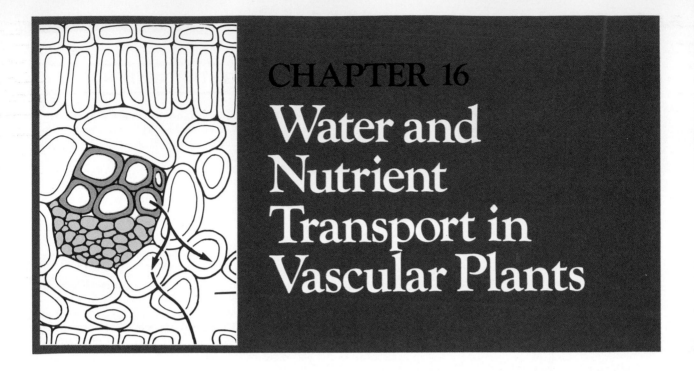

CHAPTER 16
Water and Nutrient Transport in Vascular Plants

One of the advantages of being multicellular is that different groups of cells can be specialized in form and function to carry out different activities, incorporating features in their design that render them most effective in their specific tasks. For example, land plants have leaves well suited for carrying out photosynthesis, and the structure of a root is well-designed for its roles of anchorage and absorption. But specialization also demands interdependence. No matter what specialized activity a given cell conducts in a multicellular organism, it depends on the activities of other cells to survive. The leaves must obtain water and mineral nutrients to remain healthy, and they "count on" the roots to absorb these vital substances. Conversely, the roots require carbohydrate provided by the photosynthetic leaves. Thus, substances produced or obtained in one region must be transported to other regions. This transport is the main function of the vascular tissues in ferns, gymnosperms, and angiosperms.

The vascular system in plants is not an active circulatory system like that found in higher animals. There are no hearts, no pumps, and no contracting muscles to push sap from one place to another. Nevertheless, plants are able to transport materials rapidly over great distances, often against gravity. For example, water moves up a redwood tree from roots below the ground to leaves 100 meters (about 330 feet) above it. In this chapter we will apply your knowledge of plant structure (from Chapter 15) and cellular water relations (from Chapter 4) to examine the pathways

and mechanisms by which vascular plants take up and transport water and minerals, and transport sugars.

WATER TRANSPORT

Water is essential to the life of a plant. As anyone who has forgotten to water their plants can testify, a constant supply of water is necessary to maintain normal plant shape. Without sufficient water, plant leaves and nonwoody stems wilt due to a loss in cell turgor. Wilting also causes the stomates to close, which greatly retards the diffusion of carbon dioxide into the leaf to support photosynthesis. Water is also a chemical reactant in several types of metabolic processes in plants, including photosynthesis. Another important role of water involves plant growth: the uptake of water into a plant cell accounts for most of its increase in size. Finally, and most importantly, liquid water is an absolute requirement for cellular metabolism. Water gives shape to enzymes, nucleic acids, and other macromolecules necessary not only to maintain the structural integrity of cellular components, but also to permit chemical reactions necessary for life to take place. Dry enzymes catalyze no reactions, dry phospholipids form no membranes; in short, a dehydrated cell does not "live."

As important as water is to the life of a plant, you might be surprised to learn that up to 97% of the water absorbed by a plant's root system may be lost by evaporation from its leaves. This evaporative loss of water is called **transpiration.** It occurs because leaves have a large surface area, numerous stomates, and a vast system of internal air spaces—an architecture ideally suited for rapid gas exchange necessary to achieve high rates of photosynthesis, but one that also favors high rates of evaporation. A single corn plant may lose more than 2 liters (about 2 quarts) of water per day by transpiration during the peak growing period; an acre of corn will transpire as much as 1,000,000 liters (roughly 265,000 gallons) over a four-month growing season (Table 16–1). This helps explain why agriculture uses approximately half of the water consumed in the United States each year.

Although transpiration is generally regarded as a necessary evil—the price the plant pays to obtain carbon dioxide for photosynthesis—it also provides some benefits for the plant. In the tropics, where high air temperatures and intense sunlight can produce

TABLE 16–1
Transpiration Rates.

Plant		Rate in Liters*
Per day, midsummer:		
3.5-meter cactus		0.02
Ragweed		5.5–6.5
3-meter apple tree		9.5–19
Coconut palm		65–75
Date palm		380–470
Per growing season:		
Tomato	100 days	115
Sunflower	90 days	475
Apple tree	188 days	6,800
Coconut	365 days	16,000
Date palm	365 days	133,000

*1 liter = 1.06 quarts

very high leaf temperatures, transpiration has a cooling effect on leaves, just as the evaporation of perspiration cools our bodies when we're hot. More significantly, leaf transpiration is the major driving force for water and mineral transport from the roots to the shoot. We will look closely at the mechanism of water transport shortly, but first let us trace the pathway of water as it moves from the roots to the leaves of the shoot.

The Pathway of Water Transport

The root tissue in direct contact with the soil is the epidermis, so our discussion of the pathway of water transport in plants must begin here. Water is first absorbed by the epidermis, then moves through the cortex, across the endodermis and pericycle, and into the conducting cells of the root xylem (see Figure 15–6 for a review of root anatomy). From the root xylem the water is conducted upward via the stem xylem to the veins of the leaves. This description is merely a tissue by tissue account of the water transport pathway, and serves as an overview for the more detailed discussion that follows. Let us return to the epidermis.

When soil water contacts a root, it is absorbed by the outer cell walls of epidermal cells such as root hairs. These cell walls, like those of all other plant cells, are composed of polar macromolecules (such as cellulose, pectins, and proteins) that actually attract water. Once in an epidermal cell wall, the water can move into the root via two possible micropathways: (1) it may wend its way through the cell walls of the epidermal cells into the adjoining walls (and intercellular spaces) of the cortex, or (2) the water may enter the epidermal cell cytoplasm by osmosis (Figure 16–1). In the cell

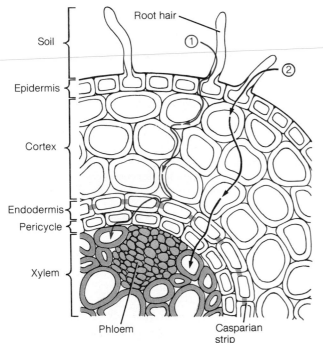

FIGURE 16–1
Micropathways of Water Movement in the Root. As indicated by the arrows, water can move through the root epidermis and cortex through (1) the system of adjoining cell walls and intercellular spaces, or (2) via the cytoplasmic route, flowing from cell to cell through plasmodesmata. Because of the Casparian strip at the endodermis, water must enter the cytoplasmic micropathway to cross this sheath of cells.

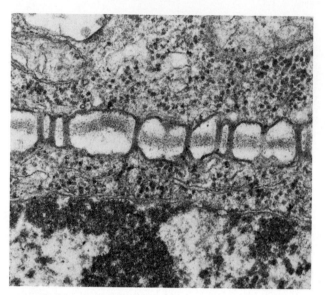

FIGURE 16–2
Plasmodesmata. This electron micrograph shows the membrane-lined plasmodesmata extending across the cell walls of two adjoining cells. Water and small molecules can pass through these cytoplasmic bridges. (Magnification: 53,000×.)

wall pathway, the water can move through the epidermis and the entire cortex without ever crossing a cell membrane, much like water moves across a blotter. Alternatively, if the water enters the epidermal cell cytoplasm, it can then flow into the cytoplasm of a neighboring cortical cell through plasmodesmata. **Plasmodesmata** are tiny bridges of cytoplasm that extend across the cell walls of adjoining cells (Figure 16–2). Since all the cortical cells are interconnected to one another by plasmodesmata, the water can continue inward toward the vascular cylinder via the network of interconnected cortical cell cytoplasms.

The two micropathways of water movement—the cell wall and cytoplasmic routes—are not mutually exclusive. Water moving through the system of cell walls may enter any cell's cytoplasm along the way, and cytoplasmic water may diffuse out into the cell wall space. This is true for the root epidermis and cortex, but the cell wall pathway ends abruptly at the endodermis, the sheath of cells encasing the vascular cylinder. Recall from Chapter 15 that the cell walls of the endodermis are impregnated with a water-impermeable wax, forming a barrier to water and mineral movement known as the Casparian strip. Any water that gets through the endodermis must go via the cytoplasmic route at this point (Figure 16–3), a requirement that has more bearing on mineral transport than on water movement (see section entitled

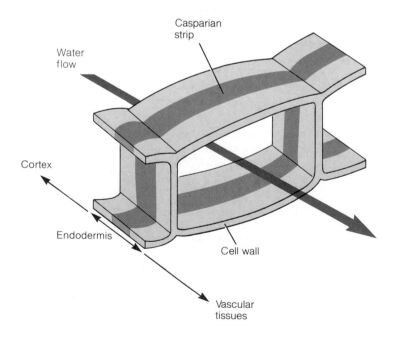

FIGURE 16–3
The Casparian Strip. This cutaway view of an endodermal cell shows the location of the Casparian strip, a collar of wax that in effect permits only cytoplasmic water and minerals to move through the endodermis.

"Transport of Mineral Nutrients" later in this chapter). Having negotiated the endodermis, the water can continue inward through the pericycle cytoplasm or cell walls, and finally enter the xylem (see Figure 16–1).

The conducting cells of the xylem—tracheids and vessel elements—are devoid of any protoplasm (see Figure 15–8). Each of these cells is filled with water, and because the tracheids or vessel elements are arranged end to end in long chains that extend from the roots through the stems and into the leaves, there are continuous columns of water that extend from the root xylem to the veins in the leaves. As we shall see in the next section, these uninterrupted columns of water in the xylem conducting elements are important to the actual mechanism that underlies water transport.

From the conducting cells of the leaf xylem, water moves into a layer of bundle sheath cells, and from there into the mesophyll (Figure 16–4). Both micropathways of water movement are operational here. Ultimately, most of the leaf water evaporates from the mesophyll cell walls into the surrounding air spaces. These air spaces are typically saturated with water vapor, and when the stomates are open, the vapor diffuses into the less humid air outside the leaf.

The Mechanism of Water Transport

We have traced the basic path of water movement in vascular plants, and have seen that water exists in continuous liquid columns in the xylem from the roots to the leaves. We are now ready to examine the mechanical aspects of water transport. In other words, what are the forces involved in the long-distance movement of water? Is water pushed up from the roots or pulled up from the leaves? Under most circumstances, the water is *pulled* up the plant, and the pulling force is created in the leaves by evaporation. Recall that water tends to move from a region of higher water potential to one of lower water potential (see Chapter 4). On a typical day, the concentration of water in the air surrounding the leaves is lower than it is in the air spaces of leaves, creating a water potential gradient that favors the diffusion of water vapor out of the leaves. This loss of vapor in turn creates a water potential gradient between the leaf's air spaces and the liquid water in the mesophyll walls, causing evaporation from the cell walls. As the mesophyll walls become slightly drier, their water potential drops, triggering water movement from neighboring cells. In fact, the water adhering to the mesophyll cell walls is continuous with liquid water present in the other leaf cells, including the water-conducting cells of the leaf xylem. As water moves to replace that lost by evaporation, a pulling force is created across this film of water that extends back to the leaf xylem. And because the water in the leaf xylem is continuous with water in the stem and root xylem, the pull is felt all the way back to the root tissues. Thus, evaporation from the leaf mesophyll cells sets up a chain of pulling forces that extends from leaf to root, so that when transpiration is occurring,

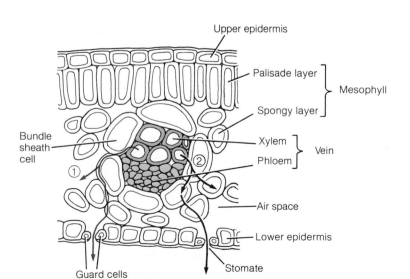

FIGURE 16–4

Micropathways of Water Movement in the Leaf. Water moving out of the leaf xylem passes into the cell walls of bundle sheath cells, a collar of photosynthetic parenchyma cells that encloses the vein. (1) Most of this water continues into the cell walls of the mesophyll; from here it evaporates. (2) Some of the water enters the cytoplasms of the bundle sheath and mesophyll cells by osmosis. Evaporation from the cell walls fills the leaf air spaces with water vapor, which escapes to the drier atmosphere when the stomates are open.

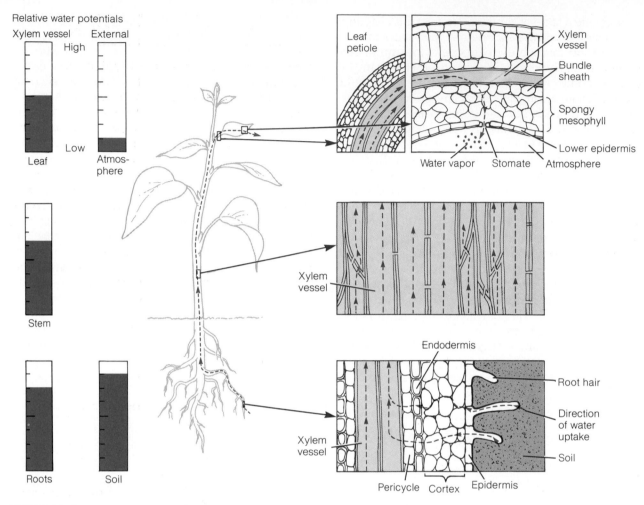

FIGURE 16–5
Evaporation-Driven Water Potential Gradient in a Plant. Evaporation from the leaf surface creates a pull on the liquid column of water in the leaf that extends back to the roots, creating a water potential gradient (bars at the left) that favors the upward movement of water along each step of the transpiration pathway.

water from the soil is actually pulled into the roots (Figure 16–5).

The beauty of this evaporation-driven mode of water transport in plants is that it is entirely passive. The plant does not expend any energy to transport water against gravity and the frictional forces encountered along the pathway. The main requirement for this mechanism to operate is that the air outside the leaf is drier (has a lower water potential) than inside the leaf, a requirement that is usually met during the day, unless it is raining or very foggy. A second requirement is that the liquid water columns in the xylem are intact and extend continuously from the leaves to the roots. Two physical properties of water, **adhesion** and **cohesion,** ensure that this is the case (see Essay 4–1). Water adheres (sticks) to polar molecules in the plant cell walls (for example, the walls of tracheids and vessel elements), and it also has a high degree of cohesion—the tendency for water to stick to other water molecules by hydrogen bonding. Therefore, as water molecules evaporate from the walls of the mesophyll cells, the force of adhesion attracts more liquid water toward these cell walls. The water molecules back in the leaf xylem cells are then pulled along by cohesion, and this movement in turn draws water up the continuous column stretching back to the roots.

When warm temperatures and/or low air humidity lead to rapid transpiration rates, you might think that the strong pull on the water column could cause the water molecules to break apart from each other. However, this does not happen because of the strong cohesion between water molecules in the column, and their adhesion to the walls of the conducting elements and other cells along the pathway. These two forces acting within the small diameter of the vessel or tracheid prevent the column of water from breaking. If you are skeptical about the capacity of liquid water to withstand strong pulling, place two wet glass plates together and then try to pull them apart. The cohesion between the water molecules and adhesion of the water molecules to the glass surfaces will bind the plates together, making it quite difficult to separate them.

MINERAL NUTRITION

Green plants have very simple nutritional needs: water, carbon dioxide (for photosynthesis), oxygen (for respiration), and a few inorganic nutrients, or minerals, obtained from the soil. We have already considered how plants obtain carbon dioxide, oxygen, and water. What remains, then, is to consider which mineral nutrients are essential for normal plant development, and how plants obtain and use them.

The mineral nutrients required by plants fall into two general categories: soil macronutrients and micronutrients. **Soil macronutrients** are those elements plants require in relatively large amounts. They include nitrogen, potassium, calcium, magnesium, phosphorus, and sulfur (Table 16–2). The plant absorbs these minerals from the soil water. Some of the key roles of these elements are listed in Table 16–2.

TABLE 16–2
Essential Plant Nutrients.

Element and Symbol	Major Form(s) Taken up by Plants	Approximate Percent of Plant Dry Matter	Major Roles in Plants
Carbon (C)	CO_2	45	Constituent of all organic compounds
Oxygen (O)	CO_2, H_2O	45	Constituent of most organic compounds
Hydrogen (H)	H_2O	6	Constituent of all organic compounds
Soil macronutrients:			
Nitrogen (N)	NH_3, NH_4^+, NO_3^-	1.5	Constituent of amino acids, nucleotide bases, vitamins, hormones, etc.
Potassium (K)	K^+	1.0	Major osmotic ion in cells for regulation of water potential
Calcium (Ca)	Ca^{+2}	0.5	Stabilizes membrane structure; forms cross-links between cell wall polysaccharides to provide rigidity
Magnesium (Mg)	Mg^{+2}	0.2	Constituent of chlorophyll; activates phosphate transfer reactions
Phosphorus (P)	$H_2PO_4^-$, HPO_4^{-2}	0.2	Constituent of nucleotides, phospholipids, sugar phosphates
Sulfur (S)	SO_4^{-2}	0.1	Constituent of two amino acids (cysteine and methionine) and certain vitamins
Soil micronutrients:			
Iron (Fe)	Fe^{+2}, Fe^{+3}	0.01	Constituent of electron transport proteins
Chlorine (Cl)	Cl^-	0.01	Required for the water-splitting reaction of photosynthesis
Manganese (Mn)	Mn^{+2}	0.005	Activates certain enzymes; also facilitates water-splitting reaction of photosynthesis
Zinc (Zn)	Zn^{+2}	0.002	Activates certain enzymes
Boron (B)	H_3BO_3	0.002	May facilitate sugar transport
Copper (Cu)	Cu^{+2}, Cu^{+3}	0.0006	Constituent of electron transport proteins
Molybdenum (Mo)	Mo^{+5}, Mo^{+6}	0.00001	Required for nitrate reduction

Soil micronutrients, or **trace elements,** are required by most plants in very small amounts. They include iron, chlorine, manganese, zinc, boron, copper, and molybdenum (see Table 16–2). Some plants also require silicon and/or sodium.

When you add a general chemical fertilizer to your lawn or houseplants, you are adding a form of nitrogen (usually nitrate or ammonium salts), potassium (usually potash), and phosphorus (phosphate salts). Some fertilizers also contain lesser amounts of iron, sulfur, and/or calcium. Such nutrients help your plants grow normally.

How do we know which mineral elements are required by plants? The answer has come through careful experiments using a system for growing plants called **hydroponics.** Typically, plants are grown to the seedling stage in washed sand or vermiculite (crushed mica that lacks soluble minerals), and are then transferred to jars containing a solution of known mineral elements (Figure 16–6). Each jar must be well aerated to provide the roots with O_2 for respiration. To determine whether a plant needs a given mineral, a control jar is set up containing all the minerals suspected of being necessary for normal plant growth—a complete mineral diet. Test jars contain all the necessary minerals except the one being tested. For example, if nitrogen is withheld from a test jar, in a week or so the plant will show symptoms of nitrogen deficiency—stunted growth, yellowing leaves, and dead spots on the leaves. While the control plant continues to grow normally, the nitrogen-starved plant will soon wither and die. For a given plant species, the deficiency symptoms for each required mineral are unique. With the aid of a book or a knowledgeable advisor, farmers can examine their crops to determine whether the soil contains too little of a particular mineral, and they can then take corrective action by applying the appropriate fertilizer.

Under natural conditions, mineral ions in the soil originate from weathered bedrock and decomposed organic matter. They dissolve in the water within the soil and plants absorb them into their roots along with the water (see Essay 16–1). The root cortex then plays a vital role in selecting which minerals are to be transported upward to the shoot system. Powered by ATP energy, cortical cell membranes actively transport certain mineral ions inward and keep nonessential and toxic substances out, often against large concentration gradients (see Chapter 4). Once inside the cells of the cortex, the minerals flow toward the root xylem along with the water in which they are dissolved. That is, they proceed from one cortical cell to another via the plasmodesmata interconnecting the root tissues. Although minerals may also move with the water traveling between the cortical cells, the Casparian strip prevents them from crossing the endodermis by this route. Therefore, only those mineral ions that are selectively absorbed *into* cortical or endodermal cells can reach the xylem, where they are transported upward in the transpiration stream.

FIGURE 16–6
A Hydroponic System for Determining Mineral Requirements. Seedlings started in sand are selected for uniform size, then transferred either to (a) a control jar containing a solution of all the mineral elements believed necessary for normal plant development, or to a test jar that lacks one of those elements, as shown in (b) and (c). After several weeks, the test plants are compared with the control plants. If an essential mineral was absent from the solution, the plant will exhibit one or more symptoms of deficiency.

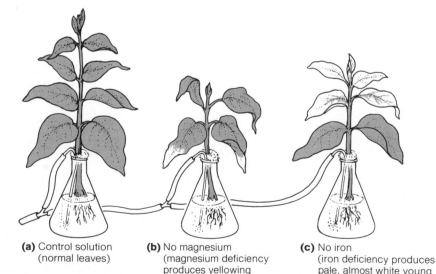

(a) Control solution (normal leaves)

(b) No magnesium (magnesium deficiency produces yellowing near tips of older leaves)

(c) No iron (iron deficiency produces pale, almost white young leaves with green veins)

ESSAY 16–1

MYCORRHIZAE

Lichens are organisms composed of an alga and a fungus—living together in a mutually beneficial relationship. Until recently, textbooks have treated the lichens as a well-known but rather isolated example of a plant–fungus partnership in which both organisms benefit from the association. All that has changed now with the discovery of **mycorrhizae** (meaning "fungus-roots"), fungi which invade the roots of plants. This relationship benefits both the plant and the fungus: the fungus obtains organic nutrients from the plant in exchange for enhancing the uptake of mineral nutrients for the plant. Mycorrhizal associations have been found in more than 90% of all families of higher plants!

Mycorrhizal fungi are usually found in the cortex of the roots, where the fungal hyphae either grow throughout the intercellular spaces or, more commonly, become embedded inside the cortical cells. In either case, the fungal hyphae extend outward and envelop the surface of the root, where they greatly increase the external surface area in contact with soil. The mycorrhizae may take over the function of root hairs and actually inhibit their formation in some plant species.

How much the various plant species depend on mycorrhizae for normal development varies greatly. Most orchids cannot exist without mycorrhizae. Many forest trees will not grow in grassland soils because mycorrhizae are absent there. However, if tree seedlings are first exposed to the appropriate fungus and then transplanted into grassland soil, they will grow normally. At the other end of the spectrum, many plants seem to thrive with or without mycorrhizae, particularly when sufficient minerals are available. In many instances, however, when the essential minerals are available only in limited amounts, plants with mycorrhizae fare much better than those without the fungus.

No one knows just how important a factor the mycorrhizal associations have been in terms of plant evolution and current plant distribution patterns. One thing is clear: we can no longer think of most plants as independent entities. Rather, they are partners in life with soil fungi.

SUGAR TRANSPORT

In addition to the essential mineral nutrients and water, all living plant cells must have a continuous supply of carbohydrates to be used as building blocks for the synthesis of other organic compounds, and as a fuel for respiration. The ultimate source of carbohydrate in the plant, of course, is the leaf (or photosynthetic stem). Because the leaves may be separated by a considerable distance from the nonphotosynthetic tissues, higher plants have a special "mass transit" system for transporting sugar from its production sites, or **sugar sources,** to areas of sugar demand, or **sugar sinks.** Recall from Chapter 15 that long-distance sugar movement takes place within the sieve tubes of the phloem, which consist of elongated cells linked end to end (see Figure 15–9). The sieve tubes, like the vessels or tracheids of the xylem, form an extensive network that extends into all the plant organs.

A comparison of the characteristics of sap movement in the xylem and phloem indicates that these two processes are quite different (Table 16–3). The different rates of movement (sugar transport is much slower), directions of transport, and physical mechanisms involved in sap transport are largely attributable to the contrasting sap compositions of these conducting tissues. Whereas xylem cells are filled with a dilute solution of minerals, the phloem sap contains a thick, syrupy mixture of sugars and small amounts of other organic compounds (see Table 16–3), the exact composition of which has been determined largely through the use of aphids (Essay 16–2). The high sugar concentration in the sieve tube favors osmotic entry of water into the tube, leading to high turgor pressures inside. As we shall see shortly, these high pressures are important to the actual mechanism of sugar transport.

The direction of phloem transport is always from

TABLE 16–3
Comparison of Phloem and Xylem Transport Characteristics.

Point of Comparison	Phloem	Xylem
Conducting cell	Sieve tube member	Vessel element or tracheid
Rate of sap movement	Up to 1 m/hr (3.3 ft/hr)	Up to 60 m/hr (roughly 200 ft/hr)
Sap composition	10–30% sugar, some amino acids, some hormones	Dilute solution of mineral ions
Direction of transport	From sugar sources to sugar sinks	From roots to leaves
Status of mature conducting cells	Alive; cytoplasmic structures greatly reduced	Dead; no cytoplasm

ESSAY 16–2

APHIDS AS LAB TECHNICIANS

The sap flowing in a sieve tube is a syrupy solution of sugars with small amounts of amino acids and hormones, and perhaps some other minor components as well. How have scientists come up with this information? The simplest and most direct approach, obviously, would be to cut through the outer tissues of a stem to the phloem. The sap in the sieve tube would push through the cut and it could be collected for chemical analysis. However, there are flaws in this technique. First, cutting through the plant's outer tissues would mix the contents of the outer cells with the otherwise pure phloem sap. Second, the loss of sap through the cut would cause the turgor pressure to drop at that part of the sieve tube. Water from adjacent cells would then move into the exuding sap by osmosis, diluting its contents and ruining the experiment.

Scientists have been able to sidestep these problems by using aphids as sap collectors. This common garden pest has a needlelike mouth part called a stylet that it inserts into a single sieve tube cell inside a stem. Because the stylet is extremely narrow, it causes little if any damage to cells. No manmade needle is as narrow or sharp as the aphid's stylet. Once the aphid has tapped into the sieve tube, sap pressure forces the sieve tube contents into the insect's digestive tract and out the other end, forming a tiny droplet of "honeydew" that can be collected and analyzed chemically.

A real fussy scientist could argue that the honeydew droplet is not pure but rather "processed" phloem sap. That is, as the sap passes through the aphid's gut, its composition will certainly be altered by the digestive process. To skirt this objection, the experimenter can simply anesthetize the aphid and then carefully sever its stylet. Pure sap will exude from the cut end for days, and can easily be collected for study.

By measuring the rate of sap flow out of the stylet, scientists have used the aphid technique to estimate rates of phloem transport. Thus, this pesky nuisance to the home gardener has been a useful tool to the plant physiologist.

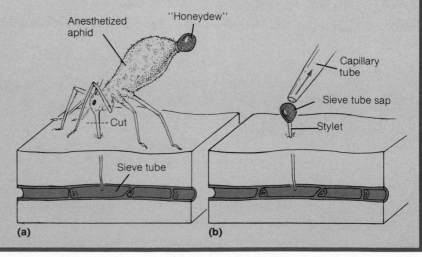

(a)   (b)

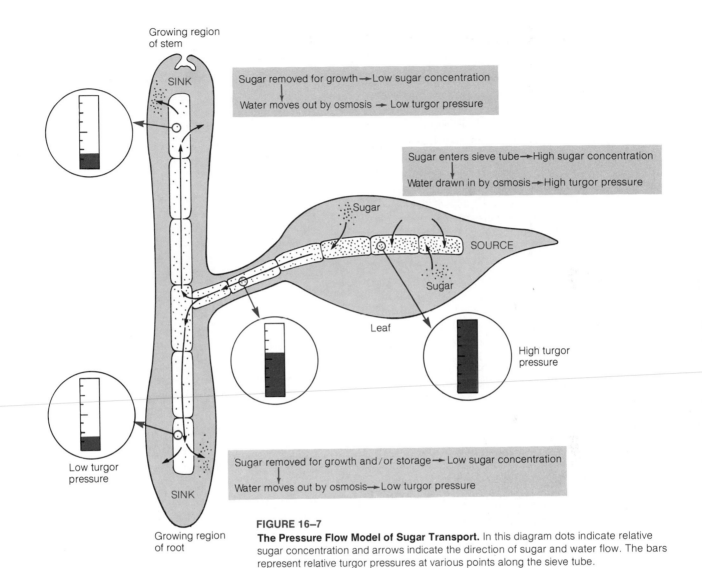

FIGURE 16–7
The Pressure Flow Model of Sugar Transport. In this diagram dots indicate relative sugar concentration and arrows indicate the direction of sugar and water flow. The bars represent relative turgor pressures at various points along the sieve tube.

a sugar source to a sugar sink. This usually means that transport takes place from the leaves to the roots or to the growing regions of the shoot. There are circumstances, however, in which the sugar sap flows from the roots to the leaves. For example, when a deciduous tree resprouts new leaves in the springtime, sugars are mobilized from storage tissues in the roots and transported upward to the young, expanding leaves. Note that the direction of sap movement is still from a sugar source (root) to a site where the sugars are used (the growing leaves).

The most widely accepted explanation for the mechanics of sugar transport is the **pressure flow model.** According to this model, sugar is actively transported from sugar-producing cells (such as leaf mesophyll) into the sieve tube cells of the phloem

(see Figure 16–4). The accumulation of sugar at the source end of the sieve tube causes water to move in from nearby cells, quickly raising the turgor pressure in these sieve tube cells. This pressure causes the sap to move toward sugar sinks in the plant, where the sugar is removed for utilization in metabolic growth processes or it is converted to starch. As sugar moves out, so does water by osmosis. Consequently, the turgor pressure drops at the sink end of the tube. The overall result is a turgor pressure gradient over the length of the sieve tube, with high turgor pressure at the sugar source end and low turgor pressure at the sugar sink end. The pressure gradient—and hence the flow of sap—is maintained as long as sugar is fed into the sieve tube at the source and removed for metabolism or storage at the sink (Figure 16–7).

SUMMARY

Water, minerals, and organic nutrients absorbed or manufactured in one part of the plant are transported to other parts in the xylem and phloem. The major driving force for water transport from the roots to the leaves is transpiration, the evaporative loss of water from leaves and sometimes from stem surfaces. Because the water in the leaf xylem is continuous with the water columns in the root xylem, its evaporation from leaves pulls water upward against gravity and frictional resistance. This pull creates substantial tensions on the water columns, but the columns remain intact because of water's strong adhesion to the xylem cell walls and cohesion between water molecules.

Minerals dissolved in the soil water enter the root and are selectively absorbed into the root cortical cells. Once inside the cortical cells, the water and minerals continue inward through the endodermis and pericycle via plamodesmata. Water and minerals can also move through the cell walls of the epidermis and cortex, but they cannot get past the Casparian strip of the endodermis via this micropathway. Thus, only the minerals able to penetrate the cell membranes of cortical or endodermal cells can gain access to the root xylem for upward transport.

All plants require certain minerals from the soil for normal growth and development. Soil macronutrients are mineral elements required in relatively large quantities; micronutrients need only be available in trace concentrations.

Sugar molecules are transported over long distances in the phloem sieve tubes from the tissues where they are produced (sugar sources) to the tissues where they are utilized (sugar sinks). The currently favored theory explaining the mechanics of sugar transport is the pressure flow model. According to this model, sugar is actively transported into the sieve tube at the source end and is actively removed at the sink end. As a result, water enters the sieve tube at the sugar source, creating a high turgor pressure there, and it tends to flow out of the tube at the sink end, leading to a reduced turgor pressure there. This turgor pressure gradient from source to sink in effect *pushes* the phloem sap through the sieve tube, resulting in a net transfer of sugar from areas of high sugar concentration to areas of low sugar concentration.

STUDY QUESTIONS

1. Define transpiration, adhesion, cohesion, water potential gradient, and turgor pressure.

2. What are the two micropathways of water transport in the root tissues? Why is only one of these pathways available at the endodermis?

3. Explain the mechanism of water transport in vascular plants.

4. Name the macronutrients and micronutrients required by a plant.

5. Design an experiment to determine the symptoms caused by a particular mineral deficiency. Be sure to include a control plant.

6. Describe the role of root cortical cells in the uptake of minerals.

7. Outline the pressure flow model of sugar transport.

8. In what important ways do long distance water transport and sugar transport in vascular plants differ?

SUGGESTED READINGS

Epstein, E. "Roots." *Scientific American* 228 (1973):48–55. Discusses water and nutrient uptake as it relates to agricultural problems.

Evert, R. F. "Sieve-Tube Structure in Relation to Function." *Bioscience* 32 (1982):789–795. Discusses the evidence for the mass flow mechanism of sugar transport and critically examines some alternative mechanisms.

Harris, D. *Hydroponics: Growing Plants Without Soils,* revised ed. Vancouver and London: David and Charles Holdings Ltd., 1974. A practical account of mineral nutrition as it applies to growing plants by hydroponic techniques.

Hewitt, E. J., and T. A. Smith. *Plant Mineral Nutrition.* New York: Wiley, 1975. This advanced text has a good discussion of mineral deficiency symptoms and diseases.

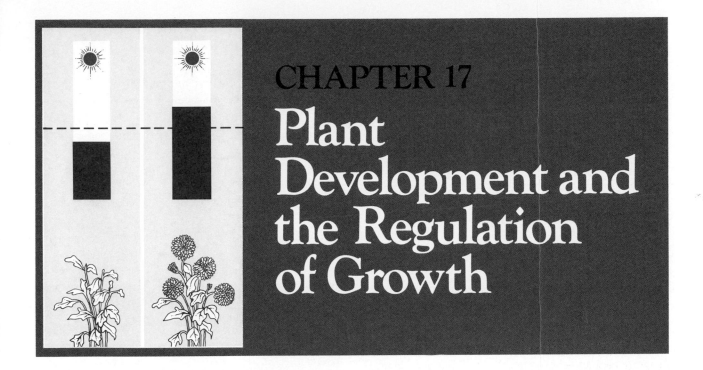

CHAPTER 17
Plant Development and the Regulation of Growth

In 1967 a group of scientists discovered lupine seeds buried in the permanently frozen silt of the Yukon basin in Alaska. According to the geological stratum in which they were found, the seeds were at least 10,000 years old. Yet when they were placed in a moist, warm environment, they germinated and grew into mature lupines. Although this is an extreme example, many kinds of seeds can exist in a state of suspended animation (alive but metabolically sluggish) for long periods, and then under suitable conditions germinate into actively growing seedlings (Table 17–1). We will begin our study of plant development with this transition from the resting seed state to the actively growing seedling.

TABLE 17–1
Seed Longevity. Listed below are the average life spans of selected seeds. Life spans vary somewhat, depending on storage conditions.

Species	Average Life Span of Seeds
Sugar maple	Less than 1 week
English elm	6 months
Sugar cane	1 year
Alfalfa	6 years
Foxtail	10 years
Red clover	30 years
Locoweed	150 years
Indian lotus	1000 years
Arctic lupine	10,000 years or more

SEEDS AND GERMINATION

The seeds of flowering plants are composed of three basic parts (Figure 17–1). The **embryo** consists of a rudimentary shoot and root and one or two seed leaves called cotyledons (monocot seeds have one cotyledon; dicots have two). Also present in developing seeds is the **endosperm,** a food-storing tissue. Finally, surrounding both the embryo and endosperm is a protective covering called the **seed coat.** Although all developing seeds start out with these three basic parts, beans, peanuts, and most other dicots transfer their accumulated food reserves from the endosperm to the expanding cotyledons during seed development. At maturity, these seeds have no endosperm, and most of the seed volume is occupied by the greatly expanded pair of cotyledons (Figure 17–1a). When you eat a peanut (a dicot seed), you are eating two cotyledons (the two halves of the seed), which are held together by a small "nub"—the embryonic

271

FIGURE 17–1
Seed Structure. These cutaway diagrams illustrate two major variations in seed structure. (a) Typical of dicot seeds, the bean's root, shoot, and two food-storing cotyledons (only one shown) comprise the embryo, which is enclosed in a seed coat. (b) The wheat seed typifies monocot seeds, which have a single cotyledon (modified for absorption of nutrients) and a large, food-storing endosperm.

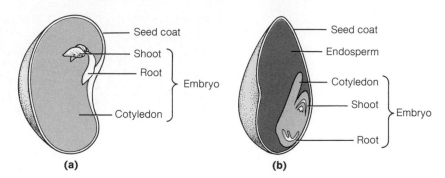

root and stem. By contrast, the seeds of monocots, such as wheat and other cereal grains, retain their endosperm when fully developed (Figure 17–1b). The single cotyledon functions in absorbing and transporting nutrients from the digested endosperm to the growing root and shoot after the seed has germinated.

As long as a seed stays relatively dry (most mature seeds are only 5–20% water by weight), its growth is suspended and its metabolism is barely perceptible. Many seed types require only water, oxygen, and a suitable temperature (5–40°C) to germinate. After a few hours to a few days under these conditions, the embryonic root pushes through the seed coat (Figure 17–2), and somewhat later the shoot system emerges. Not all seeds germinate so easily, however. The seeds of stone fruits such as peach and cherry require several weeks or more of cold temperatures (0–10°C) before they can germinate. Other types of seeds are enclosed in extremely tough or waxy seed coats, which mechanically impede growth of the embryo or restrict the uptake of water and/or oxygen. Such seeds germinate readily once the seed coat is nicked or worn away, which in natural situations can occur in various ways— by bumping into rocks along a swift-moving stream, digestion by microorganisms, or even partial burning of the seed coat (Figure 17–3). Seeds that require some special environmental condition or action before they will germinate in a moist, warm, and aerobic environment are said to be **dormant.**

In nature, seeds normally germinate under the soil, where no light is available for the photosynthetic production of carbohydrates. Thus, one of the earliest activities of a germinating seed is the mobilization of stored food in the cotyledons (most dicots) or endosperm (most monocots). Most plants have evolved sophisticated mechanisms to accomplish this; a good example is the barley seed, a monocot. When a barley seed takes up water, the embryo begins producing a

FIGURE 17–2
Seed Germination. The embryonic root is the first structure to emerge from these germinating radish seeds.

chemical messenger substance known as **gibberellin** (Figure 17–4). Minute amounts of this substance diffuse from the embryo to a specialized part of the endosperm called the **aleurone layer.** The aleurone layer completely surrounds the large starch-storing region of the endosperm, the seed's major site of food reserves. In response to gibberellin, the cells of the aleurone layer synthesize and then secrete several digestive enzymes into the starchy endosperm. One of these enzymes breaks down the starch to yield sugars, which are absorbed by the cotyledon and transported to the growing seedling. Thus the embryo, by communicating with the aleurone layer, initiates the mobilization of the food reserves it needs to support its early growth.

FIGURE 17–3
Seed Dormancy and Fire. Fire occurs periodically on southern California and Baja California hillsides. Many species of the native vegetation have adapted to fire by evolving seeds that actually require fire before they will germinate. These lovely wildflowers seen among charred shrubs grew from seeds that lay dormant in the soil since the previous fire, perhaps 10, 20, or even 50 years ago. The seeds produced by these plants will rest quietly beneath the shrubs until they are burned again.

FIGURE 17–4
The Hormonal Messenger System in Barley Seeds. This series of diagrams illustrates nutrient mobilization in the barley seed during early seedling development. The seed is planted on day 1. In response to water uptake, the embryo begins to produce gibberellin, which diffuses to the aleurone layer. On day 2 the aleurone layer begins to produce and secrete enzymes that digest the starchy endosperm (color). By day 3, the embryo has grown considerably, using the nutrients obtained from digestion of the endosperm. The endosperm has been completely digested by day 6; the plant is now producing carbohydrates by photosynthesis.

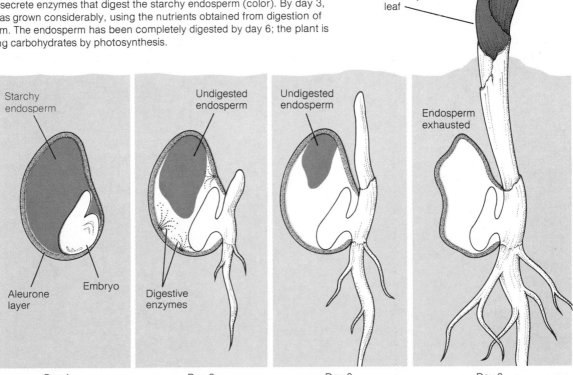

Photosynthetic leaf

Starchy endosperm

Aleurone layer

Embryo

Undigested endosperm

Digestive enzymes

Undigested endosperm

Endosperm exhausted

Day 1

Day 2

Day 3

Day 6

HORMONES AND PLANT DEVELOPMENT

The gibberellin produced by the barley embryo as it germinates is only one type of chemical messenger plants use to regulate their development. There are several other types of chemical messengers produced by plants that are collectively known as hormones. A **hormone** is a substance produced by cells in one part of an organism that can cause a physiological change in another part. In the case of the barley seed, gibberellin produced in the embryo causes a chemical change—the synthesis and secretion of digestive enzymes—in the aleurone layer of the endosperm. From germination to death, many growth and developmental events in plants are triggered by hormones, the presence and concentration of which are in turn regulated by various external factors (such as light, daylength, temperature, and gravity) and internal conditions. Thus, much of our knowledge concerning plant development has come indirectly from studies involving the actions of the plant hormones. We will consider the four major types of plant hormones here in this section: auxin, the gibberellins, the cytokinins, and ethylene.

Auxin

Auxin (from the Greek word *auxein,* to increase) was the first of the plant hormones to be studied and identified. Its presence in plants was originally inferred from experiments begun over a century ago by Charles Darwin.

PHOTOTROPISM. Charles Darwin, most famous for his formulation of evolutionary theory (see Chapter 28), also performed some of the first experiments on plant growth regulation. Darwin systematically studied the tendency of certain seedlings to bend toward light, a phenomenon called **phototropism** (*photo* = light, *tropism* = bending toward or away from). Under most natural conditions, the shoots of seedlings grow straight upward, toward the sun, but you have probably seen a houseplant that appears to be leaning toward a window, apparently "reaching" for light. Darwin noticed this, and decided to study phototropism in plants. His subjects were **coleoptiles** of grass seedlings—sheathlike structures which enclose the first leaves. Darwin found that when light was directed at the coleoptiles from the side, they bent until their tips pointed toward the light source (Figure 17–5a). This bending resulted from an uneven growth of the region below the tip—the cells on the dark

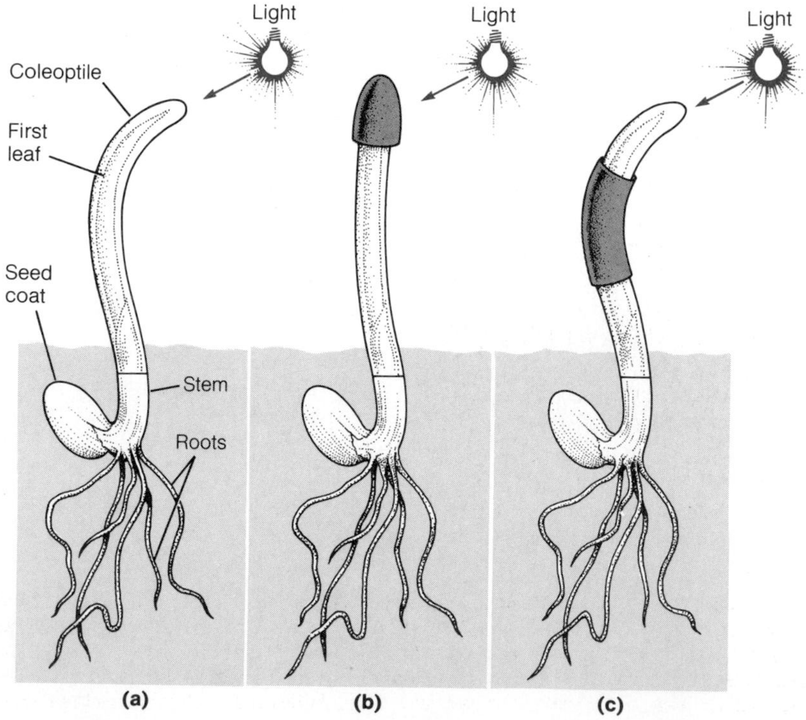

FIGURE 17–5
Phototropism. (a) Darwin found that grass coleoptiles receiving light from only one side grow toward the light source. (b) If he prevented light from reaching the tip, however, no bending occurred. (c) A light-proof sheath placed around the growth region itself did not prevent bending. Darwin concluded that something from the tip "influenced" growth in the region below the tip.

Labels: Coleoptile, First leaf, Seed coat, Stem, Roots, Light

(a)   (b)   (c)

side of this region grew faster than those on the lighted side, causing the coleoptile to bend toward the light. If, however, Darwin shielded the coleoptile tip from light with a metal foil cap, the bending response was prevented (Figure 17–5b). Furthermore, if he covered the growing region of the coleoptile with foil but left the tip exposed, the coleoptile bent toward the light (Figure 17–5c). From these experiments Darwin concluded that light causes the tip of the coleoptile to transmit some kind of "influence" to the growth region below it. There, the "influence" causes the uneven growth pattern and hence, bending.

In 1926, the Dutch plant physiologist Frits Went showed that the "influence" Darwin spoke of was a chemical substance. Went cut off tips of oat coleoptiles and placed them on small blocks of agar, a gelatinlike material (Figure 17–6a, b). When he placed one of these agar blocks on top of one side of a tipless oat coleoptile in the dark, the seedling began to bend toward the opposite side (Figure 17–6c). Apparently some growth-promoting substance had diffused from the coleoptile tips into the agar, and then from the agar into the cells of the growth region. Went called this substance auxin. Several years later, auxin was identified as indole-3-acetic acid. Extremely small

amounts of this hormone can cause bending in coleoptiles and stems; the amount of curvature depends on the amount of auxin.

These and other experiments led investigators to the current explanation for phototropism. Apparently pigments in the coleoptile tip absorb the light to which a seedling is exposed. If light is coming from only one side, relatively more pigment molecules on the lighted side become photoactivated. The unequal absorption of light, in some manner not yet understood, causes auxin in the coleoptile tip to be transported to the shaded side. Auxin accumulation on the shaded side of the growth zone causes the cells there to elongate more rapidly than the auxin-depleted cells on the lighted side. This asymmetric growth results in the bending response toward the light source.

Darwin's and Went's experiments centered on grass coleoptiles, but the principles of phototropism apply to all higher plant shoots. Auxin is produced by shoot meristems and young leaves and is transported downward, where it regulates the rate of cell growth in the zone of elongation of the stem or coleoptile. If auxin is evenly distributed throughout the growing region, all the cells there will grow at equal rates. If some environmental factor causes an

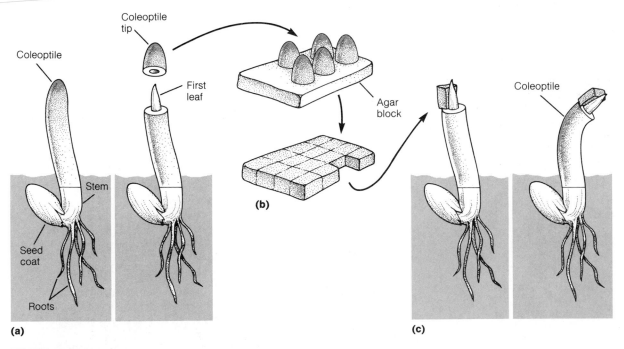

(a) **(b)** **(c)**

FIGURE 17–6

Auxin and Growth. (a) Frits Went cut coleoptile tips from oat seedlings and (b) transferred them to a thin slab of agar for about an hour. During this time a growth-promoting substance (auxin) diffused from the tips into the agar. (c) When a small section of the agar block was placed on one side of a decapitated coleoptile, auxin diffused into that side of the coleoptile and bending resulted. These experiments were performed in the dark so a light source could not influence the results.

unequal distribution of auxin, however, growth will be asymmetric and the shoot will bend.

GRAVITROPISM.   Another kind of environmentally directed growth is the gravity-oriented response called **gravitropism.**

If a young tomato seedling is placed in a horizontal position, within a few hours the root tip will bend downward and the stem tip will point upward. Like phototropism, gravitropism results from an unequal pattern of cell elongation. Let us first consider the upward bending of the stem. When a stem is oriented horizontally, auxin produced in the shoot tip begins to accumulate on the lower side. This accelerates growth on the lower side of the stem's zone of cell elongation, causing the stem to bend upward (Figure 17–7). Once the shoot is in the vertical position, auxin becomes evenly distributed around the stem; the amount of growth on all sides becomes equal, and the shoot continues to grow vertically.

Obviously, root gravitropism must be different because roots grow *toward* the source of gravity. The gravitropic curvature of roots does not appear to be regulated directly by auxin but rather by an inhibitor of cell elongation produced in the root cap. When a root is placed in a horizontal position, this inhibiting substance accumulates on the lower side of the root cap. From there it is transported toward the root's zone of cell elongation. The higher concentration of inhibitor in the cells of the lower side of the growing zone causes those cells to elongate more slowly than the cells on the upper side. The result is that the root bends downward until it is growing vertically. At that point the inhibitor concentrations and growth rates are equal on all sides of the root.

OTHER EFFECTS OF AUXIN.   In addition to controlling the rate at which cells elongate, auxin plays other regulatory roles within plants. For example, in many plants, auxin moving downward from the tips of the shoots prevents lateral buds from growing. If you remove a shoot's tip (the source of auxin), some of the lateral buds will begin developing into branches. This is the principle behind "pinching back" plants: removing the tips of the leading shoots gives rise to a much fuller, more branched plant.

Auxin is also involved in the initiation of roots on stem cuttings in many species. When a cutting is made, auxin produced in the young leaves moves down the stem, accumulates at the cut end, and triggers the formation of new roots. We take advantage of this property of auxin, as the rooting of stem cuttings is a major means of propagating many ornamental and crop plants. Synthetic forms of auxin are often used to promote root initiation, particularly in difficult-to-root species. Certain synthetic auxins called herbicides are also used to selectively kill plants (Essay 17–1).

Gibberellins

In addition to their role in the mobilization of food reserves in barley and other cereal seeds, the class of

(**a**) Zero time   (**b**) 3 hours   (**c**) 12 hours

FIGURE 17–7
Gravitropism. (a) When a stem is oriented horizontally, auxin produced in the tip accumulates on the lower side of the stem. (b) The increased auxin concentration on the lower side enhances the growth rate of the cells there, causing the stem to bend upward. (c) Once the stem is vertical, auxin becomes evenly distributed around it and the growth rate is equal on all sides.

ESSAY 17-1

HERBICIDES: BOON OR BANE?

Herbicides are plant killers. They are manufactured chemicals that are widely used to control weeds in agricultural fields, suburban lawns, and along roadsides. The discovery of the first herbicides in the mid-1940s grew out of earlier work on auxins. With the identification of indoleacetic acid as the naturally-occurring auxin in 1935, plant scientists began looking for synthetic compounds that would mimic the powerful regulatory effects of this plant hormone. It was not long before a host of auxinlike substances had been synthesized and tested, compounds that could promote growth of stems, enhance root formation on stem cuttings, and cause other effects attributable to auxin. All of these effects were brought about by fairly low concentrations of the synthetic auxins. When researchers tumbled to the fact that these same compounds at higher concentrations could kill plants, the United States military establishment threw a cloak of top secrecy over further research in this area. World War II was in full swing and these synthetic auxins, now recognized as herbicides, had potential as agents of chemical warfare. In 1944, one of the most potent of the auxinlike herbicides, 2,4-dichlorophenoxyacetic acid (2,4-D), was discovered.

The herbicides were never used in World War II, and when the war was over, the chemical warfare division lifted the restrictions on research and development of the herbicides. Since then over 180 different compounds have been discovered and used for the control of weeds. Some of the herbicides, like 2,4-D, are more effective against dicots than monocots, and this is very useful to farmers who grow monocot crops, such as wheat, corn, and rice. It means they can spray their fields to kill all the dicot weeds without damaging the crop plants. Suburbanites can spray their grass lawns (bermuda, blue grass, rye grass and others are monocots) with these herbicides to selectively eliminate dicots such as dandelion, oxalis, and clover. There are also herbicides that act selectively against monocots and spare the dicot crops. Overall, the herbicides have been a boon to agriculture by increasing productivity and decreasing costs. When used appropriately, they pose little problem to environmental quality because they are destroyed by microorganisms in the soil.

The herbicides got their test as a chemical warfare agent during the Vietnam conflict in the 1960s. In an effort to eliminate many of the food crops used by the North Vietnamese troops and to expose their supply routes hidden by the dense foliage, the United States Army undertook a massive herbicide spraying program, using concentrations of herbicides that kill *all* plants. The substance sprayed was the controversial **agent orange,** a 50:50 mixture of 2,4-D and its close relative, 2,4,5-trichloro-phenoxyacetic acid (2,4,5-T). Although agent orange proved to be a very effective defoliant, it caused some very serious side effects. The herbicide mixture contained trace amounts of **dioxin,** a very toxic chemical that is generated as a contaminant during the manufacture of 2,4-D and 2,4,5-T. Tests indicate that dioxin is extremely carcinogenic (cancer-causing), and when injected into pregnant laboratory animals, it causes a high rate of birth defects. The people of Vietnam who were exposed to the spray have suffered similar medical problems, and the United States Army personnel directly involved in the defoliation program are reporting a higher-than-normal rate of cancer. This has led to a very sticky legal battle with the federal government over responsibility for medical costs and personal damages accruing from this ill-conceived spraying program. But of course, the pain and suffering of those affected cannot be put in terms of dollars and cents.

Until 1979, 2,4,5-T was sprayed on forest lands in the United States to kill unwanted brush and hardwoods, thus encouraging the growth of commercial conifer trees. In 1979, however, the Environmental Protection Agency suspended use of 2,4,5-T on timber lands, citing evidence linking its use to mis-carriages in a region of Oregon. Apparently the dioxin present in the 2,4,5-T preparations is responsible for the ill effects.

Despite the problems associated with some herbicides, it is important to keep in mind that herbicides have benefited agriculture greatly, providing more food for a hungry world. We cannot ignore the benefits of a herbicide application program that is carefully planned and carried out.

hormones known as the gibberellins influence many other physiological functions in plants. Like auxin, the level of gibberellins in a seedling can regulate the plant's growth. For example, the growth effects of gibberellins can be quite dramatic when they are applied to certain dwarf varieties of plants, such as corn (Figure 17–8). Dwarf corn is a genetic mutant that lacks an enzyme necessary for the biosynthesis of gibberellins. Without the ability to synthesize gibberellins, stem growth in these dwarfs is severely retarded. If the dwarfs are treated with a commercial preparation of gibberellins, however, they begin to grow rapidly and soon attain the height of normal corn plants.

You might wonder from results like these whether applying auxin or gibberellins would be a useful way to increase the size of crop plants. In most cases, giving intact plants more auxin does not increase their growth. This is because meristems and young leaves already produce as much auxin as a plant's growing tissues can use. Additional auxin is only effective when the tissues that produce it are removed.

Applying gibberellins, on the other hand, does enhance growth of some intact plants. Unfortunately for agricultural purposes, much of the enhanced growth is due to increased water uptake, which adds little or no food value to the plant. Furthermore, treated plants usually have other undesirable characteristics such as top-heaviness. In some cases, however, gibberellin applications are beneficial. Gibberellin-treated grape bunches, for example, elongate dramatically, which makes the grapes less tightly packed and thus less susceptible to spoilage by fungi. In addition, the grapes are sweeter and larger. Gibberellins can also induce certain types of plants to flower at an earlier stage than they would otherwise, which is valuable to plant breeders as well as to commercial flower growers.

Cytokinins

In the 1940s, about 20 years after the discovery of auxin, several research groups became interested in growing isolated plant tissues on a nutrient medium. In the early attempts, the medium contained sugars, vitamins, various salts, and auxin. When a piece of stem tissue was placed on this medium, the tissue enlarged somewhat but did not grow by cell division. Apparently the medium lacked some factor essential for plant cell division.

FIGURE 17–8
Gibberellin and Dwarf Corn. (Left) If dwarf corn plants are treated with gibberellin (GA), they grow as tall as normal corn. (Right) Normal corn apparently produces enough gibberellin for maximum growth—the application of gibberellin to these plants has no effect on growth.

One of the first clues to what might be missing was the discovery that when diluted coconut milk was added to a nutrient medium, the plant tissue divided and grew into a large, undifferentiated mass of cells. What was the active factor in the coconut milk that caused cell division? Researchers seeking the answer found a second clue: the nucleic acid base adenine was found to be a component of coconut milk, and when it was added to the nutrient medium in place of the coconut milk, it caused a notable increase in cell division. With this slim lead, Carlos Miller in 1953 began testing various sources of adenine from the laboratory shelf. Realizing that DNA contained adenine, he took down an old bottle of herring sperm DNA and added some of it to his nutrient medium. Surprisingly, the tobacco stem tissue proliferated at an enormous rate, much better than with adenine itself. Miller knew he had stumbled onto a major discovery,

FIGURE 17–9
Cytokinins. The steps taken to culture plant material on a nutrient medium are shown here. (a) A small piece of plant tissue (usually from the stem) is removed under sterile conditions and transferred to (b) a flask containing agar with nutrients (sugars and various minerals), vitamins, auxin, and a cytokinin. (c) After about 3 weeks, the original piece of transplanted tissue grows into a large, unorganized mass of cells which, under special conditions, can give rise to new plants.

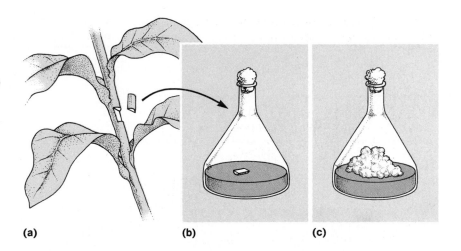

(a)   (b)   (c)

so he eagerly ordered some new bottles of herring sperm DNA from a nearby chemical supply house. You can imagine his frustration when the fresh DNA preparations were totally inactive in stimulating cell division of his cultured tobacco tissue. What went wrong? Miller soon realized that the only difference between the original DNA preparation and the fresh material was age. When he rapidly aged the new herring sperm DNA by heating it, Miller found it to be just as active in promoting cell division as the original, slowly-aged DNA. Shortly thereafter, Miller isolated and identified the active ingredient in the aged DNA—it was a breakdown product of adenosine, which he named **kinetin.**

Although kinetin itself has never been found in plants, a group of closely related substances, all derivatives of adenine, do exist in small amounts in plants. These substances form the class of plant hormones called **cytokinins** (from the term *cytokinesis,* meaning "cell-splitting"). They are produced in both roots and shoots and are most abundant in actively dividing tissues, where they promote cell division. Kinetin and other synthetic cytokinins are widely used in the commercial propagation of certain plant species (most notably orchids) by tissue culture (Figure 17–9).

Ethylene

Over 70 years ago citrus growers discovered that their lemons would ripen faster if they were stored in a room heated with kerosene stoves. Most growers believed this effect was due to the warmth of the stoves, and understandably many switched over to electric heaters when they became available. Imagine the growers' surprise and embarrassment (particularly financial) when the electric heaters failed to ripen the lemons! Years passed before the problem was explained. It turned out that what affected the lemons was not heat but a volatile hydrocarbon called **ethylene,** one of the by-products of burning kerosene.

Ethylene was soon identified as a natural product of fruits and other plant tissues as well as of kerosene. Like all hormones, plants produce it in very small amounts, but nonetheless it can have potent effects. As the fruits on a citrus tree approach their mature size, for example, ethylene production in the fruits rises, and ethylene triggers the final stages of the ripening process (change in coloration, firmness, and sweetness). Commercial fruit growers and shippers have taken advantage of this phenomenon in several ways. When fruit has to be shipped long distances to market, cool, ethylene-free air is circulated through the storage containers to prevent any buildup of naturally produced ethylene. Synthetically produced ethylene is added at the final destination to trigger ripening. Tomatoes and many other fruits you purchase in the supermarket today have been picked unripe and then treated with ethylene just before arrival at the market, allowing growers to harvest a whole field at once rather than waiting for individual fruits to ripen at their own speed.

In addition to its major role in ripening fruit, ethylene has other physiological effects on plants. For example, ethylene can cause leaves to drop, flowers to fade and die, and plant stems to swell.

The major effects of ethylene and the other plant hormones on plant development are summarized in Table 17–2.

LIGHT AND PLANT DEVELOPMENT

As we have seen in the cases of phototropism and gravitropism, plant growth and development are often tied to external environmental cues. One of the most influential environmental factors in this regard is light. The presence of light can break the dormancy of certain kinds of seeds and guide the early development of young seedlings. In temperate regions, the seasonal changes in light duration (daylength) affect various aspects of plant development. As we shall see, many of these light-mediated phenomena involve a rather remarkable pigment called **phytochrome** (*phyto* = plant, *chrome* = coloring agent).

Photodormancy and the Discovery of Phytochrome

In the late 1940s a research group at the United States Department of Agriculture's research station in Beltsville, Maryland, discovered a curious phenomenon. The group was working with a **photodormant** variety of lettuce, which germinated only if exposed to light (most varieties do not require light to germinate). By exposing seeds to different wavelengths of light, they found that the lettuce seeds germinated most readily under red light (the most effective wavelength was about 660 nanometers), and that only a minute or so of light was necessary. Surprisingly, if the seeds were exposed to far-red light (about 730 nanometers in wavelength) just after the red light exposure, they did not germinate. When additional flashes of red or far-red light were tested, the researchers found that the total number of flashes made little difference, but the wavelength of the *last* flash was critical. If the last flash was red light, most of the seeds germinated; if it was far-red, most did not (Table 17–3).

To explain these results, the Beltsville group proposed that the embryo of the lettuce seed must contain a rather unusual pigment, which they named phytochrome. They hypothesized that phytochrome can exist in two forms, and each form can be converted to the other when it absorbs the appropriate wavelength of light. One form, called P_{660}, absorbs

TABLE 17–2
Plant Hormones and Some of Their Effects on Plants.

Hormones	Effects
Auxin	Promotes stem elongation; involved in phototropic and gravitropic bending responses of stems.
	Inhibits lateral bud growth in many species.
	Promotes root initiation in stem cuttings.
	Promotes differentiation of tracheids and vessel elements in xylem.
Gibberellins	Promotes mobilization of food reserves in the endosperm of cereal seeds.
	Promotes stem elongation.
	Promotes breaking of seed dormancy in some species.
	Promotes flowering in some species.
Cytokinins	Promotes cell division in tissue culture cells and apical meristems of plants.
	Promotes lateral bud outgrowth.
	Inhibits senescence (aging) of leaves.
Ethylene	Promotes fruit ripening.
	Inhibits stem and root elongation in favor of lateral swelling.
	Promotes flowering in some species.
	Promotes leaf fall.

best at 660 nanometers (red light). When it absorbs red light, it is converted to the other form, called P_{730}, which absorbs mainly far-red light with an absorption peak at 730 nanometers. When P_{730} absorbs far-red light, it is converted back to P_{660}. Furthermore, the research group proposed that P_{730} slowly changes to P_{660} in darkness. These various phytochrome interconversions can be summarized as follows:

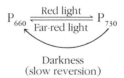

$$P_{660} \xrightleftharpoons[\text{Far-red light}]{\text{Red light}} P_{730}$$

Darkness
(slow reversion)

Let us consider how these transformations of phytochrome can help explain the results of the lettuce seed experiments. When seeds are in total darkness, most of their phytochrome molecules will be in the P_{660} form. (Any P_{730} initially present would slowly revert to P_{660}.) Because these seeds do not germinate, we assume P_{660} must be the inactive form of the pigment. If the seeds are then given a flash of red light, P_{660} is converted to P_{730}, the active form of the

TABLE 17–3
**The Effects of Red and Far-Red Light on Germination
of Photodormant Lettuce Seeds.** In this experiment
carried out by a plant physiology class at San Diego
State University, photodormant lettuce seeds were
soaked for 12 hours in darkness, treated as indicated,
then returned to darkness. Each light exposure lasted
2 minutes.

Light Treatment	Percent Germination
None (continuous darkness)	7
White	76
Red	78
Far-red	9
Red, then far-red	10
Far-red, then red	75
Far-red, then red, then far-red	9

pigment, and the seeds ultimately germinate. However, if the red light is immediately followed by a
flash of far-red light, the active P_{730} is converted back
to the inactive P_{660} before the sequence of events
leading to germination can start; the seeds therefore
do not germinate. The number of flashes is unimpor-
tant; only the nature of the final flash (red or far-
red) is critical.

Photomorphogenesis

The lettuce seed studies suggested a way to test for
involvement of phytochrome in any light-mediated
event in plants: if a flash of far-red light reverses
the effects of a flash of red light, and vice versa,
phytochrome must be involved. This triggered investigators to examine other light-mediated phenomena
in plants with an eye toward the red/far-red reversi-
bility effect. These studies proved that phytochrome is
a very versatile pigment. We now know that this
photoreversible pigment is also involved in **photomorphogenesis** (*photo* = light, *morpho* = shape,
genesis = beginning), a term that encompasses several
light-triggered changes associated with seedling
development.

To illustrate the effects of light on seedling development, let us consider two bean seedlings, one that
has been grown in total darkness, and another that
has been exposed to a typical day-night regime (for
example, 12 hours of light and 12 hours of darkness).
After about a week under these two conditions, there
will be some very obvious differences in the appear-

FIGURE 17–10
**Phytochrome and Seedling Develop-
ment.** Bean seeds were germinated and
allowed to grow for 7 days (left) in total
darkness or (right) under a 12-hour light,
12-hour dark regime. Compare height,
leaf size, and the orientation of the stem
apex.

ance of the two seedlings. The dark-grown seedling
will be pale yellow in color and will have a long,
thin stem, tiny leaves, and a shoot apex that curves
downward (is "hooked") (Figure 17–10). In contrast,
the light-grown seedling will be green due to the
presence of chlorophyll in well–developed chloroplasts, its stem will be shorter and somewhat wider,
the leaves will be expanded, and its shoot apex will
point upward (Figure 17–10). If the dark-grown bean
is exposed to light for a few minutes every day, however, it will soon come to look like the light-grown
plant. Brief exposures to daylight or red light have
this same effect unless the seedlings are exposed to
far-red light immediately thereafter. Thus, both darkness and far-red light encourage the pale and spindly

appearance of the seedlings. Apparently P_{730}, the form of phytochrome that results from exposing the plant to red light, initiates the photomorphogenetic events: chloroplast development, leaf expansion, and straightening of the apex. P_{730} also inhibits stem elongation and promotes stem thickening. P_{660}, the form of phytochrome that results from exposure to far-red light or extended periods of darkness, is the inactive form, just as it is in lettuce seeds that require light to germinate.

Photoperiodism and Flower Induction

Photoperiodism refers to the regulation of a given developmental event by a seasonal change in daylength, or **photoperiod.** In plants, events triggered by the decreasing daylengths of autumn include leaf fall in deciduous plants, root thickening in radish, and tuber development in potatoes. The increasing daylengths of spring bring about developmental events such as bulb formation in onions and renewed growth of vegetative buds in deciduous trees. Probably the most spectacular event in the life cycle of angiosperms that is subject to photoperiodic control in many species is flower induction. Flowering is also the best studied of all the photoperiodic phenomena in plants, so we will restrict our discussion to this dramatic event.

To human observers, flowering is pleasing to the eye and nose; for the plant, it represents a major transition to a fundamentally different physiological state. A plant can grow as a strictly vegetative individual, and then, just as if someone threw a switch, its shoot meristems stop producing new leaves and stem tissues and change into flower buds. In many plants, as you have probably noticed, flowering is correlated with seasonal changes. Tulips blossom in the early spring, roses in the late spring and summer, and chrysanthemums in the fall. Many environmental variables change with the seasons, including temperature, water availability, light intensity, and, of course, daylength. Of these, the most predictable and reliable variable is the changing daylength. Thus, in many species, mechanisms have evolved which cue their reproductive development to a specific photoperiod, thereby ensuring that flowering will occur at the same time each year.

With respect to flower induction and photoperiod, plants fall into three general categories: short-day plants, long-day plants, and day-neutral plants. In temperate regions, **short-day plants** normally flower during the relatively short days of late summer or fall;

examples include soybeans, chrysanthemums, and poinsettias. **Long-day plants** flower in the spring and early summer; they include red clover, oats, spinach, radishes, and many others. Finally, peas, beans, corn, and most other crop plants are **day-neutral.** As the name implies, flowering in these plants is not cued to changing photoperiods, but rather to some other environmental signal—the cold of winter, for example—or simply after reaching vegetative maturity. In fact, many of our crop plants were originally photoperiodic flowerers, but this quality was lost through selective breeding. The new day-neutral strains will flower and set seed at any time of the year, so farmers can grow two or even three generations of these plants per year rather than just one.

Before plant scientists began examining the mechanism underlying photoperiodic flower induction, it was assumed that the plants were responding to changes in length of the light period (actual daylength). As it turned out, however, it is the length of the *dark period* that is crucial. Short-day plants require dark periods that exceed some critical duration in order to flower, whereas long-day plants need dark periods shorter than some critical length (Figure 17–11). Thus, short-day plants should actually be called long-night plants, and long-day plants should be called short-night plants. However, the terms "short-day" and "long-day" were well ingrained in the literature, so they have been retained.

Evidence in support of the critical dark period idea has come from studies with short-day plants. If a short-day plant is kept in darkness for regular periods that exceed its critical dark requirement (the exact length varies from species to species), it will flower, even if it is placed on a long-day, long-night schedule that exceeds the standard 24-hour day. Furthermore, if the long dark period is interrupted by so much as a flash of light, short-day plants will not flower (see Figure 17–11). Interruption of a long day with a period of darkness will not induce flowering in a short-day plant as long as the night length falls short of the critical duration.

TIME MEASUREMENT. If flower induction in photoperiodic plants is triggered by changing night lengths, then these plants must have some sort of internal clock to measure the length of the dark period. Although we know very little about the mechanism of time measurement in plants, we do know that phytochrome plays a role. What is the evidence for the involvement of phytochrome? Again, let us use a short-day plant to answer this question experimentally. Cocklebur is a popular plant for such studies because it becomes

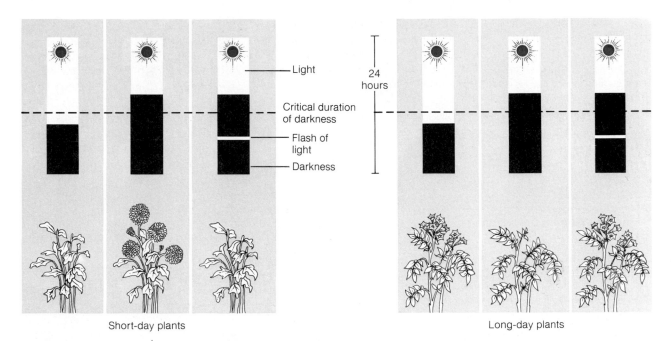

FIGURE 17-11

Flower Induction in Photoperiodic Plants. The length of the dark period (indicated by the dark vertical bars) determines whether short-day and long-day plants will flower or remain vegetative. If a short-day plant is exposed to a dark period that would normally induce flowering but is interrupted by even a brief flash of light, the plant will not flower. On the other hand, if a long-day plant is exposed to a noninductive long night interrupted by light, it will be induced to flower. Most photoperiodic plants require several inductive dark periods in succession to induce flowering.

induced to flower after receiving a single period of darkness exceeding nine hours. If a cocklebur is exposed to 10 hours of darkness which is interrupted midway by a brief flash of light, it will not flower. It is as if the light flash resets the internal clock, and the subsequent 5-hour dark period is not long enough to qualify as a long night (that is, the timer must run 9 hours in continuous darkness). Subsequent night-interruption experiments have shown that the wave-length of light flash most effective in preventing flowering in cocklebur is 660 nanometers (red). Furthermore, the inhibitory effect of a red flash can be reversed if it is immediately followed by an equally intense flash of far-red light. This result clearly impli-cates phytochrome in the process of time measure-ment. Unfortunately, we do not yet understand either the nature of the timer or how phytochrome conver-sions can reset it.

FLORIGEN: THE FLOWER-INDUCING HORMONE. As mentioned earlier, the vegetative buds of a plant be-come transformed into flower buds during flower induction. Thus, it seems logical to hypothesize that time measurement leading to flowering occurs in these

buds. To test our hypothesis, let us place a mature cocklebur plant into a growth chamber set for a 16-hour light, 8-hour-dark photoperiod (which would not induce flowering). We then cover some of the buds with foil for 16 hours to simulate an inductive long night for these buds. If indeed the buds are the site of time measurement for flowering, they should be-come induced to flower by this treatment. The results of this experiment, however, would be negative—no flowering would occur. On the other hand, if we cover the leaves with foil for 16 hours, flower induc-tion will occur. As a matter of fact, if only *one* leaf re-ceives an inductive dark period (all other leaves being kept on a noninductive, long-day photoperiod), the cocklebur will flower. Thus, time measurement for flower induction occurs in the leaves.

The fact that the initial steps of flower induction occur in the leaves and the final steps occur in the buds implies the participation of a hormone. That is, some "influence" (as Darwin would have called it) must pass from the induced leaves to the bud or buds. This hypothetical hormone has been named **florigen** ("flower-producing" substance).

Although no one has yet been able to isolate and

FIGURE 17–12
Grafting Experiments Supporting the Florigen Concept. Long days? Short days? It makes no difference when a long-day plant is grafted to a short-day plant. As long as one plant is induced to flower, florigen can flow through the graft union and induce flowering in the partner plant. Such experiments suggest that florigen is common to both long-day and short-day plants.

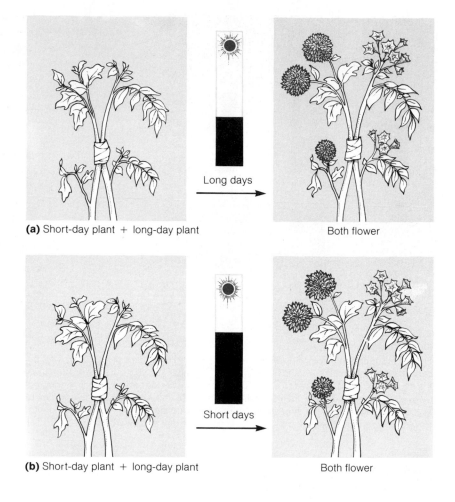

(a) Short-day plant + long-day plant

Long days

Both flower

(b) Short-day plant + long-day plant

Short days

Both flower

identify this elusive hormone, the existence of florigen is supported by several lines of evidence, the most dramatic of which comes from grafting experiments. For example, two plants of the same species can be grafted together under lighting conditions that do not induce flowering. Then, if one plant is subjected to a photoperiod that should cause it to flower, while the other is kept under the original noninductive photoperiod, both plants will flower. This indicates that a messenger compound produced in the first plant crosses the graft union and triggers flowering in the second plant. Moreover, analogous grafting experiments have been carried out in which two dissimilar species of plants were grafted together. In many cases both species flowered despite the fact that only one received an inductive photoperiod (Figure 17–12). Apparently florigen is a plant hormone common to many species.

SENESCENCE

It is fitting to end this chapter with a discussion of aging in plants, for this is the final phase of plant development. You may find it odd to think of aging, or **senescence,** as part of development. After all, isn't aging simply the deterioration of the cells and tissues of an organism? Although plant senescence certainly involves deterioration, this does not imply that plants get tired and weary as the end of their life approaches. In many cases, plant senescence is a very active, genetically predetermined series of events.

One of the most dramatic cases of programmed plant senescence is found in the **annuals**—plants that complete their life cycles during one growing season. In these plants, the formation of flowers triggers the beginning of the end. Apparently changes

FIGURE 17–13
Bristlecone Pines. The bristlecone pines are among the longest-lived creatures on earth. They grow at high elevations in the White Mountains on the California–Nevada border.

forming and do not amass large amounts of nutrients or act as strong nutrient "sinks," they still initiate senescence in the leaves. This suggests that in addition to nutrient competition, the formation of flowers triggers the processes leading to death in annuals by some other means.

The programmed senescence of annuals culminates in the death of the entire individual plant. This is similar to the situation found in higher animals, but it is not typical of plants. Many plants are **perennial** —they have a potentially indefinite life span. Indeed, one bristlecone pine near Reno, Nevada, is estimated to be nearly 5000 years old (Figure 17–13). Senescence also occurs in perennials, but its pattern is usually localized and often seasonal. For example, the above-ground parts of the tulip and daffodil wither and die in late spring, but the underground bulbs of these plants persist and produce new shoots and flowers the following spring. In sycamores and maples, seasonal senescence is restricted to the leaves, an event which clearly adds beauty to an autumn landscape. Even in the so-called evergreens (conifers, ivy, many tropical plants, and others), the leaves senesce and fall, but younger leaves remain and new leaves are constantly produced as the plant continues to grow. For many perennials, therefore, the aging pattern is one of senescence of individual organs.

SUMMARY

Plant hormones are organic substances produced in one part of a plant that influence physiological processes in another part of the plant. Together with many environmental factors, hormones play major roles in the regulation of plant growth and development. When barley seeds germinate, for example, the embryo within the seed produces gibberellins, which trigger the cells of the aleurone layer to manufacture and secrete digestive enzymes. One of these enzymes breaks down the starchy endosperm into sugars that can be used for food by the growing seedling. Gibberellins also regulate plant growth; their absence can cause dwarfism.

Another plant hormone, auxin, also influences growth by enhancing cell elongation in stems. In the case of phototropism, laterally directed light brings about an unequal distribution of auxin within the zone of cell elongation of the shoot. Greater amounts are present on the shaded side, which elongates more rapidly to produce the characteristic bending response. Asymmetrical growth is also responsible for the gravi-

take place in the plant during flower development that commit it to age and die soon after seeds are formed. Yet aging in these plants can be delayed for months to years if the flower buds are carefully removed when they appear. This observation has led some plant physiologists to propose that senescence in annuals may result from competition for nutrients between the rapidly growing reproductive structures and the older vegetative parts of the plant. That is, the flowers and developing seeds and fruits use up nutrients as they grow, thereby diverting these materials from the leaves, stems, and roots. Perhaps the vegetative plant parts become so depleted that they effectively starve to death. This sort of mechanism to account for senescence was suggested years ago and is undoubtedly partially accurate, but it still does not explain all the facts. Even when flowers are just

tropic behavior of shoots and roots. The upward bending of the shoot results from auxin accumulating on the lower side of a horizontally oriented stem. On the other hand, a horizontal root accumulates an inhibitor of cell elongation on the lower side, which in turn initiates events leading to downward bending. Auxin also plays a role in other physiological processes, such as root initiation in stem cuttings and the inhibition of lateral bud outgrowth.

Other plant hormones include ethylene, a volatile substance that controls several plant responses including fruit ripening, and cytokinins, which promote cell division.

Of the various external factors influencing plant development, none is more important than light. The presence of light stimulates the germination of light-requiring seeds and helps to regulate the development of seedlings. In these and other phenomena, light is sensed by the phytochrome pigment system. Phytochrome exists in two chemical forms: P_{660}, which absorbs red light, and P_{730}, which absorbs far-red light. In the presence of red light (and to a lesser extent, of sunlight or white light), most of the P_{660} in the plant is converted to P_{730}. In far-red light or darkness, P_{730} is transformed to P_{660}. P_{730} is the active form of phytochrome.

Phytochrome is also involved in the time measurement aspect of photoperiodic phenomena, including flower induction in short-day and long-day plants. Short-day plants normally flower in the late summer or autumn. To become induced to flower, they require one or more periods of continuous darkness (nights) that *exceed* some critical length. Long-day plants require a regime of dark periods that are *less* than some critical duration, a situation that occurs naturally during the lengthening days of late winter and spring. The critical dark period differs for each species. The length-of-the-night measurement appears to take place in the leaves, where the hypothetical flowering hormone florigen is synthesized.

The final phase of plant development is senescence. In annual plants, senescence is triggered in some unknown manner by flower formation, and it culminates in the death of the entire plant. In perennials, however, the pattern of aging is generally that of organ senescence. Persisting parts of the plant replace lost organs seasonally or continuously. Seasonal leaf fall and replacement, for example, is characteristic of deciduous plants; continuous leaf formation is characteristic of evergreens.

STUDY QUESTIONS

1. How do typical dicot and monocot seeds differ in structure?

2. What is the role of gibberellins in the germination of barley seeds? What are some of the other physiological effects of gibberellins?

3. What evidence suggests that auxin is involved in phototropism?

4. Auxin is transported from the tip of a coleoptile toward the base, never in the opposite direction. Design an experiment to verify this. How could you determine whether gravity causes this downward movement?

5. Considering the effect of ethylene on fruit ripening, what precautions can you take to slow down the ripening process in picked fruit?

6. Explain the germination of light-dependent lettuce seeds in relation to the proposed model of phytochrome interconversions.

7. How would you determine whether a particular plant species is a short-day, long-day, or day-neutral plant? Predict some hypothetical results and then explain them.

8. Describe two major patterns of senescence exhibited by plants. What physiological changes occur within the plant or plant organ that could in part be responsible for the onset of senescence?

SUGGESTED READINGS

Devlin, R. M., and F. H. Witham. *Plant Physiology*. 4th ed. Boston: Willard Grant Press, 1983. Chapters 17–21 provide an in-depth look at plant growth hormones and photoperiodism.

Luckwill, L. C. *Growth Regulators and Crop Production*. Baltimore: University Park Press, 1981. Survey of the agricultural and horticultural uses of plant hormones.

Moore, T. C. *Biochemistry and Physiology of Plant Hormones*. New York: Springer-Verlag, 1979. This advanced text reviews current knowledge of how plant hormones act at the cellular and subcellular level.

Ray, P. M., T. A. Steeves, and S. A. Fultz. *Botany*. Philadelphia: Saunders, 1983. Chapters 12 and 14 of this well-illustrated text cover the information presented here.

Thimann, K. V. *Hormone Action in the Whole Life of Plants*. Amherst, MA: University of Massachusetts Press, 1977. This book covers a broad range of topics and is especially interesting because of the personal insights provided by the author, a pioneer in the field of plant hormone research.

Animal Diversity and Physiology

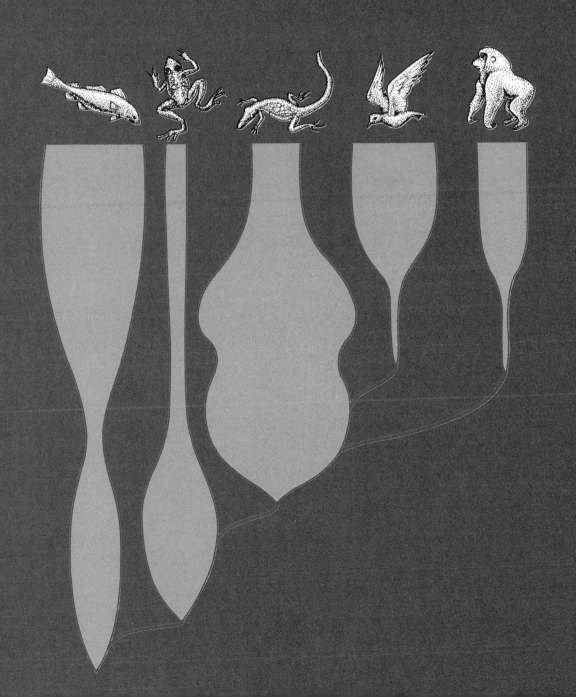

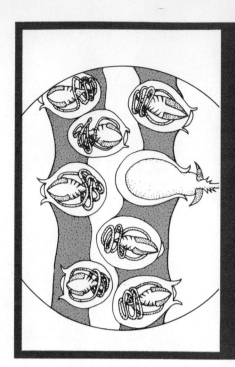

CHAPTER 18
The Animal Kingdom: Invertebrates

Of all the kingdoms of organisms on earth, most of us have the least difficulty recognizing members of the kingdom Animalia. Not only do we identify with this group as members ourselves, but many of our earliest childhood experiences dealt with recognizing animals, such as household pets and animals in a zoo. But have you ever stopped to think what it is about animals that distinguishes them from other groups of organisms? Although there is no single unique feature, animals have most if not all of the following characteristics:

1. All animals are heterotrophic and most ingest their food.
2. They are multicellular.
3. All animals reproduce sexually and have multicellular sex organs. Some of the lower animals can also reproduce asexually.
4. All animals pass through an embryonic or larval stage early in their lives.
5. Most animals are motile—they are able to move.

The animals are by far the most diverse of all the groups of organisms. There are well over one million animal species known and untold millions waiting to be discovered. To make some sense out of this tremendous diversity, most biologists divide the animals into 31 phyla. Of these, 30 include animals that have no backbone—the so-called **invertebrates.** The 31st phylum—the chordates—will be the subject of Chapter 19.

The range of variation among the invertebrates is enormous—just compare a sponge to a dragonfly! This has made the classification of these boneless creatures a rather difficult task, and no single scheme has captured the approval of all the zoologists. Nevertheless, by analyzing the fossil record (which is quite sketchy for many of the invertebrate groups), and by comparing the structural characteristics of present groups, we can organize the invertebrates in a way that reflects their most probable evolutionary history (Figure 18–1). This abbreviated "family tree" of the major animal groups takes into account several adaptive breakthroughs in body plans that occurred during the course of evolution. Many of the adaptive changes, such as new mechanisms for feeding, locomotion, or protection from prey, permitted the new groups to exploit their environments more effectively. The development of flight in the insects, a major subgroup of the arthropods, is a prime example of such an adaptive breakthrough.

In general, the fossil record gives a picture of increasing biological complexity over time. The assumption that less complex organisms are more primitive, however, must always be qualified by the possibility that simplicity in design could be an adaptation to special conditions or habitats. For example, some parasites are structurally quite simple in their mature form inside their host, yet their immature stage has all the complex structures found in their free-living relatives.

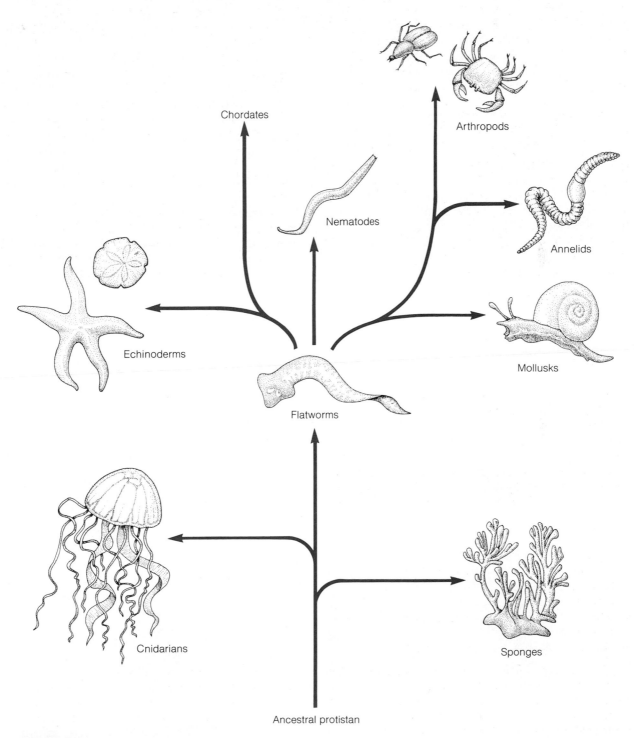

FIGURE 18–1
**Evolutionary Relationships Among the
Major Animal Phyla.** Of the 30 phyla
of invertebrates, the eight shown in this
diagram will be discussed in this chapter.

In general, simplicity is taken to indicate primitiveness only in free-living organisms that are not found in special habitats. On this basis, biologists have concluded that the sponges represent the most primitive phylum of invertebrates. Sponges, therefore, shall be our point of departure for our survey of the eight most important phyla of invertebrates.

PHYLUM PORIFERA: SPONGES

The phylum **Porifera** includes some 10,000 species of **sponges,** the simplest of the major animal phyla. Sponges are not your typical animal. Indeed, prior to 1765, sponges were classified as plants.

Although the bodies of sponges may contain several different types of cells, these cells are not arranged into complex tissues. Sponges are generally regarded as colonies of coordinated but more or less independent cells. In a few species, the individual cells composing the sponge's body can be separated by forcing the sponge through a fine sieve. Under appropriate conditions, the cells reaggregate into a whole sponge.

Despite their simplicity, sponges have made effective use of their few cell types to become the vacuum cleaners of the sea. The body cavity of a sponge is lined with **collar cells,** whose continuously beating flagella draw water into the body cavity (Figure 18–2). Minute food particles in the water adhere to the surface of the collar cells, and they are then engulfed and digested by roving cells called **amoebocytes.** The amoebocytes distribute the digested food to other cells of the sponge.

Water enters the sponge's body cavity through special **pore cells** and leaves via the **osculum,** a large opening located at one end of the body (see Figure 18–2). Carried with the departing water are metabolic waste products and sperm. The sperm cells may be drawn into the body cavities of other nearby sponges, where they fertilize nonmotile eggs. The resulting zygotes develop flagella and swim out through the osculum. Each zygote eventually settles onto the sea floor and grows into a new sponge.

Most sponges have an internal skeleton composed of either crystalline **spicules** or protein fibers called **spongin.** Some species have both. The spicules are made of silica or calcium carbonate and help maintain the animal's shape. The spongin skeletons of bath sponges are harvested and cleaned for many commercial uses, but their utility in the household has waned since the advent of the cheaper, synthetic "sponges."

Certain solitary and colonial protozoans are very similar in structure and function to the collar cells of sponges. Thus, it seems likely that a similar line of protistans gave rise to the first sponges through a process of increasing colonialization and differentiation of cell types. However, because sponges share no distinctive characteristics with any other known animal group,

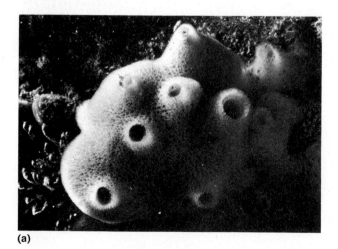

(a)

FIGURE 18–2
Anatomy of a Sponge. (a) Photograph of a sponge. (b) This diagram gives a cutaway view of a sponge with its specialized cell types; the arrows indicate the direction of water flow through its body.

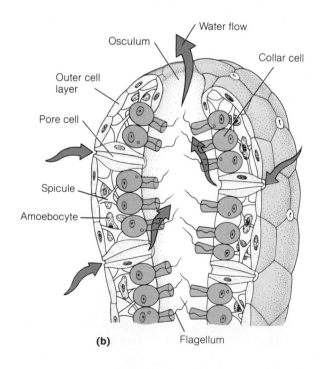

(b)

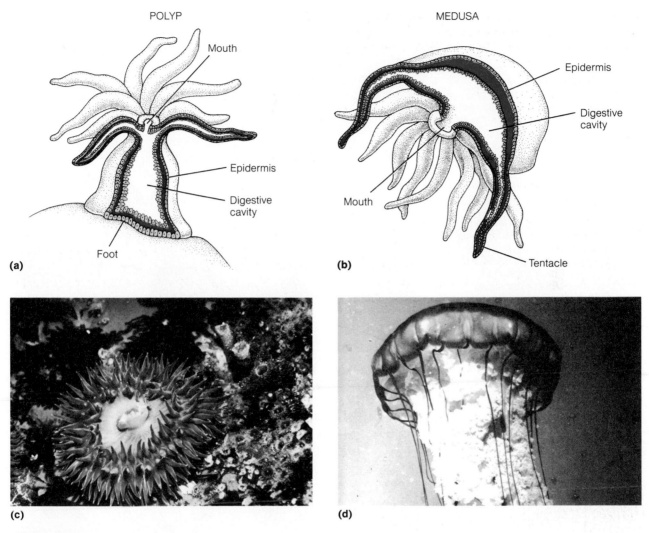

POLYP

Mouth

Epidermis

Digestive cavity

Foot

(a)

MEDUSA

Epidermis

Digestive cavity

Mouth

Tentacle

(b)

(c)

(d)

FIGURE 18–3
Cnidarian Body Forms. All cnidarians are essentially a sac with an opening surrounded by tentacles.
(a) The polyp form is erect and attached at its base to a rock or other substrate. (b) The medusa is a
free-floating form; it looks like a flattened polyp turned upside down. (c) Sea anenomes are polyps and
(d) Jellyfish have the medusa body form.

and because of their unique body organization, they are generally considered an evolutionary dead end.

PHYLUM CNIDARIA: JELLYFISH, CORALS, AND OTHERS

The phylum **Cnidaria** (formerly Coelenterata) consists of hydras, jellyfish, sea anemones, and corals. Like the sponges, the cnidarians are very simple aquatic animals with limited powers of movement. Their bodies are **radially symmetrical** (like a wheel with spokes) and consist of a sac with a single opening lined

with tentacles. In adult cnidarians, this sac may take either of two forms: an erect form attached at the base called a **polyp,** or a free-floating, umbrella-shaped form called a **medusa** (Figure 18–3). Some species exist only as polyps (see anemones and corals), some only as medusas (jellyfish), and others pass through both forms in their life cycle (hydras; Figure 18–4).

A unique feature of the cnidarians is the presence of **nematocysts,** a remarkably efficient device for self-protection and capture of prey. Nematocysts are fine threads that are formed in special cells located in the tentacles. When a nematocyst-containing cell is stimulated, either chemically or mechanically, it ejects its

FIGURE 18–4
Life Cycle of *Obelia*, a Hydrozoan.
Obelia, a marine hydrozoan related to the freshwater hydras, has both polyp and medusa forms in its life cycle. The mature vegetative body is actually a colony of individual polyps, some of which are equipped with tentacles and nematocyst-containing cells (see Figure 18–5) that function in feeding; others, termed reproductive polyps, produce buds that separate and develop into medusas. Both male and female medusas are formed, which represent the sexual stage. As they drift with the ocean currents, they release gametes (sperm and eggs) into the water. Fertilization results in a zygote, which undergoes cell division to become a solid ball of cells called the blastula. The outer cells of the blastula develop cilia; the immature organism is then called a planula, a swimming larva. The planula eventually settles onto a rock and grows into a polyp. With time, new polyps bud from the original one, forming a mature vegetative colony.

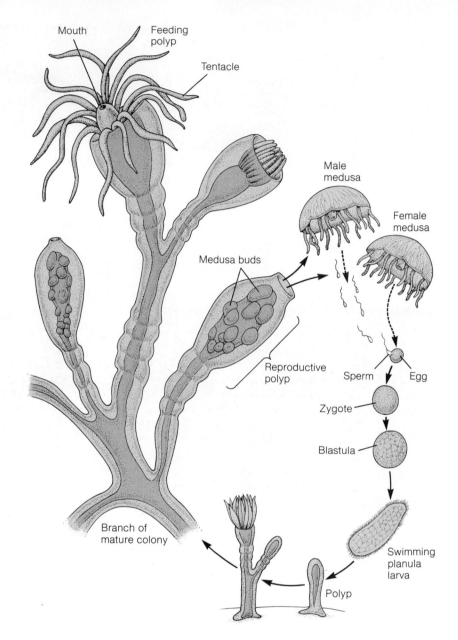

Mouth

Feeding polyp

Tentacle

Male medusa

Female medusa

Medusa buds

Reproductive polyp

Sperm

Egg

Zygote

Blastula

Branch of mature colony

Swimming planula larva

Polyp

nematocyst (Figure 18–5). In some species the nematocyst entangles the prey; in others, it penetrates the victim, injecting a toxin. The prey is then maneuvered through the mouth and into the digestive cavity, where digestion takes place. Any indigestible material is pushed back out through the mouth.

Members of the phylum Cnidaria possess rudimentary nervous systems, which represents an important evolutionary advance over the sponges and allows for some degree of coordination of body activities. In addition, cnidarians have musclelike cells that permit limited movement.

Few if any cnidarians are of direct economic importance to humans, but they are very significant components of the marine ecosystem. The corals in particular play a major ecological role, for their calcium carbonate skeletons contribute to the coral reefs that abound in the tropical and subtropical oceans. These reefs provide habitats for a wide diversity of small marine organisms that form the basis of extensive and very complex food chains (see Figure 33–4). A prime example of a coral reef is the Great Barrier Reef, which stretches 2000 kilometers (over 1200 miles) along Australia's east coast.

FIGURE 18–5
Nematocysts. Jellyfish paralyze their prey with the venomous tips of thousands of nematocysts, which are ejected from specialized cells lining the tentacles. New nematocyst-containing cells must be produced to replace those that have fired.

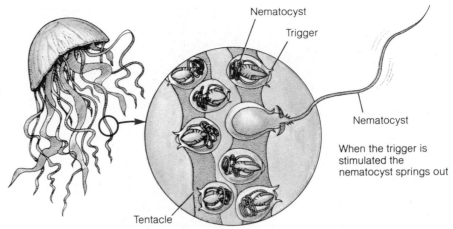

When the trigger is stimulated the nematocyst springs out

PHYLUM PLATYHELMINTHES: FLATWORMS

Animals commonly known as "worms" appear in several different phyla; the **Platyhelminthes,** or **flatworms,** are the simplest of these. The flatworms include planarians, flukes, and tapeworms. Planarians and their relatives are small, free living, mostly aquatic flatworms; flukes and tapeworms are parasites (Essay 18–1).

The flatworms display several important structural advancements over the simpler animals we have seen so far. First, they have **bilateral symmetry:** if a flatworm is divided in half along its longitudinal axis, the two halves are essentially mirror images of one another. A bilateral body plan results in distinct "head" and "tail" regions and upper (dorsal) and lower (ventral) surfaces (Figure 18–6). This is the most efficient design for locomotion, and because the "head" region usually goes first, the sensory organs can be concentrated in this region that encounters the environment first as the animal moves forward.

Another evolutionary advance in the flatworms is the presence of **organs**—units composed of two or more tissues that are closely integrated in function. Flatworms have well-developed reproductive organs and a primitive excretory system for getting rid of nitrogenous wastes. Although they have no respiratory or circulatory system (gas exchange occurs across their body wall), flatworms have a crude central nervous system consisting of several nerve cords running the length of the body. These nerve cords terminate in a nerve mass (a rudimentary "brain") in the head region. In addition, some flatworms have light-sensing eyes situated just above the nerve mass (see Figure 18–6).

Most flatworms are **hermaphroditic;** that is, each individual has both male and female reproductive

FIGURE 18–6
A Flatworm. The planarian is a free-living pond dweller and one of the most primitive animals that displays bilateral symmetry. (If this animal were split lengthwise down the middle, each half would be a mirror image of the other half.) Note the distinct "head" and "tail" regions and dorsal and ventral surfaces that result from a bilaterally symmetrical body plan.

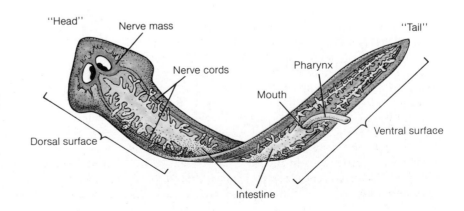

ESSAY 18–1

SCHISTOSOMIASIS

Schistosomiasis—not exactly a household word, at least in the Western world. In much of Asia and Africa, however, it is a dreaded disease that affects hundreds of millions of people. In Egypt, China, Japan, Korea, the Philippines, and about 65 other countries, schistosomiasis is well known.

Schistosomiasis is caused by the blood fluke, *Schistosoma*. This tiny parasite divides its life cycle between two hosts: humans and a freshwater snail common to parts of Africa and Asia (see figure). The adult fluke lives in the blood vessels lining the intestinal walls of humans, where it lays millions of eggs. The egg-laden capillaries eventually rupture, releasing the eggs into the intestinal cavity. The eggs are then carried out of the body with the feces. In countries having modern sewage systems, the fluke's life cycle ends here; in many parts of Africa and Asia, however, untreated human excrement often ends up in the irrigation canals, rice paddies, and other waterways. Here the eggs hatch into tiny swimming larvae that seek out a particular species of water snail. The few that actually find their aquatic host quickly bore into its soft tissues, feed for a while, then reproduce asexually to yield hordes of new larvae called cercaria. The cercaria leave the snail in search of an unsuspecting human, perhaps a rice farmer wading in his paddy. These human-seeking larvae generally bore through the skin of the feet, get into the bloodstream, and eventually settle down in the blood vessels of the intestines. The

larvae then mature into egg-laying adults to begin the cycle anew.

If the cercaria all settled into the intestinal blood vessels of their human hosts, then possibly the suffering caused by this pest would be more bearable. Unfortunately, many of the cercaria lodge in the veins of the liver, lungs, kidneys, and urinary bladder where they cause abscesses and ulcers. They also trigger an immune response wherever they are, which further complicates matters by blocking many of the vessels that feed into the affected tissues. Death may occur from extensive liver damage or kidney failure, but in most fatal cases, the victim succumbs to an unrelated disease that he or she is too weak to fight off. There are several drugs that are moderately effective when applied at an early stage of infestation, but like most parasite-caused diseases, the best medicine is prevention. Proper sanitation and effective control of the water snail population are the best preventative measures. Although nearly half a million immigrants in the United States suffer from schistosomiasis, the disease cannot be transmitted for lack of the water snail host.

It would be difficult to find redeeming value in anything so debilitating as the blood fluke, but for the people of Taiwan, there may be some. In the clash between the communist Chinese and the nationalist Chinese troops (headed by Chiang Kai-Shek) in the late 1940s, the communists got the upper hand and forced the

nationalists to retreat to the island of Formosa (now called Taiwan). Not satisfied with this state of affairs, the communists decided to mount an attack on Formosa in 1950. In preparation for the offensive, the communist soldiers began training for amphibious warfare in a swampy area near the southern coast of the mainland, an area infested with *Schistosoma*. It wasn't long before more than 30,000 troops came down with acute schistosomiasis, delaying the assault for at least six months.

In the meantime, intelligence reports of the intended assault got back to President Harry Truman. Truman ordered the Seventh Naval Fleet into the Straits of Formosa to preserve neutrality, a deployment that apparently dissuaded the communist Chinese government from going ahead with the attack. With their political integrity intact, the nationalist Chinese formed the Republic of China (Taiwan).

There are many who speculate that the outbreak of schistosomiasis among the training communist troops saved the nationalist Chinese from being overrun; others believe that Chiang Kai-Shek's troops would have repelled the attack. What's more, many believe that had the attack taken place as planned, the United States would have been drawn in on the side of Chiang Kai-Shek's forces. This alliance would have precipitated Russian support for the communist forces, and possibly World War III. Could it be that the tiny blood fluke forestalled Armageddon? We'll never know.

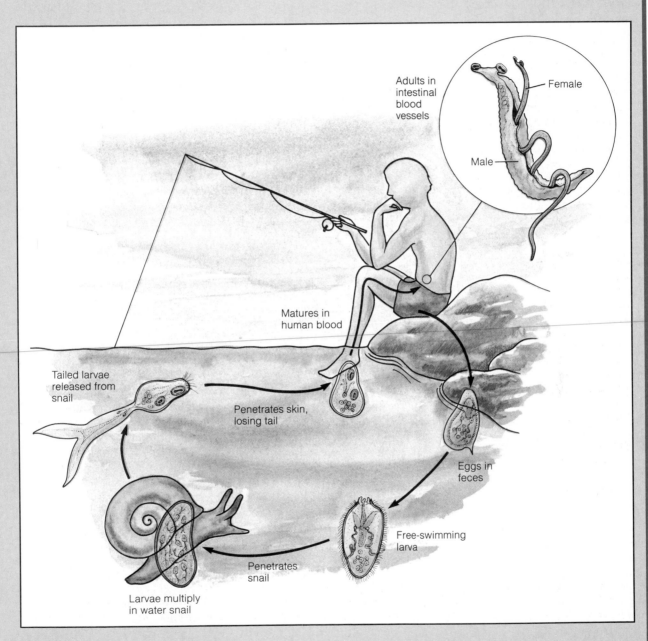

The life cycle of the blood fluke that causes schistosomiasis.

organs that produce sperm and eggs. Thus, any two individuals are capable of mating. Flatworms are also capable of reproducing asexually by fission. In some species, the fission plane forms directly behind the head region. As the worm squirms around, the head and tail halves split apart, and each half then regenerates the missing half.

The flukes and tapeworms are internal parasites of other animals. Most of them have very complicated life cycles that require two to four different hosts. In some cases, humans serve as intermediary hosts. For example, the blood fluke requires two hosts—a human and a water snail—to complete its life cycle. In humans, the blood fluke causes a disease called **schistosomiasis,** a debilitating and sometimes fatal infection that affects about 200 million people. It ranks second only to malaria as the most common parasite-inflicted disease in humans. The life cycle of this pest and some of the problems it has caused are discussed in Essay 18–1.

CLASS NEMATODA: ROUNDWORMS

The phylum Aschelminthes includes about 14,000 species of wormlike creatures, of which 12,000 species belong to the Class Nematoda. The **nematodes,** or **roundworms,** are extremely abundant in nature, occupying virtually every habitat imaginable. They are found in fresh water, salt water, and soil, ranging from hot springs to icy polar regions, and from ocean depths to mountain tops. There are both free-living and parasitic species; the parasitic nematodes spare neither plants nor animals, including humans. They are so abundant in some hosts that if the host's tissues could be made to disappear without disturbing the nematodes, a ghostly likeness of the original organs and tissues would remain visible in the form of the nematodes.

Part of the success of the nematodes may be due to their mode of locomotion, which is surprisingly effective in a variety of environments. Roundworms have a tough external covering, or **cuticle,** which can change shape by means of special longitudinal muscles under the skin (Figure 18–7). This allows nematodes to swim through water or push themselves through soil in a series of S curves.

Another significant advance over previous phyla is the roundworm's tubular digestive system that has two ends, a mouth and an anus. Unlike the flatworms,

which have a simple blind sac with a single opening for a digestive cavity, the roundworm's digestive tract permits food to enter at one end (the mouth) and undigested wastes to exit from the other end (the anus). Because the food passes through the gut in only one direction, the evolutionary door to greater specialization of the digestive system swung open. That is, the tubular digestive tract was an adaptive breakthrough paving the way for further specialization of the food processing activities—pulverization, digestion, absorption, and waste removal. Each activity could occur in its proper turn as food passed through the specialized section designed to carry it out.

Another evolutionary advance seen in the nematodes is the presence of an enclosed body cavity, or **pseudocoelom,** situated between the intestinal wall and the muscle tissue of the external body wall (see Figure 18–7). The sponges, cnidarians, and flatworms (the acoelomate phyla) do not have such a body cavity, but most of the more advanced invertebrates and all the vertebrates have a **true coelom.** (The difference between a pseudocoelom and true coelom is not important to our discussion here; both are body cavities located between the digestive tract and body wall.) The advent of the coelom (pseudocoelom) has permitted the evolutionary development of specialized internal organs and tissues not possible in the acoelomate organisms.

About 50 species of nematodes are parasites of humans. These include (1) hookworms, tiny bloodsucking worms that inhabit the intestines; (2) pinworms, intestinal worms that commonly infect children; (3) *Trichinella spiralis,* a muscle-infesting nematode usually obtained from undercooked pork, which causes the disease trichinosis; and (4) microfilaria, microscopic worms passed to humans by mosquitos. The microfilaria worms lodge in lymph vessels where they reproduce in such numbers that they block the flow of lymph fluid. Lymph blockages often result in enormous swelling (elephantiasis) of the affected regions.

PHYLUM MOLLUSCA: SNAILS, CLAMS, SQUIDS, AND OTHERS

The phylum **Mollusca** (meaning "the soft ones") contains about 100,000 living species, making it the second largest phylum in the animal kingdom. Among the most familiar mollusks are snails, slugs, clams, mussels, scallops, oysters, squids, and octopuses.

FIGURE 18–7
***Ascaris*, a Parasitic Nematode.** Ascarid worms are intestinal parasites of humans, pigs, and horses. (a) They have a complete digestive system with a mouth and anus. (b) The system of longitudinal muscles just inside the tough cuticle is responsible for the wriggling motion of roundworms.

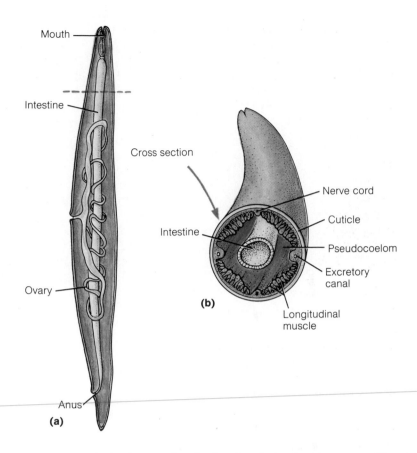

While the mollusks display a tremendous diversity in outward appearance, all share a body plan that consists of four basic parts: (1) a **head region** that houses the brain and sense organs; (2) a muscular **foot** that most mollusks use in locomotion; (3) a **visceral mass** located above the foot, which contains the internal organs involved in digestion, excretion, circulation, and reproduction; and (4) a **mantle,** which covers the visceral mass and in many species secretes a hard shell (Figure 18–8).

The mollusks show several significant advances over the less complex animal groups we have already discussed. They have a circulatory system that includes a chambered heart, which has permitted some species to attain relatively large sizes (some squids are up to 15 meters, or about 50 feet, in length). Advances in the nervous system, including a complex brain and well-developed sense organs, have permitted the mollusks to fend for themselves quite well as free-living organisms. Their digestive system is a complete two-ended tube, as in roundworms, but it is more complex: it includes a mouth, pharynx, esophagus, stomach, intestine, and anus (see Figure 18–8). Most mollusks (except for the bivalves) also have a tonguelike rasping organ called a **radula,** which is used to rasp food off of rocks or pulverize prey. Finally, all mollusks except the land snails have gills that enhance the exchange of respiratory gases—oxygen and carbon dioxide. In some groups the gills have also become adapted to carry out other functions, such as filter feeding in clams and their relatives (see page 299).

The Mollusca are generally divided into six classes, of which the classes Gastropoda, Bivalvia, and Cephalopoda contain the most familiar representatives. These classes are also the most diverse of the mollusks—the gastropods alone are about 80,000 species strong. We will consider each of these three classes separately.

Class Gastropoda: Snails, Slugs, and Their Relatives

Most **gastropods** occupy marine habitats and produce a coiled shell, although in some (abalone, for instance) the coiling is not very pronounced. The intricate and often colorful gastropod shells are prized collector

FIGURE 18–8
Structure of a Chiton, a Mollusk.
(a) Photograph of *Chaetopleuro apiculata,* a chiton common on the eastern coastline of the United States. (b) This lateral section through a chiton shows the major structures found in mollusks.

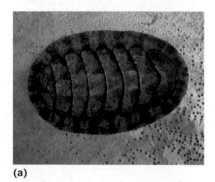

(a)

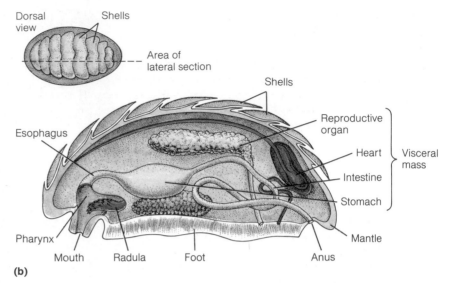

Dorsal view

Shells

Area of lateral section

Shells

Esophagus

Reproductive organ

Heart

Intestine

Stomach

Visceral mass

Mantle

Pharynx

Mouth

Radula

Foot

Anus

(b)

items for many beachcombers. Some gastropods, such as nudibranchs and slugs, do not produce a shell (Figure 18–9a).

The common land snail, one of the few mollusks that is fully terrestrial, has a conspicuous foot that secretes a slimy material as it moves along. (The slime aids in locomotion.) Transverse waves of muscle contractions moving from the back to the front of the foot enable the snail to glide over a surface. Like most gastropods, the snail has a well-developed head with two eyes and two antennae (Figure 18–9b). The snail's visceral mass lies above its head inside the shell. Enough space remains inside the shell, however, for the snail to withdraw its head and foot entirely into the shell at the first sign of danger and during inactive periods. The snail's filelike radula rips through plant tissue at a devastating pace, much to the dismay of home gardeners. It seems fitting, therefore, that millions of garden snails find their way onto hors d'oeuvre plates in French restaurants.

FIGURE 18–9
Gastropods. (a) The marine nudibranch is a gastropod that lacks a shell. (b) The common garden snail.

(a)

(b)

Class Bivalvia: Clams, Oysters, and Their Relatives

Class Bivalvia, or **bivalves,** are so known because of their two-part shell. The two parts, or valves, are hinged on one side by large muscles, enabling these creatures to open and shut their shells (Figure 18–10). Besides offering protection from predators, the opening and closing of the valves is used by scallops to propel themselves through the water.

The bivalves have no radula; they obtain organic nutrients by straining tiny food particles from the water that flows through their gills, a process called **filter feeding.** The animals take in water through an incurrent siphon and pass it through the ciliated, mucus-secreting gills (Figure 18–10b). Tiny bits of food stick to the mucus and are swept toward the mouth by the coordinated beating of the cilia. The filtered water then flows out through the excurrent siphon. Clams and oysters can filter up to several liters of water per hour.

Most bivalves have a large, hatchet-shaped foot they use for burrowing. The mantle covers the visceral mass and secretes the components of the shell. If a grain of sand or other irritant lodges between the mantle and the shell, the mantle cells secrete calcium carbonate in layers around it, resulting in a pearl.

Class Cephalopoda: Squids, Octopuses, and Their Relatives

By now you may have an image of mollusks as sedentary, or at best slow-moving, creatures. However, our third group of mollusks, the **cephalopods,** are neither. This group includes the squids and octopuses (Figure 18–11), which have a jet propulsion mechanism that propels them away from danger or toward prey. Their modified mantle cavity fills with water; then, by squirting the water out through a small opening, the squid or octopus "rockets" through its watery habitat. Once it catches up to a fish, shrimp, or other tasty morsel, the cephalopod grasps its prey with its long tentacles (equipped with suckers) and moves it into its mouth. The mouth has two tearing beaks and a radula to rip apart the unfortunate captive.

Cephalopods are equipped with many features typically present in vertebrates (see Chapter 19). They have internal cartilaginous supports similar to the vertebrate skeleton and a brain covering analogous to

(a)

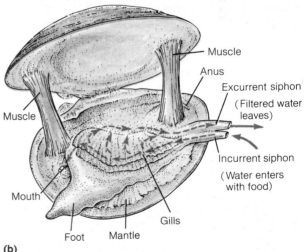

(b)

FIGURE 18–10
Bivalves. (a) This photograph of a scallop shows the two valves in the open position. Note the prominent gills used for filter feeding. (b) Internal view of a clam emphasizing the feeding mechanism.

FIGURE 18–11
The Octopus, a Cephalopod.
Octopuses live in small burrows between rocks. This one is on a hunting expedition. If confronted by one of its own predators, the octopus (and its cousin the squid) can eject a cloud of dark inklike material in its wake.

the vertebrate skull. Cephalopods also have a well-developed nervous system, including a complex brain and two image-forming eyes very much like ours. These similarities should not be taken as evidence of a close ancestral link between cephalopods and vertebrates (they are not close at all), but rather as an interesting example of convergent evolution—both groups evolved similar structures along separate evolutionary lines. Their similar adaptations reflect the similar, very active lifestyles of the cephalopods and vertebrates.

PHYLUM ANNELIDA: SEGMENTED WORMS

The **annelids,** represented by nearly 9000 species, form the third and most advanced phylum of worms. Phylum Annelida includes earthworms, leeches, and polychaetes (marine worms). Their most distinctive feature is a segmented body. Considered an evolutionary advance, body segmentation made possible the evolution of structural specializations that are characteristic of the more complex animals.

Segmented worms are the simplest group of animals to have all the principal features of the more advanced animals, except for jointed appendages and a skeleton. The digestive system of annelids is more elaborate than that of the mollusks—there is a **crop** for storing food and a **gizzard** for grinding it (Figure 18–12a). The gut is suspended in a rather large true coelom. Also within the coelom is the circulatory system, consisting of blood vessels and five paired hearts. The nervous system of the annelid includes a ganglion (a brainlike nerve mass), a nerve cord that runs the length of the body, and special sensory cells sensitive to light, taste, and touch. There is no respiratory system; gas exchange in segmented worms occurs across their body surfaces.

Reproduction in annelids is exclusively sexual. Both earthworms and leeches are hermaphroditic—each individual has both ovaries and testes. However, two individuals are required for reproduction; they cannot fertilize themselves. Mating earthworms lie side by side, head to tail, each depositing sperm in its partner's sperm receptacles (Figure 18–12b). After the worms separate, a membranous cocoon is secreted from the **clitellum,** a thickened band of tissue overlying several of the middle segments. The earthworm then pushes the cocoon toward its head end, collecting eggs on its way. Once it is shrugged off, the egg-containing cocoon provides a moist chamber for the early development of the young worms.

In marine polychaetes, the sexes are separate. Male and female worms release their sperm and eggs into the water, where fertilization takes place. In many

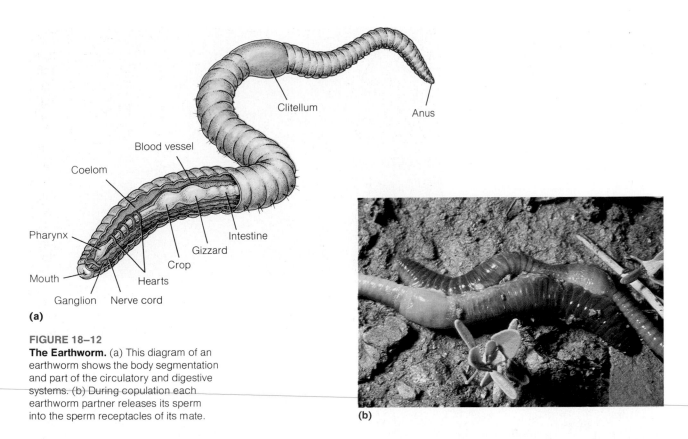

Clitellum

Anus

Blood vessel

Coelom

Pharynx

Intestine

Gizzard

Mouth

Crop

Hearts

Ganglion Nerve cord

(a)

FIGURE 18–12
The Earthworm. (a) This diagram of an
earthworm shows the body segmentation
and part of the circulatory and digestive
systems. (b) During copulation each
earthworm partner releases its sperm
into the sperm receptacles of its mate.

(b)

species of polychaetes, the zygote develops into a
distinctive larva before becoming an adult. This larva
is very similar to the larval forms of many mollusks,
which argues strongly for a close evolutionary link
between the mollusks and the annelids.

PHYLUM ARTHROPODA: SPIDERS, INSECTS, AND OTHERS

The phylum **Arthropoda** (literally "jointed legs") is
composed of over a million species—more than three
times as many as all other animal species combined!
Although the **arthropods** include the crabs, lobsters,
centipedes, millipedes, scorpions, spiders, mites, and
many other creatures, most of its members (indeed,
most animals) constitute a single arthropod class, the
insects. Representatives of the five major classes of living arthropods are shown in Figure 18–13, and the

major characteristics of each class are outlined below:

1. **Class Chilopoda:** Centipedes with one pair of legs
   per body segment. All are carnivores (meat-eaters),
   with the first pair of legs modified as poisonous
   fangs. 3000 species.
2. **Class Diplopoda:** Millipedes with two pairs of legs
   per body segment. They feed primarily on decaying plant material. 8000 species.
3. **Class Crustacea:** Crabs, shrimp, lobsters, crayfish,
   pill bugs, barnacles, and others. These vary in form
   and type of appendages but almost always have
   two pairs of antennae. 26,000 species.
4. **Class Arachnida:** Spiders, scorpions, ticks, mites,
   daddy longlegs, and others. Each individual has one
   pair of pinchers or fangs, one pair of feelers (palps),
   and four pairs of walking legs. 55,000 species.
5. **Class Insecta:** Flies, fleas, butterflies, beetles, bees,
   ants, true bugs, grasshoppers, cockroaches, termites, and many others. Insects have three pairs
   of walking legs, three pairs of mouthparts, and usually wings. 900,000 species.

FIGURE 18–13
Representatives of the Five Major Classes of Arthropods. (a) A centipede. (b) A millipede. (c) These lobsters are crustaceans. (d) A spider, a member of the arachnids. (e) A beetle, an example of an insect.

(a)

(b)

(c)

(d)

(e)

FIGURE 18–14
The Compound Eyes of an Insect. This frontal view of a Mediterranean fruit fly shows its eyes to be composed of many tiny light-sensitive units. Although such compound eyes can detect movement and form images of objects very close, they cannot be focused. Thus, most of the insect's visual world is a blur.

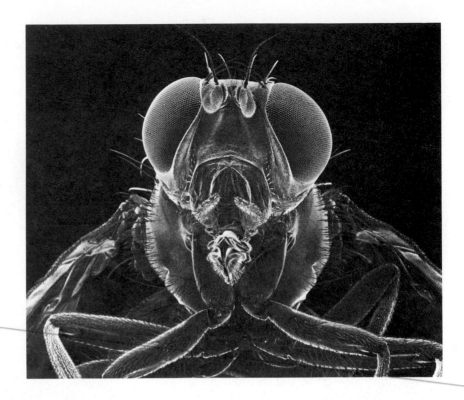

General Characteristics of Arthropods

The unparalleled success of arthropods almost certainly stems from the development of a segmented **exoskeleton** (a tough outer covering) and jointed legs. The exoskeleton is made up of a tough polysaccharide called **chitin,** and in crustaceans, the chitin is impregnated with calcium salts to make it even tougher. The exoskeleton completely covers the body, even the lens of the eye, and offers protection against predators and water loss. The exoskeleton also provides points of attachment for an elaborate system of muscles that give the arthropods such an extensive repertoire of movements. But having a hard exoskeleton has one major drawback. It must be shed periodically to allow the animal to grow, leaving it vulnerable for a while. The shedding of the exoskeleton is called **molting.** A lobster may molt up to seven times during its first growing season.

The jointed appendages characteristic of the arthropods are clearly an advance over the annelids for a terrestrial life-style. With legs strong enough to support their bodies above the ground, the arthropods are much better equipped to move from place to place. In many species some of the appendages have been modified for functions other than locomotion. Included among such accessories are jaws, antennae, fangs, pincers, claws, sucking tubes, and many others.

The arthropods show certain aspects of body design that suggest an evolutionary link to the annelids. For example, the digestive systems of these two groups are similar. Food is mechanically disrupted and temporarily stored in a **foregut,** then digested in the **midgut** where nutrient absorption also occurs; water is reabsorbed from the forming feces in the **hindgut.** In addition, the arthropod body represents a specialization of the segmentation pattern so prominent in annelids. The major trends in specialization were: (1) a reduction in the number of body segments; and (2) fusion of individual segments to form distinct body regions, such as the head, thorax, and abdomen of insects (see the following section on insects).

Unlike the annelids, arthropods have an open circulatory system in which blood circulates between vessels and large cavities. The blood in the cavities bathes the internal organs and facilitates the exchange of nutrients, respiratory gases, and metabolic wastes, then collects in the vessels leading to a single heart, which keeps the blood flowing throughout the body.

Another feature that sets the arthropods apart from their annelid cousins is the presence of very keen sense organs, such as the extraordinary compound eyes of insects (Figure 18–14). These creatures can thus sense very subtle changes in their environment and respond to them quickly. You have to be very quick indeed to catch a fly.

FIGURE 18–15

The Body Plan of Insects. As indicated in this diagram of a grasshopper, the trunk of adult insects is divided into a head, thorax, and abdomen. Most adult insects have 3 pairs of walking legs (the hind pair in the grasshopper is specialized for leaping) and 2 pairs of wings attached to the thorax; the head bears various jointed appendages modified for sensory and eating functions.

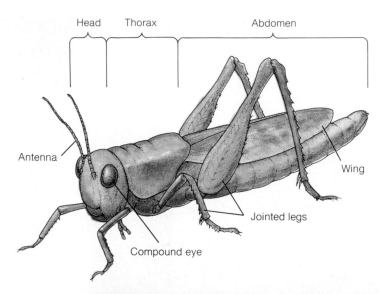

Insects

Of the five major classes of arthropods, we choose to discuss the insects in depth for one simple reason: insects cannot be ignored. They occupy virtually every available habitat on earth, and their diversity and sheer numbers boggle the mind. Indeed, if we judged biological success by numbers alone, insects would rank number one over all the animal groups that have ever lived. The overwhelming success of this class is undoubtedly due to their distinctive body design, the specialization of individual species, and the fact that they are the only invertebrates able to fly.

The body plan of insects consists of three basic parts: a **head,** consisting of six fused segments, that bears the antennae and mouthparts; a **thorax** of three fused segments bearing three pairs of legs and usually two pairs of wings; and an **abdomen** of up to eleven segments, which almost always lacks appendages (Figure 18–15). This is the familiar adult insect design; the immature form may be quite different. In fact, most insects pass through four developmental stages: the fertilized **egg,** the **larva,** the **pupa,** and the **adult** (Figure 18–16). Each stage is typically quite different from the others in appearance; similar stages also vary in form from species to species. The larval stage of a butterfly, for instance, is a caterpillar; the larva of a fly is a maggot. After a period of extensive feeding and growth, the larva becomes a pupa (for example, the cocoon of a butterfly), from which it eventually metamorphoses into an adult.

One other curious feature of certain insects is worth mentioning. Some species are highly social, forming colonies that function as a biological unit (see Chapter 30). For example, a honeybee colony usually consists of a single queen, several thousand sterile female workers, and a hundred or so males (drones). The workers are further specialized: young adults are usually restricted to the hive and perform housekeeping activities, such as tending the larvae and preparing new wax cells, whereas the older workers venture into the field in search of nectar. Thus, in bee colonies as in other insect societies (such as ants, wasps, and termites), there is a specialization of functions or roles above the level of the individual organism. The population has in effect become an organism in itself. Such complex societal structures are not found in any other invertebrate phylum.

FIGURE 18–16

Stages in the Life of an Anise Swallow-tail Butterfly. (a) Fertilized egg.
(b) Caterpillar—the larval stage.
(c) Cocoon building—the start of the pupal stage. (d) Early pupal stage.
(e) Swallowtail emerging from pupa.
(f) The anise swallowtail butterfly.

(a)

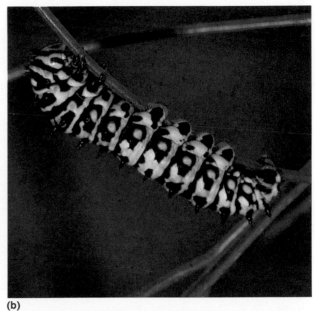

(b)

(c)

(d)

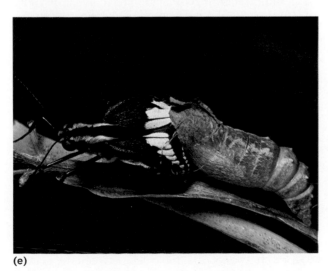

(e)

(f)

FIGURE 18–17
Echinoderms. The radial symmetry of echinoderms is illustrated in these photos of (a) sea urchin and (b) starfish.

(a)

(b)

PHYLUM ECHINODERMATA: STARFISH, SEA URCHINS, AND OTHERS

The phylum **Echinodermata** ("spiny-skinned" animals) includes starfish, sea urchins, sea cucumbers, sand dollars, and similar creatures. The echinoderms are exclusively marine animals; most live a rather sedentary life on the ocean floor and intertidal rocks. They are at the same time among the most distinctive and the most puzzling of the invertebrate groups. All exhibit radial symmetry, with five (or multiples of five) arms or other structures radiating from a central area (Figure 18–17). Their radial symmetry is especially puzzling because echinoderm larvae have bilateral symmetry. Echinoderms are also unusual because they have an extensive system of calcium-containing plates within their body wall that functions as an internal skeleton, or **endoskeleton.** Outward projections rising from these plates make up the bumps on the surface of starfish and the rather nasty spines of the sea urchin. Echinoderms also have a unique system of water-filled tubes and canals called the **water vascular system,** which plays an important role in locomotion. Existing as extensions of the system of canals are tiny tube feet that protrude from the skin on the lower surface. The tube feet can be retracted or extended, enabling the animal to move.

In many respects, the echinoderm body design is strikingly simple. Most of these creatures have a spacious coelom, a very simple circulatory system, no respiratory or excretory system, and a radially organized nervous system consisting of nerve cords but no brain. In short, if a biologist were to design an organism with a set of structural and functional characteristics geared for evolutionary success, he or she would not come up with an echinoderm. Yet the phylum has prospered—perhaps because these creatures' endoskeletons and relative lack of tissue makes them unappetizing to predators.

The most familiar example of an echinoderm is the common starfish. Most starfish have five arms radiating from a central disk. Centered on the lower surface of the disk is the mouth, and on the upper surface is the anus. Beneath its rough or spiny skin is the starfish's endoskeleton (Figure 18–18).

The starfish's digestive system begins with the mouth, which is connected by a short esophagus to a large stomach occupying most of the interior of the central disk. The stomach in turn is connected to five pairs of digestive glands, one in the coelomic cavity of each arm. To feed, the starfish turns its stomach outward through its mouth directly onto its prey. Enzymes produced in the digestive glands flow out onto the food; after a while, the stomach with the partially digested food retracts. Most of the nutrients are then absorbed by the digestive glands. A starfish is one of the few animals that can prey on clams. It slips its stomach through the thin crack between the clam's two valves, and digests the clam's tissues.

Starfish have separate sexes. Both males and fe-

FIGURE 18–18
Anatomy of a Starfish.

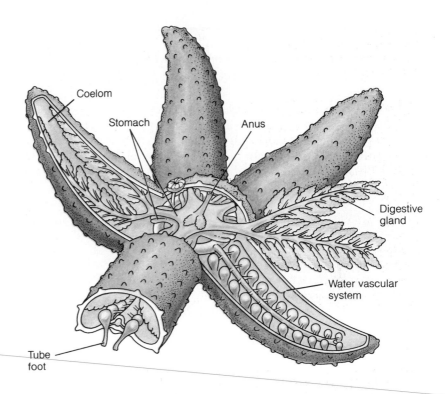

Coelom

Stomach

Anus

Digestive gland

Water vascular system

Tube foot

males shed their sperm and eggs into the water, where fertilization takes place. A fertilized egg develops into a distinctive swimming larva, which eventually undergoes a complex metamorphosis during the process of becoming an adult.

Although outward appearances would not suggest it, most biologists believe that the ancestral roots of the vertebrates lie near the echinoderms. In the chapter that follows, we will trace the evolution of these backboned animals and look at the evidence for their ancestral kinship with echinoderms.

Table 18–1 presents the major characteristics of the eight invertebrate phyla we have studied in this chapter, highlighting the evolutionary advances.

TABLE 18–1
Major Characteristics of the Invertebrate Groups. Features in **boldface** represent evolutionary advances when they first appeared.

		Adult Characteristics				
Major Animal Groups	Body Symmetry	Highest Level of Multicellular Specialization	Body Cavity	Digestive System	Skeletal System	Other Noteworthy Characteristics
Sponges	Asymmetrical	Few cell types	None	None	Spicules/spongin	
Cnidarians	**Radial**	**Tissues**	None	None	Calcium carbonate in corals	
Flatworms	**Bilateral**	**Organs**	None	**Blind sac**	None	
Nematodes	Bilateral	Organs	**Pseudocoelom**	**Tubular**	None	
Mollusks	Bilateral	Organs	**Coelom**	Tubular	Shells in most	
Annelids	Bilateral	Organs	Coelom	Tubular	None	**Body segmentation**
Arthropods	Bilateral	Organs	Coelom	Tubular	Chitinous **exoskeleton**	**Jointed appendages**
Echinoderms	Radial	Organs	Coelom	Tubular	**Endoskeleton**	Water vascular system

SUMMARY

Animals are motile, multicellular heterotrophs that reproduce almost exclusively by sexual means. The overwhelming majority of animals are invertebrates, spineless creatures that comprise all but one of the 31 currently recognized animal phyla. Biologists have used many criteria to evaluate the evolutionary relationships among this diverse group of animals. This chapter has presented eight major invertebrate phyla, roughly in order of their evolutionary development, noting the major structural advances that characterize each group.

Phylum Porifera (sponges) are judged to be the most primitive of all animals. Their bodies, usually asymmetrical, are composed of several types of cells but no complex tissues. Sponges feed on minute food particles in the water, which is filtered through their porous bodies.

Cnidarians, which include hydras, jellyfish, sea anemones, and corals, have radial symmetry. True tissues are present, including primitive nervous and muscle tissues. In addition, these animals have unique stinging threads called nematocysts, which aid in capturing prey. Two basic body forms include the sedentary polyp and the free-floating medusa, both of which occur at alternate stages in the life cycles of some species.

Like all higher invertebrates except the echinoderms, platyhelminthes (flatworms) have bilateral symmetry. Their digestive system is a blind sac with a single opening. Flatworms can reproduce both sexually and asexually. Most are parasitic.

Another group of worms is Nematoda, the roundworms, whose many free-living species inhabit virtually all life-supporting regions of the world. The parasitic forms are also extremely abundant, living in the bodies of almost all species of plants and animals. Roundworms have a tubular digestive tract, a more efficient design than the blind sac characteristic of the flatworms. The digestive tube is suspended in a body cavity called a pseudocoelom.

The soft-bodied (but often shelled) members of the phylum Mollusca have a true coelom. The true coelom permitted the evolution of the more complex organ systems characteristic of these and higher animals. The mollusks include the gastropods (such as snails, nudibranchs, and slugs), the filter-feeding bivalves (such as clams, mussels, scallops, and oysters), and the large, fast-moving cephalopods (squids and octopuses). All of these creatures have the same basic body plan, which includes a head region, foot, mantle, and visceral mass.

The annelids, or segmented worms, include the earthworms, leeches, and marine polychaetes. These worms have a closed circulatory system and reproduce sexually. The segmented body form of the annelids has been retained in various modified forms in the arthropods.

The largest animal phylum, Arthropoda, contains many familiar organisms, all of which have an exoskeleton and jointed appendages. Because the rigid exoskeleton does not expand as the animal grows, it must be periodically molted and replaced. Arthropods have highly developed nervous and sensory systems and an open circulatory system. The insects, which comprise only a single class of arthropods, include more species than all other groups of animals combined. Many insects have life cycles that include metamorphoses through egg, larva, pupa, and adult stages.

Finally, the spiny-skinned members of Echinodermata (starfish, sea urchins, sea cucumbers, and others) all have an endoskeleton. These organisms exhibit radial symmetry as adults; the larvae, however, are bilaterally symmetrical. Although the echinoderms are structurally simpler than many of the other invertebrates, they are apparently closely related to the higher animals, the vertebrates.

STUDY QUESTIONS

1. Name five general characteristics of animals.

2. Which features of the sponges suggest that this group is the most primitive of all animals?

3. Nematocysts are characteristic of which animal phylum? What is their function?

4. Name the most primitive animal phylum to exhibit bilateral symmetry.

5. Name the two animal groups that include the largest numbers of human parasites.

6. What are the advantages of a tubular digestive system over a blind sac system?

7. Which phylum includes the shelled, filter-feeding animals?

8. Which class of mollusks includes organisms with the most vertebratelike structures?

9. The segmented worms belong to which phylum?

10. What does *molting* mean?

11. What are the two most distinguishing structural characteristics shared by all arthropods?

12. Name the three basic body regions of insects.

13. Arrange the following terms in their proper developmental sequence: pupa, egg, adult, larva.

14. What are the major characteristics of echinoderms?

15. Which group of invertebrates is most closely related to the vertebrates?

SUGGESTED READINGS

Barnes, R. D. *Invertebrate Zoology*. 4th ed. Philadelphia: Saunders, 1980. This is a classic advanced text on the invertebrates.

Buchsbaum, R. *Animals without Backbones*. 2nd ed. Chicago: University of Chicago Press, 1976. An excellent introductory text on the invertebrates.

Eisner, T., and E. O. Wilson, eds. *The Insects*. San Francisco: W. H. Freeman, 1977. This collection of readings from *Scientific American* traces 25 years of research on insects.

Lane, F. W. *Kingdom of the Octopus*. New York: E. P. Dutton, 1968. This is a fascinating book about cephalopods. Especially interesting is an account of how an amateur naturalist, the Reverend Moses Harvey, came to acquire one of the largest squids ever captured.

Schneirla, T. C. *Army Ants*. San Francisco: W. H. Freeman, 1971. This book focuses on the behavioral patterns of army ants from both an environmental and genetic perspective.

CHAPTER 19

The Animal Kingdom: Lower Chordates and Vertebrates

You and I are vertebrates. So are fishes, frogs, lizards, birds, and horses. All these animals are vertebrates because they possess a **vertebral column,** or **backbone,** that surrounds and protects the spinal cord, and a **skull** that encases the brain. The backbone also provides a strong but flexible framework for the attachment of muscles.

We and the other vertebrates belong to a larger group of animals—the **phylum Chordata.** Besides vertebrates, the chordates include two other less familiar groups of animals collectively known as the **lower chordates.** Let us begin by examining the general features of chordates that distinguish them from the invertebrate phyla discussed in Chapter 18.

CHARACTERISTICS AND ORIGIN OF CHORDATES

Members of phylum Chordata exhibit the more advanced features of the higher invertebrate phyla: a segmented body in most forms, well-developed organ systems, and a tube-within-a-tube body plan (that is, a gut suspended in a coelom). What distinguishes the chordates from other animal phyla is the presence of three additional structures that are at least present in the chordate embryo if not always in the adult (Figure 19–1). They are:

1. A **notochord** (hence the name chordate). A firm, rod-shaped structure, the notochord runs longitudinally just above the digestive tract; it provides skeletal support. In most vertebrates, the notochord is partially or entirely replaced by a bony or cartilaginous backbone during embryonic development.
2. A **dorsal hollow nerve cord.** In vertebrates, the embryonic nerve cord develops into the brain and spinal cord.
3. **Pharyngeal gill slits.** The gill slits (or gill pouches) are located on either side of the pharynx (throat) between the mouth and esophagus. In fishes they become the gills; in higher vertebrates they disappear or become modified for other purposes during embryonic development. For example, in humans the uppermost gill pouches develop into the cavities of the middle ear and the eustachian tubes that connect the throat to the ears.

Although the fossil record does not tell us how the chordates got their start, we can make some reason-

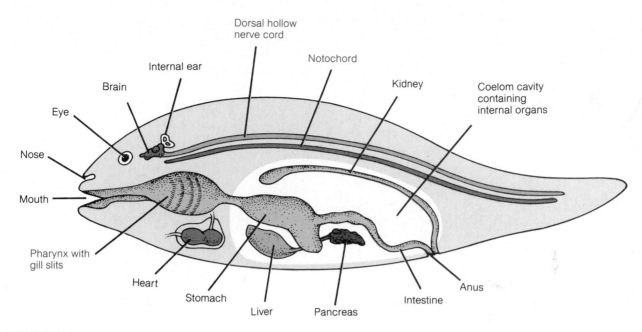

FIGURE 19–1
Generalized Body Plan of Chordates.
This idealized diagram shows the major
structures of chordates. Note particularly
the pharyngeal gill slits, notochord, and
dorsal hollow nerve cord, structures that
occur in all chordates at some time
during their life cycle.

able deductions on this matter. To begin with, we
know that all the invertebrate phyla (see Chapter 18)
appeared on earth before the chordates, so it is logical
to seek our chordate "roots" in these groups. Of these,
Echinodermata is the most promising candidate for
several reasons. First, embryo development in echinoderms and chordates is similar in several respects, one
of which has to do with the formation of the digestive
tract. In both groups the anus develops first, then the
mouth; in all the other invertebrate groups (arthropods, annelids, and lower phyla), the mouth develops
first. Second, the larvae of some echinoderms bear a
striking resemblance to the larvae of **hemichordates,**
a small phylum of advanced wormlike creatures that
have both pharyngeal gill slits and a dorsal nerve cord
that is hollow in some species (Figure 19–2). Thus,
the hemichordates have some echinoderm features
and two very prominent chordate characteristics, and
this has caught the attention of zoologists who view
them as a transitional group between the ancient
echinoderms and the lower chordates.

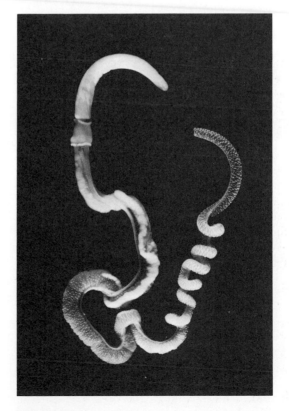

FIGURE 19–2
An Adult Hemichordate.

THE LOWER CHORDATES: TUNICATES AND LANCELETS

The lower chordates include about 2000 species of **tunicates** (subphylum Urochordata) and 30 species of **lancelets** (subphylum Cephalochordata), all of which are marine. The tunicates are small, sedentary organisms usually found attached to subtidal rocks (Figure 19–3a–c). Most are filter feeders—they take in water through a mouth, channel microscopic bits of food through the gill slits into the gut, then pass the filtered water out a pore called the excurrent siphon. Although the adult tunicate has pharyngeal gill slits, it lacks the other major chordate structures. However, the free-swimming larval tunicate, which resembles a tadpole, does have a notochord and dorsal hollow nerve cord. These two structures disappear during metamorphosis

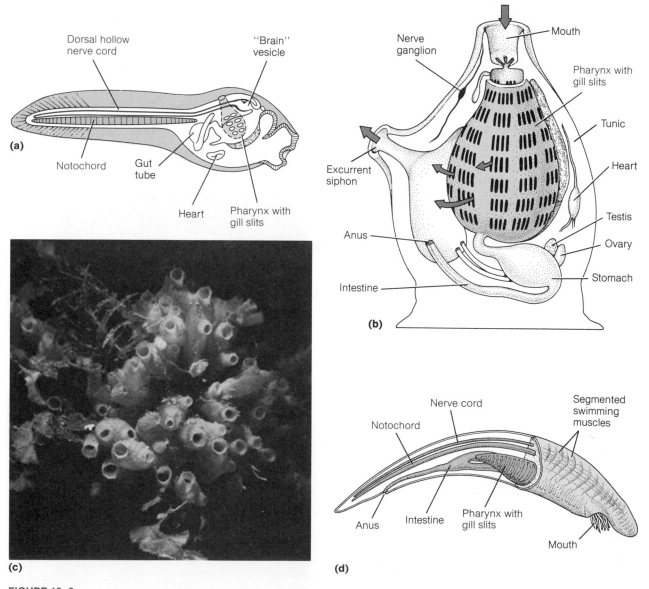

FIGURE 19–3
Lower Chordates. (a)–(c) Tunicates. (a) The larval tunicate (not shown to scale) has the three chordate characteristics (color). (b,c) In adult tunicates, the body is encased in a tough outer sheath called the tunic. (d) An adult lancelet. In addition to the three chordate structures (color), lancelets exhibit a distinct segmentation pattern, a vertebrate characteristic.

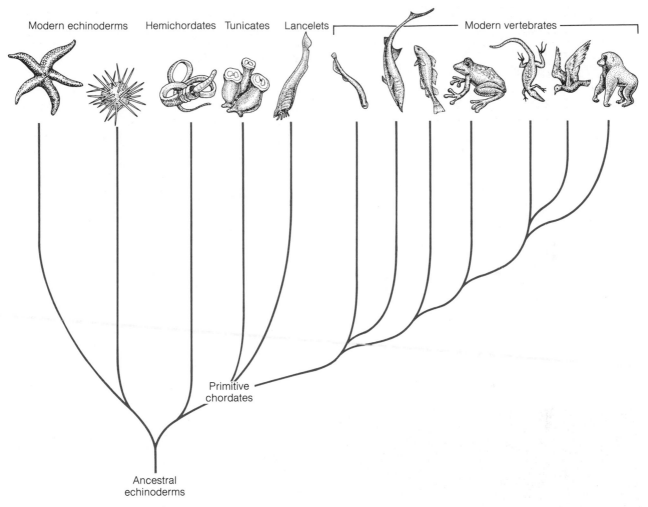

Modern echinoderms   Hemichordates   Tunicates   Lancelets   Modern vertebrates

Primitive
chordates

Ancestral
echinoderms

FIGURE 19–4
Vertebrate Ancestry. The vertebrates probably stemmed from primitive chordates that resembled the modern tunicates and lancelets. The primitive chordates presumably arose from an ancient group of hemichordates, which derived from early echinoderms.

of the larva into the adult form. Thus, the tunicates are true chordates by virtue of their larvae.

The lancelets are fishlike animals about 5 cm (2 inches) long that inhabit shallow coastal waters (Figure 19–3d). There they spend most of their time partially buried in the sand. Only their mouths stick up into the water, where they filter feed. In addition to the three major chordate characteristics, lancelets have a segmented body. This segmentation is most pronounced in the swimming muscles that are attached to the notochord.

The tunicates and lancelets probably arose from an ancient line of hemichordates, but their ancestral relationship to the first vertebrates has divided zoologists into two schools of thought. Because lancelets are fishlike and have a segmented body (a vertebrate characteristic), some zoologists believe that an ancestral form of this group gave rise to the first vertebrate fishes. On the other hand, the larval tadpoles of tunicates also bear a strong resemblance to the primitive fishes, and so others favor the tunicate-to-vertebrate theory. There is still no solid evidence to decide between these two possible lines of descent. Figure 19–4 summarizes our current "best guess" of vertebrate ancestry.

Although the fossil record for the vertebrate forerunners is very scanty, the vertebrates themselves are quite well represented in the rocks. This is largely because vertebrates have bones and/or cartilage that resist decomposition, and thus leave distinct impres-

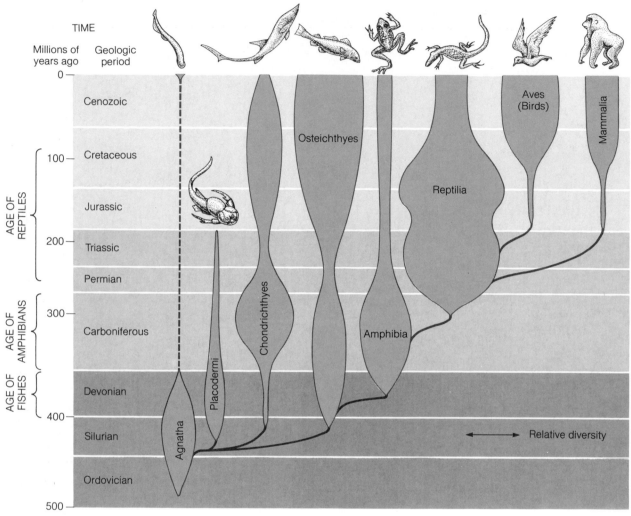

FIGURE 19–5
Evolutionary Relationships among the Vertebrate Classes. The width of each column indicates the
relative diversity at any given time. The dotted line for Agnatha indicates an absence of fossils for this
group from the Carboniferous period to recent times. Only the placoderms lack modern representatives.

sions in sedimentary rocks. By studying the fossilized
remains of earlier vertebrates, paleontologists have
been able to piece together a fairly comprehensive picture of vertebrate evolution—a story that begins some
500 million years ago and culminates in the appearance
of a backboned animal able to contemplate its own
ancestry.

VERTEBRATES

The subphylum **Vertebrata** is the most familiar group
of chordates as well as the most diverse—there are
approximately 43,000 living species of vertebrates.
Vertebrata includes the largest animals ever to inhabit
the earth. If you are thinking of the dinosaurs, you're
right in that these giant reptiles were vertebrates, but
they were not the largest. That distinction goes to the
present-day blue whale, which can attain a length of
over 100 feet and a weight of nearly 200 tons. Although
the largest dinosaur fossil (a beast named *Diplodocus;*
see Figure 19–14) nearly matches the blue whale in
length, estimates have it weighing in at a mere 32 tons.

The vertebrates are divided into seven living
classes and one extinct class. These classes and their
evolutionary relationships are shown in Figure 19–5.
As you examine this chart, you will see that the relative diversity (number of species) of each class has
changed considerably over time. For example, we
know from the fossil record that there was great diver-

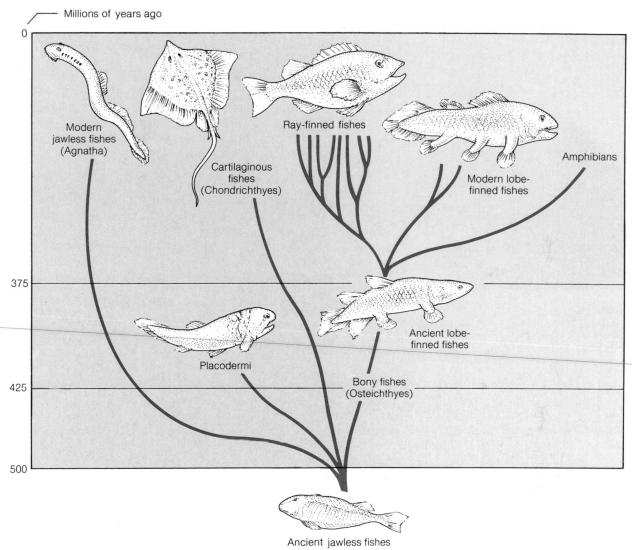

— Millions of years ago

0

375

425

500

Modern
jawless fishes
(Agnatha)

Cartilaginous
fishes
(Chondrichthyes)

Ray-finned fishes

Modern lobe-
finned fishes

Amphibians

Ancient lobe-
finned fishes

Placodermi

Bony fishes
(Osteichthyes)

Ancient jawless fishes

FIGURE 19–6
Evolutionary Relationships Among the Fishes. The oldest vertebrates, the jawless fishes (Agnatha),
gave rise about 425 million years ago to three separate lines: the placoderms, the cartilaginous fishes,
and the bony fishes. The first bony fishes had lungs and fleshy, lobed fins; a few species of lobe-finned
fishes with lungs still exist today. The ancestral lobe-finned fishes also gave rise to the amphibians and
the ray-finned fishes. The ray-finned fishes make up the vast majority of bony fishes today.

sity among amphibians about 325 million years ago,
but their numbers dwindled at about the time the
reptiles began their explosive diversification. Today,
the bony fishes (Osteichthyes) are the most diverse
vertebrate group: some 26,000 species are known at
this time, and just as many may remain undiscovered.

Fishes

Fishes are aquatic animals with a streamlined shape
that allows them to move rapidly through water. Their

fins, which provide thrust and stability against rolling,
are also used for turning and changing depth. Fishes
exchange gases with their environment through gills
located in gill pouches. In addition, a few species have
lungs that supplement the gills for gas exchange.

The earliest vertebrates were fishes without jaws.
This class, called **Agnatha,** first appeared about 500
million years ago (Figure 19–6). Then, about 425 million years ago, this group gave rise to three other
classes of fishes, all of which had jaws: (1) **Placodermi**
(all species are extinct); (2) **Chondrichthyes,** the

FIGURE 19–7
Lampreys. These jawless remnants of
the most ancient vertebrates have a
round sucker mouth and seven external
gill openings. In this photograph two
lampreys can be seen attached to rocks.

cartilaginous fishes that include sharks, skates, rays, and their relatives; and (3) **Osteichthyes,** the bony fishes. The modern bony fishes include the ray-finned fishes and the less familiar lobe-finned fishes. We will begin our discussion of the fishes with the most primitive group—Agnatha.

CLASS AGNATHA: THE JAWLESS FISHES. There are about 50 species of agnathans living today, which include the lampreys and hagfishes. These jawless fishes feed by sucking blood and other body fluids from their prey, usually other fish. The lamprey (Figure 19–7) attaches its round sucker mouth to its victim, and with it horny, toothlike mouth parts, it bores a hole through the skin. Hagfishes attack dead or dying fish in much the same way.

The lampreys and hagfishes retain a notochord throughout their life cycles but they lack true teeth, jaws, scales, and bones. No bones? Why, then, are they classified as vertebrates—animals with a backbone? The reason is quite simple—they have a vertebral column and skull, but these structures are made of cartilage, not bone.

CLASS PLACODERMI: FIRST FISHES WITH JAWS. The long extinct placoderms, earliest known fishes with jaws,

stemmed from the agnathans. The placoderms were freshwater bottom-dwellers with bony armor plates and an endoskeleton that was at least partially bony. Their most distinctive feature—one that also characterizes the later vertebrates—was jaws. The advent of jaws was a major breakthrough for vertebrates and changed their lifestyle from sifters and scavengers to ferocious predators. With jaws and their associated teeth, the placoderms could grasp prey and tear it into pieces small enough to ingest.

Although the placoderms were clearly transitional between the jawless fishes and the more advanced fishes with jaws, it seems that the latter groups arose separately from jawless ancestors. In other words, the placoderms were an evolutionary dead end. The more advanced jawed fishes—the cartilaginous fishes and the bony fishes—appear in the fossil record shortly after the placoderms, but unlike their extinct cousins, they have persisted.

CLASS CHONDRICHTHYES: THE CARTILAGINOUS FISHES. The cartilaginous, jawed fishes of today include about 600 species of sharks, skates, and rays. Many more species existed 300 million years ago (see Figure 19–5). As their name implies, these fishes have an endoskeleton made entirely of cartilage. They also have 5–7 prominent gill openings, which are visible as vertical slits on either side of the throat (Figure 19–8).

The sharks are the most notorious of the cartilaginous fishes. These carnivores tear into their prey (usually fish) with rows of razor-sharp teeth. Although shark attacks on humans have been overdramatized

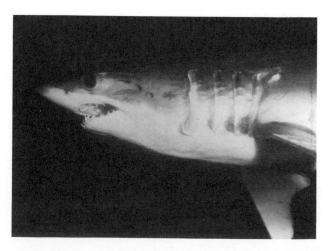

FIGURE 19–8
A Shark. Note the gill slits and teeth.

FIGURE 19–9
A Cutaway View of a Ray-Finned Fish.
Note the position of the swim bladder, a ballast organ that evolved from the lung of their predecessors, the ancient lobe-finned fishes.

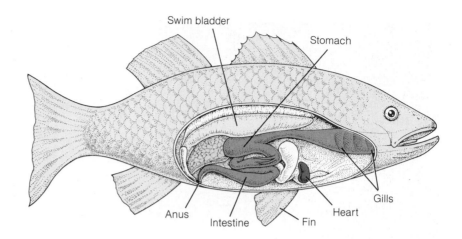

by Hollywood filmmakers, they do occur. Deep-sea divers still await the development of a truly effective shark repellent.

CLASS OSTEICHTHYES: THE BONY FISHES. The bony fishes include most of the familiar fishes living today, both in the oceans (herring, tuna, mackerel, and so on) and in fresh water (trout, striped bass, goldfish, and others). This group began their tremendous proliferation during the Devonian (400 million years ago), a period often called the Age of Fishes. Since then, they have remained the dominant form of aquatic vertebrates.

The first bony fishes were freshwater inhabitants that had, in addition to gills, lungs or lunglike air sacs. They also had pairs of fleshy pectoral and pelvic fins (see Figure 19–6). With their air-breathing lungs and lobed fins, these fishes could probably move about sluggishly on land. Several relict species of **lobe-finned fishes** still inhabit muddy freshwater ponds and lakes in parts of Africa, South America, and Australia, and they do indeed have limited mobility on land. During the dry season they bury themselves in mud and breathe air until the rains refill the ponds. Some species can survive for extended periods outside of water, and at least one species will drown if held under water.

One major line of descent from the early lobe-finned fishes led to the **ray-finned fishes,** the predominant type of bony fish living today (see Figure 19–6). Early on in their evolution, the ray-finned fishes lost the two traits that made their ancestors partially adapted to land. First, the fleshy fins became modified into ray fins—webs of skin supported by thin ribs of bone or cartilage. Second, the gas-exchanging lung evolved into a swim bladder, a gas-filled sac that lies

between the gut and backbone (Figure 19–9). By regulating the flow of gases into or out of its swim bladder, the fish can change its buoyancy, and hence, its depth in the water.

The first ray-finned fishes occupied freshwater habitats, but they soon spread into the oceans. Certain present-day species, such as salmon and many types of trout, spend different parts of their life cycles in fresh water and in salt water. However, all fishes, freshwater and marine alike, must contend with the problem of keeping a proper balance of water and solutes in their body fluids that is different from that of their aquatic surroundings. The way fishes maintain a constant internal water and salt concentration is explored in detail in Chapter 23.

Amphibians

About 375 million years ago, while one group of ancestral lobe-finned fishes was giving rise to the ray-finned bony fishes, another line was becoming more **amphibious**—that is, able to live both on land and in water. Some 15 million years later there were recognizable **amphibians** (Figure 19–10). The amphibians were most widespread during the Carboniferous period, also known as the Age of Amphibians (see Figure 19–5). Today there are relatively few species left (about 2500), which include the frogs, toads, salamanders, and the wormlike apodes.

What circumstances could have led an ancient line of lobe-finned fishes to leave behind their comfortable pond habitat for the relatively inhospitable shoreline? Perhaps a dwindling food supply in the ponds forced desperate individuals to launch brief forays onto land. At that time, the land offered rich

(b)

(c)

(a)

FIGURE 19–10
Amphibians. (a) *Diplovertebron*, an early amphibian from the Carboniferous period. Two modern amphibians: (b) a spotted salamander and (c) a poison arrow frog from Colombia.

vegetation and an abundance of flightless insects that would have been easy prey. Furthermore, few if any land invertebrates were predators of fish. Alternatively, in regions where ponds and streams would dry up periodically, the local populations of lobe-finned fishes may have been forced to seek other water holes or die. Those whose fleshy fins gave them some mobility on land could wriggle over to an adjacent larger pond and complete their life cycle. For whatever reason, the fleshy fins gradually gave way to legs, an evolutionary advance that involved many structural changes in the bones and muscles of these limbs (Figure 19–11).

Based on fossil reconstructions of the early amphibians (see Figure 19–10a), these land pioneers were at best sluggish walkers. Without the buoyant support of water, dragging their bodies across the mud must have required considerably more energy than swimming. Nevertheless, the land offered some advantages with respect to meeting the extra energy demands. Aerobic respiration (and hence, biological energy production) is often limited by the rate at which molecular oxygen can be supplied to the respiring cells. Compared to aquatic animals, air-breathing animals have an easier time meeting their oxygen demands because: (1) gases diffuse faster in air than in water; (2) the concentration of oxygen in air is much greater than it is in water; and (3) because air is much lighter than water, moving air through a respiratory system requires less energy.

Most living amphibians have retained the air-breathing lungs of their predecessors, the lobe-finned

fishes. Besides using lungs, however, modern amphibians also exchange respiratory gases through their skin and the inner lining of their mouths. Gas exchange is also enhanced by a more efficient circulatory system for moving oxygen from the skin and lungs to the rest of the body cells (see Chapter 20). With these modifications of the respiratory and circulatory systems, amphibians became adapted to meet the rigorous energy demands of a terrestrial existence.

Even with the adaptations just noted, the amphibians never became totally terrestrial. The same skin that permits the exchange of oxygen and carbon dioxide also permits rapid losses of water by evaporation. To avoid dehydration, amphibians must stay in moist habitats. Furthermore, the amphibian egg is also permeable to water; if it is not laid in water or very moist soil, it will quickly dry up. Even the sperm are subject to dehydration, because instead of being deposited inside the female, they are shed externally on or near the egg mass. Finally, the tadpoles that emerge from the fertilized eggs are totally aquatic, exchanging gases with gills and feeding on small freshwater algae and animallike protistans. Thus, the amphibians are "chained" to aquatic habitats, and this is probably why their overall success as land dwellers has been limited. After their heyday some 300 million years ago, the amphibians declined in diversity. They were soon overshadowed by a group of vertebrates that are truly terrestrial—the reptiles.

Reptiles

The 6500 or so species of living **reptiles** are only a small remnant group of a vast and glorious past. This class dominated the earth for over 150 million years during the Age of Reptiles—a period of time extending from the middle of the Permian through most of the Cretaceous (see Figure 19–5). The survivors include turtles, lizards, snakes, and the crocodilians (alligators and crocodiles) (Figure 19–12).

FIGURE 19–11
Fins to Limbs. Compare the bone structure of (a) a pectoral fin of a lobe-finned fish to (b) the forelimb of a fossil amphibian. The three major bones—ulna, humerus, and radius—are recognizably similar.

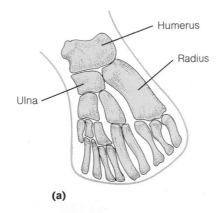

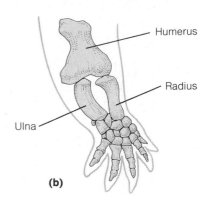

(a)

(b)

(a)

(b)

(c)

(d)

FIGURE 19–12
Representatives of the Modern Reptiles. (a) A Galapagos tortoise. (b) Two Galapagos land iguanas engaged in a dispute. (c) A coiled smooth green snake. (d) A crocodilian.

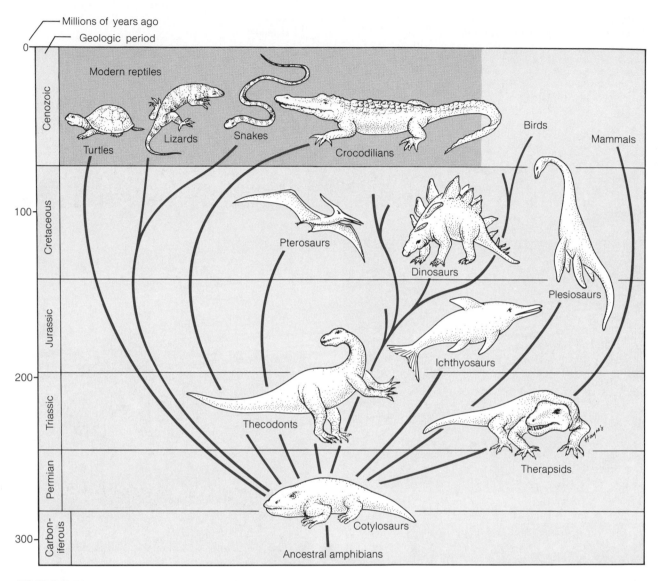

FIGURE 19–13
Evolution of the Major Reptilian Groups. The ancient cotylosaurs, or stem reptiles, gave rise to many reptilian lines of decent, including the therapsids (forerunners to the mammals) and the thecodonts (forerunners to the crocodilians, giant pterosaurs, and the dinosaurs).

All of the modern reptiles except the crocodilians are direct descendants of an ancient stem group of reptiles known as the **cotylosaurs** (Figure 19–13). The cotylosaurs diverged from the amphibians about 300 million years ago and soon branched into many reptilian lineages. These include two groups of aquatic reptiles, the **ichthyosaurs** and the **plesiosaurs;** the **therapsids,** forerunners of the mammals; the **turtles;** a stem line to the **lizards** and **snakes;** and the **thecodonts.** The thecodonts in turn gave rise to flying

reptiles such as the **pterosaurs,** some of which had wingspans exceeding 9 meters (30 feet). Thecodonts were also ancestral to the crocodilians and the infamous **dinosaurs** (Figure 19–14). Birds are believed to have descended from a branch of dinosaurs; accordingly, birds are more closely related to these extinct giants than are the modern reptiles.

Although they were more abundant in times past, the reptiles still must be regarded as successful land dwellers. They display a number of adaptations that

FIGURE 19–14
Dinosaurs. A Jurassic landscape some 135 million years ago. Shown from left to right are *Allosaurus, Camptosaurus,* and *Diplodocus* (the largest land animal that ever existed).

permitted them a true conquest of land, leaving their amphibian ancestors behind at the water holes. One successful adaptation is their dry, scaly skin, which effectively retards water loss. Terrestrial reptiles such as snakes and lizards lose water from their skin only about 3% as rapidly as amphibians. Those reptiles that have limbs also have a structural advantage over the amphibians. The legs of lizards, for example, are attached directly under the body rather than at the sides like those of the amphibious salamanders (compare Figures 19–10b and 19–12b). This provides better support and allows reptiles to move more quickly on land.

In addition, unlike their amphibian predecessors, fertilization in reptiles occurs internally. Because the male delivers sperm directly into the body of the female, there is no need for external water to effect fertilization. The fertilized egg then develops for awhile inside the female. Inside the egg, the growing embryo is surrounded by a system of membranes, which in turn is enclosed within a hard, water-resistant shell (Figure 19–15). This so-called **amniote egg** is eventually laid on dry land; even aquatic reptiles (turtles and crocodilians) deposit their eggs on land. The amniote egg was an innovation just as important in the adaptation to land as was the development of legs.

Other reptilian advances occurred in the respiratory and circulatory systems. The lungs of reptiles are more convoluted than those of amphibians, yielding a greater surface area exposed to each breath of air. This increase in lung surface area was necessary because of the evolutionary conversion of the moist, gas-exchanging skin of amphibians to the scaly, water-retarding reptilian skin. The circulatory system also changed in a way that reflects on the reptile's sole dependence on its lungs for gas exchange. A three-chambered heart effectively keeps oxygenated and deoxygenated blood separate (see Figure 20–3b).

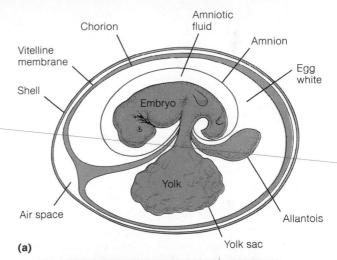

(a)

(b)

FIGURE 19–15
The Amniote Egg. (a) This diagram of a very early stage in the development of an amniote egg (reptiles and birds) shows the young embryo and a series of five extraembryonic membranes: vitelline membrane, chorion, amnion, allantois, and yolk sac. The yolk inside the yolk sac and the egg white are a reserve supply of food and water for the growing embryo; the allantois encloses a storage space for solid wastes excreted by the embryo; and the amniotic fluid bathes the embryo, acting as a substitute for the ancestral pond. The hard shell restricts water loss. Many of the membranes shown here are also present during mammalian embryo development, including humans. (b) A young king snake hatching from its amniote egg.

FIGURE 19–16
Archaeopteryx. (a) From fossil remains, (b) scientists have reconstructed this image of the oldest known bird. Note the presence of feathers and other bird-like features, despite the reptilian teeth and mouth (birds have a horny beak and no teeth), claws at the wing edges, and the long jointed tail. The reconstruction reflects the position of this animal's head in the fossil remains, not its normal position.

(a)

(b)

Birds

The oldest known fossil bird, *Archaeopteryx* (Figure 19–16), was an exciting find for paleontologists. With its reptilian characteristics—teeth, claws, and long, jointed tail—together with its birdlike feathers and wings, *Archaeopteryx* represents a clear fossil link between the ancient reptiles and the first true **birds.**

Many of the structural and physiological features of birds are clearly adaptations for flight. First, consider the overall body plan of a typical bird. The body is streamlined to minimize air resistance during flight. For example, birds do not have earlobes or external genitalia. Aerodynamic considerations also limit the size range of the 10,000 species of modern birds. (The exceptionally large birds, such as emus and ostriches, are flightless.) Birds also have a relatively light body for their size. The bones making up the endoskeleton are light and hollow, and many are fused together to provide compact strength capable of withstanding the jarring landings (Figure 19–17). For example, the ribs are fused to the massive sternum (breastbone) and the vertebrae are fused to make a very rigid backbone. The pelvic girdle is also modified to permit passage of eggs. Birds did not evolve the characteristic of giving birth to live young as the mammals did, presumably because carrying their young for the entire course of embryonic development would add too much body weight to the female.

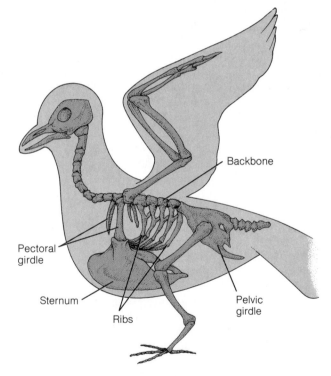

FIGURE 19–17
Anatomy of a Bird (Pigeon). The skeleton of birds is light but very strong. The pectoral and pelvic girdles are fused to the rigid backbone, which consists of fused vertebrae. The ribs are fused to the sternum, making the rib cage rigid as well. The bone fusions provide compact strength to withstand the jolt of landing.

Another obvious adaptation to flying is the presence of feathers in birds (Figure 19–18). These lightweight yet strong appendages are evolutionary derivatives of reptilian scales. Feathers also insulate well against the loss of body heat and they retard water loss. Beneath the outer layer of flight feathers is a fluffy underlayer called down. Goose down is such an effective light-weight insulator that it is commonly used in mountaineering jackets and sleeping bags.

In contrast to the modern reptiles, birds are warm-blooded **(homeothermic),** a characteristic they share with the mammals. By maintaining a constant and warm body temperature, birds can remain active re-

FIGURE 19–19
A Hummingbird. When hovering, a hummingbird can expend more than 300 Calories of food energy per minute.

gardless of fluctuating environmental temperature. The maintenance of internal temperature is aided by the insulating feathers and a four-chambered heart that ensures that the respiring cells (heat-generators) receive an undiluted source of oxygen from the blood. Since birds do not store much body fat (it would weigh them down), they must eat more or less continuously to support their high energy demands (Figure 19–19).

Mammals

Mammals are hairy homeotherms that nourish their young with milk. The presence of body hair (or fur) coincides with their warm-blooded nature, and like feathers in birds, hair helps insulate the body against extremes in external temperature. Certain aquatic mammals (whales, sea lions, walruses, and others) have insulating layers of blubber instead of fur; other body coverings also exist, such as armor plates in armadillos and quills in porcupines. But the most distinctive feature of mammals, and the one for which they are named, is mammary glands that produce milk. After birth, the young mammal is nurtured by its mother's milk for weeks to years, depending on the species.

As mentioned earlier, mammals arose from the therapsids, an ancient group of reptiles that lived over 200 million years ago (see Figure 19–13). For the next 100 million years the mammals remained rather inconspicuous while the dinosaurs underwent their explosive diversification. The fossil record is very scanty on

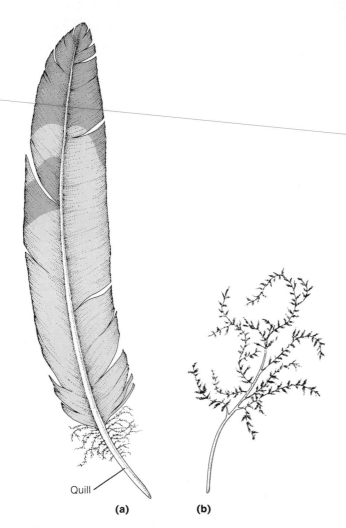

Quill

(a)   **(b)**

FIGURE 19–18
Feathers. Most birds have two types of feathers: (a) contour or flight feathers, and (b) down feathers.

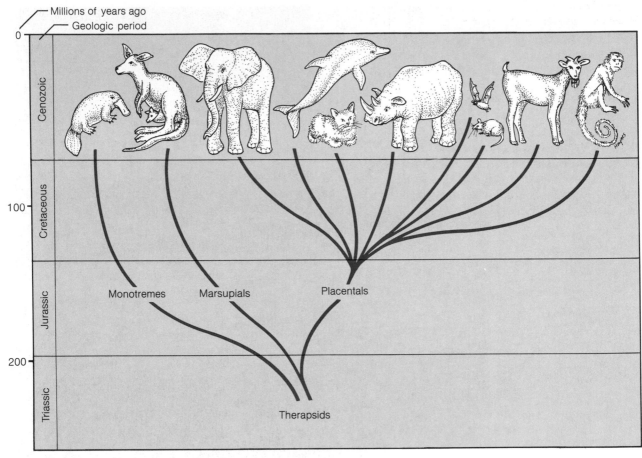

FIGURE 19–20
Evolutionary Relationships Among the Mammals. The therapsids, an ancient group of reptiles,
diverged into the three stem lines of mammals—monotremes, marsupials, and placentals—before the
dinosaurs were in their heyday. The placental mammals are the most diverse of the three groups today.

mammals during this period, and paleontologists have not been able to pin down precisely when the mammallike therapsids became the reptilelike mammals. We do know, however, that relatively early in mammalian evolution, several separate lines split off from the ancestral therapsids (Figure 19–20). One of these lines of descent was a group called the **monotremes**—egg-laying mammals that produce milk to nourish their hatchlings. Only two species of monotremes have survived to the present: the duck-billed platypus and the spiny anteater, both found in Australia. Apparently, these odd creatures became geographically isolated from the rest of the mammals when the continents of Australia and Antarctica separated about 65 million years ago (see Figure 29–15). This isolation may have saved them from extinction.

A second evolutionary line gave rise to the modern **marsupials,** represented today by the kangaroo, wombat, oppossum, koala, and quite a few others that are native to Australia and nearby islands. Although the marsupials do not lay eggs, their embryos develop within the mother's uterus (womb) for only a short time. After birth, the still-immature babies (an opossum newborn is no bigger than a horsefly) crawl over to the mother's abdominal pouch, where they attach by their mouths to the milk-secreting nipples. There they complete their development within the protected environment of the pouch.

The third group of mammals—the **placentals**—includes most of the more familiar furry creatures: mice, rabbits, deer, cats, dogs, bears, bats, seals, monkeys, and of course, humans (see Figure 19–21). These

FIGURE 19–21

Placental Mammals. The placentals shown here illustrate the wide variety of adaptations that have led to the remarkable success of this group. (a) The mule deer belongs to a group known as the running herbivores (herbivores are plant-eaters). Members of this group have elongated limbs and walk or run on their toenails that take the form of hooves. Their teeth consist of two front rows of incisors designed for nipping vegetation, and behind them, molars for grinding. (b) The cougar chasing this bighorn sheep is a running carnivore (meat-eater). Members of this group have toe-nails modified into sharp claws. Their teeth are designed for grasping and tearing flesh, and their stomachs can accommodate a huge meal. After a kill, these animals gorge themselves, then may not eat for several days. When not eating or on the prowl, carnivores are usually asleep, a habit that aids the digestive process and reduces energy demand. (c) Aquatic mammals such as these sea lions have forelimbs that are modified as paddles or fins. The hind limbs are either modified for swimming or absent altogether. Most aquatic mammals also have a tail used as a strong swimming fin, which moves up and down rather than side to side like those of fishes. (d) Bats are the only true fliers of the mammals. Bat wings are modified forelimbs consisting mostly of skin stretched across elongated, fingerlike bones. Most bats fly at night; some have very keen vision while others navigate and find food by means of echolocation, a type of aerial sonar system. The frequency of sounds they emit are too high for our ears to detect.

(a)

(b)

(c)

(d)

mammals have filled virtually all of the ecological slots vacated by the dinosaurs. Their name derives from the presence of a **placenta,** a nutritive connection between the embryo and the mother's uterine wall. In contrast to the marsupials and monotremes, placental embryos complete their development inside the uterus.

Mammals, particularly the placentals, are diverse not only in species but in habitat. You can find them in oceans, lakes and streams, in deserts and on ice masses, underground, on the ground, in trees, and even in the air (Figure 19–21). No other vertebrate group, not even the reptiles in their heyday, has spread over so many habitats or geographic regions on earth. Part of this success must certainly be attributed to the mammalian advances in structure, physiology, and behavior.

A major reason why mammals exist virtually everywhere across the globe is because they are homeothermic. The capacity to maintain their bodies at a warm and fairly constant temperature is aided by structural features, such as insulating body coverings (hair, blubber, and so on) and a four-chambered heart (as described for birds). Homeothermy is also aided by certain behaviors displayed by mammals. When the external temperature is very warm, for example, mammals employ a number of devices to keep cool—some as simple as finding a shady spot or taking a dip in a nearby pond or stream; others a bit more complex, such as sweating or panting. When the cold weather comes, many mammals grow extra thick fur; many aquatic mammals migrate to warmer waters; and some simply avoid the cold by hibernating for the season. Shivering, another behavioral response to external cold, is only a temporary means of staying warm.

Some of the most significant advances that have contributed to the success of mammals have come in the way they bear and rear young. Fertilization in mammals is internal, as it is for reptiles and birds; but in mammals (except for the egg-laying monotremes), the young embryo develops within the protected environment of the mother's uterus. Here all the normal functions of life, such as nutrition, waste removal, temperature regulation, immunological defense, and other vital activities are provided by the mother (see Chapter 25). The babies are born live, yet the intimate relationship between mother and young does not end at birth. In addition to nourishing their young with milk produced in mammary glands, mothers (and in some cases, the fathers too) tend to be very protective of their babies until they are strong enough to fend for themselves. The often prolonged prenatal and postnatal care seems to be correlated with the mammalian advances in the nervous system, especially the brain.

The nervous system has reached its highest degree of sophistication in the mammals. These creatures' senses of smell and hearing are especially acute compared to those of other animals. Information a mammal gathers by its senses is rapidly processed by its central nervous system, and then stored or used to evoke an appropriate response. It is the storage function of the mammalian brain that truly distinguishes mammals from other animals. No other group relies so heavily on memory and learning to guide its activities (see Chapter 27). These faculties give mammals a wide latitude in selecting appropriate responses to different environmental situations. Their capacity to analyze and draw from past experience has played a major role in their biological success, particularly for primates, the group to which humans belong.

PRIMATES AND HUMAN EVOLUTION

Primates are an advanced group of placental mammals that are arboreal—they are adapted to living in trees. They also get special recognition because they include one very advanced species, which happens not to be a tree dweller: *Homo sapiens*, the human.

Primatologists recognize two major groups of primates. The **prosimians**, with a fossil record that extends back 50–60 million years, were the first primates. They are represented today by the lemurs, lorises, aye-ayes, tarsiers, and a few others (Figure 19–22a). The second group, the **anthropoids**, are more recent in origin and include the Old World monkeys, New World monkeys, and hominoids (Figure 19–22 b,c,d). The hominoids are in turn divided into the apes (gibbons, gorillas, orangutans, and chimpanzees) and the hominids (humans and their extinct relatives). Of all the primates, only the hominids are (were) capable of prolonged **bipedal locomotion** (walking or running upright on two legs).

Since humans are clearly primates, we will begin the journey into our ancestral past with a discussion of those characteristics that distinguish primates from other mammals.

The Characteristics of Primates

1. Unlike other mammals that have paws or hooves, primates have hands and feet with flattened nails, not

(a) (b) (c)

(d)

FIGURE 19–22
Examples of Living Primates. (a) The prosimians, represented here by a lemur, are the most ancient group of primates. (b–d) Anthropoids. (b) A spider monkey, a New World monkey, from South America. (c) A rhesus, an Old World monkey, from Africa. (d) A chimpanzee, one of the hominoids.

claws. The digits (fingers and toes) operate independently, and one digit (the thumb or large toe) apposes the others. This permits primates to grasp objects such as tree limbs, as well as to carry their young.

2. The movements of the forelimbs (arms) and to a lesser extent the hindlimbs (legs) are not restricted to a single plane parallel to the body as in most mammals. The ability to rotate the wrist, elbow, shoulder, and hip provides the great flexibility needed for swinging from branch to branch. The wide repertoire of forelimb movements coupled with an extraordinary manual dexterity is also important in primate feeding behavior, grooming, and care of the young.

3. Primates usually give birth to only one offspring per pregnancy. For any creature of size, multiple births is simply not feasible in an arboreal habitat. The infant nurses on two nipples located on the chest rather than the abdomen, which permits the mother to hold the nursing infant while moving. Also, face to face association during nursing is believed to increase bonding between mother and infant with important social consequences.

4. The eyes of primates are directed forward rather than to the sides of the face like most other mammals. This limits the total field of sight but it permits stereoscopic vision and, hence, depth perception. The ability to perceive depth is crucial for animals that must accurately judge distance between one tree limb and the next. This increased visual acuity was accompanied by a decline in the sense of smell, which is not particularly useful in trees where odors are blown around by the wind.

5. One of the most significant trends in primate evolution has been an increase in brain size. Most of this increase has occurred in the cerebral cortex, the part of the brain associated with learned behavior. More than any other group of animals, the primates rely on learning for survival. Among the early hominids, the capacity to learn must have provided the foundation for reasoning, language, and culture.

As you can see, many of the general features of primates are clearly adaptations for tree dwelling. This is not to say that the primate characteristics are totally inconsistent with a terrestrial existence—the baboons and gorillas actually spend most of their time on the ground. But even these rather large primates are no match for the big cats that roam the African savannahs, and so they must take refuge in the trees to escape these swift predators. In addition, baboons and gorillas obtain most of their food from trees. Only one group of primates—the hominids—has completely cut the umbilical cord to an arboreal life.

Human Evolution

In the forests of eastern Africa some 18–25 million years ago lived *Dryopithecus*. This small anthropoid spent most of its time in the trees and used all four limbs when moving on the ground, much like the apes of today. According to most interpretations of the fossil record, *Dryopithecus* gave rise to several lines of descent, including those that culminated in the modern apes. Of particular interest to us, however, is an evolutionary line that led to *Ramapithecus* (Figure 19–23).

We do not have many fossil remnants of the 12-million-year-old *Ramapithecus*, but its teeth and jaw tell a revealing story. They are unlike those of both *Dryopithecus* and the modern apes, and the teeth in particular bear some resemblance to human teeth. They suggest that *Ramapithecus* lived on a diet of seeds and fibrous material gathered from the open grasslands rather than the soft fruits and leaves abundant in the forests. Thus, *Ramapithecus* may represent a transition from a strictly arboreal existence to a more terrestrial one.

Ramapithecus and *Dryopithecus* may have been capable of semi-erect posture for brief periods, but both apparently walked on all fours. The upright posture and bipedal locomotion so characteristic of humans and their immediate ancestors (all hominids) appeared

first in members of the genus *Australopithecus,* which apparently descended from the ramapithecine line. The oldest members of this group, about 3.7 million years old, were found by Donald Johanson and his associates in the Afar region of Ethiopia. In what must be described as an archaeological bonanza, Johanson found the remains of as many as 65 individuals in one location. Dubbed *Australopithecus afarensis* and nicknamed First Family, these fossils put to rest the popular notion that brain enlargement must have co-evolved with an erect posture, theoretically because the ability to use tools required a larger brain and free hands. Although *A. afarensis* clearly had an erect posture, its brain was only 400 cubic centimeters (cc), about the same size as a chimpanzee brain (Figure 19–24). As a point of comparison, the human brain averages 1400 cc.

According to one view, *A. afarensis* represents a stem group of hominids that gave rise to two separate branches of descent about 3 million years ago (see Figure 19–23). One line yielded several later species of the genus *Australopithecus* (*A. africanus*, *A. robustus*, and others), which were somewhat taller and heavier than the First Family. They had protruding jaws, heavy brow ridges, and a brain size of 500–600 cc. The australopithecine line persisted up to 1.5 million years ago, then died out. Their demise was probably related to the presence of another hominid—the genus *Homo*.

THE RISE OF *HOMO*. The other line stemming from *A. afarensis* was *Homo*. (Some anthropologists contend that *A. africanus*, not *A. afarensis*, should be accorded the ancestral position to *Homo*.) The criteria for separating *Australopithecus* from *Homo* is a controversial subject among archaeologists, but in general the fossils designated *Homo* had larger brains and are generally associated with cultural artifacts.

The oldest known representatives of *Homo* appeared sometime between 2 and 3 million years ago. Designated *Homo habilis* ("handy man") by their finders, Louis and Mary Leakey, they had slightly larger brains (600–800 cc) than their australopithecine contemporaries. The two most remarkable features of *H. habilis*, however, are its humanlike teeth and hands. With this fossil we have the first sign of humanlike manual dexterity, and crude stone tools have been found in some of the *H. habilis* excavations. Did "handy man" take the next great step and fashion weapons for hunting? This question remains unanswered, but if he did have weapons, they would have been quite useful in skirmishes against other hominids, most notably the australopithecines. *Homo* and *Australopithecus* over-

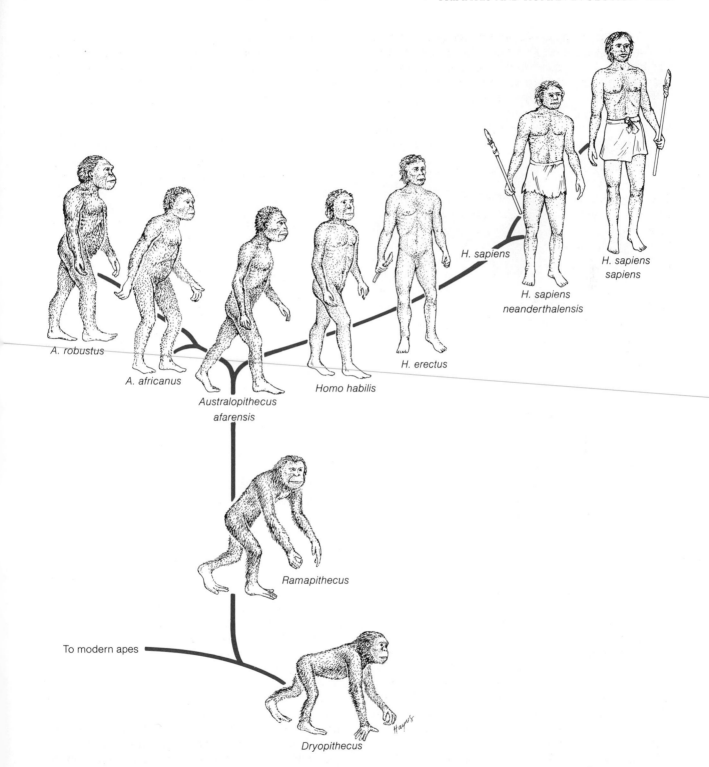

FIGURE 19–23

Evolutionary Tree of the Hominids. Note that modern humans did *not* descend from the apes; rather, both humans and apes evolved from a common ancestor, *Dryopithecus.* The arrangement of human ancestors shown here represents only one possible interpretation of the fossil record.

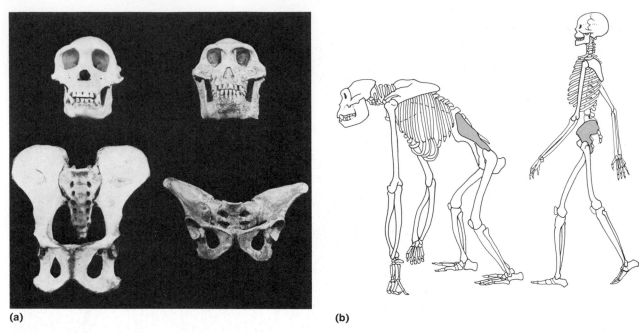

(a) (b)

FIGURE 19–24

The Pelvis and Posture. (a) A comparison of a skull and pelvis of a chimpanzee (*left*) and *Australopithecus afarensis* (*right*). Note that the *Australopithecus* pelvis is shaped more like a flat dish to support internal organs and permit bipedal locomotion. However, the two skulls are about the same size. (b) These drawings indicate the position and shape of the pelvis (color) in the chimp (*left*) and human (*right*).

lapped by at least a half-million years, and fossils of both genera have been found in some of the same archaeological sites. It is quite possible, then, that the smarter, tool-bearing *Homo* killed off his australopithecine cousins.

The next great step in human evolution is manifested in *Homo erectus* (formerly known as Java Man, or *Pithecanthropus erectus*). About 1.4 million years old, this early human walked fully upright with a stature of 1.5 meters (about 5 feet) (see Figure 19–23). The brain of *Homo erectus* (950–1100 cc) was larger than any of its predecessors, approaching the size range of modern humans (1200–2000 cc). Many features of its skeleton are similar to our own with the exception of the skull, which was very primitive. The teeth were massive and prominent brow ridges extended over the eyes.

In contrast to the australopithecines, whose range was limited to Africa, *Homo erectus* has been found in Indonesia, China, Germany, Hungary, Morocco, and Tanzania. In each location they left clear evidence of a material culture based on stone tools, such as

axes, choppers, scrapers, anvils, and awls. They were hunters and gatherers, and these activities must have been supported by a well-developed social structure and language. Evidence obtained in the Hungary and China sites indicates *Homo erectus* was using fire at least 500,000 years ago.

***HOMO SAPIENS* — THE CULMINATION.** *Homo erectus* disappears from the fossil record about 300,000 years ago, just about the time that *Homo sapiens* appears. These events are not unrelated, as *H. erectus* is presumed to be ancestral to *H. sapiens*. The definitive fossil transition is still lacking, however. The early *Homo sapiens* fossils, ones that date back to 250,000 years ago, are clearly human. These people had larger brains (1400 cc), smaller teeth and jaws, a more flattened face, and a more rounded top to the skull than *H. erectus* (Figure 19–25). In contrast to modern humans, they retained the prominent brow ridges and had a receding chin. The early *H. sapiens* also made extensive use of tools, but the artifacts did not become abundant until about 100,000 years ago. The bones of this period

FIGURE 19–25

A Comparison of the Heads of Early Humans. The top panel shows restored skulls; the bottom panel shows an artist's reconstruction of possible facial features. From left to right: *Homo erectus, Homo sapiens neanderthalensis* (Neanderthal people), and Cro-Magnon. Cro-Magnon is indistinguishable from modern humans and therefore is classified *Homo sapiens sapiens*.

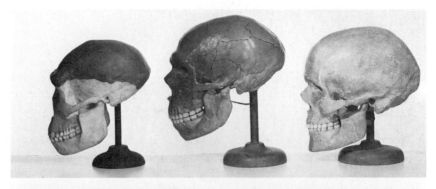

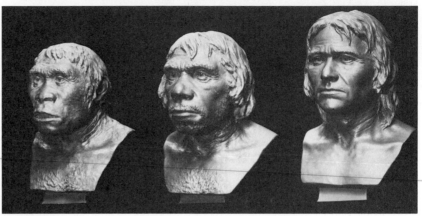

are designated *Homo sapiens neanderthalensis*, or Neanderthal people (neanderthalensis is a "grade" used to distinguish this group from modern humans, *Homo sapiens sapiens*). Neanderthal bones have been found extensively in Europe, Israel, Iraq, China, and Indonesia, but it is their cultural artifacts that tell us most about their way of life.

The Neanderthals were skilled hunters capable of killing animals many times their size (reindeer and mammoths) by ingeniously driving the beasts over cliffs or into pits. To coordinate these hunts we can imagine that some kind of language was used, though of course there was nothing like writing yet. There is also evidence that Neanderthals had an appreciation and capacity for symbolic meaning, for they apparently performed rituals and had a form of religion. They buried their dead, and in at least one instance surrounded the corpse with flowers. They had good mastery of fire, constructed shelters, and wore the skins of the animals they killed.

About 40,000 years ago the Neanderthal fossils fade out and a new group designated Cro-Magnon becomes evident. In every skeletal detail Cro-Magnon

is indistinguishable from modern humans, *Homo sapiens sapiens* (see Figure 19–25). They had an advanced material culture with tools and weapons far superior than those fashioned by Neanderthal. Indeed, their development of a stone tool sharpened on both sides, forming a blade, stands as the crowning achievement of stone tool manufacture.

Perhaps the most significant creations of the Cro-Magnon people were their sculptures and cave paintings (Figure 19–26). From these it is clear that information and attitudes were being communicated from generation to generation, and language and symbol had major roles in their lives.

All of the accomplishments of the Cro-Magnon people, even those of their modern descendants, were made possible by the evolutionary advancements we have traced through their ancestors. The primate features that permitted our forerunners to live in trees still serve us. But it was the development of an erect posture and an intelligent brain that truly distinguished the hominid line, for these attributes paved the way for the development of a uniquely human characteristic—culture.

FIGURE 19–26
Cro-Magnon Art. Cro-Magnon people painted the animals important to their lives on the walls of caves. These paintings are arresting in their beauty and vitality. They attest to the visual acuity, memory, and sense for form of these people.

SUMMARY

Members of the phylum Chordata, clearly the most advanced group of animals, are characterized by having a notochord, a hollow dorsal nerve cord, and pharyngeal gill slits, at least at the embryonic stage. Vertebrates can be further distinguished from the other chordate subphyla (tunicates and lancelets) by their cartilaginous or bony endoskeleton.

The modern vertebrates include fishes, amphibians, reptiles, birds, and mammals. Probably the most primitive vertebrates are the fishes, a group which includes the agnathans, or jawless fishes (such as lampreys), the placoderms (all extinct), the cartilaginous fishes (such as sharks, skates, and rays), and the bony fishes. Of these, the bony fishes include the greatest number of species today. They have a lung or swim bladder, and gills.

Amphibians reproduce in water by means of external fertilization. The tadpole larvae are totally aquatic, but most of the adult forms spend part of their time in water and part in moist terrestrial habitats. To meet the higher energy demands associated with life on land, amphibians have a three-chambered heart, and many use their moist skins as well as their lungs for gas exchange.

Unlike the amphibians, most reptiles are not dependent on a moist environment for survival. Some of their adaptations to a terrestrial existence include their thick, scaly skin, internal fertilization, and eggs in which the embryo is protected by a series of membranes and a shell. The reptilian dinosaurs were the largest terrestrial animals ever to exist.

Birds have wings, feathers, and a completely divided, four-chambered heart. They reproduce by internal fertilization and the production of eggs. Birds are homeothermic. Except for the handful of flightless birds, virtually everything about a bird's structure and physiology is an adaptation for flight. The body is streamlined, moderate in size, and light in weight. The bones are hollow and many are fused to provide compact strength. Birds' feathers, which are primarily adaptations for flight, also help maintain body heat. The efficient circulatory system of birds helps meet the increased energy demands of flight and temperature regulation.

Mammals consist of three major subgroups: monotremes, marsupials, and placentals (the most diverse group). Their most notable features are body hair and mammary glands. In addition, they are homeothermic, they have a four-chambered heart, and in placentals, embryo development is completed within the mother's uterus. In terms of species and habitat diversity, they are among the most successful animal groups ever to appear on earth.

Table 19–1 summarizes the various characteristics of the vertebrate groups, with an indication of the characteristics considered to be adaptations to survive on land.

Primates, a subcategory of mammals, includes the prosimians (lemurs, tarsiers, etc.) and the anthropoids (monkeys, apes, and humans). Their most distinctive characteristics include hands and feet with independent digits and an apposable thumb and large toe, flexible joints in the arms and legs capable of rotation movements, keen stereoscopic vision, single fetus pregnancies, and a large brain. All of the features are considered adaptations to life in the trees.

The descent of hominids (humans and their humanlike ancestors) is traced in the fossil record, which includes (in their order of appearance) *Dryopithecus, Ramapithecus, Australopithecus,* and finally *Homo.* The fossils representing these genera reflect some major trends in human evolution: gradual progressions toward erect posture, bipedal locomotion, a grassland habitat, enlarged brains, and an increased dependence on material culture. The genus *Homo* appeared between 2 and 3 million years ago, but the first clear

TABLE 19–1
Major Vertebrate Characteristics. Features in **boldface** represent adaptations to a terrestrial lifestyle as they first appeared in a land-dwelling vertebrate.

Vertebrate Group	Characteristics						
	Endoskeleton	Respiratory Structure	Fertilization	Eggs	Body Covering	Temperature Regulation	Limbs
Cartilaginous fishes	Cartilage	Gills	Internal	Not amniote	Scales	Not internally regulated	Fins
Bony fishes	Mostly bone	Gills	External (in most)	Not amniote	Scales	Not internally regulated	Fins
Amphibians	**Mostly bone**	**Lungs** (in adults)	External (in most)	Not amniote	Smooth skin	Not internally regulated	**Legs**
Reptiles	Mostly bone	**Lungs**	**Internal**	**Amniote**	**Scales**	Not internally regulated	Legs
Birds	Mostly bone	Lungs	Internal	Amniote	**Feathers**	**Homeothermic**	Legs, **wings**
Mammals	Mostly bone	Lungs	Internal	Amniote	**Hair**	Homeothermic	Legs

evidence of language and culture did not appear until 100,000 years ago with the Neanderthal people. Modern humans in the form of Cro–Magnon first appeared about 40,000 years ago.

STUDY QUESTIONS

1. Describe three major structures that distinguish the chordates.

2. Which features of the tunicates, and which of the lancelets, suggest an evolutionary link to the vertebrates?

3. For each major class of vertebrates, list the characteristics distinguishing its members from those of other groups.

4. What features of the lobe-finned fishes make them partially adapted to survive on land?

5. What problems confront organisms living in a terrestrial environment?

6. How have the amphibians met the challenges associated with living on land?

7. Describe the advances made by the reptiles in adapting to land.

8. Explain how birds are particularly well-adapted for flying.

9. What factors have contributed to the success of mammals in a great diversity of environments?

10. Describe three characteristics of primates that make them well adapted to an arboreal life style.

11. Beginning with *Dryopithecus,* trace the evolutionary line of descent that led to Cro-Magnon.

SUGGESTED READINGS

Gordon, M. S. *Animal Physiology: Principles and Adaptations.* 3rd ed. New York: Macmillan, 1977. This is an advanced text which emphasizes function as it relates to the survival of organisms in their natural environments.

Leakey, R. E. and R. Lewin. *People of the Lake: Mankind and Its Beginnnings.* Garden City, New York: Doubleday, 1978. An engaging account of early human evolution.

Neill, W. T. *Reptiles and Amphibians in the Service of Man.* New York: Pegasus, 1974. This nontechnical paperback provides some interesting insights into the interactions of these animals with humans.

Welty, J. C. *The Life of Birds.* 2nd ed. Philadelphia: Saunders, 1975. This is a comprehensive text covering all aspects of the biology of birds.

Wessels, N. K. (Introduction) *Vertebrates: Physiology.* San Francisco: Freeman, 1980. This collection of readings from *Scientific American* concentrates on the properties and control of vertebrate systems. A number of very instructive illustrations are included.

CHAPTER 20
The Circulatory and Immune Systems

AN OVERVIEW OF ORGAN SYSTEMS

Every living organism must exchange matter with its environment. Small multicellular organisms have no need for specialized structures to carry out these exchanges because all their cells are either in direct contact with their external surroundings or are very near their surroundings. Such is not the case for large multicellular animals, however. With large volumes relative to their external surface areas, these animals depend on special organ systems to carry out the exchange of materials. Of the 11 organ systems found in vertebrates (Table 20–1), four of them—the respiratory, digestive, excretory, and circulatory systems—are directly involved in moving materials into and out of the animal (Figure 20–1). These systems essentially bring the external environment into the body.

The respiratory systems of animals, from the simple surface gills of invertebrates to the complex lungs of mammals, provide an extensive surface area for gas exchange between the organism and its surroundings.

At these surfaces, molecular oxygen used in cellular respiration is taken into the body and carbon dioxide is released. To obtain nutrients, most animals have a digestive tract, a long, usually folded tube that is actually an extension of the external environment. As food passes through this tube, it is digested into small molecules that are absorbed into the body. A third system, the excretory system, removes nitrogenous wastes and excess water and salts from the body. We will examine each of these environment-interfacing systems in Chapters 21–23.

Because every living cell in an animal's body must receive oxygen and nutrients and dispose of cellular waste products, large animals must have a transport system for moving these materials between the exchange surfaces and the body cells. The circulatory system provides this vital transport function. In addition, the circulatory system and the associated lymphatic system house the components of the immune system, which defend the body against foreign cells or substances. In this chapter we will look at these three systems and their major roles: transport and defense.

TABLE 20–1

The Organ Systems of Vertebrates. The bodies of vertebrates are composed of trillions of cells, most of which are arranged into **tissues**—groups of similar-looking cells that perform similar functions. In many cases, tissues are arranged into larger structural units called **organs**, such as the stomach, pancreas, heart, and many others. Organs in turn may cooperate in the performance of a general body process, and thus are grouped into larger categories called **organ systems**. Listed below are the major organ systems of vertebrates (including humans), selected examples of organs and tissues that comprise them, and their general function.

Organ Systems	Examples of Associated Organs and Tissues	General Function
Circulatory	Heart, veins, arteries, capillaries, blood	Transport of nutrients, wastes, and gases
Immune	Spleen, thymus, lymph vessels, lymph nodes	Defense against foreign cells and molecules
Respiratory	Lungs or gills, trachea, bronchi	Exchange of respiratory gases with external environment
Digestive	Stomach, pancreas, liver, intestines	Food disruption and digestion; nutrient absorption
Excretory	Kidneys or gills, urinary bladder	Excretion of metabolic wastes; salt and water balance
Nervous	Brain, spinal cord, sense organs, nerves	Perception of and response to external stimuli; regulation of body processes
Muscular	Skeletal, smooth, and cardiac muscle tissues	Movement
Skeletal	Bones, cartilage	Support; attachment sites of muscles
Integumentary	Skin, hair, sweat glands	Protection against infection, water loss, heat loss
Endocrine	Thyroid, pituitary gland, adrenal glands, pancreas	Internal regulation
Reproductive	Testes, ovaries, uterus, mammary glands	Reproduction and nurture of young

FIGURE 20–1

Organ Systems. Of the 11 organ systems in the vertebrate body, these four— respiratory, circulatory, digestive, and excretory—move materials into and out of the body.

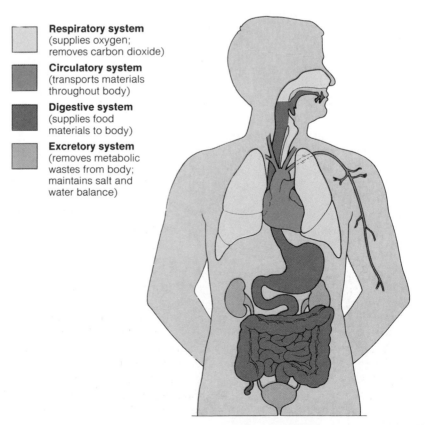

Respiratory system (supplies oxygen; removes carbon dioxide)

Circulatory system (transports materials throughout body)

Digestive system (supplies food materials to body)

Excretory system (removes metabolic wastes from body; maintains salt and water balance)

INTERNAL TRANSPORT SYSTEMS IN ANIMALS

The most primitive invertebrates, sponges and cnidarians, do not have circulatory systems. With their large surface area-to-volume ratios, these simple aquatic animals exchange materials directly across their body walls. They also employ cilia or flagella to sweep currents of water past their bodies, which facilitate the diffusion of materials into and out of their tissues. All higher animals have internal transport systems that circulate body fluids.

Open versus Closed Circulatory Systems

Most mollusks (such as clams, mussels, and snails) and all arthropods (such as spiders, insects, and lobsters) have an **open circulatory system.** In their bodies, blood moves freely within large cavities, where it bathes the body tissues directly (Figure 20–2). The blood is kept circulating through these cavities by a pulsating, tubular heart.

Annelids, echinoderms, and vertebrates all have **closed circulatory systems** in which the blood is entirely confined within a system of vessels. Blood movement in vertebrates is powered by a single muscular heart, which contracts rhythmically to propel blood through a system of vessels called **arteries.** The arteries branch into smaller and smaller vessels called **arterioles,** which direct the blood into the tiny **capillaries.** It is in the capillaries that the exchange of vital materials between blood and body cells occurs. The blood then collects in **venules** that coalesce into larger **veins** for return to the heart.

Evolution of the Vertebrate Heart

During the course of vertebrate evolution, many changes took place in the structure of the heart. These changes reflect a transition from an aquatic lifestyle (fishes) to a progressively more terrestrial one (amphibians, reptiles, birds, and mammals).

In the fishes, the heart consists of two chambers: an **atrium,** which collects oxygen-poor blood from the veins, and a **ventricle,** which propels blood through arteries toward the gills (Figure 20–3a). In the gill capillaries the blood picks up oxygen from water passing across the gills and releases carbon dioxide into the same water. The oxygen-enriched blood then moves through more arteries from the gills to the capillaries of the gut, liver, kidneys, muscles, and all the other body tissues of the fish. This type of blood circulation is called **single circulation** because the blood passes through the heart only once during its entire circuit through the body.

In most amphibians and all reptiles, birds, and mammals, blood moves through a **double circulation** pathway. In these creatures, the heart pumps blood first to the lungs, where it absorbs oxygen and releases carbon dioxide. Then the oxygen-rich blood returns directly to the heart to be pumped a second time to the capillary beds of all the other body tissues (Figure 20–3b,c). Thus, for each complete circuit through the body, blood passes through the heart twice. Double circulation is necessary for active, terrestrial animals with lungs. If oxygenated blood proceeded directly from narrow lung capillaries to the rest of the body, its flow rate would be too slow to meet the high oxygen demands of respiring body cells. By returning to the heart, oxygenated blood is forcefully pumped toward oxygen-demanding tissues, thereby

FIGURE 20–2
An Open Circulatory System. In an open circulatory system, as shown here in a grasshopper, blood bathes the internal organs directly. Arrows indicate the general direction of blood flow.

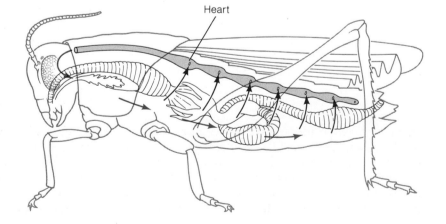

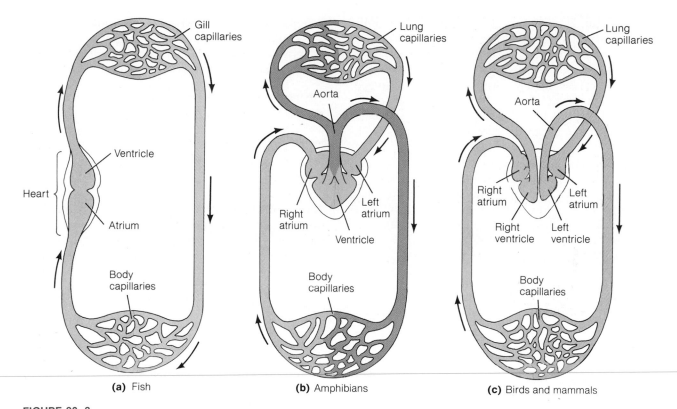

FIGURE 20–3
The Three Main Types of Vertebrate Circulatory Systems. (a) The two-chambered fish heart consists of a single atrium and a single ventricle. In this single circulation system, blood pumped from the ventricle is oxygenated in the gills, then moves directly to the rest of the body. (Oxygenated blood is indicated by color.) (b) In the three-chambered hearts of amphibians, blood passes through the heart twice during each complete circuit through the body. Blood from both the lungs and the rest of the body are mixed in the single ventricle, then pumped to the lungs and other organs via a forked aorta. The hearts of most reptiles have two atria and a partially divided ventricle. (c) Birds and mammals have a four-chambered heart, the most efficient design for a double circulation system. Oxygenated and deoxygenated blood are kept separate in the left and right sides of the heart, respectively. Thus, fully oxygenated blood is pumped directly to the oxygen-demanding tissues of the body.

ensuring a high rate of blood flow and oxygen de-livery throughout the body.

Double circulation in amphibians is aided by a three-chambered heart (see Figure 20–3b). The atrium is divided into two chambers, both of which empty blood into a single ventricle. The left atrium collects blood returning from the lungs, and the right atrium receives blood returning from all other body tissues. The single ventricle pushes blood through a forked **aorta** (the major artery leaving the heart) back to the lungs and other tissues. At first glance, the am-phibian heart appears inefficient because oxygenated blood from the lungs mixes with blood from the rest of the body in the ventricle. Amphibians, however, take oxygen in through their skin as well as their lungs. Blood returning from the skin capillaries to the right atrium may be as fully oxygenated as the blood coming from the lungs. Thus, the mixing of blood in the ventricle is of no particular disadvantage to amphibians.

Most reptiles also have a three-chambered heart, but the single ventricle is partially divided into a right and left half. Oxygen-poor blood entering the right atrium spills into the right side of the ventricle. From there the blood is pushed to the lungs, where it picks up oxygen. The oxygen-rich blood now returns to the left atrium of the heart, empties into the left side of the ventricle, and is then forced out the aorta toward the body tissues.

Although the reptilian ventricle is only a partially

divided structure, physiological tests have shown that oxygenated blood entering the left side of the ventricle remains separated from the venous (oxygen-poor) blood in the right side, so that only fully oxygenated blood enters the aorta for distribution to the oxygen-demanding tissues. This evolutionary division of the ventricle was crucial for reptiles and higher animals, which have a single gas exchange site, the lungs. Without this division, blood pumped to the body's tissues would contain only half the normal amount of oxygen.

Evolutionary advancement in heart design reached its climax in the birds and mammals. In these animals, the separation of the ventricle into two chambers is structurally complete, resulting in a four-chambered heart (see Figure 20–3c). Thus, there is no chance for the mixing of oxygen-rich and oxygen-poor blood in the heart. In addition, the aorta is not forked as it is

in reptiles and amphibians. This means that blood flows through the aorta under greater pressure, thereby permitting a higher rate of flow toward the capillary beds.

THE HUMAN CIRCULATORY SYSTEM

The Pump

The human heart is four-chambered and roughly the size of a person's clenched fist. Venous blood enters the right atrium, then moves into the right ventricle from where it is pumped to the lungs (Figure 20–4). The fully oxygenated blood then returns to the heart

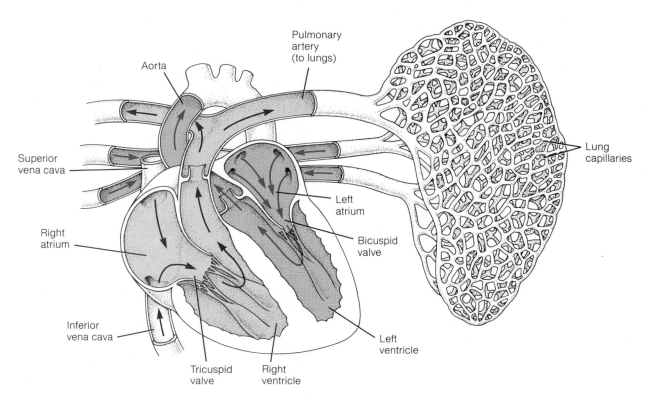

FIGURE 20–4
The Human Heart. Deoxygenated blood (gray) returns from the body tissues and enters the right atrium through large veins called the inferior vena cava and superior vena cava. Blood then flows through the tricuspid valve into the right ventricle. The ventricle then contracts, creating pressure that snaps this valve shut and forces blood into the pulmonary arteries that lead to the lungs. Oxygenated blood (color) returns from the lungs and enters the left atrium. The blood then flows through the bicuspid valve into the left ventricle, and finally leaves the heart under pressure via the aorta. The aorta branches into the major arteries, distributing blood to all body tissues except the lungs.

by entering the left atrium. The left atrium squeezes the blood into the left ventricle, which then pumps it through the aorta for distribution to the rest of the body.

The walls of the atria and ventricles are made up of cardiac (heart) muscle cells that undergo precisely coordinated waves of contraction. First, a wave of contraction spreads across both atrial walls, forcing blood from each atrium into its adjoining ventricle. A split-second later both ventricles contract, pushing blood toward the lungs (from the right ventricle) and through the aorta (from the left ventricle). Each sequence of atrial and ventricular contractions is a single heartbeat, and this sequence repeats about 70 times a minute in a typical adult at rest. Although subject to regulation by nerves and hormones, the heartbeat is initiated and propagated by the heart muscle cells themselves. If heart muscle tissue is isolated and placed in the proper artificial environment, it will keep on beating for hours and possibly days.

What triggers the heartbeat? The heart muscle contractions are initiated in the **sinoatrial node** (also known as the **pacemaker**), a small mass of specialized muscle fibers located at the upper end of the right atrial wall (Figure 20–5). These fibers initiate an electrical impulse that spreads rapidly across the walls of both atria, causing them to contract almost simultaneously. During this contraction, the impulse reaches a second mass of muscle fibers, the **atrioventricular node,** located in the heart wall just above the ventricles. This node sends out a wave of impulses across the walls of the right and left ventricles, triggering their simultaneous contractions. Following ventricular contraction, there is a brief rest period (about 0.3 second) before the sinoatrial node initiates the next heartbeat.

During the contraction of the ventricles, blood is specifically directed into the pulmonary artery and aorta, not back into the atria. Backflow into the atria is prevented by the presence of special **heart valves** situated between each atrium and ventricle pair (see Figure 20–4). During atrial contraction the valves remain open and permit the blood to empty into the ventricles. The blood pressure created by the subsequent ventricular contraction, however, snaps these valves shut and prevents backflow. Similar types of valves are located in the pulmonary artery and aorta, which remain open during ventricular contraction, then close to prevent backflow into the ventricles. In some people, particularly those who have had rheumatic fever, one or more of these valves may be damaged and unable to shut completely. A certain amount of backflow may occur, producing a muffled

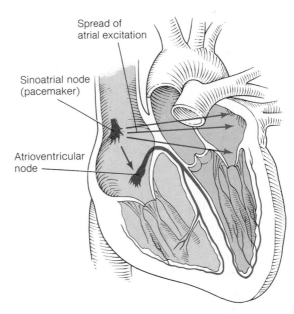

FIGURE 20–5
Anatomy of a Heartbeat. The heartbeat cycle begins when the pacemaker at the upper end of the right atrium sends out electrical impulses to the atrial muscles, causing both atria to contract simultaneously. When the impulse reaches the atrioventricular node, this fiber mass sends out a wave of electrical excitation to the ventricular muscles, causing ventricular contraction. The timing of these impulses is such that the ventricles do not contract until the atrial muscles have relaxed.

hissing sound, or **heart murmur,** as heard through a stethoscope. Many murmurs produce no ill effects, but in some cases the heart must work harder to maintain a normal rate of blood flow through the body.

When a doctor listens to your heartbeat through a stethoscope, he or she hears a long, low-pitched "lub" followed by a shorter, high-pitched "dup." The lub is the sound of the two heart valves closing as the ventricles contract. When the ventricles relax, the closing of the arterial valves (in the pulmonary artery and aorta) produces the dup sound. Although certain types of cardiac problems can be detected with a stethoscope, a more precise and sensitive diagnostic tool is the **electrocardiograph (EKG).** Electrodes attached to the skin monitor the electrical impulses generated by the heart. Irregularities in the heartbeat cycle can be seen on an oscilloscope or on a graphic printout called an electrocardiogram (Figure 20–6).

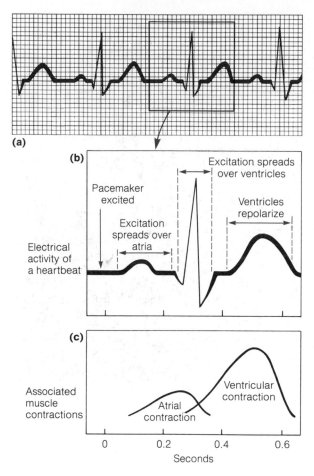

(a)

(b)

Pacemaker
excited

Excitation spreads
over ventricles

Excitation
spreads over
atria

Ventricles
repolarize

Electrical
activity of
a heartbeat

(c)

Associated
muscle
contractions

Atrial
contraction

Ventricular
contraction

0   0.2   0.4   0.6
Seconds

FIGURE 20–6
Electrocardiogram. (a) This electrocardiogram of a normal heart traces the
heart's electrical activity through four
heartbeat cycles. (b, c) Analysis of a
single heartbeat showing the succession
of (b) electrical events and (c) associated
heart muscle contractions.

Blood Pressure

Blood pressure is a measure of the force with which
blood pushes up against blood vessel walls. It is the
force that keeps blood flowing from the heart to all
the capillary networks in the body. Because it is generated by the contracting ventricles, blood pressure
is greatest in the arteries nearest the heart, then progressively decreases as the pulse of blood encounters
frictional resistance in the arteries and arterioles feeding into the capillaries (Figure 20–7). Resistance to
flow in the capillaries and veins further reduces the
pressure, which becomes nil by the time blood reenters the heart.

Blood pressure is greatest in the arteries during
systole, the phase of the heartbeat cycle when the
ventricles are contracting (see Figure 20–7). In between heartbeats, a period called **diastole,** the blood
pressure at a given point in an artery reaches its lowest
value. For humans, systolic and diastolic blood pressure measurements are conventionally taken on an
artery located in the upper arm (Figure 20–8). These
pressures are generally expressed in terms of how high
(in millimeters) they can push a column of mercury
(Hg). For a young adult at rest, the values are typically
about 120 mm Hg for systolic pressure and 80 mm Hg
for diastolic pressure, or 120/80. Abnormally high
blood pressures (such as 200/120), often a result of

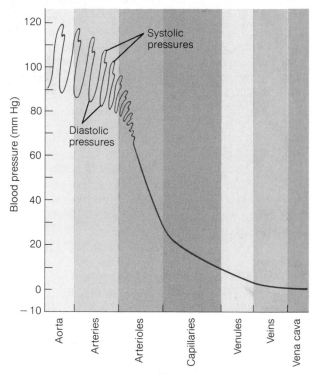

Systolic
pressures

Diastolic
pressures

FIGURE 20–7
**Blood Pressures in Various Parts of the
Human Circulatory System.** The range
indicated by the wavy line at the upper left
represents vacillations in blood pressure
(most notable in the aorta and arteries)
caused by the intermittent ventricular
contractions of the heart. In the arterioles,
blood pressure drops dramatically
because of the elasticity of these vessels'
walls, as well as an increased resistance
to blood flow caused by their small
diameters.

FIGURE 20–8
Measuring Blood Pressure. Human blood pressure is commonly measured by a device known as a sphygmomanometer, which consists of a rubber cuff, a pressure gauge, and a rubber bulb that pumps air into the cuff. The cuff is wrapped around the upper arm, then inflated with air from the bulb until the cuff pressure stops the flow of arterial blood down the arm. A stethoscope is then placed over one of the major arteries just below the cuff. At this point, no pulse can be heard in the stethoscope. When the air pressure in the cuff is slowly released, a point is reached at which a small amount of blood can spurt through the cuffed artery. This spurt sounds like a slight pulsating sound in the stethoscope. The air pressure registered on the gauge when the first sound is heard is noted as the systolic pressure. As the cuff pressure is reduced even further, the flow of blood down the artery increases. When the blood surges have reduced to a continuous flow of blood, a muffled sound can be heard in the stethoscope. At this point, the air pressure on the gauge is noted as the diastolic pressure.

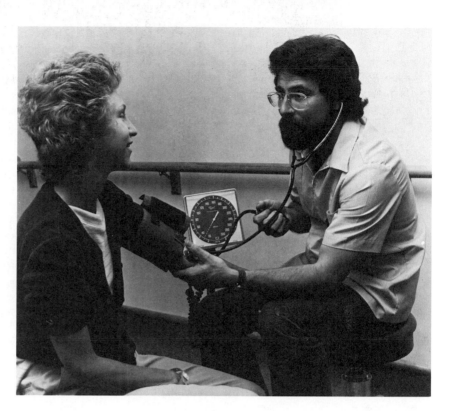

declining elasticity of arteries, increases the chances of cerebral hemorrhage, the breaking of an artery in the brain.

Blood Vessels

The average human body contains about 6 liters (13 pints) of blood, which course through some 60,000 miles of arteries, capillaries, and veins (Figure 20–9).

FIGURE 20–9
The Human Circulatory System. This simplified diagram shows the major internal organs with their capillary beds. Blood pumped by the left ventricle enters the aorta and is diverted to the head, upper limbs, and lower body. Major arteries branch toward the various capillary beds, delivering oxygen and nutrients to all the body tissues. Deoxygenated blood collects in the veins and returns to the heart in the inferior vena cava (from the lower body) and the superior vena cava (from the upper body). Color indicates oxygenated blood.

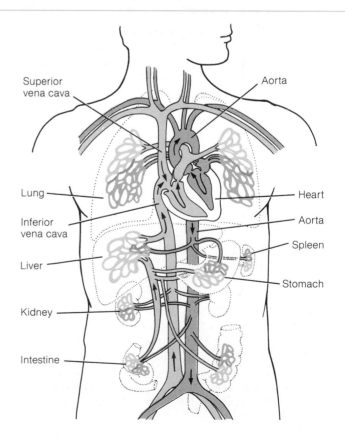

During periods of rest, the heart pumps about 5 liters of blood per minute. At that rate, the body's entire volume of blood recirculates through its system of vessels every 70 seconds. Let's take a closer look at this remarkable circulatory network.

Arteries are lined with a single layer of thin, flattened cells that compose the **endothelium** (Figure 20–10). This inner lining is surrounded by smooth

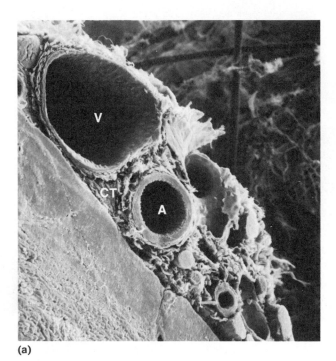

(a)

muscle (see Chapter 26), which in turn is surrounded by connective tissue containing elastic fibers. The connective tissue layer gives arteries their characteristic elasticity, and the smooth muscle controls their diameter. The elasticity of arteries generally decreases with age, a condition made worse by the cardiovascular disease known as atherosclerosis (Essay 20–1).

Arteries branch into narrower **arterioles,** whose walls are composed almost entirely of smooth muscle. The contraction and relaxation of smooth muscle in these vessels controls the amount of blood flow to any given tissue or organ. For example, the arterioles leading to the capillary networks of the gut become fully dilated (expanded) after a meal, causing an increased flow of blood toward the intestinal capillaries for nutrient absorption. If you exercise vigorously right after you eat, much of the blood will be diverted away from the intestines and toward the contracting skeletal muscles. As a result, neither may receive an adequate blood supply, which can cause muscle cramping and poor nutrient absorption. The constriction and dilation of these vessels is under the control of the nervous and endocrine systems.

The capillaries are the narrowest vessels of the circulatory system. Many are barely wider than the blood cells that pass through them, so in some places the cells must move along in single file. The walls of the capillary consist of a single layer of endothelium only (see Figure 20–10), thus permitting gases and nutrients to diffuse across them fairly easily. This is important because it is within the capillary beds that the real work of the circulatory system goes on. We

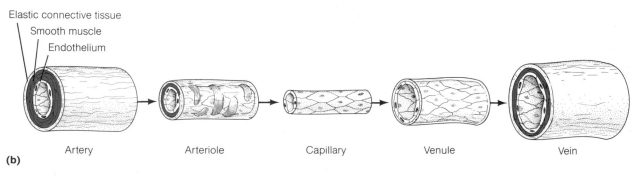

Elastic connective tissue
Smooth muscle
Endothelium

Artery   Arteriole   Capillary   Venule   Vein

(b)

FIGURE 20–10
The Structure of Blood Vessels. (a) This scanning electron micrograph shows cross-sectional views of a vein (V) and an artery (A) embedded in connective tissue (CT). Note the rather thick layer of muscle in the wall of the artery. (Magnification: 822×.) (b) Cutaway diagrams of the various vessels. Capillaries consist of a single layer of endothelium only. (Photo from *Tissues and Organs: A Text Atlas of Scanning Electron Microscopy* by Richard G. Kessel and Randy H. Kardon, W. H. Freeman & Co., © 1979.)

have seen that blood pressure created by the beating heart declines as the blood passes through the arteries and arterioles. However, enough pressure reaches the capillaries to force blood fluid (though not blood cells) through the capillary walls into the extracellular spaces between the body cells (Figure 20–11). Carried with this fluid are oxygen, nutrients (such as sugars, amino acids, and a variety of mineral ions), and a little protein. (Most of the protein molecules in the blood are too large to cross the capillary walls.) Once across the capillary wall, the oxygen and nutrients are free to move into the body cells by diffusion or by carrier-assisted transport. Meanwhile, carbon dioxide and nitrogenous wastes (urea) diffuse down their concentration gradients from the body cells into the capillaries. Virtually all of the water originally squeezed out of the capillary returns to it by osmosis.

From the capillary system, the oxygen-depleted blood collects in small venules that eventually feed into veins. The walls of veins are similar to those of arteries in that they are composed of endothelium, smooth muscle, and connective tissue, but they are much thinner and less rigid than the arterial walls (see Figure 20–10). They change shape easily when skeletal muscles press against them. Many veins also have one-way valves that operate in conjunction with nearby skeletal muscles to help move the blood toward the heart (Figure 20–12). These features are necessary because the residual blood pressure in veins, particularly those in the lower extremities, is virtually nil. The contraction of skeletal muscles pressing on the veins pushes venous blood toward the heart (like

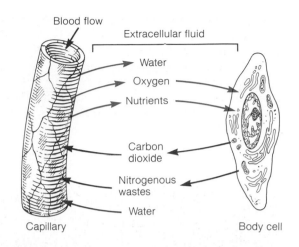

FIGURE 20–11

The Exchange of Gases and Nutrients in the Capillaries. As blood enters a capillary, the blood pressure forces water and dissolved substances through the capillary wall into the extracellular fluid surrounding the tissue cells. Oxygen, nutrients, carbon dioxide, and nitrogenous wastes move between the tissue cells and the capillary according to their individual concentration gradients or by active transport. Most of the water that was forced out of the capillary by pressure returns to the blood by osmosis.

FIGURE 20–12

One-Way Valves in the Veins.
(a) Pressure created by the contractions of nearby skeletal muscles forces blood upward through the valve flaps. (b) The weight of the blood above the valve forces it shut, preventing blood from flowing back to the capillaries.

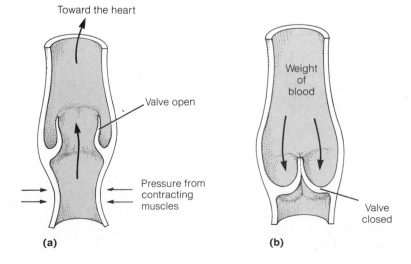

ESSAY 20–1

HEART DISEASE AND THE ONSET OF THE BIONIC ERA

Sometimes it comes quietly but more often with jolting force. One moment you're feeling fine and in the next you are weak, nauseous, dizzy, and breaking out in a cold sweat. A great pressure bears down on your chest and pain radiates out to your shoulders, arms, and jaw. These are the symptoms of a heart attack. In the United States, heart attacks strike 4000 people each day and kill more people than all other diseases combined, including cancer.

The list of our transgressions against our hearts is long and familiar. We don't exercise enough. We eat too many hamburgers, steaks, and french fries that are rich in cholesterol and saturated fats. We consume too much salt. We smoke too much (those who smoke more than 20 cigarettes a day are three times more likely to have a heart attack than nonsmokers). And many of us get caught up in a high-pressure, fast-paced lifestyle that produces too much stress. All of these factors add up to a greater risk of developing heart disease, particularly for those of us living in industrial societies.

The leading cause of heart attacks is obstruction (blockage) of the coronary arteries, vessels that deliver oxygen and nutrients to the heart muscles and connective tissue (see figure). If a blood clot lodges in one of these arteries or their tributaries, the heart tissue that lies downstream from the blocked point soon starves for oxygen and dies. The seriousness of the attack depends on where the blockage occurs and how much heart tissue is damaged. By age 40, most people have experienced minor attacks

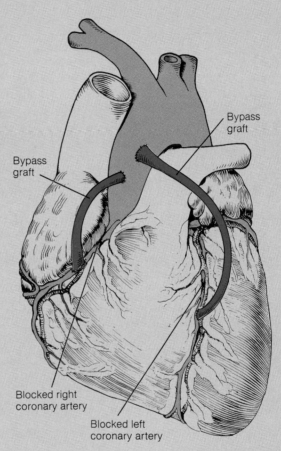

Bypass graft

Bypass graft

Blocked right coronary artery

Blocked left coronary artery

Human heart showing two blocked coronary arteries (arrows) and two bypass grafts.

without even knowing it, mistaking the slight discomfort for indigestion. A major heart attack, however, generally involves extensive heart tissue damage. If the damaged area interferes with the normal spread of electrical impulses across the heart walls, death by fibrillation (uncoordinated heart spasms) may be swift. Even if therapy, drugs, or surgery

restore activity to the heart, some muscle tissue remains permanently damaged, requiring an altered lifestyle for the victim.

The formation of blood clots is normal and they rarely cause a problem for normal, healthy arteries. The chances of obstruction rise dramatically, however, for individuals with an advanced stage of

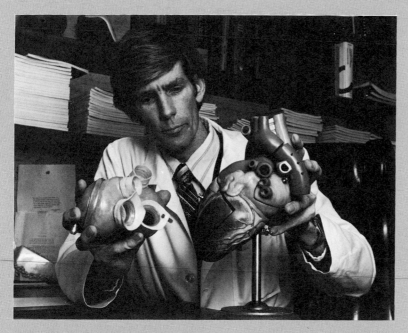

William C. DeVries, M.D., compares the Jarvik-7 artificial heart with an instructional model of a human heart. Dr. DeVries headed the surgical team that implanted a Jarvik-7 heart in Barney Clark, the first human recipient of an artificial heart.

atherosclerosis, a progressive condition that narrows and hardens arteries (see Figure 2–10). This is where the cholesterol and saturated fats come in. If excess levels of these lipids build up in the blood, they are deposited along the inside walls of the coronary arteries, decreasing their internal diameter. Calcium ions then bind to the lipids, forming a hardened, bone-like material called plaque that greatly reduces the elasticity of the arteries. With the passageway for blood narrowed and unexpandable, the atherosclerotic artery becomes a disaster waiting to happen.

In one of the major advances in coronary medicine, obstructed coronary arteries can be replaced in a procedure known as bypass surgery. First performed in 1967, this operation is now done on 125,000 Americans every year. A portion of a vein from the patient's leg is removed and grafted to the healthy parts of the coronary arteries, detouring blood around blockages and resupplying the heart with needed nourishment (see figure). Since the vein comes from the patient's own body, it presents no challenge to the immune system, and rejection of the transplanted tissue is not a problem.

For badly damaged hearts, a heart transplant may be the only recourse. The problem is to find a suitable donor in time. Also, in all tissue or organ transplants, the recipient's immune system will recognize the transplant as foreign and mount an attack that can kill the patient. Even in cases where the transplanted heart is a close match to the patient's own tissue type, powerful drugs must be used to suppress the immune response, which of course makes the patient particularly susceptible to infections. Nevertheless, 40% of heart-transplant patients at Stanford Medical Center survive at least 5 years after the operation, and the development of new drugs to combat rejection has led to increasing optimism.

In December of 1982, Barney Clark became the first recipient of a purely mechanical heart (see photo). The world watched with guarded optimism as this brave man showed us hope for the future. For 112 days Mr. Clark's blood was pushed through his circulatory system by a permanently implanted artificial heart that was connected to an air compressor at his bedside. This is at best a clumsy system that imposes severe restrictions on the patient's activities. Biomedical engineers are now working on artificial hearts that are operated by an internal motor that is implanted with the heart. The only outside connection would be to a system of batteries that could be worn on a belt. It seems the bionic era has begun.

squeezing a tube of toothpaste), and the one-way valves prevent blood from falling back toward the capillaries. If you stand still in one spot for long periods, your feet will begin to swell as they collect pools of blood. Without the contractions of the leg skeletal muscles, which occur when you walk or move about, the blood in your legs cannot move upward easily.

In mammals, venous blood returning from the veins in the lower body collects in a large vein called the **inferior vena cava;** blood returning from the upper body collects in the **superior vena cava.** Both vena cavae join with the right atrium of the heart (see Figure 20–9). The inferior vena cava, with an internal diameter of 2.5 cm (1 inch), is the largest blood vessel in the human body.

The Lymphatic System

The **lymphatic system** is a transport network of vessels that conducts an extracellular fluid called **lymph** from all the various body tissues to major blood veins in the upper chest (Figure 20–13). The narrowest lymphatic vessels, the **lymph capillaries,** are similar in construction to the blood capillaries except that their single layer of endothelium is much more permeable to proteins. Thus, any proteins that were squeezed out of the blood capillaries can cross into the lymph capillaries and eventually return to the bloodstream. The lymph capillaries coalesce into larger **lymph veins,** some of which are interrupted at various points throughout the body by **lymph nodes**— nodules of connective tissue that filter out debris (such as the dead cell fragments and the infectious microorganisms).

In addition to its role in recirculating body fluids, the lymphatic system plays a major part in the body's system of defenses. Most of the battles waged against foreign cells, viruses, and toxins take place here. The active components that help rid the body of foreign matter are also present in the blood, however. Let us now identify some of these components (and others) by examining the composition of blood.

Blood

Blood is a liquid tissue. It is composed of several types of cells suspended in a fluid called **plasma** (Table 20–2). The plasma constitutes slightly more than half the volume of whole blood; the other half of the

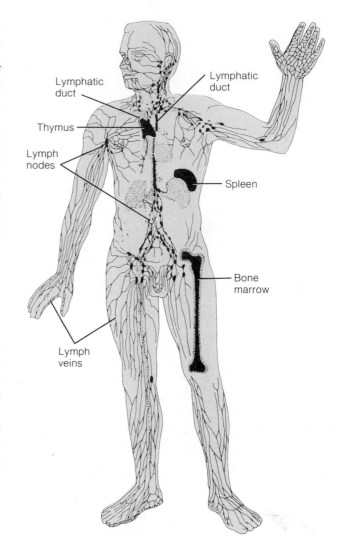

FIGURE 20–13
Human Lymphatic System. Lymph fluid flows from body tissues in lymph veins, ultimately reentering the bloodstream through two large lymphatic ducts in the upper chest. The lymph nodes, especially abundant in the abdomen, filter the lymph fluid. The nodes, spleen, and thymus are storage sites for leukocytes, cells that defend the body against foreign cells and substances.

volume is taken up by cells. Plasma is about 90% water; the remaining 10% is dissolved and suspended substances, including gases, inorganic ions, organic nutrients, various cellular waste products, hormones, and plasma proteins. Various plasma proteins are asso-

TABLE 20–2
Human Blood Composition.

Blood Component	Functions
Cells:	
Erythrocytes (red blood cells)	Oxygen transport
Leukocytes (white blood cells)	Destruction of foreign cells and molecules; production of antibodies
Platelets	Blood clot formation
Plasma:	
Dissolved substances:	
Inorganic ions	Osmotic balance, nutrition
Small organic substances (sugars, amino acids, fatty acids)	Nutrition
Nitrogenous wastes (ammonia, urea)	Metabolic wastes (to be excreted)
Gases (oxygen, carbon dioxide, nitrogen)	Oxygen for cellular respiration
Hormones	Regulation of metabolic activities
Suspended proteins:	
Blood-clotting proteins	Blood clot formation
Antibodies	Binding and inactivation of foreign cells and molecules

ciated with the blood clotting process (discussed later in this section), the transport of water-insoluble substances (lipids, certain vitamins, and the steroid hormones), and the inactivation of invading microorganisms. Some of the plasma proteins are enzymes.

Human blood (and the blood of all other vertebrates as well) contains three general types of cells: platelets, erythrocytes, and leukocytes. The **platelets** initiate the chain of events that ultimately produces a blood clot. **Erythrocytes,** or red blood cells, manufacture and store hemoglobin, the oxygen-transporting protein in blood. Finally, the **leukocytes** (white blood cells) and the proteins some of them produce are the main weapons of the immune system (discussed later in this chapter). Let's have a closer look at each of these important blood cells.

PLATELETS AND BLOOD CLOTTING. Platelets are not actually cells but rather disk-shaped fragments of cells present in the bone marrow. The tiny platelets are quite stable in the blood, but when they encounter a wound, they set in motion a complex series of events. One of these events is the conversion of **fibrinogen,** a soluble plasma protein, into its insoluble form, **fibrin.** The fibrin threads enmesh red blood cells and other platelets in the area of the damaged tissue, ultimately forming a blood clot (Figure 20–14). The clot serves as a temporary seal to prevent bleeding until the damaged tissue can be repaired.

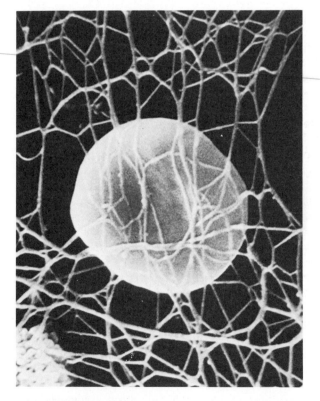

FIGURE 20–14
A Blood Clot. In this scanning electron micrograph, strands of fibrin have enmeshed a red blood cell.

FIGURE 20–15
Blood Cells. This scanning electron micrograph shows the inside of a blood vessel containing biconcave erythrocytes (E) (red blood cells) and spherical leukocytes (L) (white blood cells). (Magnification: 2030×.) (From *Tissues and Organs: A Text Atlas of Scanning Electron Microscopy* by Richard G. Kessel and Randy H. Kardon, W. H. Freeman & Co., © 1979.)

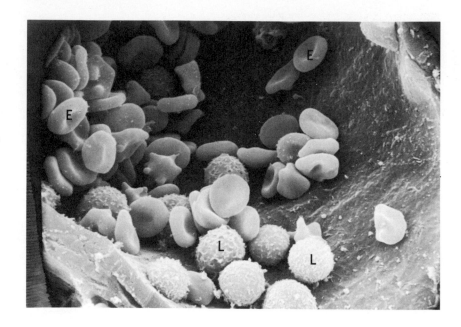

ERYTHROCYTES (RED BLOOD CELLS). Erythrocytes, which are produced in the bone marrow, are disk-shaped, biconcave (indented on both sides) cells (Figure 20–15). They are by far the most numerous cell type in the blood. Each drop of blood contains approximately 200,000,000 red blood cells compared to about 10,000,000 platelets and 300,000 leukocytes. Two million erythrocytes are formed every second in the human body to replace the old and damaged ones. The average life span of a human red blood cell is four months.

The mature red blood cell is specialized for a single role: the transport of oxygen. During its early development in the bone marrow, the immature erythrocyte has a nucleus and a full set of cytoplasmic organelles. As development proceeds, however, its biosynthetic machinery becomes programmed to synthesize one major protein—hemoglobin. Eventually even this capacity is lost as the nucleus and organelles disintegrate, leaving behind little more than a tiny sac of hemoglobin.

The rate of red blood cell production in the bone marrow is precisely regulated to ensure a nearly constant level of these vital transport cells in the blood. Under certain conditions, such as the increased bodily demand for oxygen brought about by a vigorous exercise program, the bone marrow will step up erythrocyte production, resulting in an elevated level of these cells in the blood. A similar effect results when the human body adjusts to high altitudes, where the oxygen concentration in air is substantially less than at sea level.

LEUKOCYTES (WHITE BLOOD CELLS). Like the other blood cells, leukocytes (see Figure 20–15) originate from cells within the bone marrow, then migrate to various positions throughout the body. There are two general types of leukocytes—phagocytes and lymphocytes—that play different but often complementary roles in the body's defense mechanisms. **Phagocytes** are literally "cell-eaters." They engulf and digest invading microorganisms (such as bacteria and viruses) and damaged or otherwise aberrant body cells. Included in this group are the **granulocytes,** which are found primarily in the blood, and the larger **macrophages** (Figure 20–16), which are more common in the lymph. Both types of phagocytes are capable of amoeboid movement and can squeeze through the thin walls of the blood and lymph capillaries. They are chemically attracted to regions of tissue injury and infection where they destroy the damaged cells and any intruding microorganisms.

The **lymphocytes,** as their name implies, are found predominantly in the lymphatic system. They consist of **B cells** (**B** for **b**one marrow), which develop in the bone marrow but take up final residence in the lymph nodes, and **T cells** (**T** for **t**hymus), which undergo their final maturation in the thymus. Both B and T cells have unique functions in the vertebrate immune system, which we will now examine.

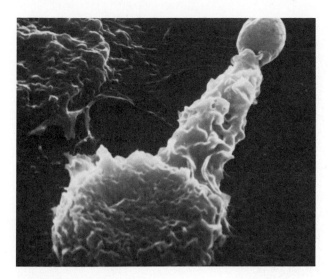

FIGURE 20–16
A Macrophage in Action. This macrophage (larger cell) is starting to attack a yeast cell (upper right). A pseudopod of the macrophage is attached to the yeast cell. (Magnification: 3100×.)

THE IMMUNE SYSTEM

All animals have at least two lines of defense against infection. First, they have physical mechanisms to keep microorganisms out. The tightly-packed cells of the skin and the hard exoskeletons of arthropods provide natural barriers that shield the animal's body from the outside environment. In addition, many vulnerable areas of the body are protected by antimicrobial secretions. In humans, for example, mucus in the nasal passages and earwax in the ears entrap microorganisms and kill certain bacteria. Chemicals present in tears and saliva, and the strong acidity of the stomach juices also have bactericidal effects.

Although the physical barriers work well to block out infectious agents, they are not impregnable. A small cut or scratch may let in hundreds of bacteria, many of which would find the internal fluids of the animal very favorable for growth. If left unchecked, the invaders could multiply and spread rapidly throughout the body. Fortunately, animals have a second line of defense—the phagocytes. Most of the body's phagocytes have fixed positions within the vessels of the lymph nodes, spleen, liver, and bone marrow, where they "lie in wait" for any microorganism that may be carried by in the bloodstream or lymph. There are also circulating phagocytes that can be chemically summoned to any region of tissue damage in the body. Here they destroy not only foreign cells but also the damaged body cells, activities that often result in an inflammation (redness and puffiness).

In addition to the physical barriers and phagocytes characteristic of all animals, the vertebrates have a third defensive mechanism—an immune system. The components of the immune system include the lymphocytes (B and T cells) and the special class of proteins they produce called **antibodies.** Antibodies have the unique ability to recognize and bind to antigens. **Antigens** are large molecules (such as proteins) or cells (such as bacteria) that are foreign to the animal. This binding immobilizes the alien substance or cell and sets in motion events that ultimately cause its destruction. Let us examine this very important defensive mechanism in detail.

Antibody-Mediated Responses

Antibodies are manufactured in B cells, then secreted into the lymph and blood where they circulate freely. Each antibody protein consists of four polypeptide chains: two light chains and two heavy chains (Figure 20–17a). At the ends of the light chains are two identical antigen-binding sites, each capable of binding to the same antigen. If the antigen has more than one **antigenic determinant** (the part of the antigen recognized by the antibody), then large antigen-antibody complexes will form (Figure 20–17b). Such complexes are easy prey for phagocytes, which engulf and digest them.

Every animal that can manufacture antibodies has an enormous variety of B cells. Mammals have hundreds of thousands of different B cell varieties, and in humans the number may exceed a million. Each variety, or **clone,** of B cells has the potential to produce one antibody **idiotype,** which recognizes and binds to one specific molecular antigen. For example, a foreign protein is a moleculer antigen that may activate a single clone of B cells to produce a single idiotype of antibody molecules against it. On the other hand, an alien bacterial cell may have different types of molecular antigens that stimulate as many clones of B cells in the host to manufacture their specific antibodies. Thus, each molecular antigen, whether it is a large molecule existing independently or part

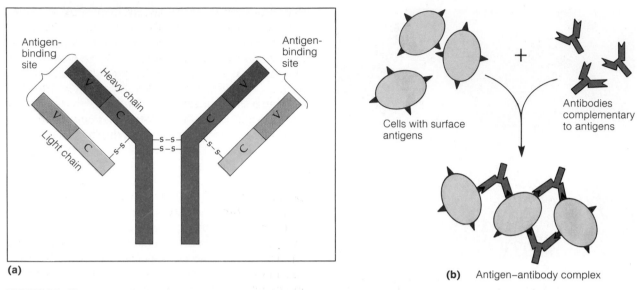

FIGURE 20–17
Antibodies. (a) An antibody molecule consists of four polypeptide chains—two identical light chains and two identical heavy chains—linked by disulfide (—S—S—) bridges. Variable amino acid sequences (V) in the light chains and upper regions of the heavy chains determine which antigen will bind to that particular antibody. Constant amino acid sequences (C) are the same for all the antibodies in one class (there are five classes of antibodies). (b) Large antigen-antibody complexes will form if there are multiple copies of the antigenic molecule on the foreign cell's surface.

of a virus or foreign cell, will prompt the appropriate clone of B cells to manufacture antibodies that attack it.

B CELL ACTIVATION. The production of any specific antibody occurs *after* the particular B cell clone has been exposed to the specific antigen. For most antigens, antibody production entails a complex series of events known as **B cell activation.** This activation process must necessarily begin with recognition. Each clone of B cells displays a set of identical receptor proteins that extend outward from their cell membranes (Figure 20–18). The structure of the receptor is clonal specific—the receptor recognizes and binds to only one type of molecular antigen. When this binding occurs (a reaction that usually requires the participation of specific T cells as well), the B cells become activated to divide. Many successive cell divisions yield a large population of daughter cells, most of which differentiate into **plasma cells.** The new plasma cells manufacture and release large quantities of the antibody against the antigen, which, in conjunction with the phagocytes, eventually destroy the invader.

Once the battle is won, the amount of plasma cells and antibody slowly decline to pre-infection levels (Figure 20–19).

In addition to the plasma cells that are produced during B cell activation, some of the daughter cells develop into **memory cells.** The memory cells are like their B cell forerunners in that they have surface receptors that recognize the same antigen. They remain in the spleen and lymph nodes long after the initial infection, and if enough of them are produced during the initial exposure to the antigen, they may persist for the lifetime of the animal. As their name implies, the memory cells represent the part of the immune system that "remembers" a previous antigen. In the event of a later infection by the same invader, the persistent memory cells respond quickly by undergoing cell divisions, yielding more plasma and memory cells specifically targeted against the returning intruder. The new plasma cells release antibodies into the bloodstream within 24 hours of reinfection, stamping out the invaders before they have a chance to multiply and cause a problem (see Figure 20–19).

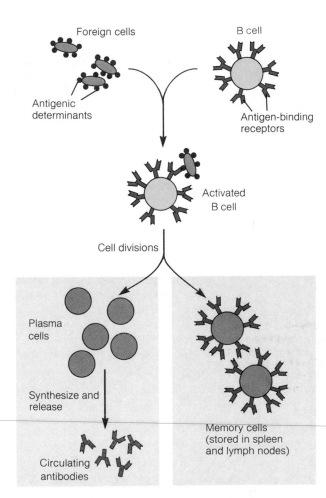

FIGURE 20–18
B Cell Activation. The binding of an antigen to the appropriate B cell activates that cell to divide, yielding daughter cells that differentiate into plasma cells or memory cells. The plasma cells produce large quantities of circulating antibodies against the antigen and the memory cells persist in the body, conferring immunity against future attacks by the same antigen.

This so-called **secondary immune response** is so quick that the host is usually unaware of the infection. Thus, if the initial encounter with an antigen triggers B cell activation, the animal will gain **immunity** against future encounters with the same antigen.

VACCINES. The production of memory cells and the immunity they confer is the underlying mechanism of vaccine action. Most vaccines consist of a nonvirulent (harmless) form of a live bacterium or virus. Although they do not cause disease, they are antigenically similar to a related, disease-causing microbe. For example, the polio vaccines available contain nonvirulent, mutant strains of the poliovirus. These mutants are unable to cause polio, but they have at least one antigenic determinant in common with the virulent poliovirus, and thus activate the production of memory cells that recognize and respond to the crippling strain. Similarly, vaccines have been developed that build up our immunity against diseases such as whooping cough, diptheria, rabies, measles, influenza, and tetanus. Smallpox, a major killer through the ages, has been totally eradicated through worldwide vaccination programs— a first in medical history. An early report (June 1983) from a British team of researchers indicates that a vaccine against genital herpes (see Essay 13–1) has been developed and might be available to the public in several years. Future candidates for control through vaccination include schistosomiasis (see Essay 18–1), malaria, gonorrhea, and syphilis (see Chapter 25).

ANTISERA. Another weapon in the arsenal of immunological medicine is **antisera.** Specific antisera are

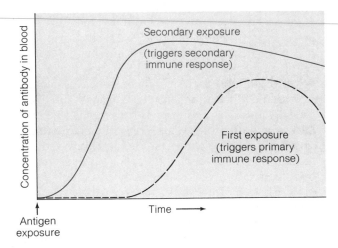

FIGURE 20–19
Primary and Secondary Immune Responses. Upon initial exposure to a given antigen, the immune system takes several days to generate a detectable level of specific antibodies against the intruder. After the battle is won, the antibody level gradually declines. In the event of a later (second) exposure to the same antigen, antibody production is quick, large, and long-lasting. The returning invader is overcome before it can cause disease.

used today to combat active infections of tetanus, infectious hepatitis, measles, and rabies; snakebite venom can also be nullified with an injection of a specific antivenom serum. In all these cases the same principle applies. Blood serum taken from another individual or animal that has been recently exposed to a particular pathogen (or snakebite venom) will have a high concentration of antibodies targeted against that antigen. When the partially purified serum containing the appropriate antibody is injected into a patient's bloodstream, the "borrowed" antibodies immediately attack the pathogen, forming antigen–antibody complexes that are gobbled up by the patient's phagocytes. Thus, the patient is spared the complications (and possibly, death) caused by the infection or venom while his immune system is gearing up to repel the invader.

There are some risks associated with the use of antisera, however. Because the antibody-laden serum itself is foreign to the recipient, the patient's immune system will ultimately respond by manufacturing antibodies against the serum antigens. In most cases this response against the serum takes days to develop and does not interfere with the immediate action of the antiserum against the pathogen. Occasionally, however, the treated patient may experience a rapid and life-threatening allergic reaction to the antiserum. Since the usual culprit in such cases is the serum, not the specific antibody, the ultimate solution to this problem will be to obtain pure supplies of the disease-fighting antibodies. This has been technically unfeasible in the past, but recent breakthroughs in cloning specific antibody-producing B cells in tissue culture has opened the door to the large-scale production of specific, pure (monoclonal) antibodies. Many scientists feel that monoclonal antibodies may be the key for curing many forms of cancer (Essay 20–2).

Cell-Mediated Responses

For certain types of cellular pathogens (such as cancer cells, cells in tissue transplants, and the tuberculosis bacterium), specialized T cells take over the role of the circulating antibodies in the fight. T cells become activated when their surface-bound receptors recognize a specific cellular antigen. These antigen-bound T cells become transformed into killer cells, which divide rapidly, migrate throughout the body, and attack any cell with the appropriate surface antigens. The killer cells release digestive enzymes into the alien cell to inactivate it, and then phagocytes complete the job by engulfing the cell remnants.

Recognizing Self

Recall that the fundamental purpose of the immune system is to distinguish between the animal's own cells and large molecules and those from another source, and then to destroy the latter. Apparently this ability develops early in embryonic growth; any new antigen introduced into the body after this critical stage will be recognized as non-self. This recognition system does not always work to the organism's advantage, however. In some instances the immune recognition system breaks down (or is tricked) and begins attacking the body's own cells; in other cases it can be too discriminating and attack a beneficial invader, such as a transplanted organ from a human donor.

One type of immune system malfunction is the so-called **autoimmune diseases.** Myasthenia gravis (a disease of the muscles), rheumatoid arthritis, rheumatic fever, and some forms of diabetes are all suspected to be autoimmune diseases. For example, cases of rheumatic fever can usually be traced to a previous streptococcus infection, such as the type that causes strep throat. The strep bacteria stimulate the production of antibodies against themselves, but the same antibodies also recognize antigenically similar human proteins on the surface of heart muscle cells. Heart muscle cells are attacked by these antibodies, which damage the heart and may cause death. (For this reason, strep infections are best treated immediately with antibiotics.)

As we mentioned, the immune system can be an obstacle to the success of a tissue or organ transplant, such as those involving a kidney, the heart, or skin tissue. Unless the transplanted tissue is a close match to the tissues of the recipient immunologically, the transplant will be rejected. The culprits in transplant rejections are T cells that respond to non-self antigens present on the surface of the transplanted cells.

One of the most talked-about health problems in recent years is **acquired immune deficiency syndrome,** or **AIDS.** This mysterious killer first appeared in homosexual men, then spread to drug addicts and hemophiliacs who require periodic blood transfusions. Even children are dying from this disease, and at the time of this writing, no one knows its cause. Most medical researchers now (1983) favor the virus theory, but until the pathogen is isolated and identified, producing a vaccine or antiserum is not possible. AIDS is so named because individuals that are affected suffer a deficiency in their immune systems that weakens their ability to fight off other diseases. The malfunction has been traced to an insufficient supply of the T cells that participate in B cell activation. As a result, certain infectious agents can multiply within the body, unchallenged by the population of plasma cells that are normally formed to repel the invader. Thus, the AIDS victim often succumbs to a bacterial disease or cancer that, under normal circumstances, the immune system can overcome.

SUMMARY

The circulatory system of animals is usually powered by a muscular heart that moves blood through a complex network of vessels. This system delivers gases, nutrients, and metabolic wastes to and from all body cells, and serves as a transport link between the body cells and the respiratory, digestive, and excretory systems that exchange materials with the external environment.

Certain invertebrates and all vertebrates have closed circulatory systems. The vertebrate system consists of arteries, veins, capillaries, and a single heart. Fishes have a two-chambered heart and a single circulation system in which the blood pumped to the gills flows from there to the rest of the body. Amphibians and reptiles have a three-chambered heart and a double circulation system: blood pumped to the lungs returns to the heart and is then pumped to the other body tissues. Birds and mammals also have a double circulation system, but they employ a four-chambered heart.

Humans have a typical mammalian heart. Venous blood enters the right atrium, where the contraction of the atrial wall forces it into the right ventricle. Ventricular contraction forces this blood into the pulmonary arteries leading to the lungs. Gas exchange occurs in the lung capillaries, and the oxygenated blood returns to the left atrium of the heart. From there it enters the left ventricle, to be pumped via the aorta to the branching system of arteries. After nutrients and waste materials are exchanged between blood and body tissues in the capillary beds, the oxygen-depleted blood returns to the heart through the veins. Blood flow through the veins is aided by muscle contractions in the vein walls and in nearby skeletal muscles, and by one-way valves that prevent backflow.

Blood pressure is greatest in the arteries during systole (ventricular contraction) and reaches a minimum during diastole (the heart's relaxation phase between ventricular contractions). Blood pressure drops dramatically in the arterioles and capillaries, but there is sufficient pressure to force water and small dissolved substances through the capillary walls into the lymph surrounding the cells. Most of this water reenters the capillaries by diffusion; excess water and some plasma proteins forced out of the blood return to the circulatory system via the lymphatic vessels.

The lymphatic system as a transport network serves as an adjunct to the circulatory system. Lymph contained within its vessels is filtered through lymph nodes, passes through various lymphoid tissues (such as the spleen and the thymus), and eventually enters the circulatory system. The lymphatic system is also the major site of operations for the immune system.

Blood is composed of plasma (a solution of dissolved substances and plasma proteins) and three types of cells: platelets, which play a role in the blood-clotting process; erythrocytes, which contain hemoglobin and are specialized for oxygen transport; and leukocytes, phagocytotic and antibody-producing cells that defend the body from foreign intruders.

The vertebrate immune system defends the body against foreign substances and cells, supplementing the physical barriers and phagocytotic cells found in all animals. The immune response involves the formation of antibodies that bind foreign antigens. Antibody production and its regulation are governed by white blood cells called lymphocytes (B-cells and T-cells). A given B cell synthesizes and releases into the blood or lymph one type of antibody that recognizes and binds one type of antigen. The formation of an antigen-antibody complex renders the invader susceptible to destruction by the circulating phagocytes (granulocytes and macrophages). During a major infection specific B cells in the spleen and lymph nodes become activated to divide, producing plasma cells and memory cells. The plasma cells release large quantities of circulating antibodies to fight off the present danger, and the memory cells serve as a future source of rapidly deployable antibodies should the same foreign cell return. The stimulation of memory cell production by a non-toxic form of an infectious agent is the basis of vaccination. Antiserum treatment for certain pathogens is more direct in that antibodies produced by another individual (or animal) are injected into the bloodstream to do battle with the disease-causing agent.

The immune responses to cancer cells and foreign-tissue transplants are believed to involve T cell activation. Abnormalities in T cell responses may be the bases for autoimmune diseases, the unchecked proliferation of cancer cells, and AIDS.

ESSAY 20–2

NEW STRATEGIES IN THE WAR AGAINST CANCER

Science fiction becomes science fact? (© 1960, King Features Syndicate, Inc. World rights reserved.)

Ted grabbed breakfast at 7:00 and dashed to the University for Friday morning classes. After lunch he played soccer, then picked up Jill. After a pizza they attended a movie, and Ted returned home at 11:30. During the day, Ted's body produced cancer cells capable of killing him in 12 months.

The formation of cancer cells is now believed to be a regular event in the human body. In daily life, agents that turn normal cells cancerous are encountered frequently: toxic chemicals, certain viruses, radiation. In most cases the new cancer cells, scientist believe, are soon recognized as foreign by the body's immune system and destroyed. Ted's immune system worked well and destroyed all of Friday's cancer cells. Only if his immune response failed would a single cancer cell be able to grow to become life threatening.

Exactly how certain agents change a normal cell into a cancer cell is unknown, but the result of such a change is unregulated growth by repeated cell divisions. This produces a mass of cells (a tumor) that may prevent normal functioning of vital organs and cause death. Surgery can remove tumors and accomplish a cure, unless the cancer has spread. When cancer cells break off from the tumor and initiate cancer growth in other parts of the body, surgery is no longer effective. Chemotherapy (drugs), radiation, or both are then used to combat the cancer cells. Although these methods of treatment are remarkably effective for certain types of cancer, they have the serious drawback of killing normal healthy cells as well as cancer cells. Clearly we need a "magic bullet" that will destroy cancer cells without harming normal cells. Two promising new approaches involve

natural components of the body's defenses, interferon and antibodies.

Interferon

Interferon is a protein—actually a class of proteins—normally produced in minute quantities by white blood cells. One of its important roles in the body is to defend against viral diseases (it *interferes* with viral replication in host cells). Extracted from cells and administered as an experimental drug, interferon has also proved effective against certain cancers.

Genetic engineering has now made available larger amounts of several types of interferon. After isolation of the interferon gene from human cells, it is transplanted into the bacterium *E. coli* by recombinant DNA methods (see Figures 11–11 and 11–12). Large scale laboratory growth of *E. coli* then produces large amounts of interferon identical to that from the human cells. The potential usefulness of interferon treatment will only be established after further tests determine its side effects, specificity, and range of action.

Monoclonal Antibodies

Because of their great specificity for target antigens, antibodies have long been a prime candidate for cancer therapy. The problem has been to devise a way to produce large amounts of pure antibody directed specifically against cancer cells. A recently discovered solution involves the construction of special cells called hybridomas, which can produce large amounts of a single species of antibody, as described below.

In this new technique, some of the patient's cancer cells are injected into a healthy mouse (see figure). The mouse's immune system responds by generating several populations of B cells. Each population makes antibodies against one type of antigen on the cancer cell surface. Since cancer cells have some surface antigens not found on normal cells (the basis of their "foreign" status), some of these B cells will make cancer-specific antibodies. Next, the mouse's spleen is removed and B cells from it are fused (hybridized) with fast-growing cells from a mouse myeloma, a type of tumor. This cell fusion combines in one cytoplasm the properties of both cells. The hybrid, called a **hybridoma,** multiplies indefinitely like the myeloma cell and makes antibodies like the spleen B cell.

A single hybridoma cell can be isolated in the laboratory and then allowed to multiply to form a clone of identical cells. All cells in the clone will secrete a single type of antibody into the growth medium. These **monoclonal antibodies** (antibodies from one clone of cells) can be harvested and tested against

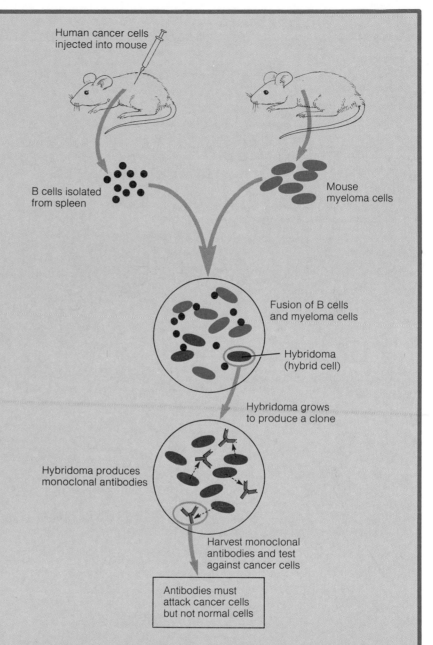

Human cancer cells injected into mouse

B cells isolated from spleen

Mouse myeloma cells

Fusion of B cells and myeloma cells

Hybridoma (hybrid cell)

Hybridoma grows to produce a clone

Hybridoma produces monoclonal antibodies

Harvest monoclonal antibodies and test against cancer cells

Antibodies must attack cancer cells but not normal cells

the donor patient's cells. If they react with cancer cells but leave similar normal cells untouched, they can be administered to the patient.

Will the monoclonal antibodies destroy the patient's cancer cells? Clinical tests thus far have had only limited success, perhaps because the binding of circulating antibodies to cells may not always kill them. One

way to overcome this problem may be to attach a radioactive material or toxic drug to the antibody molecules for direct, lethal delivery to the cancer cells. Another, more long-term approach may be to seek ways to mimic or enhance the roles played by other parts of the immune system (such as killer T cells) in defending the body against cancer.

STUDY QUESTIONS

1. What are the differences between an open and closed circulatory system? Between single and double circulation?

2. Name and describe the general function of the three major types of cells found in blood. Which of these cell types is found predominantly in the lymph?

3. Contrast the general structure of the amphibian and mammalian hearts. How are the differences related to these animals' respiratory systems?

4. What is the function of the layer of smooth muscle in vertebrate arteries and arterioles?

5. What exchanges take place between blood and body cells in the capillary beds?

6. What is an electrocardiogram?

7. What force causes blood pressure? What forces reduce it?

8. Name the four types of leukocytes and briefly describe their functions.

9. What are the three major lines of defense that protect the vertebrate body against infection?

10. What is the relationship between B-cell activation and immunity?

11. What is an autoimmune disease?

SUGGESTED READINGS

Hood, L. E., I. L. Weissman, J. H. Wilson, and W. B. Wood. *Immunology.* 2nd ed. Menlo Park: Benjamin/Cummings, 1984. This up-to-date text is written by leaders in the field. Chapter 1 provides a very nice overview of immunology.

Jarvik, R. K. "The Totally Artificial Heart." *Scientific American* 244(1981):74–80. Explains the development of the world's first artificial heart for humans.

Milstein, C. "Monoclonal Antibodies." *Scientific American* 243(1980):66–74. A very readable account of how hybrid cell clones have been used to generate large quantities of a single type of antibody.

Spence, A. P., and E. B. Mason. *Human Anatomy and Physiology.* 2nd ed. Menlo Park: Benjamin/Cummings, 1983. Chapters 17–20 of this nicely illustrated text provide an indepth look at the human circulatory and immune systems.

Zucker, M. B. "The Functioning of Blood Platelets." *Scientific American* 242(1980):86–103. Explains how these small cellular fragments are involved in wound healing, a very complex but interesting process.

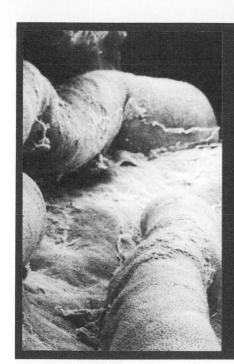

CHAPTER 21
Nutrition and the Digestive System

Heterotrophic organisms, as you may recall, are those that cannot make their own food but must obtain organic nutrients from their environment. This is accomplished by a variety of feeding mechanisms, each adapted to the nature of the food source. Fungi and heterotrophic bacteria, for example, feed by absorbing small organic compounds from the material they grow on, which may be anything from a rotting log to an animal's large intestine. Many of these microorganisms secrete enzymes directly onto their food source. These enzymes break down the protein and polysaccharide components into amino acids and sugars that can be absorbed. If the organisms on which they are feeding are dead, the feeders are termed **saprophytes** (Figure 21–1a). The activities of saprophytic fungi and bacteria are extremely important for all life on earth because the decomposition of organic matter returns minerals to the soil or water for use by plants (see Chapter 31). Heterotrophs that live and feed within other living organisms are termed **parasites** (Figure 21–1b).

The way most animals obtain nutrients is more complex. They feed by ingesting chunks of food (Figure 21–1c) into a digestive cavity or tract, where extracellular digestion takes place. The digestion products are then absorbed by cells lining the cavity or tract.

The nutrients that animals obtain in their food serve two basic purposes. First, they provide chemical energy to fuel the regeneration of ATP needed by all living cells in the body. Second, they are the molecular building blocks used in the construction of more complex compounds in the cells. However, animals vary somewhat as to which nutrients must be available in their diets to serve these metabolic roles. Since it would be beyond the scope of this book to describe the nutritional requirements of all animals, we will discuss only human nutrition.

HUMAN NUTRITIONAL REQUIREMENTS

Humans (and most other animals) require four basic types of nutrients: energy foods, essential organic precursors (building blocks of larger molecules), vitamins, and minerals. Let us examine each of these nutrient categories individually to understand what they are and why they are needed.

357

(a)

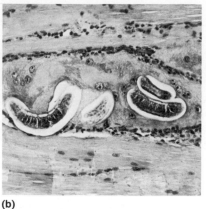

(b)

(c)

FIGURE 21–1
Feeding Strategies of Heterotrophs. (a) Saprophytes obtain food from dead organisms. Here, a dead beetle serves as a food source for a water mold. (b) Parasites live and feed within other living organisms. These *Trichinella* larvae are feeding on muscle tissue. In humans, *Trichinella* is often obtained from eating undercooked pork. (c) Most animals ingest their food into a digestive cavity or tract, as this baboon is preparing to do.

Energy Foods

The three major energy foods are carbohydrates, fats, and proteins. These compounds can be chemically dismantled within the body cells and respired to provide energy for ATP formation (see Chapter 5). The energy released during this process is measured in **Calories.** Ounce for ounce, proteins and carbohydrates in foods have about the same caloric value, whereas fats have roughly twice the energy value.

An adult human weighing 150 pounds must consume about 1800 Calories per day just to sustain his or her vital bodily functions (heartbeat, breathing, and so on). Leading the fairly sedentary life of a typical college professor might boost the daily demand up to 2500 Calories, whereas breaking rocks on a chain gang might generate a need for up to 6000 Calories per day. The number of Calories you need each day depends on your weight and the rate at which you **metabolize,** or "burn up," food through physical activity. If you want to lose weight, all you have to do is take in fewer Calories than you burn. This sounds simple enough, but we all know that it is not easy to lose weight and, in fact, crash diets may be dangerous to your health (Essay 21–1). The idea behind a weight-reduction diet is to get rid of some of the reserves of stored food (fats) within the body. When a person's metabolism outpaces his or her intake of Calories, the immediate result is a drop in the blood sugar level. Next to go are the carbohydrates stored in the liver. After that, fats are gleaned from various storage areas in the body, places that probably need no mention. The problem is that when the sugar level in the blood drops, we get hungry.

In many areas of the world, people take in too few Calories not by choice but because sufficient food is not available. As a result, these people are generally frail and listless, a condition known as **undernutrition.** Millions of people die of starvation each year, although the immediate cause of death is often pneumonia or some other infectious disease that takes advantage of their weakened condition.

A more widespread form of starvation—**malnutrition**—is related to food quality, not total Calories. For example, many people in parts of Southeast Asia and Africa subsist on diets that are almost exclusively starch in content (such as polished rice). Eating two or three bowls of white rice can certainly meet the energy needs of people, but purely starchy foods are deficient in certain other nutrients the body demands. Let us now examine these other human dietary requirements.

Essential Organic Precursors

As the name suggests, **organic precursors** are substances that become units of larger molecules. The essential ones must be obtained in the diet as preformed compounds because the animal in question lacks the metabolic pathways to synthesize them from other compounds. In humans this applies to several amino acids (precursors to proteins) and linoleic acid (an unsaturated fatty acid precursor to fats).

Of the 20 different kinds of amino acids typically found in proteins, humans can synthesize 11 from simple organic compounds that contain nitrogen. The others, designated the **nine essential amino acids,** must be obtained in the food we eat. They are histidine, isoleucine, leucine, lysine, methionine, phenylalanine, threonine, tryptophan, and valine. If our diet is deficient in one or more of these essential amino acids, our ability to synthesize proteins will be greatly impaired.

Because proteins are long chains of amino acids, it stands to reason that the best source of the nine essential amino acids is protein in the diet. However, the proteins in foods vary considerably in amino acid composition. For humans, meat and animal products have the highest quality protein because animal proteins are generally similar to human proteins in amino acid composition. Digesting animal protein thus gives a person an appropriate balance of all 20 amino acids. Plant proteins are less ideal because they often are low in one or more of the essential amino acids. People can still meet their nutritional requirements on an exclusively vegetarian diet, but to do so they must take in more total protein, and include more than one source of that protein in their daily diet. For instance, the proteins in cereal grains (rice, wheat, barley, and others) are typically low in lysine but are good sources of methionine; legumes (such as beans and peas) are deficient in methionine but have adequate amounts of lysine. Thus, vegetarians whose main staple is cereal grains should supplement their diet with legumes or other plant foods rich in lysine to ensure an adequate intake of all the essential amino acids.

The most common form of malnutrition worldwide is protein deficiency, which really means a deficiency in one or more of the essential amino acids. Protein malnutrition is prevalent in Southeast Asia, Africa, and parts of South America, where it is particularly serious in young children—it can cause stunted growth, tissue swelling, brain damage, and even death. In West Africa, these symptoms are known as **kwashiorkor,** which literally means "the sickness a child develops when another baby is born" (Figure 21–2). The kwashiorkor child does well until it is weaned from its mother's milk and begins eating the standard protein-deficient, starchy diet of the region.

Besides the essential amino acids, humans also need a small amount of unsaturated fat in their diet. **Linoleic acid,** a type of unsaturated fatty acid, is a

FIGURE 21–2

Child with Kwashiorkor, a Protein Deficiency Disease. The bloated abdomen and altered texture of the hair are characteristic of this condition. Other symptoms include listlessness and mental retardation.

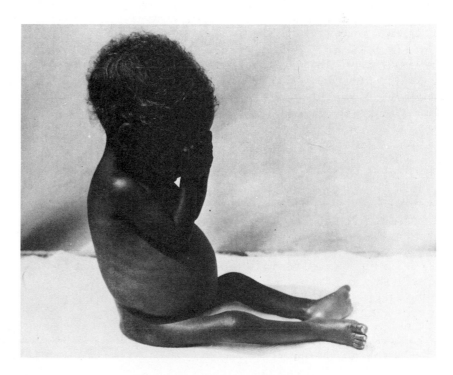

ESSAY 21-1

THE DIETER'S DILEMMA

Many of us, at one time or another, wage the battle of the bulge. In this fight we try to lose weight, even though our bodies scream out, "Feed me!"

Excessive weight, considered unattractive and unhealthy by most of our society, is the enemy. So, by various means, we try to reduce. Weight loss is accomplished by decreasing caloric intake and/or increasing energy expenditure so that the body reserves of fat are burned up. Certain foods like sugar and alcohol are high in calories and have little nutritional value aside from supplying energy. Other foods are high in calories, but they also contain essential nutrients. It is important to remember that nutritional requirements do not change even when caloric intake is decreased to lose weight.

Some people who are overweight yield to the enemy without a fight. They may defend their surrender with various reasons (which may be valid) such as "my metabolism is too slow," "it's genetic," or "I don't have the willpower to give up the food I enjoy." Others do fight—sort of. These people make compromises such as drinking diet soda with their fried chicken or skim milk with their chocolate cake. Still others decide to enter the foray with vigor, pushed on by the image of a thinner body.

Some warriors fight the enemy from the front lines, advancing slowly but surely. Their strategy is to reduce their caloric intake by eliminating sweets; cutting down on the quantity of fatty, fried foods; and eating well-balanced but smaller meals. Armed with substantiated nutritional recommendations for weight control, these fighters not only lose weight but usually stay healthy in the process.

Others engage in guerrilla warfare, resorting to hit and run tactics such as crash diets and diet pills to gain quick results. Always looking for the "quick fix," they are easy targets for the many, occasionally bizarre diet plans developed by individuals who often lack medical or nutritional expertise. In their impatience to lose weight rapidly, they may find themselves caught in the "lose now, pay later" trap. By participating in fad diets or using diet aids that promise beauty and slenderness in a short period of time, these individuals may be at risk for possible long-term effects of vitamin or mineral deficiencies, gastrointestinal problems, and cardiovascular disease.

Let's look at some of the tactics the guerrilla fighter may resort to:

High-Protein, Low-Carbohydrate Diets

These diets prescribe eating foods containing large amounts of animal protein (eggs, lean meat, poultry, fish). Foods containing carbohydrates are permitted but in extremely low amounts. These diets produce weight loss by reduced caloric intake as well as the development of **ketosis,** an abnormal metabolic condition in which partially oxidized fat residues (ketones) are excreted in the urine. Ketosis can also produce a loss of appetite, which promotes weight loss. However, another side effect of ketosis is elevated levels of cholesterol, triglycerides, and uric acid, which increase the risk of cardiovascular disease and gout.

High-Carbohydrate Low-Protein Diets

These diets are based on the ingestion of large amounts of complex carbohydrates (whole grain foods, fruit, and vegetables), and little fat or animal protein is allowed. The Pritikin diet, the most well-known and controversial of the high-carbohydrate, low-protein diets, promises not only weight loss but also a reduction in coronary heart disease risk. In its favor, this diet does incorporate many of the recommendations made by experts to prevent heart disease. But the diet is so austere that it may result in deficiencies of vitamins B, D, and E, and it borders on protein insufficiency for large men.

Other Fad Diets

Other diets such as the Beverly Hills diet (all fruit), the grapefruit diet, the high-fiber diet, and so on are based on the ingestion of primarily one type of food for a limited period. Caloric intake is reduced but so is nutritional quality, since only those nutrients present in the one type of food are supplied. Those participating in these diets may find themselves not only bored but deficient in certain essential nutrients.

Diet Aids

An incredible array of diet pills, powders, and liquids are available to the person who needs assistance in the battle against obesity. Some of these products are prescription drugs; others are available over the counter. Many of the diet pills are appetite suppressants. By acting on the central nervous system, especially the hypothalamus, which is involved in regulating appetite, appetite suppressants inhibit that physiological feeling of discomfort or pain called hunger. Amphetamines have been the most commonly prescribed appetite suppressants, but their use has fallen into disfavor because of the potential for drug abuse.

Until recently, starch-blocking diet pills were part of the dieter's arsenal. Made from kidney bean extract, starch blockers inhibit the digestion of food starches, which then pass, undigested, through the large intestine. However, after users reported nausea, vomiting, abdominal cramps, and diarrhea, starch blockers were banned from sale by the Food and Drug Administration.

Ideal Weight.

Height (without shoes)	Weight (without clothing)		
Men	Light build	Medium build	Heavy build
5 ft. 3 in.	118	129	141
5 ft. 4 in.	122	133	145
5 ft. 5 in.	126	137	149
5 ft. 6 in.	130	142	155
5 ft. 7 in.	134	147	161
5 ft. 8 in.	139	151	166
5 ft. 9 in.	143	155	170
5 ft. 10 in.	147	159	174
5 ft. 11 in.	150	163	178
6 ft.	154	167	183
6 ft. 1 in.	158	171	188
6 ft. 2 in.	162	175	192
6 ft. 3 in.	165	178	195
Women	Light build	Medium build	Heavy build
5 ft.	100	109	118
5 ft. 1 in.	104	112	121
5 ft. 2 in.	107	115	125
5 ft. 3 in.	110	118	128
5 ft. 4 in.	113	122	132
5 ft. 5 in.	116	125	135
5 ft. 6 in.	120	129	139
5 ft. 7 in.	123	132	142
5 ft. 8 in.	126	136	146
5 ft. 9 in.	130	140	151
5 ft. 10 in.	133	144	156
5 ft. 11 in.	137	148	161
6 ft.	141	152	166

Source: U.S. Department of Agriculture

What many people tend to forget is that the battle of the bulge is just one of many in the war for good health. Healthy people don't just look good; they feel good too.

Feeling good is an indicator of physical and emotional fitness, and a nutritionally balanced diet contributes to a person's overall fitness.

component of certain lipids vital to cell membrane function. Although humans cannot synthesize it, it is a common component of vegetable oils and fish, so a dietary deficiency of linoleic acid is extremely rare. The "vitamin F" sold in health food stores is not actually a vitamin, but rather a source of unsaturated fatty acids. Taken as a dietary supplement, vitamin F merely supplies excess lipids to the diet.

Vitamins

Vitamins are organic compounds that humans (and other animals) need in minute amounts in the diet. Two major criteria define vitamins: they are not synthesized by the animal in question, and they have a known metabolic function critical to the animal's normal health. Most vitamins are precursors of **coenzymes,** small molecules that are required for various enzyme-catalyzed reactions.

The first recorded indication of a human vitamin requirement dates back to the mid-eighteenth century. In the 1750s it was discovered that eating fresh fruit could prevent **scurvy,** a disease characterized by swollen joints, bleeding gums, loosening of the teeth, and anemia. Scurvy was common among sailors, who spent long periods at sea without fresh fruit and vegetables. The British navy was the first to include limes and lemon juice as part of the ship's rations, a practice for which the British sailors were nicknamed "limeys." Not until the twentieth century was ascorbic acid, or vitamin C, identified as the scurvy-preventing factor.

In the late nineteenth century, another dietary deficiency syndrome, **beriberi,** was plaguing Dutch troops stationed in the East Indies. The soldiers suffered from muscle deterioration, paralysis, and loss of mental acuity. The nutritional basis of beriberi surfaced when it was learned that adding whole-grain rice to the usual polished rice diet of the troops prevented these symptoms. In 1912, Casimir Funk identified the anti-beriberi agent in whole-grain rice as an organic amine known today as thiamin (vitamin B_1). (An amine is an organic compound with an amino group, $-NH_2$.) Since this amine was vital to normal health, Funk coined the term *vitamine* ("vital amine"). When it was later discovered that many of these vital dietary factors were not amines, the *e* was dropped to yield *vitamin.*

We currently know of 13 vitamins required by humans (Table 21–1). They are generally split into two classes on the basis of their solubility: Four are fat-soluble and nine are water-soluble.

FAT-SOLUBLE VITAMINS. Vitamins A, D, E, and K, all required by vertebrates, are virtually insoluble in water. They tend to dissolve in the body's fatty tissues, not the blood, so they are not very accessible

TABLE 21–1
Vitamins Required by Humans.

Vitamin	Major Dietary Sources	Deficiency Symptoms
A (retinol)	Butter, cheese, liver, egg yolk, green and yellow vegetables	Night blindness, degeneration of cornea, dry skin
B_1 (thiamin)	Whole grain cereals, liver	Beriberi: muscle weakness, heart failure, mental dysfunction
B_2 (riboflavin)	Milk, liver, eggs, whole grains	Cracked lips, sore tongue, stunted growth
Niacin	Milk, eggs, meat, whole grains	Pellagra: dermatitis, diarrhea, nervous disorders
B_6 (pyridoxine)	Eggs, fresh meat, whole grains, fresh vegetables	Dermatitis, convulsions
Pantothenic acid	Eggs, fresh meat and vegetables, whole grains	Dermatitis, disorders of the digestive tract
Biotin	Liver, egg yolk, fresh vegetables	Nausea, muscle pain, dermatitis
Folic acid	Liver, leafy vegetables	Anemia
B_{12} (cobalamin)	Liver, meat, eggs, milk	Pernicious anemia: malformation of red blood cells, weakness, numbness
C (ascorbic acid)	Citrus fruits, tomatoes, leafy green vegetables	Scurvy: anemia, bleeding gums, swollen joints, improper wound healing
D (calciferol)	Fish liver oil, sunlight on skin; added to milk	Rickets: weak bones, defective bone formation in children
E (tocopherol)	Leafy green vegetables, eggs, vegetable oils	Anemia in children (suspected)
K	Leafy green vegetables	Excessive bleeding, internal hemorrhaging

to the excretory system. As a result, a large dose of one of these vitamins in a single meal might meet the animal's requirement for months. In fact, an increasing problem among health food enthusiasts is getting an overdose of such vitamins, which can lead to a condition called **hypervitaminosis.**

Vitamin A is part of the visual pigment rhodopsin present in the eyes of vertebrates. A deficiency of vitamin A can cause a hardening of certain eye tissues, and a prolonged deficiency might lead to total blindness. Vitamin A deficiency is common in certain parts of the Middle East and Asia, where it is the leading cause of blindness among children. A less extreme symptom is night blindness, in which the affected individual has difficulty seeing objects under dim light.

Vitamin D is commonly called the sunshine vitamin because it is formed in the outer skin layers when they are exposed to the ultraviolet rays of sunlight. However, because animals can synthesize vitamin D, technically it is not a vitamin. Nevertheless, it has long been recognized as a vitamin in the clinical sense because symptoms stemming from its deficiency can be overcome or prevented by the appropriate diet supplements. At one time, vitamin D deficiency was fairly common, particularly among children living in northern temperate regions. Remaining mostly indoors during the long, cold winter, and heavily clothed during their brief outdoor excursions, many children did not get enough sunlight exposure to produce adequate levels of vitamin D. The result was often a bone malformation disease known as **rickets** (Figure 21–3). Vitamin D is now added to milk, which has virtually eliminated rickets among children in developed nations.

The most mysterious of the fat-soluble vitamins is **vitamin E.** It has proven to be a nutritional requirement in a variety of experimental animals, but the case is less clear for humans. Although many claims have been made for vitamin E, ranging from enhanced sexual potency to the prevention of heart disease, no one has yet discovered its exact role in humans.

Vitamin K, named for the Danish word *koagulation,* is required for normal blood clotting. It is not included in most commercially-available vitamin supplements because bacteria present in the colon synthesize enough of it to meet normal human needs. However, newborn babies are routinely given an injection of vitamin K because their colons lack sufficient levels of the bacteria that produce it.

WATER-SOLUBLE VITAMINS. The water-soluble vitamins include vitamin C and the B-complex vitamins. Because of their solubility in water, these vitamins

FIGURE 21–3
Child Suffering from Rickets. A deficiency of vitamin D has resulted in soft, malformed bones, causing this young girl's legs to bow out.

are constantly excreted in the urine. To assure adequate levels of these vitamins in the body, they must be provided in the diet on a daily basis.

Vitamin C is a required nutrient for several vertebrates, including monkeys and humans. Although its specific metabolic function remains a mystery, we do know that vitamin C is important for the maintenance of healthy connective tissue and the prevention of scurvy. In addition, there are many who believe that megadoses of vitamin C help prevent (and reduce the severity of) the common cold. It has also been claimed to bolster the immune system against cancer. So far, however, there is no definitive clinical evidence for either claim.

The **B-complex vitamins** include eight different compounds that, in their chemically active forms, serve as coenzymes. (Each B vitamin is converted to its active form in animals' cells.) Half of the B vitamins—niacin, riboflavin, thiamin, and pantothenic acid—are precursors of coenzymes that function in the Krebs cycle of respiration. Therefore, a deficiency of any one of these vitamins will seriously curtail the cells' ability

TABLE 21–2
Minerals Required by Humans.

Minerals	Major Roles
Macrominerals:	
Calcium	Constituent of bones and teeth; involved in nerve and muscle cell action
Phosphorus	Constituent of many organic compounds such as nucleotides, sugar phosphates, phospholipids; necessary for bone formation
Magnesium	Required for various enzyme-catalyzed reactions, particularly those involving ATP; constituent of bones and teeth
Sodium Potassium Chloride	Used to maintain osmotic balance in cells; involved in nerve and muscle cell action
Microminerals:	
Iron	Constituent of oxygen-carrying hemoglobin and electron-transporting cytochrome molecules
Iodine	Component of hormones produced by thyroid gland
Copper Zinc Manganese Molybdenum Chromium	Required by a variety of enzymes for catalytic activity

to produce ATP. As you might imagine, such deficiencies can have grave consequences.

Like vitamin K, five of the B vitamins (niacin, pantothenic acid, biotin, folic acid, and vitamin B_{12}) are synthesized in varying amounts by intestinal bacteria. Under normal conditions these bacteria supply enough biotin and pantothenic acid to meet our needs. However, someone taking large doses of antibiotics to fight a bacterial infection may need a dietary supplement of these vitamins to make up for the loss of intestinal bacteria—innocent victims of the antibiotic treatment.

Interestingly, the two vitamins for which a specific biochemical role is not known—E and C—have received the greatest acclaim for a myriad of health-promoting effects. Nutritionists and biochemists are eager to discover the real functions of these vitamins, if for no other reason than to dispel the myths generated by popular articles, books, and rumors.

Minerals

All organisms require certain inorganic substances we call **minerals.** Plants normally absorb minerals from the soil; animals obtain them in the food and water they ingest. The minerals required by humans are shown in Table 21–2. Note that carbon, hydrogen, oxygen, nitrogen, and sulfur are not listed. These five elements must be provided in organic form, whereas the substances designated as minerals are usually assimilated as inorganic ions. As the table shows, the mineral nutrients humans need are divided into two categories based on the daily amount required. Macrominerals are required in larger quantities than microminerals.

DIGESTIVE SYSTEMS

As you recall, most animals feed by ingestion—food is taken into an internal body cavity for processing. In sponges, cnidarians, and flatworms, food is digested in a **gastrovascular cavity,** which has a single opening to the outside. For example, the cnidarian *Hydra* captures prey with its tentacles, then maneuvers its catch through the mouth into the gastrovascular cavity (Figure 21–4). Cells lining the cavity then secrete digestive enzymes to break down the food. Any undigested morsels must be expelled through the same mouth. Such a relatively simple digestive system places strict limitations on what *Hydra* can call food.

Annelids, mollusks, arthropods, echinoderms, and chordates have a **tubular digestive tract,** with a mouth for taking in food and an anus for eliminating the undigested material. Food passes one way through the tract, encountering specialized compartments that store, pulverize, digest, and absorb it along the way.

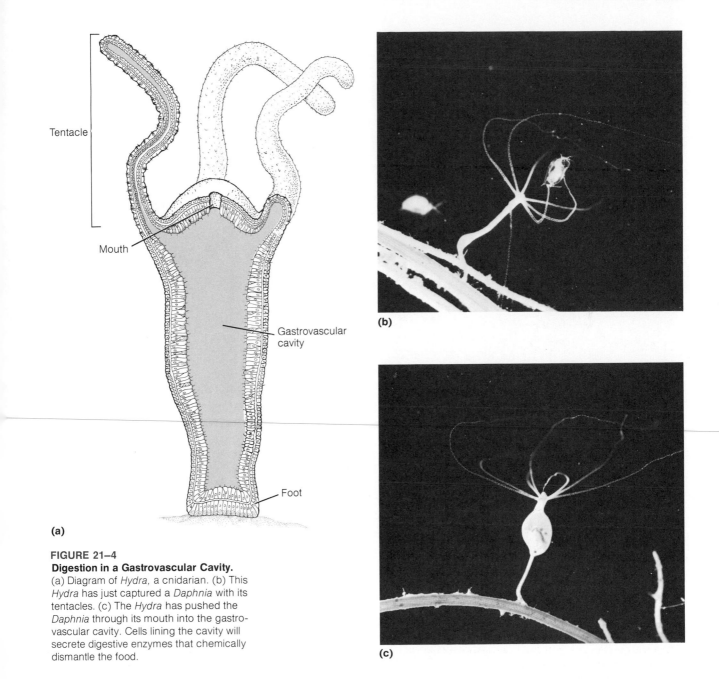

(a)

(b)

(c)

FIGURE 21–4
Digestion in a Gastrovascular Cavity.
(a) Diagram of *Hydra,* a cnidarian. (b) This *Hydra* has just captured a *Daphnia* with its tentacles. (c) The *Hydra* has pushed the *Daphnia* through its mouth into the gastro-vascular cavity. Cells lining the cavity will secrete digestive enzymes that chemically dismantle the food.

Each of the food processing activities occurs in sequence because of the *dis*assembly line arrangement of the compartments.

During the course of evolution, many variations in the food processing compartments in animals have appeared, each an adaptation for a specific feeding strategy. For instance, earthworms and birds have a **crop,** an expanded region of the foregut where food can be temporarily stored (Figure 21–5). The crop enables these creatures to take a meal when it is available or safe to do so, then process the food gradually during periods of nonfeeding. They also have a muscular **gizzard** for grinding food into smaller, digestible bits.

Mammals do not have crops or gizzards; they have other specializations that perform the functions of food storage (stomachs) and grinding (teeth). The main features of the mammalian digestive tract are embodied in our own system, so let us now turn to the human food processing machine.

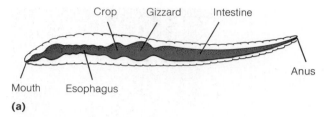

(a)

FIGURE 21–5
Tubular Digestive Tracts. Most animals have a tubular digestive tract, in which food passes in one direction from the mouth to the anus. Both (a) earthworms and (b) chickens have a crop for food storage and a muscular gizzard for grinding and pulverizing food. The intestine is the major site of digestion and nutrient absorption.

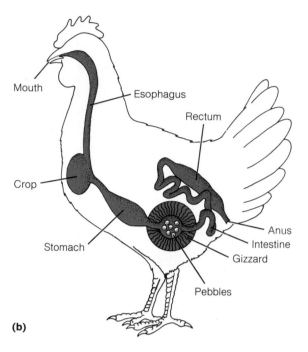

(b)

THE HUMAN DIGESTIVE SYSTEM

The human digestive system consists of the digestive tract and associated structures and organs that participate in the mechanical and chemical processing of food (Figure 21–6). As you munch on your potato chips or other snack, let us trace the fate of this food as it moves through your system.

Oral Cavity, Pharynx, and Esophagus

The human digestive tract begins with the **oral cavity**—the mouth and associated structures, including the teeth, tongue, and salivary glands (see Figure 21–6). One of the most important activities which takes place here is food selection. The food must pass smell and taste tests before further processing, although most of us have occasionally had to suppress this basic activity under pressure. ("You can't leave the table until you eat your spinach.") Once we take a bite, the food is broken up into small pieces by chewing. The teeth shear and grind the food, increasing its surface area for attack by the digestive juices. Thus, chewing your food well not only decreases the chances of getting it stuck in your throat, but aids the digestive process as well.

As the food is being chewed, it is mixed with saliva secreted by three pairs of **salivary glands** (see Figure 21–6). Saliva acts as a lubricant to ease the passage of food toward the stomach. It also contains **salivary amylase,** an enzyme that begins breaking down starches and glycogen. The bulk of starch digestion occurs later, however, in the small intestine.

The tongue is a muscular organ that positions food for chewing, forms it into a small mass, then pushes the mass back into the pharynx for swallowing. The **pharynx** is the short tube we commonly call the throat; it doubles as an air passage to the lungs. The base of the pharynx branches into the trachea (windpipe) and the esophagus (see Figure 21–6). As food enters the pharynx, the swallowing reflex closes off the openings into the trachea and nasal passages, thereby directing the food mass into the esophagus. If food becomes lodged in the pharynx or gets into the trachea by mistake, it triggers the coughing reflex.

The **esophagus** is a tube about 30 cm (12 inches) long that extends from the pharynx to the stomach. Circular bands of muscles in the esophageal wall constrict in a wavelike fashion behind the food mass, squeezing it along in the proper direction. This coordinated series of muscle constrictions is called **peristalsis;** similar peristaltic contractions take place in the intestines. A special band of muscle tissue called a **sphincter** closes off the entrance to the stomach. When food approaches, the sphincter muscles relax, allowing the food to enter the stomach.

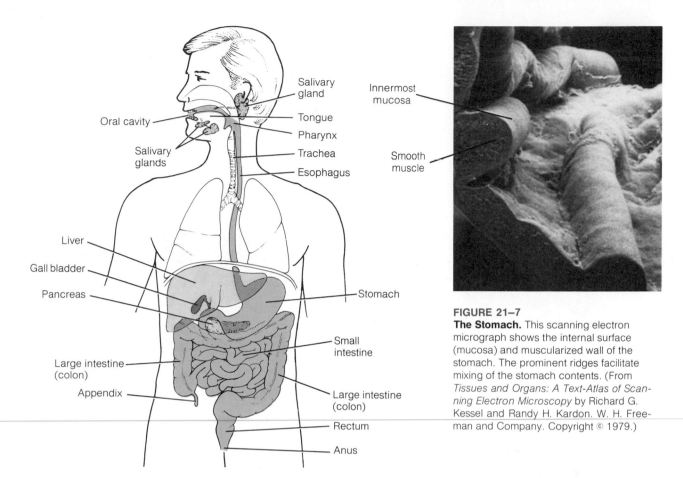

FIGURE 21–6
The Human Digestive System.

FIGURE 21–7
The Stomach. This scanning electron micrograph shows the internal surface (mucosa) and muscularized wall of the stomach. The prominent ridges facilitate mixing of the stomach contents. (From *Tissues and Organs: A Text-Atlas of Scanning Electron Microscopy* by Richard G. Kessel and Randy H. Kardon. W. H. Freeman and Company. Copyright © 1979.)

Stomach

The **stomach** is a large muscular sac lying just below the lower ribs. Its walls consist of an inner lining of tissue called the **mucosa,** which lies over a thick layer of smooth muscle (Figure 21–7). The presence of food in the stomach triggers involuntary muscle contractions that churn the stomach contents and physically break up the food mass. The violent agitations help mix the food with gastric juices in the stomach cavity.

Besides its role in physically disrupting food, the stomach is also a major site of digestion. Several types of glandular cells in the mucosa secrete substances that collectively make up the gastric juice. One type of cell secretes an inactive form of **pepsin,** an enzyme that digests proteins into smaller peptides. Another type of glandular cell releases hydrochloric acid, which has several important effects. This strong acid converts the inactive form of pepsin to the active form and provides the very acidic environment needed to optimize pepsin's catalytic activity. Hydrochloric acid also kills some microorganisms that are carried in with the food, and helps separate the cells of ingested tissues.

A third type of glandular cell produces and secretes mucus. The mucus coats the stomach lining and helps protect it from being digested by the hydrochloric acid–pepsin mixture. Occasionally the protection mechanism breaks down and the gastric juices wear away part of the lining. The result is a weakened area, or **ulcer,** in the stomach wall or adjoining section of the small intestine. Although there seems to be a correlation between ulcers and emotional stress, we still do not understand exactly how these very painful sores are caused.

Small Intestine

The partially digested, soupy mixture of food in the stomach cavity is gradually released into the **small intestine,** where the bulk of digestion and nutrient

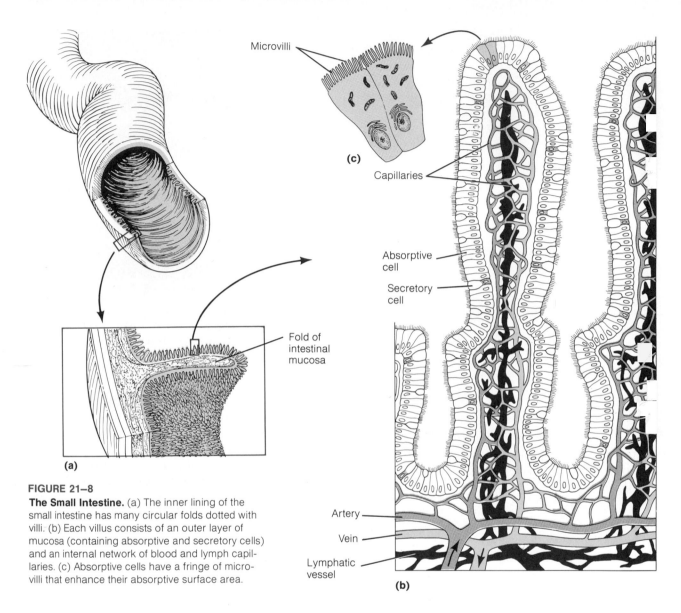

Microvilli

Capillaries

(c)

Absorptive cell

Secretory cell

Fold of intestinal mucosa

(a)

Artery

Vein

Lymphatic vessel

(b)

FIGURE 21–8
The Small Intestine. (a) The inner lining of the small intestine has many circular folds dotted with villi. (b) Each villus consists of an outer layer of mucosa (containing absorptive and secretory cells) and an internal network of blood and lymph capillaries. (c) Absorptive cells have a fringe of microvilli that enhance their absorptive surface area.

absorption take place. The small intestine is a highly folded tube about 2.5 cm (1 inch) in diameter and up to 7.5 m (25 feet) in length. Its inner lining contains many circular folds that are dotted with thousands of tiny, fingerlike projections called **villi** (Figure 21–8). Even the villi have microscopic projections termed **microvilli.** All of these folds and projections create an enormous surface area (more than 500 square meters) for both the secretion of digestive enzymes into the intestinal cavity and the absorption of nutrients from it.

Digestion within the small intestine is aided by secretions from three different sources: the pancreas, the intestinal glands, and the liver. The **pancreas,** a large organ that lies just below the stomach (see Figure 21–6), releases a mixture of digestive enzymes via the pancreatic duct into the first section of the small intestine. Glandular cells interspersed in the intestinal mucosa also secrete a battery of digestive enzymes that complement the actions of the pancreatic enzymes. **Bile,** a complex mixture of bile salts, bile pigments, and cholesterol produced in the liver and stored in the gall bladder (see Figure 21–6), also empties into the small intestine via the bile duct. All of these secretions, which may total about two liters of fluid per day, are alkaline. They neutralize the acidic gastric juices entering the small intestine, which is important because the digestive enzymes

TABLE 21–3
Sources and Actions of the Major Digestive Enzymes.

Source	Enzyme	Action
Salivary glands	Salivary amylase	Starch and glycogen → Maltose
Stomach	Pepsin	Protein → Peptides
Pancreas	Trypsin	Protein → Peptides
	Chymotrypsin	Protein → Peptides
	Carboxypeptidase	Peptides → Amino acids
	Pancreatic amylase	Starch and glycogen → Maltose
	Lipase	Fats → Glycerol and fatty acids
	Nuclease	Nucleic acids → Nucleotides
Small intestine	Maltase	Maltose → Glucose
	Sucrase	Sucrose → Glucose and fructose
	Lactase	Lactose → Glucose and galactose
	Aminopeptidase	Peptides → Amino acids
	Dipeptidase	Dipeptides → Amino acids
	Nuclease	Nucleic acids → Nucleotides

operating in the small intestine are most active under slightly alkaline conditions.

The digestive enzymes released into the small intestine perform virtually every conceivable digestive process (Table 21–3). Proteins are broken down into peptides by the actions of pancreatic **trypsin** and **chymotrypsin,** then the peptides are further degraded into amino acids by various **peptidases.** The digestion of starch and glycogen begun by salivary amylase in the mouth is completed by **pancreatic amylase,** generating maltose (a disaccharide) as the major product. Maltose and other disaccharides are cleaved into their corresponding monosaccharide units by specific enzymes released from the intestinal glands (see Table 21–3). **Nucleases** originating from both the pancreas and intestinal glands digest nucleic acids (DNA and RNA) into their component nucleotides, and **pancreatic lipase** breaks down fats to fatty acids and glycerol.

Of all the digestible macromolecules present in food, fats present the greatest challenge. Recall that fats are very nonpolar molecules that do not dissolve in water. Thus, they tend to clump together in the watery fluid of the gut, making them difficult targets for the enzyme lipase. The digestion of fats is greatly aided, however, by the bile salts produced in the liver. These salts act as detergents that break up the fatty clumps into tiny fat droplets, just as dish detergent breaks up grease on pots and pans. This increases the surface area of the fats and makes them more accessible to lipase.

The ultimate products of digestion—sugars, amino acids, nucleotides, fatty acids, and glycerol—are ab-sorbed into the mucosa of the small intestine, along with dissolved mineral ions. The nutrients then enter nearby blood capillaries and lymph vessels (see Figure 21–8) for distribution throughout the body.

Large Intestine

As food moves through the small intestine by intermittent peristaltic contractions, most of the liberated nutrients and some of the water are absorbed by mucosal cells. The undigested matter, bile, and digestive juices pass on to the **large intestine,** or **colon.** The large intestine reabsorbs most of the bile salts, returning them to the liver via the bloodstream; it also absorbs most of the water, forming compact feces in the process. If the lining of the small or large intestine becomes irritated by certain types of protozoans, bacteria, or influenza viruses, peristaltic contractions increase, and the intestinal contents move through too rapidly to permit adequate water reabsorption. The result is frequent defecation and watery feces, a condition commonly known as **diarrhea.** Alternatively, **constipation** results when the contents of the large intestine move too slowly, so that too much water is reabsorbed and the undigested matter becomes dry and difficult to move. An effective way to avoid constipation is to eat plenty of fiber (indigestible cellulose), which gives bulk to the colon contents. Leafy vegetables, celery stalks, carrots, and bran flakes are a few examples of high-fiber foods.

Large populations of bacteria (such as *E. coli*) inhabit the colon and derive their nutrition from the waste material that passes by. As mentioned earlier, humans benefit from this association by obtaining certain vitamins the bacteria produce. More than half of the fecal mass is composed of these bacteria.

At the end of the large intestine is a muscular region called the **rectum,** which terminates in an opening to the outside called the **anus.** The urge to defecate is brought about by the filling of the rectum and the distention of its muscular walls. Because the entry of food into the stomach normally triggers the emptying of the large intestine, the desire to defecate often comes just after a meal.

The movement of material through the entire digestive tract normally takes 24 to 36 hours. Depending on the size of the meal, food is acted upon by the stomach in 1 to 4 hours. It then makes its way through the small intestine in about 8 hours, and passage through the large intestine takes another 12 to 24 hours.

SUMMARY

All animals are heterotrophic and most feed by ingestion. They rely on organic compounds present in their food to meet their needs for chemical energy and molecular building blocks.

Humans and most other animals require four basic types of nutrients: energy foods, certain organic precursors, vitamins, and minerals. Energy foods, such as carbohydrates, fats, and proteins, can be broken down during respiration to provide energy for ATP production. Their energy value is measured in Calories. Individuals who do not get enough Calories in their diet are said to be undernourished.

The human requirement for organic precursors includes the nine essential amino acids (normally obtained from protein in the diet) and linoleic acid (an abundant component of vegetable oils). Dietary deficiencies in linoleic acid are extremely rare, but protein deficiencies are a leading cause of malnutrition worldwide.

Vitamins are relatively small organic compounds that function in a variety of metabolic processes. The fat-soluble vitamins (A, D, E, and K) tend to be stored in the fatty tissues of the body, where they are not very accessible to excretion. The water-soluble vitamins (C and the B-complex), on the other hand, are readily excreted in the urine and therefore must be taken in every day. If the diet lacks sufficient quantities of a particular vitamin, specific deficiency symptoms soon appear.

All organisms require certain minerals—inorganic nutrients that serve various metabolic functions. Macrominerals are required in relatively large amounts daily; microminerals are needed in smaller quantities.

The digestive systems of animals range from the simple gastrovascular cavities of sponges, cnidarians, and flatworms to the complex tubular tracts of higher animals. In animals with a tubular digestive tract, food enters a mouth, passes through a sequence of compartments specialized to store, pulverize, digest, and absorb nutrients, and the undigested wastes are eliminated through an anus. Through evolutionary time, many variations on the structure of the digestive tract have appeared as adaptations to different feeding mechanisms.

The processing of food in humans and other mammals begins in the oral cavity, where it is chewed up by the teeth, lubricated by saliva secreted by the salivary glands, and then positioned for swallowing by the tongue. As a food mass enters the pharynx, the swallowing mechanism closes off the passages to the nasal cavities and trachea, thereby directing the food into the esophagus. Circular bands of muscles in the esophageal wall undergo waves of constrictions (peristalsis), pushing the food toward the stomach.

The stomach is a muscular sac capable of violent agitation for mechanical disruption of the food mass. Special glandular cells of the gastric mucosa secrete hydrochloric acid and pepsin that initiate the digestion of proteins. Other mucosal cells release mucus that helps shield the stomach lining from self-digestion.

The partially digested food is gradually released from the stomach into the small intestine, where most of the digestion and absorption of digestion products take place. Batteries of enzymes produced by the pancreas and glandular cells of the small intestine are responsible for the digestion of starches, proteins, nucleic acids, and fats. Bile manufactured in the liver passes via the gall bladder into the small intestine, where it aids in the dispersion of fats into tiny droplets. The digestion products are absorbed into the mucosa of the intestinal wall, then passed into the blood and lymph capillaries for transport to the rest of the body. The villi and microvilli of the intestinal mucosa provide an enormous surface area for the absorption of nutrients.

Any material not absorbed in the small intestine (such as indigestible cellulose, bile, and the digestive juices) passes on to the large intestine (colon), where the bile salts and most of the water are reabsorbed into the bloodstream. Intestinal bacteria are abundant here, feeding on wastes and, as a benefit to their host, producing vitamins. At the end of the large intestine is the rectum, a muscular region that compacts the feces and expels them through the anus.

STUDY QUESTIONS

1. Name the four general classes of nutrients required by humans and give a specific example for each.

2. Distinguish between undernutrition and malnutrition.

3. Why must strict vegetarians (with no meat or animal products in their diets) eat a variety of plant foods?

4. It is more common for individuals to exhibit symptoms of hypervitaminosis from taking excessive amounts of vitamin A, D, or E than vitamin C or the B-complex vitamins. Explain.

5. What is the difference between a gastrovascular cavity and a tubular digestive tract?

6. What part(s) of the digestive tract is specialized for grinding food in a bird? In humans?

7. What are the roles of hydrochloric acid in the stomach? How is the stomach lining normally protected from this acid?

8. Explain the roles of the pancreas and liver in digestion.

9. How is it possible that the small intestine, a tube roughly 2.5 cm in diameter and 7.5 meters long, has an internal surface area greater than 500 square meters?

10. What are the major roles of the large intestine? The rectum?

SUGGESTED READINGS

Brody, J. E. *Jane Brody's Nutrition Book: A Lifetime Guide to Good Eating for Better Health and Weight Control.* New York: W. W. Norton, 1981. Jane Brody is a personal health columnist for the *New York Times* and her book is both very readable and nutritionally sound.

Combs, B. J., D. R. Hales, and B. K. Williams. *An Invitation to Health.* 2nd ed. Menlo Park, CA: Benjamin/Cummings, 1983. Chapter 13 of this book has a good practical discussion of nutrition, food controversies, and diet.

Lappe, F. M. *Diet for a Small Planet.* 2nd ed. New York: Ballantine Books, 1981. This book, with recipes, describes nonmeat foods that produce a high-grade protein diet.

Mason, E. B. *Human Physiology.* Menlo Park, CA: Benjamin/Cummings, 1983. Chapter 16 of this well-illustrated text has a good discussion of the digestive system.

Moog, F. "The Lining of the Small Intestine." *Scientific American* 245(1981):154–176. This article discusses the structure and transport functions of the cells composing the lining of the small intestine.

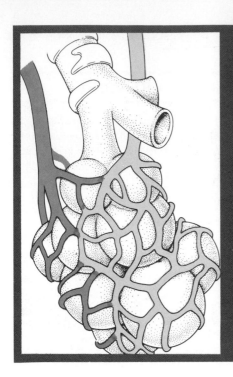

CHAPTER 22

Gas Exchange and the Respiratory System

Humans can survive for weeks without food and for days without water, but we will die in about six minutes if deprived of oxygen. Oxygen is necessary for aerobic respiration, the means by which about 90% of cellular ATP is generated (see Chapter 5), and our cells require a continuous supply of ATP to carry out the energy-demanding processes essential to life. Because of their high metabolic rates, human brain cells have especially high ATP (energy) demands. Death by suffocation is actually due to ATP starvation in vital brain cells.

Most animals have **respiratory systems** that facilitate the exchange of respiratory gases (oxygen and carbon dioxide) between the organism and the external environment. Let us begin by considering the variety of ways animals meet their gas exchange needs.

TYPES OF RESPIRATORY SYSTEMS

In all organisms, the exchange of gases between a cell and its immediate environment takes place by diffusion. A respiring cell consumes oxygen (O_2) and produces carbon dioxide (CO_2), thereby maintaining diffusion gradients that favor oxygen uptake and carbon dioxide release relative to the cell's surroundings. The relatively small and simple animals (sponges, cnidarians, flatworms, nematodes, and most annelids) have large surface areas in contact with the environmental water. No cell is far removed from this external source of oxygen, and gas exchange occurs directly across the body surfaces (Figure 22–1a). Many semiterrestrial amphibians also exchange gases across their moist skins.

For large animals, however, size adds complications to the gas exchange process. Most of a large animal's body cells are tucked far away from the external source of oxygen, yet they need this vital gas to support cellular respiration. This dilemma has been solved in higher animals by the coevolution of respiratory structures for external gas exchange and circulatory systems to transport gases between the respiratory surfaces and the body cells.

Two general types of respiratory structures are found among the animals. The first is the **evaginated** type, which consists of feathery or platelike extensions from the body surface. The gills of the large aquatic invertebrates and the fishes are examples of evaginated structures adapted for gas exchange in water (Figure

FIGURE 22–1
Animal Respiratory Systems. (a) Small aquatic animals and land animals that maintain a moist skin (e.g., amphibians) exchange gases directly across their skins. (b) Larger aquatic animals (certain annelids, mollusks, arthropods, echinoderms, and most fishes) have evaginated structures called gills that provide a large surface area in contact with water. (c) Tracheae are invaginated systems of air pores and channels characteristic of insects and a few other terrestrial arthropods. (d) Most large, terrestrial animals (certain amphibians and all reptiles, birds, and mammals) have large invaginated air sacs called lungs. Gas exchange across skin and gill surfaces are aquatic adaptations; tracheae and lungs are found predominantly in air-breathing animals. Interestingly, the semiaquatic frog utilizes three of these general respiratory surfaces: gills in the tadpole stage, and skin and lungs as an adult.

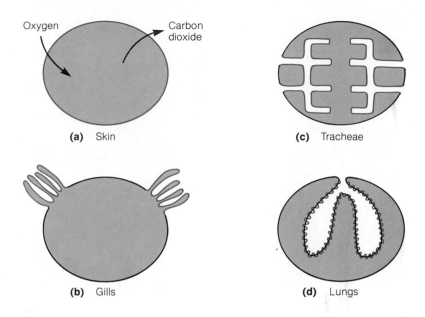

(a) Skin

(c) Tracheae

(b) Gills

(d) Lungs

22–1b). The second type of respiratory structure is **invaginated,** forming tubes or cavities inside the body. The tracheae of insects and the lungs of the land-dwelling vertebrates are invaginated systems that are suited for gas exchange in air (Figure 22–1c and d). Let us examine these respiratory structures in more detail.

Gills

The most advanced type of gill system is found in the fishes (Figure 22–2). Fish **gills** are essentially complex networks of finely branched blood capillaries embedded in thin cellular plates called gill lamellae. The gill capillaries are separated from the external water by only a single layer of cells, so oxygen dissolved in the water crosses only one gill cell and the capillary wall to enter the fish's bloodstream. As water passes through the gills, oxygen molecules diffuse into the blood, where they bind to hemoglobin for transport to the rest of the body. At the same time, carbon dioxide produced within the body cells diffuses into the blood, eventually reaching the gills, where it diffuses outward into the water.

The spaces between the rows of gills extend into the oral cavity of a fish, so water taken into the

mouth passes through the gills and out the other side. Fast-swimming fish such as the tuna require large amounts of oxygen to keep pace with their high metabolic rates. Thus, they keep their mouths open while swimming, thereby facilitating the flow of oxygen-containing water through the gills. Many less active fish obtain enough oxygen by opening and closing their mouths while stationary and flexing their bony gill covers to draw water past the gills.

Tracheae

The respiratory systems of most terrestrial arthropods (such as spiders and insects) consist of branched tubes called **tracheae** that extend from many surface pores to the internal body tissues (Figure 22–3). Through such tubes air contacts virtually all of the individual body cells. Although some larger insect species can increase air flow through their tracheae by muscular contractions, the tracheal systems of most species are entirely passive.

The scattered system of pores and tubes in arthropods clearly meets the gas exchange needs of these relatively small creatures. However, the appearance of larger land dwellers hinged on the evolution of a centralized internal gas exchange organ that was intimately linked to a circulatory system.

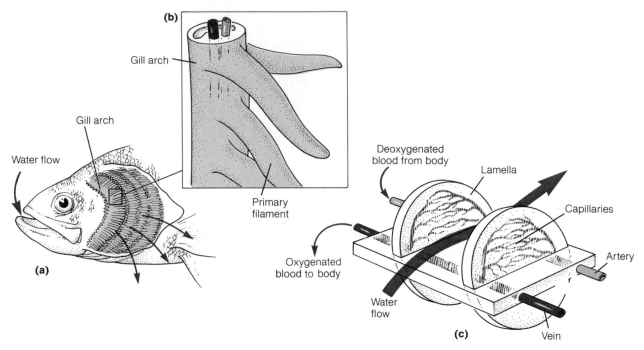

FIGURE 22–2
Structure of a Fish Gill. (a) On each side of a fish's head are two or more rows of gills covered by a bony gill cover (removed in diagram to expose gills). (b) Each gill consists of a gill arch from which two rows of primary filaments extend outward. (c) Each primary filament in turn is composed of a stack of gill lamellae, very thin plates which are rich in capillaries. Blood moves down one side of the primary filament through an artery and enters the capillary beds of the lamellae. As water flows past the capillary beds, gas exchange takes place: oxygen enters the blood from the water and carbon dioxide diffuses into the water from the blood. The oxygenated blood collects in the vein on the other side of the filament and is then transported to the rest of the body. Note how the design of the gill maximizes the tissue surface area in contact with water.

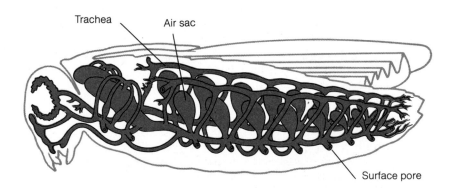

FIGURE 22–3
The Tracheal System of an Insect. Air moves in and out of the tubular tracheae and air sacs through surface pores.

Lungs

In their most rudimentary form, **lungs** are simply inpocketings of the body wall, with a passageway leading to the exterior, and numerous blood vessels near their internal surface. Simple lungs are found in certain marine snails that breathe air only occasionally (when exposed to low tides). As lungs evolved into more complex forms, the inpocketings became more highly branched and folded, thereby exposing a greater surface area to air. The trend toward increasing lung surface area in animals reached its climax in the mammals. Human lungs, for example, have a total surface area of about 90 square meters, approximately 50 times greater than the skin's surface area. Land snails, most amphibians, and all reptiles, birds, and mammals have well-developed lungs and a mechanism to ventilate them (breathing).

THE HUMAN RESPIRATORY SYSTEM

Anatomy

The respiratory system of humans, one of the most sophisticated gas exchange systems known, begins with the nose (Figure 22–4). Air breathed in through the nostrils enters the **nasal cavity,** where it is warmed and humidified. The ciliated epithelial cells lining this cavity secrete mucus, which traps inhaled dust and debris. The beating cilia move the mucus and entrapped particles toward the throat, where they are either swallowed or spat out. The moistened, filtered air then passes into the **pharynx** (throat) and on to the **larynx** (better known as the Adam's apple).

Just below the larynx is the **trachea,** a tube whose thick walls are reinforced with rings of cartilage to prevent collapse during inhalation. (You can feel these rings by moving your fingers along the front of your trachea below your Adam's apple.) Lining the inner

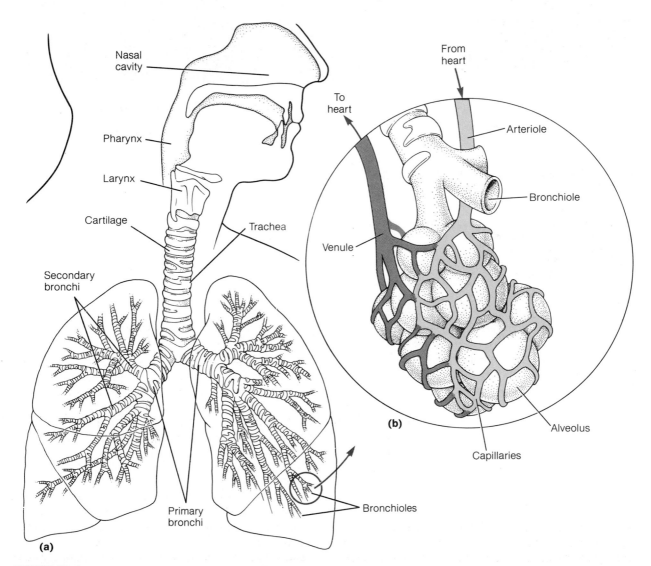

FIGURE 22–4
The Human Respiratory System. (a) Inhaled air is directed through the trachea, which branches into the primary bronchi leading to the right and left lungs. Air continues through smaller and smaller bronchi, which branch into the very narrow bronchioles. (b) Each bronchiole terminates in a cluster of alveoli, which are in close association with blood capillaries for gas exchange.

surface of the trachea are millions more cilia that also sweep mucus and debris into the pharynx (Figure 22–5). Prolonged cigarette smoking tends to kill these ciliated cells lining the respiratory passageway, so heavy smokers must often rely on coughing to keep the air tubes clear.

About midway into the chest cavity, the trachea branches into two **primary bronchi** (singular **bronchus**) (see Figure 22–4), which direct air into the left and right lungs. The bronchi also have cartilaginous rings reinforcing their walls and are lined with ciliated epithelial cells. Each primary bronchus branches into smaller and smaller bronchi, which in turn branch into **bronchioles.** The bronchioles terminate in clusters of tiny air sacs known as **alveoli** (singular **alveolus**). The human lungs consist of about 300 million alveoli (see Figure 22–4b).

The Movement of Oxygen and Carbon Dioxide

The alveoli have extremely thin walls surrounded by a dense network of capillaries (see Figure 22–4b). It is here that blood meets air. Incoming oxygen dissolves in the film of water on the inner surface of each alveolus, then diffuses across the alveolar and capillary walls into the blood. Once in the blood, most of the oxygen binds to hemoglobin in the red blood cells. Meanwhile, carbon dioxide dissolved in the blood plasma diffuses into the alveolar air to be exhaled. The oxygen-enriched blood moves from the lung capillaries into venules and eventually into the pulmonary veins for return to the heart (see Figure 20–4). With each heartbeat an entirely new supply of oxygen-poor blood moves into the lungs to replace the oxygenated blood returning to the heart. If your heart rate is 80 beats per minute, then a red blood cell remains in your lung capillaries for an average of only three-quarters of a second.

Fresh air is 20% oxygen and 0.4% carbon dioxide by volume. When air is inhaled, the atmosphere within the lung air spaces is thus relatively high in oxygen and low in carbon dioxide concentration. In contrast, blood arriving at the lung capillaries is relatively low in oxygen and high in carbon dioxide concentration. Because of these concentration differences, oxygen diffuses into the blood and carbon dioxide diffuses out into the air spaces of the alveoli.

By the time the blood moves from the capillaries to the venules of the alveoli, however, it actually contains a higher oxygen concentration than is present

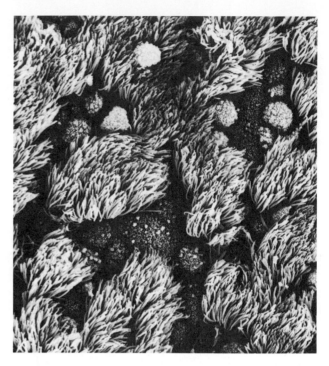

FIGURE 22–5
Ciliated Internal Surface of the Trachea.
Each cell may have more than 200 cilia.

in the air spaces. Clearly something beyond simple diffusion has taken place. The explanation is that oxygen in the blood binds to hemoglobin in the red blood cells, lowering the concentration of free oxygen in the blood plasma. This permits more oxygen to diffuse into the blood. Therefore, the combination of free and hemoglobin-bound oxygen in the blood adds up to a greater oxygen concentration than that in the alveolar air spaces. Researchers have estimated that without hemoglobin, the blood could carry only 2% as much oxygen as it actually does.

How does hemoglobin carry oxygen? Hemoglobin is composed of four polypeptide subunits (two alpha and two beta chains), each containing a heme group bound to an iron atom (Figure 22–6). Each heme iron can bind loosely to a single oxygen molecule. Thus, one hemoglobin molecule can carry up to four oxygen molecules, though the number bound depends on the oxygen concentration in the blood plasma. In the lungs where oxygen is abundant, the blood becomes nearly saturated with oxygen, so most of the hemoglobin molecules carry a full load of four oxygen molecules. When this oxygenated blood reaches the oxygen-depleted tissues elsewhere in the body, the free oxygen in the blood plasma rapidly diffuses out,

FIGURE 22–6

Hemoglobin. Each of hemoglobin's four polypeptide chains has a heme group bound to an iron atom. Each heme iron can bind a single oxygen molecule.

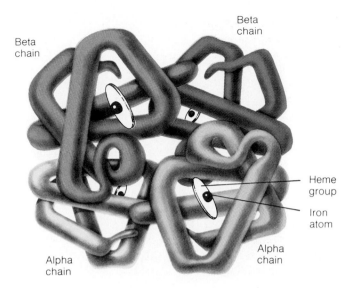

Beta chain

Beta chain

Alpha chain

Alpha chain

Heme group

Iron atom

creating a concentration gradient favoring the unloading of the hemoglobin. Consequently, hemoglobin in venous blood contains on the average less than one oxygen molecule per hemoglobin molecule.

Carbon dioxide, a waste product of cellular respiration, diffuses down its concentration gradient from the body cells into the blood. This gas binds only weakly to hemoglobin, so most of it is carried to the lungs as a dissolved component of the blood plasma.

BREATHING MECHANISMS

If two balloons, one filled with oxygen and the other filled with carbon dioxide, were connected by a pipe, the rates of diffusion of these gases between the balloons would depend on the length and cross-sectional area of the pipe. A short, wide pipe would permit faster gas diffusion rates than a long, narrow pipe. The bronchioles, bronchi, and trachea represent a long, narrow system of pipes that connect the lungs to the outside air. They are bottlenecks to air flow, and if air movement through them were entirely passive, the exchange of oxygen and carbon dioxide would occur too slowly to be of any use to the organism. Thus, with the evolution of lungs came the active means to ventilate them—namely, breathing mechanisms.

The Mechanics of Breathing

Breathing entails the periodic and forceful replacement of "used" air in the lungs with fresh air from outside the body. Frogs and toads use a force-pump method of breathing. They fill their mouth cavity and pharynx with fresh air, then close off the nostrils and force the air into the lungs by muscular contraction of the mouth–pharyngeal cavity. This mechanism of *pushing* air into the lungs is limited to the amphibians and lungfish; all reptiles, birds, and mammals breathe by *suction.* Let us examine the suction pump method of breathing by considering the human example.

Our lungs are suspended inside the **pleural cavity,** surrounded and protected by the rib cage. Below the lungs, separating the pleural and abdominal cavities, is a sheet of muscle tissue called the **diaphragm** (Figure 22–7). The inward rush of air, called **inhalation,** is caused by the concerted action of the rib muscles and the diaphragm. As the contraction of the rib muscles draws the rib cage up and outward, the diaphragm pulls downward (Figure 22–7a). These movements expand the pleural cavity, causing the air pressure inside to fall below atmospheric pressure. Consequently, the higher atmospheric pressure forces air into the lungs until the atmospheric and lung pressures equalize. You should note that the lungs are not attached to the inner walls of the pleural cavity, but rather, suspended like two balloons in an expandable jar (Figure 22–8). During inhalation, the lungs inflate because of a drop in pressure created by the expanding pleural cavity; the lungs themselves have no musculature.

Breathing out, or **exhalation,** is caused by the relaxation of the diaphragm and rib muscles. The ribs sink down and inward, and the diaphragm returns to its relaxed, upward position (Figure 22–7b). The resultant decrease in pleural cavity volume increases

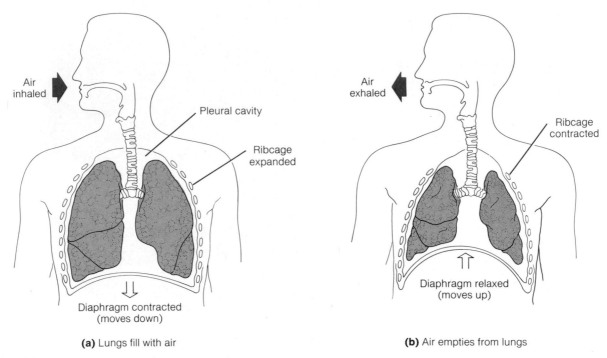

Air
inhaled

Pleural cavity

Ribcage
expanded

⇓⇓

Diaphragm contracted
(moves down)

(a) Lungs fill with air

Air
exhaled

Ribcage
contracted

⇑⇑

Diaphragm relaxed
(moves up)

(b) Air empties from lungs

FIGURE 22–7
Inhalation and Exhalation. Note how the position of the rib cage and diaphragm affect the volume
(and hence, pressure) of the pleural cavity. Lung inflation and deflation hinge on changes in pleural
cavity pressure.

FIGURE 22–8
**A Mechanical Model of the Suction
Pump Method of Breathing.**

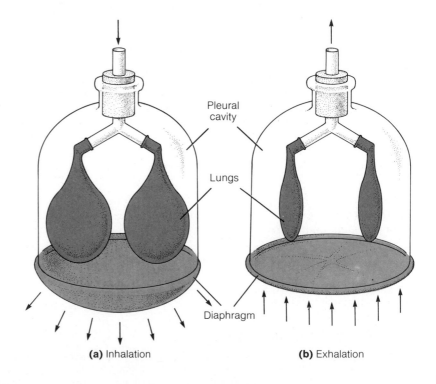

Pleural
cavity

Lungs

Diaphragm

(a) Inhalation

(b) Exhalation

ESSAY 22–1

HIGH ALTITUDE ADVENTURE

Imagine that you are in a jetliner cruising at 30,000 feet. Suddenly, something goes wrong and there is a loss of cabin pressure. Almost immediately, an oxygen mask drops from above and you'd be well advised to use it. At that altitude, the oxygen concentration of the outside air is less than a third of what it is at sea level. Were it not for the emergency oxygen, you and the other unlucky passengers would lose consciousness in a matter of minutes, lapse into comas and probably die before the plane could land.

When the oxygen level in the brain cells falls so low as to cause mental disturbances, a person is said to be suffering from hypoxia. Many of us have experienced a mild form of hypoxia when traveling from sea level to the mountains. Dizziness and nausea are the earliest signs of "mountain sickness." As the body begins to adjust to the low oxygen levels, the unpleasant effects of hypoxia gradually disappear, usually within a few days.

Many people live their entire lives in the high mountains with little or no threat to their health or well-being. High in the Andes, for example, Peruvian Indians live and work at almost 19,000 feet above sea level. On the other side of the world, but just as high up, are the Sherpas inhabiting the slopes of the Himalayas. Physiologists have noted that the typical native highlander has a larger and stronger heart, a greater lung capacity, and a more extensive system of blood vessels containing more red blood cells than his counterpart living at lower altitudes. Equipped with these permanent changes, people like the Sherpas are

better able to pick up oxygen from the air and deliver it to their respiring tissues. This has made them particularly well-suited for the numerous assaults made on Mt. Everest, including the first successful climb (recorded) to the 29,028 foot peak. Most people have heard of Sir Edmund Hillary who, in 1953, reached Everest's lofty summit. Few are aware, however, that beside him stood the Sherpa Tenzing Norgay.

As historic as the first ascent to the top of Everest was, the recent solo climb by Reinhold Messner is perhaps even more remarkable. It took the German-born climber two and a half days to complete the final "dash" to the top from High Base Camp at 21,325 feet. In spite of dehy-

dration, occasional hallucinations, and extreme fatigue, he managed the feat without using the bottled oxygen other climbers find absolutely essential.

How could such a climb be possible if, at nearly the same altitude, the passengers in our ill-fated jetliner blacked out in minutes? The answer lies in the body's ability to adjust to high altitudes, a process called acclimatization. Messner, like the Sherpas, had become acclimatized to the air of high altitudes after spending more than 30 years in the high Alps. His cardiovascular system was primed for this high adventure, but the "dash" to the top of Everest was not without its cost. Messner spent two months in the hospital recovering from his success.

the internal air pressure, forcing air out of the lungs through the mouth and nostrils.

The average adult human breathes in about half a liter of air with each normal breath, even though the lung capacity is about 4 liters (roughly the volume of a basketball). In other words, only about 12% of the air in the lungs is actually exchanged with each breath. (One deep breath, however, can sweep in up to 3 liters of air.) Furthermore, of each half-liter of fresh air taken in during a normal breath, approximately 30% fills dead space in the trachea and bronchi. This air will be the first to leave during the next exhalation, and its gas composition will be similar to that of fresh air. This is why air exhaled into the lungs of a person receiving mouth-to-mouth resuscitation contains enough oxygen for the technique to be potentially life-saving.

The Regulation of Breathing

In all birds and mammals, breathing is precisely regulated by the nervous system. Any increase in physical activity increases the cellular consumption of oxygen and the production of carbon dioxide; the nervous system responds automatically by stepping up the breathing rate, the volume of air inhaled with each breath, or both. How does the nervous system know the body cells are running out of oxygen? It seems reasonable that the concentration of oxygen in the blood would influence the brain's control of breathing, so that when exercise lowers the blood oxygen level, the breathing rate increases to compensate. In fact, however, laboratory experiments have verified that the carbon dioxide, not the oxygen, concentration is the chief regulatory factor. When animals are exposed to air containing an increasing concentration of carbon dioxide while the oxygen concentration is held constant, their breathing rate increases. Conversely, if the oxygen level is varied while the carbon dioxide concentration is kept constant, the breathing rate remains the same, even when the animal is not getting adequate amounts of oxygen. Swimmers and skin divers take advantage of this regulatory mechanism when they wish to remain submerged for extended periods. By taking a series of deep breaths (hyperventilating) just before jumping into the water, they greatly diminish the level of carbon dioxide in their lungs and blood, which delays for a time the urge to breathe. This practice can be dangerous, however. As a person swims under water, cellular respiration reduces the oxygen level in the blood, but it takes quite a while for the

carbon dioxide level to reach a point where the urge to breathe is extreme. As the brain cells are deprived of oxygen, the swimmer may lose consciousness without even being aware of the danger.

Breathing intensity is largely an involuntary response controlled by a respiratory center in the brain. Nerve signals sent to the various rib and diaphragm muscles regulate both inhalation and exhalation. The respiratory center responds primarily to the amount of carbon dioxide reaching its cells, but other factors may also exert secondary influences. For example, we know that the conscious part of the brain can exercise control over the respiratory center, for we can control our breathing voluntarily. The voluntary control is limited, however. No one has ever committed suicide by holding his or her breath, despite all the threats made by toddlers who didn't get their way. It simply cannot be done: involuntary control takes over when the person becomes unconscious. Involuntary control also ensures that breathing continues normally while we are asleep. However, the respiratory center is sensitive to nervous system depressants such as ether, morphine, barbiturates, and alcohol. Taking too much of a depressant or taking a dangerous combination of depressants can interfere with the normal function of the respiratory center, and breathing can stop forever.

SUMMARY

Relatively small and simple animals have sufficient surface area exposed to the external environment to exchange gases rapidly enough for their needs. Large, multicellular animals, however, have relatively few body cells in direct contact with the surrounding air or water. In these organisms, gas exchange (oxygen uptake and carbon dioxide release) is facilitated by specialized structures whose large surface area exposes the gas-carrying blood of the circulatory system to the environment. In fishes, these gas exchange structures are called gills, and extend outward from the body; most terrestrial animals rely on inward-extending systems of tubes (tracheae) or air sacs (lungs) to carry out gas exchange.

The human respiratory system consists of a pair of lungs and associated passageways which bring oxygen-rich air in contact with oxygen-poor blood. Inhaled air, warmed and humidified in the nasal cavity, passes through the pharynx and larynx to enter the trachea. The air then moves into the lungs through bronchi, which branch into smaller bronchioles. The trachea and lung passages are lined with ciliated, mucus-secreting epithelial cells that help keep the passages clear of foreign matter. Clusters of tiny alveoli at the ends of the bronchioles are embedded in a network of capillaries, and are the sites of actual gas exchange between organism and environment. Oxygen diffuses into the capillaries, and oxygen-enriched blood then returns to the heart to be pumped throughout the body. Carbon dioxide diffuses from the capillaries into the alveoli for exhalation.

Hemoglobin molecules in the red blood cells bind to oxygen molecules, increasing the oxygen-carrying capacity of the blood. When the oxygen concentration is high, as in alveoli containing freshly inhaled air, hemoglobin becomes nearly saturated with oxygen molecules. In the vicinity of the respiring body cells, where the oxygen concentration is low, the weakly bound oxygen molecules readily dissociate from hemoglobin and diffuse down their concentration gradient into the cells.

Higher animals have a muscle-aided breathing mechanism to facilitate the movement of air into and out of the lungs. In most vertebrates, including humans, simultaneous contractions of the rib muscles and diaphragm expand the pleural cavity. The resulting decrease in internal air pressure surrounding the lungs causes them to inflate with incoming air. Relaxing the rib muscles and the diaphragm leads to exhalation.

The rate and depth of breathing are controlled by the carbon dioxide concentration in the blood, which is detected by receptors in the brain's respiratory center. Nerves connect this center to the muscles that control breathing.

STUDY QUESTIONS

1. Explain why the evolution of large, multicellular animals must have been accompanied by the evolution of centralized respiratory organs.

2. Why do fast-swimming fish such as tuna open their mouths while swimming?

3. Trace the pathway of an oxygen molecule from the human nasal cavity to an alveolus in one of the lungs.

4. What is the function of the bands of cartilage in the linings of the human trachea and bronchi?

5. Which parts of a hemoglobin molecule bind oxygen molecules? What factor determines how many oxygen molecules bind to a given hemoglobin molecule?

6. Explain the role of the diaphragm in breathing.

7. Why is hyperventilation potentially dangerous to skin divers?

SUGGESTED READINGS

Avery, M. E., N-S. Wang, and H. W. Taeusch. "Lungs of the Newborn Infant." *Scientific American* 228(1973):75–85. Good description of the physiological changes that take place prior to birth, and how respiratory distress in premature infants can be reduced.

Comroe, J. H. "The Lung." *Scientific American* 214(1966): 57–68. Provides a detailed look at the relationship between structure and function in the lung.

Gordon, M. S., G. A. Bartholomew, A. D. Grinnell, C. B. Jorgensen, and F. N. White. *Animal Function: Principles and Adaptations.* New York: Macmillan, 1982. Chapter 5 of this advanced text covers respiratory mechanisms in a variety of animals.

Spence, A. P., and E. B. Mason. *Human Anatomy and Physiology.* 2nd ed. Menlo Park, CA: Benjamin/ Cummings, 1983. Chapter 21 provides in-depth coverage of the human respiratory system.

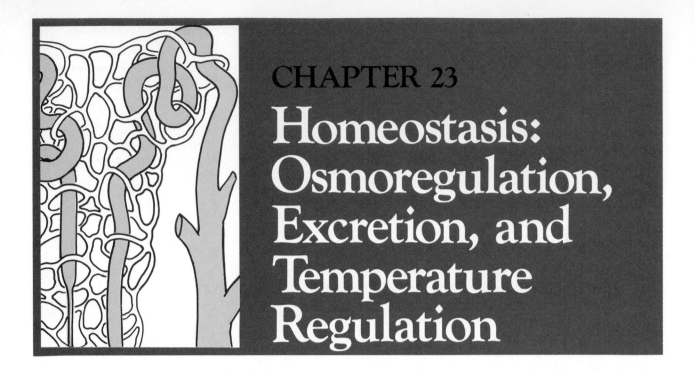

CHAPTER 23
Homeostasis: Osmoregulation, Excretion, and Temperature Regulation

During the first 3 billion years of life on earth, virtually all organisms lived in the oceans. The sea was a logical place for life to begin and first develop because of all the environments on earth, it is the most stable. The oceans are so vast that their physical properties, such as temperature and salt concentration, fluctuate very little. And because oceans do not dry up periodically, the availability of water, the substance so vital to life, is never a problem.

Many of the creatures inhabiting the seas are reflections of their external surroundings. Their body temperatures and internal solute concentrations are similar, if not the same, as the sea's. This is not to say that the body fluid of a clam is seawater. The solute *composition* of marine animals (which includes sugars, amino acids, proteins, and various inorganic salts) is different than that of seawater, even though the overall solute *concentrations* may be similar. These animals must expend energy not only to maintain this difference in solute composition, but also to keep their internal environments relatively constant. The maintenance of internal constancy, called **homeostasis** (literally, *staying the same*), encompasses a broad range of activities, including the regulation of temperature, acidity, solutes, and internal levels of nutrients and metabolic wastes.

The nonmarine environments present even greater challenges to biological homeostasis. Freshwater lakes and rivers are certainly more changeable than the oceans, particularly with respect to temperature and the concentration of salts. The terrestrial environment is even more changeable, and the scarcity of water adds to its challenge. Nevertheless, life forms obviously adapted to these less hospitable environments, and some of the mechanisms by which they and their marine counterparts achieve homeostasis will be the subject of this chapter. In particular, we will consider first how marine, freshwater, and terrestrial animals maintain a suitable balance of water and solutes, then take up the related topic of excretion. Finally, we will close with a discussion of how animals regulate their body temperatures.

OSMOREGULATION: ENVIRONMENTAL ADAPTATIONS

All organisms maintain a certain concentration of solutes in their cells. Animals above the flatworms also regulate the solute concentration of their extracellular body fluids. As you may recall from Chapter 4, **hypertonic** cells have a *greater* solute concentration than their surroundings, and thus tend to lose solutes by diffusion and gain water by osmosis (Figure 23–1, left). In the reverse situation, **hypotonic** cells with *lower*

FIGURE 23–1
Hypertonic and Hypotonic Cells. The hypertonic cell has a higher solute concentration than the external environment; it tends to lose solutes by diffusion and gain water by osmosis. In the opposite case, a hypotonic cell tends to lose water and gain solutes.

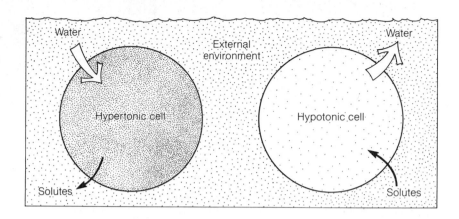

solute concentrations relative to their surroundings tend to gain solutes and lose water (Figure 23–1, right). Only the **isotonic** cell (*iso* = equal) is in osmotic equilibrium with its external environment. Regardless of the **osmotic status** (hyper-, hypo-, or isotonic) of the cell or organism, most creatures must perform work to maintain their particular status. The homeostatic process of maintaining a constant osmotic status is called **osmoregulation.** Let us now examine some of the osmoregulatory mechanisms employed by different animals to regulate the movement of water and salts into and out of their bodies.

Marine Invertebrates

Most of the ocean-dwelling invertebrates have no specialized osmoregulatory system; their cells and body fluids remain isotonic with their external environment. For example, if a starfish finds itself in a slightly more dilute seawater environment than usual (such as in brackish water), its body fluids will become more dilute to match the more dilute salt concentration of the new environment. That is, the starfish simply loses some internal salts and gains some water (Figure 23–2). Such an organism is called an **osmoconformer**

FIGURE 23–2
Osmoconformers and Osmoregulators among Marine Invertebrates. The starfish is an osmoconformer. In an environment more dilute than seawater, it loses salts and gains water until it becomes isotonic with its new environment (note increase in size due to water uptake). The shore crab is an osmoregulator. It maintains a nearly constant internal solute concentration regardless of the salt concentration of its external environment.

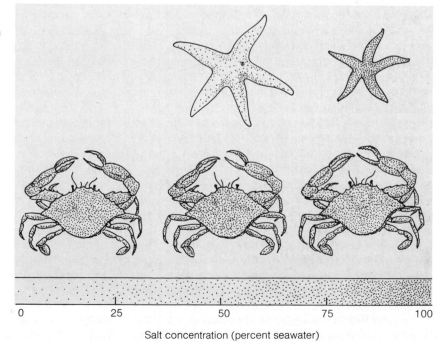

Salt concentration (percent seawater)

because its osmotic status conforms to the osmotic conditions of its liquid surroundings. There are limits to the capacity for osmoconformation, however. If placed in fresh or very dilute salt water, the starfish will soon die because its internal fluids will be fatally diluted by the rapid gain of water and loss of body salts.

Many marine invertebrates live in estuaries where rivers empty into the ocean. Here salt water and fresh water alternate as tides fill the river mouth and then recede. To survive in this constantly changing environment, some invertebrates, such as mussels and clams, simply close their shells against the fresh water as the tide ebbs. They remain inactive until the tide brings back the salt water. Others, such as the shore crab, are partial **osmoregulators.** Their body fluids are isotonic with seawater, but when the water becomes brackish (15–85% fresh water), their gills actively transport salts into the body, while their excretory system eliminates the excess water that has entered by osmosis.

Freshwater Animals

When certain ancestral marine animals made their evolutionary trek into freshwater ponds and rivers, presumably by way of estuaries, some major physiological changes were clearly necessary. The concentration of salts in fresh water is less than 2% of the concentration in seawater. In such a dilute environment, attempting to maintain an osmotic status equivalent to seawater would have overloaded their osmoregulatory systems. And switching to a cellular solute concentration equivalent to fresh water would have been even more unrealistic because the metabolic activities of cells cannot operate under such dilute conditions. The problem was solved with a compromise—most freshwater animals today maintain an internal solute concentration equivalent to approximately 30% seawater. Even so, they are still quite hypertonic to the fresh water around them. They constantly face the problem of losing salts and swelling up with water.

How do the freshwater fishes cope with the relentless influx of water? Part of the answer is water avoidance. This may sound strange for an animal that lives in water, but these fish (and their marine cousins) have scales covering most of their body that prevent passage of water and salts. Also, they rarely drink. (You may see a goldfish opening and closing its mouth, but it is merely circulating water past its gills, not swallowing it.) However, the water shield breaks down at the gills, highly permeable structures with large surface areas. To meet their oxygen demands, freshwater

fishes must expose their gills constantly to the water, and this leads to excess water uptake into the fish's tissues. To compensate for this water gain, freshwater fishes excrete large amounts of very dilute urine, passing the excess water out almost as fast as it comes in (Figure 23–3a). A 12-pound trout may excrete up to 4 pounds of water in a single day.

In addition to the constant influx of water, a freshwater fish must contend with the continuous loss of body salts, which leak into the urine and out from the gills. The loss of body salts is balanced to some degree by the intake of salts present in food. These fish also gain a large share of their salts by active transport, which occurs in specialized cells in the gills. Of course, the active uptake of salts and the excretion of hypotonic urine involves moving substances against concentration gradients. Therefore, osmoregulation in freshwater fishes has a substantial energy cost.

Marine Vertebrates

The marine bony fishes evolved from freshwater ancestors (see Chapter 19), and they retained an internal solute concentration equivalent to about 30–40% seawater. Consequently, they are faced with an osmotic problem just opposite to that of their freshwater cousins: they tend to lose water by osmosis and gain excess salts by diffusion. To gain water and maintain their hypotonic status, these fishes drink seawater continuously; excess salts are excreted from their gills (Figure 23–3b). Although they have kidneys, relatively little urine is produced. Like the excess salts, most of the nitrogenous wastes are excreted at the gills.

Marine cartilaginous fishes, such as sharks and rays, solve the water balance problem in a different way. They maintain an internal salt concentration roughly equivalent to that of the marine bony fishes, but their blood also contains a high concentration of the nitrogenous waste, urea. Apparently, the cartilaginous fishes can tolerate levels of urea that would poison other vertebrates. (Most shark meat must be soaked to rid it of urea before it can be eaten by humans.) By accumulating high levels of urea in the blood, the total solute concentration in these fishes is equal to or slightly greater than that of seawater. Thus, they have no problem with osmotic water loss. Nevertheless, because the cartilaginous fish maintains an internal salt concentration considerably less than that of seawater, salts leak into its body through the gills. Apparently, sharks and rays lack an active transport mechanism in their gills to extrude salt. Instead, they excrete excess salts

FIGURE 23-3
Osmoregulation in Bony Fishes. (a) A freshwater fish must contend with a constant influx of water, mainly into the gills. These fish osmoregulate by excreting copious amounts of dilute urine. The loss of salts both in the urine and by diffusion from the gills is compensated partially by food intake and mainly by active accumulation in the gills. (b) A marine bony fish drinks substantial amounts of seawater to counteract the relentless loss of water from its gills and urine into the salty environment. The excess salts it ingests are actively removed by the gills and kidneys.

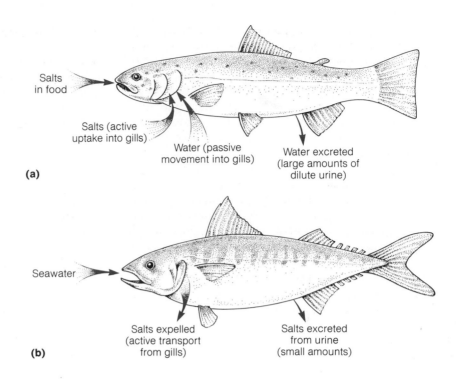

from special glands located near the rectum.

Marine mammals (porpoises, whales, seals, etc.) maintain a blood osmotic concentration similar to that of the bony fishes (about 30% seawater), so they too are hypotonic to their environment. Unlike the bony fishes, however, these air-breathing mammals have no gills, and hence, have less water- and salt-permeable surface area exposed. Most never drink; rather, they obtain water from the food they eat. Excess salts in the body are removed by the kidneys, which produce a strongly hypertonic (salty) urine.

Terrestrial Animals

When animals first ventured onto land, they gained the benefit of air for rapid gas exchange, but at the same time they took on the danger of dehydration. This danger has always existed for land animals: the major environmental factor influencing the distribution and density of terrestrial organisms today is water availability. Not only are animals' bodies exposed to air, but the gas exchange surfaces lose considerable water by evaporation (Figure 23-4). Terrestrial creatures must

FIGURE 23-4
Water Budget of a Pig. This diagram illustrates the pathways by which water enters and leaves a typical terrestrial mammal. Most of the water is gained by eating and drinking, but a significant amount is also produced metabolically in cellular respiration. The water loss pathways include evaporation from the general body surface (25% of the total water lost) and from the lungs (15%), elimination from the digestive tract (fecal water, 5%), and excretion from the kidneys (urine, 55%).

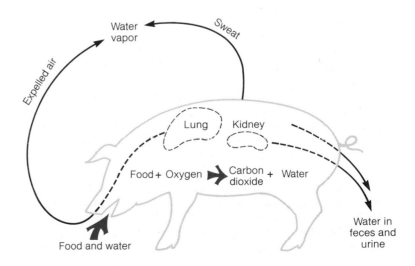

also sacrifice some of their precious water for excretion and fecal elimination. Thus, animals are continually replacing lost water by drinking and eating food containing water.

Animals also generate water within their body cells as a by-product of cellular respiration (organic nutrients + oxygen → carbon dioxide + water). For most animals, metabolism is only a minor source of water; but for some adapted to desert habitats, such as the kangaroo rat, it is the primary source (Figure 23–5).

Humans are quite inefficient water users. A typical adult takes in about 2300 ml (about 2.5 quarts) of water each day in his or her food and drink, and gains another 300 ml from cellular respiration. Of the 2600 ml of daily water gain, more than half is lost by urination, and the rest escapes by evaporation and fecal elimination. Of course, the amount of internal water lost to evaporation varies with temperature, humidity, and physical activity. On a warm day, a marathon runner may lose up to 8000 ml of water during a race by evaporation and sweating. As water uptake and loss change with environmental conditions and physiological activity, humans, like most terrestrial animals, regulate the amount of water in their bodies through urine excretion. A high internal water content can be adjusted downward by frequent excretions of dilute urine (a homeostatic mechanism familiar to beer drinkers). When it becomes necessary to conserve water, we produce a smaller volume of concentrated urine. Salt balance is handled in a similar manner. After an especially salty meal, such as a ham dinner, the kidneys compensate by producing urine with a high salt content.

EXCRETION

Whereas osmoregulation refers to the processes by which organisms maintain an appropriate *balance* of

FIGURE 23–5
The Kangaroo Rat, a Water Conserver. The kangaroo rat of the southwestern deserts of the United States is a well-adapted water miser. It avoids the hot, dry days by remaining inactive until nightfall. It also excretes relatively small amounts of a very concentrated urine and eliminates very dry feces. In fact, the kangaroo rat is so efficient at retaining water that it may never take a drink. It produces most of its own water from cellular respiration.

TABLE 23–1
Embryo Habitats and Nitrogenous Wastes.

Animal Group	Adult Habitat	Embryo Habitat	Major Nitrogenous Waste Product
Aquatic invertebrates Bony fishes Larval amphibians	Aquatic	Aquatic	Ammonia
Cartilaginous fishes Adult amphibians Mammals	Aquatic Semiaquatic Terrestrial	Aquatic	Urea
Insects Reptiles Birds	Terrestrial	Egg	Uric acid

water and salts in their body fluids, **excretion** encompasses those processes by which substances (including water and salts) originally present in cells, tissue fluids, or blood are *removed* from the body. Excretion differs from fecal elimination in that feces contain material that never entered the body cells.

When animal cells metabolize amino acids and nucleotides, they generate **ammonia** (NH_3) as a nitrogenous waste product. Ammonia is highly toxic to cells and must be removed quickly to avoid cell poisoning. Fortunately, it is also quite soluble in water and will diffuse rapidly from an area of high concentration to one of low concentration. Aquatic marine invertebrates, with their large body surface areas, take advantage of its solubility and release ammonia into the surrounding seawater by simple diffusion. Freshwater fishes produce copious amounts of a dilute urine containing ammonia, and the marine bony fishes release ammonia both from their gills, and to a lesser extent, in the small amount of urine they excrete. Terrestrial animals, however, are not surrounded by a large body of water to sweep away their nitrogenous wastes. Moreover, excreting large amounts of a dilute urine is a luxury they can ill afford in an environment where water is at a premium. Therefore, terrestrial animals have solved their nitrogenous waste problem by converting ammonia into less toxic nitrogenous compounds, **urea** and **uric acid.** These products can be collected, stored, and excreted in more concentrated amounts in the urine, thereby minimizing the loss of body water.

Among the land-dwelling animals, mammals excrete urea; insects, birds, and reptiles excrete uric acid, the semi-solid paste that adorns many statues and park benches. The reason for this difference is probably related to the environments in which these animals' embryos develop. The mammalian embryo grows within the liquid environment of the uterus, where waste products are removed as dissolved substances in the maternal bloodstream. Urea, which is very water-soluble, is thus a fitting excretory product for mammals (Table 23–1). On the other hand, insect, reptilian, and bird embryos develop within eggs whose shells are permeable only to oxygen and carbon dioxide. Although the shell keeps water and nutrients from leaking out of the egg, it is also a barrier to the removal of nitrogenous wastes. Consequently, if urea were produced by the developing embryo, it would accumulate inside the egg and eventually poison the embryo. Uric acid, on the other hand, has a low solubility in water. It can be deposited as a solid paste inside the egg in a location removed from the embryo.

In order for urea or uric acid to be excreted, it must first be collected from the blood, concentrated, and then expelled from the body. In vertebrates, the collection and concentration functions are carried out by the kidneys. We will focus our attention on the human kidney.

The Human Kidney

The human **kidneys** are two bean-shaped organs that lie behind the stomach and liver (Figure 23–6). In an adult male weighing 160 pounds, the kidneys process about 190 liters (50 gallons) of blood each day, filtering out urea and maintaining an appropriate osmotic status for the body. If the kidneys fail and these vital functions are not carried out, death will occur quickly.

FIGURE 23–6

The Human Excretory System. Blood enters the kidneys through the renal arteries, which branch from the aorta. After chemical processing takes place in the kidneys, the filtered blood returns to the inferior vena cava by way of the renal veins. The urine produced within the kidneys is piped through the ureters to the urinary bladder for temporary storage, and is ultimately released from the body via the urethra.

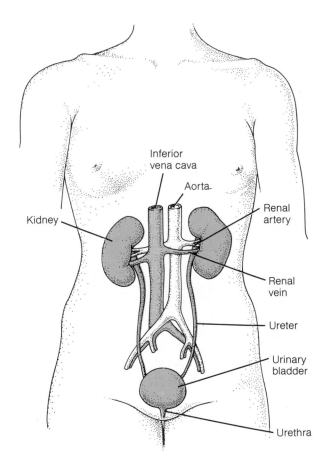

When disease or injury interferes with normal kidney function, dialysis therapy or a kidney transplant must be performed (Essay 23–1).

Blood enters the kidneys through the **renal arteries,** major vessels which branch from the aorta (see Figure 23–6). These arteries divert about 20% of the aortic blood into the kidneys with each beat of the heart. The filtered blood leaves the kidneys by way of the **renal veins** and empties into the inferior vena cava for return to the heart. While the blood is in the kidneys, urea and other substances are removed from the blood, a process we will return to in a moment. The extracted substances are collected in the central region of the kidney as a solution called **urine.** The urine is then transported to the **urinary bladder** through a long tube called the **ureter.** The bladder stores the urine until it is discharged from the body through the **urethra.**

THE NEPHRON.   The actual filtering unit of the kidney is the **nephron,** of which there are about a million buried inside each kidney (Figure 23–7). Each nephron consists of a long, twisted tubule that runs a U-shaped course through the midregion (**medulla**) of the kidney. At one end of the tubule is a hollow, cup-shaped structure called **Bowman's capsule.** This capsule surrounds a knot of capillaries—the **glomerulus**—that originates from a branch of the renal artery (see Figure 23–7). This is where the *filtration* of blood takes place. Blood enters the glomerulus under pressure, which forces some of the blood plasma through tiny pores in the capillary walls into Bowman's capsule. Included in this blood filtrate are water and small dissolved substances, such as urea, salts (mainly sodium chloride), glucose, and amino acids; plasma proteins and blood cells are too large to pass through the pores. The filtrate then passes through the **proximal convoluted**

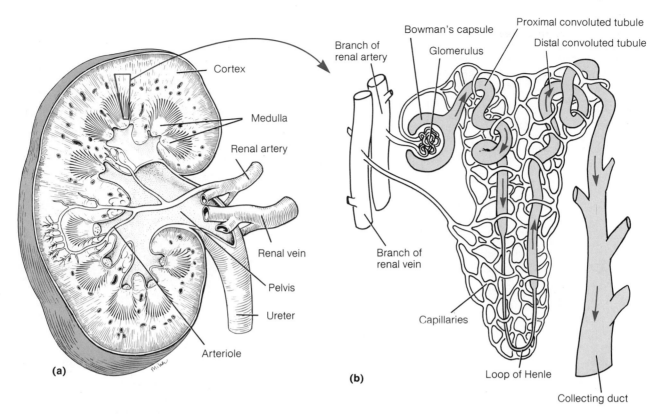

FIGURE 23–7
The Human Kidney and Nephron. (a) This drawing shows the three main regions of a human kidney:
the outer cortex, the central medulla, and the pelvis. The nephrons are located in the cortex and
medulla, which are amply supplied with blood vessels. The urine collects in the pelvis for drainage
out of the kidney via the ureter. (b) Filtration of fluid from the blood takes place in the million or
so nephrons found in each kidney. Blood is directed toward each nephron through an arteriole,
passes through the glomerulus, then circulates through a larger capillary network downstream before
collecting in a venule. The blood filtrate that is squeezed out of the glomerulus enters Bowman's
capsule of the nephron, passes through the proximal convoluted tubule, loop of Henle, and distal
convoluted tubule before entering a collecting duct that is shared by other nearby nephrons. The
collecting duct drains urine into the pelvis of the kidney.

tubule into the **loop of Henle,** then the **distal convoluted tubule,** and finally into a **collecting duct** (see Figure 23–7). Along the way, the nutrients and most of the water and salts are returned to the blood, but urea remains within the tubule. In this most important phase of nephron action, called **selective reabsorption,** the blood filtrate is transformed into urine, and crucial nutrients, water, and salts are conserved.

FORMATION OF URINE.   As the filtrate makes its way through the proximal convoluted tubule of the nephron, glucose, amino acids, and sodium chloride (NaCl) are actively reabsorbed from the filtrate into the surrounding blood vessels (Figure 23–8a). As a result, the blood becomes hypertonic to the filtrate, causing the passive movement of some water out of the tubule.

More water leaves the filtrate (now called urine) from the descending arm of the loop of Henle, because this section of the nephron is embedded in the medulla, which has a high concentration of sodium chloride (Figure 23–8b). This salt in the medulla tissue is generated by the active transport of NaCl out of the ascending limb of the loop, which also has the effect of diluting the urine as it moves up toward the distal convoluted tubule. At this stage, the urine is quite dilute, not because of water reentry (the ascending portion of

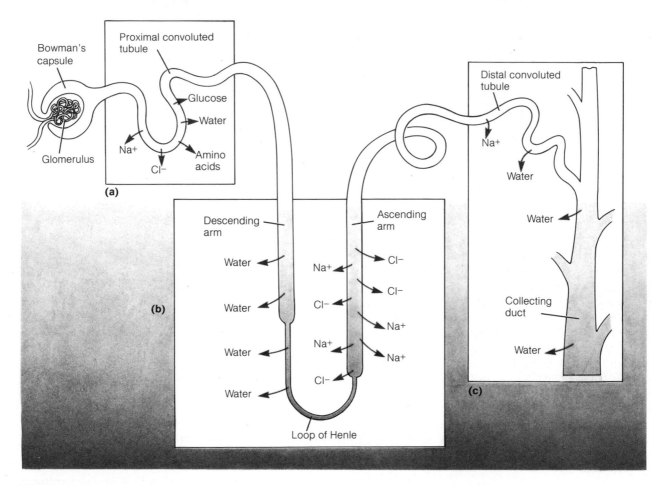

FIGURE 23–8
Processes Occurring in the Nephron. (a) In the proximal convoluted tubule, glucose, amino acids,
sodium ions, and chloride ions are actively transported out of the filtrate, and water follows by
osmosis. (b) The loop of Henle is surrounded by the very salty tissue of the medulla (shaded dark),
a condition created by the active removal of sodium and chloride ions from the ascending arm of
the loop. The salt gradient causes water to move out of the loop and descending arm; the ascending
arm is impermeable to water. (c) Final processing of the urine occurs in the distal convoluted tubule
and collecting duct. If the body is low in Na+, this ion is actively reabsorbed from the distal convoluted tubule; water follows osmotically. The collecting duct passes through the salty environment of
the medulla, causing more water to move out and yielding a concentrated urine. (Note: the nephron
is grossly distorted in this diagram.)

the loop of Henle is impermeable to water), but due to
the loss of salt.

Final processing of the urine occurs in the distal
convoluted tubule and the collecting duct (Figure
23–8c). The concentration of sodium ions in the body
is adjusted at the distal convoluted tubule. If the body
is low in sodium ions (Na+), the adrenal glands (located on top of the kidneys) increase their production
of **aldosterone.** This hormone increases removal of
Na+ from the distal convoluted tubule. High levels
of Na+ in the blood trigger reduced levels of aldo-

sterone, and hence, less reabsorption of Na+ (greater
Na+ excretion).

The collecting duct is permeable to water, and
since it passes through the very salty tissue of the medulla, more water flows out, leaving a concentrated
urine. Each collecting duct receives urine from several
nephron tubules and directs it to the central cavity
(**pelvis**) of the kidney (see Figure 23–7a). Here the
urine empties into the ureter for passage to the bladder.

Overall, selective reabsorption results in the conservation of water, nutrients, and salts, and the elimina-

TABLE 23-2
Concentrations of Various Substances in the Blood Plasma, Glomerular Filtrate, and Urine.

Substance	Concentration (grams/liter)		
	Blood Plasma	Glomerular Filtrate	Urine
Glucose	1.0	1.0	None
Amino acids	0.3	0.3	None
Protein	70.0	0.2	None
Urea	0.3	0.3	20.0
Sodium ions	3.0	3.0	3.0
Chloride ions	4.0	4.0	6.0

tion of urea. Virtually all of the urea that enters Bowman's capsule ends up inside the collecting ducts for excretion, but 99% of the water and salts and 100% of the nutrients are reabsorbed into the blood. Thus, although the glomerulus is relatively nondiscriminatory in its filtration of the blood (it excludes only blood cells and plasma proteins), selective reabsorption occurring along the length of the nephron creates a urine that is quite different in chemical composition than either the blood plasma or glomerular filtrate (Table 23–2).

WATER BALANCE. Of the 190 liters of glomerular filtrate that our 160-pound adult male processes each day, only about 1%, or about 2 liters, is excreted as urine. This amount will vary somewhat depending on the water status of the body. Therefore it is appropriate to ask how the kidneys "know" when to remove excess water and when to conserve water if the body's water content is low.

The excretion of excess water or its conservation by the kidneys is controlled indirectly by the hypothalamus region of the brain (see Chapter 24). A below-normal water content in the body (a 1% water deficit is enough) is sensed by specialized nerve cells in the hypothalamus called **osmoreceptors.** The osmoreceptors stimulate other nerve cells nearby to release small quantities of the hormone **vasopressin** (also known as antidiuretic hormone) into the blood. Vasopressin is transported through the bloodstream to the kidneys, where it increases the water permeability of both the distal convoluted tubule and the collecting duct. Thus, more water can be reabsorbed into the blood, leaving behind a more concentrated urine. Producing less urine enables the body to conserve water. When the water level in the body is back to normal, vasopressin release is stopped, and the kid-

neys excrete a more dilute urine. Incidentally, alcohol depresses the release of vasopressin, even when the body's water level is low. A high blood alcohol content, therefore, can lead to excessive excretion of water. Generally this does nothing more serious than contribute to a hangover, but in more severe cases it can cause fatal dehydration.

The sensation of thirst is also regulated by the hypothalamus. When the body's water content drops by 1%, the hypothalamus sends out impulses to the salivary glands, causing a reduction in saliva production. The mouth and throat consequently become drier, a condition that can be relieved only by taking in water.

TEMPERATURE REGULATION

Every organism must live within a rather narrow temperature range, primarily because enzymes, the catalysts of cellular metabolism, are quite sensitive to temperature (see Figure 5–7). However, the earth's surface offers a far greater range in temperatures, both geographically and seasonally, than what is optimal for life. Yet we find creatures on the polar ice caps and in the deserts, and they survive the frigid winters and blistering summers. Faced with the wide fluctuations in external temperature, it is not surprising that animals have evolved ways to keep their internal temperature within the narrow range suitable for life's processes.

Ectotherms

All organisms generate heat as a by-product of metabolism. However, most creatures lose this metabolic

ESSAY 23–1

DEFERRING DEATH: DIALYSIS

Kidney failure, the end result of many different diseases and injuries, is an automatic death sentence. Cause of death—the unrelenting build-up of compounds that are toxic to the body's cells. The only way death can be averted is the frequent removal of these toxic substances by **dialysis** therapy and/or the replacement of a dysfunctioning kidney with a healthy one.

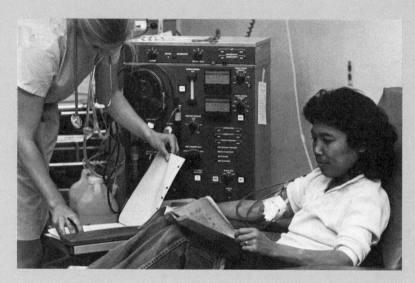

Dialysis is based on the process of diffusion: small molecules pass through a semipermeable membrane down a concentration gradient. In dialysis therapy, the solution on one side of the porous membrane is the patient's blood, which contains urea and other toxic substances normally filtered out by the kidneys. On the other side of the membrane is the **dialysate**, a salt solution isotonic with blood. Substances in the blood that are not present in the dialysate will pass through the membrane into the dialysate. Conversely, if the dialysate contains a higher concentration of a substance than the blood, then it will diffuse across the membrane into the blood.

There are two forms of dialysis: peritoneal dialysis and hemodialysis. The major difference between these techniques is the type of filtering membrane used. Peritoneal dialysis uses the patient's own periosteum (the membrane lining the abdominal cavity). Hemodialysis uses an artificial membrane.

In peritoneal dialysis, the dialysate is delivered through a catheter into the abdominal cavity. Time is allowed for the toxic substances to be exchanged and the dialysate, now containing the waste products, is removed. Traditional peritoneal dialysis is time-consuming and inconvenient for the patient, requiring the patient to be immobile 10–12 hours per treatment three to five times a week. But recently, an alternative has been developed: Continuous Ambulatory Peritoneal Dialysis (CAPD). In this technique, a plastic bag containing dialysate is connected to a catheter implanted in the patient's abdominal cavity. The bag is raised to shoulder level, allowing the dialysate solution to drain down into the abdominal cavity. The empty bag is then rolled up and hidden in the patient's clothing. For the next 6 hours, the patient is free to move about as he pleases, reassured by the knowledge that inside his body, toxic substances are passing from the blood into the dialysate. Then the patient unrolls the bag and places it below his abdomen so that the waste-ridden dialysate can drain back into the bag, which is then discarded. A new bag filled with fresh dialysate is then attached to the catheter, and the patient is ready to continue his activities.

In hemodialysis, a patient is connected to a machine by means of small tubes inserted in the forearm,

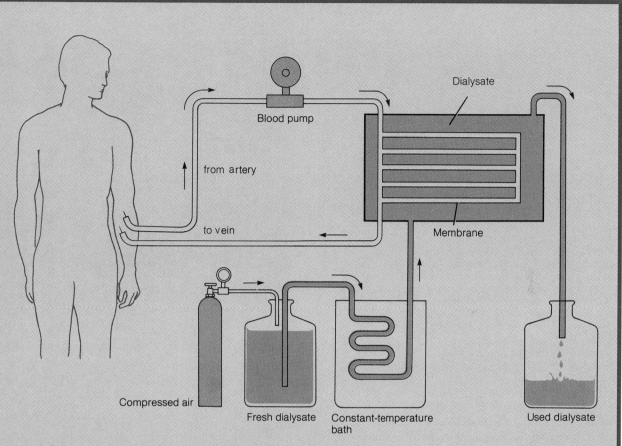

Blood pump

Dialysate

from artery

to vein

Membrane

Compressed air

Fresh dialysate

Constant-temperature bath

Used dialysate

Hemodialysis.

one tube in an artery and one in a vein (see figure). The patient's blood is then pumped from the artery into a long, porous tube made of synthetic membrane. The tube coils through a bath filled with dialysate, where the toxic substances in the blood diffuse through the membrane into the bath. The cleansed blood is then returned to the patient's body through the vein. The bath solution is replaced every two hours to avoid the return of any toxic substances to the blood.

Treatment schedules of those undergoing hemodialysis vary, de-pending on medical need and personal preferences. The patient may be dialyzed three days a week for 6 to 10 hours per treatment, or the patient may be dialyzed daily for 2 hours, which allows for a more normal daily routine.

Although hemodialysis is usually performed in hospitals or clinics, self-hemodialysis in the patient's home is becoming more popular. It is also much less expensive. Other modes of self-hemodialysis that allow the patient more control and independence include the wearable artificial kidney and the "kidney in a suitcase," small portable machines that the patient can take almost anywhere.

Although many individuals who suffer irreversible kidney failure have had their lives prolonged by dialysis therapy, spending many hours connected to a machine takes its psychological toll. With the continuing efforts to develop more convenient, less immobilizing dialysis procedures, these patients can look forward to major improvements in the quality of their lives.

(a)

FIGURE 23–9
Ectotherms. (a) Most amphibians regulate their body temperatures by moving in and out of water. On warm days water provides relief from the heat; on cool days it is a source of heat. (b) Terrestrial reptiles also move around to regulate body temperature. Basking in the sun raises their temperature, and they move to a shady spot to cool off.

(b)

heat so rapidly that its contribution to body temperature is insignificant compared to the heat they acquire from their surroundings. Such organisms are said to be **ectothermic**—they obtain most of their body heat from the outside. All invertebrates and most fishes, amphibians, and reptiles are ectothermic. Let's look at some of the ways ectothermic vertebrates regulate their body temperatures.

Apart from a few fast-swimming species such as tuna, which generate considerable amounts of heat internally, the body temperatures of fishes are the same or very similar to that of the water around them. They can withstand small, gradual changes in water temperature, but a rapid change of even a few degrees can be fatal—as anyone who has left a goldfish bowl exposed to the summer sun may have witnessed. When water temperatures do change, such as the seasonal fluctuations that occur in oceans of temperate regions, many of the marine fishes migrate, seeking warmer waters in the winter and cooler waters in the summer.

Water can gain or lose relatively large amounts of heat without changing much in temperature (see Essay 4–1). In contrast, air temperatures can vary widely from day to night, season to season, and place to place. In addition, air is much more transparent to solar radiation than water, which screens out most of the sun's rays. Thus, although terrestrial animals are faced with greater variability in outside temperatures, they can take better advantage of the sun's warmth.

For the typical amphibian, the key to regulating body temperature is its semiaquatic lifestyle. In a recent study, the daytime body temperatures of the southern California bullfrog were found to range between 26° and 33°C. On warm, sunny days, the bullfrog raised its temperature by basking in the sun. As its body temperature neared 33°C, the frog typically took a quick dip in a nearby pond to cool off. On chilly, overcast days, the pond water became the heat source. Selecting appropriate surroundings is the major way bullfrogs and most other amphibians adjust their body temperatures (Figure 23–9).

Except for the aquatic turtles, crocodiles, and alligators, reptiles must face the more severe temperature fluctuations associated with a completely terrestrial lifestyle. Nevertheless, dry-land reptiles such as lizards and snakes regulate their body temperatures in much

(a)

(b)

FIGURE 23–10
Endotherms. (a) The blubber of this elephant seal helps her conserve the heat generated by metabolism. (b) This dog cools its body by panting.

the same way as the bullfrog: they move around. By shuttling back and forth between sunny and shady spots, and orienting their body position relative to the sun, they can keep their body temperatures within an optimal range (Figure 23–9b).

Endotherms

In many ways, terrestrial ectotherms are at a disadvantage. Because they depend on the environment to attain a suitable body temperature, amphibians and reptiles are limited as to where they can live and when they can be active. Birds and mammals are not restricted in this way because they are **endothermic.** Their primary source of body heat is metabolism. Birds and mammals are also **homeothermic**—they maintain a constant body temperature (usually about 37°C) no matter what the outside temperature is.

To be a successful endotherm, an animal must be sufficiently large. As you may recall from Chapter 3, the ratio of surface area to volume of an object decreases as the size of the object increases (see Figure 3–5). Thus, a large animal has a relatively small, heat-dissipating surface area in relation to its heat-generating volume, as required for an endothermic way of life. The smallest endotherms are hummingbirds and shrews, animals that must work hard to stay warm.

For an animal smaller than a shrew to maintain a constant, warm body temperature, it would have to eat around the clock to "fuel the internal furnace."

To generate sufficient heat, endotherms characteristically have high metabolic rates. But heat generation is only part of the story. In order to maintain a constant body temperature in a changing climate, the homeotherm must be capable of conserving or dissipating body heat as conditions warrant. The feathers of birds and the fur, hair, or blubber of mammals are good insulators against excessive heat loss (Figure 23–10a). These animals also have an internal thermostat—the hypothalamus—that senses very small changes in body temperature and initiates behavioral or physiological mechanisms to adjust it. When exposed to cool external temperatures, for example, many animals curl up in a ball, resulting in less surface exposed to the cold air. Other responses include the involuntary constriction of blood vessels near the skin to reduce heat loss, and shivering to generate more heat from muscle tissues.

When the mercury in the thermometer climbs, many homeotherms seek shady retreats, cut down on their heat-producing physical activities, and expose as much of their body surface as possible to the air or cooler ground. Many also dissipate excess body heat through evaporation of water; panting in dogs and perspiring in horses are two examples (Figure 23–10b).

FIGURE 23–11
A Hibernating Chipmunk.

For every gram of liquid water turned to vapor, more than 500 calories of heat escape.

A few birds and mammals have developed some rather elaborate behavioral strategies to cope with extreme climates. For example, female Emperor penguins of the Antarctic lay their eggs on an ice floe in the dead of winter (as if it's not cold enough in summer), then leave them for the males to incubate. If left alone, the body temperature of the male penguin would fall too low to keep the egg sufficiently warm, so the brooding males huddle together in close formation to conserve body heat. Other animals, such as some rodents and certain bats, evade the winter chill by hibernating (Figure 23–11). **Hibernation** involves resetting the hypothalamic thermostat to a temperature far below normal; the rates of breathing, heartbeat, and cellular metabolism are also severely depressed. At this very low level of activity, the animal can "sleep" through winter, using its energy reserves very slowly.

SUMMARY

Homeostasis encompasses all the processes by which an organism maintains a constant internal environment in the face of a different and/or changeable external environment. Osmoregulation is one type of homeostatic process utilized by animals to regulate their water and salt balance. Most marine invertebrates are osmoconformers—their internal solute concentration changes with the osmotic status of their aquatic environment. All other animals are osmoregulators that actively maintain a constant osmotic value.

Freshwater fishes are hypertonic to their surroundings. They actively absorb salts (accomplished by the gills) and excrete copious amounts of dilute urine to compensate for the relentless outward leakage of salts and osmotic gain in water. Conversely, hypotonic marine fishes must drink seawater and excrete excess salts. Terrestrial vertebrates regulate their internal osmotic status through urine production.

Ammonia is a toxic breakdown product of cellular metabolism that must be removed from the body. In most aquatic animals, ammonia diffuses across the body wall (most invertebrates) or from the gills and urine (most vertebrates) into the surrounding water. Terrestrial animals convert ammonia to less toxic nitrogenous wastes, urea or uric acid. The egg-laying insects, reptiles, and birds excrete the solid uric acid; mammals expel the water-soluble urea in urine.

Osmoregulation and excretion in humans is handled by the kidneys, each of which contains about a million filtering units called nephrons. The nephron receives blood fluid containing nutrients, salts, and urea. As this fluid moves through the long, twisted tubule of the nephron, all of the nutrients and about 99% of the salts and water are reabsorbed into the blood. Urea is retained in the urine. The urine formed in each nephron empties into a collecting duct for passage via the ureter to the urinary bladder.

Sodium ion concentration and water level of the body are regulated by the actions of the hormones aldosterone and vasopressin, respectively, on the nephron.

Temperature regulation, a key feature of animals, is handled in different ways by the different groups of animals. Most animals are ectothermic, deriving most of their body heat from the external environment. Their body temperatures tend to conform to the outside temperature, but most employ various behavioral tricks to warm up or cool off.

Birds and mammals are endothermic—they get most of their body heat from internal metabolism. They are also homeothermic, since they maintain a constant body temperature.

STUDY QUESTIONS

1. If an organism is hypotonic to its surroundings, does water tend to enter or leave its tissues? What about solutes?

2. The starfish is an osmoconformer. Explain what this means in terms of solutes and water flow when the starfish is transferred to a slightly hypertonic medium.

3. How do marine bony fishes deal with the constant tendency of their tissues to lose water and gain salts? How do freshwater fishes deal with the reverse problem?

4. Mammals typically excrete their nitrogenous wastes in the form of urea, whereas birds excrete uric acid. How does this relate to the environments of the embryos of these two groups?

5. What types of substances pass from the blood into Bowman's capsule in a nephron? Which blood substances are generally excluded?

6. Explain how the filtration and selective reabsorption functions of the nephron determine which substances appear in the urine.

7. Explain the role of vasopressin in regulating water balance in a vertebrate.

8. Differentiate between ectothermic and endothermic temperature regulation mechanisms. What is the difference between endothermy and homeothermy?

9. Describe some of the ways homeotherms maintain a constant body temperature when exposed to cold external temperatures.

SUGGESTED READINGS

Dantzler, W. H. "Renal Adaptations of Desert Vertebrates." *Bioscience* 32(1982):108–113. This readable article discusses the wide variety of water-conserving strategies utilized by desert vertebrates.

Degabriele, R. "The Physiology of the Koala." *Scientific American* 243(1980):110–117. About half of this article deals with how the koala achieves water balance.

Gilles, R., ed. *Animals and Environmental Fitness.* New York: Springer-Verlag, 1978. Several chapters in this book deal with osmoregulation.

Smith, H. W. *From Fish to Philosopher.* New York: Doubleday, 1961. A lively treatment of vertebrate evolution as viewed from the perspective of excretory mechanisms and osmoregulation.

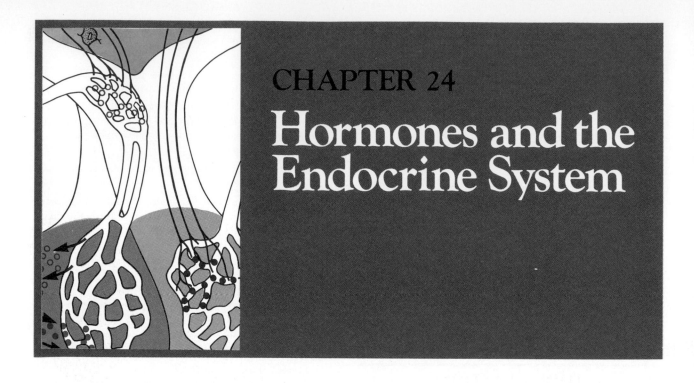

CHAPTER 24
Hormones and the Endocrine System

As organisms evolved into increasingly complex forms, their various internal activities became more and more specialized and divided among different regions of the multicellular body. Functional specialization would have served no advantage, however, had the organisms not also developed the means to harmonize the various activities into an integrated whole. Thus, internal communication systems evolved which integrate and coordinate the specialized activities.

Whereas multicellular plants coordinate their internal activities and process external information exclusively by chemical messenger systems, animals above the sponges supplement chemical controls to varying extents with a nervous system. The structure and function of the nervous system will be covered in Chapters 26 and 27. For now, let us look into the ways animals regulate their activities and achieve homeostasis through chemical control systems.

HORMONES

Hormones are chemical messengers that are produced in minute quantities at specific sites in the body, then transported to other sites where they cause some physiological change. In animals, hormones are transported in the blood and thus come in contact with virtually every cell in the body. However, they usually affect only specific **target cells** that recognize and respond to a particular hormone. Target cells have specific receptors that bind to the hormone, forming a **hormone-receptor complex.** This active complex then sets in motion a series of biochemical events in the target cell that culminates in some physiological action.

Insect Hormones

Among the invertebrates, hormones have been found in mollusks, annelids, arthropods, and echinoderms. Although our knowledge is still fragmentary, the hormones of insects have been studied more extensively than those of any other invertebrate group. We will limit our coverage of chemical messenger systems in invertebrates to an example taken from the insects.

Many animal hormones are involved in the regulation of growth and development. Two of the insect hormones, ecdysone and juvenile hormone, play such a role. As we noted in Chapter 18, the rigid exoskeleton

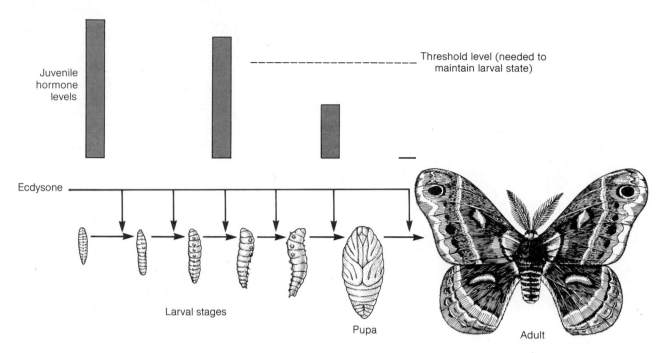

FIGURE 24–1
Hormonal Regulation of Development in an Insect. Moltings of the silkworm larva are triggered by
bursts in ecdysone level. Each molt produces another larval stage as long as juvenile hormone
is above a threshold level. The final larval molt yields a pupa when the juvenile hormone level drops
below threshold, and the pupa develops into the adult moth when juvenile hormone is absent.

of insects must be shed periodically (a process called molting) to accommodate body growth. This molting is apparently triggered by **ecdysone,** which is produced by a small gland located in the head region. Experiments have shown that surgical removal of this gland prevents molting in insect larvae because they are unable to produce ecdysone. When the glandless larvae are treated with synthetic ecdysone, they undergo a normal molt.

As its name suggests, **juvenile hormone** acts to maintain the juvenile (larval) stage of insect development. When high levels of juvenile hormone are present in the larva, each successive molt (triggered by a surge in ecdysone) yields another larval stage (Figure 24–1). This pattern continues until the level of juvenile hormone drops below some critical concentration. When it does, the larva molts into a pupa. Final metamorphosis of the pupa into an adult can only proceed in the absence of juvenile hormone.

With knowledge of how insect hormones regulate development, and the ability to synthesize these hormones in the laboratory, pest control scientists are looking toward hormone applications as a way to keep insect pests in check. For example, many insects are serious crop pests as adults, but their larvae are not particularly destructive. It is theoretically possible to keep these insects in their larval forms by periodically spraying the fields with synthetic juvenile hormone, at least until the crop is harvested. There are serious drawbacks to the indiscriminate use of "hormonal warfare," however. One major problem is that the insect hormones are not species specific. The life cycles of nearly all insect species would be interrupted by juvenile hormone spraying, including those beneficial insects that prey on the pests or that pollinate the crops one is trying to protect.

Vertebrate Hormones

Vertebrates produce many different hormones, and the number seems to grow every year with new discoveries. These hormones fall into three general classes based on their chemical structure: (1) steroid

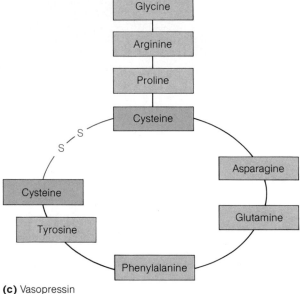

(a) Testosterone

(b) Thyroxine

(c) Vasopressin

FIGURE 24–2
Steroid, Amine, and Peptide Hormones.
(a) Testosterone, one of the male sex hormones, is a steroid hormone; note the four
rings characteristic of steroids. Because
they are nonpolar lipids, the steroid hormones are insoluble in water. (b) Thyroxine
is an amine hormone produced by the
thyroid gland. Note the presence of four
iodine (I) atoms, which is why iodine is an
essential mineral for humans (we get it
primarily from iodized salt). (c) Representing the peptide class of hormones, vasopressin consists of nine amino acids, six
of which are held in a loop by the disulfide
bridge (-S-S-) between the cysteine units.
Vasopressin stimulates water reabsorption
from urine forming in the kidneys.

hormones, (2) amine hormones, and (3) peptide hormones (Figure 24–2). The **steroid hormones** are
fat-soluble compounds made from cholesterol. They
include the androgens (male sex hormones) and estrogens (female sex hormones). The **amine hormones,**
such as thyroxine, are small, water-soluble compounds
related to the amino acid tyrosine. The **peptide hormones** exhibit the greatest diversity, both in number
and size. They vary from small peptides consisting of as
few as three amino acids, to very large polypeptides
with over 200 amino acid units.

The steroid hormones differ from the other hormone classes in general mode of action. Because of
their fat-solubility, the steroids readily pass through the
plasma membrane of their target cells, then bind to
protein receptors in the cytoplasm (Figure 24–3). In
most cases that have been studied, the hormone–
receptor complex moves into the nucleus, where it
regulates the transcription of specific genes into messenger RNA. As a result, the target cells begin to manufacture specific proteins (usually enzymes) that cause
some specific physiological change.

Most of the amine and peptide hormones act by
a fundamentally different mechanism. These water-
soluble hormones do not enter their target cells, but
instead bind to receptors extending from the cell surface. This external binding transmits a signal across
the plasma membrane that (in most cases examined)
activates an internal membrane-bound enzyme. This
enzyme converts ATP into cyclic adenosine monophosphate (cyclic AMP). Cyclic AMP is called a **second
messenger,** since it conveys the message of the
externally-bound hormone (the "first messenger") to
the inside of the cell. Cyclic AMP then initiates a chain
of enzyme activations that culminates in a particular
end effect. For example, one of the effects of adrenaline, an amine hormone produced by the adrenal
glands, is to enhance the breakdown of glycogen into
glucose in target liver cells. This is accomplished by a
cascading system of enzyme activations in the liver
cells. These activations are set off by adrenaline-
mediated cyclic AMP production (Figure 24–4). The
net result is that large amounts of glucose enter the
bloodstream to support the "fight or flight" response
associated with adrenaline action. (Your cells may
need extra glucose if you encounter a grizzly bear!)

Now that you have an idea of how hormones
work, let us turn to the orchestra of glands that produce hormones. We will also examine the regulatory
functions that hormones exert to create the biological
symphony of orderly growth, development, and homeostasis. We will emphasize the human system.

FIGURE 24–3

A General Model for Steroid Hormone Action. Steroid hormones penetrate the plasma membrane of their target cells and bind to a specific protein receptor. The hormone–receptor complex activates transcription of certain genes in the nucleus, and the new messenger RNAs are translated into proteins in the cytoplasm. The new proteins are responsible for the biochemical (and physiological) changes that occur in response to those hormones. Different hormones result in the production of different proteins.

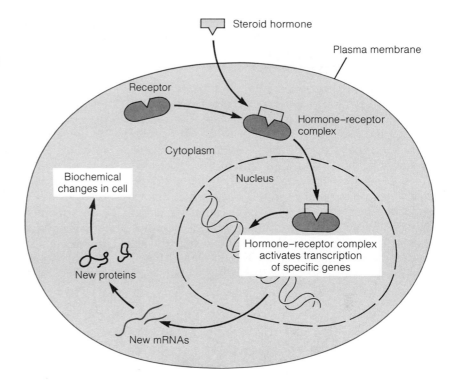

FIGURE 24–4

The Second Messenger Model for Adrenaline Action. Adrenaline acts on liver and muscle cells to promote the breakdown of glycogen into glucose units. Like other amine (and peptide) hormones, adrenaline binds to an external receptor on the plasma membrane, which activates an internal membrane-bound enzyme (adenyl cyclase) to convert ATP into cyclic AMP. Cyclic AMP, the second messenger, initiates a cascade of enzyme activations in the cytoplasm that culminates in the increased catalytic activity of phosphorylase, the enzyme that breaks down glycogen. The overall result is an increase in glucose level in the blood.

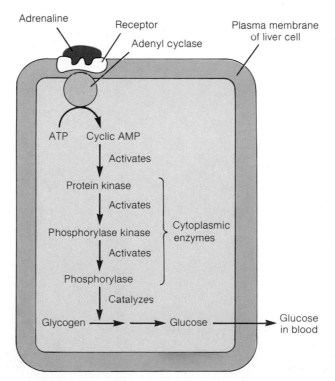

THE HUMAN ENDOCRINE SYSTEM

In many animals, hormones are made and released by specialized secretory organs called **endocrine glands.** The endocrine glands in humans are distributed throughout the head, neck, and torso regions of the body, and collectively make up the **endocrine system** (Figure 24–5). The hormones produced by this system affect nearly every bodily function (Table 24–1). To gain an understanding of how the various endocrine glands contribute to the chemical regulation of body functions, we will examine some of the major glands (and the hormones they produce) individually. In keeping with our musical metaphor, let us begin with the "conductor," the pituitary gland.

The Anterior Pituitary

The **pituitary gland** in humans is a round body about the size of a pea, located in the head just beneath the brain (see Figure 24–5). The pituitary is divided into two regions: the anterior (front) lobe and the posterior (rear) lobe. The **anterior pituitary** produces and releases at least six different hormones, four of which exert their regulatory actions on other endocrine glands (hence, the anterior pituitary is sometimes called the "master gland"). Because they stimulate activities in other endocrine glands, these four are called **tropic hormones.** They are **thyrotropin** (stimulates the thyroid gland), **adrenocorticotropic hormone (ACTH)** (stimulates the adrenal cortex), and two **gonadotropic hormones** (stimulate the gonads—testes or ovaries). The tropic hormones will

FIGURE 24–5
The Human Endocrine System.

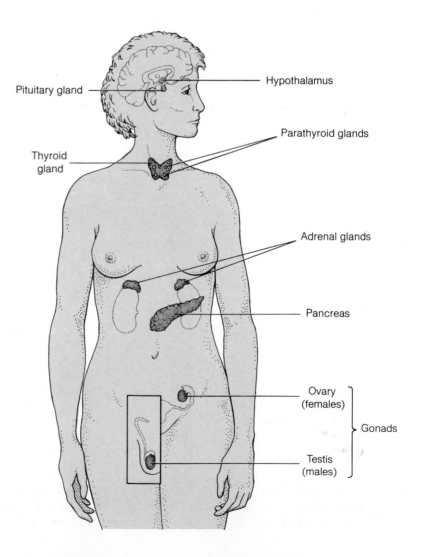

Hypothalamus

Pituitary gland

Parathyroid glands

Thyroid gland

Adrenal glands

Pancreas

Ovary (females)

Gonads

Testis (males)

TABLE 24–1
The Endocrine Glands and Their Hormones.

Endocrine Glands and Hormones They Release	Hormone Structure	Major Function(s)
Anterior Pituitary:		
Thyrotropin	Polypeptide	Stimulates thyroid gland to produce thyroxine
Adrenocorticotropic hormone (ACTH)	Polypeptide	Stimulates adrenal cortex to produce and release corticosteroids
Gonadotropic hormones	Polypeptides	Stimulate gonads (testes and ovaries) to produce sex hormones
Growth hormone (somatotropin)	Polypeptide	Promotes metabolism of carbohydrates and fats; stimulates growth of bones and muscles
Prolactin	Polypeptide	Stimulates milk production
Posterior Pituitary:		
Oxytocin	Peptide	Stimulates milk secretion; promotes contraction of uterus during childbirth
Vasopressin (antidiuretic hormone)	Peptide	Enhances water reabsorption by kidneys; promotes constriction of small arteries
Hypothalamus:		
Various releasing and inhibiting hormones (see Table 24–2)	Peptides	Stimulate or inhibit anterior pituitary to release tropic hormones
Thyroid:		
Thyroxine	Amine (contains iodine)	Regulates cellular respiration in metabolically-active body cells
Parathyroid:		
Parathyroid hormone	Polypeptide	Increases blood calcium level through release from bones
Pancreas:		
Insulin	Polypeptide	Decreases blood sugar level by promoting glucose uptake into body cells; promotes glycogen synthesis in liver
Glucagon	Polypeptide	Increases blood sugar level by promoting breakdown of liver glycogen
Adrenal Medulla:		
Adrenaline and norepinephrine	Amines	Increase heart rate, blood sugar level; dilate certain arteries and air passages to lungs; create "fight or flight" body status
Adrenal Cortex:		
Glucocorticoids	Steroids	Regulate metabolism of carbohydrates, fats, and proteins; reduce inflammation
Mineralocorticoids	Steroids	Regulate amounts of sodium and other mineral ions in the body fluids
Sex hormones	Steroids	Stimulate secondary sex characteristics
Gonads:		
Androgens (in testes)	Steroids	Stimulate development and maintenance of male secondary sex characteristics; stimulate sperm production
Estrogens (in ovaries)	Steroids	Stimulate development and maintenance of female secondary sex characteristics
Progesterone (in ovaries)	Steroid	Promotes preparation of uterus for embryo; maintains pregnancy

be discussed later in this chapter and in Chapter 25, when we discuss the endocrine glands they act on. The other two hormones that originate from the anterior pituitary, growth hormone and prolactin, have nonendocrine target tissues, and will be discussed now.

GROWTH HORMONE. Growth hormone, also known as somatotropin, is a large polypeptide. It has many target tissues in the body, where it promotes the breakdown of stored carbohydrates and fats into building blocks and energy units needed for growth. It also promotes the specific growth of bones and muscles, particularly during the childhood years. This function is most evident when abnormal amounts of the hormone are produced. A child whose anterior pituitary produces too much growth hormone may grow to an adult height of $2\frac{1}{2}$ meters (8 feet) or more, a condition known as **gigantism.** Too little growth hormone during the developmental years leads to

dwarfism (Figure 24–6). Pituitary dwarfism has been potentially treatable for years, but until recently, the only source of human growth hormone has been cadavers (growth hormone produced by other animals is ineffective in humans). Now, with the recent advances made in recombinant DNA technology (see Chapter 11), human growth hormone is being synthesized in large quantities by bacteria grown in the laboratory. The early clinical tests of this synthetic hormone are very encouraging, and before long, pituitary dwarfism may be a thing of the past.

In spite of their abnormal size, pituitary giants and dwarfs usually have normal body proportions. If excessive amounts of growth hormone are produced after adolescence, however, only those parts of the body still susceptible to the growth-promoting effect of this hormone respond. Such individuals suffer from **acromegaly,** a fairly rare condition in which the bones of the hands, feet, and face enlarge, and the tongue, lips, nose, and ears thicken (Figure 24–7).

PROLACTIN. The other nontropic hormone produced in the anterior pituitary, prolactin, is also a polypeptide. In conjunction with oxytocin, a hormone released from the posterior pituitary (see next section), prolactin stimulates milk production. When a woman gives birth, her pituitary gland begins to release prolactin, which triggers glands in her breasts to produce milk. Milk is not actually delivered, however, until

FIGURE 24–6
A Pituitary Dwarf and Giant Backstage.
Pituitary dwarfism and gigantism are caused by under- or overproduction of growth hormone during childhood. The man in the center is of normal height.

(a)   (b)   (c)

FIGURE 24–7
Acromegaly. Acromegaly is caused by overproduction of growth hormone in an adult. Note the progressive coarsening of the facial features and enlargement of the jaws and hands.

the infant begins suckling. This physical stimulation triggers the release of **oxytocin** into the bloodstream, causing milk to flow from the glands into the ducts that empty out into the nipple.

The Posterior Pituitary

The posterior lobe of the pituitary gland stores and releases two hormones: oxytocin and vasopressin. In addition to its effect on milk secretion from the breasts, oxytocin promotes the uterine contractions associated with childbirth. **Vasopressin,** or antidiuretic hormone (ADH), facilitates water reabsorption in the kidneys (see Chapter 23). Factors known to enhance the secretion of vasopressin include stress and nicotine, which thus have the effect of reducing the volume of urine excreted. Alcohol inhibits the release of vasopressin, thereby retarding water reabsorption and enhancing urine production. Besides its effect on the nephrons of the kidneys, vasopressin in high concentrations also causes constriction of the small arteries, which contributes to an elevation of blood pressure.

Oxytocin and vasopressin, though stored and secreted from the posterior pituitary, are not produced there. Rather, they are synthesized in the hypothalamus. The hypothalamus also regulates the release of hormones from the anterior pituitary. Metaphorically, the hypothalamus "conducts the conductor," or is "master over the master gland." Let's see how this interaction operates.

The Hypothalamus

The **hypothalamus** is a small region of the brain located just above the pituitary gland. It receives information from the nervous system on the internal status of the body (such as temperature and the concentration of various substances in the blood). The hypothalamus then acts as a switching station between the nervous and endocrine systems, relaying information to the pituitary. As you might suspect, there is an intimate structural and functional relationship between the hypothalamus and the pituitary (Figure 24–8).

FIGURE 24–8
The Hypothalamus–Pituitary Connection. Nerve cells in the hypothalamus produce and secrete a variety of hormones. One of the nerve clusters synthesizes oxytocin and vasopressin, then stores them in nerve endings located in the posterior pituitary. Upon proper stimulation from the brain, oxytocin and vasopressin are released into the blood supply of the posterior pituitary. Other nerve clusters in the hypothalamus produce and secrete a battery of releasing and inhibiting hormones, which are carried by the blood to the anterior pituitary. There they regulate the secretion of various tropic hormones, growth hormone, and prolactin manufactured by the anterior pituitary cells.

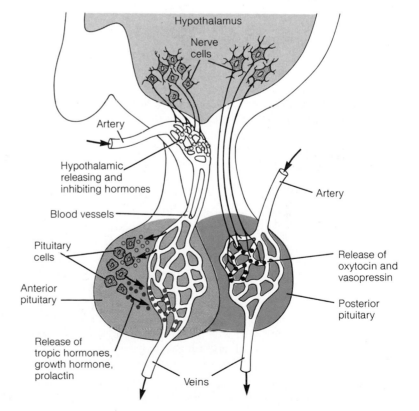

Hypothalamus
Nerve cells
Artery
Hypothalamic releasing and inhibiting hormones
Blood vessels
Pituitary cells
Anterior pituitary
Release of tropic hormones, growth hormone, prolactin
Veins
Artery
Release of oxytocin and vasopressin
Posterior pituitary

Specific nerve cells originating in one part of the hypothalamus produce oxytocin and vasopressin. These nerve cells extend into the posterior lobe of the pituitary, where their swollen endings store the hormones (see Figure 24–8). Upon proper stimulation of the hypothalamus, the nerves release their hormonal cargo into the blood vessels of the posterior pituitary for distribution to other parts of the body.

Another region of the hypothalamus exerts control over the anterior pituitary. Here, special nerve cells produce a battery of small peptide hormones that either stimulate (**releasing hormones**) or inhibit (**inhibiting hormones**) the release of the various tropic hormones from the pituitary (Table 24–2). The hypothalamic hormones reach their pituitary target via a system of blood vessels that stretch between these two regions (see Figure 24–8). We will examine how one of these small peptides (thyrotropin releasing hormone) plays its role in the section that follows.

The Thyroid Gland

The **thyroid gland** extends around both sides of the trachea just below the larynx (see Figure 24–5). Its primary function is the synthesis and release of three hormones, of which thyroxine is the most important. **Thyroxine** is an amine hormone that stimulates the rate of cellular respiration in body cells, particularly in the liver and skeletal muscles.

The regulation of thyroxine synthesis is a good example of the checks and balances that frequently operate in the endocrine system. The production of thyroxine in the thyroid gland is triggered by thyrotropin, one of the tropic hormones produced in the anterior pituitary. The release of thyrotropin into the bloodstream, however, requires the presence of **thyrotropin releasing hormone (TRH)** made in the hypothalamus (Figure 24–9a). Thus, thyroxine cannot be produced unless the anterior pituitary and hypothalamus "say so." High levels of thyroxine in the blood inhibit the release of both thyrotropin and TRH, so further synthesis of thyroxine does not occur (Figure 24–9b). This type of homeostatic mechanism is called a **negative feedback system,** and it is similar in principle to a thermostat, which controls the temperature of a house. The on-and-off thyroxine regulatory mechanism ensures a nearly constant level of thyroxine in the blood.

TABLE 24–2
Hypothalamus Releasing and Inhibiting Hormones.
All the hormones listed either stimulate or inhibit the release of specific hormones produced in the anterior pituitary, as suggested by their names.

Thyrotropin releasing hormone (TRH)
Adrenocorticotropin releasing hormone
Gonadotropin releasing hormones
Growth hormone releasing hormone
Growth hormone inhibiting hormone
Prolactin releasing hormone
Prolactin inhibiting hormone

Hyperthyroidism, or the overproduction of thyroxine, is an abnormal condition characterizd by nervousness, high blood pressure, and an inability to gain weight. At the other extreme, **hypothyroidism** (too little thyroxine) in adults produces dry skin, loss of hair, an intolerance to cold, and general lethargy. **Infant hypothyroidism,** or **cretinism,** is more serious, causing mental retardation and dwarfism. The most common cause of hypothyroidism in the past stemmed from dietary iodine deficiencies (each thyroxine molecule contains four iodine atoms; see Figure 24–2b). Iodine deficiencies in adults often led to **goiter,** a swelling of the thyroid gland. This deformity has been virtually eliminated in developed nations since manufacturers began including iodine in table salt. Synthetic thyroxine is available for individuals with a hypothyroid condition stemming from other causes.

The Pancreas

As a secretory organ, the pancreas has double duty. Not only does it produce and release digestive enzymes into the small intestine (see Chapter 21), it also secretes several hormones into the bloodstream. These hormones, which include insulin and glucagon, are produced in specialized clusters of pancreatic cells called the **islets of Langerhans.**

INSULIN. The primary roles of insulin are to promote the uptake of glucose into all metabolically active body cells, and, in the liver, to stimulate the formation of glycogen from glucose units. Both these actions have the effect of reducing blood glucose levels. As

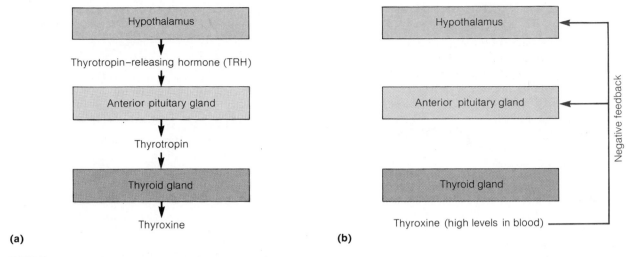

FIGURE 24–9
Feedback Regulation of Thyroxine Levels. (a) The synthesis of thyroxine in the thyroid is promoted by the release of thyrotropin from the anterior pituitary, which in turn depends on TRH released from the hypothalamus. (b) High levels of thyroxine inhibit both thyrotropin and TRH secretion, causing a reduction of thyroxine synthesis. This negative feedback system keeps the thyroxine in the blood adjusted to an appropriate level.

you might suspect, high blood glucose concentration is the trigger for insulin production and release from the pancreas.

Diabetes mellitus, the most common form of diabetes, is caused by a deficiency in insulin production. This disorder leads to excessive sugar levels in the blood after a meal, overtaxing the kidney's ability to reabsorb glucose from the urine. (Sugary urine is a common diagnostic symptom of diabetes.) The abnormally high levels of glucose in the nephron tubules in turn prevent proper water reabsorption, resulting in excessive water loss through urination. The untreated diabetic complains of constant thirst and may suffer from dehydration. Prolonged diabetes causes weight loss, blindness, heart disease, kidney damage, and, if left untreated, coma and death. These symptoms can be averted (or at least forestalled) by taking daily injections of insulin and restricting intake of carbohydrates and salts. Most diabetics live active, normal lives with proper treatment.

Oversecretion of insulin can lead to **hypoglycemia** (low blood sugar). Individuals prone to hypoglycemia tend to release too much insulin in response to an elevation in blood sugar (such as after a meal), and the excess of insulin drops the blood sugar to abnormally low levels. This brings on visible symptoms

such as a high pulse rate, profuse sweating, weakness, hunger, and loss of skin color. Long-term treatment usually involves reducing the amount of carbohydrates in the diet. Candy bars and regular soda pop are definitely off limits for the hypoglycemic and the diabetic alike.

GLUCAGON. Glucagon opposes the effect of insulin in the liver by stimulating liver cells to break down glycogen and release glucose into the blood. At the molecular level, the action of glucagon is like that of adrenaline, which also promotes glycogen breakdown in liver cells (see Figure 24–4). The pancreatic cells that produce glucagon are extremely sensitive to glucose concentration. If the blood glucose drops sufficiently, these cells release glucagon, which stimulates the liver to release more glucose into the blood.

As we have seen, a high blood glucose level triggers the release of insulin, which acts to lower the sugar level; conversely, low blood sugar stimulates the release of glucagon, which raises the blood sugar level. Thus, we have all the ingredients for blood sugar homeostasis—the maintenance of a fairly constant glucose concentration in the blood (Figure 24–10). Through this sensitive hormonal regulation system, the body cells are ensured of a steady diet of glucose.

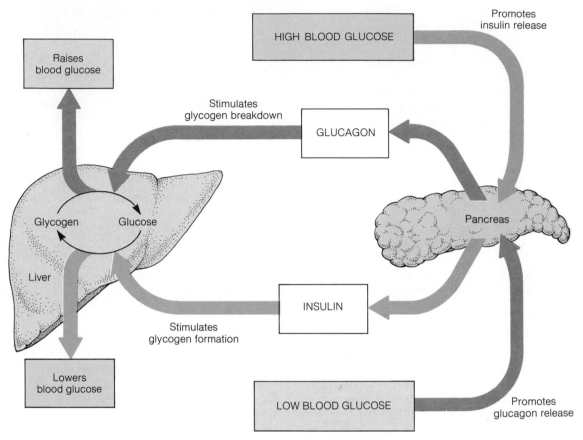

FIGURE 24–10

Insulin and Glucagon Regulate Blood Sugar Level. Special secretory cells in the pancreas monitor the concentration of glucose in the blood, then respond to an abnormal concentration by releasing the appropriate hormone. Glucagon and insulin have opposing effects on the glycogen–glucose balance in the liver.

The Adrenal Glands

The two **adrenal glands** sit like caps on the upper ends of the kidneys (see Figure 24–5). Each adrenal gland has two distinct regions that synthesize and secrete different sets of hormones: the central core of the gland, called the **adrenal medulla,** and the outer region, called the **adrenal cortex** (Figure 24–11).

ADRENAL MEDULLA. The adrenal medulla is composed of nervelike cells that secrete two closely related hormones, **adrenaline** (also called **epinephrine**) and **norepinephrine.** The secretion of adrenaline and norepinephrine is activated under situations of stress, anxiety, fear, and other conditions requiring the body

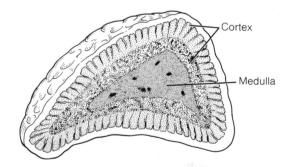

FIGURE 24–11

An Adrenal Gland. The medulla (central core) of the adrenal gland produces adrenaline and norepinephrine. The cortex (outer layers) produces more than 50 corticosteroid hormones.

to prepare for unusual exertion or perceptual activity. These hormones increase the heart rate, blood glucose level, and the diameters of arteries leading to the muscles and heart; the air passages to the lungs also dilate, permitting greater air flow to the lungs. The overall feeling that results from the outpouring of these hormones is familiar to all of us and lives up to its popular name—the **fight-or-flight reaction.** The emergency reactions triggered by adrenaline and norepinephrine are quickly neutralized, however, because the blood contains enzymes that inactivate these hormones.

ADRENAL CORTEX. Under the influence of ACTH from the pituitary, the adrenal cortex produces over 50 different steroid hormones, collectively referred to as the **corticosteroids.** All of these steroids are closely related in chemical structure, but have rather divergent hormonal actions. The **glucocorticoids** comprise one family of corticosteroids that regulate various aspects of carbohydrate, fat, and protein metabolism. They also reduce inflammation of tissues, and synthetic forms (such as cortisone) have been used to reduce tissue swelling and other symptoms brought on by injury or allergic reactions. Another group, the **mineralocorticoids,** regulate the levels of mineral ions in the extracellular body fluids. For example, aldosterone promotes sodium reabsorption during the processing of urine in the kidneys (see Chapter 23). Finally, the adrenal cortex produces small amounts of the male and female **sex hormones.** Larger quantities of these hormones are formed in the gonads, so we will discuss them in the section on the gonads.

People whose adrenal cortex fails to produce enough corticosteroids (a condition called Addison's disease) generally have a low blood sugar level and are weak and lethargic. Often their skin has a bronze tone. During the presidential campaign of 1960 it was revealed that John Kennedy suffered from a mild case of Addison's disease, which was partly responsible for his year-round tan. From time to time he had to take oral doses of corticosteroids.

The Gonads

As endocrine glands, the **gonads** (testes or ovaries) are the major production sites of the sex hormones. The male's testes produce **androgens,** of which **testosterone** is the most abundant. The female's ovaries produce a class of hormones called **estrogens** and the hormone **progesterone.** The characteristics that distinguish males from females are due largely to differences in the relative amounts of androgens or estrogens generated by their respective gonads.

Because the gonads are absent in the earliest stages of human embryo development, it is impossible in the first six weeks to distinguish anatomically between male and female embryos. Then, at about six weeks of age, embryos destined to become males (XY chromosomes) begin to produce a substance known as **H-Y antigen,** which by some unknown mechanism initiates testicular development. Shortly thereafter, the developing testes begin to produce androgens. In normal female (XX) embryos, the gonads remain undifferentiated until about the eighth week. Apparently, the development of the ovaries is triggered not by the production of a special molecule or factor, as with males, but by the *absence* of the H-Y antigen and androgens. In any case, as the testes and ovaries begin to form, they start producing sex hormones, which influence the sexuality of an individual throughout his or her lifetime.

The period in human development when the sex hormones exert their most dramatic effects is puberty, when the individual begins to mature sexually. In males, the transition from youth to sexual maturity is gradual and begins at about the age of 10 to 13. The voice deepens; body growth increases; the testes and penis enlarge; facial, underarm, and pubic hair begin to form; and sperm production begins. These changes result primarily from a substantial rise in testosterone production in the testes.

The regulation of testosterone production (at puberty and throughout the male's lifetime) is under the direct control of the anterior pituitary. One of the pituitary gonadotropin hormones, called luteinizing hormone (LH), stimulates testosterone production. Adding to the complexity, LH release requires the presence of LH releasing hormone secreted by the hypothalamus, and its production is subject to negative feedback control by testosterone. As we have seen for other hormonal systems, this feedback loop serves to regulate the levels of testosterone in the body. However, this does not explain why testosterone levels increase at the onset of puberty; that question remains unanswered.

In females, puberty generally begins between the ages of 11 and 13. As the ovaries increase their production and release of estrogens and progesterone, the breasts and uterus enlarge, the hips broaden, long-bone growth ceases, pubic and underarm hair develop, and ovulation and the menstrual cycle begin. We will have more to say about ovulation, the menstrual cycle, and pregnancy—and the regulatory roles played by the sex hormones in each of these—in the next chapter.

SUMMARY

Chemical control systems in animals serve the important roles of coordinating the activities of specialized tissues and organs in the body, and maintaining internal constancy. The most important character in chemical control systems is the hormone, a messenger substance produced in one region that acts on target cells elsewhere in the body.

The most extensively studied hormones among invertebrates are those that regulate development in insects. Ecdysone, the molting hormone, acts in conjunction with juvenile hormone to control metamorphosis and the shedding of the old exoskeleton.

The vertebrate hormones are of three general classes: the steroid hormones, the amine hormones, and the peptide hormones. The steroids enter their specific target cells and bind to a cytoplasmic receptor. The hormone–receptor complex then activates the transcription of specific nuclear genes that ultimately results in the synthesis of new proteins. Amine and peptide hormones bind to receptors on the surface of the plasma membrane. This binding triggers the production of cyclic AMP in the target cell's cytoplasm. Cyclic AMP acts as a "second messenger," activating certain cellular enzymes that alter that cell's metabolic functions. The result is a change in cell function that is characteristic for the particular hormone.

Human endocrine glands produce hormones that are transported in the bloodstream. The hormones then interact with their specific target cells to regulate various metabolic and developmental processes.

The anterior pituitary is often called the "master gland" of the endocrine system because it produces tropic hormones that regulate the activities of other endocrine glands. The tropic hormones include thyrotropin (regulates thyroxine synthesis in the thyroid gland), adrenocorticotropic hormone (regulates production of the corticosteroids in the adrenal cortex), and the gonadotropic hormones (regulate activities in the ovaries and testes, including sex hormone synthesis). The anterior pituitary also produces growth hormone (regulates growth of bones and muscles) and prolactin (stimulates milk production after childbirth).

The posterior pituitary is the site of release of oxytocin and vasopressin, two hormones manufactured in special nerve cells of the hypothalamus.

The hypothalamus is a region of the brain located just above the pituitary. It receives signals from various parts of the body and other regions of the brain, and responds by transmitting specific releasing or inhibiting hormones to the anterior pituitary. These hormones stimulate or inhibit the release of the anterior pituitary hormones into the bloodstream.

The thyroid gland produces thyroxine, a hormone that stimulates the rate of cellular respiration in body cells. Thyroxine levels are under negative feedback control. The synthesis of thyroxine is promoted by secretions of the anterior pituitary and the hypothalamus, both of which are inhibited by high levels of thyroxine.

The endocrine function of the pancreas is attributable to special cells in the islets of Langerhans that produce and secrete insulin and glucagon. Insulin acts to reduce blood sugar level by promoting glucose uptake into body cells and stimulating the conversion of glucose into glycogen in the liver. The inability to produce enough insulin causes diabetes. Glucagon promotes the elevation of glucose concentration in the blood by stimulating glycogen breakdown in the liver.

The adrenal glands, located at the upper ends of both kidneys, are actually two glands in one. The adrenal medulla (inner core of the gland) produces adrenaline and norepinephrine, two hormones that prepare the body for "fight or flight" responses. The adrenal cortex (outer region) synthesizes a vast array of corticosteroid hormones, including the glucocorticoids (regulate metabolism of carbohydrates, fats, and proteins), mineralocorticoids (regulate ion levels in the extracellular fluids), and the sex hormones (determine male and female characteristics).

The sex hormones are produced in greatest quantity by the gonads. The ovaries synthesize estrogens and progesterone (female sex hormones), and the testes produce the androgens (male sex hormones, including testosterone). The sex hormones not only influence the development and functions of the sex organs, but also stimulate the secondary sex characteristics that develop during puberty. The gonadal production of these hormones is regulated by gonadotropic hormones secreted by the anterior pituitary, the release of which is controlled by the hypothalamus.

STUDY QUESTIONS

1. How does a target cell "recognize" a hormone that affects it?

2. What is the effect of juvenile hormone on insect development?

3. Describe the general mechanism by which steroid hormones act in their target cells.

4. Describe the second messenger model of hormone action. Which classes of endocrine hormones operate via a second messenger?

5. What is the nature of the relationship between the hypothalamus and the pituitary gland?

6. Often the roles of hormones have been discovered from the effects of over- or underproduction in individuals. Describe the effects of abnormally high and low amounts of (a) growth hormone and (b) insulin.

7. What is a negative feedback system? Describe the system by which the level of thyroxine is regulated in humans.

8. Which two hormones control the level of sugar in the blood? How do they work?

9. Name the three families of steroid hormones produced in the adrenal cortex?

10. Explain the effects of the sex hormones during the onset of puberty in human males and females.

SUGGESTED READINGS

Arnold, C. *Sex Hormones: Why Males and Females Are Different.* New York: William Morrow, 1981. Sexual differences influenced by hormones are explained for animals and humans, including the role of hormones in animal behavior.

Bloom, F. E. "Neuropeptides." *Scientific American* 245 (1981):148–168. Discusses how the discovery of neuropeptides—substances that can act as hormones as well as neurotransmitters—blurs the classical distinctions between the endocrine and nervous systems.

Mason, S. *Human Physiology.* Menlo Park: Benjamin/Cummings, 1983. Chapter 10 provides an in-depth yet understandable account of the endocrine system.

O'Malley, B. W., and W. T. Schrader. "The Receptors of Steroid Hormones." *Scientific American* 234(1976): 32–43. Discusses how the steroid–receptor complex may interact with nuclear DNA to initiate a hormonal response.

Witzmann, R. F. *Steroids: Keys to Life.* New York: Van Nostrand-Reinhold, 1981. The role of steroids in life is described in an enjoyable and informative manner. The work of key scientists involved in steroid research is also discussed.

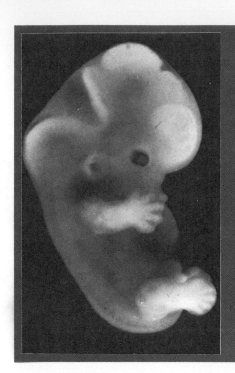

CHAPTER 25
Human Reproduction and Development

Everyone around you looks, behaves, and thinks in unique ways. Even members of your family are different from you, unless you have an identical twin. This diversity among members of our own species also extends to members of other species. Even though two bullfrogs may look exactly alike to us, they are genetically unique individuals.

The diversity among individuals is largely attributable to the processes and events underlying sexual reproduction, the means by which humans and most other animals produce more of their kind. The primary cellular events associated with sexual reproduction are gametogenesis and fertilization. **Gametogenesis** is the formation of **gametes**—haploid cells capable of cell fusion. In all animals except sponges, the formation of gametes occurs by meiosis in specialized reproductive organs. In human males, sperm are produced in the testes; in human females, eggs are produced in the ovaries.

Fertilization is the fusion of a sperm with an egg, forming a **zygote** that contains the combined genetic material of the gametes. The result is a new individual having a unique combination of genes contributed by both parents.

In terms of evolution, sexual reproduction has advantages over purely asexual modes of propagation because it contributes to the genetic variability in a population. You may recall that during meiosis, crossover and the independent assortment of chromosomes shuffle the nuclear genes into many possible combinations (see Chapter 8). Thus, no two gametes formed by an individual are likely to be the same; their genetic makeup varies. As a result, the union of gametes invariably produces offspring that are genetically unique. And when sexual reproduction involves combining genes from two different parents, as it does in most animals, genetic variation among the offspring is further enhanced. As we will explore in Chapter 28, genetic variation is a crucial component of the evolutionary process—it underlies the adaptations of organisms to new and changing environments.

The mechanics of gametogenesis and fertilization vary little from one animal to another. What *does* vary is the means by which sperm meet eggs. In most aquatic animals, fertilization occurs externally: both eggs and sperm are released into the water. For many terrestrial invertebrates and all reptiles, birds, and mammals, fertilization is internal. The eggs are retained in the female, and the male deposits sperm into the female's reproductive tract, where fertilization occurs.

In most mammals (except humans and a few other primates), mating or copulation occurs only at specific times when the female is physically and behaviorally receptive. This sexually receptive period is called **estrus.** How often estrus occurs and how long it lasts depend on the animal. Female dogs usually come into estrus ("heat") twice a year, with each receptive period lasting about a week. Cats generally are receptive three times during the year; mice may go into estrus as often as every four days. In most species, the male is alerted to the female's receptivity by her behavior, by odors resulting from vaginal discharge, or by both. Regardless of how often estrus occurs, each receptive period is characterized by a heightened sexual desire on the part of the female that coincides with **ovulation**—the release of an egg or eggs from the ovary—and a thickening of the lining of the uterus. In most mammals that have distinct periods of estrus, if fertilization of the egg or eggs does not occur, the uterine lining is reabsorbed into the body.

The reproductive cycle of human females differs in a fundamental way from the estrus cycles of most other mammals. Although a woman's ability to conceive is cyclical, her sexual receptiveness is not. A woman's menstrual cycle lasts about 28 days, and she can become pregnant only during a day or so within this period when a mature egg is available for fertilization. She may enjoy sexual activity, however, at any time during the cycle. Thus, unlike other animals, copulation (or sexual activity in general) is not strictly tied to a reproductive function in humans.

THE FEMALE REPRODUCTIVE SYSTEM

The reproductive system of the human female is relatively complex (Figure 25–1). This is not surprising, given that it makes an egg available for fertilization and prepares an environment for embryo development once every 28 days or so for some 30 to 40 years. The source of the eggs is the **ovaries,** two almond-shaped structures about 4 cm ($1\frac{1}{2}$ inches) long lying on each side of the pelvic cavity. The ovaries are both egg-producing organs and endocrine glands that secrete sex hormones (see Chapter 24). Once every menstrual cycle, one of the ovaries discharges an egg into one of the **fallopian tubes,** ducts which extend between the ovaries and the uterus. It is in the fallopian tubes that the egg normally is fertilized. These tubes are lined with cilia and mucus that slowly sweep the egg, fertilized or not, toward the uterus.

The **uterus** is a hollow, muscular organ about the size and shape of a small pear. Its inner lining, the **endometrium,** is where the fertilized egg implants and develops into a fetus. Situated at the lower end of the uterus is the muscular **cervix,** which opens into the vagina. The **vagina** is a tubelike canal about 10–12 cm (4–5 inches) long that leads to the exterior of the body. The vaginal opening is partially protected by folds of skin on either side known as the labia. The outermost folds, the **labia majora,** are thicker and more fleshy than the innermost folds, the **labia minora.** The opening of the urethra lies above the vaginal opening and below the **clitoris,** a tissue richly imbued with nerves whose sole purpose is to heighten sexual arousal. The labia and clitoris, collectively referred to as the **vulva,** constitute the external genitals.

The Menstrual Cycle

The **menstrual cycle** refers to an overlapping series of events that includes egg development, ovulation, and growth of the uterine endometrium. The first cycle, or **menarche,** occurs at puberty. From then on, this series repeats roughly every 28 days (except during pregnancy) until it ceases when the woman reaches **menopause,** sometime between the ages of 45 and 52. The cyclical occurrence of events coordinates the release of an egg from an ovary with the time at which the endometrium is fully developed and receptive to implantation of the fertilized egg. If the released egg is not fertilized, part of the endometrium degenerates and is sloughed off through the cervix and vagina in the menstrual flow. This is called **menstruation.** Menstruation marks the end of one cycle and the beginning of the next, when another egg in one of the ovaries starts to develop.

OOGENESIS AND OVULATION. Human eggs (ova) are ultimately derived from immature sex cells called **oogonia** within the ovaries through a process called **oogenesis.** When the human female is still a fetus, the diploid oogonia proliferate by mitosis and begin to develop into **primary oocytes.** By the time of birth, 2–4 million primary oocytes have already begun meiosis and have reached the first meiotic prophase. Further meiotic development is suspended until the

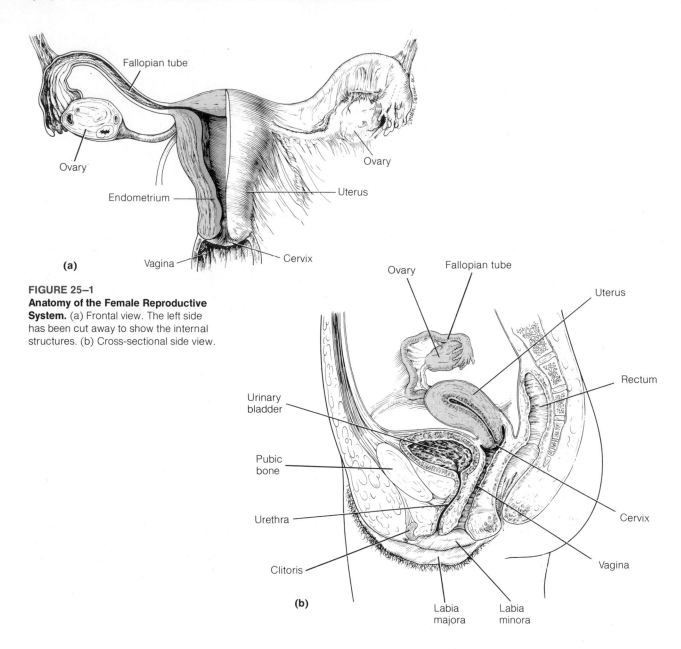

(a)

FIGURE 25–1
Anatomy of the Female Reproductive System. (a) Frontal view. The left side has been cut away to show the internal structures. (b) Cross-sectional side view.

(b)

girl reaches puberty. Then some unknown stimulus triggers menarche, and, except during pregnancy, one primary oocyte matures every 28 days or so for the next 30 to 40 years of the woman's life.

At the onset of each menstrual cycle (taken as the first day of menstrual bleeding), several primary oocytes in each ovary become activated to resume development, each surrounded by a spherical group of cells known as the **primary follicle** (Figure 25–2a). After several days, however, all but one of the primary follicles cease development, and the lone survivor

enlarges to become a **Graafian follicle.** (Occasionally, more than one follicle matures, particularly if fertility drugs are taken, which can cause multiple births.) In the meantime, the primary oocyte enclosed within the enlarging follicle resumes meiosis. It completes the first meiotic division at the Graaffian follicle stage, yielding a **secondary oocyte** and a **polar body** (Figure 25–2b). The homologous chromosomes are segregated equally into these two cells, but the secondary oocyte retains nearly all the cytoplasm; the polar body is little more than a nucleus.

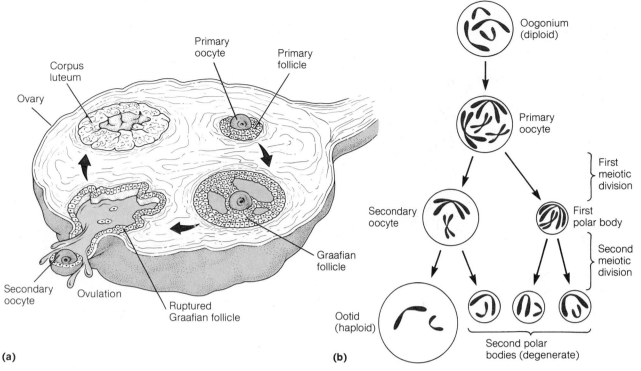

FIGURE 25–2
Oogenesis and Ovulation. (a) Cross section of an ovary showing the progressive development of a follicle and oocyte.
(b) Diagram of oogenesis.

At about day 14 of the cycle, the Graafian follicle ruptures and releases the secondary oocyte and polar body through the ovary wall into the fallopian tube (ovulation). The remnants of the ruptured follicle then reassemble into a yellow-colored body called the **corpus luteum** (see Figure 25–2a).

The expelled secondary oocyte and its attached polar body are swept slowly toward the uterus by the cilia lining the fallopian tube. If perchance it encounters sperm in the tube, a single sperm penetrates the oocyte, triggering both the secondary oocyte and the polar body to undergo the second meiotic division. The completion of meiosis results in four haploid cells: three tiny polar bodies (which soon degenerate) and one large **ootid** (see Figure 25–2b). (The unequal divisions of cytoplasm following both divisions of meiosis ensure that the egg cell will have sufficient cytoplasm and stored food to support its development after fertilization.) The ootid then quickly matures into an **ovum** (egg), and the sperm nucleus fuses with the egg nucleus to form the zygote.

If the migrating oocyte is not penetrated by a sperm within 6–24 hours after ovulation, it will not complete meiosis and soon degenerates. About 10 days later, the corpus luteum in the ovary also withers, a prelude to the onset of a new cycle.

DEVELOPMENT OF THE ENDOMETRIUM. The events occurring in the ovary are closely tied to events taking place in the uterus. After menstrual flow stops (usually three to five days after it starts), the endometrial lining of the uterus is quite thin. Then, over the next three weeks, the endometrium gradually grows in thickness. Many new blood vessels are formed, providing oxygen and nutrients to this burgeoning tissue. Some of the nutrients are tucked away into food-storing glands buried in the endometrium. These changes prepare the endometrium to receive and nurture an embryo should the egg become fertilized. If pregnancy does occur, the embryo **implants** in the endometrium and, through a hormonal mechanism, signals the corpus luteum in the ovary to remain intact and active. The corpus luteum in turn produces hormones (see next section) that maintain the integrity of the endometrium (prevent menstruation). If pregnancy does not occur, no positive signal reaches the corpus luteum to keep it active. It degenerates, and a few days later the endometrium begins to slough off.

FIGURE 25–3

The Menstrual Cycle. This composite diagram illustrates how the levels of FSH, LH, estrogens, and progesterone correlate with developmental events taking place in the ovary and uterus.

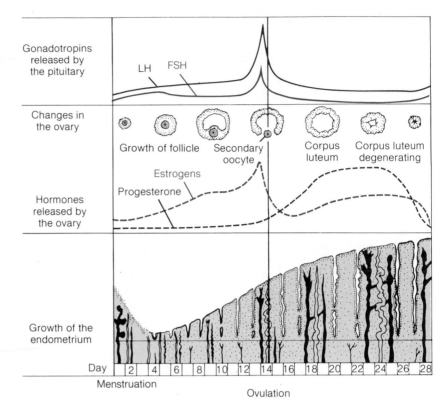

Gonadotropins released by the pituitary

LH   FSH

Changes in the ovary

Growth of follicle   Secondary oocyte   Corpus luteum   Corpus luteum degenerating

Estrogens

Progesterone

Hormones released by the ovary

Growth of the endometrium

Day 2 4 6 8 10 12 14 16 18 20 22 24 26 28

Menstruation

Ovulation

HORMONAL REGULATION OF THE MENSTRUAL CYCLE. The menstrual cycle is regulated and coordinated by a rather complex interplay of endocrine hormones. The processes occurring in the ovary are triggered by rising and falling levels of two gonadotropic hormones released from the anterior pituitary: **follicle-stimulating hormone (FSH)** and **luteinizing hormone (LH)** (Figure 25–3). The secretion of these hormones is in turn triggered by specific releasing hormones produced in the hypothalamus.

Under the influence of FSH and LH, the developing follicle synthesizes and secretes **estrogens,** which stimulate development of the endometrium. At about the 12th day of the cycle, the rising level of estrogens stimulates the hypothalamus to produce more of the gonadotropic releasing hormones, which cause a burst in FSH and LH release from the pituitary. The sharp rise in LH level induces ovulation (day 14), and shortly thereafter the FSH and LH levels fall off.

Following ovulation, the corpus luteum produces both progesterone and estrogens. These hormones promote the continued development of the uterine lining in preparation for embryo implantation. In addition, the combination of high progesterone and estrogen levels has a negative feedback effect on the gonadotropic releasing hormone centers of the hypothalamus, resulting in a drop in FSH and LH release from the pituitary. Eventually the LH level becomes too low to sustain the corpus luteum. With the degeneration of the corpus luteum, the levels of progesterone and estrogens drop, and shortly thereafter, the thickened endometrium also degenerates. This marks the end of one cycle.

Pregnancy has the effect of short-circuiting the cyclical fluctuation of hormone levels, and thereby suspending the menstrual cycle. We will examine this effect later in the chapter in the section on pregnancy.

THE MALE REPRODUCTIVE SYSTEM

The male reproductive system is somewhat less complicated than the female system (Figure 25–4). The external genitals consist of two **testes** (singular *testis*) held within a sac called the **scrotum,** and a **penis.** The testes, also referred to as the **testicles,** produce

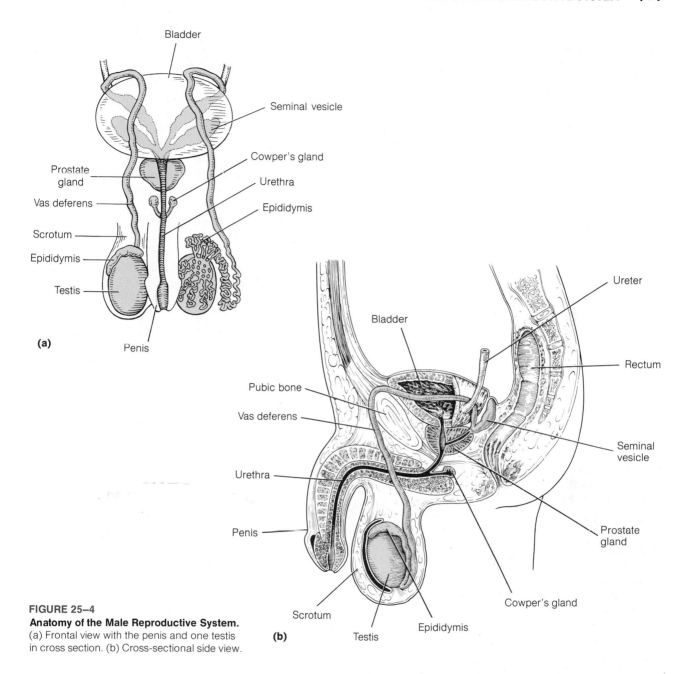

FIGURE 25—4
Anatomy of the Male Reproductive System.
(a) Frontal view with the penis and one testis
in cross section. (b) Cross-sectional side view.

sperm and the male sex hormones. The penis, consisting of blood vessels, fibrous tissue, and spongy tissue, doubles as a urinary and copulatory organ.

The sperm produced in each testis collect in the **epididymis** for storage before they are transported through a long duct called the **vas deferens.** Both vas deferens extend upward into the abdominal cavity above the urinary bladder, then downward to join with the urethra. Various glands (including the pros-

tate, seminal vesicles, and Cowper's glands) near this juncture secrete seminal fluids into the sperm tract to form the **semen** (sperm plus seminal fluids). The semen is discharged through the urethra, which runs the length of the penis. Since the male urinary and reproductive systems have the urethra in common, these two systems are collectively referred to as the **urogenital system.** In females the urinary and genital systems are separate.

Spermatogenesis

Sperm production (**spermatogenesis**) in males begins at puberty and, under normal circumstances, continues throughout their lives. Unlike females, there is no cyclic variation in this gamete-producing activity. The sperm are formed in highly convoluted structures within the testes called the **seminiferous tubules** (Figure 25–5a). **Spermatogonia,** specialized diploid cells located within these tubules, propagate continuously by mitotic cell divisions and eventually differentiate into **primary spermatocytes** (Figure 25–5c). Each primary spermatocyte yields two **secondary spermatocytes** after the first meiotic division, and each secondary spermatocyte yields two **spermatids** after the second meiotic division. Thus, one diploid primary spermatocyte produces four haploid spermatids (Figure 25–5d). Each spermatid then differentiates

into a sperm, which consists of a head region containing the chromosomes and an enzyme-laden cap called the **acrosome;** a middle piece housing numerous mitochondria; and a flagellar tail for propulsion (Figure 25–6). The development of spermatogonia into sperm takes about two months. Each day several hundred million sperm complete this developmental sequence.

During sexual stimulation, the sperm move through the vas deferens into the urethra, where they mix with fluid secretions from the seminal vesicles, the prostate gland, and the Cowper's glands (see Figure 25–4). These secretions not only provide a liquid suspension for the sperm cells, but also contain the sugar fructose to nourish them. The secretions are also alkaline and thus help neutralize the acidic fluids in the female reproductive tract, which otherwise would inactivate the sperm.

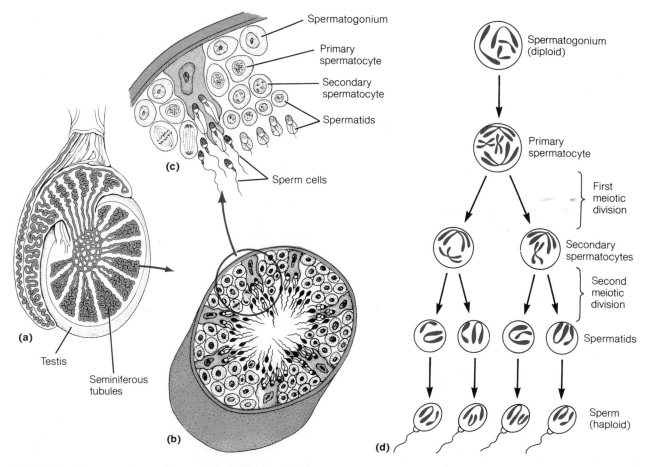

FIGURE 25–5

Spermatogenesis. (a) Cross section of a testis showing the arrangement of the seminiferous tubules. (b) Cross section of a seminiferous tubule. (c) A closer view of the cells undergoing spermatogenesis. (d) Diagram of spermatogenesis.

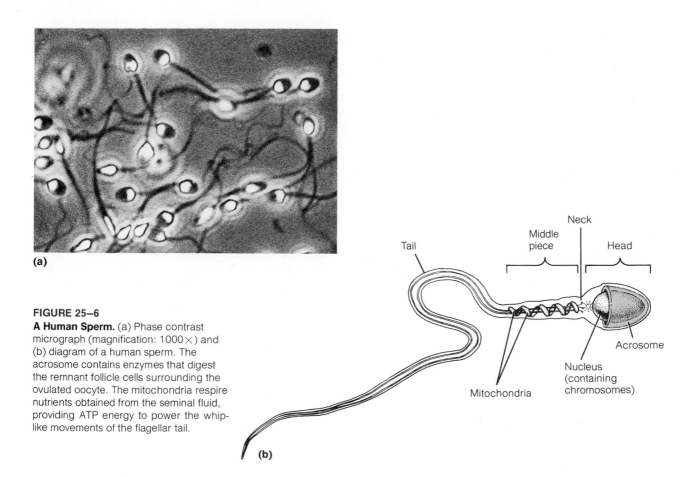

(a)

(b)

FIGURE 25–6
A Human Sperm. (a) Phase contrast micrograph (magnification: 1000×) and (b) diagram of a human sperm. The acrosome contains enzymes that digest the remnant follicle cells surrounding the ovulated oocyte. The mitochondria respire nutrients obtained from the seminal fluid, providing ATP energy to power the whip-like movements of the flagellar tail.

Tail

Middle piece

Neck

Head

Acrosome

Nucleus (containing chromosomes)

Mitochondria

SEXUAL INTERCOURSE

The penis is a highly vascularized copulatory organ well suited for transmitting sperm to the female reproductive tract, but this function first requires that the penis become erect to facilitate its entry into the vagina. Erection normally results from sexual stimulation, which increases the flow of blood through arterioles leading to the penis. The blood fills sinuses (cavities) within the spongy tissue of the penis. These blood-filled sinuses increase the fluid pressure within the penis and partially close off the veins that carry blood back out. With the blood flow increased into the penis and the outward flow partially restricted, the penis enlarges and becomes erect.

In women, foreplay and sexual stimulation also result in increased blood flow to the breasts and genitals, causing the nipples to become firm and the vulva to enlarge. In addition, sexual stimulation causes the vagina to secrete a lubricating fluid that facilitates

entry of an erect penis. The actual insertion of the penis into the vagina marks the beginning of **sexual intercourse** (also referred to as **coitus**). Subsequent rhythmic motions and the repeated thrusting of the penis within the vagina further enhances the sexual arousal of both partners, causing increases in pulse rate, blood pressure, and breathing rate.

The culmination of this sexual arousal is **orgasm.** During orgasm in males, the muscles surrounding the vas deferens and urethra go through a series of strong contractions, forcefully expelling the semen, an event termed **ejaculation.** Consequently, orgasm in men is not only physically well defined, it is also clearly an adaptation to facilitate fertilization. Soon after ejaculation, the penis returns to its shorter, flaccid condition. Although orgasm in women can also occur at the height of arousal, it is not necessary for **conception**— fertilization leading to pregnancy. In general, orgasm in women is preceded by an increased pelvic awareness followed by a series of strong contractions of the muscles surrounding the vagina and uterus. Unlike

FIGURE 25–7
Sperm Meets Egg. Before fertilization can occur, a sperm must first pass through the external covering of the oocyte. These scanning electron micrographs of a sea urchin egg and sperm show the sperm (a) contacting the external covering of the egg, called the vitelline membrane, then (b) penetrating it. As this occurs, changes take place in the vitelline membrane that prevent other sperm from penetrating it.

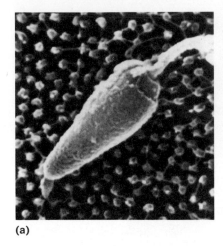

(a)

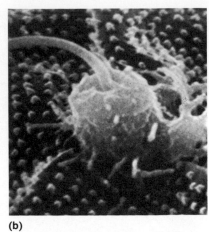

(b)

men, many women can experience multiple orgasms over a relatively short period of time. For both sexes, orgasm is followed by a decline in the breathing rate, pulse rate, and blood pressure, and usually accompanied by a general feeling of relaxation and well-being.

CONCEPTION

Ejaculation deposits about 3–4 ml of semen containing some 200–400 million sperm in the vagina. Given such a vast number of sperm, you might think it virtually certain that at least one sperm will survive the journey to the fallopian tube, encounter the egg, and fertilize it. In fact, however, this often is not the case. The vagina is a relatively hostile environment for sperm because of its high acidity. This condition helps protect the female reproductive tract from fungal and bacterial infections but also kills sperm cells. Furthermore, those sperm that do survive must navigate through the narrow opening in the cervix into the uterus, across the uterine cavity, and up one of the fallopian tubes to reach the egg. In addition to the chemical barriers and the distance the sperm must travel, the timing of sexual intercourse is important. Recall that after ovulation, the secondary oocyte's receptivity to sperm only lasts up to 24 hours. Sperm can live about two to three days in the female reproductive tract. Therefore, intercourse can result in conception only during a 4-day period of the 28-day cycle (3 days before and 1 day after ovulation). (Remember, however, that menstrual cycles vary from woman to woman. Although this formula is used by some individuals in family planning, the margin for

error is great. See the section on contraception later in this chapter.)

Once a sperm does encounter a viable oocyte, the sperm and oocyte membranes unite, and the sperm nucleus enters the oocyte cytoplasm (Figure 25–7). This fusion causes a number of rapid changes in the oocyte. First, its cell membrane changes in a way that prevents other sperm from penetrating. If this did not occur, more than one sperm nucleus could fuse with the egg nucleus, resulting in a zygote with too many chromosomes. Second, the secondary oocyte is triggered to go through its second division of meiosis, ultimately yielding the egg. A few hours later, the sperm and egg nuclei shed their nuclear membranes and fuse to form the nucleus of the zygote.

PREGNANCY

From the moment of fertilization to the birth of a baby takes on the average 266 days (approximately 9 months), or 280 days from the onset of the woman's last menstrual period (assuming that the last ovulation was on day 14 of the woman's cycle). During this nine-month period, referred to as **pregnancy,** the single-celled zygote develops into a multicellular, complex organism.

Embryo Development

After the ovum is fertilized, it travels through the fallopian tube toward the uterus—a journey that takes about a week to complete. During this trip, the zygote

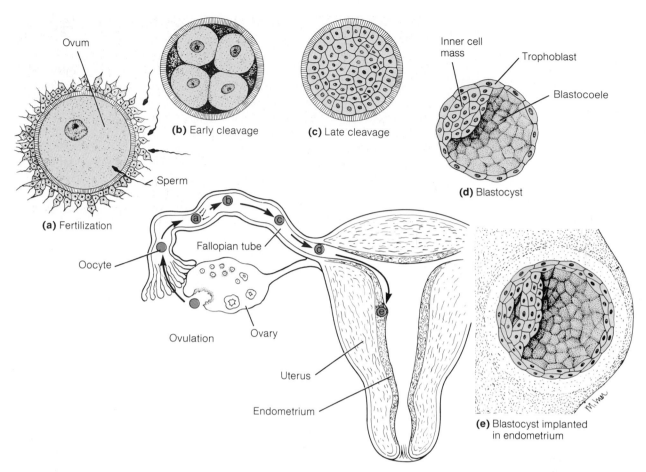

Ovum

(b) Early cleavage

(c) Late cleavage

Inner cell mass

Trophoblast

Blastocoele

(d) Blastocyst

Sperm

(a) Fertilization

Oocyte

Fallopian tube

Ovulation

Ovary

Uterus

Endometrium

(e) Blastocyst implanted in endometrium

FIGURE 25–8
Ovulation to Implantation. The sequence of events depicted in this diagram takes about one week.

divides by mitosis into two cells, then four, eight, and so on. This series of cell divisions is called **cleavage** (Figure 25–8a–c). Since the individual cells do not enlarge after each cell division, the total volume of cells created by cleavage is about the same as the original zygote. By the time it reaches the uterus, the zygote has developed into a semi-hollow ball consisting of an outer layer of cells called the **trophoblast,** a thickened **inner cell mass** at one pole, and a central, fluid-filled cavity called the **blastocoele** (Figure 25–8d). The entire structure is called a **blastocyst.**

Between the seventh and ninth day after fertilization, the blastocyst **implants** in the endometrium. Upon contacting the uterine wall, the trophoblast secretes digestive enzymes that eat away part of the endometrium. This creates a pocket in which the blastocyst settles, and the digestion products, which are absorbed across the trophoblast, serve as the source

of nutrients. Within a few days, the blastocyst is completely surrounded by uterine tissue (Figure 25–8e). The trophoblast then grows extensions into the surrounding endometrium that enhance nutrient absorption.

In addition to its roles in implantation and nutrient procurement, the trophoblast has an endocrine function. It signals its existence to the now pregnant woman by producing a hormone called **human chorionic gonadotropin (HCG).** The level of HCG in the maternal bloodstream begins to rise during the second week of pregnancy, reaches a peak by the beginning of the third month, and then rapidly declines (Figure 25–9). (Tests to confirm pregnancy are based on the high level of HCG present in the blood and urine of a woman during the first three months of pregnancy.) With properties similar to LH, HCG apparently prevents degeneration of the corpus luteum, which con-

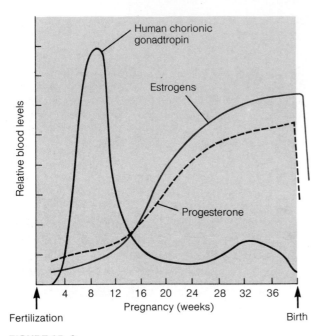

FIGURE 25–9
Hormone Levels in the Maternal Bloodstream During Pregnancy. The production of human chorionic gonadotropin by the trophoblast stimulates the corpus luteum to remain active and produce estrogens and progesterone. After about the eighth week of embryo development, the placenta becomes the primary source of the sex hormones.

tinues to produce estrogens and progesterone. This is crucial to the survival of the embryo, for without the ovarian source of sex hormones, the endometrium would slough off about two weeks after conception, expelling the embryo in the menstrual flow. Menstruation is prevented by the release of the sex hormones initially from the corpus luteum, and later from the placenta (discussed later in this section). Furthermore, the maintenance of high sex hormone levels in the mother blocks the hypothalamus-stimulated release of FSH and LH from the pituitary, so no new follicles develop during the entire term of pregnancy.

Shortly after implantation, a process called **gastrulation** begins. The inner cell mass, the forerunner of the embryo, pulls away from the trophoblast and then splits from inside to form two cavities, the **yolk sac cavity** and the **amniotic cavity** (Figure 25–10). The yolk sac cavity has little importance in the development of the human embryo and is considered to be a vestige of an evolutionary past when it may have functioned in the storage of food (yolk). In humans, the yolk sac cavity ultimately becomes part of the embryo's gut. On the other hand, the amniotic cavity is very important. Surrounded by a layer of embryonic cells called the **amnion,** this fluid-filled cavity expands and eventually surrounds the embryo, cushioning it from physical injury. Meanwhile, the trophoblast gives rise to the **chorion,** which encloses the embryo and its amniotic and yolk sac cavities (see Figure 25–10).

As gastrulation continues, two layers of cells, the ectoderm and endoderm, form between the amniotic and yolk sac cavities (see Figure 25–10). The layer

FIGURE 25–10
Gastrulation. Shortly after implantation, the inner cell mass forms the amniotic and yolk sac cavities, and the central disk of cells differentiates into the ectoderm, endoderm, and mesoderm—all forerunners of the embryo. The trophoblast becomes the chorion with its nutrient-procuring extensions, the chorionic villi.

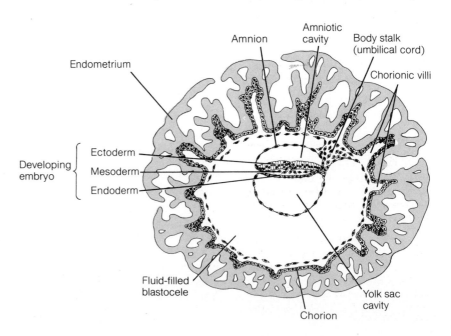

nearest the amniotic cavity, the **ectoderm,** ultimately gives rise to the body's outermost covering (skin epidermis, nails, and hair) and the entire nervous system. The layer adjoining the yolk sac cavity is the **endoderm,** which gives rise to the inner linings of the digestive tract, liver, pancreas, respiratory tract, urinary bladder, and several endocrine glands. After about two weeks of development, a third layer, the **mesoderm,** forms between the endoderm and ectoderm (see Figure 25–10). The mesoderm is the forerunner of muscles, blood, bones, connective tissues, and many internal organs.

With the formation of the ectoderm, endoderm, and mesoderm, gastrulation is complete and the process of **differentiation** begins. Cells that started out alike in all respects begin to take on specialized characteristics and eventually become organized into tissues and organs with specialized functions. At the cellular level, differentiation usually results from differ-

ential gene expression, whereby individual cells become specialized by virtue of producing a select group of proteins. Since proteins determine cell function, and genes encode the information to make proteins, then the specialized activities performed by a cell— activities that differentiate one cell type from another— are determined largely by which genes are being expressed into protein products. Not much is known about the regulation of gene expression during embryonic development, but we do know that it involves very complex chemical and physical interactions between the differentiating cell and its immediate environment. One of the most well-studied types of such interactions is called **embryonic induction.** In this process, one group of embryonic cells becomes signaled (induced) to differentiate into a specific tissue by chemical substances released from a nearby group of cells. An example of embryonic induction is described in Figure 25–11.

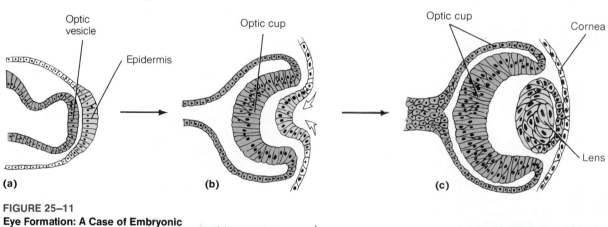

(a)   (b)   (c)

FIGURE 25–11

Eye Formation: A Case of Embryonic Induction. (a) At the three-week stage of human embryo development, two bulges of the brain called the optic vesicles grow toward the overlying epidermis. (b) Each optic vesicle then folds inwardly to form the optic cup, which becomes the retina of the eye. The optic vesicle induces the overlying epidermis to fold inward and thicken. (c) The epidermal tissue pinches off and develops into the lens of the eye, and the nearby epidermis grows over it to become the cornea. (d) Embryonic eye of a chick.

In experiments with frogs, optic vesicles transplanted to other parts of the frog embryo induce the nearby epidermis to develop into a lens and cornea. Thus, some chemical factor produced in the optic vesicle directs the developmental fate of the overlying epidermis.

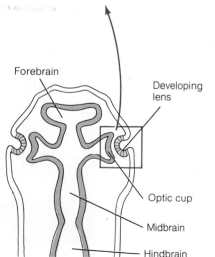

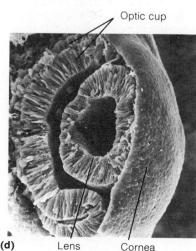

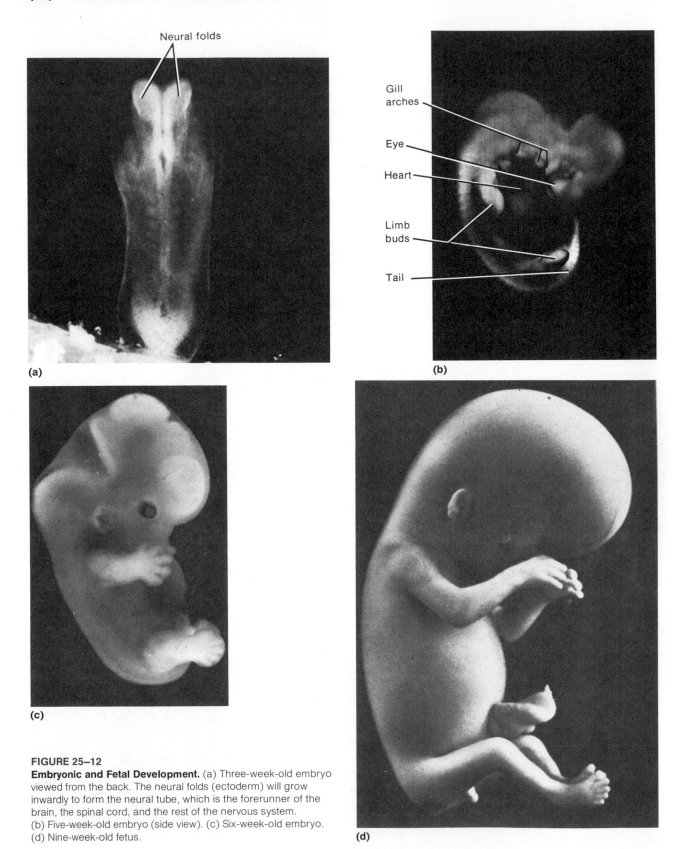

FIGURE 25–12
Embryonic and Fetal Development. (a) Three-week-old embryo viewed from the back. The neural folds (ectoderm) will grow inwardly to form the neural tube, which is the forerunner of the brain, the spinal cord, and the rest of the nervous system.
(b) Five-week-old embryo (side view). (c) Six-week-old embryo.
(d) Nine-week-old fetus.

As the embryo completes its third week of development, it has a cylindrical shape with a clearly defined head and tail region (Figure 25–12a). The nervous system and heart have already begun to develop.

In the meantime, the chorionic (trophoblastic) villi have grown extensively into the surrounding endometrium, which has also grown. Together, these tissues continue to grow to become the **placenta.** The placenta constitutes the transport network between the embryo and the mother. Blood from the mother's circulatory system bathes the placental capillaries of the embryo (Figure 25–13). Maternal blood and embryonic blood do not mix, however. Oxygen and dissolved nutrients in the maternal pool of blood are transported (or diffuse) into the embryo's capillaries. (So are many harmful substances if present in the mother's bloodstream, such as carbon monoxide from cigarette smoke, lead, insecticides, tranquilizers, heroin, caffeine, and alcohol. These substances increase the risks of miscarriage and the birth of underweight and/or deformed babies.) Wastes, including carbon dioxide and urea, move from the embryo's capillaries into the maternal blood, eventually to be expelled by the mother's respiratory and excretory systems. Materials move between the placenta and the embryo via blood vessels in the **umbilical cord,** the embryo's lifeline to the placenta (see Figure 25–13).

By the beginning of the fifth week, the embryo has assumed a C-shaped appearance. Limb buds—the future arms and legs—begin to grow out, and the head region enlarges substantially (Figure 25–12b and c). At this stage the embryo has gill arches and a tail, vestiges of the ancestral past.

Fetal Development

Even after eight weeks of development, the embryo is quite small—about 3 cm (1 inch) long and weighing less than a gram (1/35 ounce). Its appearance, however, is fairly humanlike (Figure 25–12d). Its major organ systems are in place, and its developing circulatory system is pumping blood through the umbilical vessels. At this point, its sex becomes apparent, as the previously uncommitted gonads begin to develop into ovaries or testes. Fingers and toes also become recognizable. The embryo is now somewhat arbitrarily referred to as a **fetus.**

The development of the fetus's internal organs continues into the third and fourth months, as the

FIGURE 25–13

The Placenta. Gases, nutrients, and wastes are exchanged between the maternal blood and the blood flowing through the embryonic capillaries. The arrows indicate the direction of blood flow.

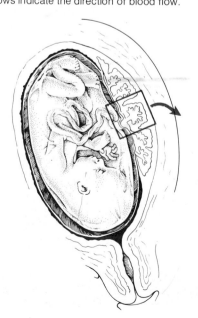

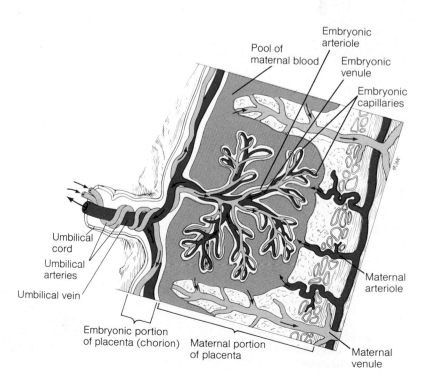

Pool of maternal blood

Embryonic arteriole

Embryonic venule

Embryonic capillaries

Umbilical cord

Umbilical arteries

Umbilical vein

Embryonic portion of placenta (chorion)

Maternal portion of placenta

Maternal arteriole

Maternal venule

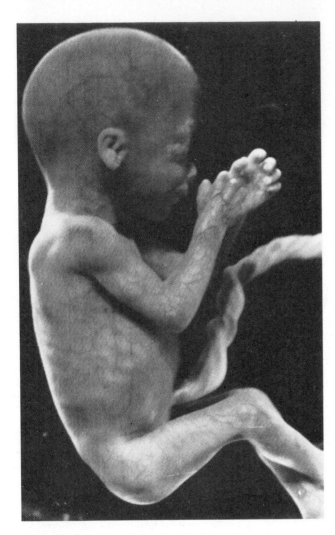

FIGURE 25–14
A Human Fetus at 28 Weeks.

in the sixth or seventh month, the fetus has some chance of surviving with medical assistance (Figure 25–14). During the eighth and ninth months of development, the growth rate slows, and the fetus's internal organs, such as the lungs, undergo the final stages of development in preparation for an independent existence. A few weeks before birth, the fetus normally turns so that its head is in the lower part of the uterus, the optimal position for a safe delivery (Figure 25–15).

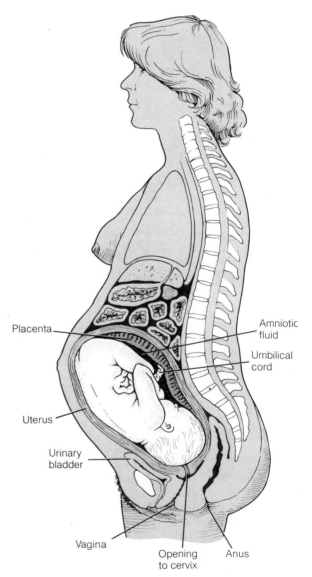

FIGURE 25–15
The Fetus at Nine Months. Shortly before birth, the fetus's head is normally positioned toward the lower part of the uterus.

skeleton and muscles become recognizable. The fetus also becomes sensitive to touch. Movements such as stretching and kicking begin in the fourth month, but they are not easily felt by the mother until the fifth month, when the fetus approaches 20 cm (8 inches) in length and weighs about 225 grams (8 ounces). Near the end of the fourth month, if conditions warrant it, a physician can perform **amniocentesis.** In this procedure, some of the amniotic fluid containing fetal cells is removed with a syringe, and the cells are transferred to a sterile tissue culture medium (Essay 25–1). After the fetal cells have proliferated sufficiently, they can be tested for various genetic defects that may be carried by the fetus.

Fetal development progresses quite rapidly in the fifth and sixth months, so that if premature birth occurs

BIRTH

Curiously, we do not know precisely what triggers **labor,** the rhythmic contractions of the uterine and abdominal muscles that ultimately expel the fetus from the uterus. The initial contractions and labor pains associated with them occur about every 30 minutes, but they become stronger and more frequent as labor progresses. When the contractions are one to three minutes apart, birth is imminent. During this **first stage of labor,** the contractions push the fetus toward the cervix, causing the cervix to dilate considerably (about 10 cm, or 4 inches) (Figure 25–16a). The contractions also cause the amniotic sac to rupture, an event commonly known as the "breaking of the waters."

Once the cervix is fully dilated, the infant's head drops into the vagina, marking the beginning of the **second stage of labor,** or expulsion (Figure 25–16b). In a series of strong contractions that lasts several minutes to perhaps two hours, the infant is pushed through the birth canal. Passage is usually aided by the mother's conscious use of her abdominal muscles to push the baby along. Soon after the infant has emerged, the umbilical cord is tied off and cut. With the connection to the placenta severed, the infant's blood quickly builds up in carbon dioxide level. This is one factor that triggers the brain to initiate the first gasping breath.

Soon after birth, the uterus contracts dramatically in size, causing the placenta to dislodge from the uterine wall and eventually be expelled as the **afterbirth** (Figure 25–16c). Placental separation and expulsion usually occurs within 30 minutes after birth, and is referred to as the **third stage of labor.**

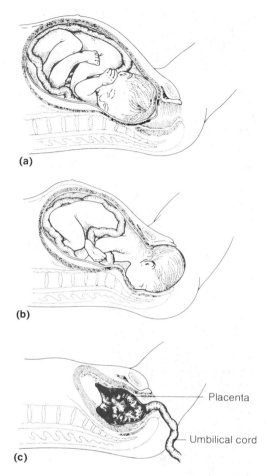

FIGURE 25–16
The Birth of a Baby. (a) First stage of labor. (b) Second stage of labor. Contractions of the mother's abdominal and uterine muscles push the baby through the dilated cervix and vagina. (c) After the baby is delivered, the placenta is expelled in the third stage of labor.

Twins

Single baby births account for 98.8% of all deliveries. Of the remaining 1.2% of multiple birth cases, the overwhelming majority are twins. **Fraternal twins** (nonidentical) result when two eggs are released during the same menstrual cycle, and each is fertilized by a different sperm (Figure 25–17a). Fraternal twins have different genetic makeups (genetically different eggs and sperm) and so bear only the same family resemblance to each other as siblings born at different times. They may be the same or opposite sex. About 70% of all twins are fraternal.

Identical twins, on the other hand, arise from a single zygote that has split at some early stage of development into two independent embryos (Figure 25–17b). (In rare cases, identical twins remain joined at some point along their bodies, resulting in Siamese twins.) The separated embryos complete their development side by side in the uterus (as do fraternal twins). Because identical twins arise from the same zygote, they have identical genetic constitutions and thus appear identical in every regard, including sex.

There is an inheritable tendency in some women to release two eggs during a menstrual cycle, so fraternal twins tend to "run in the family." Producing identical twins, on the other hand, does not appear to be an inheritable trait.

ESSAY 25–1

AMNIOCENTESIS

Not too many years ago, a pregnant woman who wondered about the health of her fetus had no recourse but to go on wondering until its birth. For a couple with a genetic disorder in either his or her family history, or for a woman who suffered rubella (German measles) during her pregnancy, the months were often filled with anxiety and doubts. This was particularly true for older women who became pregnant. One of the more common genetic disorders is Down syndrome, a chromosomal abnormality resulting in mental retardation and various physical deformities. The chances of having a baby with Down syndrome increase dramatically with the age of the mother (see Figure 8–10).

Many of the doubts concerning the genotype of a human fetus can now be eliminated by a relatively simple procedure called amniocentesis. With this technique, an obstetrician can determine whether a fetus has genetic abnormalities, in-

cluding defects in chromosomes and deficiencies in the ability to synthesize certain critical enzymes. If an abnormality is discovered, the expectant parents are notified so that they can decide whether to terminate the pregnancy with an abortion or to allow it to continue.

Amniocentesis is performed at about the sixteenth week of pregnancy and requires only a local anesthetic. With the aid of a sonar probe to locate the exact position of the fetus, the physician inserts a long, slender needle through the woman's abdominal and uterine walls into the amniotic sac. Some of the amniotic fluid containing freely suspended fetal cells is withdrawn, and the cells are grown in tissue culture for several weeks until enough of them accumulate for medical testing purposes. At that time some of the cells are stained to visualize their chromosomes under the microscope; others may be analyzed for the presence or absence of one or more critical enzymes.

Physicians and genetic counselors generally recommend that amniocentesis be seriously considered by women over 35 years of age, as well as by women who have a family history of genetic disorders. For others, the choice may be less clear-cut because the procedure does pose some risk (although a very small one) to the fetus. In other words, if you are merely curious to know the sex of your child at 16 weeks of age, it may be wiser to live with the uncertainty rather than risk fetal injury.

Amniocentesis is also performed to evaluate the ability of the fetus to live outside the uterus in cases where an infant's well-being may depend on early delivery. The amounts of certain substances in the amniotic fluid indicate whether the lungs of the fetus are mature enough to begin the process of respiration. Amniocentesis is performed weekly during the last two months of pregnancy until maturity is confirmed. Then labor can be induced or cesarean delivery performed.

FIGURE 25–17

Twins. (a) Fraternal twins result from the union of two different sperm with two different eggs. Note that each fetus has a separate placenta. (b) Identical twins are formed from a single zygote that splits in half at an early stage of embryo development. Identical twins share the same placenta.

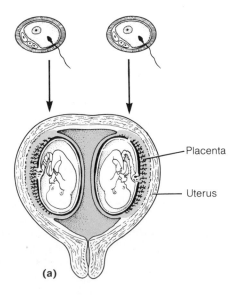
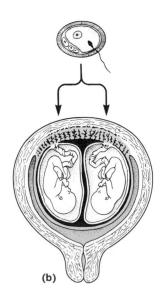

(a) (b)

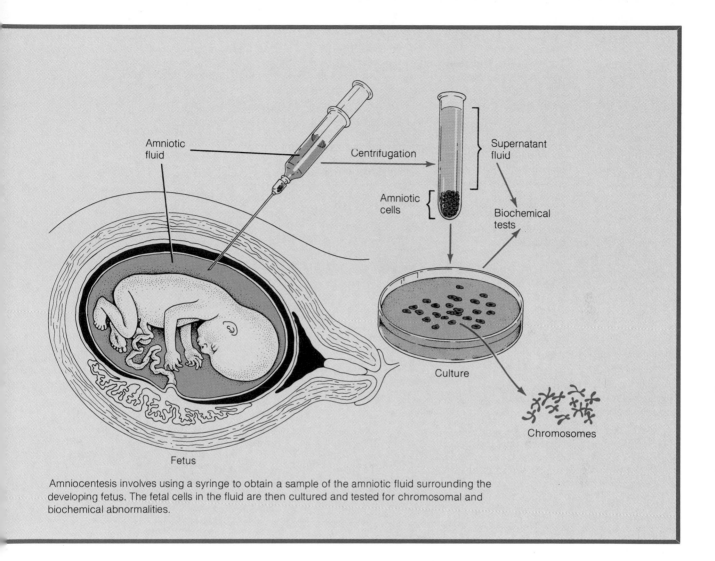

Amniocentesis involves using a syringe to obtain a sample of the amniotic fluid surrounding the developing fetus. The fetal cells in the fluid are then cultured and tested for chromosomal and biochemical abnormalities.

CONTRACEPTION

Contraception, or birth control, refers to the willful prevention of conception that might otherwise result from sexual intercourse. There are several general ways to effect contraception, including:

1. Restricting intercourse to periods of the menstrual cycle when the egg is not available for fertilization;
2. Preventing the sperm from reaching the egg;
3. Preventing ovulation;
4. Post-fertilization interference with embryo implantation or development;
5. Sterilization.

Table 25–1 summarizes the effectiveness, the inconveniences, and the possible side effects of the most popular contraceptive techniques now in use. A more complete description of some of these methods follows.

Methods Based on the Menstrual Cycle

Recall that normally one egg is shed from an ovary into the fallopian tube on about day 14 of the menstrual cycle, where it remains viable for approximately 24 hours. Sperm can remain viable for up to 3 days in the female reproductive tract. Theoretically then,

TABLE 25–1
The Effectiveness and Potential Side Effects of Various Contraceptive Techniques. The effectiveness of various birth control methods is expressed here in pregnancies per 100 woman-years of use. For example, a failure rate of 10 indicates that each year 10 women out of every 100 using a given method will become pregnant. Implicit in the figures are faults in the method itself as well as human error, such as positioning a diaphragm incorrectly or forgetting to take a pill.

Contraceptive Method	Rate of Failure per 100 Woman-Years	Potential Side Effects
Tubal ligation	0	Usually irreversible sterility
Vasectomy	0	Usually irreversible sterility
Oral contraceptives (birth control pills)	2–5	Increased risk of headache, nausea, swelling of breasts, blood clots, heart disease, and possibly cancer
Intrauterine device (IUD)	2–8	Can cause irritation and excessive menstrual flow; in rare cases, sterility (perforation of the uterus)
Condom	3–10	Occasional mild irritation and diminished sensitivity in males; may break or leak
Diaphragm (with foams, creams, or jellies)	3–17	Inconvenient to insert
Spermicidal foams and jellies	5–30	Sometimes minor irritation
Coitus interruptus (withdrawal of penis just before ejaculation)	10–25	Frustration in males and females
Natural planning methods	15–25	Requires abstinence for 8–12 days during monthly cycle
Douche	34–60	Infection, irritation

intercourse will not result in pregnancy except during those few days before and one day after ovulation.

The **natural family planning methods** (also referred to as **fertility awareness methods**) are based on **abstinence** from sexual intercourse during the period of high risk, usually about 8–12 days during the midphase of the woman's cycle. Each of the natural family planning methods differs in how the ovulatory period is determined.

With the **calendar method,** also called the **rhythm method,** the woman estimates her fertile period by counting the number of days in her menstrual cycle for about six months and using an averaging formula to determine her high-risk days. The **basal body temperature method** is based on temperature fluctuations before and after ovulation. Generally, the body temperature drops slightly on the day of ovulation and then rises the following day to a level that is maintained until the end of the cycle.

With the **ovulation method,** also known as the **mucus method,** the woman learns to interpret the changes that occur in the cervical mucus during the menstrual cycle. When the normal vaginal secretions are yellow or white and sticky, the fertile period has begun. At ovulation, the mucus is clear, stringy, and stretchy.

The advantages of the natural family planning methods include the absence of harmful side effects and little or no expense. Major disadvantages include lack of reliability and long periods of abstinence from intercourse. Reliability suffers because menstrual cycles can vary from month to month, and a number of physiological and psychological factors can affect the timing of ovulation.

Physical and Chemical Barriers to Sperm Movement

A variety of contraceptive techniques are based on preventing viable sperm from reaching the fallopian tubes. A **condom** (rubber) is a sheath usually made of thin surgical latex or sheep membrane that is worn over the penis during intercourse (Figure 25–18a). Ejaculated sperm cannot get through the condom and are thus prevented from entering the vagina. Some men complain that a condom reduces the sensitivity of the penis during intercourse, and taking time to put one on at the height of arousal is distracting and inconvenient. In addition, it may break or leak, permitting semen to escape. On the plus side, using a condom reduces the risk of spreading or contracting certain sexually transmitted diseases because most microorganisms cannot penetrate the sheath.

A **diaphragm** is a thin, flexible dome that fits over the woman's cervix and prevents ejaculated sperm from moving into the uterus (Figure 25–18b). Diaphragms are generally used in conjunction with spermicidal (sperm-killing) foams, creams, or jellies to maximize their effectiveness. The diaphragm, which is fitted to the wearer by a qualified medical practitioner, is inserted by the woman prior to intercourse. It must then be kept in place at least six hours after intercourse.

Diaphragms are effective if used properly, but they are bothersome for the woman. Occasionally they develop holes, allowing some sperm to get through. In addition, diaphragms must be refitted after childbirth, which irreversibly stretches the cervical opening, and after a major weight gain or loss.

Cervical caps are similar to the diaphragm in that they are placed over the cervix and must be fitted to the wearer (Figure 25–18c). The cervical cap stays in place by suction, however, and can remain in the woman's body for one or more days. This method of birth control has not been approved for use in the United States, although it is popular in Europe.

The **vaginal sponge** has recently been approved for use in the United States. The sponge is made of natural collagen (a fibrous protein), and has a spermicidal chemical within its matrix. The small, pillow-shaped sponge is inserted into the vagina prior to intercourse and placed near the cervix. Its effectiveness rate is about the same as the condom and diaphragm.

Spermicidal foam or jelly can be used alone as a contraceptive agent, but it is not as effective as when used with a diaphragm or other physical barrier. The spermicide is inserted into the vagina just before intercourse.

Coitus interruptus, withdrawal of the penis just before ejaculation, is a very ineffective form of birth control. Besides requiring great control on the male's part, sperm-containing secretions may "leak" from the penis prior to ejaculation.

Preventing Ovulation

The most commonly-used oral contraceptive in the United States today is the **combination birth control pill.** This pill contains a mixture of two synthetic hormones similar in structure and action to estrogens and progesterone. Generally, a woman takes one pill each day between day 5 and day 25 of the menstrual cycle. (Pills marketed in 28-day packets consist of seven nonhormone pills taken on the first four and last three days of the cycle, thus eliminating the need to remember the start-up day of the next cycle—one pill is taken *every* day.)

The combination pill works in three ways. First, the synthetic estrogen acts on the hypothalamus to block the pituitary from releasing FSH and LH. In the absence of these gonadotropins, follicle development is suspended, so ovulation does not occur. Second, progestin (a synthetic hormone similar to progesterone) provides additional contraceptive protection by causing a thickening and chemical alteration of the cervical mucus. This makes passage of viable sperm into the uterus more difficult. Third, progestin also leads to changes in the uterine lining that make implantation less likely.

A more recent addition to the oral contraceptive market is the **progestin-only pill**, often called the mini-pill. The mini-pill actually contains less progestin than the combination pill, and lacks estrogens entirely. Women taking the mini-pill continue to ovulate at least occasionally, but they derive the same contra-

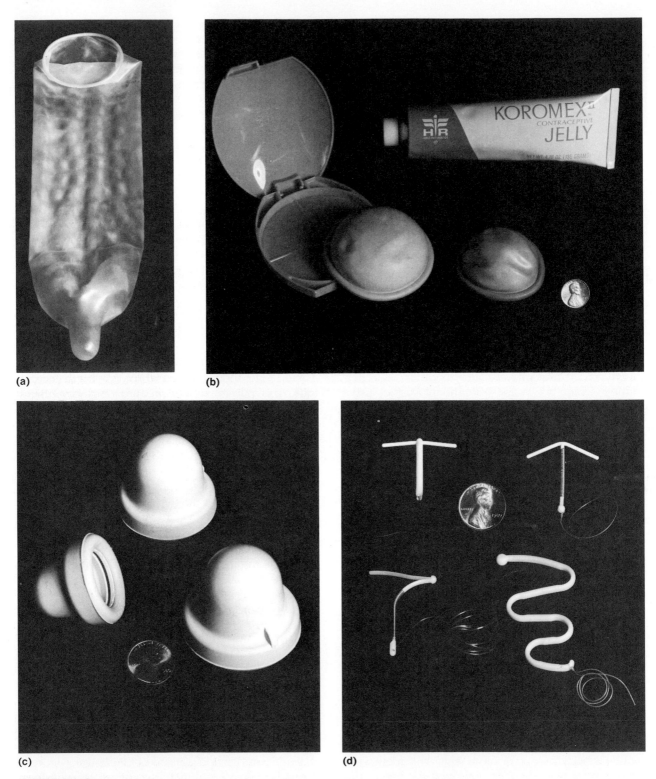

FIGURE 25-18
Contraceptive Devices. (a) Unrolled condom with reservoir tip. (b) Diaphragm and contraceptive jelly. (c) Cervical caps. (d) Various types of intrauterine devices (IUDs).

ceptive effects of the combination pill with regard to the changes in the cervical mucus and uterine lining.

Both types of pills are very effective birth control agents and over 50 million women throughout the world use them. There are some drawbacks, however. In mimicking the contraceptive aspect of pregnancy, the combination pill also mimics many of pregnancy's side effects, including weight gain, headaches, nausea, depression, and tenderness of the breasts. Furthermore, users over the age of 35, especially smokers, run a greater risk of heart attack, stroke, blood clots in the legs and pelvis, and possibly certain forms of cancer. Interestingly, taking the combination pill reduces the risk of getting cervical or uterine cancer. Long-term studies on the potential health hazards of the mini-pill are incomplete. In terms of serious side effects, however, most physicians feel the mini-pill is safer than the combination pill. Even so, users of the mini-pill may experience changes in weight, depression, gastrointestinal problems, and breast tenderness.

Prevention of Implantation or Embryo Development

Intrauterine devices (IUDs) are made of metal or plastic and come in a variety of shapes (Figure 25–18d). An IUD is inserted into the uterus by a qualified medical practitioner, and then requires only occasional checking by the user. While in the uterus, the IUD causes mild irritation of the endometrium, which apparently prevents implantation of a blastocyst. It is a highly effective contraceptive, although it may be spontaneously expelled without the wearer realizing it.

About 25% of all women cannot use an IUD because it causes excessive (and sometimes midcycle) menstrual flow, severe cramps, and/or an allergic reaction. Recent evidence indicates that IUD users may be more prone to vaginal and tubal infections. There is also a risk of uterine perforation.

There are several post-intercourse measures that are designed to prevent implantation or disrupt embryo development. For example, taking high levels of synthetic hormones, such as progestin, estrogen–progestin combinations, or diethylstilbestrol (DES), within 72 hours after intercourse prevents implantation. Interestingly, DES, also known as the **morning-after pill,** was once prescribed to reduce the chances of miscarriages (Essay 25–2). **Abortion,** the removal of the fetus by mechanical or chemical means, is a more drastic (and controversial) measure to interrupt fetal development.

Sterilization

Sterilization for the purpose of contraception involves blocking or removing some part of the reproductive system so that a person is no longer fertile. For men, the most common method is **vasectomy.** This simple operation requires only a local anesthetic and entails removing a short piece of each vas deferens, then tying off the loose ends (Figure 25–19a). Thus, the sperm are blocked from moving into the urethra during ejaculation, but the seminal fluids are still emitted. Sperm production continues, but the sperm are absorbed into the body and destroyed by phagocytotic white blood cells. A vasectomy does not affect a man's sex drive or his hormone production. The operation is generally considered irreversible, although new surgical procedures have improved the chances that a severed vas deferens can be reconnected.

Female sterilization is also relatively safe, simple, and inexpensive. **Tubal ligation** involves severing and tying off or cauterizing the fallopian tubes (Figure 25–19b). The woman continues to ovulate, but the severed tubes prevent any possibility of fertilization. This operation can be accomplished using small incisions and local or general anesthesia. The reversal rate of female sterilization is greater than that for vasectomies, but in general, all sterilization procedures should be considered irreversible.

SEXUALLY TRANSMITTED DISEASES

The human reproductive system is susceptible to a variety of infectious diseases. Many are spread primarily through sexual contact; hence, they are designated **sexually transmitted diseases (STDs).** Certain STDs, such as gonorrhea and syphilis, are subclassified as venereal diseases because they are transmitted almost exclusively by sexual contact; others, such as yeast infections and AIDS (acquired immune deficiency syndrome), may be contracted by other means in addition to sexual intimacy.

STDs are a major health problem in the United States and elsewhere. Millions of new cases of gonorrhea and genital herpes are reported in the United States each year, and probably untold millions go unreported. What is particularly alarming is that the rate of occurrence of these and other STDs has been increasing lately. Major contributing factors to this trend seem to be the increased sexual activity among

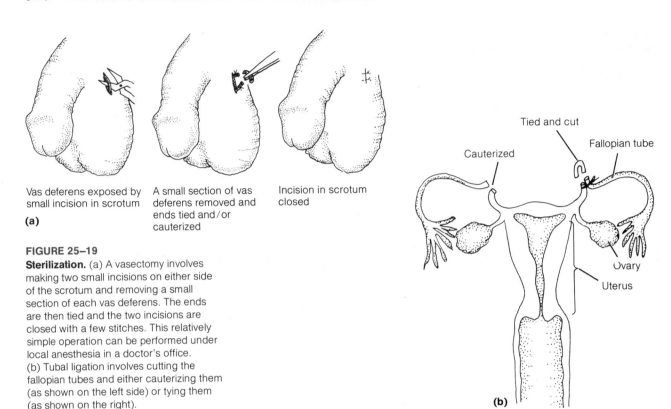

Vas deferens exposed by small incision in scrotum

(a)

A small section of vas deferens removed and ends tied and/or cauterized

Incision in scrotum closed

Cauterized

Tied and cut

Fallopian tube

Ovary

Uterus

(b)

FIGURE 25–19
Sterilization. (a) A vasectomy involves making two small incisions on either side of the scrotum and removing a small section of each vas deferens. The ends are then tied and the two incisions are closed with a few stitches. This relatively simple operation can be performed under local anesthesia in a doctor's office. (b) Tubal ligation involves cutting the fallopian tubes and either cauterizing them (as shown on the left side) or tying them (as shown on the right).

young people, and an increased tendency toward having multiple sex partners. Another factor may be simply a greater incidence of reported cases as a result of an increased willingness to seek medical treatment for STDs. Whatever the case might be, the ultimate solution to the epidemic of STDs is for sexually active individuals to take the responsibility for their health and those with whom they are intimate. This involves learning to recognize the symptoms of STDs, seeking the appropriate medical assistance when necessary, and refraining from sexual contact during periods of active infection (Table 25–2). All but two STDs— herpes and AIDS—are presently curable, and the cure for herpes seems close.

SUMMARY

Human reproduction differs from that of most other mammals in that female sexual receptivity is not restricted to periods coinciding with ovulation. Thus, humans have a unique ability to engage in sexual intercourse that is not exclusively reproductive in function.

The human female's reproductive system is composed of several organs, each specialized for part of the reproductive process. The ovaries produce and release eggs, which then travel down the fallopian tubes (normally one egg per month), where fertilization can occur. The fertilized egg implants in the endometrium, the highly vascularized lining of the uterus, where the embryo develops until birth. Ovulation and endometrium development are coordinated by the regulated release of hormones: follicle-stimulating and luteinizing hormones produced in the pituitary, and estrogens and progesterone released from the ovaries. In the absence of pregnancy, these events repeat roughly every 28 days and are referred to as the menstrual cycle.

The male reproductive system is specialized for the production and ejaculation of sperm. Hundreds of millions of sperm are produced daily in the seminiferous tubules (highly coiled tubes located in each testis) and collect in the epididymis. During ejaculation, the sperm move through the pair of vas deferens into the urethra, where they mix with nutrient-laden fluids secreted by the seminal vesicles, prostate gland, and Cowper's glands. The sperm deposited in the

TABLE 25—2
Common Sexually Transmitted Diseases (STDs).

STD	Transmission	Symptoms	Treatment
Trichomoniasis	*Trichomonas vaginalis,* a single-celled parasite, is passed through genital sexual contact; or less frequently by towels, toilet seats, or bathtubs used by an infected person.	White or yellow vaginal discharge that has an unpleasant odor; vulva is sore and irritated.	Metronidazole (Flagyl), a prescription drug.
Moniliasis (yeast infection)	The *Candida albicans* fungus may accelerate growth when the chemical balance of the vagina is disturbed; it may also be transmitted through sexual interaction.	White, "cheesey" discharge; irritation of vaginal and vulvar tissue.	Vaginal suppositories of Mycostatin or candicidin (fungicides)
Gonorrhea	*Neisseria gonorrhoeae* ("gonococcus") bacteria is spread through genital, oral-genital, or genital-anal contact.	Most common symptoms in men are a cloudy discharge from the penis and burning sensations during urination. If untreated, complications may include inflammation of scrotal skin and swelling at the base of testicle. In women, some green or yellowish discharge is produced. At a later stage, pelvic inflammatory disease may develop.	Penicillin, tetracycline, or erythromycin (antibiotics)
Syphilis	*Treponema pallidum* (a bacterium) is transmitted from open lesions during genital, oral-genital, or genital-anal contact.	*Primary stage:* A painless chancre appears at the site where bacterium entered the body. *Secondary stage:* The chancre disappears and a generalized skin rash develops. *Tertiary stage:* Heart failure, blindness, mental disturbance, and many other symptoms. Death may result.	Penicillin, tetracycline, or erythromycin.
Acquired immune deficiency syndrome (AIDS)	Agent believed to be a virus passed through homosexual contact between men. May also be transmitted via hypodermics used for blood transfusions and drugs.	Impaired immune response, making victim susceptible to other diseases. Usually fatal.	No cure; treatment is for associated diseases
Herpes	Genital herpes appears to be transmitted primarily by vaginal, oral-genital, or anal sexual intercourse. Oral herpes is transmitted primarily by kissing.	One or more small red, painful bumps (papules) appear in the region of the genitals (genital herpes) or mouth (oral herpes). The papules develop into painful blisters that eventually rupture to form wet, open sores.	A variety of treatments may reduce symptoms, but no cure is presently available.
Genital warts (venereal warts)	Primarily spread through genital, anal, or oral-genital interaction.	Warts are hard and yellow-gray on dry skin areas; soft, pinkish-red, and cauliflower-like on moist areas.	Surface applications of podophyllin; large warts may require surgical removal.

ESSAY 25-2

DES AND ITS TRAGIC LEGACY

In recent years the incidence of a rare and often fatal form of vaginal cancer, called clear-cell adeno-carcinoma, has risen dramatically among young women. More cases have been reported in the last 15 years than previously reported in the world's medical history. After an exhaustive search into these women's medical histories, one common feature emerged: their mothers had all taken diethylstil-bestrol (DES), a synthetic estrogen, during their pregnancies.

Between 1940 and 1971, DES was prescribed for pregnant women who had histories of miscarriages. Lacking substantiating studies, drug companies and medical authorities claimed that DES improved the chances for a safe, full-term pregnancy. In 1971, after the association between DES and clear-cell adeno-carcinoma was discovered, its use for preventing miscarriages was discontinued. Ironically, DES has been shown to be ineffective in preventing miscarriages.

DES daughters and their tragic inheritance have received much attention. But at least as many sons were born to women who took DES during pregnancy, and evidence is growing that they too are at risk for serious health problems. What is the legacy of DES?

Although clear-cell adenocarcinoma in DES daughters is potentially the most dangerous of the DES-linked medical problems, there are others. In women who took DES during pregnancy, studies suggest an increase in the incidence of breast and gynecological cancers, particularly if other types of synthetic estrogens (such as those in oral contraceptives) were taken at the same time. Daughters exposed to DES in utero (as fetuses in a mother taking DES) display an unusually high incidence of vaginal adenosis, an abnormal but noncancerous growth of glandular tissue that is considered a possible forerunner of vaginal cancer. In addition, a number of structural abnormalities in the reproductive organs of individuals exposed to DES during early fetal development have been recorded in both males and females. The abnormalities in males are often associated with sterility and increased risk of testicular cancer.

Although much has been learned about DES over the years, the search for victims continues, and the legal controversy over DES usage rages on. To compound the problem, DES is used as a "morning-after" pill (a post-intercourse contraceptive) and as a lactation suppressant for women who have just given birth. These practices are currently under scrutiny by the Food and Drug Administration (FDA). In a related matter, the FDA banned the use of DES as a food additive used to fatten cattle and poultry before slaughter. The ban was issued in 1973 (and upheld in the courts in 1979), when significant levels of DES were detected in the food products derived from these animals.

The prognosis for those people exposed to DES is still uncertain, as medical researchers continue to study its long-term effects. For the rest of us, the lesson is clear: both medical practitioners and medical consumers must exercise caution in the use of any synthetic hormone. Hormones are potent regulators of vital bodily processes, but if present in excessive amounts, they can have the opposite effect— unregulated and abnormal cellular activities.

---

woman's vagina during sexual intercourse must swim through the uterus and into one of the fallopian tubes to fertilize an egg.

Pregnancy in humans normally lasts about 266 days (nine months). The zygote begins dividing during its journey to the uterus, and by the time it reaches the uterus, the week-old embryo consists of about 100 cells shaped into a hollow ball called the blastocyst. The outer layer of cells—the trophoblast—invades the endometrium during implantation and also produces human chorionic gonadotropin, a hormone that stimulates the ovarian corpus luteum to persist and continue to synthesize estrogens and progesterone. This prevents menstruation and the development of a new follicle. The trophoblast eventually develops into the chorion (the fetal portion of the placenta), an organ through which maternal and fetal blood exchange nutrients, gases, and wastes.

In the meantime, the inner cell mass within the trophoblast begins to differentiate into the ectoderm, endoderm, and, later, the mesoderm. These will ultimately give rise to all the tissues and organs of the new individual. Also during this second week of embryonic growth, the yolk sac and amniotic cavity are formed; the latter eventually surrounds the embryo and fills with amniotic fluid to cushion the fetus.

By the end of the second month, the embryo has developed a heart and the major blood vessels, the limbs have grown out, the major organ systems are in place, and the sex organs have begun to form— all of this in an embryo that is now about the size of an almond.

The next five months of development are marked by a large increase in size and the development of the skeleton, muscles, and nervous system. In the eighth and ninth months, the growth rate slows and final preparations are made for expulsion from the womb.

The birth process is characterized by three stages of labor: a period of strong and increasingly more frequent contractions of abdominal and uterine muscles, the expulsion of the infant, and the expulsion of the placenta. Shortly after passage of the baby, the umbilical cord is tied off and cut, which helps trigger the infant's first independent act: taking its first breath.

People use various contraceptive techniques to minimize the chance of conception when pregnancy is not desired. Some methods attempt to coordinate abstinence from intercourse with the stage in the woman's menstrual cycle when ovulation is most likely to occur. Physical and chemical barriers such as the condom, diaphragm, cervical cap, vaginal sponge, and spermicidal agents prevent the sperm from reaching an egg. The most popular method of contraception is female oral contraceptives (birth control pills). Consisting of various combinations of synthetic female sex hormones (estrogen and/or progestin), the pills prevent ovulation. Intrauterine devices or taking a large dose of hormones such as diethylstilbestrol after intercourse may be used to prevent a fertilized egg from implanting in the endometrium. An increasingly popular contraceptive technique, and probably the safest and most effective other than total abstinence, is sterilization. In males, sterilization involves cutting and tying off the vas deferens (vasectomy), thereby preventing sperm from entering the ejaculated seminal fluids. To sterilize females, the fallopian tubes are similarly cut and tied off, a procedure called tubal ligation.

Sexual contact can result in the passing of infectious agents that may cause uncomfortable and potentially serious diseases. Sexually transmitted diseases (STDs) include gonorrhea, syphilis, yeast infections, herpes, and others. If not treated promptly, some STDs can cause extensive damage to the reproductive and other organ systems.

STUDY QUESTIONS

1. How does human reproductive behavior differ from that of other mammals with estrus cycles?

2. What are the roles of FSH and LH in the menstrual cycle?

3. How does fertilization prevent the completion of the menstrual cycle?

4. Describe the pathway of a sperm cell from its site of production in a testis to its release during ejaculation.

5. Why can only one sperm fertilize an egg?

6. What are the functions of the embryonic trophoblast?

7. List the key events in the first, second, and third stages of labor.

8. How do fraternal twins differ from identical twins?

9. If a released egg can remain viable in the fallopian tube for only 24 hours, why does the rhythm method recommend that abstinence begin on day 10 and last through day 18 of the woman's menstrual cycle?

10. How does the combination pill prevent ovulation?

SUGGESTED READINGS

Castleman, M. *Sexual Solutions.* New York: Simon and Schuster, 1981. Provides frank, practical information on male sexuality.

Combs, B. J., D. R. Hales, and B. K. Williams. *An Invitation to Health.* 2nd ed. Menlo Park: Benjamin/Cummings, 1983. This inclusive health guide includes informative chapters on sexuality, reproduction, pregnancy, childbirth, and parenting.

Crooks, R., and K. Baur. *Our Sexuality.* 2nd ed. Menlo Park: Benjamin/Cummings, 1983. This informative text deals with both the social and biological components of human sexuality. Chapters 4, 5, 11, and 12 are especially pertinent to the material presented here.

Djerassi, C. *Politics of Contraception.* New York: Norton, 1979. An informative look at how the pill was developed, including the legal and political issues it generated.

Epel, D. "The Program of Fertilization." *Scientific American* 237(1977):128–138. A number of outstanding electron micrographs illustrate interactions between sperm and egg that lead to fertilization.

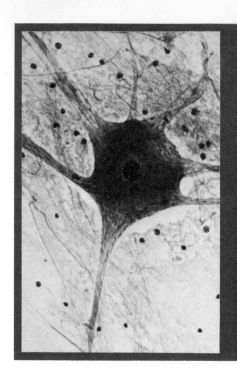

CHAPTER 26

Neurons, Sensory Reception, and Muscle Response

Who has not witnessed the uncanny ability of a house-fly to escape the swatter? Or wondered how a seem-ingly snoozing dog senses the mail carrier approaching and instantly springs to the door, barking loudly? These common examples point out two of the most obvious features of animals: they are motile, and they can sense and respond to changes in the environment. **Motility,** or the ability to move, is made possible by systems of muscles. But the activities of muscles must be coordinated in some manner if directed move-ments, rather than random ones, are to be made. This coordination is the function of the nervous sys-tem. Various parts of the nervous system *collect infor-mation* from the environment, *integrate and process* the information received, and, finally, *act* upon the information by coordinating the actions of muscles (or glands) to bring about appropriate responses. Thus, nerves and muscles work together to give animals their wide repertoires of physical activities, including escaping the fly swatter and sensing an approaching mail carrier.

All animals except sponges have systems of nerves and muscles, but the complexity of these systems varies greatly. For example, the nervous system of sea urchins consists of only a sparse collection of nerve fibers, whereas the human nervous system is exceedingly complex. This variation, however, is primarily a reflection of the number of nerve cells and their degree of organization and centralization. When we examine the nervous systems of sea urchins and humans at the cellular level, the distinction be-tween the two becomes less obvious. Let us begin our discussion then by describing the structure and function of the basic unit of all animal nervous sys-tems—the nerve cell, or **neuron.**

NEURONS

All neurons share two basic properties: **excitability,** the ability to sense and respond to a stimulus, and **conductivity,** the capacity to carry electrical mes-sages, or impulses, from one end of the cell to the other. It is tempting to picture neurons as the cellular equivalents of telephone wires, but such a view is too simplistic. Neurons are living cells; they contain cytoplasm, a nucleus, and organelles, and they carry out the same bioenergetic processes as other living cells. What distinguishes neurons from other cells is their highly specialized design for receiving, processing, and transmitting information.

Types of Neurons

Neurons can be classified into three categories based on their function, location, and the direction in which they conduct impulses. One type of neuron is the **sensory neuron,** which carries impulses *toward* the central nervous system. (In vertebrates, the central nervous system consists of the brain and spinal cord; see Chapter 27.) Some sensory neurons are activated directly by a stimulus, such as touch; others receive stimuli indirectly from specialized sensory receptors that detect light, odors, sound waves, or other stimuli. The second type of neuron is the **motor neuron,** which conducts impulses *from* the central nervous system to a muscle or gland. There, the motor neuron triggers some form of physical activity, such as muscle contraction or glandular secretion.

    Interneurons, the third category of nerve cells, are found only in the spinal cord and brain. Interneurons usually form the connecting link between sensory neurons and motor neurons. In the case of a **reflex arc,** one or more interneurons in the spinal cord connect a sensory neuron to one or more motor neurons. If you accidentally step on a tack, you don't have to think about pulling back; the reflex arc takes over before you realize what happened (Figure 26–1). In most cases, however, many interneurons are involved in the transmission of information between the sensory and motor systems. The vast complex of interneurons in the brains of higher animals integrates and processes most incoming signals, and is responsible for such complicated behavior as memory, learning, and emotions (see Chapter 27). In humans, interneurons are by far the most abundant type of neuron, comprising more than 90% of the 100 billion or so nerve cells in the body.

Neuron Structure

The two functional characteristics of all neurons— excitability and conductivity—have their bases in nerve cell structure. This becomes apparent when we examine a typical vertebrate neuron. Every neuron has four distinct regions: a cell body, dendrites, an

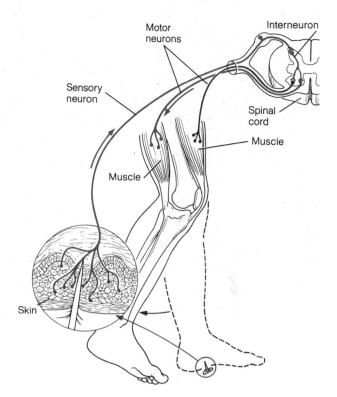

FIGURE 26–1
The Reflex Arc. Reflex arcs such as the one diagrammed here permit extremely quick responses to external stimuli. If you step on a tack, the signal is detected by a sensory neuron, transmitted to an interneuron in the spinal cord, and finally sent to the appropriate leg muscles via motor neurons. The contraction of the leg muscles raises your foot.

axon, and terminal branches (Figure 26–2). The **cell body** contains the nucleus and most of the organelles involved in cell maintenance and growth. The **dendrites** are cytoplasmic extensions of the cell body that are thin, brushy, and usually short. Their function is to receive incoming signals. The **axon** is also an extension of the cell body, but it is generally less branched and much longer than a dendrite. Axons conduct impulses. They end in **terminal branches** that bear tiny swollen tips called **synaptic knobs** (Figure 26–2c). As we shall see later in this chapter, these knobs are important in the relay of chemical information between neurons, or from a motor neuron to a muscle or gland cell.

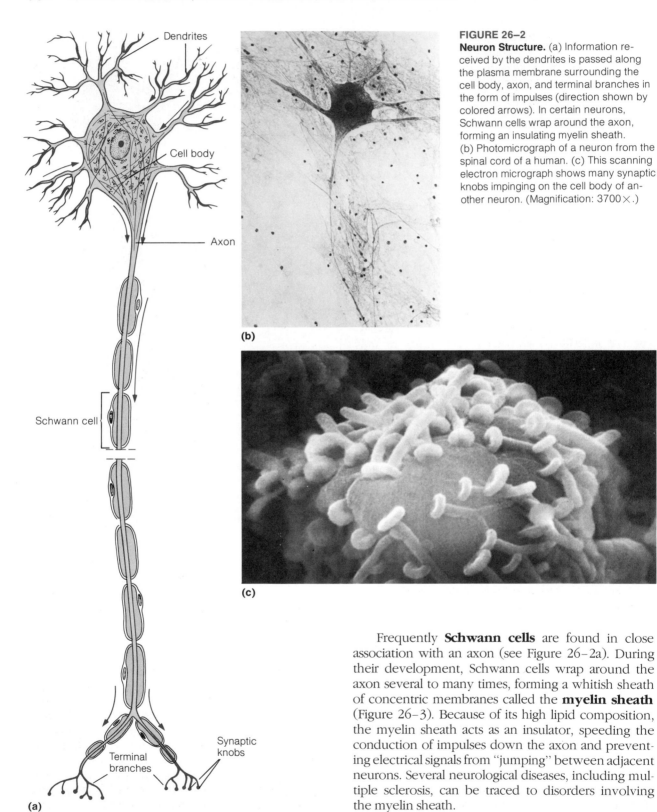

Dendrites

Cell body

Axon

Schwann cell

Terminal branches

Synaptic knobs

(a)

(b)

(c)

FIGURE 26–2
Neuron Structure. (a) Information received by the dendrites is passed along the plasma membrane surrounding the cell body, axon, and terminal branches in the form of impulses (direction shown by colored arrows). In certain neurons, Schwann cells wrap around the axon, forming an insulating myelin sheath. (b) Photomicrograph of a neuron from the spinal cord of a human. (c) This scanning electron micrograph shows many synaptic knobs impinging on the cell body of another neuron. (Magnification: 3700×.)

Frequently **Schwann cells** are found in close association with an axon (see Figure 26–2a). During their development, Schwann cells wrap around the axon several to many times, forming a whitish sheath of concentric membranes called the **myelin sheath** (Figure 26–3). Because of its high lipid composition, the myelin sheath acts as an insulator, speeding the conduction of impulses down the axon and preventing electrical signals from "jumping" between adjacent neurons. Several neurological diseases, including multiple sclerosis, can be traced to disorders involving the myelin sheath.

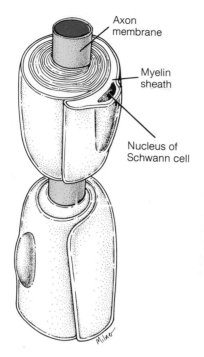

(a)

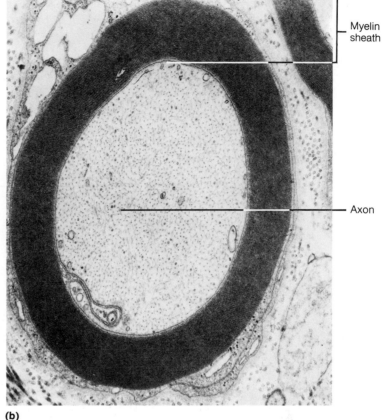

(b)

FIGURE 26–3
Myelin Sheath. (a) The myelin sheath forms as the plasma membranes of Schwann cells wrap around the axon. (b) This electron micrograph shows the extensive layers of membrane composing the myelin sheath. (Magnification: 23,000×.)

The Nature of an Impulse

How do neurons transmit information? All the evidence to date says that a neuron transmits messages in the form of **impulses,** short-lived electrical changes at the plasma membrane. We can draw an analogy to the Morse code system of relaying messages down a telegraph wire: each impulse traveling down a neuron is equivalent to a "dot," and the pause between impulses is a "space." Thus, as complex as nervous systems are, their language is quite simple. Let us now investigate this neural "dot"—the impulse.

Perhaps the best way to describe the nature of an impulse is by carrying out an experiment on an isolated motor neuron. The giant axon of squids is a favorite choice of neurophysiologists, partly because of its large size (up to 1 mm in diameter). Let us begin by positioning two electrodes (small probes that detect electrical changes) on the surface of a squid axon near its midregion (Figure 26–4a). (Another advantage of the squid axon is that it lacks the myelin

sheath insulation. Thus, electrical changes occurring on the axon's surface can be detected with electrodes.) We then apply a very mild stimulus to one end of the axon, such as barely touching the axon with a fine hair. At such a low stimulus intensity, the electrodes record no electrical change whatsoever (Figure 26–4b). Next, we increase the intensity of the stimulus gradually by pressing harder with the hair. Again, the electrodes record no change *until* we reach the **threshold stimulus**—the stimulus intensity at which an impulse is triggered. The surface electrodes detect this impulse first at the electrode nearest the point of stimulation, and then, a fraction of a second later, at the second electrode (Figure 26–4c).

Now we continue our experiment by increasing the intensity of the stimulus above the threshold level. Again our electrodes detect an impulse, but they record the same amount of electrical change at the membrane surface. Furthermore, the rate at which the impulse moves down the axon (the time it takes the impulse to travel from the nearest electrode to

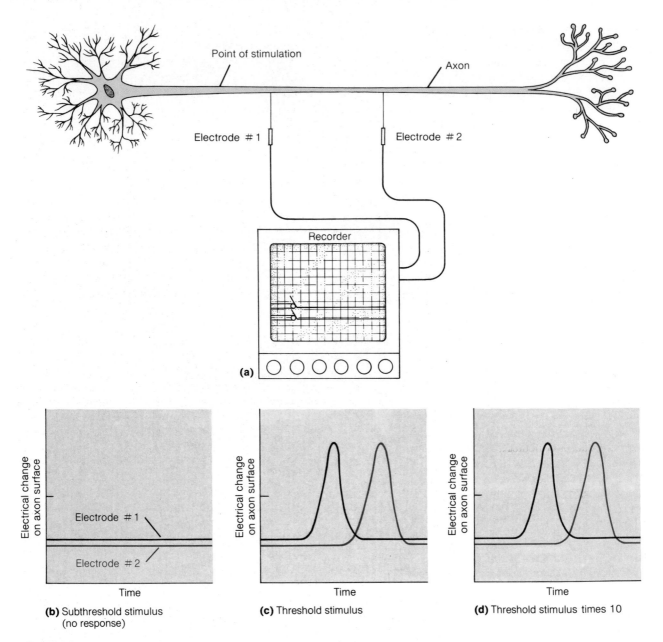

FIGURE 26–4
Detecting an Impulse. (a) To measure electrical changes occurring at the membrane of an axon, two electrodes connected to a recording device are placed on the surface of the neuron. (b) If a subthreshold stimulus is applied, no electrical change is detected. (c) If a stimulus above the threshold is applied, both electrodes detect an identical electrical change. (d) Even if the stimulus is ten times greater than threshold value, the amount of the electrical change remains the same. The time necessary for the impulse to travel from electrode #1 to electrode #2 also stays the same.

the next one) is also the same. This phenomenon is called the **all-or-none response** of a neuron—a stimulus either triggers an impulse in a neuron or it doesn't. There are no intermediate levels of response. That is, all stimuli at or above threshold level produce identical impulses in a given neuron. These impulses all involve the same amount of electrical change in the axon's membrane, and they all travel down the axon at the same speed.

The results of our experiment suggest that the language of the neuron is electrical, and it consists of only two "words"—*impulse* and *no impulse* (or, in keeping with our earlier analogy, "dots" and "spaces"). This is true of nearly all neurons regardless of the nature of the stimulus. The sensory neurons in your big toe that become activated when you step on a tiny pebble fire off impulses in the same manner as they do when your toe finds a sharp tack. But this raises an important question: If the language of a nerve cell consists of only two "words," how do animals distinguish among the wide range of different stimuli? For instance, how do you sense the difference between stepping on a pebble and stepping on a tack? The answer has two parts. First, although the *nature* of the impulse carried by a neuron does not change with the intensity of the stimulus, the *frequency* of impulses can. That is, an intense stimulus generally causes a neuron to fire not just once but over and over. The more intense the stimulus, the more frequently impulses travel down the affected neurons. Second, some neurons have higher stimulus thresholds than others. Since various regions of the body contain both low and high threshold neurons, a weak stimulus (such as stepping on a tiny pebble) will activate only a few low threshold neurons, whereas an intense stimulus (the tack) will trigger both low and high threshold neurons. The information an animal receives about the intensity of any given stimulus depends on the number of neurons that receive a stimulus at threshold or higher, and the frequency of impulses traveling along each individual neuron.

RESTING POTENTIAL. So far we have described the impulse as an electrical change that occurs at the surface membrane of a neuron. To understand the actual nature of this change, we must introduce the concept of electrical potential. **Electrical potential** is a measure of the capacity to do electrical work. It represents a type of stored energy, which is manifested as a separation of charges across a barrier. In the case of the neuron, the charges are positive and

negative ions and the charge-separating barrier is the plasma membrane.

Physiologists refer to the electrical potential that exists across a cell membrane as the **membrane potential.** For most living cells, the membrane potential lies in the range of −40 millivolts (−40 mV, or −0.04 volts) to −90 mV. The minus sign indicates that the inside of the cell is negative relative to the outside of the cell. That is, the cytoplasm has more negative ions and/or fewer positive ions than the cell's external environment. The greater the net charge difference is, the more negative the membrane potential will be.

A typical neuron at rest (when no impulses are traveling along it) has a membrane potential of approximately −70 mV (Figure 26–5a). This is called the **resting potential.** Three major factors contribute to the negative resting potential. First, all living cells, including neurons, contain many organic compounds (such as proteins, organic acids, and so forth) that, on balance, have more negative than positive charges associated with them. Since the extracellular fluid has few if any of these organic substances, the excess of negative organic charges inside the cell contributes partly to the negative resting potential. Second, all neurons have very active **sodium–potassium pumps** located in their cell membranes (Figure 26–5b). Driven by the splitting of ATP, these pumps transport Na^+ out and K^+ into the cell, both against their respective concentration gradients. The exchange of Na^+ for K^+ is unequal, however. For every two K^+ that are actively transported inward, three Na^+ are pumped out. As a result, more positive ions are going out than coming in, so the operation of these pumps tends to drive the membrane potential downward (make it more negative).

The third factor has to do with K^+ leakage from the neuron. The cell membrane is virtually impermeable to all ions except K^+. Now, recall that the sodium–potassium pump transports K^+ inward against its concentration gradient (a neuron at rest has about 20 times more K^+ inside than exists outside the cell). This concentration gradient favors the outward diffusion of K^+, and since the membrane is slightly permeable to K^+, some of it leaks out of the cell. The loss of this positive ion from the neuron by diffusion accounts for a large measure of the negative resting potential.

ACTION POTENTIAL. A resting potential is characteristic of all nerve cells in their unstimulated state, but this should not imply that these cells are "at rest."

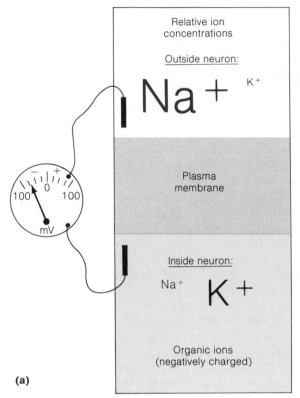

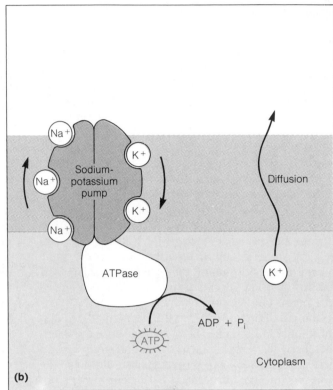

FIGURE 26–5

Resting Potential. (a) In the unstimulated state, a neuron has a membrane potential of approximately −70 mV. The relative concentrations of the principal ions inside and outside the neuron are indicated by the sizes of the chemical symbols (Na+ = sodium ion, K+ = potassium ion). (b) Two of the major processes that contribute to the negative resting potential are the active exchange of Na+ for K+, and the outward diffusion of K+. The sodium–potassium pump actively transports Na+ out and K+ into the cell, and is powered by the splitting of ATP by an associated enzyme, ATPase.

On the contrary, the neuron is constantly expending metabolic energy to maintain the −70 mV charge difference across its plasma membrane. It is more accurate to consider the unstimulated neuron as being in a high-energy state poised for action, much like a stretched rubber band. When a neuron receives a threshold stimulus, this poised energy is instantly converted into action in the form of an impulse. The changes that occur at the membrane during an impulse are called an **action potential.**

The initial stimulation of a neuron causes the plasma membrane at the point of stimulation to become highly permeable to Na+. Exactly how this happens is not known, but most neurophysiologists believe the increased permeability is due to the opening of specific pores in the membrane termed **sodium gates.** When these gates open, sodium ions rush into the neuron by diffusion. Within less than a millisecond (thousandth of a second), the influx of positively charged Na+ **depolarizes** the membrane (decreases its charge), and before the Na+ inrush is over, the membrane potential is actually reversed, reaching a value of +35 mV (Figure 26–6). At this point in time, the stimulated region of the neuronal membrane has an excess of positive ions at its internal surface.

At the peak of depolarization (+35 mV), the sodium gates close and the membrane once again becomes impermeable to Na+. Now the permeability to K+ increases dramatically, and these ions diffuse down their concentration gradient outward. With the rapid loss of K+ from the cell, the membrane quickly **repolarizes** to about −80 mV (see Figure 26–6b). Although the outward diffusion of K+ restores the negative membrane potential, the membrane is not

FIGURE 26–6

Action Potential. (a) When a neuron is stimulated, the cell membrane at the point of stimulation undergoes a momentary reversal in charge (dark color) called an action potential. For perhaps a millisecond, the inside of the membrane becomes positive relative to the outside. (b) Sequence of membrane potential changes associated with an action potential: (1) resting potential; (2) sodium gates open and Na+ diffuses into the cell, causing a depolarization of the membrane; (3) sodium gates close and potassium gates open; (4) K+ diffuses out, causing a repolarization of the membrane; (5) sodium–potassium pump restores original ion gradients and resting potential (recovery). Steps (2)–(5) take a mere 2–3 milliseconds.

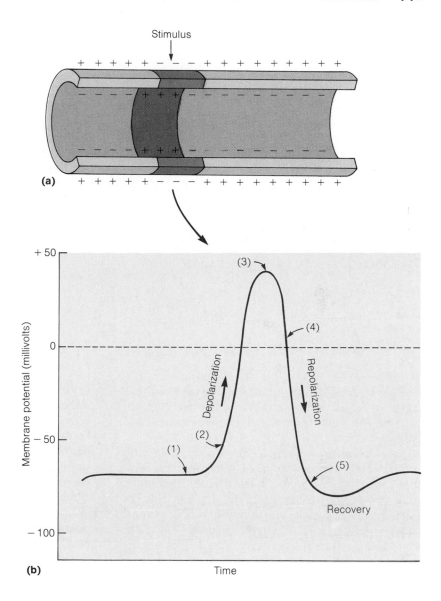

yet "at rest." There is still too much Na+ and too little K+ inside the membrane. The sodium–potassium pump quickly restores the original ion gradients and resting potential by pumping Na+ out and K+ in. This is called the **recovery phase** of the action potential.

Propagation of the Impulse

Initially, an action potential is limited to the region of the membrane at the point of stimulation. However, it quickly spreads down the length of the entire neuron. The change in membrane potential (depolar-

ization) at one spot of the membrane triggers an action potential in the membrane region immediately next to it. This occurs rapidly and repeatedly at the leading edge of the action potential, resulting in a wave of depolarization (the impulse) that moves quickly along the neuron (Figure 26–7). However, when a region of the membrane is just recovering from an action potential, it is **refractory** to further stimulation—that is, it cannot be triggered to undergo another action potential until it has fully recovered its resting potential. Thus, the moving impulse cannot retrace its path because of the refractory period in the region of the membrane where it has most recently occurred. This

FIGURE 26–7
Propagation of an Action Potential.
When a region of a neuron's cell membrane is undergoing an action potential (dark color), sodium ions rush in and diffuse to nearby regions of the cell membrane (arrows). The positively charged sodium ions cause a slight depolarization of the membrane potential in these regions. If the direction of impulse propagation is to the right, then the depolarization of membrane just to the right of the action potential triggers an ensuing action potential there. The depolarization of the adjacent region on the left has no effect because this membrane region is refractory—it is still recovering from the action potential that occurred there 3 milliseconds ago.

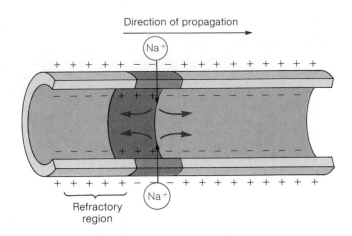

feature ensures that the impulse is propagated from its point of origin (usually a dendrite) toward the other end of the neuron (terminal branch of the axon), where it stops.

As complex as the molecular events of an action potential are, they occur almost instantaneously. From the initial opening of the sodium gates to complete recovery of the resting potential requires only 2–3 milliseconds. This means that the action potential travels very rapidly down the neuron. The fastest rate of impulse conduction ever recorded for a human neuron is 120 meters per second (269 miles per hour)!

Communication Between Neurons

Recall that the simplest of nerve systems, the reflex arc, involves at least three neurons. Information received by a sensory neuron is passed to an interneuron, and then on to a motor neuron to evoke a response. You now know how this message (the impulse) is conducted down a single neuron, but in most cases, action potentials cannot jump from one neuron to the next in line. Rather, the message is transmitted across junctions called **synapses** in the form of a chemical messenger.

Remember that the terminal branches of a nerve cell have swollen tips called synaptic knobs (see Figure 26–2). Inside each synaptic knob are numerous membranous sacs called **synaptic vesicles,** which contain a chemical substance known as a **neurotransmitter** (Figure 26–8a). When an impulse reaches a synaptic knob, some of the synaptic vesicles fuse with the cell membrane of the knob (**presynaptic membrane**), causing the release of their neurotransmitter cargo into the **synaptic cleft** (Figure 26–8b). The neurotransmitter molecules diffuse across the cleft and bind to receptor molecules extending from the surface of the **postsynaptic membrane** of the next neuron. This binding triggers a depolarization in the postsynaptic membrane, which is then propagated down the length of this second neuron. Thus, at each neuron-to-neuron junction, an electrical message is transformed into a chemical message, which triggers another electrical message.

NEUROTRANSMITTERS.   Many different kinds of neurotransmitters are known, including acetylcholine, adrenaline, norepinephrine, serotonin, and dopamine. Acetylcholine is the main transmitter for synapses that lie outside the central nervous system; most of the others are involved in synaptic transmission within the brain and spinal cord.

Not all neurotransmitters are stimulatory. Some act as inhibitors that suppress other incoming signals. This is especially true for interneurons, which may receive both stimulatory and inhibitory neurotransmissions simultaneously from different presynaptic

FIGURE 26–8
Communication Across a Synapse.
When an impulse reaches a synaptic knob, synaptic vesicles within fuse with the presynaptic membrane, causing the release of neurotransmitter molecules into the synaptic cleft. The neurotransmitter molecules bind to receptors on the postsynaptic membrane, triggering an action potential in the postsynaptic neuron.

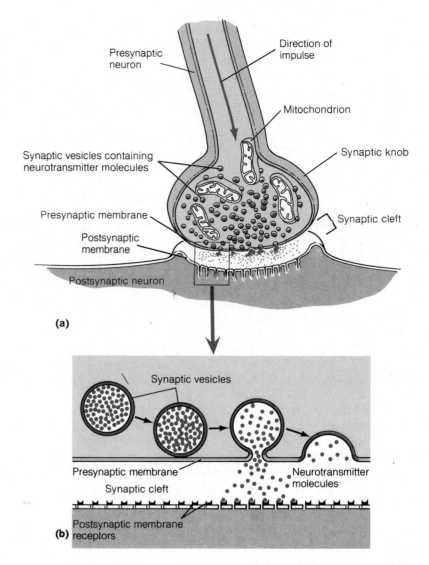

neurons. The postsynaptic neuron essentially sums up the opposing signals to either generate a response (action potential) or have them cancel one another for no response at all.

CLEARING THE SYNAPSE. Given that release of neurotransmitters can trigger impulses in postsynaptic neurons, an obvious question arises. Once a transmitter initiates an impulse, why doesn't the substance continue to stimulate the postsynaptic neuron indefinitely? The answer is that the neurotransmitter is removed from the synapse almost as quickly as it appears. Depending on its nature, the transmitter may be broken down or modified chemically, or reabsorbed by the presynaptic neuron. For example, quickly after its release at a synapse, acetylcholine is broken down by an enzyme present in the synaptic cleft. Other transmitters are destroyed or altered by enzymes associated with the postsynaptic membrane. Still others are quickly reabsorbed into the synaptic knobs, where they are either destroyed or repackaged into synaptic vesicles.

The importance of clearing the synapse of neurotransmitters is easily understood when we consider what happens when such removal mechanisms are blocked. For example, organophosphates are compounds found in several common insecticides; in the past, similar compounds were used in nerve gas. When taken into the body, organophosphates inactivate the enzyme that breaks down acetylcholine. As a result, acetylcholine molecules discharged into millions of synaptic clefts continue to fire off postsynaptic neurons, leading to uncontrollable spasms that often cause death.

SENSORY RECEPTION

Up to this point we have considered how neural messages are transmitted along neurons and between them. But how does information enter the nervous system in the first place? The answer is through sensory reception. Obviously, the ability to sense changes in their environment is vital for animals who must forage for food, escape predators, and find mates. All aspects of animal behavior, from the simple blinking of an eye to elaborate courtship rituals, depend strongly on sensory reception. We will first describe the various types of sensory receptors and how they detect stimuli, then examine the specialized senses of vision and hearing.

Sensory Receptors

All sensory receptors are designed to gather specific types of information about the external or internal environment, then transmit it to the central nervous system. Most sensory receptors are neurons or modified neurons. Some are clearly not neurons, but rather highly specialized cells. These nonneuronal receptors pass on information to nearby sensory neurons via a synapse.

Sensory receptors are classified by the type of information they gather. **Photoreceptors** respond to light; **chemoreceptors,** such as those involved in the sensations of taste or smell, are sensitive to specific types of chemicals; **thermoreceptors** detect hot and cold temperatures; and **mechanoreceptors** perceive changes in pressure and position. These general categories of receptors are further subdivided according to even more specialized functions. For example, the human skin has two different kinds of thermoreceptors—one to detect cold and another to detect heat. Also in the skin are mechanoreceptors that sense touch, pressure, and pain (Figure 26–9). In short, animals are equipped to detect a variety of external stimuli.

Equally important are sensory receptors that monitor internal changes. **Stretch receptors** (a type of mechanoreceptor) located in muscles inform the brain of the position of our limbs in space and the degree

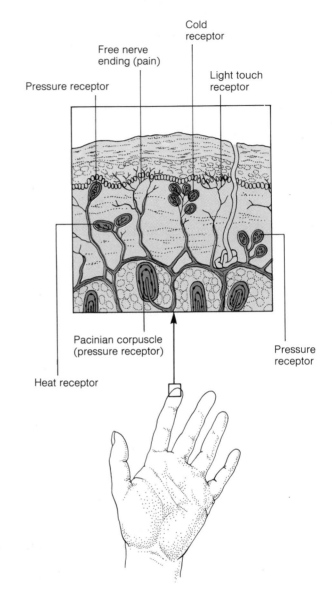

FIGURE 26–9
Sensory Receptors of the Skin. The sensations of touch, pressure, heat, cold, and pain are detected by modified sensory neurons having naked nerve endings (touch and pain receptors) or specialized cellular capsules (pressure, hot, and cold receptors).

of muscle stretch. Also, chemoreceptors in the walls of the aorta and carotid arteries (major vessels in the neck that deliver blood to the head) detect changes in the blood's oxygen concentration to help the brain regulate breathing rate. Other chemoreceptors monitor blood glucose levels; if glucose is in low supply,

TABLE 26–1
Sensory Receptors.

Type of Receptor	Sensation	Body Location
Mechanoreceptors	Touch	Skin
	Pressure	Skin, intestines, bladder
	Stretch	Muscles, tendons, lungs
	Hearing	Inner ear (cochlea)
	Balance	Inner ear (semicircular canals)
Chemoreceptors	Taste	Taste buds on tongue
	Smell	Nasal cavity
	Solute concentration in blood	Hypothalamus
	CO_2 concentration in blood	Brain, carotid artery
	O_2 concentration in blood	Aorta, carotid artery
Thermoreceptors	Hot	Skin
	Cold	Skin
Photoreceptors	Sight	Retina of eye
Nocireceptors	Pain	Skin, internal organs

the chemoreceptors send out messages that ultimately trigger the sensation of hunger. These and other sensory receptors inside the body keep a constant vigil on the internal status of the organism, relaying information to the brain, which makes any necessary adjustments. Table 26–1 lists the major types of receptors and their functions.

Sensory Cell Activation

Although the various sensory receptors are activated by different stimuli, all transmit information to the central nervous system by the same mechanism—the nerve impulse. The initial events that lead to the impulse, however, are somewhat different than the all-or-none response we have described for axons. There is also variation between receptors that are modified sensory neurons and the nonneuronal receptors.

When the nerve endings of a neuronal sensory receptor (such as a pain receptor) are stimulated, the cell membrane at the point of stimulation becomes permeable to all ions. The diffusion of ions across the membrane causes a partial depolarization called a **generator potential** (Figure 26–10). Unlike the action potential of axons, however, the magnitude of the generator potential varies with the intensity of the stimulus—the more intense the stimulus is, the greater the generator potential (depolarization) will be (see Figure 26–10). Furthermore, repeated

stimulation of the nerve endings leads to a summation of the individual generator potentials, producing a larger depolarization. The generator potential is then propagated to the axon of the sensory neuron where, if it exceeds the required threshold, triggers an action potential. Large generator potentials produced by high intensity or repeated stimulation set off many successive action potentials.

Stimulation of a nonneuronal sensory receptor triggers a receptor potential. Like the generator potential, the **receptor potential** is a graded depolarization of the cell membrane—its magnitude varies with stimulus intensity and repetition. The receptor potential causes the release of a neurotransmitter into the synaptic cleft that lies between the receptor cell and its associated sensory neuron. If the stimulus is above threshold, enough neurotransmitter is secreted to trigger an action potential in the postsynaptic sensory neuron.

Photoreception and the Human Eye

Photoreception, or the ability to sense light, is widespread among living things. Many single-celled organisms move in response to the direction of light, a phenomenon called **phototaxis.** Even the chloroplasts in the cells of plants are phototactic—they can become reoriented for maximum exposure of their photosynthetic pigments to sunlight. And the photo-

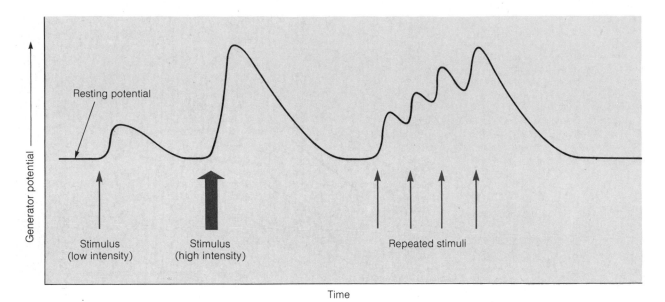

FIGURE 26–10
Generator Potential. When stimulated, a neuronal sensory receptor's membrane undergoes a
depolarization (generator potential), the magnitude of which depends on the intensity of the stimulus.
A rapid succession of stimuli can result in a summation of generator potentials.

tropic bending of plant shoots toward a light source involves detection of light by specialized photoreceptive pigments in the shoot tip. These types of photoreception, however, are primitive when compared to the light-sensing phenomenon of vision.

Vision is the ability to form an image in the brain of objects that reflect light. The eye is the most precise of all visual organs. Several groups of invertebrates, including mollusks (snails, squids, octopuses, and others), insects, and crustaceans (lobsters, shrimp, crabs, and others) have eyes (Figure 26–11). Almost all vertebrates have eyes that form images much like cameras do. Let us look into the eye of humans.

The human eyeball is a sphere about 1 inch in diameter that is surrounded by a three-layered wall (Figure 26–12). The outermost layer, the **sclera,** is a fibrous protective coating that houses numerous blood vessels. Muscles attached to the sclera along the sides of the eye aid in eye movement. The visible portion of the sclera is white with one important exception— the perfectly transparent **cornea.** Often called the window to the eye, the cornea helps to focus light rays toward the back of the eye.

The middle layer of the wall is called the **choroid;** it contains light-shielding pigments that reduce light

scattering inside the eye. The front section of the choroid is the **iris,** an adjustable light shield containing the pigments that determine eye color. In bright daylight, the iris is maximally constricted, permitting a narrow beam of light to enter the eye through the small dark opening known as the **pupil.** When you move to a dimly lit room, however, the iris recedes slowly (the pupil dilates), permitting more light into the eye. This adjustment, called **dark adaptation,** may take 30 seconds to complete, as most of us realize when we turn off the bedroom light and stumble over the bed.

Light entering the pupil is focused by the crystal-clear **lens** onto the innermost layer of the eye covering called the **retina.** The retina, which lies over two-thirds of the back section of the choroid, is composed of several kinds of neural elements, a back layer of black pigmented cells that reduce light scattering inside the eye, and two major types of photoreceptors, the rods and cones (Figure 26–13).

Rod cells are extremely sensitive to light, but they cannot form clear images of objects nor can they discriminate colors. They are concentrated toward the outer edge of the retina where they come into play mainly under darkened conditions when the pupil

FIGURE 26–11
The Eyes of Some Invertebrates.
(a) Most spiders, such as this rabid wolf
spider, have clusters of simple eyes that
can detect movement and form crude
images at close range. (b) This scanning
electron micrograph of a Mediterranean
fruit fly shows the compound eyes typical
of insects. (c) The eye of an octopus,
like those of vertebrates, forms images
in much the same manner as photographic cameras.

(a)

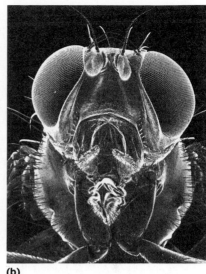

(b)

(c)

is fully dilated. The retinas of nocturnal animals such
as owls and many bats are made up exclusively of
rods. Although they can see well at night, their vision
is apparently restricted to shades to gray.

The light-absorbing pigment in rod cells is a
protein that is attached to **retinal,** a light-harvesting
unit derived from vitamin A. When light is absorbed
by retinal, a chemical change occurs in the pigment
that causes a permeability change in the rod cell
membrane. Ultimately, this information is transmitted
to the brain in the form of impulses. Since these
events cannot take place without retinal, which is
synthesized from vitamin A, a prolonged deficiency
of vitamin A can lead to night blindness, or worse,
total blindness.

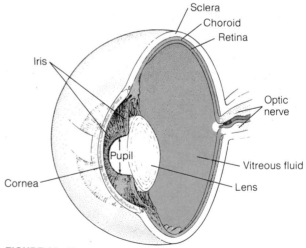

FIGURE 26–12
The Human Eye. Light enters the eye
through the transparent cornea, then
travels through the pupil (an opening in
the iris). The light is focused by the lens
onto the light-sensitive retina. Photoreceptors in the retina convert the light
signals into nerve impulses, which are
carried to the image-forming centers of
the brain by the optic nerve.

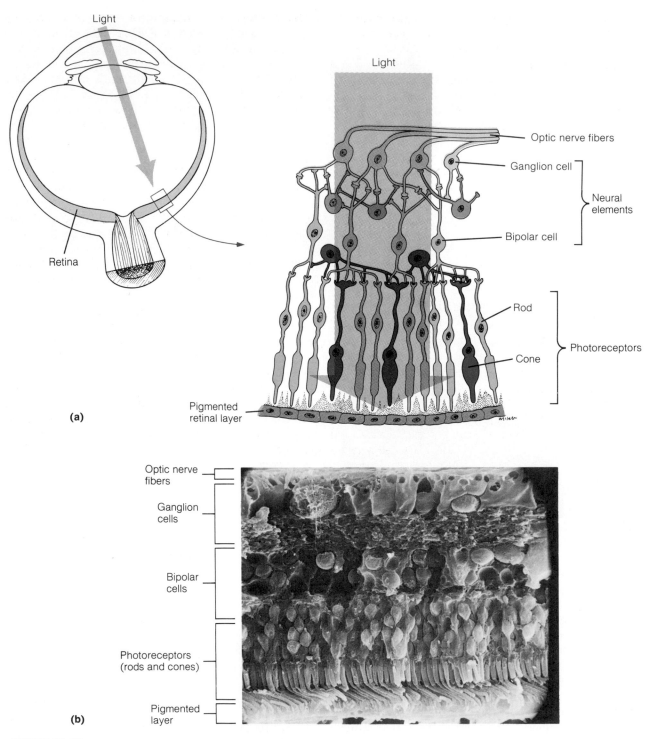

FIGURE 26–13

Retina. (a) This greatly simplified diagram shows the arrangement of the pigmented cells (which reduce light scatter), the photoreceptors (rods and cones), and the neural elements that transmit the light-activated impulses via the optic nerve to the brain. Note the position of these cells relative to the direction of incoming light. (b) Scanning electron micrograph of a cross-section through part of a retina. (Magnification: 650×.) (From *Tissues and Organs: A Text-Atlas of Scanning Electron Microscopy* by Richard G. Kessel and Randy H. Kardon, W. H. Freeman and Company, Copyright © 1979.)

Cone cells (see Figure 26–13) are less sensitive to light, but they do yield very precise images and discern colors. They are more abundant in the central region of the retina where high-intensity light is focused. Color perception involves three types of cones, each having a slightly different pigment (all contain retinal, but the protein unit varies). Each pigment absorbs mainly blue, green, or red light. For instance, because the cones containing the red-absorbing pigment are preferentially stimulated by red-colored objects, we see red. We can also see in-between shades of color. For example, yellow color lies between green and red in the spectrum, and is "seen" by both the green- and red-sensitive cones. The brain receives impulses from both of these two types of cone cells, yielding the yellow image.

Various indirect tests suggest that humans are not alone in having color vision. Besides other primates, many birds, fish, reptiles, and insects respond to (and thus apparently perceive) colors. Interestingly, nonprimate mammals apparently cannot see colors, indicating that color vision has evolved independently in different groups.

Regardless of the type of photoreceptor cell, the conversion of light energy into the electrical-chemical language of neurons is the same basic process for all animals with vision. Chemical changes in the pigment molecules, which are activated by light, lead to the development of a receptor potential in these sensory receptors. The magnitude of the receptor potential depends on the number of pigments activated, a function of light intensity. The receptor potential in turn triggers impulses in the adjacent neural elements called **bipolar cells** (see Figure 26–13). These transmit impulses to a second set of elements in the retina called **ganglion cells.** The axons of ganglion cells come together at the back of the eye to form the large **optic nerve.** The neurons of the optic nerve synapse with interneurons in the thalamus region of the brain, which in turn lead to other visual centers of the brain. (See Chapter 27 for further discussion of the brain.)

Hearing and the Human Ear

Sounds are detected in humans and other vertebrates by special mechanoreceptors in the ears. Although this may seem strange, the appropriateness of touch-sensitive mechanoreceptors for hearing will become clear after we describe briefly what sound is.

Sounds are actually compressed air waves, much like the ripples that form when you toss a stone into a pond. One of the characteristics of sound is **frequency,** or the number of sound waves passing a given point per second. **Pitch** is determined by the frequency of sound waves. The human ear can detect sound frequencies ranging from 16 cycles per second (very low pitch) up to 20,000 (extremely high pitch). Although this may seem like a broad range, dogs can detect sound frequencies at 30,000 cycles/sec, and bats can hear frequencies up to 100,000 cycles/sec.

Another characteristic of sound waves is **amplitude,** or the height of the individual waves. The amplitude determines **volume,** or loudness of the sound, and is measured in **decibels.** Volumes above 120 decibels are painful to the human ear and can cause deafness with prolonged exposure. Table 26–2 lists the decibel values of some common sounds.

The ear is designed to detect sound waves, and much like the senses of vision and smell, hearing enables an animal to gather information about its surroundings. The human ear has three main regions: the **outer ear,** the **middle ear,** and the **inner ear** (Figure 26–14). Sound waves enter the **auditory canal** and bump into the **tympanic membrane,** or **eardrum,** located at the back of the outer ear. The sound vibrations cause the eardrum to vibrate at the same frequency as the sound waves, and these vibrations are amplified by a series of three small bones located in the middle ear. The innermost bone (the **stirrup**) rests against the **oval window,** a membrane-covered opening in the back wall of the middle ear. Vibrations transmitted to the oval window create pressure waves

TABLE 26–2
Loudness of Some Common Sounds.

Sound	Decibels*
10 ft away from a jackhammer	130
Motorcycle (no muffler)	115
2000 ft from a jet taking off	105
50 ft from a moving heavy truck	90
50 ft from a highway	70
Noisy office	65
Quiet conversation	60
Quiet office	45
Quiet rural night	30
Whisper	30

*The decibel scale is logarithmic, like the Richter scale for measuring the intensities of earthquakes. A sound measuring 70 decibels is ten times louder than a 60-decibel sound. A quiet conversation (60 decibels) is 1000 times louder than a whisper (30 decibels).

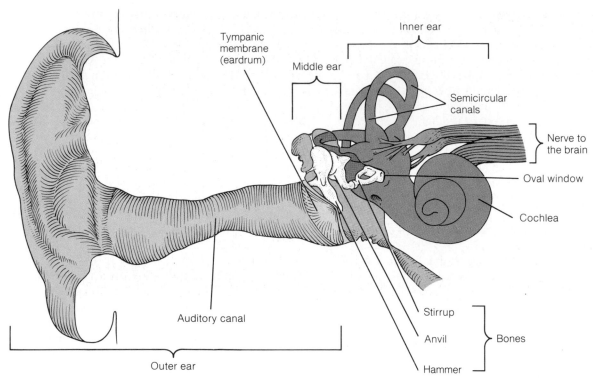

FIGURE 26–14
The Human Ear. Sound waves entering the auditory canal cause vibrations of the eardrum, which are amplified by a series of three bones in the middle ear. The stirrup bone transmits the vibrations to the oval window, which initiates pressure waves in the fluid filling the cochlea. The pressure waves displace hair cell receptors in the cochlea, causing a specific pattern of impulses that is carried to the brain via a nerve.

in the fluid-filled canals of the **cochlea** in the inner ear, waves that have the same frequency as the original sound waves that entered the ear. These pressure waves cause a certain pattern of **hair cells** located in the cochlea to become stimulated physically (Figure 26–15). Each time a hair cell becomes disturbed, an impulse is sent through its associated sensory neuron to the brain. The brain thus is informed as to which hair cells are stimulated, and reads the incoming signals as pitch.

Louder sounds create larger pressure waves in the cochlea, and hence, greater stimulation of the corresponding hair cells. This presumably increases sensory neuron firing, which the brain interprets as loudness.

In addition to the cochlea, the inner ear contains three fluid-filled **semicircular canals,** which provide the brain with information about the position of the head with respect to gravity. Any movement of the head causes a movement of fluid and calcium carbonate crystals within the semicircular canals. The fluid and crystals exert a pressure on the sensory hair cells lining the canals, stimulating them to initiate impulses in their corresponding sensory neurons. These sensory neurons then carry information to the brain concerning the position of the head. Similar mechanisms underlie the related senses of balance and acceleration.

MUSCLE RESPONSE

We have seen how various stimuli are perceived by specialized sensory receptors and how this information is transmitted between neurons toward the central nervous system. Two major functions of the nervous system remain: (1) the processing and integration of

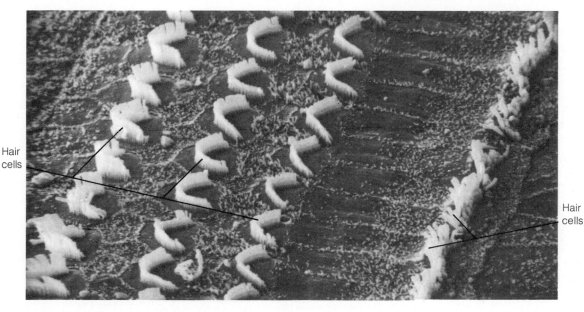

Hair
cells

Hair
cells

FIGURE 26–15
Scanning Electron Micrograph of Hair Cells in the Cochlea. (Magnification: 2420×.) (From
Tissues and Organs: A Text-Atlas of Scanning Electron Microscopy by Richard G. Kessel and
Randy H. Kardon, W. H. Freeman and Company. Copyright © 1979.)

information entering the central nervous system, and
(2) the generation of an appropriate response, if any,
to this information. The processing function of the
central nervous system is important enough that we
have reserved the entire next chapter for it. The
remainder of the present chapter will be devoted to
a major type of nervous response—muscle activity.

Movement in multicellular animals is achieved
largely through the coordinated actions of tissues
specialized for contraction—**muscles.** With the exception of sponges, all animals have a system of muscles,
even the lowly sea anemone. This creature has two
layers of muscle in its body wall, a circular layer and a
longitudinal layer. When the circularly arranged muscle
cells contract, the body becomes longer and thinner.
When they relax and the longitudinal muscles contract,
the body becomes shorter and fatter (Figure 26–16).
Because of their different orientations in the body wall,
these two systems of muscle cells operate **antago-
nistically;** that is, the action of one opposes the action
of the other.

For muscles to operate effectively in causing
movement, the force of contraction must be applied
against a supporting framework. In the sea anemone,
the muscles contract against the fluid in the animal's
central cavity. Arthropods have a rigid exoskeleton to
which the muscles responsible for locomotion (and

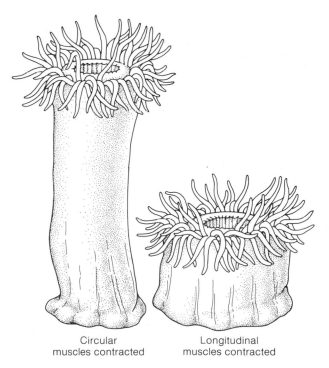

Circular
muscles contracted

Longitudinal
muscles contracted

FIGURE 26–16
Movement in a Sea Anemone. The different shapes of the sea anemone are
produced by two layers of antagonistic
muscles in its body wall.

FIGURE 26–17
The Human Skeleton and Associated Muscles.

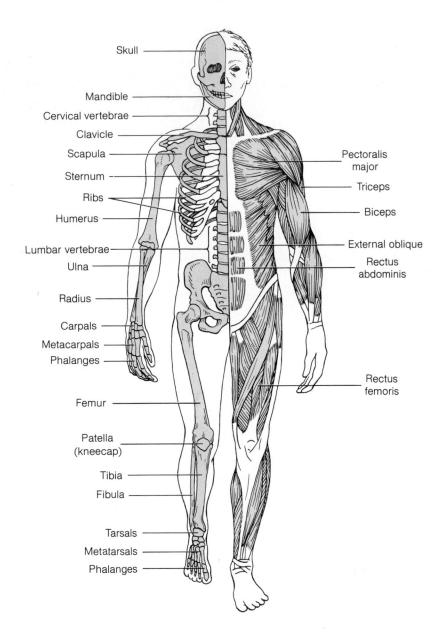

Skull

Mandible

Cervical vertebrae

Clavicle

Scapula

Sternum

Ribs

Humerus

Lumbar vertebrae

Ulna

Radius

Carpals

Metacarpals

Phalanges

Femur

Patella (kneecap)

Tibia

Fibula

Tarsals

Metatarsals

Phalanges

Pectoralis major

Triceps

Biceps

External oblique

Rectus abdominis

Rectus femoris

flight in winged insects) are attached. The cartilaginous and bony endoskeletons of vertebrates serve as the muscle attachment framework for these animals. The human skeleton and some of the major muscles are shown in Figure 26–17.

Muscles that are attached at their ends to the skeleton are called **skeletal muscles.** In vertebrates, a given skeletal muscle is usually attached to two different bones. Thus, its contraction causes the movement of those bones relative to one another, and hence, movement of that body part. For example, the movement of your forearm is under the control of the **biceps** and **triceps,** a pair of antagonistic muscles (Figure 26–18). By contracting your biceps, your forearm bends upward at the elbow. Your arm will return to its straight position when you simultaneously relax the biceps and contract the triceps muscle.

Besides skeletal muscles, vertebrates have cardiac and smooth muscle (Figure 26–19). **Cardiac muscle** is found only in the heart. Although similar to skeletal muscle in many respects, cardiac muscle has unique properties that permit it to contract spontaneously

FIGURE 26–18
Arm Movement. Contraction of the biceps muscle causes the arm to bend at the elbow. Relaxation of the biceps and contraction of the triceps causes the arm to straighten.

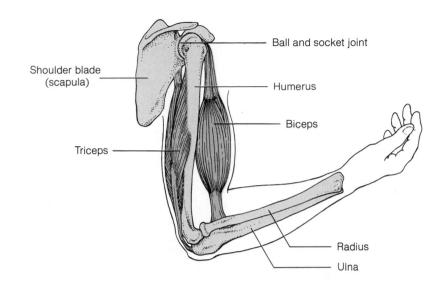

Shoulder blade (scapula)

Triceps

Ball and socket joint

Humerus

Biceps

Radius

Ulna

and rhythmically without direct nervous system control. **Smooth muscle** is found in the walls lining the major blood vessels and various internal organs, such as the stomach, small intestine, and urinary bladder. Most of these muscles are under involuntary control of the central nervous system, but certain smooth muscles contract spontaneously, like cardiac muscle. Smooth muscle is so named because its cells lack the striations that characterize skeletal muscle cells (see Figure 26–20).

Structure and Contraction of Skeletal Muscles

To appreciate how skeletal muscles work, we must first have a look at the anatomy of the structural unit of such muscles—the **muscle fiber.** A muscle fiber is a single muscle cell. Each fiber is composed of elongated subunits called **myofibrils** that extend the length of the cell and occupy most of the cell's volume (Figure 26–20). Organelles such as mitochondria

(a)

(b)

FIGURE 26–19
Muscle. (a) Cardiac (heart) muscle. (b) Smooth muscle.

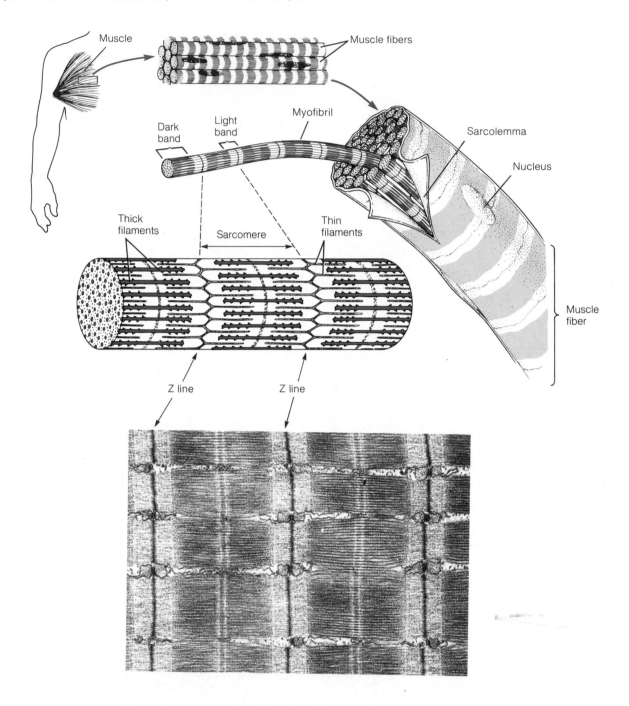

FIGURE 26–20

Substructure of a Skeletal Muscle. Skeletal muscles are made up of many individual muscle fibers (cells). Muscle fibers contain many myofibrils, each of which consists of sarcomeres, the smallest contractile unit of a muscle. The arrangement of the thick and thin filaments in sarcomeres gives skeletal muscle its characteristic striated appearance.

are usually found near the periphery of the cytoplasm, next to the cell's plasma membrane. In a muscle fiber, the plasma membrane is called the **sarcolemma.** Each myofibril consists of repeating units of light and dark bands called **sarcomeres,** and each sarcomere is composed of alternating thick and thin filaments that are bounded on either side by a dark, vertical **Z line.** The thick filaments are made up of the protein **myosin;** the thin filaments consist primarily of the protein **actin.** The dark bands of each sarcomere are regions where thick and thin filaments overlap; the light bands on both sides of each Z line contain only thin filaments. As we shall see, the physical interaction between these thick and thin filaments is responsible for muscle fiber contraction.

When a muscle contracts, it shortens in length. This is the cumulative result of the shortening of many individual sarcomeres in the myofibrils of many muscle fibers. The whole process is triggered by impulses traveling down one or more motor neurons that impinge on the muscle fibers. Let us begin with the electrical events that lead to muscle cell contraction.

ELECTRICAL EVENTS. The mechanism of communication between a motor neuron and a muscle fiber is similar to the process that occurs at a synapse between two neurons. An impulse traveling down a motor neuron triggers the release of acetylcholine from its terminal synaptic knobs. Each knob is closely associated with an individual muscle fiber by means of a **neuromuscular junction** (analogous to a synapse). The acetylcholine molecules bind to receptors on the sarcolemma, triggering an action potential there. The action potential quickly runs the length of the sarcolemma, triggering similar membrane depolarizations in the **transverse tubules** (membranes) that weave around the myofibrils (Figure 26–21). These tubules in turn link up with a vast system of membranous sacs called the **sarcoplasmic reticulum (SR).** Although structurally similar to endoplasmic reticulum, the SR is modified for one primary function: the storage of calcium ions. When stimulated by electrical changes moving along the transverse tubules, the SR suddenly releases calcium ions into the spaces surrounding the thick and thin filaments. The flood of calcium ions in turn initiates biochemical changes that result directly in the contraction of the sarcomeres.

BIOCHEMICAL EVENTS. In its resting state (unstimulated) the sarcomere is at its maximum length, with the degree of overlap between thick (myosin) and thin (actin) filaments at a minimum. Although they are physically close to one another, the actin and myosin filaments cannot bind to each other in the absence of calcium ions. Thus, a relaxed muscle is easy to stretch. Upon stimulation, muscles not only contract but also become tense and hard to stretch. This suggests that the release of calcium ions from the sarcoplasmic

FIGURE 26–21

Internal Membrane System of a Muscle Fiber. Upon neural stimulation, an action potential runs the length of the sarcolemma, but is also propagated inwardly along each transverse tubule. The electrical changes in the transverse tubules activate the sarcoplasmic reticulum to release its stored calcium ions into the myofibrils. The calcium ions trigger contraction.

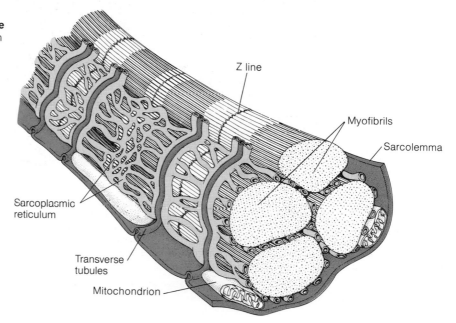

Z line

Myofibrils

Sarcolemma

Sarcoplasmic reticulum

Transverse tubules

Mitochondrion

FIGURE 26–22
Sliding Filament Hypothesis of Muscle Contraction. (a) Sarcomere in its relaxed and contracted state. Note that the thick and thin filaments must slide past one another during contraction. (b) Model of the interaction between myosin and actin during contraction. Calcium ions must bind to specific components of the actin filament to permit the binding of the myosin head to actin subunits (not shown).

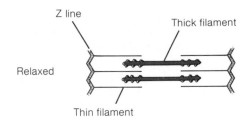

(a)

1. Myosin head bound to actin subunit

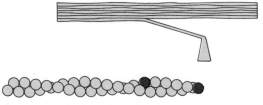

2. ATP split; myosin head swivels and filaments slide past each other

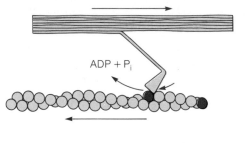

3. Myosin head releases from actin subunit

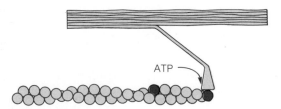

4. Head attaches to new actin subunit one "notch" ahead

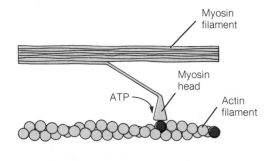

(b)

reticulum causes an interlocking between the thick and thin filaments of each sarcomere, thereby causing the muscle to stiffen. But how does the muscle fiber contract?

The currently popular model of muscle fiber contraction is called the **sliding filament hypothesis.** According to this model, the release of calcium ions from the sarcoplasmic reticulum causes a reorientation of certain components in the actin thin filaments, permitting them to bind with extensions (heads) from the myosin thick filaments (Figure 26–22). Each myosin head then binds and splits an ATP molecule, and the energy released powers the head forward to the next binding component on the actin filament. As this occurs, the actin filament moves one "notch" past the myosin filament. As long as calcium ions and ATP are available in the cytoplasm, the myosin heads continue to "crawl" along the actin filaments, thereby contracting the sarcomere (and muscle).

RECOVERY: MUSCLE FIBER RELAXATION. When the electrical impulses reaching a muscle fiber cease, the sarcoplasmic reticulum begins to reaccumulate the calcium ions by active transport. Once most of the calcium is sequestered in the sarcoplasmic reticulum sacs, which takes only milliseconds, the binding between the myosin heads and the actin filaments can no longer occur. As a result, the thick and thin filaments slide past one another, returning to their relaxed state of minimal overlap. The sarcomeres (and muscle fibers) once again achieve their maximal length and stretchability.

Control of Muscle Contraction

The contraction of a muscle fiber is normally an all-or-none phenomenon. Once it is stimulated, a muscle fiber will contract to a set length, regardless of intensity of the stimulus above the threshold level. The question then arises: If fiber contraction is an all-or-none phenomenon, how do we manage the fine control of muscular activity that permits us to lift a pencil on one occasion and a bowling ball on another?

Part of the answer has to do with the physical relationship between motor neurons and muscle fibers. The axon of a motor neuron has many branches, each branch terminating at a single muscle fiber. Thus, depending on how many branches it has, one neuron can stimulate several to many different muscle fibers. All the muscle fibers triggered by a single neuron contract simultaneously as a single **motor unit** (Figure 26–23). Since a particular muscle may consist of many motor units, the total amount of muscle contraction depends on the number of motor neurons conducting impulses to their motor units in that muscle. If many neurons carry impulses at once, many motor units within the muscle will contract. This causes a stronger overall contraction of the muscle than if only a few motor units are activated.

The degree of fine motor control also depends on the sizes of the individual motor units that make up the muscle. In the muscles controlling human eye movements, for example, each motor unit consists of relatively few muscle fibers, whereas some of the motor units in the skeletal muscles of the leg consist

FIGURE 26–23
Motor Units. Each motor unit consists of one motor neuron and several muscle fibers. An impulse traveling down a motor neuron activates all the muscle fibers connected to it to contract simultaneously.

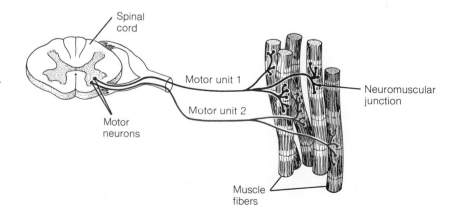

ESSAY 26-1

HOW FAST? HOW HIGH? HOW FAR?

His lungs ballooning and his heart furiously pumping almost twice the normal blood volume, English track star Sebastian Coe broke the tape in August, 1981 in a mile run at Zurich in 3 minutes 48.53 seconds. It was a new world record, beating the 1980 mark—3:48.80—set by Coe's countryman Steve Ovett. One week later, in Koblenz, West Germany, Ovett returned the compliment, running the mile in 3:48.40 and recapturing the championship. Only two days elapsed before Coe got back on the track. In a stunning performance in Brussels, the 24-year-old running machine blazed yet another mile record, this time in 3:47.33, or 1.07 seconds better than Ovett's still fresh mark. That incredible two-man, ten-day international spectacle surprised fans and raised some provocative questions among sports physiologists. How much faster can mere mortals run the mile? How much higher can they leap? How much more weight can they lift? How much farther can they jump, or hurl a javelin, or put a shot? Are there limits to athletic performance? And if so, what are they, and when will they be reached?

Since the golden days of Greece, when athletes paid homage to their sporting deities in the Olympic games, man has sought to outrace the wind, outleap the deer, outdo his opponents—and outfox his own body. With competitive spirit, strict discipline, and hard training, dedicated athletes have shown their disdain for limits. Today, in virtually every kind of sport, they continue to pit themselves against both competitors and the apparent constraints of their own muscles and organs in order to break record after record.

These continuing assaults on the record books have confounded the physiologists who try to understand the limits of human physical endeavor. No sooner do they believe that they have seen the impossible in athletic performance than the sports pages send them back to the laboratory to review their data.

Much of the improvement in performance can be traced to improvements in training techniques, nutrition, and equipment. Moreover, renewed interest in physical fitness has led to greater participation in all sports, and consequently to a greater pool of talent from which record breakers will naturally emerge.

Still, scientists remain convinced that the body *does* have its limits, and they wonder how far human muscle, bones, and sinew can be pushed. When does the advantage of having a great coach no longer matter, and when do the ultimate biological limitations of muscle, the skeletal structure, and the cardiovascular system come into play? In a handful of laboratories across the country, scientists have reached some tentative conclusions suggesting barriers that body and will cannot surmount. Weight lifters in the lightest body-weight categories, for example, may reach their limits some day soon—unable, finally, to set new records; high jumpers, milers, and certain other track athletes will also have to accept the fact that their recordbreaking days are numbered.

To assess the factors that explain athletic prowess and define its possibilities, scientists analyze human muscle fibers, study enzymes, and even crush bones to test resilience. Living specimens—top athletes—are run on treadmills, worked out on

Nautilus machines, wired, tapped, probed, and poked from toe to top.

Whether the athlete is strapped into a padded chair or is flipping dolphin-like over the pole-vault bar, the physiologist must study the body as a machine that operates according to mechanical and chemical laws. The human body requires a certain amount of energy to move, and the efficiency with which it can convert food into energy is limited by its own internal mechanisms.

That conversion system involves two distinct yet interacting processes—aerobic and anaerobic—that produce energy for the muscles. In aerobic metabolism, the muscle cells rely on oxygen, delivered to them by the blood, to convert fats and carbohydrates into energy at a relatively slow rate. In anaerobic metabolism, the muscle cells use no oxygen but make the conversion from carbohydrates much faster. Different physical activities make different demands on the two processes. During marathon running, in which an athlete must pace himself to run for hours, the muscles depend almost entirely on the slow, steady form of aerobic conversion. By contrast, a weight lifter who heaves more than 500 pounds overhead for only a few seconds needs the quick energy that the anaerobic process provides. Other sports, like swimming and running, use both processes in roughly equal proportions. Says Robert Fitts, a muscle physiologist at Marquette University in Milwaukee, "The body knows just when to depend on one or the other."

In past decades, scientists have gauged the relative efficiency of the two metabolic systems and have discovered how the body switches them on and off. They have found that the anaerobic process, though speedy, is extremely inefficient. Muscle cells that function anaerobically discard carbohydrates after only a small percentage of the available energy has been extracted from them. Moreover, the cells tend to rebel in these circumstances, and, through a complex mechanism, will release hydrogen ions into the muscle tissue, causing fatigue.

At that point, the body's aerobic machinery steps in to take up the slack. With the help of oxygen, the muscle cells turn the by-products of anaerobic metabolism, along with additional fats snatched from pockets in obscure corners of the body, into carbon dioxide, water, and finally energy. The oxygen delivered by the blood helps to clear out the fatigue-producing hydrogen ions embedded in the muscle cells.

Although the general picture of the body's power-train system is understood, scientists must also learn how much energy can be spent before the body collapses from exhaustion. They consider the microcosmic details of enzyme reactions as well as the macrocosm of different muscle types and body proportions. Certain features, like long legs, are genetically determined. Other factors, like the amount of blood the heart can pump, change with training, and here researchers try to extrapolate the limits by working athletes to utter exhaustion.

Physiologists David Costill and William Fink, at Ball State University in Muncie, Indiana, have calculated the maximum volume of oxygen that certain athletes are able to consume per pound per minute; this would be an approximate measure

of the best aerobic performance that can be expected from those athletes. Costill experiments with champion runners; he plugs up their noses, stuffs a tube in their mouths, and runs them relentlessly on treadmills. He measures the volume of air that is breathed in and out, as well as the ratio of carbon dioxide to oxygen in the expired air. The percentage of carbon dioxide tells him how well the body metabolizes its deposits of fat and carbohydrates, a process that depends upon the ability of the heart to deliver oxygen and the ability of the muscles to use that oxygen. Fink reckons that top

runners like Bill Rodgers use roughly 80 milliliters (4.8 cubic inches) of oxygen per 2.2 pounds of body weight per minute (an average male Sunday jogger uses about 45 milliliters, he thinks). He has now concluded that "the upper limit is probably in the eighties or nineties," because the heart could be incapable of delivering any additional oxygen.

In another experiment, Swedish researchers injected radioactive dyes into the bloodstream of a cross-country ski champion who had an exceptionally large build and heart, and monitored him at rest. They

---

ESSAY 26-1   HOW FAST? HOW HIGH? HOW FAR? (continued)

discovered that his heart could pump about 36 quarts of blood per minute. (The heart of a typical man delivers about 18 quarts.) It is doubtful whether any athlete could surpass 36 quarts: the hearts of the average-sized Coe and Ovett pump about 30.

Another figure that is not likely to improve, and one that affects sprinters and weight lifters, is the amount of pressure that a human femur can withstand. At the University of California at San Diego, orthopaedic surgeons subjected the thigh bone of a human cadaver to the jaws of a compression device. They found that the femur can tolerate 1,600 pounds per square inch before it splinters. At present speeds, the leg bones of a 160-pound sprinter in action withstand forces up to five times his body weight, or 800 pounds. As sprinters get faster and weight lifters stronger, the forces will increase accordingly and approach the breaking point.

Another element in athletic performance is the genetically determined quality of muscle fiber, which sports scientists divide into two

types, depending on the speed with which muscles move. So-called fast-twitch fibers are stronger and larger and are most suited for anaerobic activity; slow twitchers are best for aerobic work. Physiologists Costill and Fink have examined these factors in athletes. After anaesthetizing volunteers, they take out a bit of muscle, "anywhere from the size of a grain of rice to a pea," says Fink. The fibers are then examined under microscopes, and compared with those of other athletes. Fink has found that a Rodgers-class marathoner seems to possess 80 to 90% slow-twitch fibers; top men and women sprinters like AAU champions Carl Lewis and Evelyn Ashford have about 70% fast twitch. But physiologists do not rule out the possibility that evolution could produce marathoners with far greater potential. Fink sees no reason why a genetic fluke could not turn out somebody with 90 + % of either fast- or slow-twitch muscles; so equipped, a person might surpass currently accepted limits on performance.

As the physiologists probe the limits, they must also consider the

occasional superhuman performance that goes off their charts. In the 1968 Mexico City Olympics, America's Bob Beamon shattered credulity as well as the world record with an unbelievable long-jump feat of 29 feet 2½ inches; the previous record was 27 feet 4¾ inches. Based on a century-long trend of records set before 1968, Beamon's great leap should not have occurred until the twenty-first century.

It is perhaps just as well that sports science was unprepared for Beamon. As Ernst Jokl, a neurologist at the University of Kentucky, says, "The wonderful thing about sport is that it is unpredictable. The arrival of geniuses like Bach and Mozart in the world of music could never have been predicted. And the same is true for genius in athletics. Unpredictability is an element of sport at its best." And, he might have added, the best sort of challenge to science.

(Adapted with permission from *Discover Magazine,* November 1981, pp. 24–30, © 1981, Time, Inc. Article by Natalie Angier.)

---

of hundreds of fibers. Consequently, an impulse coming from a single motor neuron in the eye stimulates fewer muscle cells, resulting in more subtle movements than is possible in the leg.

Although individual muscle fibers exhibit the all-or-none response, there are situations that can lead to greater-than-normal muscle fiber contractions. When an isolated whole muscle is stimulated sufficiently to activate all the fibers composing it, a simple **twitch** contraction results (Figure 26–24). If the same stimulus is given a few seconds later (after the muscle has had time to relax completely), the same twitch response appears. However, if this muscle is stimulated re-

peatedly in rapid succession so that the muscle cannot relax between stimuli, it will contract to a shorter length than that of a simple twitch, and it will remain contracted until its supply of ATP runs out. This sustained contraction is called **tetanus** (not to be confused with lockjaw, a symptom of a particular bacterial infection). Tetanus is a normal response for any muscle that is subjected to sustained stress, such as holding up a bowling ball. With prolonged stimulation, the muscle will eventually become **fatigued** (ATP supply becomes low). For example, the arm that is holding the bowling ball will eventually have to set the ball down or drop it.

FIGURE 26–24
Muscle Twitch, Tetanus, and Fatigue.
When a single threshold stimulus is applied to a muscle, a twitch contraction and rapid relaxation response are noted. When stimulated repeatedly, the muscle undergoes a tetanus contraction—the muscle remains contracted at a shorter length than that observed for a twitch. With prolonged tetanus, the muscle cells eventually deplete their store of ATP, resulting in fatigue.

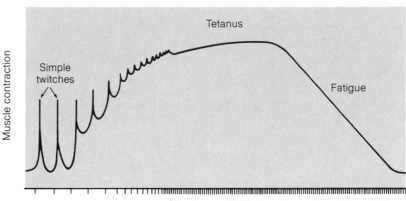

SUMMARY

The ability of animals to sense and respond to environmental stimuli is related to their highly specialized nervous and muscular systems. The functional unit of the nervous system is the neuron, a type of cell that has the properties of excitability and conductivity. Upon excitation, sensory neurons conduct impulses (electrical messages) *to* the central nervous system. Motor neurons transmit impulses *from* the central nervous system to muscles or glands to initiate a response. Sensory and motor neurons are linked functionally by one to many interneurons present in the brain or spinal cord.

A typical neuron consists of a cell body, which contains the nucleus and most of the organelles; a system of highly branched dendrites specialized to receive messages from other neurons (or, in the case of sensory neurons, from sensory receptors or direct physical stimulation); and an axon that conducts impulses toward terminal branches that end in neurotransmitter-storing synaptic knobs.

All neurons transmit information in the form of impulses. When a neuron is in its unstimulated state, its plasma membrane has an electrical potential of approximately −70 millivolts; that is, the inside of the membrane has more negative (less positive) ions than the outside. This resting potential is established partly by the activity of an ATP-driven sodium–potassium pump, which transports three sodium ions out of the cell for each two potassium ions in, and to a greater extent by the diffusion of potassium ions outward. Upon receiving a threshold stimulus, the membrane quickly depolarizes (inside of the membrane becomes more positive than outside) at the point of stimulation, and this electrical charge reversal (action potential) becomes the stimulus to trigger the adjacent segment

of the membrane to depolarize. Thus, from the point of initial stimulation, an impulse moves down the neuron as the action potential is generated in successive regions of the membrane. Behind this wave of depolarization the membrane quickly returns to its resting potential, as the neuron membrane regains its original permeability properties.

An axonal impulse does not vary with the strength of a stimulus—it is all or none. Animals sense the relative intensity of a stimulus instead from the frequency of the impulses transmitted during a given period (how often a given neuron fires) and from how many neurons are stimulated to threshold level or higher. When the action potential reaches the end of an axon, a neural transmitter substance is usually released across the presynaptic membrane into the synaptic cleft. This substance then binds to receptor molecules on the postsynaptic membrane of the next neuron. Neural transmitters ordinarily are either destroyed or reabsorbed across the presynaptic membrane almost immediately after release, preventing uncontrolled firing of the postsynaptic neuron (or muscle cell).

Information enters an animal's nervous system through specialized sensory receptors (or modified neurons), which translate an environmental stimulus, such as light or sound waves, into membrane depolarization. The receptor cells in the human eye are rods, which are very sensitive to light intensity (thus responsible for night vision), and cones, which are highly precise image and color detectors. In the human ear, sound vibrations are transmitted inward to the cochlea where they are detected by hair receptors. Sensory hair cells in the semicircular canals of the inner ear are responsible for the senses of head position, balance, and acceleration. Ultimately, all information from sensory receptors and the neurons associated with them is transmitted to the brain, which interprets these impulses as images, sounds, or whatever.

Muscles are contractile tissues responsible for movement in animals. The three types of muscle tissue are skeletal, cardiac, and smooth. The latter two types are associated with internal organs and either display spontaneous contractile properties or are under involuntary control of the nervous system. As the name implies, skeletal muscles are attached to the skeletal framework of animals (bone or cartilage in vertebrates), usually occur in antagonistic pairs, and are subject to voluntary control by the brain.

As is true of neurons, the function of skeletal muscles is closely related to the structure of their cellular units, the muscle fiber. Each fiber is composed of elongate, striated units called myofibrils. The myofibrils in turn are made up of sarcomeres, contractile units consisting of thick filaments (myosin) and thin filaments (mostly actin). The myofibrils are surrounded by a system of membranous sacs which store calcium ions (sarcoplasmic reticulum) and transverse tubules.

When a muscle cell becomes stimulated by acetylcholine released from a stimulated motor neuron, the muscle fiber sarcolemma (plasma membrane) conducts an action potential that spreads into the transverse tubules, causing the release of calcium ions from the sarcoplasmic reticulum. Calcium ions bind to the specific components of the thin filaments, which permits actin to bind to myosin units extending from the thick filaments. With a supply of energy from the splitting of ATP molecules, the actin and myosin filaments slide past one another, causing the sarcomere to shorten (contract). Muscle relaxation occurs when the calcium ions become sequestered once again in the sarcoplasmic reticulum.

The amount of total muscle contraction depends on how many individual muscle fibers become stimulated to contract, which in turn hinges on the number of impulse-conducting motor neurons reaching the muscle. In addition, rapid and repetitive muscle stimulation will cause each muscle fiber to contract to a length shorter than that triggered by a single stimulus.

STUDY QUESTIONS

1. How do dendrites and axons differ functionally? Which is often associated with Schwann cells?

2. What is the relationship between threshold stimulus and the all-or-none response of neurons?

3. How does the brain "know" the difference between stepping on a pebble and stepping on a tack?

4. Why is an unstimulated neuron not really "resting"?

5. What ion movement event associated with the action potential actually leads to the depolarization of the membrane?

6. Which part of the information transfer process between neurons is a secretion event?

7. What general type of sensory receptor is involved in the sensation of taste? Vision? Pressure?

8. Which type of photoreceptor in the human retina is most sensitive to light? To colors?

9. What part of the ear is responsible for amplifying sound waves? For maintaining balance?

10. Name the three major types of muscles. Which of these would be found in the arm? In the walls of blood vessels?

11. What accounts for the striated appearance of skeletal muscle fibers?

12. What stimulates the release of calcium ions from the sarcoplasmic reticulum of muscle fibers?

13. Is the contraction of a sarcomere unit correlated with the greatest or least amount of overlap between thin and thick filaments?

14. Why is eye movement subject to finer motor control than leg movement?

SUGGESTED READINGS

Cohen, C. "The Protein Switch of Muscle Contraction." *Scientific American* 233(1975):36–45. This article discusses in detail how calcium ions trigger events leading to muscle contraction.

Mason, E. B. *Human Physiology.* Menlo Park, CA: Benjamin/Cummings, 1983. Chapters 8 and 9 of this intermediate-level text provide an indepth look at the senses and nervous system.

Morell, P., and W. T. Norton. "Myelin." *Scientific American* 242(1980):88–116. This article describes the molecular composition of myelin, and its assembly into the sheath, as well as its role in electrical insulation.

Newman, E. A., and P. H. Hartline. "The Infrared Vision of Snakes." *Scientific American* 246(1982):116–127. This interesting article focuses on a special organ in pit vipers which can act as an infrared "eye." Information from the pit organ plus visible-light information gives these snakes a unique view of their environment.

Nicholls, J. G., and D. Van Essen. "The Nervous System of the Leech." *Scientific American* 230(1974):38–48. The relatively simple nervous system of the leech makes it an ideal experimental animal in many ways. This article shows how the study of neuron connections can help explain certain behavioral reflexes.

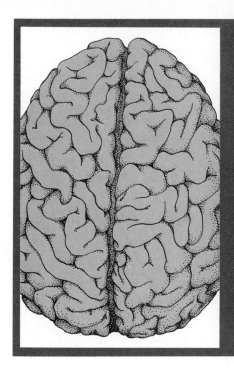

CHAPTER 27
Nervous Systems and the Brain

Nervous systems in animals collect many types of information, integrate the inputs, then evoke appropriate responses. All of the steps, from input to response, often take no more than a few milliseconds, and this feature of quickness distinguishes the nervous system from its slower-acting partner, the endocrine system (discussed in Chapter 24).

In the previous chapter we examined the cellular unit of nervous systems, the neuron, and how it conducts and transmits neural information. We also described how neurons communicate with one another across synapses, how sensory receptors gather information, and how motor neurons trigger responses such as muscle contraction. These cellular processes are carried out in essentially the same manner in all animals with nervous systems. A neuron is a neuron whether it is from a sea slug or an elephant. What really differs among the nervous systems of different animals is the pattern in which the neurons are arranged and functionally organized. Animals capable of complex behaviors have correspondingly complex organizations of their nervous systems.

In this chapter we will begin with a brief comparison of nervous systems in selected invertebrates, then describe the general organization of the vertebrate nervous system, and conclude by examining the most sophisticated information processor on earth—the human brain.

NERVOUS SYSTEMS OF INVERTEBRATES

The simplest nervous systems are found among members of the phylum Cnidaria (sea anemones, corals, jellyfish, and hydrozoans). *Hydra,* for example, has a diffuse system of interconnected neurons known as a **nerve net** (Figure 27–1a). The net of neurons links sensory receptors present on *Hydra*'s internal and external surfaces to nearby muscle cells. There are no clusters of neurons that collect and integrate the incoming signals; rather, impulses generated in one region of the body simply radiate in all directions. The magnitude of the initial stimulus determines how far the impulse will travel and thus, how many muscle cells will be stimulated to contract.

A planarian (a flatworm, phylum Platyhelminthes), has a slightly more complex nervous system than *Hydra* (Figure 27–1b). Part of the nerve net is condensed into two **nerve cords** (parallel arrays of nerve cells) that run the length of the body, each terminating in a ganglion in the head region. These **ganglia** are clusters of neuronal cell bodies (perhaps 500 to 5000 per ganglion) that represent rudimentary "switching stations" for the integration of incoming signals. Containing both excitatory and inhibitory synapses (see Chapter 26), a ganglion may receive a single impulse

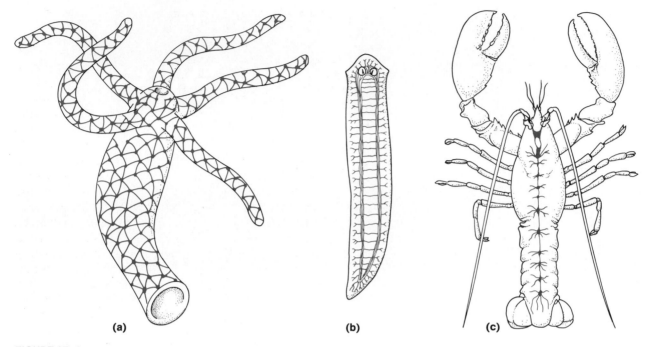

(a) (b) (c)

FIGURE 27–1
Nervous Systems of Invertebrates. (a) Nerve net of *Hydra,* a cnidarian. (b) Flatworms have a nerve net that condenses into two nerve cords running the length of the body. The nerve cords terminate in the head region as two distinct ganglia—a rudimentary brain. (c) Arthropods such as this lobster have a single nerve cord with a major ganglion in the head (brain) and many secondary ganglia downstream.

FIGURE 27–2
A Motor Ganglion. This highly simplified diagram of a motor ganglion shows how a single impulse (colored arrows) entering a ganglion along a presynaptic neuron can lead to the coordinated actions of two different muscles. As the impulse reaches an inhibitory synapse, inhibitory neurotransmitters are released that prevent the firing of the postsynaptic neuron (*top*) and its associated muscle. At the same time, the impulse is transmitted across an excitatory synapse and along another postsynaptic neuron (*bottom*) that stimulates its associated muscle to contract.

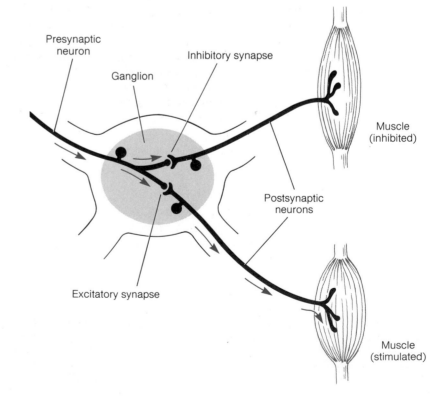

Presynaptic neuron

Inhibitory synapse

Ganglion

Muscle (inhibited)

Postsynaptic neurons

Excitatory synapse

Muscle (stimulated)

that ultimately triggers the contraction of one muscle and inhibits another (Figure 27–2). With its two integrating ganglia, a planarian has a limited capacity to coordinate responses to external stimuli.

The flatworms, such as planarians, which are accorded an ancestral position to all higher invertebrates (see Figure 18–1), apparently started an evolutionary trend in the organization of nervous systems. They represent the beginnings of **cephalization**—the gradual centralization of the major sense organs and ganglia in the anterior (head) region of animals. Over time the anterior ganglia increased in size and relative importance, and in higher animals they are recognized as brains. The lobster (an arthropod) exhibits an intermediate stage of cephalization. Its single nerve cord terminates in a rather small brain, but there is also a chain of ganglia that extends along the length of the nerve cord (Figure 27–1c). Although the brain is the dominating neural center in lobsters, the other ganglia exert direct control over many important activities, such as the movements of the legs and tail.

THE VERTEBRATE NERVOUS SYSTEM

Cephalization reached its climax in the vertebrates. The sense organs responsible for vision, hearing, smell, and taste are all located in the head, and with the exception of reflex arcs (see Figure 26–1), all information processed by the vertebrate nervous system goes through a single brain. The development of this "master computer," the brain, made possible the highly complex behaviors so characteristic of these backboned animals. Let's now have a closer look at the organization of the vertebrate nervous system.

Because of its complexity, the vertebrate nervous system is split into distinct divisions based on the function and location of their parts. Although many of these divisions are themselves referred to as nervous systems, you should keep in mind that a given animal has only one nervous system, and that each division thereof is not distinct, but rather an integral part of the whole.

The vertebrate nervous system has two major divisions: (1) the **central nervous system (CNS),** which includes the brain and spinal cord, and (2) the **peripheral nervous system (PNS),** which is composed of all the neurons that lie outside of the brain and spinal cord. We will consider first the peripheral nervous system.

The Peripheral Nervous System

The PNS is divided into **sensory** and **motor divisions,** which consist of sensory neurons and motor neurons, respectively (Figure 27–3). The motor division is in turn divided into the somatic nervous system and the autonomic nervous system. The **somatic nervous system,** also known as the voluntary system, consists of motor neurons that lead from the CNS to the body's skeletal muscles, which are generally under conscious (voluntary) control. The **autonomic nervous system,** or involuntary system, regulates the activities of involuntary muscles, such as cardiac muscles of the heart and the smooth muscles that line the blood vessels and many internal organs. The autonomic nervous system also controls the secretion of substances (such as hormones) from certain glands. The major divisions of the nervous system and their functional interrelationships are diagrammed in Figure 27–3.

In terms of overall function, the peripheral nervous system links sensory reception to motor response by way of the central nervous system. Its sensory division of neurons gathers information from the internal and external environments, either directly with modified sensory neurons or indirectly from specialized sensory receptors, then relays the messages to the CNS. The motor division carries information from the CNS to the various muscles and glands to initiate specific responses. Even very simple reflexes involve both the PNS (sensory and motor divisions) and CNS (spinal cord) (see Figure 26–1).

The axons of the PNS neurons exist in bundles called **nerves** (Figure 27–4). Most nerves are mixed; that is, they contain both sensory and motor axons. Nevertheless, impulse "jumping" between axons in a nerve does not take place because each axon (also called **nerve fiber**) is insulated from the others by a myelin sheath and some connective tissue. Humans have 43 pairs of nerves: 12 pairs of **cranial nerves,** which originate in the brain (such as the optic nerve shown in Figure 26–12), and 31 pairs of **spinal nerves,** which arise from the spinal cord. These nerves branch repeatedly into smaller bundles of nerve fibers that extend toward the outlying regions of the body.

Most of the cell bodies of the PNS neurons come together in ganglia that lie just outside the spinal cord. Some of the motor neurons have their cell bodies actually inside the spinal cord. (As you can see, the anatomical distinction between the CNS and PNS often becomes fuzzy.) Unlike the nerves, however, each

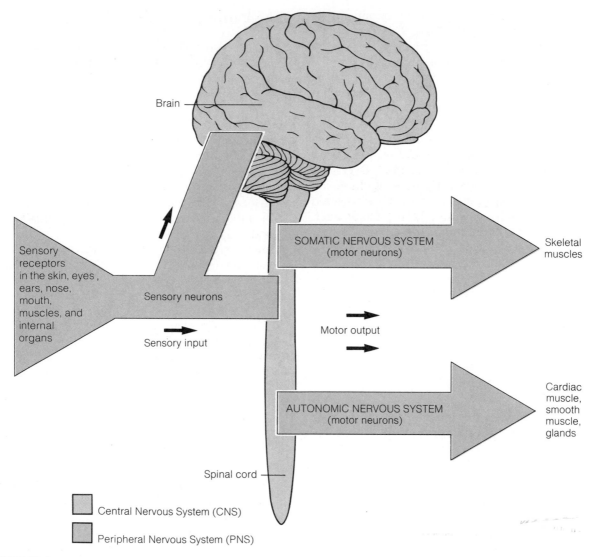

FIGURE 27–3
General Organization of the Vertebrate Nervous System. Information gathered by sensory receptors is transmitted to the central nervous system via sensory neurons of the peripheral nervous system. Motor responses are initiated in the central nervous system and transmitted to the appropriate skeletal muscles via the somatic nervous system, or to the appropriate cardiac muscles, smooth muscles, or glands via the autonomic nervous system. The somatic and autonomic nervous systems are motor divisions of the peripheral nervous system.

ganglion contains cell bodies from *either* sensory *or* motor neurons. Thus, in contrast to the situation in arthropods (see Figure 27–1c), vertebrate ganglia cannot exert independent control over any body activity, but must pass messages to the central nervous system (in the case of sensory ganglia) or receive messages from the CNS (in the case of motor ganglia).

THE SOMATIC NERVOUS SYSTEM. The motor neurons of the somatic nervous system run directly from the CNS to the various skeletal muscles, where their impulses trigger muscle contraction. They are generally involved in conscious activities, stimulating muscles that help you walk, run, talk, write, eat, and many other voluntary responses. An exception is the group

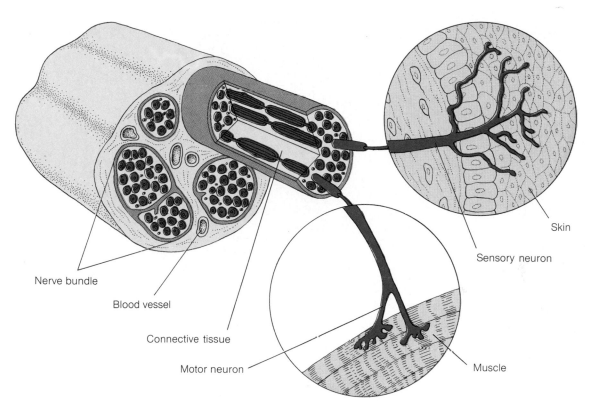

FIGURE 27–4
A Nerve. Nerves are bundles of axons (fibers), which are myelinated and embedded in connective tissue. Mixed nerves (such as the one diagrammed) contain both sensory and motor fibers.

of somatic motor neurons involved in reflexes, which are not controlled consciously.

THE AUTONOMIC NERVOUS SYSTEM. The autonomic motor pathways relay messages from the unconscious centers of the brain to their target muscles and glands. The autonomic nervous system differs from the somatic nervous system in three major ways:

1. Instead of skeletal muscle targets, the autonomic nervous system regulates the actions of heart muscle, smooth muscle, and glands. All of the regulated responses (such as muscle contraction and glandular secretion) are involuntary—we don't have to think about increasing or decreasing our heart rate, or moving food through the small intestine by peristaltic contractions. Nevertheless, a few activities such as breathing are subject to control by both the somatic and autonomic systems. Furthermore, certain biofeedback practices (yoga and other forms of meditation) enable individuals to

exert some conscious control over what were previously thought to be strictly involuntary muscles, such as those that produce the heartbeat and control the diameter of arteries.

2. Whereas each somatic motor pathway consists of one neuron that runs from the CNS to its target muscle, each autonomic motor pathway consists of two neurons. The first, or **preganglionic**, neuron (fiber) originates in the brain or spinal cord and synapses with a second, or **postganglionic**, fiber in a ganglion (see Figure 27–5).

3. The somatic system can communicate only a single message to its target muscles: "contract." On the other hand, the autonomic system has both fibers that stimulate muscle contraction or glandular secretion, and fibers that inhibit such responses.

The autonomic nervous system itself has two structurally and functionally distinct divisions: the **sympathetic division** and the **parasympathetic division.** Anatomically, the sympathetic preganglionic

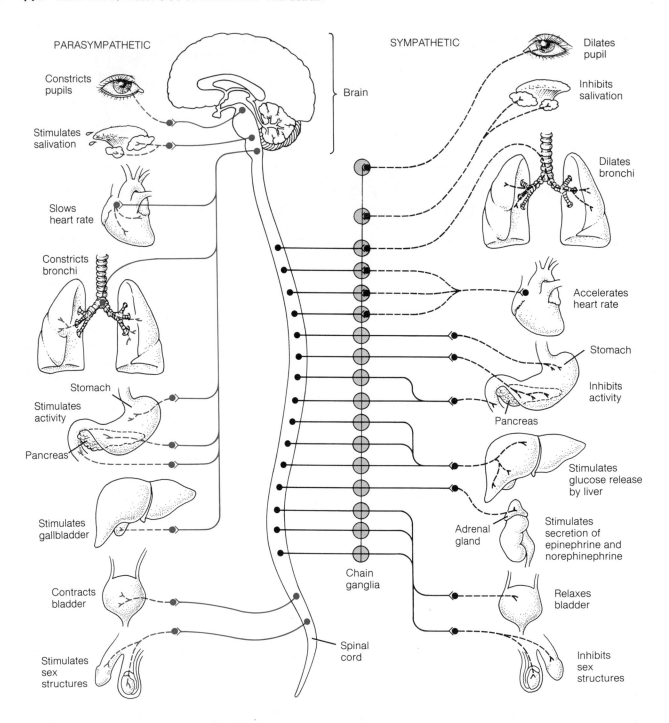

FIGURE 27–5

The Autonomic Nervous System. This diagram shows some of the major nerves of the parasym-pathetic division (*left*, in color) and sympathetic division (*right*, black), and their effects on various organs. Impulses traveling through the parasympathetic nerves tend to put the body in a relaxed state and stimulate digestion. Sympathetic stimulation prepares the body for strenuous activity and inhibits digestive activities. Note that the postganglionic fibers (dashed lines) of the parasympathetic system are shorter than those of the sympathetic system.

fibers arise from the middle segments of the spinal cord, are relatively short in length, and terminate in ganglia located near the spine (Figure 27–5). The postganglionic fibers extend from the ganglia to their target muscles or glands. In contrast, the preganglionic fibers of the parasympathetic division originate in the brain and lowermost section of the spinal cord and run to ganglia located in or near their target tissues. In this case, it is the postganglionic fibers that are quite short.

For target organs that have both sympathetic and parasympathetic connections, the impulses coming from the two systems tend to have opposite effects. For example, the rate of heartbeat is increased by sympathetic signals and decreased by parasympathetic impulses. Broadly speaking, sympathetic stimulation tends to prepare the body for strenuous physical activity, such as the "fight or flight" responses. Anger and fear bring on enhanced levels of sympathetic stimulation. On the other hand, parasympathetic stimulation predominates when the body is relaxed and external conditions are more tranquil. These two systems work cooperatively to alert or calm the body as external conditions warrant, thereby providing an effective homeostatic mechanism to regulate body functions.

The Central Nervous System

The brain and spinal cord form the central nervous system. Both are enclosed by membranes and encased in bones. The brain is protected by the skull, and the spinal cord is surrounded by the vertebrae. Cerebrospinal fluid, which is similar in composition to blood plasma, bathes the neurons of these structures and cushions them against the bumps and jolts of everyday living.

In terms of function, the spinal cord is simply an extension of the brain. It consists of areas of gray and white matter. The **gray matter** occupies the central region of the spinal cord and contains the cell bodies and unmyelinated processes (dendrites, axons, and terminal branches) of interneurons (Figure 27–6). Also present are the cell bodies of many motor neurons and the axons of many sensory neurons, both of which synapse with interneurons and leave the spinal cord to form the spinal nerves. The gray matter is surrounded by the **white matter,** which consists primarily of myelinated axons of interneurons and sensory neurons (the myelin sheaths impart the whitish appearance of these cells). Their function is to *conduct* sensory impulses to the brain and motor impulses from the brain to the motor fibers of the spinal nerves.

The vertebrate brain is undoubtedly the most spectacular accomplishment of biological evolution. The human brain in particular contains hundreds of billions of neurons that form trillions of interneuronal synapses. Some individual brain neurons may have up to 75,000 connections with other brain cells. The complexity of the brain is staggering, and researchers have only begun to sort out some of its most basic functions. Often referred to as the last great frontier of biology, brain research has provided many new insights into the mysteries of this amazing organ. Yet, one wonders if the human brain will prove intelligent enough to actually understand itself. Will emotions, ideas, and dreams eventually surrender to the scientist's analytical "scal-

FIGURE 27–6
Spinal Cord. A segment of the spinal cord showing two paired spinal nerves. The gray matter consists of unmyelinated fibers of interneurons and sensory neurons, and includes the cell bodies of motor neurons. Myelinated fibers of interneurons and sensory neurons make up the white matter.

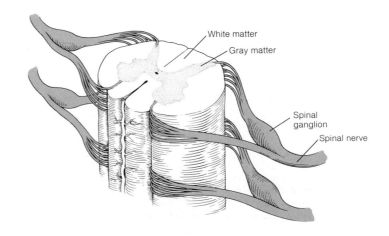

pel" and become explicable in terms of molecules, ions, and energy? Time will tell.

THE HUMAN BRAIN

The brains of all vertebrates, including humans, are divided into three major regions: the **hindbrain,** the **midbrain,** and the **forebrain** (Figure 27–7). The **medulla** is the most basal portion of the hindbrain; it serves as a communication link between the spinal cord and other regions of the brain. It also exerts some direct control over several autonomic functions, such as heart rate, breathing, blood vessel diameter, and the coughing and swallowing reflexes. The **cerebellum,** a dorsal (toward the back side) outgrowth of the medulla, is concerned with balance and skeletal muscle coordination. An individual whose cerebellum has been injured may experience difficulty in walking and standing erect. The third section of the hindbrain, the **pons,** appears as a bulge on the ventral (toward the front side) surface of the medulla. It contains neural fibers that connect both the cerebellum and upper regions of the brain to the medulla.

The midbrain of humans is a rather small, constricted region that lies between the hindbrain and forebrain. It controls involuntary movements of certain eye muscles that dilate and constrict the pupils (bright light triggers constriction) and other muscles that move the eyelids (blinking, squinting). It also serves as a relay station for auditory (ear) impulses entering the brain. In fishes and amphibians, the midbrain houses the optic lobes, where visual images are formed. In mammals, the visual centers are located in the forebrain.

Along with the evolutionary reduction of the midbrain in higher vertebrates came a dramatic increase in the size and complexity of the forebrain, particularly the region called the cerebrum (to be discussed shortly). The human forebrain is especially impressive, and its activities and capabilities distinguish our species from all the rest. The forebrain is divided into three major regions: thalamus, hypothalamus, and cerebrum (see Figure 27–7).

The Thalamus

The **thalamus** is located just above the midbrain. It is the major sensory switching station of the brain, relaying all sensory inputs (except those of smell) to the appropriate sensory areas of the cerebrum, and in certain instances, to the nearby hypothalamus.

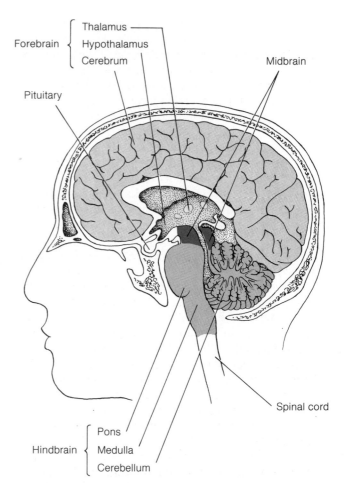

FIGURE 27–7
The Human Brain. Side view of the brain in section.

The Hypothalamus

As its name suggests, the **hypothalamus** lies below (and to the front of) the thalamus (see Figure 27–7). It controls many involuntary aspects of behavior, including appetite, thirst, body temperature, sexual responses, and even emotions such as rage, fear, and pleasure. Many of these regulatory activities are channeled through the autonomic nervous system. Others derive from the influence of the hypothalamus on the endocrine system. Recall from Chapter 24 that the hypothalamus produces a group of hormones that stimulate or inhibit the release of other hormones made in the anterior pituitary gland. Many of the pituitary hormones in turn affect the activities of other endocrine glands. Thus, the tiny hypothalamus has many far-reaching effects on important, hormone-regulated bodily functions.

The Cerebrum

The largest part of the human brain (in fact, all mammalian brains) is the **cerebrum** (see Figure 27–7). It consists of two halves, or **cerebral hemispheres** (Figure 27–8), which are joined by a bundle of nerve fibers called the **corpus callosum.** Covering the surface of the cerebrum is a thin layer of gray matter called the **cerebral cortex,** which has a highly convoluted appearance in primates (see Figure 27–8). The cerebral cortex processes much of the sensory input to the brain and initiates many motor responses. It is also the center for many of the more complex functions of the brain, including learning and memory, abstract thinking, and personality.

MOTOR AND SENSORY FUNCTIONS. Because it lies just under the skull, the cerebral cortex has been the object of many experimental studies, both in laboratory animals and humans. In humans, such studies have been carried out mainly on epileptics who have had major sections of the skull removed for explorative surgery. As the patient is usually conscious during brain surgery (the brain has no pain receptors), the surgeon can electrically stimulate specific areas of the cortex, then monitor the effect on the patient's sensory or motor functions. From these types of studies, scientists have located specific regions of the cortex associated with speech, hearing, vision, and many other sensory-motor functions (Figure 27–9a). For example, when the occipital lobe located at the very back of the cortex is stimulated electrically, the patient sees flashes of light. (The same result can occur with a sharp blow to the back of the head—we see "stars.") If the occipital lobe is seriously damaged by injury, the result is blindness.

Situated in the middle region of the cerebral cortex are two strips of gray matter called the **primary sensory area** and **primary motor area** (see Figure 27–9a). Each of these areas has been mapped experimentally with regard to the specific areas of the body from which they receive impulses (sensory strip) or to which they send out impulses (motor strip) (Figure 27–9b). For example, stimulating one spot in the motor area will cause the wrist to bend; triggering another nearby spot moves the thumb.

Most of the specific sensory and motor areas of the cerebral cortex are present in both hemispheres. Thus, a region associated with arm movement in the left hemisphere has a corresponding region in the right hemisphere. Interestingly, however, each hemisphere receives sensory stimuli from and sends motor impulses to the opposite side of the body. That is, the

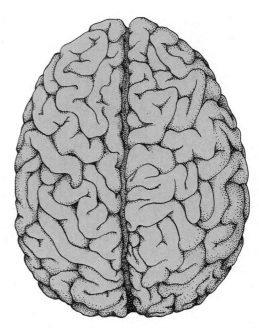

FIGURE 27–8
The Human Brain. Top view, showing the two cerebral hemispheres.

left hemisphere receives and sends impulses to the right side of the body, and the right hemisphere receives and sends impulses to the left side. This crossover of motor and sensory pathways from one side of the body to the other side of the cerebral cortex occurs near the medulla.

ASSOCIATION AREAS. Only about 25% of the cortical surface controls well-defined motor and sensory functions. Stimulation of any region within the remaining 75% of the cortical surface elicits no sensations or visible motor response. These so-called "silent" areas, or **association areas,** have other important functions.

The association areas are involved in the most complex (and least understood) activities of the brain. It is in these largely unmapped regions of the cortex that we are able to associate new sensory information with past experience, then respond appropriately. For example, the smell of bacon frying may trigger fond memories of a past camping trip, which might start one thinking about and planning for the next trip. In other words, the association areas are responsible for learning, memory, and integrating stored information with new experiences (thinking). In primates, particularly humans, most of the cerebral cortex is devoted to such functions.

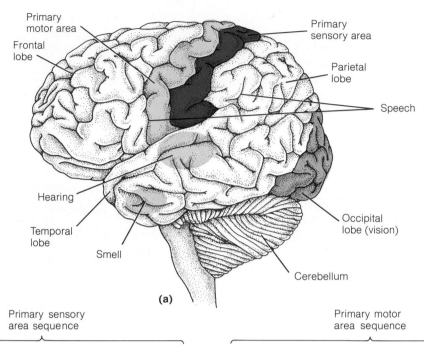

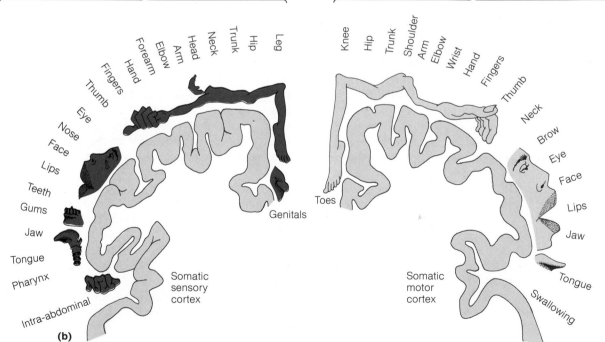

FIGURE 27–9

Sensory and Motor Areas of the Cerebral Cortex. (a) Each hemisphere of the cerebral cortex consists of four lobes as indicated. Most visual stimuli are processed within the occipital lobe, and the major hearing center is found in the temporal lobe. Many of the sensory and motor functions involving outlying regions of the body are processed by neurons occupying two strips of the cortex. The primary motor area lies at the back of the frontal lobe, and the primary sensory area is the frontal strip of the parietal lobe. (b) Electrical stimulation of various spots along the surface of the primary sensory area (*left*) causes the sensation of touch in specific regions of the body. Stimulation of localized regions in the primary motor area (*right*) causes muscle twitches to occur in the indicated body regions.

The association areas of the right and left hemispheres are not identical in the kinds of activities they carry out. In the human brain, the left cerebral hemisphere appears to house the majority of neural pathways related to language ability, logical thought, and the capacity to perform routine mathematical tasks. On the other hand, the right hemisphere seems to be concerned more directly with creative activities, such as art and music composition, and spatial relations (picturing three-dimensional objects in space). This discovery has led to such colloquialisms as "left brain" and "right brain" individuals; that is, we tend to think of scientists and accountants as "left brain" people, and Picasso and John Lennon as "right brain" people. Educators today are becoming more aware of the need to stimulate learning in both sides of the cerebral cortex.

The Limbic System

The **limbic system** is a wishbone-shaped collection of neurons that loops around the central region of the brain, linking the hypothalamus with the cerebral

cortex (Figure 27–10). Its chief role seems to be that of mediator between reason and emotion, permitting conscious (cerebral) control over passions and drives emanating from the hypothalamus. The limbic system relays neural messages from the cerebral cortex that allows us to overcome fears, fight back tears, or suppress the urge to eat or drink. Occasionally, the balance is upset and anger, fear, or another strong emotion takes over. For example, stimulation of one region of a cat's limbic system generated aggressive behavior, including hissing, clawing, and retraction of the ears. In other experimental studies, removing a small part of the limbic system in monkeys rendered them docile and unfearful of situations that normally produce extreme fright. Curiously, emotions relating to sexual activity were intensified.

The Reticular Formation

Like many aspects of brain activity, the **reticular formation** (also known as the reticular activating system) is defined more by function than by anatomical location. This diffuse net of neurons runs from the

FIGURE 27–10
The Limbic System. The limbic system is a complex network of nerve fibers located at the base of the cerebrum. It is the seat of emotions. The olfactory bulb, where sensations of smell are processed, is also part of the limbic system.

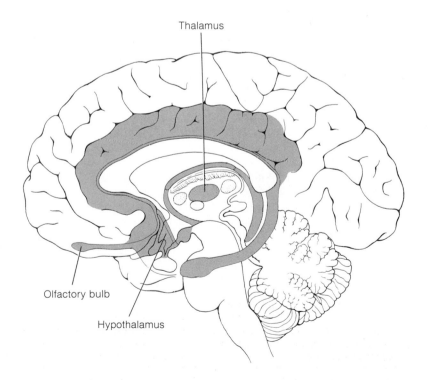

Thalamus

Olfactory bulb

Hypothalamus

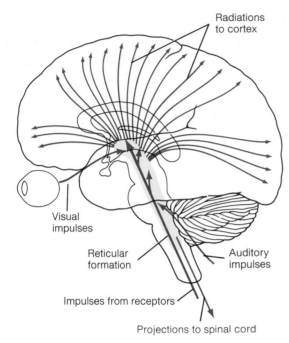

Radiations to cortex

Visual impulses

Reticular formation

Auditory impulses

Impulses from receptors

Projections to spinal cord

FIGURE 27–11
The Reticular Formation. The reticular formation is a diffuse array of nerve fibers that extend through the central regions of the hindbrain and midbrain, and radiate into the cerebral cortex. These fibers filter incoming sensory stimuli and are associated with arousal and sleep.

medulla, through the midbrain, and into the base of the cerebrum (Figure 27–11). All the sensory systems have fibers that feed through this net, which filters incoming signals and discriminates between those that deserve attention and those that should be ignored. Thus, the reticular formation is concerned with attention and arousal.

A few examples might help. As you are reading these words, your reticular formation is permitting visual stimuli to arouse centers in your cerebral cortex that enable you to understand the words. In the meantime, all the noises around you (traffic, crickets, a nearby conversation) are screened out—that is, until we mentioned them. Now, if we can have your attention again, it is your reticular formation that allows you to focus your senses, thoughts, and motor responses on a specific object or activity at the exclusion of other more peripheral ones. That nearby conversation will probably not "register" unless something is said that piques your curiosity, such as your name. If that

occurs, you will undoubtedly screen out other sensory stimuli in favor of the conversation and quickly become "all ears."

The reticular formation can also become conditioned to sensory stimuli. For example, people who live near the beach are not awakened by the pounding surf, but while visiting relatives in rural Iowa, may be instantly alerted by the first crowing of a rooster. The farmer vacationing near the ocean may have the opposite problem. And everyone knows of those individuals with gifted reticular formations—the sports enthusiast who agrees with everything you say ("Ah-huh") when watching a football game, or the student who can study biology with the TV or stereo blaring.

SLEEP AND DREAMS.   In addition to filtering information and controlling our level of awareness, the reticular formation also controls wakefulness and sleep. Recent studies indicate that being awake or asleep is not a simple on or off condition. Rather, there are various levels of alertness (or sleep), and each has a characteristic pattern of brain waves—oscillating electrical activity recorded from the brain's surface by an electroencephalograph (Figure 27–12).

On the basis of brain wave activity, we can distinguish between three clearly different active states: wakefulness, light sleep, and deep sleep (see Figure 27–12b). Light sleep is characterized by slow, symmetrical brain waves. The muscles become relaxed but still retain some tension. A person in light sleep is easily aroused, and when awakened, may say, "I was just resting my eyes."

During a regular sleep interval, periods of light sleep alternate with three or four episodes of deep sleep, each lasting 5 to 20 minutes. During deep sleep, or **REM** (rapid-eye-movement) sleep, the brain waves "come alive" and are very similar to the wakefulness pattern (see Figure 27–12b). Despite the increased brain cell activity, the muscles are totally relaxed and all sensory information is screened out (it takes a loud noise or vigorous shaking to wake the REM sleeper). The notable exceptions to total muscle relaxation include a rapid back-and-forth movement of the eyeballs (hence the term rapid-eye-movement) (Figure 27–13), and an occasional flailing of the limbs. These spontaneous movements appear to be associated with dreaming, an activity that occurs primarily during REM sleep. Some believe that the eye movement is in response to the brain scanning the visual image of the dream; the limb movement apparently results from the dreamer taking an active role in the dream.

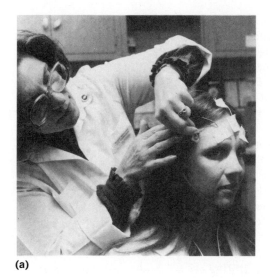

(a)

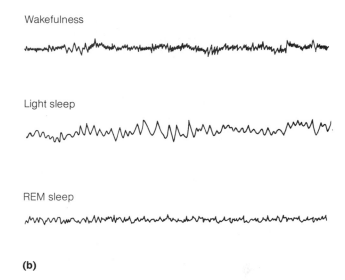

Wakefulness

Light sleep

REM sleep

(b)

FIGURE 27–12
Brain Waves. (a) This subject is being wired to an electroencephalograph, an instrument that records electrical activity of the brain. (b) Brain wave patterns of the three major states of consciousness. Note the similarity in the patterns for wakefulness and deep (REM) sleep.

FIGURE 27–13
Rapid Eye Movements. Time-lapse photography has captured this individual during REM sleep. Note the different positions of the eyes as seen through the closed eyelids.

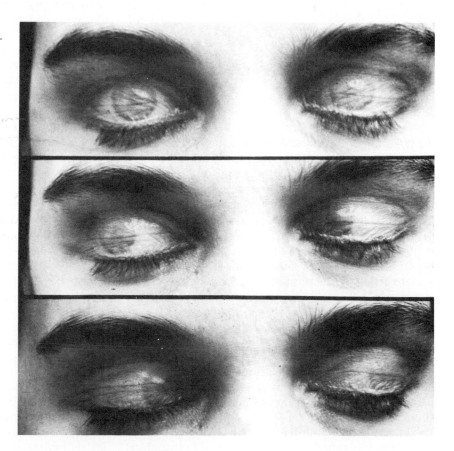

CHEMICAL ACTIVITIES OF THE BRAIN

When we described brain studies as possibly the "last great frontier in biology," we had in mind the chemistry of the brain. Some of the most exciting research in recent times has been in the area of brain chemistry. We are just beginning to understand how the myriad of brain neurotransmitters work, and as a medical spinoff, how analgesics and behavior-modifying drugs exert their actions. As progress continues on this front, we will undoubtedly see the human brain become less "psychological" and more "chemical."

Neurotransmitters

Recall that the cell-to-cell relay of neural messages is chemical. When an impulse reaches a presynaptic knob, it causes the release of a neurotransmitter into the synaptic cleft, which either triggers the postsynaptic neuron to fire, or inhibits it from becoming excited by other neurons (see Figure 26–8). The type of action (excitation or inhibition) depends on the nature of the neurotransmitter.

Scientists have uncovered nearly 30 different kinds of neurotransmitters in the vertebrate central nervous system. They fall into three general categories based on chemical structure: amino acids, neuropeptides, and amino acid derivatives. Among the amino acid transmitters, aspartic acid and glutamic acid are excitatory; glycine and gamma-aminobutyric acid (GABA) are inhibitory. **Huntington's chorea,** a genetic disorder characterized by uncontrolled spasmodic twitching, is apparently caused by the inability to produce sufficient levels of GABA. Lacking enough of this inhibitory transmitter in certain parts of the brain leads to unsuppressed neural "chatter," which results in excessive motor neuron firing.

Small chains of amino acids called neuropeptides also act as neurotransmitters within the CNS. One interesting neuropeptide, called substance P, is released from sensory neurons within the spinal cord. Here it excites spinal nerves that relay to the brain information associated with pain. Two other classes of neuropeptides, the endorphins and enkephalins, are also associated with our perception of pain. We shall have more to say about these interesting substances shortly.

In addition to amino acids and neuropeptides, a wide variety of other compounds derived from amino acids function as neurotransmitters. For example, norepinephrine is found in relatively high concentrations in the reticular formation, limbic system, and hypothalamus. It is thought to be involved in the maintenance of arousal and the regulation of mood. Serotonin is a neurotransmitter that has been implicated in a wide range of functions including temperature regulation, sensory perception, sleep, and emotions. Dopamine plays a major role in the control of movement. **Parkinson's disease** (sometimes called shaking palsy) is related to the degeneration of neurons that secrete dopamine. A closely related compound, L-dopa, is often prescribed to reduce the uncontrolled shaking by partially making up for the dopamine deficiency.

Natural Opiates

Everyone is aware of the existence of pain, some people more than others. Actually, pain has value because it makes us aware of potentially dangerous underlying causes. Of course, knowing this does not make pain any easier to bear. It is not surprising, then, that humans have sought and experimented with a variety of painkilling substances (**analgesics**). Opiate drugs, such as heroin and morphine, have been used for centuries to kill pain (Figure 27–14). Recent studies on how these painkillers work have led to the discovery of opiate receptors in two regions of the central nervous system: (1) the point of entry of nerves into the spinal cord, where the identification of pain sources occurs, and (2) the limbic system, where emotions are controlled. The opiate drugs apparently bind to receptor sites in both regions, inhibiting the transmission of pain-signaling messages.

The human body does not produce morphine, so why are there opiate receptors in the CNS? They must serve some purpose. Does the body produce its own painkillers? Indeed it does. Since 1975, neurophysiologists have isolated a host of opiatelike neuropeptides collectively known as the **enkephalins** and **endorphins.** Both of these groups of natural substances resemble morphine in chemical structure and function. Enkephalins, found in various areas of the brain and spinal cord, are effective for only a few seconds as pain suppressors. Endorphins, which are produced in the brain and stored and released by the pituitary, are effective for up to three hours. Individual differences in the amount of pain perceived is most likely due to differences in enkephalin and endorphin levels. Some researchers believe that acupuncture blocks pain by stimulating key neural pathways that trigger the release of enkephalins (Essay 27–1).

ESSAY 27-1

ACUPUNCTURE

Acupuncture is an ancient Chinese medical technique based on the premise that needles, when used to stimulate specific points on the body, can cure disease, alleviate pain, and induce anesthesia. When China opened its doors in the early 1970s, Westerners caught a glimpse of how trained doctors practice this 5000-year-old healing art in every major hospital, right alongside Western-style medicine. Dramatic accounts of surgery performed on wide-awake patients, using only needle manipulation to numb the pain, have spurred interest and inquiry among medical practitioners throughout the world.

Until recently, the varied effects of acupuncture were explained only according to the metaphysical philosophy of Oriental medicine. This view holds that health is a balance between two naturally opposing forces within the body (Yin and Yang) and that energy circles the body along hypothetical channels called meridian lines, each associated with a particular organ system. Pain and disease are thought to result from a disruption of this vital energy flow and an imbalance in the body's equilibrium. By insertion and calculated manipulation of a needle at certain surface points along the meridian line, the flow of energy may be released and the body's harmony restored.

Today, both Chinese and non-Chinese investigators are seeking a more physiological explanation for the phenomenon of acupuncture. But, since acupuncture is a tool which performs a number of functions, no single mechanism has yet been identified. Research is yielding some valuable clues, however. Variations in the skin's electrical resistance, cell structure, and temperature

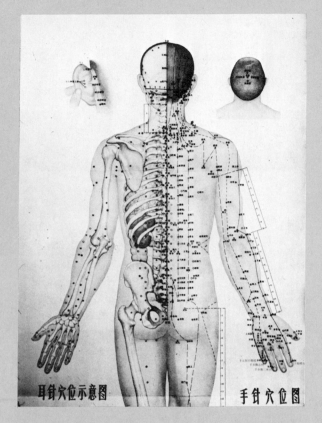

have been traced along the meridians and the 365 acupuncture points described in ancient Chinese textbooks. This evidence appears to lend validity to the existence of hypothetical meridian lines. Furthermore, since acupuncture points lie close to clusters of nerve endings, needle insertion may trigger impulses which cause the brain to release enkephalins.

It has also been demonstrated that acupuncture has regulatory effects on blood pressure, pulse rate, perspiration, temperature, and other autonomic functions. Thus, the curative effects of acupuncture on internal disorders may occur through its action on the autonomic nervous system. (Coincidentally, the two divisions of the autonomic nervous system, the sympathetic and parasympathetic systems, must be bal-

anced for optimum health. In many ways they are analogous to the Chinese concept of Yin and Yang.)

Acupuncture now stands at the crossroads between ancient Chinese philosophy and traditional Western medicine. As researchers continue to relate the empirical practice to modern knowledge, it appears that these divergent views may be drawing closer together. Many Chinese physicians have modified their technique by choosing acupuncture points on the basis of neuropathways and by employing electrical needle stimulation to replace manual manipulation. In other parts of the world, this ancient Oriental technique is gaining prominence, and because of the absence of hazardous side effects, it may prove to have valuable applications as a form of anesthesia.

FIGURE 27–14
Opium and the Opium Poppy. Opium is a general term for the solidified sap that exudes from cuts made in the unripe seed pod of the poppy *Papaver somniferum*. The most abundant active agent in opium is morphine. Heroin is produced by chemically modifying morphine in the laboratory so that it becomes more soluble in lipids. Because of this increase in lipid solubility, heroin enters the brain more rapidly than morphine. Once in the brain, however, cellular enzymes convert heroin back to morphine.

Because of their high concentration in the limbic system, enkephalins may also regulate mood. Physicians who have administered endorphins to patients suffering from severe depression, schizophrenia, and other mental disorders report many successes in relieving symptoms and improving moods. The "natural high" experienced by athletes and others after vigorous exercise may result from elevated levels of enkephalins. In fact, exercise enthusiasts may even become addicted to these internal opiates and experience a form of withdrawal and mood depression when regular exercise is not possible.

Drugs

Most drugs that affect the nervous system exert their actions at the synapse. In general, they either stimulate or inhibit synaptic transmission of impulses, but the exact mechanism of action varies with the drug (Table 27–1). For example, amphetamines ("uppers") enhance the release or synthesis of excitatory transmitters (norepinephrine and serotonin) in the brain, causing a general increase in neuron activity. For this reason, amphetamines are used clinically as stimulants to elevate alertness and mood, and to reduce fatigue. Because they also act on the hypothalamus to reduce the sensation of hunger, amphetamines have been used as appetite suppressants. Barbiturates ("downers"), on the other hand, have the opposite effect. They interfere with the synthesis or release of norepinephrine and serotonin, particularly in the reticular formation. As a result of this interference, barbiturates induce drowsiness and sleep.

Knowing the specific actions of drugs makes them useful in the treatment of many disorders of the nervous system. For example, the drug prostigmine inhibits the activity of acetylcholinesterase, the enzyme responsible for breaking down acetylcholine shortly after its release at a synapse. Prostigmine has been useful in the treatment of myasthenia gravis, a neurologic disorder in which individuals lose voluntary muscle control because their motor neurons release too little acetylcholine at neuromuscular junctions. Prostigmine amplifies neuromuscular signals by slowing down the destruction of the subnormal levels of acetylcholine.

The medical benefits of drugs are often overshadowed by instances of drug abuse. In the case of amphetamines, for example, individuals who take these substances to get "high" or lose weight often suffer from lack of sleep, which may ultimately trigger hallucinations. In extreme cases, as with an overdose, schizophrenic behavior (confusion of reality and fantasy) and paranoid behavior (increased distrust and delusions of persecution or grandeur) can result. Unfortunately, no known drug creates a feeling of well-being without also interfering with our ability to sense and respond to stimuli properly. Occasionally, as we read in the newspapers, the interference can be total and permanent.

TABLE 27–1
Some Drugs That Alter Behavior.

Drug Type	Mechanism of Action	Physiological Effect(s)
Stimulants:		
Amphetamines (Benzedrine, Methedrine, etc.)	Stimulate the synthesis or release of excitatory transmitters (norepinephrine and serotonin) in the brain	Lessen fatigue and depression; reduce appetite; addictive and can lead to mental disorders
Caffeine (coffee, tea, cola)	Facilitates synaptic transmission and enhances activity of neurotransmitters	Lessens fatigue and promotes alertness; can produce insomnia and tenseness
Nicotine	Mimics the effect of acetylcholine	Mild stimulant in low doses from smoking
Tranquilizers:		
Barbiturates (Nembutal, Seconal)	Inhibits production and release of norepinephrine and serotonin	Can induce sleep, euphoria, weight loss; highly addictive; withdrawal can produce convulsions and hallucinations
Chloropromazine (Thorazine, Compazine)	Blocks normal receptor sites for epinephrine and acetylcholine and, most important, dopamine	Depresses activity of reticular formation; suppresses hallucinations and delusions; used in treatment of schizophrenia
Anesthetics:		
Ether	Inhibits transmission of impulses	General anesthetic
Alcohol	Central nervous system depressant but exact mechanism of action unknown	Sedative causing loss of coordination and alertness
Novocaine	Prevents transmission of impulses to central nervous system	Local anesthetic used by dentists; numbs treated area
Cocaine	Prevents or reduces uptake of norepinephrine into presynaptic neuron terminals	Produces euphoria and alertness; addictive
Narcotics and hallucinogens:		
Opium and derivatives (heroin, morphine)	Depressants of central nervous system activity	Causes euphoria, drowsiness, loss of appetite; addictive; used clinically to treat pain and diarrhea
Mescaline and psilocybin	May alter (enhance) the effect of serotonin	Can cause hallucinations and psychotic behavior; increase sensory awareness
LSD	May antagonize or alter the action of serotonin	Increases sensory awareness; produces vivid hallucinations
Marijuana and hashish	Unknown	Mild stimulants and euphoric in low doses; can cause loss of appetite and seem to reduce pain in cancer patients

SUMMARY

The comparison of nervous systems of invertebrates provides a model for visualizing the evolutionary trends that led to the vertebrate nervous system. The cnidarians display the simplest of all nervous systems, a diffuse array of interconnected neurons called a nerve net. In the more advanced flatworms, part of the nerve net is condensed into two nerve cords that terminate in two anterior ganglia. This pattern represents the first sign of cephalization, a trend toward concentration of the major sense organs and ganglia in the head region of the body. The arthropods have a single nerve cord with several ganglia located along its length; the most anterior ganglion is the brain. In vertebrates, the single nerve cord (spinal cord) terminates in a single, very complex brain at its anterior end.

Considering the wide range of activities carried out by the vertebrates, the organization of the vertebrate nervous system is understandably complex. It consists of the central nervous system (CNS), which includes the brain and spinal cord, and the peripheral nervous system (PNS). The PNS refers to all the ganglia, nerves, and individual neurons that lie outside of the

CNS, and it is divided into sensory and motor divisions. The motor division is in turn divided into the somatic nervous system and the autonomic nervous system. The somatic nervous system directs motor impulses to the system of skeletal muscles, which are primarily under voluntary (conscious) control of the brain. The autonomic nervous system exerts involuntary motor control over the heart muscles, smooth muscles associated with the internal organs and blood vessels, and various glands. The sympathetic division of the autonomic nervous system directs impulses that prepare the body for strenuous physical activity (increase heart rate, dilate blood vessels, etc.), whereas stimulation of the parasympathetic division tends to have a calming, restorative effect.

The spinal cord contains a central region of gray matter (mainly unmyelinated nerve fibers), which is surrounded by white matter (mostly myelinated axons of interneurons and sensory neurons). The spinal cord conducts messages in both directions between the brain and outlying regions of the body.

The human brain is clearly the most sophisticated accomplishment of biological evolution. It consists of the hindbrain (medulla, cerebellum, and pons), midbrain, and forebrain (thalamus, hypothalamus, and cerebrum). In humans the cerebrum is the largest and most distinctive part of the brain. It is divided into two cerebral hemispheres, each of which is covered by a thin layer of gray matter called the cerebral cortex. The cortex processes most of the sensory input to the brain and initiates many motor responses. It also controls many of the more complex functions of the brain, including learning and memory, abstract thinking, and personality.

Many regions of the cerebral cortex have been mapped as to function. The primary sensory area contains neurons that receive sensations from, and the primary motor area initiates responses to, the various parts of the body. Other regions called association areas are responsible for memory, learning, and integrating stored information with new experiences.

The limbic system is located at the base of the cerebrum, linking the cerebral cortex with the hypothalamus. It is involved in the interplay between reason and passion, permitting conscious control over emotions, such as anger, fear, sexual desire, appetite, and thirst.

The reticular formation consists of a diffuse set of neurons that runs through the medulla and midbrain, and conveys signals up to the cerebrum. It filters incoming sensory information and permits the brain to concentrate on the most important sensory input or motor response at a given time. The reticular formation also controls the states of consciousness associated with wakefulness and sleep. By monitoring brain wave patterns recorded on an electroencephalograph, scientists have discerned three major states of electrical activity of the brain: wakefulness, light sleep, and REM sleep. During the sleep interval, short periods (5–20 minutes) of REM sleep alternate with light sleep. Dreams occur primarily during REM sleep.

Some of the most exciting finds in brain research have come in the area of chemical activity of the brain. Nearly 30 different kinds of neurotransmitters have been isolated from the central nervous system. Unique neuropeptides, the enkephalins and endorphins, are substances that act like opiates to kill pain and regulate moods. These internal opiates appear to be responsible for the anesthetic effects of acupuncture and the feeling of well-being experienced after strenuous exercise.

Drugs that affect the nervous system act primarily at the level of the neuronal synapse, stimulating (such as the effect of amphetamines) or inhibiting (such as the effects of barbiturates and anesthetics) synaptic transmission of impulses. Drugs have many clinical applications in the treatment of nervous system disorders, but drug abuse can create disorders.

STUDY QUESTIONS

1. What trend in the organization of animal nervous systems was apparently started by the flatworms?

2. Outline the subdivisions of the peripheral nervous system and indicate how the two systems at each division level differ from one another.

3. What are the three major regions of the hindbrain? Which is involved in balance?

4. List the three major regions of the forebrain. Which of these exhibits an extraordinary level of development in primates?

5. Describe the functional relationship between the hypothalamus and the endocrine system.

6. What are association areas of the brain and how do they differ from the primary sensory and motor areas?

7. What types of nervous behavior are associated with the limbic system? The reticular formation?

8. Name three characteristics that distinguish light sleep from REM sleep.

9. In which parts of the central nervous system do the opiate drugs act?

10. What are natural opiates?

SUGGESTED READINGS

Iversen, L. L. "The Chemistry of the Brain." *Scientific American* 241(1979):134–149. The relationship between chemical transmitters and drugs is covered in depth in this article.

Kety, S. S. "Disorders of the Human Brain." *Scientific American* 241(1979):202–214. The relationship between genetic and environmental factors in mental illness is but one of the topics covered in this article.

Naute, W. J. H., and M. Feirtag. "The Organization of the Brain." *Scientific American* 241(1979):88–111. A nicely illustrated article that discusses overall brain anatomy as it relates to function.

Van Dyke, C., and R. Byck. "Cocaine." *Scientific American* 246(1982):128–141. This article discusses all aspects of this drug from its preparation to mode of action.

Wurtman, R. J. "Nutrients That Modify Brain Function." *Scientific American* 246(1982):50–59. This article describes how three nutrients in pure form or ingested in food (tryptophan, tyrosine, and choline) can modify brain function.

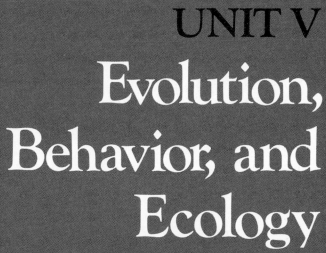

UNIT V
Evolution, Behavior, and Ecology

CHAPTER 28

The Idea of Evolution

According to the theory of evolution, existing biological species are derived from previous ones by descent; that is, all organisms, past and present, share a common ancestry. So much evidence for this theory has accumulated over the years that most scientists no longer speak of biological evolution as a highly probable occurrence, but as an actuality—as we have done in this book. Today, evolution is a unifying theme in biology. It provides a framework for the organization of diverse and seemingly unrelated observations into logical patterns. At the same time, data obtained from a variety of fields—paleontology, comparative anatomy, comparative embryology, genetics, and biochemistry—provide an overwhelming mass of evidence consistent with the theory of evolution.

There are those who do not support the tenets of evolutionary theory. After all, no one has witnessed the evolution of amphibians from fishlike forerunners or humans from apelike ancestors. Rather, evolution is based almost entirely on circumstantial evidence. But we must keep in mind that circumstantial evidence combined with logical thought is the heart of science; it is the way we learn about things in the laboratory, field, and in everyday life. It is also true

that no one has ever witnessed the planets revolving around the sun, but it would be very difficult to fit all the observations regarding our solar system into a different interpretation. The astronomers have provided the circumstantial evidence, and we (most of us?) accept their model that the planets orbit the sun. Let us now have a closer look at the case for evolution.

EVIDENCE FOR EVOLUTION

The nearest thing we have to a written record of evolution comes from the field of **paleontology.** Paleontologists find, reconstruct, and interpret ancient life forms which have been preserved as **fossils** (from Latin, "something dug up"). By comparing similar-looking fossils of different ages, paleontologists try to reconstruct the past histories of organisms. The evolution of the horse is a well documented case in point (Figure 28–1). By comparing the bone structures and sizes of ancestral horses from different periods, scientists have been able to trace the evolution of the modern horse from its ancient predecessor, *Eohippus* ("dawn horse").

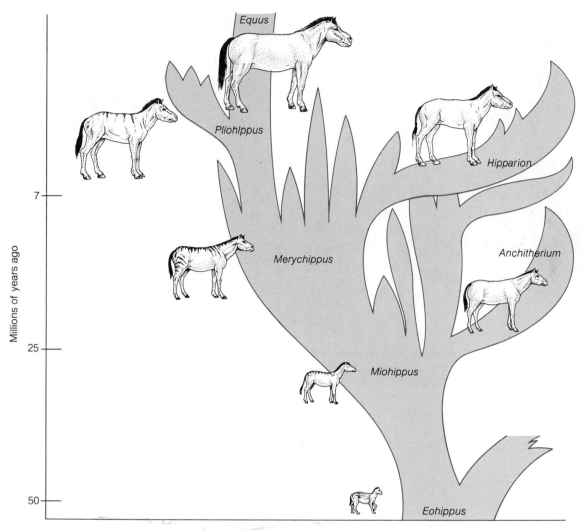

Millions of years ago

7

25

50

Equus

Pliohippus

Hipparion

Merychippus

Anchitherium

Miohippus

Eohippus

FIGURE 28–1

Evolution of the Horse. The extensive fossil record of earlier horselike animals allows paleontologists to construct an evolutionary history of the horse, such as the one shown here.

Few creatures can be traced as thoroughly through the fossil record as the horse, however. Another source of data from which evolutionary inferences are drawn is **comparative anatomy,** the science of comparing structural features of present-day organisms. For example, the forelimbs of humans, dogs, birds, and whales all have a similar number of bones (Figure 28–2), muscles, nerves, and blood vessels. Such similarities in limb structure strongly suggest that these animals share a common evolutionary origin.

In addition, **comparative embryology** has provided some rather astounding parallels in embryonic structures which would be difficult to interpret in any terms other than evolutionary. For example, the early human embryo resembles a fish embryo with gill slits, a two-chambered fishlike heart, and a tail with musculature to wag it. Later on, the human embryo passes through an embryonic reptilian stage before taking on the distinctive characteristics shared by all mammalian embryos. The very early embryos of all vertebrates, from fish to humans, are difficult to tell apart (Figure 28–3).

The breeding of domestic animals and plants by human beings lends further support for evolution.

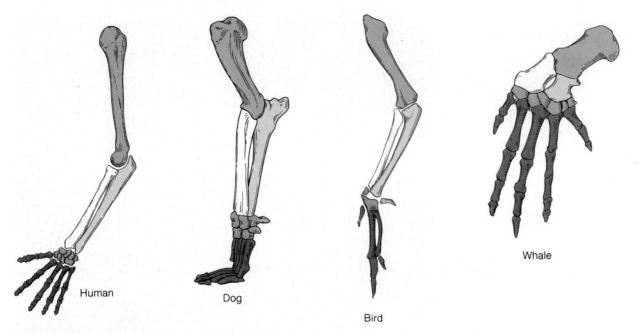

FIGURE 28–2
Comparative Anatomy of Selected Forelimbs. The arm of a human, the forelimb of a dog, the wing of a bird, and the front flipper of a whale are all modified for different purposes, but the supporting bones are very similar. *Dark color,* humerus; *light color,* ulna; *white,* radius; *gray,* carpals, metacarpals, and phalanges.

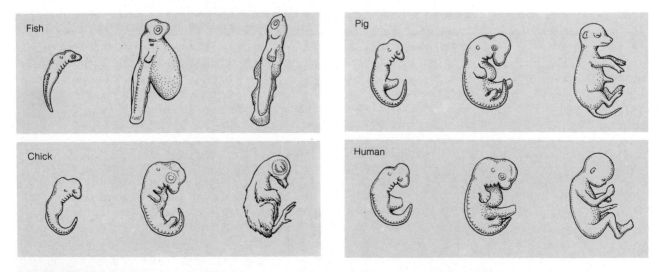

FIGURE 28–3
Comparative Vertebrate Embryology. The early stages in the development of the fish, chick, pig, and human embryos are remarkably similar in form. All show pharyngeal gill slits and a tail.

FIGURE 28–4
The Results of Artificial Selection in Dogs.

Breeders choose specimens with some desirable characteristic, such as speed in racehorses or sweetness in corn, and breed them to preserve or enhance that feature. If you were to encounter a St. Bernard and a miniature poodle for the first time, you would probably consider them two different species. Yet we know they are the same species and that these and other domestic breeds of dogs are all descended from a single ancestral species of wild dog. Thanks largely to artificial selection, dogs have evolved over the past few thousand years into many breeds (Figure 28–4), including those specialized for hunting (pointer and beagle), for herding sheep (Australian sheepdog), and for their comical appearance and good disposition (Bassett hound). By applying genetic principles, our species is constantly effecting small-scale evolution for other species. As we shall see later, Charles Darwin recognized that artificial selection in domestic animals was a special case of evolution.

The most recent and probably most profound of all evidence supporting the notion of evolution has come from the fast-paced fields of **biochemistry** and **molecular biology.** Comparative biochemical studies have shown that all organisms, from simple bacteria to humans, use similar biochemical mechanisms to trap and transform energy; they all build proteins from a similar set of amino acids, using a common molecular code of genetic information; and many of the proteins in closely related groups (such as apes and humans) are not just similar but identical (Figure 28–5). As we delve deeper and deeper into the fine structure of diverse organisms, we are struck more by the similarities than the differences. The conclusion seems unavoidable: the creatures that exist today share common structural and functional features, and therefore, a common ancestry. The basic molecular machinery of the earliest life forms is still with us, but its expression into form and function has changed over time, yielding the differences among organisms we see today.

	Human, chimp	Monkey	Pig, cow, sheep	Chicken, turkey	Turtle	Tuna fish	Yeast
Human, chimp	0						
Monkey	1	0					
Pig, cow, sheep	10	9	0				
Chicken, turkey	13	12	9	0			
Turtle	15	14	9	8	0		
Tuna fish	21	21	17	17	18	0	
Yeast	45	45	45	46	49	47	0

FIGURE 28–5
Comparative Biochemistry of Cytochrome c. Cytochrome c is an electron transport protein found in the mitochondria of all eukaryotes. Its sequence of amino acids has been determined for a wide range of organisms, including those listed here. When comparisons are made between the structures of cytochrome c molecules from two distantly related species, the number of amino acid differences is relatively high. However, these differences are small or nil when closely related species are compared. For example, by all the traditional criteria of ancestral relatedness, humans and chimpanzees are closely related. This is further supported by the fact their cytochrome c's are identical. The next closest group is the monkeys (second row) whose cytochrome c molecules differ from those found in chimps and humans (first column) by a single amino acid. Turtles (5th row) and monkeys (2nd column) are more distantly related, as evidenced by having cytochrome c's that differ by 14 amino acids.

THE BEGINNINGS OF EVOLUTIONARY THOUGHT

Of all those who have contributed to our present understanding of the evolutionary process, the name of Charles Darwin stands above the rest. But Darwin, the nineteenth century naturalist, did not invent the concept of evolution; it had been proposed in one form or another well over 2000 years ago. Rather, Darwin provided a sound explanation for how evolution works, and most importantly, he organized a wealth of data and convincing arguments in its support. Before we examine Darwin's thesis, let us look into the "evolution" of the intellectual climate that influenced Darwin's thinking.

Early Ideas on the Origins of Species

The earliest writings on evolution date from the sixth century B.C. in Greece. Several Greek philosophers speculated about the origins of various groups of organisms; some of them even recognized that fossils represent clues to the biological past. Aristotle (384–322 B.C.), recognized as the first great "scientist," contributed to evolutionary thinking by proposing a natural hierarchy of organisms spanning from the very simple to the complex, or as he called it, a "ladder of nature." He envisioned a quasievolutionary process whereby living forms progressed from a "less perfect" to a "nearly perfect" state with time.

The contributions of Aristotle and the other ancient Greeks did not gain widespread attention, however, until nearly 2000 years later. With the rise of Christianity in Europe, the Biblical doctrine of special creation held sway over Western intellectual thought in the interim. The Bible describes how God created the world and all the living things in it.

The Christian Church maintained that questions concerning biological origins and the history of the earth were answered in the Scriptures: they were not proper subjects for theorizing or investigation. The Biblical alternative to a natural evolutionary process centers on two ideas expressed in the Old Testament: special creation and the immutability of species. **Special creation** is the view that God set all creatures (Biblical word for "created beings") on earth in their present forms. **Immutability of species** refers to the idea—also supported by the authority of the Church—that the earth's creatures have remained unchanged since the day of Creation.

At the time the doctrine of special creation became popular, the sheer complexity of nature—its intricate patterns, enormous diversity, and powerful unexplained forces—seemed clear evidence of the work of a supernatural being. And because the Creation, according to one bishop's widely accepted interpretation of the Bible, had taken place in 4004 B.C. there had not been enough time for species to change from their original forms. This view of nature was not seriously challenged until the middle of the nineteenth century.

Challenges to Special Creation

With the development of modern science during the Renaissance, certain aspects of Church dogma began to come into question. Copernicus (1473–1543) and Galileo (1564–1642) disputed the long-held belief that the earth was at the center of the universe. Somewhat later, James Hutton (1726–1797), generally regarded as the father of modern geology, argued against the firmly entrenched geological principle of **catastrophism.** According to most eighteenth century scientists, the earth had been shaped by sudden and violent catastrophes, all of which had taken place within a relatively short time. Catastrophism was consistent with the Biblical reference to the Great Flood, as well as the notion that the earth was relatively young. In contrast, Hutton's **theory of uniformitarianism** spoke of a gradual, continuing process of change in which the earth had been shaped by more localized natural forces—wind and water erosion, volcanoes, minor floods—over a time frame much longer than the 6000 or so years granted by the Church.

The importance of Hutton's theory was overshadowed for a time by an artful compromise formulated by a biologist and contemporary of Hutton, Georges Cuvier (1769–1832). A noted French paleontologist, Cuvier was an expert in recreating the appearance of extinct animals from a few fossil bones. He had a good knowledge of the fossil record, and he knew of the hordes of extinct species; but he was a relentless and eloquent opponent of biological evolution. He reasoned that what had killed off the various extinct groups of animals was a series of catastrophes, such as fires and floods. After each catastrophe, a new set of organisms moved in from parts unknown to replace those wiped out. Cuvier was willing to grant Hutton his period of gradual geological change—but only for the time period since the Great Flood. The era before the Flood, Cuvier contended, was marked by catastrophes and a succession of strange, now extinct, life forms. Cuvier's compromise, which harmonized various Biblical doctrines with recent scientific findings, was widely embraced.

FIGURE 28–6
Jean Baptiste Lamarck (1744–1829).
Lamarck was the first to propose a reasonable mechanism for evolution. Although his proposal that characteristics acquired by an organism during its lifetime are inherited proved to be wrong, Lamarck's work inspired an intellectual curiosity about evolution.

Lamarck's Theory

Around the same time, the French zoologist Jean Baptiste Lamarck (1744–1829) offered the first comprehensive explanation of how evolution might operate (Figure 28–6). Like Aristotle, Lamarck envisioned a ladder of nature, a vertical hierarchy from simple single-celled organisms at the bottom to human beings at the top. But Lamarck's ladder was a dynamic one, more like an escalator: the lower organisms were continually increasing in complexity and moving up the ladder, while the void left at the bottom was filled by spontaneous generation of the simplest creatures.

Lamarck believed that all organisms unconsciously strive to evolve, and that this "evolution by will" takes place by two mechanisms. The first was **use and disuse of parts.** As environmental conditions change, Lamarck believed, organisms develop new traits to cope with the change. If the new conditions warrant greater use of a particular part of an animal's body, that part will become more efficient at carrying out its function. Similarly, if an animal no longer uses a particular body part, that part will wither and eventually disappear. The reasoning behind Lamarck's proposal is sound if you consider the effect that lifting weights has on human muscle development, for example, versus the way a bedridden patient with no opportunity for exercise quickly loses muscle tone. Second, Lamarck proposed that **acquired characteristics are inherited.** He believed that the physical alterations an organism acquired through use and disuse of parts were passed on to the next generation. If the environmental pressure for change persisted, the increments of structural change gained during each generation would eventually add up to a major alteration.

To illustrate his evolutionary mechanisms, Lamarck applied them to explain the origin of the giraffe. He theorized that the giraffe evolved from an ancestral stock of horselike creatures, which, possibly because of overgrazing, had to stretch their necks to reach leaves on the higher branches of tall shrubs and trees. Over a lifetime of stretching, each individual's neck became slightly longer, and this acquired trait was inherited by the next generation. Over many years, with each generation contributing a small gain in neck length, a long-necked giraffe evolved (Figure 28–7).

Lamarck's view of evolution included four key elements:

1. A progressive change from simpler to more complex organisms.
2. An inherent striving to adapt to altered environmental circumstances.
3. Changes in biological characteristics brought about by use and disuse of parts.
4. The inheritance of any acquired characteristics.

To Lamarck's contemporaries, the most objectionable of these four points was the notion that simpler organisms give rise to higher forms. Cuvier publicly humiliated Lamarck and his theories, as he had done earlier to Hutton, and the intellectual community became even more skeptical of an evolutionary process. Today, however, 200 years later, Lamarck has been vindicated on his idea of progressive change. The discoveries of the twentieth century clearly indicate that creatures have developed from simpler to more complex forms with time. We know too that use and disuse of body parts can alter an organism's physical char-

FIGURE 28–7
Giraffe Reaching for Vegetation. According to Lamarck, this long-necked browser evolved from a line of short-necked horses by stretching its neck over many generations. Today we recognize the horse and giraffe as having evolved along separate lines.

acteristics during the individual's lifetime. However, modern genetics has disproven Lamarck's concept that acquired characteristics can be inherited. Only genes are inheritable; environmentally altered characteristics are not. Finally, adaptation is a passive phenomenon, not the result of an individual striving to change. Contrary to Lamarck's belief, creatures do not evolve by will, whether conscious or unconscious.

DARWIN AND THE CONCEPT OF NATURAL SELECTION

Lamarck's theories on the mechanism of evolution were published in 1809. That same year, a relatively insignificant item appeared in the local newspaper of Shrewsbury, England: the birth of a son, Charles Robert, to Dr. and Mrs. Robert Darwin. Charles Darwin was born to a family of education and means—his father had a respectable medical practice and his mother was the daughter of Josiah Wedgwood, the millionaire who engineered the first mass production of table china. Charles was expected to follow his father into medicine, and so at the age of 17 he went off to the University of Edinburgh. But like many students then and now, Darwin quickly became bored with lectures and pompous professors. He also had a distaste for the surgical demonstrations (done without benefit of anesthesia). After two years he transferred to the University of Cambridge to study theology. This was not as radical a shift as it might seem, for science and religion were still intertwined during the early nine- teenth century, although Darwin's work would soon change that situation.

Although Darwin's three years at Cambridge earned him a degree in theology, it was his extracurricular activities that really prepared him for his later work. Outside the classroom, Darwin was an avid insect collector and amateur geologist. Through his friendship with a Cambridge botanist, John Henslow, he learned to identify plants. It was Henslow who recommended that Darwin fill the position of shipboard naturalist on the *Beagle,* a sailing vessel which was to embark on a five-year expedition around the world. Although the position was a nonpaying one, Darwin seized upon this opportunity to escape from the stuffy Cambridge atmosphere and pursue his favorite activity, observing nature.

Voyage of the *Beagle*

The *Beagle* sailed from Devonport, England, on De- cember 27, 1831, with the 22-year-old naturalist, Charles Darwin, on board (Figure 28–8). Although the

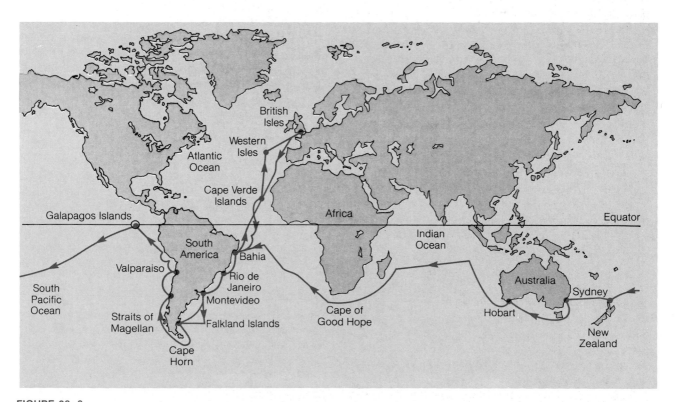

FIGURE 28–8

Voyage of the *Beagle*. This map shows the route taken by the *Beagle* during its five-year voyage around the world. By the time Darwin reached the Galapagos Islands, he was firmly convinced that only an evolutionary process could account for the diversity of life forms.

voyage completely encircled the globe, much of the five years was spent "hopping" from one anchorage to another along both coastlines of South America. While the ship's captain was constructing chart maps and looking for exploitable natural resources, Darwin was taking careful notes on the flora and fauna of each area, and the geological characteristics as well. Often he collected specimens to bring back to England for further study. During his forays onto land, Darwin was most impressed by the constantly changing varieties of organisms he observed. The species on the east coast of South America, he noted, were on the whole very different from those on the west coast.

Probably the most interesting and impressive part of the entire journey to Darwin was his visit to the Galapagos Islands, a small archipelago about 600 miles west of Ecuador (see Figure 28–8). With their numerous craters and lava beds, Darwin recognized these islands to be of recent volcanic origin. They were on the whole sparsely populated, but some of the plants and animals were like nothing Darwin had seen before—giant cacti, huge sea-going lizards which fed on algae, and of course the giant tortoises for which the islands were named (*galapagos* in Spanish means tortoise), many of which weighed over 200 pounds.

From the standpoint of Darwin's studies, the most famous of all the Galapagos fauna were the islands' 14 species of finches, now known as Darwin's finches. The major characteristics which led Darwin to recognize these birds as separate species were the different shapes and sizes of their beaks, and their distinctive eating habits (see Figure 29–11). Most ate seeds but some preferred an insect diet. Despite these differences, they were all clearly finches. These observations stuck in Darwin's mind, and he would later write: "One might fancy that from an original paucity of birds in this archipelago one species had been taken and modified for different ends."

During the journey Darwin had ample time for reading. One book in particular, a gift given to him by his friend Henslow just before the voyage began, had a tremendous impact on him. It was Charles Lyell's *Principles of Geology.* Lyell (1797–1875) described new evidence for Hutton's theory of uniformitarianism, reiterating that a slow, geological evolutionary process driven by natural forces had taken place on earth. Citing the enormous amount of geological change represented in rock strata, and coupling this to the fact that such changes are barely visible over a single lifetime, Lyell concluded that the earth must have a history spanning millions of years.

Modern geology has extended the earth's age to 4.7 billion years; but back in the 1830s, Lyell's ideas were revolutionary. His book provided Darwin with three important incentives for considering a biological evolutionary process. The first was time. If the earth is not thousands but millions of years old, then life forms would have had the time necessary for a slow evolutionary progression. Second, if the face of the earth had evolved through the action of natural forces, why shouldn't natural forces have shaped the realm of the living? Third, Lyell's advances in geology cast new doubt on a literal interpretation of the Bible's account of the Creation, and thus the doctrines of special creation and the immutability of species became suspect.

Formulating the Theory

Darwin returned to England in 1837 to an inherited fortune (Figure 28–9). Not having to work for a living, he immediately set about sifting through his notes and specimens, reading books, and discussing his observations with other scientists. In 1838 he read *An Essay on the Principle of Population* by Thomas Malthus, a British economist. In his book, Malthus warned that the size of the human population was growing so rapidly that, if left unchecked, it would soon outstrip its food supply. Darwin quickly saw how Malthus' principle could be extended to all natural populations of organisms (that is, a group of interbreeding members of the same species in a given geographical area). In countless instances during his travels, Darwin had noticed that an animal's biological capacity to leave offspring far exceeded the capacity of its environment to support such unlimited growth. For example, based on his observations of *Doris,* a sea slug which inhabits the coastal waters of the southern tip of South America, Darwin wrote:

> I was surprised to find, on counting the eggs of a large white *Doris* (this sea-slug was three and a half inches long), how extraordinarily numerous they were. From two to five eggs (each three-thousandths of an inch in diameter) were contained in a spherical little case. These were arranged two deep in transverse rows forming a ribbon. The ribbon adhered by its edge to the rock in an oval spire. One which I found, measured nearly twenty inches in length and half in breadth. By counting how many balls were contained in a tenth of an inch in the row, and how many rows in an equal length of the ribbon, on the most moderate computation there were six hundred thousand eggs. Yet this

FIGURE 28–9
Charles Darwin (1809–1882). This portrait was painted when Darwin was 31 years old, three years after he returned from his voyage on the *Beagle*. By this time Darwin had already hit upon the theory of natural selection, but he would be 50 years old before unveiling this most elegant biological principle to the public.

Doris was certainly not very common: although I was often searching under the stones, I saw only seven individuals. No fallacy is more common with naturalists, than that the numbers of an individual species depend on its powers of propagation.

Darwin recognized that, given the enormous reproductive potential of organisms, population sizes would increase at a geometric rate in the absence of environmental limiting factors. In fact, however, limiting factors such as competition for food and habitat, predation, and disease had the net result of keeping populations at a relatively constant size. Thus, although the sea slug has the reproductive capacity to cover the earth after a few generations, environmental factors keep the number of sea slugs in check.

Darwin concluded that there must be a struggle for existence among members of any population. Not all eggs develop into embryos, not all embryos become juveniles; and few juveniles survive to adulthood. The question then became: How are the survivors selected?

To address the problem of selective survival, Darwin drew from his enormous powers of observation. His insect-collecting hobby had trained his eye to recognize small, seemingly insignificant differences between individuals of the same population. Darwin reasoned that some of these variations might confer a survival advantage to those individuals fortunate enough to have them. A moth that could fly faster than its fellow moths, for example, would have a better chance of eluding predators. Darwin's recognition of the importance of variation was a crucial step toward understanding how evolution operates. It was a point entirely overlooked by previous evolutionists.

Combining the "struggle for existence" concept with the variation he observed in natural populations, Darwin had the basis for his mechanism of evolution: natural selection. Having spent a good deal of time in the English countryside, Darwin had seen how certain desirable traits in horses (such as speed) could be artificially selected by a careful breeding program. Why couldn't this same type of selection process operate in native populations of organisms, with the environment rather than the human breeders acting as the selecting agent? With this analogy in mind, the concept of **natural selection** took shape: In the struggle for existence, individuals of a population with favorable variations have the best chance of surviving and reproducing. As a result, a higher proportion of the next generation will exhibit those same favorable traits.

The environmental factors that keep population size in check—competition for food and so on—provide a route for the natural selection process. That is, the creatures weeded out by starvation, losing fights over territory, and other struggles are most likely to be those that lack the competitive edge provided by favorable variations. Furthermore, as environmental conditions change, different variations may become favorable, and thus new characteristics will be selected through differential survival and reproduction (Essay 28–1).

Publication of the Theory

In his autobiography, Darwin indicated that he had hit upon the idea of natural selection in 1838. It was not

MICROEVOLUTION OF THE PEPPERED MOTH

One of the basic tenets of Darwin's theory of evolution is that populations adapt to changing environmental conditions through natural selection. One of the most remarkable examples of this process is the case of **industrial melanism** in the English peppered moth *Biston betularia*. Before 1850, the tree trunks in most regions of England were covered with a dappled white lichen, which provided excellent camouflage for the similarly colored peppered moths (see photographs). Once in a while, however, a butterfly collector would report capturing a dark-colored peppered moth (the dark color is due to an excess of melanin content). The birds that prey on these moths would have had little trouble in spotting the dark-colored variant against a pale backdrop of lichen, making these moths less fit in the struggle for existence.

(Left) Two moths are resting on this lichen-covered tree trunk. The pale moth blends in well with its background, but the dark moth does not. (Right) On this soot-covered tree trunk, the pale moth is more conspicuous than the darker one.

In the latter part of the nineteenth century, the industrial revolution swept the Western world. England's main source of energy at that time was coal—a very dirty coal by modern pollution standards. The nation's cities soon became covered with a layer of black soot, and the tree trunks were no exception. The lichens were darkened, and now it was the pale peppered moths which became "sitting ducks" for predatory birds. Dark-colored moths, however, blended in well with the sooty backdrop. Soon the dark variant outnumbered the dappled white moth in urban areas, thanks to natural selection. A change in the environment favored greater predation of the pale moth variant, causing a shift in moth coloration, or industrial melanism.

More recently, England has tightened pollution standards and severely restricted the amount of soot released into the air. The coal now used is much "cleaner," and there has been a shift toward the use of oil and natural gas. The lichens covering the tree trunks have re-emerged in their natural pale color once again, and the dappled white moth variant has made a resurgence.

The adaptive changes experienced by the peppered moth population are termed microevolutionary, since a change in the pattern of variation in a species has taken place, but no new species has been formed. Over 100 years ago the dappled white moth gave way to the dark form, but the dark moth was not a separate species, just a variant of the same species, *Biston betularia*. And as conditions have changed again, so has the moth.

FIGURE 28–10
Alfred Russel Wallace (1823–1913).
Wallace, an English explorer and naturalist, proposed natural selection as the driving force of evolution independently of Darwin. Although both Wallace's and Darwin's papers on the subject were read before the Linnaean Society in 1858, Darwin received the lion's share of the credit because he had the data to support his proposal.

Wallace (1823–1913), an English naturalist and adventurer, had spent a number of years in the Amazon basin, and later in the East Indies, studying nature much as Darwin had done on his voyage (Figure 28–10). Like Darwin, Wallace became convinced of a natural evolutionary process, and independently came up with his own view of how evolution worked. He drafted a manuscript describing his ideas and sent it to Darwin in 1858 for comments and suggestions. Upon reading Wallace's paper, Darwin was astounded. It was a perfect description of natural selection! Darwin found himself in a quandary as to how to proceed, so he sought the advice of his closest friends, Charles Lyell and Sir Joseph Dalton Hooker. It was decided that both Darwin's and Wallace's work would be presented together at the next meeting of the Linnaean Society of London. The two papers were read on July 1, 1858, evoking only mild interest from the audience of Society members. Interestingly, Wallace was still in the East Indies at this time, totally unaware that his manuscript was being presented.

In the following year Darwin's book appeared, with the title *On the Origin of Species by Means of Natural Selection, or the Preservation of Favoured Races in the Struggle for Life.* The more popular abbreviated title is simply *On the Origin of Species.* By that time, however, the importance of the work was appreciated, and the first printing of 1250 copies sold out on the very first day. Obviously, it created quite a stir among scientists and lay people alike. But because his ideas, arguments, and supporting data were so well presented in the book, many, but by no means all, scientists accepted Darwin's theory of natural selection immediately.

Natural Selection in a Nutshell

Let us have a closer look at this most profound of biological theories. Outlined here are the major observations and conclusions of the Darwin–Wallace theory:

Observation: The reproductive potential of organisms is large. In the absence of environmental constraints, population sizes would tend to increase at a geometric rate ($2\rightarrow4\rightarrow8\rightarrow16\rightarrow32\rightarrow$ etc.).

Observation: In nature, the size of a given population of organisms tends to remain fairly constant.

until 20 years later, in 1858, that Darwin made his ideas public. During those 20 years, Darwin methodically built his case, preparing detailed arguments supported by thousands of observations, all to substantiate this new theory of evolution. He was a very thorough man who certainly realized the potential impact of his revolutionary ideas. Was this the reason he held back for so long? Perhaps he was concerned about having to defend his proposal before the hordes of anti-evolutionists. He was keenly aware of the humiliation Lamarck had suffered at the hands of Cuvier. In the end, however, the decision to present his work was in a sense forced on him by Alfred Russel Wallace.

Conclusion: Environmental constraints oppose the geometric potential of reproduction. Individuals within a population must struggle to exist.

Observation: There is considerable variation among the individuals in a population.

Conclusion: In the struggle for existence among members of a population, those individuals having variations (traits) which make them better adapted to their environment will survive and reproduce in greater numbers than those who are less well adapted. Consequently, succeeding generations will have a higher proportion of these same favorable variations, because environmental limiting forces act to weed out the less well adapted organisms and thereby eliminate their traits from the population. This process of differential survival and reproduction is natural selection, or more popularly, survival of the fittest.

Some key elements of the Darwinian mechanism clearly distinguish it from Lamarck's view of evolution. First of all, Darwin emphasized that no choice or will is involved in this process. People commonly speak of a species changing "in order" to survive in a changing environment, as if organisms recognize shifts in conditions and adjust to them accordingly. However, this is not the case, as the enormous number of extinct species attests. Rather, variations occur by chance and are selected for or against by the environment. Here again Darwin differed from Lamarck in emphasizing that the agent of evolution is not the organism but the environment. Furthermore, the environment applies its cutting edge of selective forces to the population as a whole—and so it is the population which evolves, not the individual. In the case of the giraffe, Lamarck, you recall, envisioned an ancestral population of short-necked, horselike animals, all stretching their necks to reach the higher leaves, then passing on this stretched-neck trait to their offspring. To Lamarck, it was the inheritance of acquired increments of change in neck length that ultimately yielded the giraffe. Darwin would have explained the evolution of the giraffe quite differently. He would have envisioned an ancestral population of horselike creatures, which exhibited some variation in neck length—certain individuals would have slightly longer necks than others. When faced with a shortage of grasses and small shrubs (an environmental constraint), the individuals with longer necks (a favorable variation) could still find food because they could reach the higher leaves of trees. Thus,

they survived and reproduced in greater numbers than the short-necked variants, many of which starved. According to Darwin, natural selection produced the giraffe (Figure 28–11).

If you find yourself now and then lapsing into a Lamarckian explanation for evolution, you will be in good company. Even Charles Darwin, when faced with the problem of the source of variation, fell back on the fallacy of inheritance of acquired characteristics. But Darwin had a good excuse: He was not aware of the hereditary mechanism. This is somewhat ironic because in 1865, 17 years before Darwin's death, Gregor Mendel published the results of his experiments on pea plants. It is not clear whether Darwin simply missed Mendel's paper (it was published in a rather obscure journal), or having read it, overlooked its significance. In any event, the discoveries of mutation and genetic recombination (see Chapter 12), the ultimate sources of variation in a population, were not made until the beginning of the twentieth century (Figure 28–12).

SUMMARY

The most fundamental concept of biology is that life forms have changed through time by means of a natural evolutionary process. Evidence from diverse fields of inquiry has accumulated over the years to lend overwhelming support to the idea of evolution. The fossil record, although incomplete, reveals a clear progression of biological forms from past ages to the present. Similarities in structural, developmental, and biochemical characteristics among present-day species provide further evidence that organisms share a common ancestry. Finally, the results of breeding domestic plants and animals amply demonstrate that biological evolution does occur, at least on a small scale.

Over the centuries people have proposed various answers to the question of biological origins. The Biblical notion of special creation has been defended for many years largely on the basis of the authority of religious documents. In the early 1800s, Jean Baptiste Lamarck published the first comprehensive attempt to explain the mechanism of evolution. Lamarck asserted that progressive changes in organisms take place as creatures seek to adapt to their changing environment, and the characteristics they acquire through the use and disuse of parts are then inherited by the next generation. Geneticists have since proven that acquired characteristics are not inherited.

FIGURE 28–11
Lamarckian and Darwinian Views of Evolution.

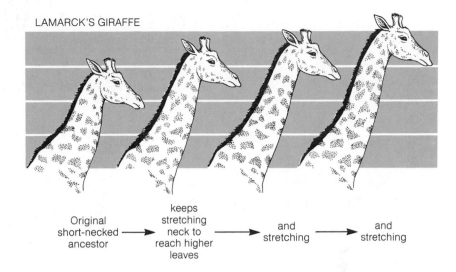

LAMARCK'S GIRAFFE

Original short-necked ancestor → keeps stretching neck to reach higher leaves → and stretching → and stretching

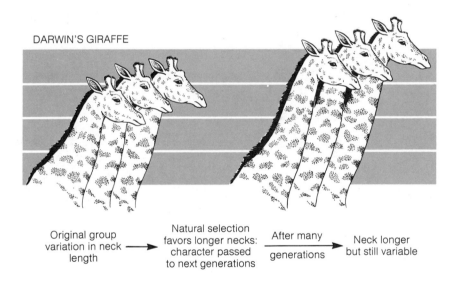

DARWIN'S GIRAFFE

Original group variation in neck length → Natural selection favors longer necks: character passed to next generations → After many generations → Neck longer but still variable

During his five-year voyage on the *Beagle* in the 1830s, Charles Darwin closely observed the flora, fauna, and physical characteristics of many areas around the world. Influenced by these observations, and by his readings on geology and human population dynamics, Darwin formulated the concept of natural selection. Alfred Russel Wallace, another English naturalist who had made similar observations in the tropical Amazon region of South America, independently reached the same conclusion: Populations evolve by means of natural selection. Although Wallace's contribution was a solid one, we generally associate Darwin's name with the theory of evolution because of his extensive contribution of published data and persuasive arguments.

In formulating the theory of natural selection, Darwin (and Wallace) recognized that while populations have the reproductive potential to expand at geometric rates, natural populations tend to remain fairly constant in size. Thus, there must be a struggle for existence among the individuals within a given

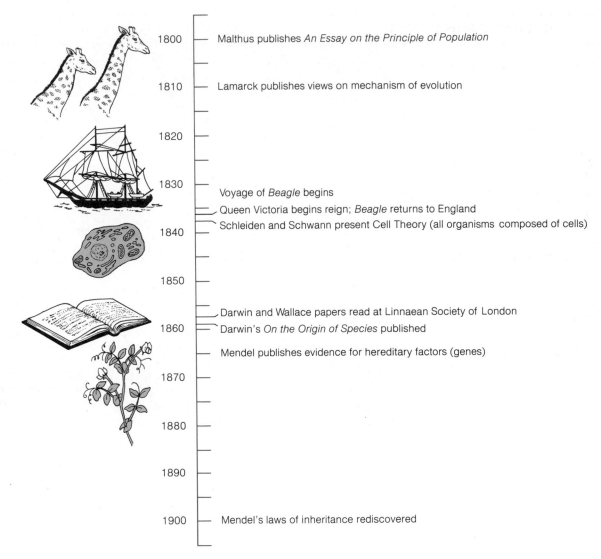

1800 — Malthus publishes *An Essay on the Principle of Population*

1810 — Lamarck publishes views on mechanism of evolution

1820

1830 — Voyage of *Beagle* begins
Queen Victoria begins reign; *Beagle* returns to England
Schleiden and Schwann present Cell Theory (all organisms composed of cells)

1840

1850

Darwin and Wallace papers read at Linnaean Society of London
1860 — Darwin's *On the Origin of Species* published
Mendel publishes evidence for hereditary factors (genes)

1870

1880

1890

1900 — Mendel's laws of inheritance rediscovered

FIGURE 28–12
Nineteenth Century Landmarks in Biology.

population. These naturalists also noticed that considerable variation exists within each population, and they reasoned that having favorable variations would give an individual a better chance of survival than its less-fit fellows. As environmental limiting factors weed out those individuals with the fewest advantageous traits, the population will change (evolve). It will contain more of the better adapted members. Thus, structural and physiological variations that are particularly compatible with the environment will increase in the population over time.

In contrast to Lamarck, both Darwin and Wallace envisioned the population, not the individual, as the biological unit that evolves in response to changing environmental circumstances. But as neither was familiar with the mechanism of trait inheritance, or the fact that hereditary units (genes) can mutate, they could not explain the source of variation within populations.

STUDY QUESTIONS

1. Name four lines of evidence that support the theory of evolution.

2. Which major findings in the physical sciences challenged the literal interpretation of the Old Testament, thereby encouraging renewed thinking on the subject of biological evolution?

3. Describe Lamarck's explanation for how evolution operated. What were the problems with his ideas?

4. As a young naturalist, Darwin developed a keen talent for observing nature. How did this talent help him in the formulation of the theory of natural selection?

5. Describe how Darwin's observations on *Doris,* the South American sea slug, helped him recognize the concept of a struggle for existence among members of a population. How did Malthus' book influence Darwin in this regard?

6. Describe and illustrate with an example the theory of natural selection.

7. How did Lamarck and Darwin differ on the point of which biological unit—individual or population—actually evolves?

SUGGESTED READINGS

Bates, M., and P. S. Humphrey, eds. *The Darwin Reader.* New York: Charles Scribner's Sons, 1956. This collection of Darwin's writings includes *The Autobiography* and excerpts from many of his other works. Darwin was a prolific writer with scientific interests spanning from movements in plants to the evolution of man.

Darwin, C. *The Origin of Species by Means of Natural Selection, or The Preservation of Favored Races in the Struggle for Life.* Garden City, NY: Doubleday & Company, 1960. This is a reprinting of Darwin's most famous work originally published in 1859. Often referred to with the abbreviated title, *On the Origin of Species,* this is Darwin's "long argument" in support of the mechanism of natural selection. The reading becomes tedious in places, but browsing through it is highly recommended.

Darwin, C. *The Voyage of the Beagle.* Garden City, NY: Natural History Press, 1962. Darwin's own account of his famous voyage around the world, this book contains many of the personal observations that led young Darwin to formulate the theory of natural selection.

Eiseley, L. *Darwin's Century.* Garden City, NY: Doubleday & Company, 1961. Written by a distinguished anthropologist and writer, this book provides an excellent account of the developments in science that paved the way for Darwin.

CHAPTER 29
Modern Evolutionary Theory

From the fossil record, we know that life forms on earth have changed continuously through time. This history of change we call evolution has spawned an incredible amount of biological diversity. How did this diversity arise? How have species changed over time, and why?

Though the groundwork for attempting to answer these questions was laid more than 100 years ago, the twentieth century, with its virtual explosion of biological discoveries, has provided some of the answers. New facts have streamed in from many corners of biology: paleontology, systematics, ecology, biogeography, biochemistry, genetics, and others. And from the new facts have grown new theories to explain them. Evolutionary biologists have not always agreed on how to put together the new data—there have been many lively debates, even heated arguments, over one point or another. However, the disagreements no longer center on whether or not evolution has occurred, but rather on *how* it has occurred.

Our current understanding of how evolution works began in 1859 with the publication of Darwin's *On the Origin of Species*. Darwin's work was clearly a piece of masterful insight, and his conception of natural selection as the major driving force of evolution

has changed little over the years. Another key piece of the puzzle was added at the beginning of the twentieth century when Mendel's principles of heredity were rediscovered and extended. Since then scientists have discovered the chemical nature of genes, how they mutate, and at least in a general way, how genes determine traits. In other words, we now know the genetic basis of variation in populations that Darwin so carefully noted. With this genetic component of the puzzle laid, we find that the genes of populations occupy a central position in our unfolding picture of the evolutionary process.

A population, you recall, is a group of interbreeding members of one species that lives in a given geographical area. From the geneticist's viewpoint, the population is a collection of genes which, like a deck of cards, gets reshuffled into new combinations with every generation. This so-called **gene pool** includes all the alleles of all the genes present in all the individuals of a population. The individual is merely a repository of genes, a temporary "warehouse" that harbors a small portion of the gene pool.

Although variations occur at the level of individuals, the evolution of a species is not simply a matter of one individual developing some new or altered characteristic. Rather, evolution involves a change in frequency of one or more characteristics in a population over time. And since the alleles of genes determine characteristics, then evolution boils down to a ***progressive change in the frequencies of alleles*** in the gene pool. As we will see, the gene pool is the medium upon which the forces of evolutionary change, such as natural selection, operate.

THE HARDY–WEINBERG LAW

One of Mendel's greatest contributions to the field of genetics was his demonstration that the alleles for any given gene become segregated and reassorted during sexual reproduction (see Chapter 9). Recall that when Mendel crossed a purebreeding red-flowered pea with a purebreeding white-flowered pea, all the F_1 progeny had red flowers (red flower color is dominant over white). When Mendel then crossed his F_1 plants, the F_2 population consisted of about three-fourths red-flowered and one-fourth white-flowered individuals.

Now let us carry this example one hypothetical step further. Suppose that Mendel had allowed this F_2 population of peas to cross-pollinate randomly, and the next generation as well, and so on for many generations. Given that the red-flower allele is dominant, would you predict that the white flower trait would eventually die out? That is, would the recessive allele specifying white flower color become less and less frequent with each passing generation, eventually reaching the point of extinction?

Similar types of questions were asked by G. H. Hardy, an English mathematician, and W. Weinberg, a German physician. In 1908 these men independently proposed what is now referred to as the **Hardy–Weinberg law:**

> In a large population of organisms in which random mating occurs, and in the absence of evolutionary forces of change, the frequencies of all the alleles in the gene pool will remain unchanged from generation to generation.

What Hardy and Weinberg realized is that the nature of an allele—whether dominant or recessive—does not affect its frequency in a population *if* the carriers of that allele have the same chance to interbreed and produce viable offspring as do the noncarriers. So long as all members of a population make equal genetic contributions to the next generation, the frequency of any given allele will stay the same, gen-

eration after generation. This principle can be demonstrated in mathematical terms, as Essay 29–1 shows.

The Hardy–Weinberg law tells us that the mere reshuffling of alleles does not alter allele frequencies in the gene pool, and thus cannot produce evolutionary change. How, then, do species change over time? The crux of the Hardy–Weinberg law lies in its assumptions for gene pool stability: large populations, random mating, and the absence of forces that can perturb allele frequencies in a population. These assumptions are theoretical only; they do not hold for real populations. In nature, many populations are quite small. The premature deaths of only a few individuals in a small population could have a significant impact on the gene pool. Even in large populations the individuals may be so spread out geographically that purely random mating is not possible. And mate preference (nonrandom mating), particularly in animal populations, is the rule rather than the exception. We also know that different alleles may be affected in different ways by the forces of evolutionary change. Thus, the Hardy–Weinberg law, although not telling us how evolutionary changes occur, does tell us where to look for them. It is in the deviations from the Hardy–Weinberg assumptions that we must look for the agents of evolution.

THE AGENTS OF EVOLUTION

Biologists recognize four principal agents of evolutionary change: mutation, gene flow, genetic drift, and natural selection. Understanding how these forces of change act on gene pools is the key to understanding how and why evolution occurs.

Mutation

Mutation, you recall, is an alteration in the structure of a gene or chromosome (see Chapter 12). Mutations are the sources of new alleles, and hence the ultimate source of all variation in a population. Biologists have determined the rates of mutations for a wide array of different genes, a few of which are listed in Table 29–1. Most of the genes examined mutate at a rate of 1 to 100 mutations per million gametes produced, although for some the rate is much higher (over a thousand mutations per million gametes) or lower (a few genes have never been observed to mutate).

Because they involve random chemical changes in DNA, many mutations are harmful, causing the altered

ESSAY 29–1

THE HARDY–WEINBERG EQUILIBRIUM

The Hardy–Weinberg law is fundamental to modern evolutionary theory because it specifies the conditions under which allele frequencies in a population will not change (hence, no evolution). We can demonstrate this principle mathematically. Let's assume a given gene has two alleles A and a, and we want to determine the frequencies of these alleles from one generation to the next under the conditions stipulated by Hardy and Weinberg: a large, randomly breeding population in which neither allele confers a selective advantage to its carrier. We can let the letter p stand for the frequency of the A allele, and q stand for the frequency of the a allele. Since A and a are the only alleles for our hypothetical gene in the population, then the frequencies of A and a must add up to 100%, or 1:

$$p + q = 1.0$$

Now suppose that in our population the frequency of A is 0.6 ($p = 0.6$), and the frequency of a is 0.4 ($q = 0.4$). If we assume that all the individuals contribute an equal share of gametes, then the frequency of gametes bearing the A allele will be 0.6 and the frequency of a-containing gametes will be 0.4. Assuming random mating, we can predict the genotype frequencies of the offspring by constructing a Punnett square:

Sperm	Eggs	
	$0.6\,A$	$0.4\,a$
$0.6\,A$	$0.36\,AA$	$0.24\,Aa$
$0.4\,a$	$0.24\,Aa$	$0.16\,aa$

The Punnett square shows that 36% of the progeny will have the homozygous AA genotype, since the chance of producing an AA individual is the product of the frequencies of A-containing sperm and eggs, or $0.6 \times 0.6 = 0.36$. Among the rest of the offspring, 48% will be heterozygous Aa ($0.24 + 0.24 = 0.48$), and 16% will be homozygous aa. Alternatively, the same calculation can be made using simple algebra. Since p represents the frequency of the A allele in both eggs and sperm, then the frequency of AA offspring will be:

$$p \times p = p^2 = 0.36$$

Similarly, the chance of producing an individual with an aa genotype is:

$$q \times q = q^2 = 0.16$$

One final calculation. The probability that an A-bearing sperm will fuse with an a-bearing egg is 0.24 (0.6×0.4). The probability that an a-containing sperm will fertilize an A-containing egg is also 0.24. Therefore, the combined chance of producing a heterozygote (Aa) by any mating is:

$$(p \times q) + (p \times q) = 2pq = 0.48$$

We also know that the frequencies of all the different genotypes produced must add up to 100%, or 1. Therefore,

$$p^2 + 2pq + q^2 = 1$$

Have the frequencies of the A and a alleles changed in this second generation? The AA individuals account for 0.36 of the population, so they have 0.36 of the population's alleles for this gene. Since they have only the A allele, the AA genotype represents an A allele frequency of 0.36. The heterozygous (Aa) individuals account for 0.48 of the alleles in the population, but only half of these are A. Thus, they contribute 0.24 A alleles to the gene pool of the second generation. By summation, the total frequency of the A allele (p) is $0.36 + 0.24 = 0.60$. And because the frequencies of the A and a alleles must add up to 1.0, then the frequency of the a allele (q) is $1.0 - 0.6 = 0.4$. As you can see, the allele frequencies (p and q) have remained unchanged. When allele frequencies in the gene pool do not change from one generation to the next, we say that the population is in **Hardy–Weinberg equilibrium.**

TABLE 29–1
Estimated Mutation Rates.

Characteristics Resulting from Gene Mutation	Mutations per 1,000,000 Gametes
Human:	
Hemophilia	32
Albinism	28
Color blindness	28
Fruit fly:	
White eyes	29
Eyeless	60
Yellow body color	120
Corn:	
Waxy (pollen lacks starch)	0.01
Sugary endosperm (seed contains sugar instead of starch)	2.4
Purple kernels	10

gene to code for a defective protein, or in some cases to lose its coding function altogether. A harmful allele that is dominant, or one that is recessive but present in the homozygous state, renders the organism less fit to survive and/or reproduce. Natural selection tends to weed out such less fit individuals, and with them, the harmful alleles they carry. Occasionally, however, a beneficial mutation occurs—a favorable variation as Darwin would have called it. The frequency of such an allele in the gene pool would tend to increase over time because individuals carrying it would have an advantage in the struggle for existence, and thus produce relatively more offspring who inherit that allele.

Finally, as a recent outgrowth of improved techniques in the analysis of proteins (the products of genes), geneticists have discovered that genes can undergo neutral changes. Neutral mutations give rise to altered genes (as measured by slight changes in the physical properties of the coded proteins), but such changes do not affect the coded proteins' functions in a way that would render organisms either more or less fit. Thus, neutral mutations are invisible to natural selection.

Although mutations are the primary source of genetic variation in a population, mutation rates in general are too low to bring about significant shifts in allele frequencies directly. Only when natural selec-

tion or another evolutionary force acts on the mutated allele can mutations have a significant impact on the gene pool.

Gene Flow

We have spoken of the population as the breeding unit of a species, a group of individuals that is geographically isolated from other groups of the same species. In many instances, however, this geographical separation is not absolute, and a few individuals from one population may come over and interbreed with members of another population. The addition of the immigrants' genes to the gene pool of the host population constitutes **gene flow,** and can lead to changes in that population's gene frequencies.

The relative contribution of gene flow to evolutionary change will depend on two factors: (1) the rate of immigration, and (2) the amount of genetic difference between the immigrants and the host population. When immigration rates are low (for example, one immigrant for every million host individuals per generation), the effect on the host's gene pool will be understandably small. Even when the immigration rates are high (such as one immigrant per hundred individuals in a host population), the effect on the host's gene pool will be minimal unless the immigrants bring in a significantly different set of alleles (Figure 29–1).

Few data exist to indicate how great an effect gene flow has on the evolution of natural populations. There is simply no direct way to measure it in the field. Nevertheless, various indirect estimates show that gene flow goes on in some species more than in others.

Genetic Drift

When the gene frequencies of a population change by accidents of chance, we attribute such changes to **genetic drift** or, more simply, drift. Evolution by drift is most likely to occur in small, well-isolated populations in which accidental death or reproductive incapacitation of a few individuals can result in a significant loss of genes from the gene pool. Changes brought about by drift are not necessarily adaptive, for it is not natural selection but rather pure chance operating in these cases. In fact, drift may even work to the detriment of a population. For example, sup-

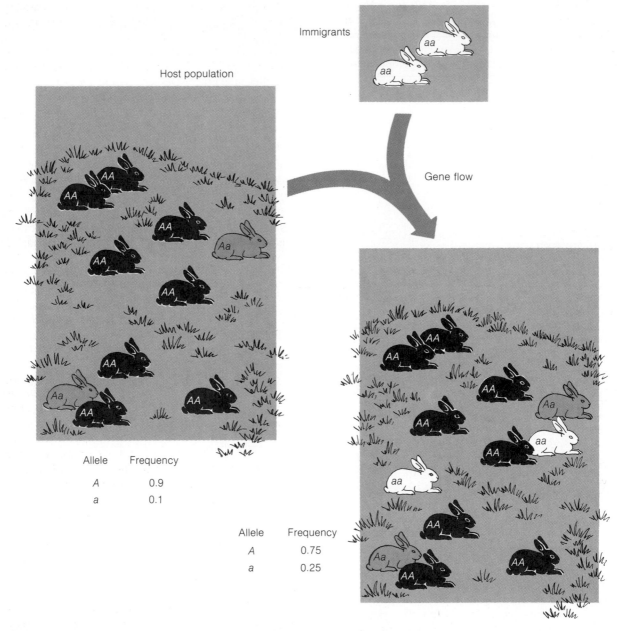

FIGURE 29–1
Gene Flow. When migration rates are high and the genetic compositions of the immigrants and host population differ significantly, gene flow can have an impact on allele frequencies in a population.

pose a small population of ten bighorn sheep migrate to a new rocky mountainside, isolated from the parent population. Let's say that three of these sheep have alleles that make them better adapted than the other seven to the new terrain (perhaps they have a narrower hoof that is less likely to get stuck between

rocks). We might expect that over succeeding generations, the frequency of the narrower-hoof alleles would increase in this new population due to differential survival and reproduction in those three sheep. Unfortunately, however, a storm triggers a rockslide that kills half of the population, including all three of the better-adapted sheep. Thus, narrow-hoof alleles have just been eliminated from the gene pool by a freak accident. The point is that genetic drift has no regard for fitness. Changes in gene frequencies caused by drift occur by sheer chance events, whether the changes are adaptive or not.

There are many indications that drift has operated in small human populations. For example, the distribution of human blood types (ABO system, see Chapter 9) is often used to assess drift because there do not seem to be any strong selection forces favoring one blood type over others. Furthermore, it is easy to obtain blood-type data from medical records or by performing simple blood tests. In most human populations, the frequency of the *A* allele (present in *AA, AO,* and *AB* genotypes) is 15–30%; the *B* allele (genotypes *BB, BO,* and *AB*) normally constitutes 5–20% of the total. In the greater Eskimo population, which extends into parts of Alaska, Greenland, and Labrador, the frequencies of the *A* allele (30%) and the *B* allele (6%) fall into these normal ranges. However, in a small tribe of Polar Eskimos living in Thule, Greenland, the frequency of the *A* allele is only 9%. In this and several other small Eskimo tribes in Labrador and Baffin Island, the *B* allele is missing altogether (Figure 29–2). As all of these tribes are presumably descended from an original stock of Asian Mongoloids, we suspect that drift may have caused these variances in blood type frequencies.

No one knows the extent to which genetic drift has played a part in the evolution of life forms, but some evolutionary biologists feel that its importance may rival that of natural selection.

Natural Selection

Each individual organism expresses its particular set of genes into a particular set of physical characteristics, or phenotype. The fitness of the phenotype is then tried by a jury of environmental circumstances. In a population, the individuals with phenotypes found most fit to survive and reproduce will contribute proportionally more offspring to the next generation than those found less fit. The result of this ongoing trial is natural selection. Most biologists today consider natural selection to be the major *driving force* of evolution.

When Darwin first described natural selection, he drew an analogy to artificial selection, the process by which breeders produce lines of domestic plants and

animals that have desirable characteristics (desirability, of course, is in the eye of the breeder). The breeder selects and mates those variants with the desired features. Future generations will thus have a high proportion of the phenotypic characteristics selected and, of course, the alleles that determine them.

In natural selection, it is not humans but rather the environment that acts as the selecting agent. Environmental circumstances select those phenotypes in the population that represent higher degrees of fitness. Relative fitness may be manifested in many ways: an individual's ability to avoid predators better than its fellows; a greater resistance to disease; or some characteristic that enables it to get more food, withstand droughts better, or in some other way compete favorably with other members of its population. Fitness can also be measured in terms of reproductive qualities, such as fertility and the ability to compete for mates. When we combine the concept of fitness with the principles of population genetics discussed earlier, we arrive at the modern version of Darwinian natural selection:

> Natural selection is the differential reproduction of individuals in a population brought about by differences in their abilities (relative fitness) to contribute to the gene pool of the next generation.

In other words, the better-adapted individuals will produce relatively more offspring that inherit their genes. Hence, the set of alleles they carry will increase in frequency in the next generation, and this, after all, is evolution.

Darwin realized that variations in a population are the raw material of evolution, the substance upon which natural selection acts. Depending on the environmental conditions, however, natural selection operates on the gene pool in different ways. Biologists today recognize three general types of natural selection: stabilizing, directional, and disruptive selection. Their effects on the distribution of variation over time are quite different, as shown diagrammatically in Figure 29–3.

STABILIZING SELECTION. If the environment is relatively static, then the selection pressures will be similar generation after generation, tending to favor an optimum phenotype (see Figure 29–3a). For example, if yellow is the most visible color to a species of insect that pollinates a particular species of plant, then yellow-flowered individuals stand the greatest chances of

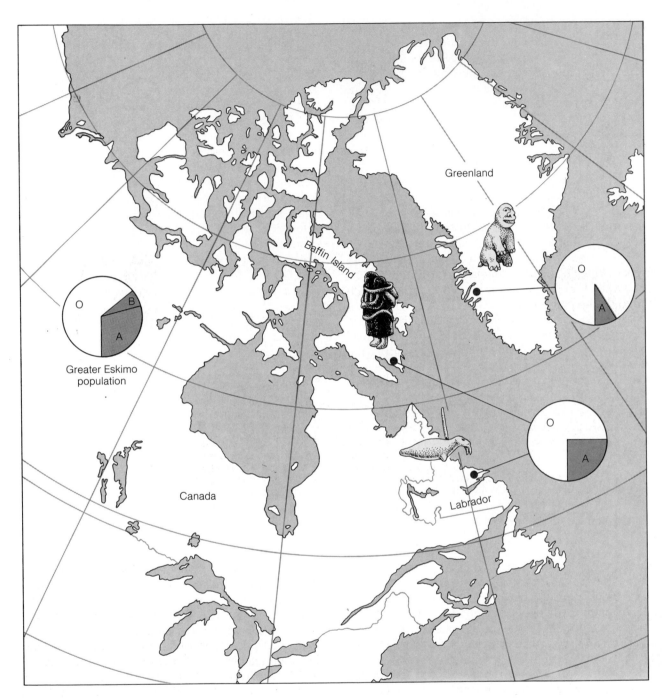

FIGURE 29–2
Genetic Drift in Eskimo Populations. Since their original migration from Asia, the Eskimos have founded many small tribes that extend from Alaska across the North American continent to Greenland. In some of these tribes—one in Thule, Greenland, and several in Labrador and Baffin Island—the frequencies of the *A* and *B* blood-type alleles are very different from those found for the larger Eskimo population. Genetic drift has probably been the major factor in causing these differences.

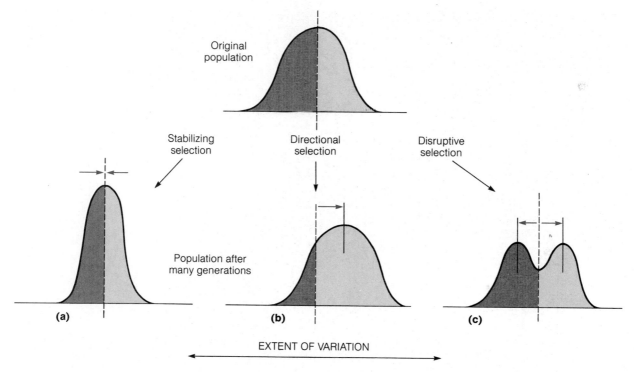

FIGURE 29–3

Modes of Natural Selection. (a) In stabilizing selection, environmental factors tend to favor the most frequent, or optimal phenotypes in a population, leading to an increase in their representation in subsequent generations at the expense of variants on either side of the optimum. Hence, the extent of variation becomes narrower. (b) Directional selection forces cause a shift in the phenotype norm to a new optimum. (c) When environmental conditions favor two or more different phenotypes in a single population, disruptive selection is operating. Vertical lines indicate the optimal phenotypes, either original (dotted lines) or new (solid lines). Colored arrows show the direction of shifts in variation.

being pollinated; red- and blue-flowered variants will be less favored because they do not have the optimum phenotype—yellow petals. In this case **stabilizing selection** forces are operating to reduce the range in variations, reducing the frequencies of phenotypic variants (red and blue petals) on either side of the optimum.

Carried to its theoretical extreme, stabilizing selection could yield a single adapted phenotype. This almost never happens in natural populations, even when only a single characteristic is being considered, because mutation and gene flow continually add new alleles to the gene pool. Thus, stabilizing selection is opposed by the processes that enhance variation, leading to a dynamic balance in the amount of variation in a population.

We can infer the workings of stabilizing selection from studies on clutch sizes (the number of eggs per brood) of birds. For Swiss starlings, the optimum clutch size appears to be five. Birds that produce fewer than five eggs win correspondingly less representation in the next generation's gene pool. On the other hand, birds with clutch sizes larger than five have difficulty in providing food for all the nestlings, so more of their offspring die. Thus, environmental pressures have operated against both small and large clutch sizes, yielding an optimum of five eggs per brood.

DIRECTIONAL SELECTION. When environmental conditions change, or when a group of individuals colonizes a new environment, the most frequent phenotype may no longer be the optimal one. Faced with new circumstances, the population's pattern of variation will generally shift toward another phenotype which is better adapted to those circumstances. This process is called **directional selection** (see Figure 29–3b).

Examples of directional selection are numerous. Industrial melanism in the peppered moth is a case in point (see Essay 28–1). Recall that the accumulation of coal dust on the lichen-covered tree trunks of England during the Industrial Revolution provided camouflage to the dark-colored form of the moth, favoring their survival from predation over the more abundant pepper-colored variant. Within a few decades the moth population shifted dramatically toward the dark color, edging out the peppered variant. Thus, directional selection can produce rapid changes in the gene pools of a population, especially when the environmental selecting force is strong.

Another example of directional selection is the development of resistance to antibiotics in many species of bacteria. Faced with the onslaught of penicillin, streptomycin, and other drugs over the past 40 years, many types of infectious bacteria have evolved resistance to these substances. Such directional selection in bacteria can be demonstrated in the laboratory in a few days. If we expose a bacterial population to an antibiotic, the overwhelming majority of cells are killed, but a few variants with the ability to resist the drug survive. These few survivors then give rise to a new population of resistant cells (Figure 29–4). In the presence of such a strong selecting agent, a population of bacteria can shift from a 99 + % nonresistant phenotype to 100% resistance virtually overnight.

To emphasize the point that environmental circumstances dictate the type of selection that acts on populations, let us consider the case of human sickle cell anemia, a trait that has been subject to both stabilizing and directional selection in different environments. You may recall that sickle cell anemia is a blood disorder that results from an abnormal form of hemoglobin, the oxygen-carrying protein of red blood cells (see Chapter 12). The homozygous recessive individuals (genotype *ss*) are severely anemic and usually die before reaching reproductive maturity. Although the carriers of the sickle cell allele (genotype *Ss*) are normally symptom-free, they too can become anemic under conditions of oxygen stress that may develop during vigorous exercise or at high altitudes. In the United States, sickle cell anemia occurs almost exclusively among American blacks whose ancestors originated from specific areas in tropical

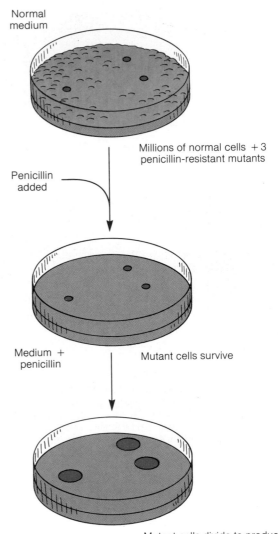

Normal medium

Millions of normal cells + 3 penicillin-resistant mutants

Penicillin added

Medium + penicillin

Mutant cells survive

Mutant cells divide to produce penicillin-resistant colonies

FIGURE 29–4
Selecting for Penicillin Resistance in Bacteria. An original population of bacteria growing on a normal medium includes three penicillin-resistant mutants. When penicillin is added to the dish, all but the three mutant cells are killed. The penicillin-resistant cells subsequently divide to form new colonies (populations), all the cells of which will be penicillin-resistant. Thus, penicillin is an environmental agent that causes directional selection toward the penicillin-resistant phenotype.

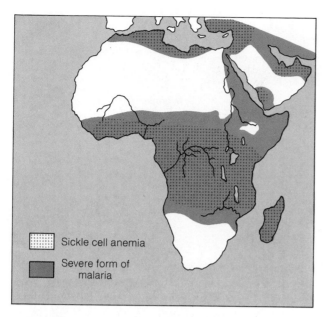

FIGURE 29–5
Coincidence of Sickle Cell Anemia and Acute Malaria. Wherever the severe form of malaria occurs in Africa and the Middle East, the frequency of the sickle cell anemia allele is high.

Sickle cell anemia

Severe form of malaria

TABLE 29–2
Estimated Frequencies of Sickle Cell Genotypes Among African Blacks, American Blacks, and American Whites.

Human Population	Geonotype Frequency (%)		
	SS	Ss	ss
Tropical Africans	68	30	2
Blacks in the United States	90	9.75	0.25
Whites in the United States	almost 100	rare	very rare

is considered rare in nature, its tendency to split a population into two or more optimal phenotypes makes it a clear candidate for the formation of new species. Let us illustrate with an example.

In the British Isles are many abandoned mines that once produced copper, zinc, and lead. The soils around these mines are contaminated with these toxic heavy metals, which make these areas unsuitable for most forms of life. Nonetheless, several species of higher plants grow in these soils, including *Agrostis tenuis* (creeping bent), a common pasture grass. Since *Agrostis* also abounds in the pastures surrounding the mines, biologists reason that the mine-dwelling variant must have originated from the nearby pasture populations. However, these mine colonizers are fundamentally different from the normal pasture inhabitants, for when pasture grasses are transplanted to the mine soil, they promptly die. On the other hand, the metal-tolerant plants do well in either type of soil, but competition with the normal variety apparently keeps their numbers low in the pastures. These two variants live in adjacent areas, exchanging genes by means of windblown pollen. But their different phenotypes for heavy metal tolerance indicate that they clearly represent two distinct norms of variation. Perhaps in time the metal-tolerant *Agrostis* will diverge in other respects to become a new species.

The case histories we have cited emphasize the importance of natural selection as an agent of change. In large populations where the effects of drift and gene flow are minimal, the combination of mutation and natural selection is the prime mover of evolution. This view, the backbone of modern evolutionary theory, extends Darwin's view on the mechanisms of evolution by recognizing the role of genes, and by identifying the sources of variation (Figure 29–6).

Africa. Its incidence in the United States has been steadily dropping due to strong directional selection against the *ss* individuals.

Selection against the *ss* phenotype is universal. Nevertheless, the frequency of the *s* allele in certain regions of tropical Africa is maintained at very high levels. How can this be? A hint to solving this apparent paradox was provided when biologists discovered a strong correlation between the geographical distributions of sickle cell anemia and malaria (Figure 29–5). As it turns out, the heterozygotes (*Ss*) are much more resistant to malaria than are homozygous normal individuals (*SS*). Thus, malaria selects against the *SS* phenotype, while severe anemia acts against the *ss* phenotype. The result is a form of stabilizing selection called heterozygote superiority that maintains a balance of *S* and *s* allele frequencies in malaria-infested regions (Table 29–2).

DISRUPTIVE SELECTION. Sometimes multiple selection forces favor more than one phenotypic optimum within a population. This is called **disruptive selection** (see Figure 29–3c). Although disruptive selection

FIGURE 29–6
The Essence of Modern Evolutionary Theory.

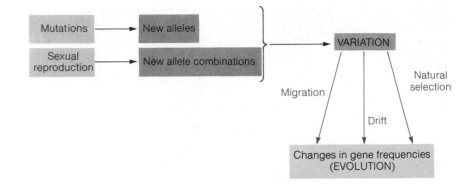

SPECIATION

Despite the title of his book, *On the Origin of Species,* Darwin never dealt directly with the problem of how a new species originates. Today we recognize several different modes of **speciation,** the process of forming a new species. Before we discuss these modes, however, we should digress for a moment to consider just what a species is.

The Concept of Species

Few other concepts in biology elude definition or incite discussion more than the notion of species. The problem, as Darwin recognized, is that drawing a clear line of demarcation between two different species is often like drawing the line between red and orange. For example, why are German shepherds and toy poodles classified as the same species (*Canis familiaris*), while the coyote (*Canis latrans*) is recognized as a separate species? (Figure 29–7) Given that species evolve continuously, Darwin saw the species as an arbitrarily defined unit, an artificial category devised by biologists with little basis in nature. Quite a few biologists today favor Darwin's view; they recognize an evolutionary continuum in which the differences among organisms blend into one another like the colors of a rainbow.

Even if we accept that dividing organisms into species is a somewhat artificial practice, the concept of species remains so useful for purposes of discussion that few if any biologists would be willing to discard it. Most biologists adhere to the **biological species concept:** Members of the same species interbreed to

yield fertile offspring, while members of different species do not. This definition focuses on the gene pool. A species is a population or group of populations whose gene pool is closed. That is, each species is reproductively isolated from all others.

Among most species, the division is obvious. No one has any difficulty recognizing dogs and horses as separate species, as they clearly cannot mate. The distinction between closely related species, however, often becomes blurred. For example, the horse and donkey can mate, but their offspring, the mule, is sterile. (Because the two sets of chromosomes in the cells of a mule cannot align properly during meiosis, the mule cannot produce viable gametes.) Biologists recognize the sterility of the hybrid mule as a form of reproductive isolation, and thus the horse and donkey are deemed different species by our definition.

Whereas reproductive isolation is the major criterion, other attributes also help biologists delineate species. Because each species has been isolated genetically from others since its inception, it has presumably undergone adaptations to its environment in ways peculiar to itself. These adaptations yield specific morphological, physiological, biochemical, and (in animals) behavioral characteristics that distinguish each species from all others.

The biological species concept works quite well when one is considering the flora and fauna of one geographical area over a relatively short span of time. But space and time play havoc with our concept of species. First, how can we know if two populations that appear to be the same species but which inhabit different regions are in fact the same species? Since they are geographically separate, we cannot know if the members of these populations would interbreed freely under natural conditions. Second, the criterion

(a)   **(b)**   **(c)**

FIGURE 29–7
An Illustration of the Species Problem. (a) A toy poodle and (b) German shepherd belong to the
same species (*Canis familiaris*), but the (c) coyote, although similar to the German shepherd in
appearance, is a different species (*Canis latrans*). The problem is compounded by the fact that
coyotes and domestic dogs occasionally interbreed.

of reproductive isolation is meaningless for distinguishing species that existed in the past. In classifying fossils, morphological characteristics of the fossilized remains are the only criteria available.

The species concept has other problems as well. How can it be applied to organisms that have no known sexual stage in their life cycles, as is true for certain fungi? Or the organisms (mainly plants) that reproduce exclusively by self-fertilization? Because the biological species concept hinges on the exchange of genes among members of a species, such organisms lie outside of our definition of species. Their designation as a species must be based entirely on phenotypic characters. Finally, there are a number of plant species that can interbreed with one another, but at much reduced rates. Gene flow between these so-called **semispecies** is restricted, but not to the point of reproductive isolation. For example, plant taxonomists recognize ten distinct species and seven semispecies of oak, making the classification of oaks a particularly difficult exercise.

As you can see, the species concept breaks down occasionally, as nature does not always conform to the

guidelines laid down by scientists. For lack of any better system, however, we continue to use the term "species," for without it there would only be more confusion.

Reproductive Isolating Mechanisms

Casting the exceptions aside for the moment, our definition of species tells us that a given pair of species, even closely related ones, are kept apart biologically by their inability to produce fertile offspring. While this reproductive isolation may have a geographical basis (obviously, two species that occupy different areas have no opportunity to interbreed), it always involves one or more forms of incompatibility at the level of sexual reproduction. The barriers to successful cross-species reproduction fall under two general headings: (1) **premating isolating mechanisms,** in which the members of the two species are prevented from mating, and (2) **postmating isolating mechanisms,** in which members of different species can mate but, for one reason or another, the results are unsuccessful

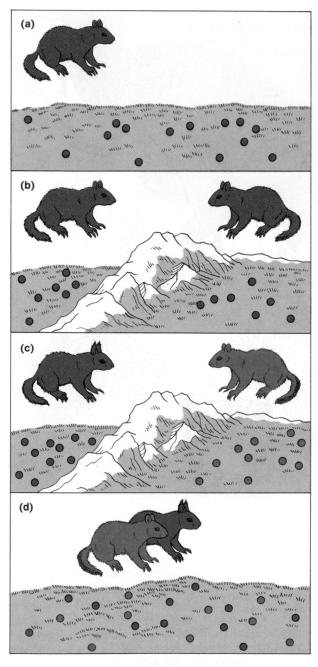

FIGURE 29–9
Speciation by Geographical Isolation.
(a) An original population becomes (b) divided by a natural barrier, preventing gene flow between the two groups. (c) Each isolated subpopulation evolves along separate lines, eventually becoming reproductively isolated. (d) Now members of different species, the two groups will not interbreed, even if the geographical barrier disappears.

(Figure 29–10). All the finches are apparently derived from an ancestral stock of finches from Ecuador some 600 miles to the east. Evidently a small number of mainland finches found their way over to the Galapagos many thousands of years ago, perhaps rafting over on some floating debris, to become established as the first birds on one of the islands. These original colonizers apparently had fairly small beaks which they used to crack small seeds for food. With little competition or predation on their island, the birds must have proliferated into a large population. Then, perhaps a few of these birds were blown over to another nearby island by strong winds. For the sake of discussion, let us assume that the major type of food on this second island was a relatively large seed (Figure 29–11). Directional selection operating on this new population of finches would have favored those birds which, because of mutations, had somewhat larger and wider beaks than normal. Such finches could gather and crush a meal of large seeds more easily than birds with small beaks. In each generation, the finches with the largest beaks would have gotten the most food, survived the longest, and reproduced in the greatest numbers, until eventually the whole population had large, broad beaks.

Finches are notoriously poor distance flyers, and the distance barrier between the two islands would have permitted these two populations to evolve separately. Continuing mutations would have led to other differences and eventually reproductive isolation. Once this had occurred, the different island populations would be different species.

Today on one of the Galapagos Islands, two species of seed-eating ground finches exist side by side: One has a large beak and gathers large seeds, and the other collects smaller seeds with its narrower, shorter beak, so they do not compete with each other for food. The origin of these two species probably occurred much as we have just described. With ten major islands in the Galapagos archipelago, it is not difficult to imagine that many geographically isolated environments were available to contribute to the diversification of the finches into 14 species.

Sympatric Speciation

While geographical isolation has undoubtedly been a forerunner to countless millions of speciations on earth, new species can also form within the same geographical range. This process is called **sympatric speciation** (*sympatric* means "in the same place").

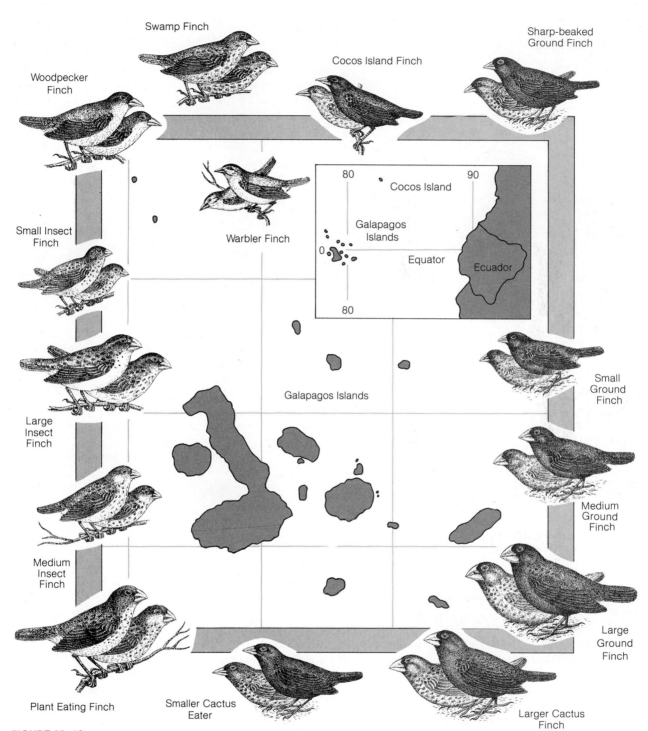

FIGURE 29–10

The Galapagos Finches. The finches of the Galapagos Islands are found nowhere else on earth, although other finch species do inhabit the mainland of South America. The Galapagos finches are not uniformly distributed over all the islands. Only the Warbler Finch is found on all the islands, while the Medium Insect Finch is found only on one island. Adaptation has given rise to different species of Galapagos finches that fill the ecological niches of hummingbirds, flycatchers, and woodpeckers elsewhere in the world.

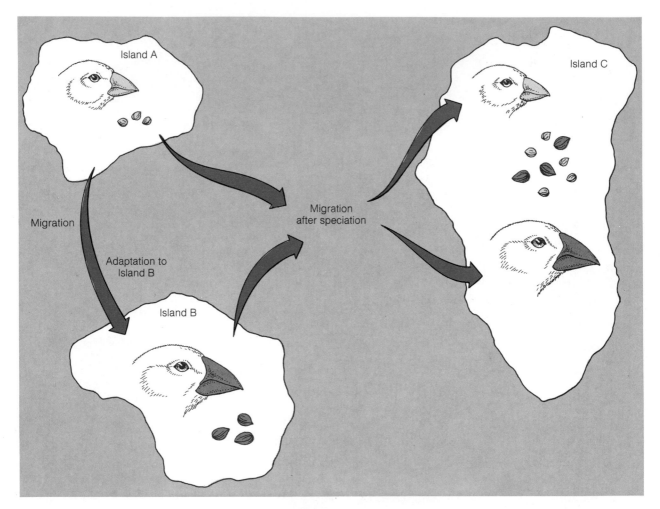

FIGURE 29–11
Possible Speciation Events for Two Species of Galapagos Finches. We can imagine that from an original population of small-beaked finches inhabiting Island A, a few migrants spread to Island B where the predominant food source is large seeds. Natural selection favors large-beaked variants among the migrants, and other changes gradually accumulate in these geographically separated populations that preclude mating between the two groups. Later, representatives of both populations migrate to Island C as different species. Because of their eating habits, the two species can coexist without competing for food.

For sympatric speciation to occur, a population must fragment into two or more reproductively isolated units, even though no geographical barriers are present to prevent gene flow between the splitting groups. Disruptive selection, as we have seen, can have a splitting effect on a population, and in theory could lead to speciation. However, disruptive selec-·tion is a poorly documented phenomenon which occurs rarely in nature, and we have no way of knowing if it has occurred in the past.

A much more common and well-documented mechanism for sympatric speciation, particularly in the plant kingdom, is polyploidy. **Polyploidy** refers to the exact duplication of the chromosome number of cells to some value greater than diploid ($2n$). For example, triploids ($3n$) are polyploids with three times the haploid (n) number of chromosomes; tetraploids ($4n$) have four times the haploid count; and so on.

Polyploidy in plants generally arises from a mistake in mitotic or meiotic cell division. For example, if identical chromatids fail to segregate in a diploid cell undergoing mitosis, one daughter cell will receive no chromosomes, and the other cell will end up with twice the diploid count to become tetraploid ($4n$). If this tetraploid cell survives and divides, it will yield more tetraploid cells and possibly, a tetraploid flower.

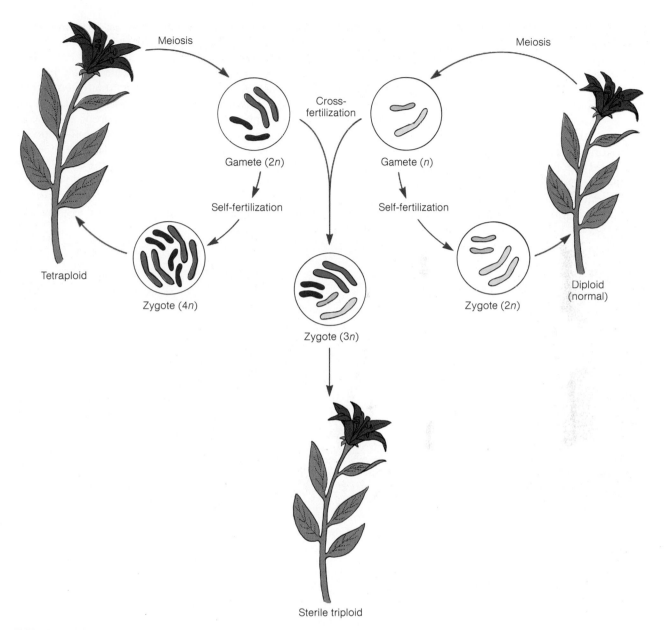

FIGURE 29–12
Speciation by Polyploidy. This diagram illustrates why a normal diploid plant species and a tetraploid derived from it by chromosomal doubling would be separate species. Cross-fertilization would
yield a triploid plant, which is sterile. If the haploid number of chromosomes is two (like the example
shown), then the triploid cells would have six chromosomes: two homologous pairs and two unmatched chromosomes (light color and gray). The presence of unpaired chromosomes generally
causes sterility due to improper meiotic segregation of these gene-bearing units. However, self-fertilization in the tetraploid plant (or fertilization among tetraploids) yields fertile tetraploid offspring; similarly, the diploid plant can produce fertile diploid offspring. Thus, the diploid and tetraploid are two
separate species.

Meiosis occurring in this flower will yield diploid
gametes. However, the union of a diploid ($2n$) gamete
with a normal haploid (n) gamete will yield a triploid
($3n$) offspring, which, although it might develop into

an adult plant, will undoubtedly be sterile due to its
inability to undergo proper meiosis (Figure 29–12).
But many plants can reproduce sexually by self-fertilization. If self-pollination takes place in our tetraploid

flower, a *2n* egg will be fertilized by a *2n* sperm. The resulting *4n* zygote can grow into a tetraploid plant capable of producing viable *2n* gametes. Any tetraploids so produced would be effectively isolated from their diploid parental stock because any interbreeding between the diploid and tetraploid plants would generate sterile triploids. But the tetraploids could go on self-pollinating or breeding with each other, forming a new species whose gene pool would be closed off from that of their ancestral diploids (see Figure 29–12).

A related and even more common form of sympatric speciation for plants is polyploidy following hybridization between two species. For example, bread wheat originated about 5000 years ago, probably in Central Europe, as a hybrid between cultivated wheat (*2n* = 28 chromosomes) and a closely related wild grass (*2n* = 14 chromosomes) that often grows as a weed in wheat fields. This hybrid started with 21 chromosomes—14 from the wheat parent and seven from the grass—and would have been sterile due to the lack of chromosome homology (Figure 29–13). But the hybrid apparently underwent a doubling in its chromosome number, becoming polyploid bread wheat with 42 chromosomes. Each of the chromosomes then had a homologous mate, and segregation

Ancient cultivated wheat (28 chromosomes)

Oat-faced grass (14 chromosomes)

Hybridization

Sterile hybrid (21 chromosomes)

Chromosome doubling (polyploidy)

Bread wheat (fertile) (42 chromosomes)

FIGURE 29–13
Origin of Bread Wheat by Hybridization and Polyploidy. Many of the kinds of wheat used in agriculture today originated as hybrids between earlier forms of wheat and oat-faced grasses. In each case, the sterility of the original hybrid was overcome by a spontaneous chromosome doubling.

of these identical homologues during meiosis was possible, making bread wheat a new, self-fertile species.

Polyploidy alone or in combination with hybridization has been an important evolutionary agent in the origin of flowering plant species. More than 100,000 species have appeared in this manner, including many important crop plants—not only wheat but also sugarcane, banana, cotton, potatoes, tobacco, and many others. Because speciation by these mechanisms is so rapid (a new species can be formed in one to several generations), chemical means have been developed to induce polyploidy in many plants. One particularly promising new hybrid is *Triticale,* an induced polyploid from a cross between wheat and rye. *Triticale* has the high grain yield of wheat and the high resistance to wheat rust (a wheat-destroying fungus) typical of rye. This and many other examples of forced hybridization and artificial selection have made humankind a powerful evolutionary agent.

EVOLUTIONARY TRENDS

Approximately 1.6 million known species of organisms inhabit earth today, and many scientists believe there are several million more species still to be discovered. This enormous diversity of flora and fauna, however, constitutes only a small fraction (less than 1% by conservative estimates) of all the species that have ever lived on earth. We must conclude therefore that speciation has been a common event. So has extinction.

When Charles Darwin was formulating his ideas on the evolutionary process, he envisioned a gradual process of change occurring over an enormous time span. Each major change in some structure, such as the development of the feathered wings of birds from scaly forelimbs of reptilian ancestors, must have been the culmination of thousands of smaller changes. And each increment of change required perhaps thousands of years to evolve by means of natural selection. Remember, life has existed on earth for 3.8 billion years, so time presents no problems for a gradualistic view of evolution.

Over the past 15 years or so, a growing number of biologists have voiced opposition to Darwin's notion of gradualism. If we take the fossil record at face value, earth's biological history has not been a smooth unfurling of new characteristics and species, but a series of evolutionary fits and starts. Periods of rapid change in life forms have interrupted relatively longer stretches of gradual transitions. The gradualists have ascribed this erratic pattern to an imperfect fossil record. They contend that the periods of rapid change only appear to be so because fossils were better preserved during those times. The long periods that seem to have been relatively static—the so-called "gaps" in the fossil record—look that way because geological conditions were not conducive to preserving the transitional forms as fossils. A new wave of evolutionary biologists, however, see no compelling reason to compromise the only available window to the past. Accordingly, they support the theory of **punctuated equilibrium:** Periods of slow rates of change with few transitional forms have been punctuated with relatively brief periods of rapid speciation (Figure 29–14).

In support of the idea of punctuated equilibrium, paleontologists have identified three major (and many minor) episodes of rapid change in the earth's fauna. First, the so-called **Cambrian explosion,** which took place some 600 million years ago, marked the origin of all the major groups of invertebrates. This phenomenal rise in diversity occurred over the relatively brief geological span of 10 million years or so. (This may seem like a long time, but if we were to condense the history of life on earth to a 24-hour day, 10 million years would be represented by less than four minutes!) Virtually all the major themes of animal body design were laid down during this relatively brief period, and subsequent animal evolution has produced only variations on these themes.

The second major change—the **Permian extinction**—occurred about 225 million years ago. More than half of the families of marine animals and 75% of the amphibians became extinct. Many explanations for this great dying have been proposed. The currently favored explanation is based on the formation of the supercontinent Pangaea during the late Permian (Figure 29–15). According to current geologic theory, the earth's crust is made up of several massive land masses called plates. Because the world's continents ride on different plates, plate movements have caused the continents to drift relative to one another. During the late Permian, continental drift caused all the continents to coalesce into one giant land mass—Pangaea. When the continents fused, many of the original oceanic shorelines disappeared, so that countless habitats occupied by the marine invertebrates were lost. This would help to explain why half of the marine invertebrate families became extinct during this period.

ESSAY 29–2

THE LATE CRETACEOUS CATACLYSM

The end of the Cretaceous period some 65 million years ago is marked by the abrupt disappearance of more than half the animal species on earth, including all the dinosaurs. As expected, this extraordinary occurrence has inspired many theories, some of which involve extraterrestrial catastrophes. There are speculations of supernovas exploding near Earth, giant comets crashing into the oceans, volcanoes on the moon hurtling massive chunks of debris toward earth, and the suggestion that a huge asteroid collided with our planet at a speed of 100,000 kilometers per hour. Where do scientists get these fantastic ideas?

Until a few years ago, all of the extraterrestrial theories were more imaginative than substantive. None had any hard facts to support it. Today, largely due to the efforts of Walter Alvarez, a geologist at the University of California at Berkeley, and his Nobel laureate father Luis, the asteroid-collision theory has gained some credibility.

While investigating a sedimentary rock formation near Gubbio in central Italy, Walter Alvarez found a peculiar 2-centimeter-thick layer of reddish clay which, based on its position in the strata, corresponded in time with the Cretaceous extinction. What time span did this 2-centimeter band of clay represent? How long did the late Cretaceous dying last? To find an answer, Walter consulted his father, a distinguished physicist.

Luis Alvarez knew that meteorites and other debris from space fall to Earth at a constant known rate and that this extraterrestrial material is chemically different in composition from the rocks on our planet. The cosmic debris contains higher amounts of certain metals in the platinum group, particularly iridium. Thus, as Luis Alvarez recalls, "It occurred to me that by measuring the abundance of iridium in the sedimentary deposits we might be able to tell how long a period of time was represented by that clay layer."

When the Berkeley group analyzed Walter's clay sample for iridium, the results were totally unexpected. The clay contained 30 times more iridium than was present in the earlier Cretaceous and more recent Tertiary rocks! Similar analyses of late Cretaceous sediments from many other regions of the world yielded similar or even higher iridium contents. So much for the assumption that meteoritic debris has fallen to earth at a constant rate throughout geological history. Our late Cretaceous planet was literally showered with extraterrestrial material. Of course, this finding shattered all hopes of obtaining a time span measurement, at least by the cosmic fallout method. But it rekindled in the Alvarez' minds visions of a giant asteroid impact.

There seems to be little doubt now that something from space bombarded the earth 65 million years ago, leaving its iridium marker in the sediments. But how is this astronomic event tied to the mass extinctions? The Alvarezes speculate that the impact of a giant asteroid would have thrown tons upon tons of matter into earth's atmosphere, perhaps thousands of times more dust than that spewed skyward by recent volcanoes. This huge dust cloud would have blocked out the sun, turning days into nights for several years and causing the death of much vegetation. Without adequate plant food, the larger animals such as the dinosaurs would have starved to death. The smaller animals (ancestors of modern reptiles and the emerging mammals) presumably eked out an existence by eating seeds, insects, and decaying vegetation until the dust cloud eventually settled to earth.

Although scientists are generally comfortable with the asteroid-collision idea, many biologists have difficulty accepting this proposed biological scenario. There are many reasons for the misgivings, but the general feeling is perhaps best summed up by William Clemens, professor of paleontology at the University of California at Berkeley, who said "It's just not very convincing to say that the animals that survived the extinctions (such as the placental mammals—moles and shrews—as well as a crocodilian reptile and perhaps even the earliest primates) ran around for five years eating seeds and insects." Clearly more work is needed before we can adequately account for this differential survival aspect of the asteroid theory. But even with its unanswered questions, the Alvarez proposal is both logical and imaginative. At the very least it has stimulated fresh thinking on a subject that has long stumped some of the greatest minds in science.

SUMMARY

With the advances made in genetics during the twentieth century, biologists turned their focus to the pool of genes in a population as the unit of evolutionary change. Thus, evolution has become synonymous with a progressive change in a population's gene pool.

The recombinations of genes through random breeding patterns does not in itself contribute to changes in allele frequencies. Hardy and Weinberg realized that allele frequencies in populations will remain unaltered from generation to generation if the population is large, mating is random, and there is no mutation or selection. In nature these conditions do not exist, and changes in allele frequencies, and hence evolution, do occur.

Mutation is the ultimate source of genetic change. However, mutations occur too infrequently to contribute alone to significant changes in allele frequencies. Mutations acted upon by natural selection or genetic drift, however, do have an impact on the evolutionary process.

Gene pools can also change as a result of the influx of genes, or gene flow, when individuals from neighboring populations migrate. The amount of change will depend on the rate of migration and the degree of genetic difference between the immigrants and the host population.

Genetic drift refers to changes in gene frequencies brought about by sheer chance events occurring in small populations. As fitness is not a factor in such cases, the changes may or may not be adaptive. Drift has probably been an important factor in effecting changes in characteristics that are relatively neutral in the face of the environment, such as the ABO blood-type alleles in humans.

Most biologists regard natural selection as the prime mover of evolutionary change. In any heterogeneous population, certain individuals will bear alleles that render them better or less fit to survive and reproduce than their fellows. The frequencies of these genes thus will increase or decrease in the next generation due to differential reproduction. This is natural selection as Darwin viewed it, but now expressed in terms of genes.

Depending on environmental circumstances, selection may take three different forms. In stabilizing selection, fairly constant environmental conditions act to reduce the range of variations in a population by weeding out extreme variants and favoring an optimal phenotype. When conditions change so that a less frequent phenotype is favored, the pattern of variation will shift in a process called directional selection. Finally, disruptive selection tends to divide a population into two or more optimal phenotypes when there are two or more opposing selecting forces operating. All three types of selection are important components of modern evolutionary theory.

A biological species is a population or group of populations which, through interbreeding, can produce fertile offspring. Due to the presence of pre- or postmating reproductive barriers, the gene pools of two different species remain distinct. Thus, the formation of a new species from an ancestral one (speciation), must involve the development of one or more reproductive isolating mechanisms.

In allopatric speciation, the reproductive isolation between the ancestral and new species is preceded by geographical isolation. A natural barrier (such as a new river or mountain) may split an original population in half and prevent any gene flow between the two groups. After perhaps thousands of years of separate evolutionary changes, the two subpopulations might have diverged sufficiently to preclude successful interbreeding, even if the geographical barrier disappears. That is, they would be separate species.

Sympatric speciation involves reproductive isolation without benefit of geographical isolation. Sympatric speciation occurs rapidly (often within a single generation time), is most common in the plant kingdom, and invariably involves changes in chromosomes. Polyploidy (an increase in chromosome number) and hybridization followed by polyploidy are two common means by which new plant species have formed.

In reconstructing the evolutionary past from the fossil record, one gets the picture of a discontinuous, often erratic series of changes. Is this pattern real, or is it the result of an imperfect fossil record? Darwin and the gradualists that followed him have adopted the latter viewpoint: Evolutionary trends have been continuous and gradual, involving many intermediate forms generated by mutation and subjected to natural selection. The absence of transitional forms in the fossil record is owing to improper fossilization or preservation during the so-called "gaps" in biological history. An alternative theory—punctuated equilibrium—maintains that the gaps merely reflect periods of evolutionary stability, when life forms changed very little over long periods of time. These relatively static periods were interrupted occasionally by comparatively short spans of rapid evolutionary transitions, such as the Cambrian explosion, and the great extinctions of the Permian and Cretaceous periods. Thus, supporters of the punctuated equilibrium theory take a more literal view of the fossil record.

STUDY QUESTIONS

1. In a population consisting of 50 *AA,* 40 *Aa,* and 10 *aa* individuals, what are the frequencies of the *A* and *a* alleles? Assuming these individuals mate randomly, what will be the frequencies of the *AA, Aa,* and *aa* genotypes in the next generation? Would the allele frequencies change?

2. What are the four major agents of evolutionary change?

3. In a population that is totally isolated from other populations of its species, which source of change in gene frequencies would not be applicable?

4. Name three types of natural selection and indicate how each affects the range of variation within a population.

5. Explain why in the United States and Africa, two different types of selection have acted on the frequencies of the sickle cell allele.

6. Describe the concept of biological species. What are some of the problems with the species concept?

7. Reproductive isolating mechanisms tend to keep species distinct. What is the basic difference between premating and postmating isolating mechanisms? Describe three specific types for each.

8. Contrast allopatric and sympatric speciation. Which generally occurs faster?

9. How do the theories of gradualism and punctuated equilibrium differ in explaining how life forms have changed over time?

10. What major geological event correlates with the massive extinctions of marine invertebrates and amphibians during the Permian? How are they related?

SUGGESTED READINGS

Eldredge, N., and S. J. Gould. "Punctuated Equilibria: An Alternative to Phyletic Gradualism." In T. J. M. Schopf, ed., *Models in Paleobiology.* San Francisco: Freeman, Cooper, and Company, 1972. A well-written article that defends the view that major evolutionary changes have been episodic and rapid.

Gould, S. J. *Ever Since Darwin.* New York: W. W. Norton, 1977. In a collection of essays reprinted and updated from his monthly column in *Natural History* magazine, Gould brings to life an assortment of interesting evolutionary topics.

Gould, S. J. *The Panda's Thumb.* New York: W. W. Norton, 1980. A continuation of the essays.

Patterson, C. *Evolution.* Ithaca, NY: Cornell University Press, 1978. A good treatment of the main topics of modern evolutionary thought. This book requires no formal training in science.

Simpson, G. G. *Splendid Isolation: The Curious History of South American Mammals.* New Haven, CT: Yale University Press, 1980. A splendid account of the adaptations of the unique mammals of South America.

CHAPTER 30
Animal Behavior

Nearly 50 years ago, Karl von Frisch placed dishes of sugar water near a hive of honeybees, then waited for the bees to discover the food. Eventually, a lone scout honeybee would discover the sugar, take a sample, and then fly away. A short time later many more bees would appear at the dish. It was apparent to von Frisch that the scout had somehow communicated her discovery to the other worker bees in the hive, who then flew to the new food source.

To study this possibility, von Frisch constructed an observation hive with glass sides so that he could watch what takes place inside the hive. When a scout bee landed at the sugar-water dish, he daubed some paint on her back for purposes of identification. Von Frisch observed that when the marked scout returned to the hive, she performed a rather elaborate dance on the surface of the honeycomb. Other bees gathered around her and became generally excited by her movements. The dancing bee stopped periodically to feed some of the watchers by regurgitating some of the sugar solution; other bees got the food scent by brushing their antennae against her. After a while, the bees left the hive and flew directly to the sugar-water.

After many years of painstaking observations and experiments, von Frisch discovered that the specific dance movements told a story. When food is found relatively close to the hive (within 85 meters), the scout bee performs a characteristic round dance. This tells the workers, "I've found some food near the hive. Follow me." The scout then flies to the food source, guiding the bees that follow her. When food is discovered at greater distances from the hive, however, the returning scout performs an entirely different routine called the "waggle" dance. The bee moves in a straight line for a short distance (called the "straight run" portion of the dance) and vigorously waggles her abdomen from side to side (Figure 30–1). She then turns, circles back to the starting position, makes the straight run again, then turns in the other direction and circles back. The dance is repeated over and over. When von Frisch changed the distance and direction of the food source, the "waggle" dance changed. To his great excitement, von Frisch discovered that the "waggle" dance gave both distance and direction. Distance to the food source was related to the number of turns made by the dancer over a given time period and direction was related to the orientation of the "straight run." If the food source was located in the direction of the sun relative to the hive, the straight run was made directly upward on the honeycomb (Figure 30–1a). A downward straight run indicated the food could be found in the opposite direction of the sun, and food located at a given angle from the hive-sun line was communicated by the cor-

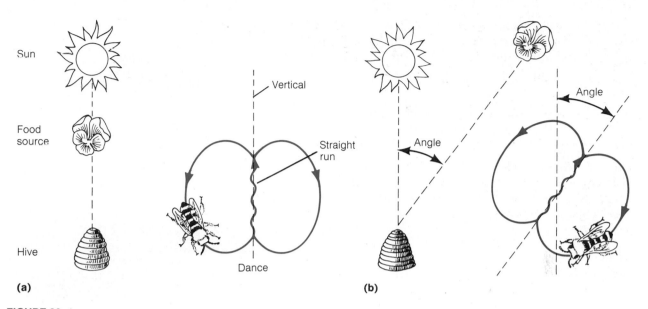

FIGURE 30–1
The Waggle Dance of Honeybees. To communicate distance and direction of a food source more than 85 meters from the hive, the scout honeybee performs a waggle dance. (a) When the food source is located in the direction of the sun, the orientation of the straight-run portion of the dance is directly upward. (b) For food sources located at a given angle to the sun-hive line, the straight run is oriented at the same angle to the vertical.

responding angle to the vertical in the straight run (Figure 30–1b).

The dance of the honeybee is an example of **behavior**—an action or sequence of actions performed by an organism in response to a stimulus. In its usual context, behavior is a characteristic normally attributable to animals, one that involves their nervous and/or endocrine systems. As the title of this chapter suggests, we will deal exclusively with the behaviors of animals. In its broader context, however, you should realize that behavior encompasses any activity of an organism that involves the components of stimulus perception and response. Thus, the directed movement of bacteria toward a food source and the bending growth of plant stems toward the sun (see Chapter 17) are certainly examples of behavior, albeit relatively simple ones.

The enormously complex nervous systems of animals permit a wide range of complicated behaviors in these creatures. This makes the study of animal behavior—**ethology**—not only interesting but also difficult and often very time-consuming. One of the inherent difficulties in behavior studies is expressing the observations in an objective way. Our language is full of words that have human connotations, so that when we observe a particular animal responding to a given stimulus, we tend to describe the behavior in terms of human experience. We say a spider weaves a web to catch flies and other insects, as if the spider gets hungry and says to himself, "Hey, I'd better build a web and snare myself a meal." Actually, we do not know if a spider can experience the sensation of hunger, and it is highly unlikely that our web-builder "knows" why he is laying down a silken net. Thus, in the design of their experiments and interpretation of results, ethologists must constantly guard against getting caught in their own "web" of anthropomorphism—the assignment of human characteristics and values to other species.

Studies in animal behavior tend to follow one of two general avenues. In one path, the scientist seeks answers regarding the physiological mechanisms underlying a particular behavior. How are the stimuli perceived? Which nerve pathways are used? Is the behavior instinctive or learned? The other path, usually a more tricky one, leads to questions concerning the adaptive value of behaviors. How does the behavior help the individual survive and reproduce? What is the "cost" to the organism? How has a specific behavior evolved? For both types of questions, there is the tacit assumption that behavioral patterns have a genetic basis—that they are phenotypic traits whose development requires genetic information. Let us begin our study by considering the relationship between genes and behavior.

GENES AND BEHAVIOR

Although traits have a genetic basis, genes determine traits only indirectly. Recall that genes code for proteins, which in turn produce traits. In the specific case of behavioral traits, the link to genes must be even further removed because animal behavior requires an intact nervous system, itself the product of the interactions of many different proteins. Exactly how the nervous system determines behavior is not known, but the indirect link between genes and behavior has been clearly demonstrated in a number of case studies, two of which we will now describe.

In a classic set of experiments performed in 1940, R. C. Tryon took a diverse group of rats and rated each individual on its ability to negotiate a maze. "Bright" rats, those that learned the maze quickly and made few mistakes, were segregated from the others and allowed to breed. "Dull" rats, those that were slow to learn the maze and made many mistakes, were similarly segregated and bred with one another. In each generation, rats were selected for their ability or inability to learn the maze, isolated, and allowed to breed as separate groups. All rats that had intermediate scores on the maze were excluded from the two breeding groups. After seven generations of artificial selection, Tryon had produced two rather distinct strains of rats: exceedingly bright ones and exceedingly dull ones (Figure 30–2). By demonstrating that it is possible to select for at least one behavioral trait (skill at negotiating a maze) in at least one species (the rat), Tryon's results suggested that behavioral differences among individuals may result from genetic (hereditary) differences. Interestingly, while Tryon's rats were labeled "bright" and "dull," these terms did not reflect overall intelligence and stupidity. When the same rats were tested in a maze that required different senses to negotiate properly, the bright rats scored no better than the dull ones!

Perhaps the clearest demonstration to date of the genetic basis of behavior is W. C. Rothenbuhler's work with two strains of honeybees. One strain he named "hygienic" because worker bees uncapped honeycomb cells in which larvae had died, then removed the corpses from the hive. Workers of the other "unhy-

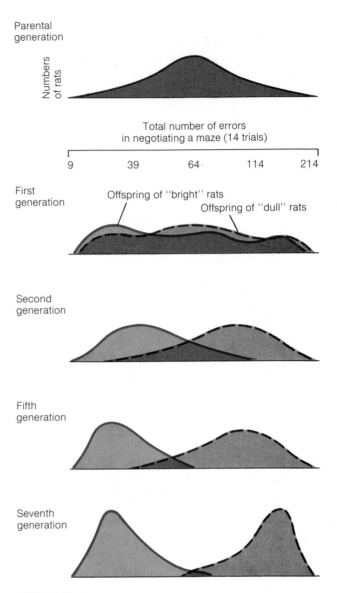

FIGURE 30–2
Artificial Selection for "Bright" and "Dull" Rats. To demonstrate that the ability of rats to negotiate mazes is a behavioral trait with a genetic basis, Tryon scored individuals from an original population with respect to the number of errors they made during the maze run. Individuals making the least number of errors (bright rats) were isolated and allowed to interbreed; those making the greatest number of errors (dull rats) were similarly isolated for interbreeding. After seven generations of artificial selection for bright and dull rats, Tryon succeeded in producing two genetically-distinct populations with regard to maze-running ability.

gienic" strain simply ignored the dead larvae. When Rothenbuhler crossed the two strains, all the F_1 hybrids were unhygienic. However, when he crossed the F_1 hybrids with the hygienic parental strain, he got roughly equal proportions of bees displaying four different behavioral patterns: (1) bees that uncapped the cells containing dead larvae and removed the corpses, (2) bees that uncapped the cells but did not remove the corpses, (3) bees that did not uncap the cells but would remove the corpses if Rothenbuhler did the uncapping, and (4) bees that neither uncapped the cells nor removed the corpses, even if Rothenbuhler uncapped the cells. These results follow simple Mendelian rules of inheritance if we assume that two independently-assorting genes control uncapping and larval removal, and the alleles specifying the unhygienic phenotypes are dominant (Figure 30–3).

The experiments carried out by Tryon and Rothenbuhler point out that behavior has a genetic component, and thus is subject to evolution by natural selection. As we shall see, the environment can also exert an influence on behavioral patterns, and the degree of this influence is high in vertebrates who rely heavily on learning. The relative contributions of genes and the environment to the behavior of organisms is best illustrated by comparing instinctive and learned behaviors.

INSTINCTIVE AND LEARNED BEHAVIORS

Instinctive (or innate) **behaviors** follow rigid, predictable patterns of responses that are triggered automatically by specific stimuli. They are not dependent on any past experiences. For example, the male European robin establishes a territory early in spring and defends it against all other male robins that may enter it. The sight of another male's red feathers sets off a pattern of instinctive behavior in the defending male, which includes the singing of specific songs, wing flapping, aggressive posture, and even an attack on the intruder (Figure 30–4). Immature male robins that lack red feathers do not evoke the threatening response, nor do birds of other species. The defending robin will perform the territory-defending behavior even if he has never seen it before. It is a purely instinctual, genetically-programmed sequence of actions evoked by the stimulus of red feathers. Thus, instincts are built into the design of the nervous system.

Learning is the development or modification of behavior in response to experience. Learned behavior may be just as stereotyped as instinctive behavior, but in learning, the organism shapes its response to a given stimulus *after* being exposed to it. Obviously there are limits to the extent that behaviors can be modified through learning. Each animal inherits its nervous and muscle systems, including well-defined neural pathways, that determine the types of behaviors it can perform. For example, no amount of training can teach a dog to speak like a human—not only are the speech centers in the brain lacking, but the dog cannot manipulate its tongue and mouth in a way to produce human sounds. Thus, we can view learning as experience-modified behavior that can occur within the realm of genetically-defined limits. Genes determine what kinds of learned behavior are possible, and the environment, through the stimuli it provides, shapes the precise character of the behavior.

To illustrate this relationship between learning and the genetic potential which bounds it, let us consider the studies made by W. H. Thorpe on the European chaffinch. When Thorpe raised young chaffinches in isolation, these birds could not sing the normal chaffinch song. When the birds were exposed to a recording of the chaffinch song, they sang like all other chaffinches. Clearly, then, the chaffinch song is learned. However, it is a little more complicated than this. If Thorpe exposed the isolated chaffinches to the recordings of similar-sounding songs of other bird species, they did not ordinarily learn them. But they did learn the chaffinch song, even if it was played backwards. Thorpe concluded that chaffinches learn the precise character of their song by listening to other chaffinches sing it (experience component), and they will only recognize and respond to the songs of their species (genetic component).

Types of Learning

There are many forms of learning. We will restrict our coverage here to the five most widely-recognized types.

HABITUATION. One of the most basic types of learning is **habituation,** or learning to ignore insignificant stimuli. Habituation generally requires repeated exposure to a particular stimulus before the animal develops a lowered response to it. For example, the first few times a young rabbit pops his head out of the rabbit hole, a

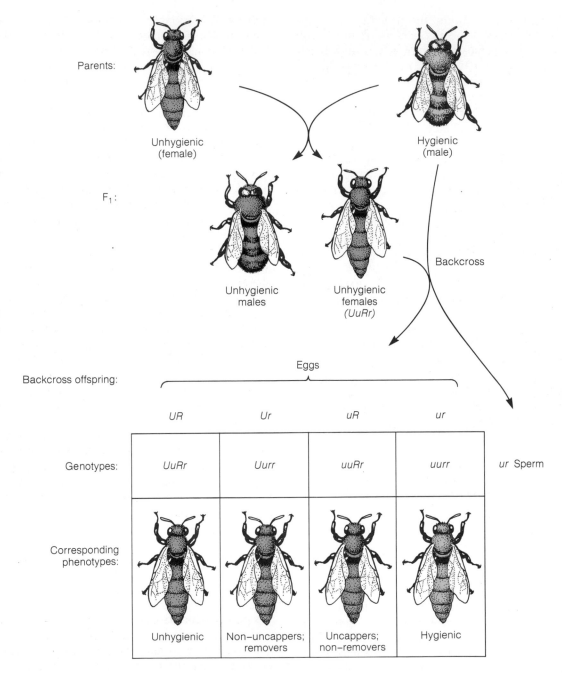

Parents:

Unhygienic
(female)

Hygienic
(male)

F₁:

Unhygienic
males

Unhygienic
females
(*UuRr*)

Backcross

Backcross offspring:

Eggs

UR   *Ur*   *uR*   *ur*

Genotypes:

UuRr   *Uurr*   *uuRr*   *uurr*   *ur* Sperm

Corresponding
phenotypes:

Unhygienic   Non–uncappers;   Uncappers;   Hygienic
              removers   non–removers

FIGURE 30–3

The Genetic Basis of Hygienic Behavior in Honeybees. Simple Mendelian rules govern the inheritance of two associated ''hygienic'' behaviors: uncapping honeycomb cells containing dead larvae, and the physical removal of dead larvae from the hive. When Rothenbuhler crossed an unhygienic strain of honeybees with a hygienic strain, all the F₁ bees were unhygienic. Thus, the hygienic traits of uncapping cells and removing corpses are apparently recessive. When Rothenbuhler backcrossed an F₁ female (genotype *UuRr*) with its hygienic male parent, he got approximately equal numbers of the four different phenotypes indicated. Conclusion: The two hygienic behaviors are determined by two separate genes.

because the "correct" response is rewarded (with food, a pat on the head, etc.) and/or "incorrect" responses are punished (no food, the crack of a whip, etc.).

The simplest type of conditioning is the conditioned reflex, first studied by the famous Russian behaviorist, Ivan Pavlov. Pavlov rang a bell each time he fed his dogs. Before long the dogs associated the bell ringing with food, and they would salivate every time they heard the bell, even if Pavlov brought no food that they could see, smell, or taste. In other words, the dogs learned to recognize a stimulus (the sound of the bell) which, under normal circumstances, would not evoke the response of salivation.

TRIAL-AND-ERROR. Perhaps the most familiar, and probably the most common, type of learning is **trial-and-error.** As in the case of conditioned learning, reinforcement is crucial to learning by trial-and-error. If an animal does something that results in a reward, it is likely to repeat the behavior. A child's first taste of candy usually evokes repeated grabs into the candy bowl. On the other hand, if a particular action goes unrewarded or results in some form of punishment, the individual will generally stop doing it. Touching a hot stove offers a punishment that teaches us not to touch a hot stove.

Trial-and-error learning also underlies improvements in behavioral performance. We learn to play musical instruments, throw darts, and play baseball from experimenting with certain actions, then learning from our mistakes and successes. Those who are very proficient at what they do have become so through many trials we call practice.

IMPRINTING. **Imprinting** is a special type of learning in which young animals, particularly birds, make a strong association with another individual, or in some cases, an object. In 1935, the Austrian zoologist Konrad Lorenz discovered that certain birds (geese, partridges, and chickens) will follow and form a bond with a moving object shortly after hatching. Under normal circumstances, the object is the mother. Thus, imprinting has survival value because it ensures that the young birds will not wander off away from their mother's protection. Under experimental conditions, however, the young hatchling will imprint on just about any moving object, particularly if the object makes sounds. It is amusing to see artificially imprinted birds following a toy train, dog, cat, or even a human being (Figure 30–5). The tendency toward imprinting and the nature of the object to which imprinting can occur are

FIGURE 30–4
Instinctive Behaviors in a Robin. When a red-breasted male robin intrudes on the territory (gray area) of another male, the defender (left) sings warning songs, assumes aggressive postures, and if necessary, will attack the trespasser as shown. The red feathers of the intruding male provoke these instinctive responses.

swaying branch may scare him back in. Gradually, the rabbit will learn that the swaying branch presents no danger and he will soon ignore it. Nevertheless, the sight of a different moving object, such as a soaring hawk, will make the rabbit scramble for cover.

CONDITIONING. In contrast to habituation, **conditioning** is behavior that is learned in association with **reinforcement,** either positive (reward) or negative (punishment). Examples of conditioned learning abound in the training of animals to respond to commands. The animals learn to associate a given verbal command, whistle, or gesture with a specific response

FIGURE 30–5
Imprinted Geese. Geese imprinted to Konrad Lorenz follow him as if he were their mother.

genetically determined; the actual formation of the bond is established through learning.

INSIGHT. Learning by **insight,** the ability to respond correctly to a stimulus never before experienced, is primarily a characteristic of primates, including humans. Insight involves the application of past, unrelated experiences to solve a new problem without having to go through trial-and-error (Figure 30–6).

ANIMAL COMMUNICATION

Communication among animals of the same species occurs under a variety of circumstances and serves a number of purposes. As we have already discussed, communication is used by male robins in defending their territory against other mature males, and by scout honeybees in directing other bees to a food source. Various means of communication are also employed to alert fellow members of a species to danger, to attract a mate, to establish dominance, and many other functions. The means of communication include visual displays, sounds, and chemical signals (odors).

Communication by Visual Displays

Visual displays are important in the courtship behaviors that precede mating in many vertebrates. Perhaps you have seen a male peacock with his tail feathers displayed like a giant fan, strutting before a female; or the antics of other male birds in spring trying to coax a female to mate. These seemingly strange displays serve some very important functions. For one, they help synchronize the sexual physiologies of the male and female, thereby ensuring coordination of gamete formation and release. Secondly, the courtship rituals for closely related species are different enough that mating invariably occurs within a particular species, not across species lines. This ensures that the reproductive effort, which usually has a considerable cost associated with it, will yield fertile offspring.

The courtship behavior of the three-spined stickleback, a small freshwater fish, is a good example of the involvement of visual displays in a complex sequence of actions. Sensing the increasing daylengths of spring, the male stickleback builds a nest on the lake bottom and undergoes a hormone-induced change in color. His underside, previously gray-green in color, becomes bright red. In the meantime, females also undergo

FIGURE 30–6
Insight Learning. (a) Incapable of insight learning, the raccoon is unable to figure out how to reach the food. (b) The chimpanzee can solve his problem even if he has never been exposed to this particular situation before.

changes leading to the production of eggs. Each female has so many eggs that her abdomen swells noticeably. When the egg-laden female spots a red-bellied male stickleback, she displays her receptivity by swimming toward the male and orienting her head upward (Figure 30–7). (Interestingly, just about any fish-shaped object with the lower half painted red will interest the female.) Upon seeing the head-up posture and swollen abdomen of the female, the male stickleback performs a zigzag dance that attracts the female to follow him to the nest. The male makes several thrusts of his head into the nest entrance, then turns on his side and finally backs away. This is the female's cue to enter the nest and deposit her eggs. Once in the nest, however, she does not release the eggs until prodded several times by the male's snout. After spawning (laying the eggs in the nest), the female swims off and the male enters the nest and releases sperm. The male stays near the nest to guard the developing young sticklebacks.

The stickleback mating ritual is so specific and regimented that no other related species of sticklebacks could gain "access to the code." Each behavioral action triggers another in an elaborate sequence of steps that synchronizes egg deposition with sperm release and ensures matings only among members of the same species.

Communication by Sounds

Relating information by sound is widespread in the animal kingdom. Because sounds can travel great distances through air and water (whales can communicate over hundreds of miles), sound is a very effective means of communication. Female mosquitos produce a familiar buzzing sound (one that we have all grown to hate) with their wings that attracts males of the same species, so the mosquito buzz is both a mating call and a species recognition signal. In more recent studies on communication in honeybees, it was learned that scout bees can also communicate direction to a food source by sounds emitted from her vibrating wings during the waggle dance. Crickets and frogs produce species-specific mating calls, and birds communicate many different kinds of messages through their songs. During the mating season, many kinds of birds (usually the male of the species) define territories with their songs. Bird songs also tell of the presence of predators, warn against intrusion of territories, attract mates, and identify members of the same species.

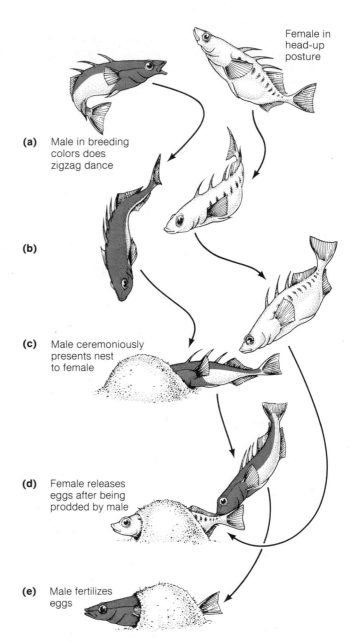

Female in head-up posture

(a) Male in breeding colors does zigzag dance

(b)

(c) Male ceremoniously presents nest to female

(d) Female releases eggs after being prodded by male

(e) Male fertilizes eggs

FIGURE 30–7
The Stickleback Mating Ritual. Each action of the male stickleback triggers a specific response from the female, which in turn triggers the next male action. This stereotyped sequence of actions ensures that the female's eggs will be fertilized immediately by sperm from a male of the same species, thereby increasing the reproductive success rate.

The most extensive user of sounds in communication is the human species. No other species even approaches the range of information that we humans can communicate through spoken language.

Communication by Chemicals (Odors)

Many animals produce and release chemicals into the environment that affect the behavior of other members of the same species. These chemicals, which are generally sensed by smelling, are called **pheromones.** Many mammals, such as dogs, release pheromones in their urine to mark territories. Female dogs and cats, as well as many other mammals, release chemicals when they enter estrus, signaling males that they are sexually receptive. Many insects utilize pheromones as sex attractants, and ants lay down chemical markers to identify the path toward a food source. These trail-marking chemicals are species-specific, so different species of ants foraging in the same area don't become confused by each other's trails.

One of the most intriguing forms of chemical communication in ants involves the so-called "death pheromone." Certain substances released from a decomposing ant corpse trigger worker ants of the same species to pick up the corpse and deposit it onto the dead ant pile outside the nest. (Ants are quite hygienic.) When investigators extracted these chemicals from dead ants and brushed them onto live ants, the workers recognized the scent, gathered up the treated ants, and dumped them onto the pile. Of course, these live ants crawled off the pile and reentered the nest, but they were promptly carried to the grave once again. Although the workers must certainly have noticed that the treated ants were alive (they put up quite a struggle), they were responding to the death pheromone in the only way they "knew."

ORIENTATION BEHAVIOR

Everyone has heard of dogs and cats finding their ways home, often over long distances. A horse taken on a long, circuitous trail ride can find its way home, even if its rider is totally lost. These animals cue on visual landmarks to find their way around, and they have a keen sense of direction. Since we humans also use visual landmarks to guide us, there is little need to discuss how this form of orientation works. Instead, we would like to discuss two more exotic types of orienting behavior: echolocation and celestial navigation.

Echolocation

Bats and dolphins are well-known for their abilities to navigate and locate objects in space using echoes. These animals emit high-frequency sounds that bounce off objects back to the animal, giving information on direction, distance, and even size of the object. This phenomenon is aptly named **echolocation.**

Bats have no trouble flying through the pitch-black caves where they often nest. So precise is the bat "radar" system that they can locate a flying insect in total darkness and catch it on the wing (Figure 30–8). As an evolutionary adaptation in their defense, many types of insects have developed the ability to detect the bat's high-frequency sounds, and thus can hear the approaching bat and take evasive action.

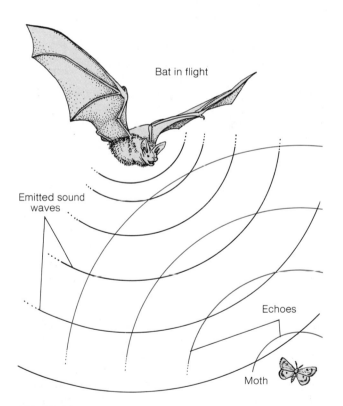

FIGURE 30–8
Echolocation in Bats. The high-frequency sound waves produced by bats reflect from nearby objects back to the bat's sensitive ears. Some bats can emit sound frequencies up to 100,000 Hertz, much too high for the human ear to detect. The upper limit for humans is a mere 20,000 Hertz.

Whales, porpoises, and dolphins all utilize "sonar" echolocation systems that are similar to the "radar" system of bats. Dolphins, which have been studied most extensively, emit high-pitched whistling sounds that they detect as reflections from any object, moving or stationary, in their vicinity. Although they have perfectly good eyes, dolphins appear to use echolocation as their primary means of navigation and finding food. For example, the dolphin can sense a clear sheet of plastic stretched through the water and avoid it. If the water is murky, or if the dolphin is blindfolded, they still navigate perfectly and can catch a swimming fish with unerring accuracy.

Celestial Navigation in Birds

Every autumn, nesting birds in northern temperate regions gather into flocks and fly south for the winter (Figure 30–9). For a given species of bird, the flight path and take-off time are the same each year, regardless of the prevailing weather conditions, availability of food, or most other variables. How can their routes and timing be so precise? Apparently, migratory birds have an internal timing mechanism, or **biological clock,** that "tells" them when to leave and in which direction. (Humans have a similar internal clock that sets various bodily activities to roughly a 24-hour cycle. This internal cycle, or rhythm, is often disturbed when we jet across time zones, resulting in "jet lag.") As the days of autumn get shorter, the bird's internal clock becomes adjusted by the decreasing daylengths, which triggers certain physiological changes (such as the accumulation of fat to meet the energy demands of the long flight) and eventually, the "itch" to fly southward.

To accomplish their annual migrations, birds must have some system of navigation. Visual landmarks would be of no use to young birds making their maiden flight over unfamiliar territory, nor to birds that fly over large bodies of water. For example, the Pacific golden plover flies nonstop from Alaska to the Hawaiian islands each autumn, using a flight path that is highlighted only by random whitecaps on the ocean (Figure 30–10). Surely these birds must be cuing on some celestial compass. Experiments begun in the early 1950s have shown that many species of migratory birds use the sun as their orienting refer-

FIGURE 30–9
A Migrating Flock of Snow Geese.

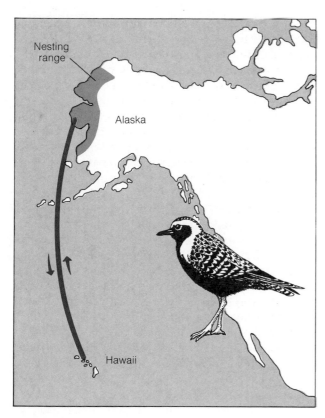

FIGURE 30–10
Migratory Route of the Pacific Golden Plover. Unable to land on water, the Pacific golden plover makes a non-stop migration from Alaska to the Hawaiian islands each autumn, and returns to Alaska in late spring. These birds are guided by celestial ''landmarks.''

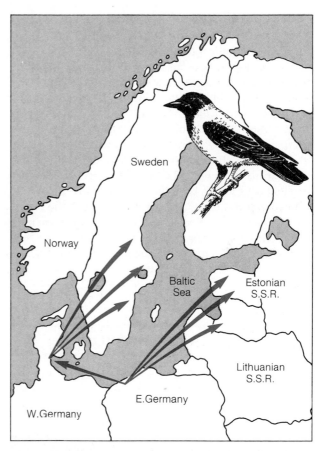

FIGURE 30–11
Sun-Oriented Migration in Hooded Crows. Just before their migration northeastward, F. Bolle transferred some hooded crows from their usual wintering site in East Germany to a location in West Germany. Cuing on the sun, the transplanted crows still migrated in the normal northeasterly direction, ending up in Sweden instead of their usual nesting territory in the Baltic countries.

ence point. They know instinctively, or in some cases through learning, the correct angle relative to the sun to head in. F. Bolle clearly demonstrated that hooded crows utilize direction rather than visual landmarks to guide their annual migrations. Just before their spring migration northeastward, a few of the hooded crows wintering in eastern Germany were captured and released in western Germany (Figure 30–11). Both groups migrated in the same northeasterly direction, but the birds transplanted to the western Germany take-off point ended up in Sweden instead of their normal nesting grounds in the Baltic countries. Since the hooded crows migrate during daylight hours, they apparently use the sun as their compass.

There is a problem with using the sun as a navigational reference point—it changes position continuously from sunrise to sunset. Again, we must invoke the bird's biological clock to explain how its direction of flight can be guided by a moving object. Presumably, the bird's internal clock (which may be set "ticking" when the sun rises) provides the time-of-day information that must be taken into account as the bird continuously adjusts its direction relative to the sun. Exactly how this works is still a mystery.

Some birds migrate at night. As you might suspect, the species that have been studied orient on the stars. Stephen T. Emlen took indigo buntings he had kept in a cage and, when they began to show signs of "migratory restlessness," released them inside a planetarium with an altered star pattern (the planetarium's dome had been rotated to yield a shifted orientation of the starry sky). The birds oriented to the south as indicated by the altered star pattern, not true south as they would have outside the planetarium. Emlen went on to show that the bunting's use of the stars as navigational cues has a learned component. When young buntings were not permitted to see the starry sky during the few weeks preceding what was to have been their first migration south, they never learned how to use the stars for navigation, even after many exposures to starry nights later on.

SOCIAL BEHAVIOR

Up to this point, our emphasis has been on the behaviors of individuals—how they learn, how they communicate, and how they find their way from place to place. But as you well know, individuals often interact with their **conspecifics** (members of the same species). These social interactions are usually cooperative, such as the mating behavior between two individuals or the herding of zebras for their mutual protection. In other instances, the interaction is definitely not cooperative, such as the aggressive behaviors often displayed by males during breeding season. These kinds of behavioral interactions between two or more conspecifics are termed **social behavior,** and the branch of biology that deals with them is called **sociobiology.**

The amount of social interaction that takes place within a group varies greatly with species and, quite often, with environmental circumstances. In some species, the individuals are basically solitary—their contact with other members of the species is limited to mating activities. In other animals, loose-knit social groups, such as flocks of birds and schools of fish, may form in which the associations are only temporary. But even such limited social interactions can be very advantageous to the individuals. For example, a solitary starling will spend approximately half of its foraging time on the ground feeding and the other half watching out for predators. In groups of ten birds, however, each starling will devote an average of 90% of its time to feeding and only 10% to surveillance.

Flocking in starlings also has advantages in getting from one place to another safely. If one of the starlings spots a falcon flying above the flock, it squawks out an alarm call that brings all the birds into a tight formation (Figure 30–12). The falcon cannot swoop down on the flock without risking injury to itself.

Depending on the circumstances, social behavior can also pose some risks. The individuals in a flock of starlings benefit from group feeding as long as the food is plentiful. When the food supply is sparse, however, the starlings will do better if they disperse over a larger region. And with respect to predators, one can imagine that a flock of birds may draw more attention to itself than would a solitary individual. Also, certain hawks prefer to attack flocks of birds, not isolated individuals. Thus, social behavior can be analyzed from the standpoint of its relative benefits and costs in terms of individual reproductive success. The underlying assumption in such analyses is that, through natural selection, those behavioral patterns that have a high benefit-to-cost ratio should come to predominate in a population through differential reproduction. What we see in living populations are the descendants of reproductively successful individuals of the past. Thus, as our working hypothesis, we will assume that traits, behavioral or otherwise, tend to help individuals reproduce successfully.

Animal Societies

The most sophisticated forms of social behavior—those that involve the most extensive communication among conspecifics—occur in **societies.** Social structures that qualify as societies are found among several groups of insects (bees, ants, wasps, and termites) and many species of vertebrates. In both insect and vertebrate societies, the social group is typically the family or, in a few cases, the extended family. Thus, its members are related genetically.

Insect societies are typically composed of a queen (the only reproductive female) and her offspring. These societies have a biologically-determined caste system featuring a clear division of labor. The various jobs include foraging, nest maintenance, guard duty, and fending off intruders. In carrying out their individual roles, the members of insect societies exhibit very rigid and stereotyped (instinctive) behaviors. Each role is so vital to the survival of the family that many biologists look upon the insect society as one supraorganism—the whole society functions more or less as a single biological unit.

FIGURE 30–12
Flocking Defense in Starlings. If flying starlings spot a falcon below them, they remain in loose formation. The falcon is not a threat when below the flock. When a falcon is sighted above the flock, one or more starlings emits an alarm call. All respond by forming a tight formation, making it difficult for the falcon to attack without risking injury to itself.

In contrast, vertebrate societies are much less rigid and structured, and the roles of individuals are not nearly as stereotyped. When they exist, caste systems (for example, the pecking order of chickens and male dominance hierarchies in baboons) are not biologically predetermined; rather, they are usually formed through social interaction (Essay 30–1). This and other forms of social interaction depend largely on the ability of individuals to recognize one another as unique individuals, a feature not found in most insect societies. Individual recognition in most vertebrate social groups is also important because all members of the society are potentially reproductive.

We will examine some of the traditional forms of social behavior in both insect and vertebrate societies, but we will do it in a nontraditional manner. Instead of probing the physiological bases of social behavior, we will focus on their adaptive qualities. That is, we will consider how and why certain social behaviors have evolved. As we mentioned at the beginning of this chapter, this is the "trickier" path of behavior studies. The questions asked are difficult and some of the answers that have come forth only recently are still quite controversial. Our context will be evolutionary theory.

Evolution and Social Behavior

Modern evolutionary theory tells us that the relative fitness of any given individual is measured against the environmental circumstances that individual faces (see Chapter 29). If we make the reasonable assumption that different patterns of social behavior, like other forms of behavior discussed previously, are due to genetic differences, then the patterns of social interaction displayed by organisms today must have evolved, just as structural and physiological traits have evolved. The question is, "How?"

To approach this evolutionary question, it is helpful to focus on genes rather than individual organisms. Remember, individuals do not evolve—they are merely transitory vehicles for genes. And it is genes (DNA), not organisms, that are passed from one generation to the next. Individuals come and go, but a gene can survive for long periods of biological history. This rather abstract view of evolution is perhaps best captured in a statement made by E. O. Wilson, a pioneering sociobiologist: "An organism is DNA's way of making more DNA."

Although this view of life tends to make some people (including biologists) nervous, it has been a particularly useful paradigm for scientists concerned with the evolution of social behavior. It permits one

ESSAY 30–1
SOCIAL LIFE OF BABOONS

Baboons are ground-dwelling Old World monkeys found throughout the African continent. Most species live in open country, such as the hamadryas baboons of the eastern deserts and *Papio anubis* (pictured here) of the central and southern savannahs. These species in particular have been popular subjects of ethological studies, as they are relatively easy to observe in their natural habitats. Such is not the case for the colorful mandrill pictured on the cover of this book. This baboon species inhabits the dense forests of West Central Africa.

Baboons are social animals. They live in groups ranging in size from small families up to large troops that may consist of 200 individuals. The hamadryas baboons of the deserts form small family units, which usually include one adult male, several adult females, and their juvenile and infant offspring. While foraging for food, however, several families may band together into a larger group, thereby affording better protection against predators. At night when they are most vulnerable to attack, two or more foraging bands may share the same general sleeping area. In a different ecological setting, the grassland-dwelling *P. anubis* has a single social unit—the large, multimale troop.

Adult male baboons are about twice the size of adult females and, understandably, the males are dominant over the females, both at the family and troop levels of social organization. The large troops of *P. anubis,* which contain many adult males, display a rather complex social structure termed **male dominance hierarchy.** This hierarchy differs from the simple pecking order of chickens and other birds in that social status

does not fall into a linear sequence of dominance. Rather, leadership among the dominant male baboons shifts depending on circumstances. One male may lead the troop on a foraging expedition; another may be responsible for choosing the sleeping site and be accorded the most desirable spot; and another may lead in the aggressive defense of the troop against a hungry lion. Occasionally, two or more dominant males form an alliance to secure an advantage over the others. Such interactions are clearly manifestations of a very complex social structure in which learning and individual recognition play large roles.

In the hierarchy, there are various levels of subordinate males below the dominant males. The status of all males in the troop is established through three general modes of aggressive behavior: play conflict, the threat of conflict, and actual conflict. In play conflict, juveniles act out mock battles they have witnessed between older males. Generally, no one gets hurt, but consistent winners may establish early dominance over their rivals. In adult males, it is the threat of conflict that is the most common means of asserting dominance. Included in the arsenal of threats are a stiff stance, a wide-eyed stare (the "look that could kill"), and the fearsome "yawn" (see photo), in which the male bares his prominent set of canines. Such threats are usually enough to deter subordinates. However, subordinates may also use threats to harass the dominant males. Actual conflict is rarely used to establish dominance, presumably because the risk of serious injury to both parties is great.

Baboons also have a repertoire of gestures that indicates submission and appeasement. When confronted by a dominant male, a subordinate may avoid conflict by crouching and grinning (the so-called "fear-grin"). Although "presenting" her genitalia to a dominant male is the way a female in estrus (at ovulation) signifies she is ready for copulation, such a gesture is also used by non-estrus females and both adult and juvenile males to show submission and vulnerability. The dominant male may acknowledge the act by reassuringly placing his hand on the other.

When a female comes into estrus, she is sought after by males. Usually the most aggressive male pursuer is successful in establishing a temporary bond with this female, but often he must contend with subordinate followers who harass him and try to sneak copulations when he and the other male followers are distracted. When the female's estrus period is over, the male and female separate and resume other social roles.

Another important social activity among baboons is grooming—the removal of insects and other parasites from the skin. Dominant males are groomed by females and juveniles, and females are groomed by juveniles and each other, especially when they are carrying infants. During their first few weeks, infants receive a lot of grooming attention from their mothers and other females of the group. Aside from its hygienic benefits, grooming is a form of attention and submission, and helps form and reestablish social bonds. Grooming is widespread among primates, and plays a social role similar to caressing in humans.

Baboon Social Behavior. (a) An infant takes refuge in an adult male's lap. (b) An adult female has greeted the mother with jaw claps and with eyes wide (an established greeting by an approaching individual). She is then allowed to touch the infant. (c) An adult male "yawns." (d) A juvenile presents to an adult male with eyes wide, and the adult touches the juvenile. (e) The dominant male threatens followers. (f) A juvenile grooms his mother and her infant.

to analyze behaviors in terms of their fitness—how they maximize the survival of genes carried by individuals. If the genes controlling the behavior of individuals confer greater fitness to those individuals, natural selection will favor the survival of such genes over time.

It is not difficult to see how natural selection can operate on the personal behavioral traits of individuals. For example, consider a hypothetical rabbit population containing two types of individuals we'll call "timid" and "courageous." At the first sight of a fox, the timid rabbits duck for cover, while the courageous ones stand and fight. Obviously, not many generations will pass before the "courageous" genes are eliminated from the rabbit's gene pool.

Can we explain the evolution of social behavior using the same type of model? No, the model is too simple. After all, social behavior involves interactions among members of a group, and recall that Darwin's natural selection operates on the fitness of individuals, not groups. Each individual struggles to exist and pass on its genes. If this were not the case, if for the good of others certain individuals sacrificed their sets of genes (including those that predispose them to such generous behavior), then it is the others' genes that will survive. The "unselfish" genes would soon become extinct. This reasoning is the foundation of the so-called **selfish gene** hypothesis: Evolution has favored genes that predispose individuals to behave in ways that maximize the likelihood of their passing such genes to the next generation. Accordingly, all forms of social behavior should be interpretable in terms of individual selfishness when it comes to reproduction.

Let us now turn to some selected cases of social behavior using the selfish gene model as the underlying theme. As we do, bear in mind that genes do not have any of the qualities we have had to use to simplify our explanation of their role in evolution. Genes are not "selfish," nor are they conscious entities striving to be passed on through reproduction. But behavior, social or otherwise, does have a genetic component. And behavior in and among individuals of a group is subject to natural selection, which in turn affects the frequencies of genes, including those related to behavior.

Territoriality

An animal's territory is defined as the geographical area an individual or group defends. In birds of prey such as the hawk, the territory may be a square mile or more, and from this area the bird will obtain all its food.

In contrast, the herring gull will defend an area of only a few square meters around its nest; it obtains no food from this space. **Territoriality,** or the defense of a territory, is a behavior common to many species of birds, fishes, reptiles, and mammals.

Although territoriality is based on the behavior of individuals, it clearly involves interaction and communication with other members of the same species. As we shall see, it has social consequences.

In defending a geographical territory, an animal must be aggressive toward its conspecifics (recall the example of the red-breasted robin). This may lead to actual fights, but in most cases the aggression never goes that far. The threat of aggression is usually enough to convey the territorial message.

To examine the adaptive value of territorial behavior, let us consider the passerine, or perching, birds during breeding season (Figure 30–13).

The passerines include the finches, sparrows, wrens, chickadees, jays, blackbirds, and a variety of other small birds. Their territories are typically large enough to contain all the essential resources, such as food and mates. During breeding season, the territories are defended by the males, an activity that has certain costs. The males spend much of their time singing, a form of territorial advertisement. And the energy they spend chasing away intruders is considerably more expensive in calories than foraging for food. Territoriality, therefore, takes away time and energy that could be spent building a nest, feeding, or carrying out other activities.

Obviously, there must be benefits to territoriality that outweigh these costs. First, because territories spread out individuals and nests, both young and adult birds may be less susceptible to predation. Furthermore, diseases and parasites are less likely to be passed among individuals when they are so widely dispersed. Another benefit of defending a territory is that limited food resources can support only a limited number of birds per region. In fact, many believe that territoriality evolved in response to food availability pressures acting on populations. In support of this hypothesis is the general relationship between territory size and food abundance. When food resources are abundant in an area, territory size tends to be smaller than when food resources are sparse. Because there is a cost to territory defense, birds simply defend less area in times of plenty.

There is yet another, more subtle benefit to territoriality for passerine birds that are monogamous (forming single-mate relationships). The relative costs of reproduction are different for males and females. Sperm are small, metabolically "inexpensive" to pro-

(a)

(b)

(c)

FIGURE 30–13
Passerine Birds. (a) The song sparrow is monogamous—both parents tend to the needs of the young. (b) The male yellow-headed blackbird is polygamous, unlike most passerine birds. He defends a large territory containing many females with which he mates, but he plays no role in the care and feeding of the young. The male spends much of his time singing to advertise his territory. (c) A female yellow-headed blackbird with young.

duce and males have the capability of inseminating many females. On the other hand, reproduction is very costly to females. Females produce fewer and larger eggs than a male does sperm. Furthermore, for animals with internal fertilization there is the metabolic cost of developing embryos. For the majority of females, the decision to mate with a particular male is one they have to live with. Most animals are seasonal breeders, restricting their breeding to a limited favorable time (such as spring). Indeed, many animals breed only once a year, or once in a lifetime. The bottom line is: females benefit by being choosy. With restricted opportunities to breed, natural selection has favored behavioral patterns which ensure that females mate with the highest quality male available. In passerine birds, the quality of the territory defended by a male is the best indication of his relative success in a given environment (the quality of his genes). Hence, females tend to select a mate based on the quality of his territory.

Most passerine birds are monogamous, so their territories contain only one mated pair. In fact, 90%

of all birds are monogamous. Because birds lay eggs, both parents are capable of caring for the young at the nest. Since newly-hatched passerines are helpless, the period of parental care is important. It appears that those males which are monogamous contribute more of their genes to future generations by helping their mate successfully raise her clutch. If they were promiscuous, they could not give full attention to all their offspring.

It is instructive to compare monogamy in nesting birds with the relatively rare cases of polygamy in birds. Polygamous passerine birds live in special ecological settings with rich and rapidly renewable food resources. With abundant resources at hand, the male is "emancipated" from his duties as a parent because the female can forage with great efficiency and meet the needs of her young alone. The male takes advantage of this by claiming large territories that will attract several females. If more than one of his mates are successful at raising their clutches, he will have contributed more genes to future generations than he would in a monogamous situation (Figure 30–13b,c).

Mating Behavior in Mammals

In mammals, unlike birds, only the mother can provide direct nourishment (milk) to developing young. Thus, care for the young is predominantly the responsibility of the female, and male mammals are generally polygamous (less than 3% of mammal species have been reported to be monogamous). Furthermore, it is not uncommon for male and female mammals to live apart for much of the year, coming together only at breeding time.

POLYGAMY. California sea lions (Figure 30–14) return each year to breed on isolated beaches on islands off the west coast of the United States. The males may winter as far north as the Aleutian Islands, while the females winter to the south. When reunited on the breeding grounds, the males establish territories by vigorous, often bloody, battles with other males. The winner remains on the territory and controls access to a group of several females. In the meantime, the females are busy giving birth to pups conceived during the preceding breeding season, and nursing them to independence. Eventually the females come into estrus and are impregnated by the male in whose territory they reside. Accepting polygamous status is not without cost for females. Males pay no attention to the presence of small pups and often crush them during territorial fights.

A similar form of behavior related to polygamy occurs in many hoofed mammals (ungulates). The elaborate antlers on deer, elk, and caribou, and horns on bighorn sheep, are weapons used to determine dominance during the fall when the females are in estrus (Figure 30–15). The strongest male generally wins these contests, and with his victory receives the opportunity to mate with the herd's females. The victor's traits, such as large antlers or horns and large body size, thus are transmitted to the next generation of males. Female offspring of these matings are likely to inherit their mother's tendency to accept males who win battles.

These polygamous mating systems based on territoriality and male aggression take their toll on the males. For example, during the estrus season, males use up most of their energy reserves. They are therefore more likely than the females to die or be heavily infected by parasites in the winter that follows.

INFANTICIDE. Infanticide, the deliberate killing of infants, seems to run counter to the tendency of animals to behave in ways likely to perpetuate their genes. On

FIGURE 30–14
California Sea Lions.

closer inspection, however, some sociobiologists interpret infanticide as one male's way of getting rid of another male's offspring and clearing the way for his own. In two species in which infanticide has been documented, African lions and Hanuman langurs, males regularly move in and take control of a new group of females. By killing infants fathered by previous males, a new male gains two advantages. First, he induces females in the group to come into estrus earlier than would happen if their infants lived. Reproduction is tightly controlled by hormone levels, and estrus is hormonally delayed during lactation, which can persist for an extended period. For example, langur mothers give birth only once every two to three years when they have infants to care for, but they generally give birth again within a year if their infants are killed. Lionesses give birth every 24 months if their cubs survive, but they may give birth within nine months after their last cub dies. Second, infanticide allows a male to avoid the costs of providing protection for or sharing resources with young that he did not father. In both these ways, males who practice

FIGURE 30–15
Caribou Bulls. Dominance among caribou is determined by battles between males. Their antlers are powerful weapons in these battles.

infanticide tend to increase the possibility of passing their genes on to future generations. Females naturally try to ward off attacks on their young, and if the male is successful they show antagonism toward him. However, the females' antagonism decreases when they come into estrus, and they eventually accept the new male's sexual advances.

Altruism

The selfish gene model faces its most severe test with **altruism,** a form of social behavior in which certain individuals jeopardize or totally relinquish future chances of reproduction to benefit others. For example, worker bees and ants are sterile, yet they slave away all their lives for the good of their societies. The female killdeer, a North American bird of the plover family, will place her life in jeopardy by feigning injury if her babies are threatened by an approaching predator. How can such unselfish servitude in social insects and life-endangering acts in killdeers be consistent with the selfish gene model? Here again, it becomes fruitless to approach this question at the level of the individual. Instead, we must seek our explanation at the level of genes. The social insects work for the benefit of their queen or future queen, both of which are genetically related to them. By helping their mother (queen) or sister (future queen), the workers increase the likelihood that other copies of their genes will be

carried forward to the next generation. Similarly, the female killdeer risks her life to protect her half of the genes present in each nestling. If she is protecting three or more young killdeers, her death in exchange for the survival of the young birds would preserve more of her genes than would the reverse situation — saving herself only to return to an empty nest. Thus, although the individuals in such cases are unselfish, maybe their genes are not!

NEPOTISM IN GROUND SQUIRRELS. Individuals live in groups for many reasons, and within these groups they apparently cooperate with one another. But why do individuals cooperate if natural selection is acting only on the likelihood of an individual passing his or her genes on to future generations? One of the most thorough investigations into this knotty problem has been conducted by Paul Sherman on Belding ground squirrels. These squirrels live on high-altitude meadows in the Sierra Nevada Mountains of California (Figure 30–16). They hibernate during the winter, and during the spring and summer they mate, raise their young, and accumulate the fat reserves they need for hibernation.

Belding ground squirrels are polygamous. Females are only sexually receptive for four to six hours once a year. During this time they run among males, inciting fights as the males compete for sexual access to them. Females watch these fights and then appear to solicit copulations with consistent fight winners and reject

FIGURE 30–16
A Belding Ground Squirrel.

mating attempts by others. Each spring relatively few males copulate frequently, while the majority never mate. As with ungulates and other mammals, these dominance battles extract their toll on males. Females generally live four to six years while males usually live only three to four years.

Of interest to us here are the dispersal patterns of males and females. Female ground squirrels are significantly more sedentary than males, and they live closer to their female relatives than to males. Males move away from their site of birth before the first winter's hibernation, while most females remain close to the burrow where they were born. Males move again following mating, with the result that they seldom live close to relatives.

With this information in mind we can now focus on one example of cooperative behavior exhibited by the Belding ground squirrels—alarm calls (a series of chattering noises) given upon the approach of predators. Alarm calls alert other members of the population to the presence of the predator. But giving an alarm call is not without cost—the individuals giving alarm calls are more likely to be detected by the predator and be its next meal! The question becomes: Why give alarm calls? Discovering which ground squirrels give these calls helps to answer this question. Alarm callers are usually reproductive, resident females with living mothers, sisters, and/or offspring. Nonreproductive, nonresident females, or females without kin seldom give warning calls. Apparently alarm calls are given only when the individual caller is likely to be surrounded by close relatives (which naturally carry many of the caller's genes). Males rarely give alarm calls, but remember that they rarely live close to relatives. Thus, their failure to give the alarm maximizes their probability of producing offspring rather than becoming someone's supper. The young females are in a similar situation. Having had little opportunity to breed, they are not surrounded by as many daughters as are the older females. Hence, more of a young female's genes are in jeopardy when an unannounced predator shows up, and so it is not to her advantage to call attention to herself.

Behaviors which benefit close relatives are referred to as **nepotism.** Clearly, the giving of alarm calls is nepotistic behavior, but it is only one of several such behaviors exhibited by Belding ground squirrels. For instance, females chase and fight each other vigorously to obtain suitable nest burrow locations each spring. Hence, females expend time and energy and take risks to establish residences. Close relatives seldom fight when establishing these nest residences, but fights are common among nonrelatives. After their young are born, females chase distantly-related and non-related individuals from the area surrounding their nest burrows, but close relatives are granted immunity. Similarly, close relatives codefend areas and cooperate directly in the defense of their young. By some un-known mechanism, the ground squirrels can apparently assess how closely related they are to other squirrels and act appropriately. There is no evidence that males ever act nepotistically.

This elaborate field study would seem to indicate that cooperative acts are not simply unselfish. The ground squirrels behave cooperatively only when they are aiding individuals with whom they share a high proportion of genes.

THE HONEYBEE SOCIETY. Invertebrate animals, especially insects, have a rich repertoire of social behaviors. The most famous and familiar example is the common honeybee.

Honeybees cooperate in caring for the young and exhibit a clear division of labor, with sterile individuals working in behalf of reproductive ones. Further, as you are undoubtedly aware, these workers have the ility to sting, a mechanism used to protect the hive against intruders. However, once a bee has used its stinger, it dies—certainly an altruistic act that represents the pinnacle of unselfish behavior, or does it? To answer this question we need to know how such a system evolved and why there are individuals that spend their lifetime (and in the case of hive defense, their life) working for others. Let us first briefly look at the structure of a honeybee society before we examine this question.

Honeybee colonies are composed of drones, workers, and a single queen bee (Figure 30–17). Queens are basically egg-laying machines. Workers are sterile females who perform all colony functions other than egg production. Drones are male bees that develop from unfertilized eggs (hence, they have a haploid number of chromosomes). Drones have a single function—they inseminate new queens. A virgin queen

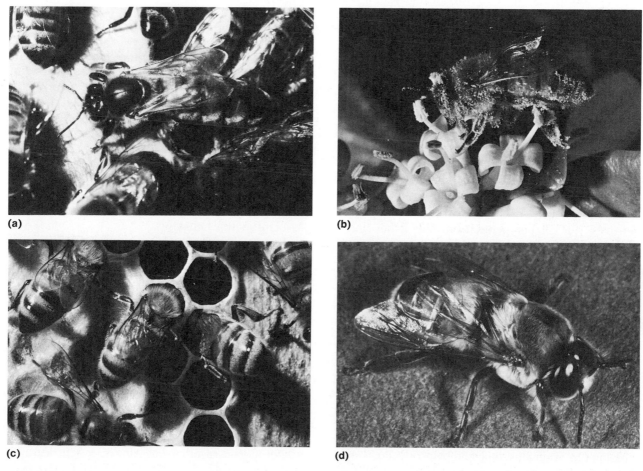

(a)

(b)

(c)

(d)

FIGURE 30–17
Honeybee Society. Honeybees show a rigid social structure with extensive division of labors among individuals. (a) The queen (center) is the only reproductive female. (b and c) Sterile females (workers) gather pollen and tend the hive. (d) The males (drones) exist only to inseminate the queen.

mates with several drones on her nuptial flight during which the queen receives enough sperm to last her entire life. Apparently the queen can control the release of sperm, because she can lay unfertilized eggs which will develop into drones as well as fertilized eggs that develop into female workers.

The queen bee begins life as an ordinary fertilized egg, but the egg is deposited in a specially constructed, larger-than-average cell of the hive, and it is supplied with a special substance known as royal jelly. There is only one queen per colony because each queen produces a chemical that prevents workers from producing more queens. If the queen dies, the lack of the inhibitory chemical signals the workers to prepare another queen by treating a larva with royal jelly.

We now have the background to hypothesize why worker bees are "unselfish." Remember that workers are normal diploid organisms with one set of chromosomes from their mother, the queen, and one set from their father, a drone. On the other hand, drones are haploid. Thus, daughters are related to each parent by $\frac{1}{2}$ because they each receive half of their genetic material from each parent. Haploid males, however, receive all of their genetic material from their mothers. Brother drones are related to each other by $\frac{1}{2}$ because for each chromosome pair of the mother they receive the same chromosome 50% of the time and a different (but homologous) chromosome the other half of the time. Sisters, on the other hand, are related by $\frac{3}{4}$ because they share all of the genes they receive from their father (since their father is haploid), and they share on the average $\frac{1}{2}$ of the genes they receive from their mother. Each sister receives $\frac{1}{2}$ of all her genes from the father and $\frac{1}{2}$ from the mother, so that the average proportion of genes shared through common descent among sisters is equal to:

$$(1 \times \tfrac{1}{2}) + (\tfrac{1}{2} \times \tfrac{1}{2}) = \tfrac{3}{4}$$

Therefore, because of the nature of sex determination in bees, sisters tend to be more genetically similar to one another ($\frac{3}{4}$ gene sharing) than they are to their mother ($\frac{1}{2}$ gene sharing) or her sons ($\frac{1}{2}$ gene sharing), and are closer genetically than are drones to one another ($\frac{1}{2}$ gene sharing). It has been hypothesized that it is because sisters are so closely related to each other that they apparently engage in their altruistic behavior. Females can propagate their genes faster by helping rear younger sisters than by producing daughters. Indirectly they end up helping the future queens, to which they are related by $\frac{3}{4}$. Thus, even though the female workers never produce offspring, they help perpetuate their own genes by protecting the queen and her offspring.

SUMMARY

Behavior is what an organism does in response to stimuli. Ethology, the study of behavior, utilizes two general approaches. In the first, the scientist attempts to understand the physiological bases of behavior—how the nervous and/or endocrine systems detect and respond to changes in the environment. The second approach attempts to understand why certain behaviors exist—how they have evolved and how they confer adaptation to organisms.

Behavior, like other traits, has a genetic basis. By artificially selecting for certain behavioral traits in rats, R. C. Tryon was able to produce two behaviorally distinct populations. This clearly shows that the differences in the selected behaviors are due to genetic differences between the subgroups. One of the most direct demonstrations of the link between genes and behavior was provided by Rothenbuhler's work with honeybees. By crossing two behaviorally different strains of honeybees, Rothenbuhler showed that the inheritance of hygienic behavior in his bees followed simple Mendelian rules.

Instinctive behaviors follow rigid and predictable patterns of responses that are triggered automatically by specific stimuli. Some territorial displays exhibited by male birds during breeding season may be examples of instinctive behaviors.

Learning is the development of modification of behavior in response to experience. The limits of what an organism can learn is determined genetically, but the precise character of a learned response is shaped by experience.

There are five general types of learning. Habituation is learning to ignore insignificant stimuli. In conditioning, a particular behavior is molded by reinforcement—either positive (reward) or negative (punishment)—doled out by a trainer. Trial-and-error learning involves reinforcement that comes in response to the animal's behavior under more natural settings. Imprinting is a special type of learning in which young animals, particularly birds, make a strong association with another individual or object. Finally, insight learning involves the application of past, unrelated experiences to solve a new problem without having to go through trial and error. The capacity for insight learning is found primarily among the primates.

Communication in animals is an extremely important component of behavior. Animals utilize a variety of means in communication, including visual displays, sounds, and chemical signals. Visual displays are crucial in the courting behavior of many birds and fish. Bats and marine mammals use a type of self-communication system (echolocation) to navigate and locate food, and many birds produce songs to advertise their territories. Chemicals called pheromones are produced by many species to define territories, mark trails, and signal sexual receptiveness.

Animals find their way from place to place by practicing various forms of orientation behavior. Most animals are guided by visual landmarks, but bats and marine mammals echolocate, and migratory birds key on celestial objects (the sun or stars) for direction.

Social behavior encompasses all forms of behavior that occur between two or more conspecifics. Social behavior involves communicative interactions between mating partners, among members of loose-knit groups (flocks of birds and schools of fish), and among individuals of more permanently structured societies. Each type of social behavior can be analyzed in terms of the costs and benefits it provides to the individual in a group. Only those behavioral patterns in which the benefits exceed the costs are preserved by natural selection.

In analyzing why various forms of social behavior have evolved, it is helpful to focus at the genetic level, for it is genes that can persist over evolutionary time. One way to view the evolution of social behavior is through the selfish gene hypothesis. This hypothesis states that evolution has favored genes that predispose individuals to behave in ways that increase the likelihood of their passing such genes on to the next generation. The selfish gene model permits us to view altruism directed at relatives as a mechanism of sacrificing oneself for genetically selfish reasons. Other cases of social behavior, such as territoriality, mating behavior, and infanticide can also be explained in terms of the selfish gene model.

STUDY QUESTIONS

1. How does a scout bee communicate to other workers the location of a *nearby* food source? How is the distance and direction of a *distant* food source communicated?

2. In 1940, R. C. Tryon performed an experiment which clearly indicated a link between genes and behavior. What was this experiment and how did it establish this link?

3. How does instinctive behavior differ from learned behavior? Cite specific examples in your answer.

4. How does imprinting in young birds increase their chance of survival?

5. Animals communicate by three major means: visual displays, sounds, and chemical signals. Provide a specific example of these three means of communication.

6. What is the signal that triggers northern temperate birds to fly south for the winter?

7. Using examples, explain how social behavior has both costs and benefits for the participating individuals.

8. From an evolutionary perspective, explain E. O. Wilson's statement: "An organism is DNA's way of making more DNA."

9. Explain why infanticide does not run counter to the tendency of animals to perpetuate their genes.

10. Explain why sister honeybees tend to be more genetically similar to one another than they are to their mother or her sons. How does this help explain their seemingly unselfish behavior?

SUGGESTED READINGS

Alcock, J. *Animal Behavior.* 2nd ed. Sunderland, MA: Sinauer Associates, 1978. For those who wish to learn more about behavior this text is an excellent way to begin.

Bekoff, M., and M. C. Wells. "The Social Ecology of Coyotes." *Scientific American* 242(1980):130–148. This article describes the variable modes of social organization characteristic of coyotes.

Gould, J. L. *Ethology: The Mechanisms and Evolution of Behavior.* New York: W. W. Norton, 1981. An advanced text for those who wish to learn more about the physiological bases of behavior.

Hall, J. "Sex Behavior Mutants in *Drosophila.*" *Bioscience* 31(1981):125–130. An interesting article that discusses some courtship-specific mutations that may help clarify the link between gene products and behavior.

Wilson, E. O. "Slavery in Ants." *Scientific American* 232(1975): 32–36. This interesting article discusses a species of ants which raids neighboring species, capturing their pupae. When hatched the captured individuals become willing slaves!

Wilson, E. O. *Sociobiology: The New Synthesis.* Cambridge, MA: Harvard University Press, 1975. This book did much to establish sociobiology as a branch of behavior—a classic.

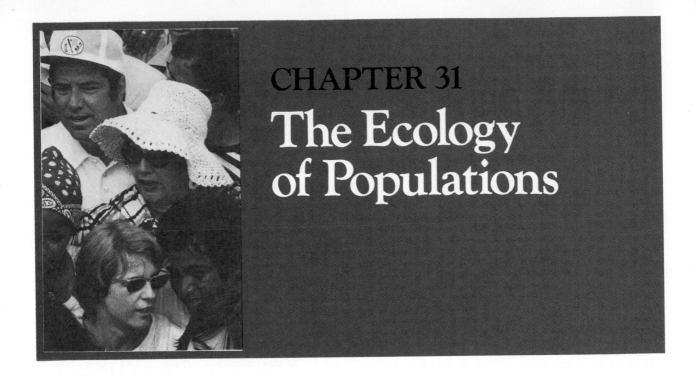

CHAPTER 31
The Ecology of Populations

Organisms face numerous challenges posed by the environments in which they live. These challenges may come in the form of dramatic changes in temperature, shortages of critical resources such as food or water, or threats such as predation from other creatures. The success of an organism and in turn of its species depends on its ability to survive and reproduce in spite of these factors. Thus, the interactions between an organism and its environment are very important in determining the range (the geographical limits) of a particular species and how many of its members inhabit a given area. The study of these interactions, which affect the distribution and abundance of organisms, is the domain of **ecology.**

Several different levels of biological organization are studied by ecologists, ranging from single populations of one species to the entire **biosphere** (all the regions of the earth occupied by living organisms). This chapter will focus on the principles of population ecology, examining the relationships that determine the population size, abundance, and distribution of particular species of organisms.

CHARACTERISTICS OF POPULATIONS

A population, you recall, is a group of individuals of the same species that lives in a given geographical area. All populations have certain statistical characteristics that are important in understanding their structure, and in more applied instances, their management. These characteristics include population size, density, range, age distribution, birth rates, and death rates. Let us begin with size.

It is possible in some cases to measure the size of a population by simply counting all the individuals comprising it. For example, the size of a population of palm trees at a small desert oasis would be relatively easy to determine. Counting all the individuals of most populations, however, is usually a laborious, if not impossible, task. Consider the difficulties in attempting to count every house sparrow in New York City, or all the Ponderosa pine trees in a mountainous region of Colorado. In such cases, ecologists must resort to calculating population density (number of individuals per unit of area or per unit volume) by randomly sampling portions of the population. By studying a 10-acre plot of New York City, we might estimate the house sparrow population there as having a density of 5 individuals per acre. Density estimates, in turn, can be used to calculate total population size. If the sparrow's population range is 10,000 acres, then the total population size could be estimated as 50,000 individuals.

Another informative characteristic of a population is its **age distribution** — the number of individuals in different age groups. An organism's ability to reproduce is largely a function of its age, so the rate at which new individuals are added to a population is determined in part by the age distribution of the population. A population that has an abundance of individuals at reproductive or prereproductive age will generally grow faster than one dominated by individuals that are beyond reproductive age. For example, the human populations of Mexico and of the United States have very different age distributions (Figure 31–1). With its population shifted much more to the younger age classes, Mexico has a considerably higher rate of population growth than the United States.

If we ignore migration of individuals into or out of a given population, then the two factors that directly affect population size will be birth rate and death rate. Obviously, if the birth rate of a given population is greater than the death rate, then the population will increase in size. If, on the other hand, the death rate exceeds the birth rate, then the size of the population will decrease. Of course, a population's size will remain constant if the birth and death rates are equal.

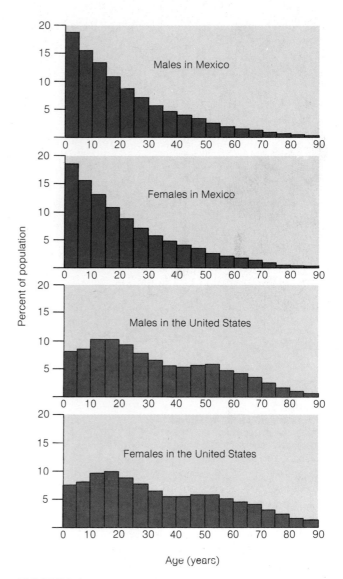

FIGURE 31–1
Age Distribution. In comparing the age distributions of the human populations of Mexico and the United States, note that the rapidly growing Mexican population has more younger individuals.

DYNAMICS OF POPULATION GROWTH

All populations have the reproductive potential to grow in size at a geometric (exponential) rate. This capacity to leave offspring, or **biotic potential,** represents the maximum rate at which a population can grow. Consider a hypothetical population of organisms in which the population's rate of increase is such that, on the average, each individual is replaced by two offspring before it dies. The size of such a population would double each generation, and a J-shaped curve would result when population size is plotted against time (generations) (Figure 31–2). This pattern of population growth is called **exponential growth.**

If our hypothetical population began with 100 individuals, the size of the population would be 1600 individuals by the 5th generation (see Figure 31–2). After 30 generations, this population would consist of more than 100 billion organisms! Such a situation is clearly unrealistic in a natural setting. For a population to grow exponentially for an indefinite period,

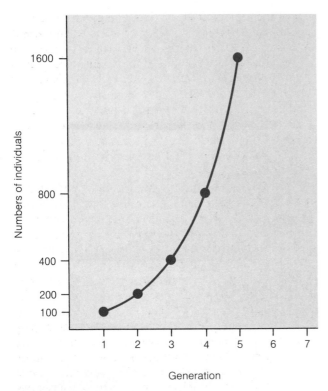

FIGURE 31–2
Exponential Population Growth. This hypothetical population doubles in size with each generation, producing a J-shaped curve.

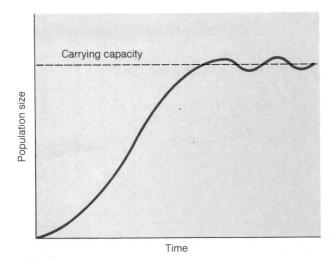

FIGURE 31–3
Population Growth and Carrying Capacity.

it would need unlimited quantities of food, space, and other resources. Real populations do not enjoy this luxury. Although they may grow exponentially for short periods, natural populations inhabit environments that have limited quantities of the resources required to sustain population growth. The maximum size of a population that can be sustained by the available resources is called the **carrying capacity** of that environment.

The concept of carrying capacity can be incorporated into our model of population growth. Our hypothetical population may grow at an exponential rate initially, but as it approaches its environment's carrying capacity, the rate of growth begins to slow down. Maybe food or shelter becomes more difficult to find, resulting in a decreased birth rate or increased death rate, or both. As the differential between the birth and death rates diminishes, the rate of population growth consequently decreases. Eventually the population reaches a relatively stable size (Figure 31–3).

REGULATION OF POPULATION SIZE

As mentioned previously, the environment imposes limits to population growth. The actual factors that limit and regulate population size are classified as either physical or biotic. Physical factors include temperature, moisture, sunlight, chemical composition of the soil or water, and natural catastrophes, such as floods and fires. Biotic factors are those that result from interactions among organisms. One common biotic factor is competition among members of the same or different species for space, food, water, light, or other resources. Predation and parasitism are also biotic factors that represent constraints exerted across species lines. We shall have a more detailed look at competition and predation later in this chapter.

Certain limiting factors tend to affect roughly the same percentage of a population regardless of the actual population size. Such factors are said to be **density-independent.** For example, consider two populations of tomato plants, one with a size of 10 individuals and the other with a size of 1000. Both are subjected to a killing frost. The small population loses 3 individuals, while the large population might lose 300. In both populations, the same proportion (30%) of individuals was affected. Thus, we can say that a killing frost affects tomato plant populations in a density-independent manner. As it turns out, most physical limiting factors are density-independent.

Biotic factors are more likely to operate in a **density-dependent** manner—the number of individuals affected depends on the population's density. For example, contagious diseases are usually density-dependent. The likelihood of an organism contracting a disease from another individual in the population depends on how often the individuals are in contact with one another. In high density populations, there will be greater contact among individuals, and hence, more opportunities to pass the disease.

INTERACTIONS AMONG POPULATIONS

While individual limiting factors can be easily identified and categorized, discerning their individual impact on natural populations can be a difficult task, because usually more than one factor is operating on a population at any given time, and interactions among different factors can further complicate matters (see Essay 31–1). Furthermore, no natural population exists in isolation. Let us now look at some of the ways populations of different species interact.

Predation and Parasitism

Predation can be defined broadly as the killing and eating of live organisms by other organisms. One species, the **predator,** obtains energy and nutrients by eating members of another species, the **prey.** We can categorize predators according to the type of prey they utilize: animals that eat plants are called **herbivores,** animals that consume both plants and animals are **omnivores,** and animals that prey on other animals are **carnivores. Parasitism** is a special form of predation in which the prey or host is not usually killed outright, but is exploited by the parasite over a period of time. Still, the parasite benefits at the expense of the host.

Predator–prey interactions are usually viewed as a feedback system, with the population size of one group affecting the population size of the other. As the density of the prey population increases, more "resources" become available to the predator population, which then increases. However, as the number of predators increases, more and more prey are eaten, and the prey population begins to decrease. Soon, not enough prey are left to support so many predators, and the predator population declines. Predator–prey interactions thus tend to oscillate, with the density of each population fluctuating in response to the other.

Oscillations in predator–prey systems have been well documented in experimental laboratory populations. Figure 31–4 illustrates this pattern for an experimental system consisting of two species of mites, one of which preys upon the other. Host–parasite populations have also exhibited the expected oscilla-

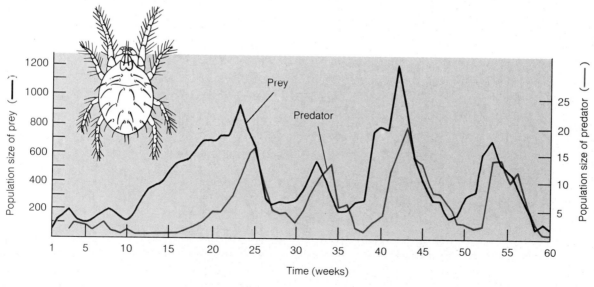

FIGURE 31–4
Population Size Fluctuations in a Predator–Prey System. In this laboratory study of two populations of mites, an increase in the size of the prey population is followed shortly by an increase in the number of predators. Both populations fall simultaneously—the prey because of increased predation, and the predators due to less food availability.

ESSAY 31-1

LEMMING CYCLES: AN ECOLOGICAL "WHODUNIT"

The efforts of scientists to obtain explanations for natural phenomena sometimes follow a course reminiscent of a mystery novel: the diligent investigator formulating plausible scenarios, doggedly amassing evidence, and occasionally being thrown off the trail by a false lead. Investigations into the "lemming cycle case" has all of these features.

Lemmings, voles, and field mice are collectively known as microtine rodents. They inhabit the higher latitudes of North America, Europe, and Asia. These rodents are very short-lived under natural conditions—few individuals live for more than a year. However, their short lifespans are compensated for by very high reproductive rates. Gestation periods as short as 21 days are not uncommon, and individuals may reach sexual maturity 30 days after they are born.

One of the most interesting aspects of microtine biology is that many species undergo periodic fluctuations, or "cycles," in population size. These cycles generally have a periodicity of 3 to 4 years, and may be synchronous over thousands of square miles. At its ebb, a popula-

tion may consist of one animal per acre. This phase is followed by an extremely rapid population growth period, with the population size increasing 10 to 15% per week. Peak densities may reach 300 individuals per acre. After roughly one year at peak density, the population tapers off to a low level, thus completing one cycle.

Outbreaks or "plagues" of microtine rodents have been recorded by humans since the dawn of written history. References to such outbreaks appear in the Old Testament and in the writings of Aristotle. Illustrations from sixteenth century Norway show lemmings falling to earth from rain clouds. The romantic myth that lemmings reduce their own numbers by embarking on suicide marches into the sea is one bit of lemming folklore that persists even today. Despite the lack of data supporting this myth, it continues to be perpetuated by film makers and cartoonists.

The first scientific studies of microtine population cycles were begun in the 1920s by Charles Elton of Oxford University. Elton was the first to demonstrate that microtine fluctuations were not random, but had a regular periodicity, and that the cycles were often synchronous over large geographical areas. At this time, most people believed that animal populations were stable in size unless disturbed by humans. Elton's initial findings challenged this "balance of nature" dogma and opened up an entirely new avenue of scientific inquiry.

Once the reality of the cycles was established, the next step was to identify the factors which generate them. Likely candidates included disease, predation, and starvation. Elton and his colleagues investigated the disease hypothesis and the solution to the cycle mystery seemed imminent when a new disease, vole tuberculosis, was discovered in the British field mouse. However, outbreaks of the disease were not always associated with changes in mouse population size, and the periodic declines were observed even in the absence of the disease.

Elton also discovered that the population densities of predators fluctuated with lemming population densities. This was taken by some as an indication that lemmings and their predators exemplified the classic predator–prey fluctuation (see Figure 31–4). However, predators have much lower reproductive rates than field mice, and they cannot produce offspring rapidly enough to check microtine population growth. It is now generally believed that predation may be an important mortality source during the decline and low phases of the cycle, but that it is not sufficient to stop lemming populations from increasing in size.

Food, or lack of it, might cause lemming populations to cycle. However, several points argue against this theory. First, animals in peak populations are larger than those at any other phase of the cycle and show no signs of starvation. Second, habitats do not show the effects of overgrazing. Third, peak populations decline even when provided with supplemental food.

Investigators soon began considering the possibility that microtine populations might possess some intrinsic mechanism which regulates population densities. One such mechanism was proposed by John Christian in 1950. Christian theorized that high levels of stress in peak populations might produce a deterioration in the general physical condition of the individuals. Unfortunately, studies of natural populations have generally failed to bring forth convincing evidence linking physiological changes from stress with lemming population cycles.

While other intrinsic mechanisms have been proposed since 1950, none to date can account for all the facts. For the time being, the lemming cycle case must still be designated as "unsolved."

(a)

FIGURE 31–5
Predation. (a) The photo shows two lampreys attached to a lake trout. (b) The introduction of the sea lamprey had a devastating impact on the fish populations of the Great Lakes. The lake trout in all three lakes were almost eliminated.

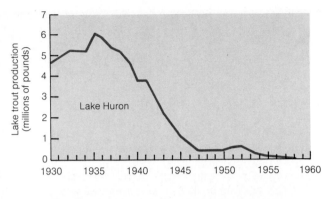

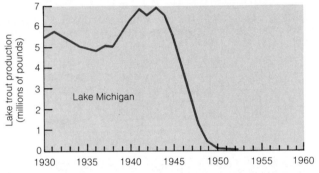

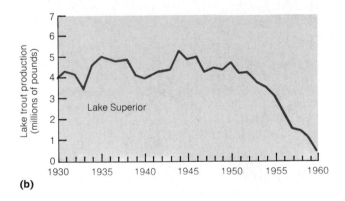

(b)

tions under laboratory conditions. However, the relatively large oscillations observed under laboratory conditions are rarely seen in nature. This presumably reflects the "fine-tuning" influence of natural selection on the evolution of predators and their prey. Large oscillations in predator–prey populations signal instability, which could lead to the extinction of both species. Natural selection has apparently favored the evolution of traits in both predator and prey populations that prevent such oscillations and thereby increase the stability of the system.

Some of the most striking examples of the imbalances in predator and prey populations have come from disturbances created by humans when new predators are introduced into an environment, or when a species is introduced into an area where there is no controlling predator. Consider the case of the marine lampreys. Lampreys, which live along the Atlantic Coast of North America and migrate into fresh waters to spawn, prey on fish. They attach their sucker mouths to the sides of fish, then suck out the fish's body fluids. Before the construction of the Welland Canal around Niagara Falls in 1829, lampreys were unable to reach the upper Great Lakes. In 1921 the first marine lamprey was found in Lake Erie; since then, lampreys have made their way into the other Great Lakes as well. The impact of lamprey predation

was dramatic—commercial lake trout catches in the Great Lakes were reduced to almost zero by 1960 (Figure 31–5). Since then the situation has improved slightly through a program of selective chemical poisoning of the lampreys. The commercial whitefish catch has steadily increased in Lake Michigan but the trout have yet to rebound sufficiently to make commercial trout fishing economically worthwhile.

The introduction of predators or parasites by humans may also be an effective means of controlling pest species. Prickly pear cactus was introduced from South America to Australia in 1839 (Figure 31–6). Without any natural predators in Australia, the cactus spread rapidly and soon became a serious pest. Australians finally managed to control the cactus by introducing its natural parasite, the cactus moth, from South America in 1926. The moth bores into the plant, and the resulting wounds eventually kill the cactus by encouraging dehydration and infection by fungi.

Competition

The size of a population, you recall, is typically limited by the amount of resources available in its environment. What happens when two populations of different species utilize one or several of the same limited resources? In such a situation, each population will limit the size of the other by an interaction called **competition.**

Competition between species is intimately related to the concept of the ecological **niche.** The ecological niche of a species is the role that species fills in relation to other species and the environmental resources. We can think of the niche as the range of environmental conditions under which the species can exist and successfully replace itself. If we think of a community as a jigsaw puzzle of resources and living creatures which are interdependent and fit together to form a whole picture, the niche of each species is the unique space it fills in the puzzle. It is not possible in practice to completely describe the niche of a species because so many variables are involved: food, temperature, shelter, and so on. We can, however, look at each individual group of variables separately, such as the range of food resources a species utilizes (its food

(a)

(b)

FIGURE 31–6

Biological Control of the Prickly Pear Cactus. Prickly pear cactus was introduced into Australia in 1839. (a) In the absence of native predators, the cactus quickly spread and became a nuisance. (b) In 1926 the cactus moth was introduced and soon brought the cactus under control.

FIGURE 31–7
Competition Between Species. (a),(b) When two species of *Paramecium* were grown separately under laboratory conditions, each population exhibited an exponential growth curve. (c) When they were grown together, however, competition for resources led to the elimination of the *P. caudatum* population.

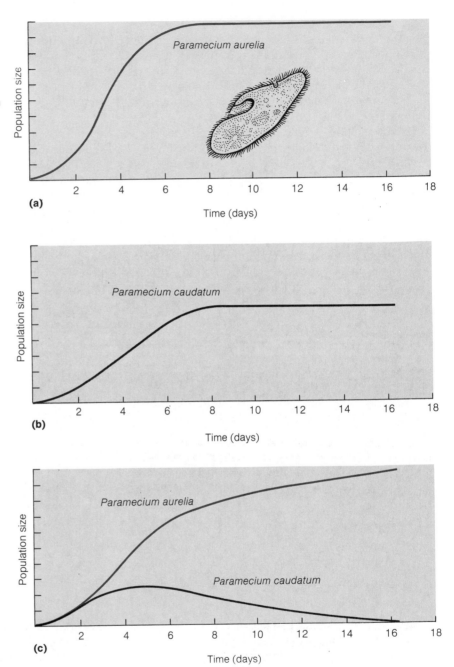

niche). Such an examination makes it clear that certain aspects of a niche relate directly to competition.

The Russian biologist G. F. Gause placed two similar species of *Paramecium* in laboratory cultures. When the two were kept separate, each species' population grew until it reached the limits of its environment. When the two species were grown in a mixed culture, however, one eventually eliminated the other (Figure 31–7). Similar results have been reported from experiments using other organisms. These findings led scientists to formulate the **competitive exclusion** principle: Two species that are ecologically identical—that share the same niche—cannot coexist in the same area for long.

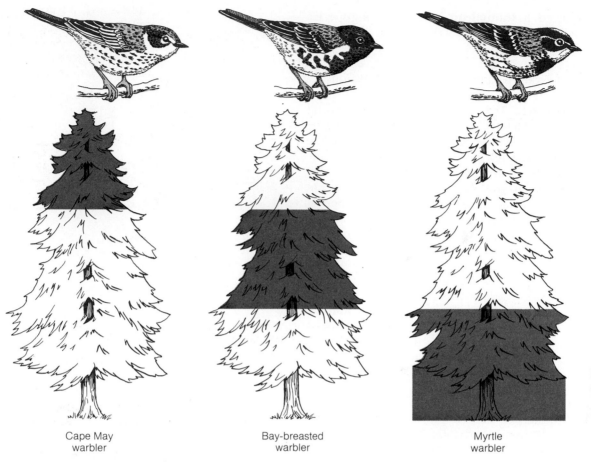

Cape May
warbler

Bay-breasted
warbler

Myrtle
warbler

FIGURE 31–8
Niches. Three different species of warblers occupy three separate niches within a single spruce tree.
The shaded areas are those parts of the tree where each species spends most of its foraging time.

How does this principle apply to natural populations? In many instances, species that appear to be ecologically identical are found in the same place. For example, five species of nocturnal, seed-eating rodents coexist in a one-acre plot in the Mojave Desert. A close study of this plot revealed that these rodents do not occupy the same niche, but display differences in the way they utilize resources. Each rodent species eats a different size of seed or forages for seeds in a different "microhabitat"—one forages under shrubs, another in bare areas, etc. Figure 31–8 illustrates how several ecologically similar species of warblers inhabiting coniferous forests in North America spatially divide the trees in which they forage for food. This niche partitioning reduces the intensity of competition among the warblers.

Symbiosis

Symbiosis (from the Greek, meaning "life together") is a relationship between two or more species that are in close and permanent contact with one another. Biologists generally recognize several forms of symbiotic relationships. **Commensalism** is a form of symbiosis in which one species benefits while the other is unaffected. In **mutualism,** both interacting species benefit. Parasitism, an interaction in which one species benefits at the expense of another, is viewed by some scientists as a type of symbiosis. We have chosen to view and discuss parasitism as a form of predation.

COMMENSALISM. True commensal relationships among species are difficult to document—it is hard to be absolutely sure that the species benefitting from the relationship does not provide its partner with some subtle benefit or weakly parasitize it. Consider the barnacles which grow on the backs of whales. At first glance it would seem that the barnacles benefit because they are provided with a home, while the whales suffer no ill effect. Closer examination, however, indicates that the barnacles may irritate the skin of the whales and initiate sites for infection. One relationship that does appear to be truly commensal involves sharks and *Remoras*. *Remoras* are small fish that attach to sharks. The *Remora* benefits from the shark by getting free transportation and it is obviously well positioned to feed on the remains of any prey the shark may kill (Figure 31-9). Its presence appears to have no effect on the shark.

MUTUALISM. In contrast to commensalism, there are literally thousands of examples of mutualism. One example, discussed in Chapter 21, is the bacteria (*E. coli*) which reside in the human gut. These bacteria

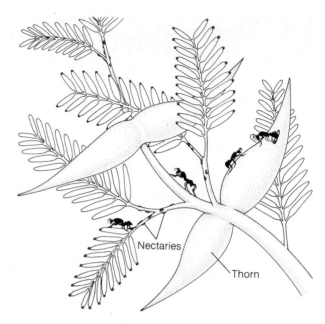

FIGURE 31–10
Mutualism. A queen ant hollows out the thorn of an acacia for her brood. As the colony grows, more and more thorns are excavated. If the ants are experimentally removed from acacias, herbivores quickly ravage the trees.

are provided with predigested food, while their human hosts benefit from the vitamins they synthesize and secrete.

Another classic example of mutualism exists between a certain species of ant and the Central American acacia plant. These acacias have large thorns which, when hollowed out, provide a protected home for ant colonies. As far as we know, the ants' destruction of the thorn's interior does not adversely affect the acacia plant—on the contrary, housing ant colonies has a distinct advantage. The ants patrol the surface of the plant, discouraging and often attacking herbivorous insects and other creatures that would normally eat acacia leaves. The ants themselves do not feed on the foliage but on pockets of nectar on the plant's surface (Figure 31-10), and on other plants nearby.

A third example of mutualism is the lichen. **Lichens** are associations composed of a fungus and an alga. The fungal component of a lichen forms the skeleton in which the algal cells are embedded (Figure 31-11). The fungi derive nourishment from the photosynthetic algal cells; the algal cells benefit because the fungal body restricts water loss from the algae and

FIGURE 31–9
Commensalism. A *Remora* attached to its host, a shark. The *Remora* benefits from the relationship and the shark is apparently unharmed.

SU

Brittingham, M
    Forest Sc
    31–35. Th
    populatio
    problem.
Colinvaux, P.A
    John Wile
    cludes an
Hazen, W.E.,
    Ecology,

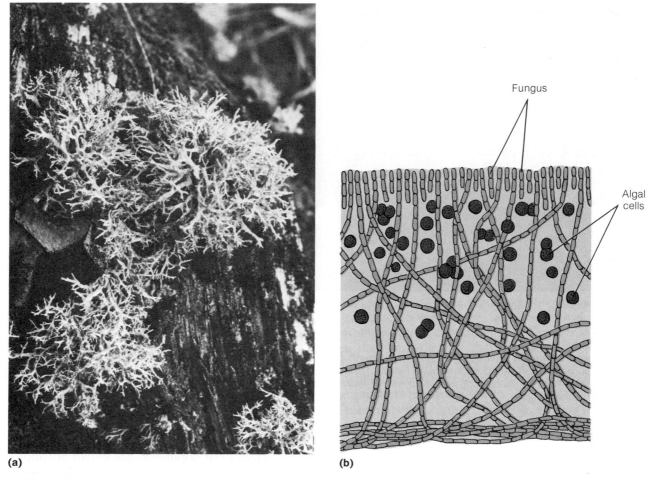

(a)

(b)

FIGURE 31–11
Lichens. (a) Frequently, lichens grow on nutrient-poor rocks or bark where few other plants could
survive. (b) In cross section, the round algal cells are embedded within a matrix of fungal filaments.

protects them from the damaging effects of intense light. Both the fungal and algal components of a lichen can be grown separately in the laboratory, but neither alone can tolerate the extreme environmental conditions to which the lichen is normally exposed.

THE HUMAN POPULATION

Now that you have a general understanding of how populations grow and what factors act to regulate population size, it seems appropriate to conclude our discussion by considering the growth, regulation, and future prospects of a population in which we humans have a vested interest — our own. Many of the problems that we face as a species stem from our own popula-

tion growth. The principles of population ecology discussed in this chapter can help us analyze these problems, and perhaps ultimately solve them.

Human Population Growth

The human population has undoubtedly been studied in more detail than any other. Yet the size of the world population is not known accurately. Even in this technological age, population data for our species are subject to substantial error. However, we can get a general picture of how the human population has changed through time and also make predictions about the future by combining estimates of population sizes before modern times with contemporary data on density, birth rates, and death rates.

It shoul
one crucial
to stabilize
of populatic
can be acc
nations of
simply wai
death rate
agents of st
and perhap
voluntarily
traceptive
spectacle c
able for h
tensive pl
of the wo
steps are
rate volur
dictate th

The scier
affect the
The basic
a group
in a give
populatic
rate at w
Und
lations g
normally
sources
support
capacity
A n
natural
ment, s
populat
such as
to regu
Pre
lating d
which
Natural
tions ir
their st
tors, a

ESSAY 32–1

OFF-ROAD VEHICLES AND ECOSYSTEMS

A Nobel Prize ought to be awarded posthumously to the genius who invented the wheel, often cited as one of the greatest technological breakthroughs. Its advantages, however, have not come without certain costs. Much of the earth's landscape is now criss-crossed with highways and their associated settlements. bill-boards, service stations, and so on. The adverse effects of automobile emissions on the quality of the air we breathe is well known, and thousands of people lose their lives or are seriously injured annually in vehicle-related accidents.

In the United States, growing numbers of people have been chan-neling money and leisure time into recreational vehicles. Sales of off-road vehicles (ORVs), such as motor-cycles, dune buggies, snowmobiles, and four-wheel drive vehicles, have skyrocketed in recent years. However, it is becoming apparent that the environmental costs associated with this form of recreation are also in-

creasing at an alarming rate. At this point, not a single natural ecosystem found in the United States has escaped ORV-related damage. Slow-healing scars can be found from the Mojave Desert to the sand dunes of Cape Cod. Not even such remote regions as the tundra of Alaska have escaped. California, with the nation's greatest number of ORVs, has suffered more damage than any other state.

ORV activity disrupts every component of the ecosystem. A recent report prepared for the Council on Environmental Quality states, "First and foremost, ORVs eat land." For example, it has been estimated that a single dirt bike traveling in the desert displaces roughly one ton of topsoil per mile. A single off-road motorcycle race can fill the air with more than 600 tons of dust in one day. This disruption of the topsoil and the plants growing on it leads to erosion by water and wind, causing an even greater loss of soil and vegetation. A single hillside in California, measuring one mile long and 200 yards wide, is estimated to have lost 11,000 metric tons of soil through erosion due to ORVs.

Because plants are the primary producers in ecosystems, their destruction by ORVs makes the environment uninhabitable for other organisms. Also, the dangers of ORV activity to animals may be more direct. Research has demonstrated that the intense noise levels created by many ORVs are responsible for hearing damage not only in those using ORVs, but in wildlife. Kangaroo rats in western deserts of the United States rely heavily on their sense of hearing to avoid nocturnal predators such as owls. Hearing loss produced by ORV noise has been documented in kangaroo rats. Since Eskimos of the Baffin Bay area in northern Canada traded in their dogsleds for snowmobiles, partial hearing loss has been reported in 85% of the adult Eskimo population there.

While the majority of ORV enthusiasts are probably unaware of the destructive impact their recreation has on environments and their inhabitants, a small segment of this group is guilty of malicious destruction of valuable natural resources. Reports of ORV drivers harassing game species are becoming increasingly common. This practice could kill animals weakened by harsh winter weather, and it may cause pregnant animals to abort. From the deserts of the Southwest come reports of dirt bike riders using the now endangered desert tortoise as "ramps" for jumping their motorcycles.

The destruction of large expanses of the environment by ORVs is a serious problem, and one that will become even more widespread as our population and the number of ORV enthusiasts increases. Many ecosystems are the product of millions of years of evolution, and they are slow to repair themselves after this kind of damage. They are finely balanced, delicate systems. The solution to this problem is complicated, and emotions on both sides run high. ORV enthusiasts will continue to demand their Constitutional right to the "pursuit of happiness." The flawed argument that there is an infinite amount of wide-open, unused space in our nation will also be employed. On the other side, backpackers, cross-country skiers, and other enthusiasts of nonmechanized outdoor activities will contend that their recreation and that of the ORV contingent are incompatible; a single dirt bike destroys the serenity and aesthetic beauty of natural areas that the former group seeks. Whether some compromise between these two factions is possible or even desirable is debatable. However, it is clear that strictly enforced laws regulating ORV use and protecting ecologically sensitive areas will be needed if we are to save our nation's valuable natural resources from irreparable damage.

the primary producers. In a terrestrial environment, these might include plant-eating insects, mice, rabbits, deer, cattle, giraffes, giant pandas—any animals that browse or graze for food. In an aquatic environment, the primary consumers include many kinds of fishes, sea urchins, manatees, and others.

The herbivores in turn are the food source for the **secondary consumers,** or carnivores—the meat-eaters. On land, these include hawks, cats, wolves, and many other predators. In an aquatic environment, larger fishes are the most conspicuous secondary consumers, but starfishes, octopuses, and many other meat-eating animals also fall into this category.

Finally, every food chain has a group of organisms known collectively as the **decomposers.** These typically include bacteria and fungi that feed on the fecal

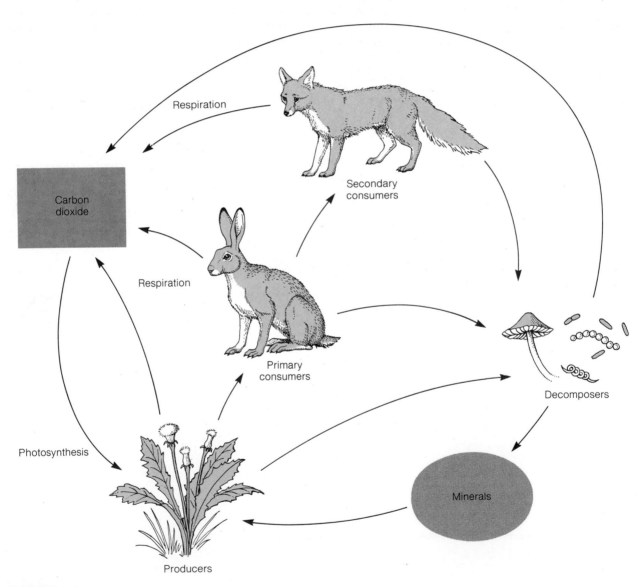

FIGURE 32–2
Trophic Levels. In this very simplified food chain, the flow of organic nutrients (colored arrows) and inorganic nutrients (black arrows) through the various trophic levels are indicated.

and metabolic wastes of animals, and the bodies of dead producers and consumers. By breaking down these materials, the decomposers speed up the rate at which nutrients are released from dead organisms and wastes, ultimately returning carbon dioxide and vital minerals to the air, soil, and water for use by the producers (Figure 32–2).

Obviously, the nutritive relationships among organisms of any ecosystem are bound to be complex. A herbivore such as a horse is likely to eat tree leaves and alfalfa as well as certain grasses; and grasses are food for cattle and sheep as well as for horses. Another complication arises with organisms that occupy more than one trophic level. The raccoon is an omnivore which feeds on both plants and animals; and although the tadpole is a herbivore, the adult frog is a carnivore, feeding on flies and other insects. Thus, the food relationships in most ecosystems are more accurately represented by **food webs** (Figure 32–3).

ENERGY FLOW THROUGH ECOSYSTEMS

All ecosystems are driven by energy, and the ultimate source of that energy is the sun. The amount of solar energy directed toward earth is enormous. About half of this radiation never reaches the surface of earth, however, because atmospheric gases, clouds, and dust particles reflect it back toward outer space (Figure 32–4). Of the radiant energy that does get through, most is either reflected back by the earth's surface or absorbed by water, rocks, and other objects, then re-emitted as heat. Only the sunlight that strikes photosynthetic structures—mainly the leaves of green plants—is potentially useful for supplying the energy needs of organisms. And when you consider that only about 0.5–3.5% of the radiation falling on leaves is actually transformed into chemical energy by photo-

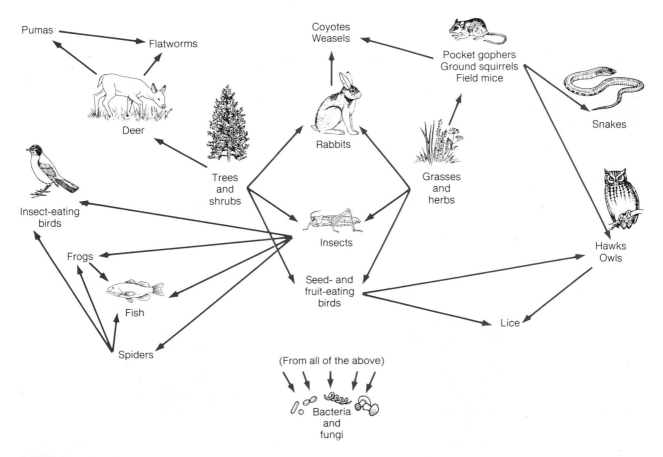

FIGURE 32–3
Food Web. Real food webs in nature are complex. The example diagrammed here shows just a few of the species of a food web studied in an American woodland.

FIGURE 32–4
The Fates of Solar Radiation.

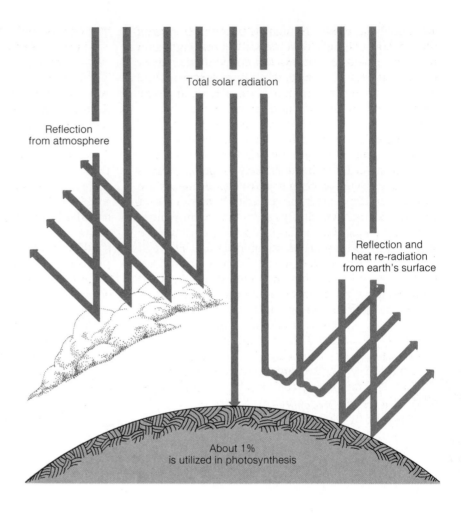

Total solar radiation

Reflection
from atmosphere

Reflection and
heat re-radiation
from earth's surface

About 1%
is utilized in photosynthesis

synthesis, it becomes clear that life on earth subsists on a tiny fraction of available sunshine.

The chemical energy produced during photosynthesis is in the form of organic compounds, such as sugars. The total amount of this energy generated by a plant (or all the plants in an ecosystem) is termed the **gross productivity** of that plant (or ecosystem). Under normal conditions, about half of the carbohydrate energy in plants is used to maintain their status quo—that is, about half of the sugar is broken down in respiration to generate the ATP each cell needs to carry out its energy-demanding activities. The remainder is either stored away as starch or fat, or used to synthesize other compounds necessary for growth. This portion of the carbohydrates from photosynthesis adds to the plant's biomass (organic bulk), and it is called the plant's **net productivity.** To summarize:

Gross productivity – Respiration = Net productivity

We can estimate the gross and net productivities of a plant or ecosystem in several ways. One way is the so-called light- and dark-bottle technique, commonly used for aquatic ecosystems. Two samples of water containing photosynthetic organisms (usually single-celled algae called phytoplankton) are collected from the same region of a lake or ocean. One bottle is transparent so that when light is available, the plankton inside will carry out both photosynthesis and respiration. Because oxygen is produced in photosynthesis and consumed in respiration, any increase in the amount of oxygen in this lighted bottle over a given time period will be a measure of the plankton's net productivity—that is, their total photosynthetic activity minus respiration. The second bottle is kept in darkness so that no photosynthesis takes place. Here the phytoplankton can only respire, consuming oxygen without generating it. The rate of oxygen loss in the dark bottle therefore gives us a measure of respiration. When we add the amount of oxygen consumed in the dark bottle (respiration value) to the amount of oxygen produced in the light bottle (net productivity value), we obtain a measure of gross productivity for the phytoplankton (Figure 32–5). Determining net and gross productivities in terrestrial ecosystems is more difficult, but the general principles are the same.

The total productivity of an ecosystem is of interest to ecologists because it indicates how much total food energy is potentially available to the heterotrophs living there. That is, because all the life in an ecosystem depends on the chemical energy produced by plants, knowing a given system's net productivity enables us to judge what range of other organisms that system could support. For this reason net productivity is also called **primary productivity,** because it describes the net energy generated by the primary producers at the most fundamental level of the system. The total productivities of a wide variety of ecosystems have been measured (Table 32–1).

Although estuaries and coastal aquatic environments often support very high productivities, terrestrial ecosystems are generally more productive than aquatic ecosystems. This is largely because mineral nutrients, particularly nitrogen, phosphorus, potassium, and iron, are usually more abundant in terrestrial ecosystems. The major factor that tends to limit productivity in terrestrial ecosystems is water supply. Consequently, tropical areas tend to be much more productive than deserts.

The primary productivity (production of plant biomass), of any ecosystem always exceeds the production of biomass by the heterotrophs it supports. Just as plants capture only a tiny fraction of the solar radiation that reaches the earth, animals do not consume every bit of plant material available. If there were enough animals to do so, the ecosystem would quickly collapse. Also, much of the plant matter they do consume is indigestible—most animals are unable to break down cellulose, a major component of all plant structures. And of the food they can digest, less than half goes toward increasing their biomass. Like plants, animals must apply much of the food energy they acquire to purely maintenance functions (repair, active transport, etc.) and the very energy-demanding processes of movement (for example, blood circulation, breathing, peristalsis, and locomotion).

Finally, consider this crucial point: During the course of every physical and chemical process that occurs in an organism, a portion of the biologically useful energy is converted to heat. The beating of a tiny cilium, the transport of a potassium ion across a

FIGURE 32–5
Measuring Productivity of an Aquatic Ecosystem. Samples of water containing phytoplankton are collected into paired light and dark bottles. The bottles are sealed and then lowered to the same depth at which the samples were collected. After 24 hours, the contents of the bottles are analyzed for changes in the amounts of oxygen. The photosynthesizing plankton in the light will cause an increase in oxygen content, a measure of the net productivity at this depth of the lake. In contrast, the sample kept in darkness will show a decline in oxygen content due to respiration activity of the plankton. Assuming that the same rate of respiration took place in both bottles, we can estimate the total photosynthetic activity (gross productivity) by adding the net changes for oxygen levels in the light and dark bottles.

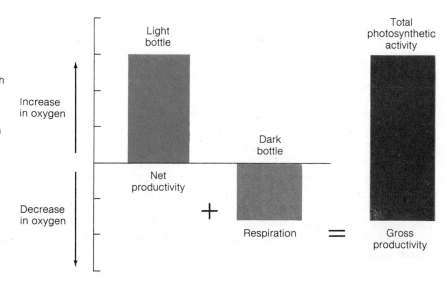

TABLE 32–1
Annual Primary Productivities of Selected Ecosystems. (Averaged from several sources.)

Ecosystem	Range of Primary Productivity (Grams of biomass/square meter/year)
Freshwater:	
Nutrient-poor areas	7–25
Average nutrient levels	75–450
Sewage treatment pond	4500
Saltwater:	
Open water	60–470
Upwelling areas*	500–3650
Shallow coastal areas	1750–7300
Estuary	3000–3300
Terrestrial:	
Desert	70–150
Tundra	100–250
Chaparral and brushlands	400–700
Coniferous forests	600–700
Grasslands	600–1300
Deciduous forests	850–1570
Tropical forests	2000–3000

*Regions where rising currents of water sweep nutrients from the ocean floor upward, often providing a rich mineral diet for the phytoplankton in these areas.

membrane, the respiration of a glucose molecule—all these processes release heat. And since organisms cannot transform internal heat into other forms of energy useful to them, it represents lost energy. Heat dissipates into space, never to return.

In summary, energy flows one way through an ecosystem. It enters as light energy, it is trapped briefly as chemical energy in organic compounds, and finally it is lost as heat. Energy is not recycled (Figure 32–6).

NUTRIENT FLOW AND BIOGEOCHEMICAL CYCLES IN ECOSYSTEMS

In contrast to the one-way flow of energy, nutrients within ecosystems move in cycles. Atoms may exist in a variety of chemical combinations as they are exchanged among organisms or between organisms and their physical environment. For example, a carbon atom in a carbon dioxide molecule can become incorporated into a sugar molecule by photosynthesis taking place in a corn leaf. The sugar may be transported to a developing corn seed which is later eaten by a crow. The crow may respire the sugar molecule,

returning the carbon atom to its original form of carbon dioxide. The same rule holds true for all elements used in biological systems: Matter is never gained or lost, only recycled. Thus, a nitrogen atom in your big toe may have been part of a protein molecule in one of the first living cells that existed 3.8 billion years ago. Understanding the nature of these cyclical nutrient transformations is a crucial part of understanding ecosystems.

The flow of nutrients from one trophic level to another is only part of the pattern of nutrient movement in an ecosystem. Remember that in addition to the living organisms themselves, an ecosystem includes a physical environment. Some of the elements that make up an animal come from the other organisms it ingested; but all the elements that exist in biochemical forms in living creatures also exist in geochemical forms in the external environment. Materials are constantly being exchanged between the living and nonliving components of an ecosystem; the circuit of exchanges for a given element is called a **biogeochemical cycle.**

The biological components of a biogeochemical cycle are the organisms themselves. Carbon, hydrogen, oxygen, nitrogen, and a few other elements are found in the organic molecules of all creatures. An

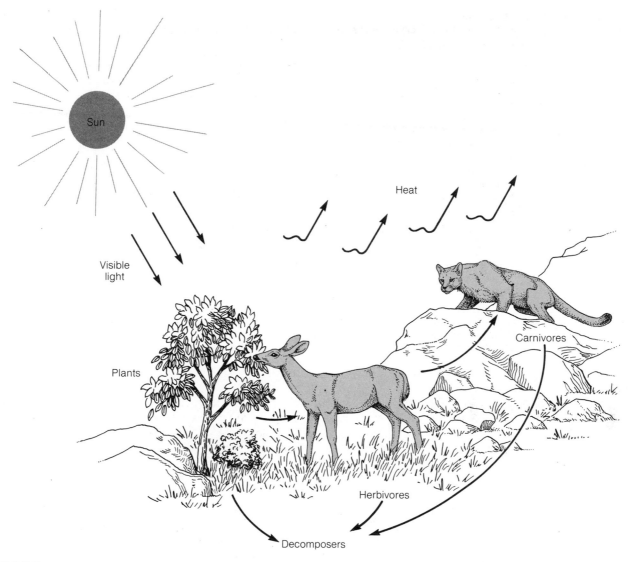

FIGURE 32–6
Energy Flow in Ecosystems. Energy enters the biological component of ecosystems as visible light trapped by the photosynthesizers, who convert it to chemical energy. The chemical energy of plants becomes transformed into the chemical energy of other organisms. All organisms lose energy to their physical surroundings in the form of heat. Energy flow is thus one-way: visible light→chemical energy→ heat.

inadequate supply of any one of these elements will severely restrict an organism's ability to grow and develop. For example, the low concentration of carbon dioxide in the air (0.04%) often limits photosynthesis, and the rate of photosynthesis in many plants can be doubled by artificially increasing the concentration of

this gas in its environment (see Chapter 6). Similarly, the increased plant growth one observes when soil is enriched with nitrogen and other fertilizers suggests that a scarcity of these nutrients may limit plant growth as well.

The nonliving or geochemical segment of all biogeochemical cycles includes two distinct parts (Figure 32–7). The first part, called the **reservoir,** consists of elements in a form that cannot be used directly by organisms. For example, nitrogen gas (N_2) comprises about 78% of our atmosphere, but plants cannot use nitrogen in this form. The gas must first be converted to ammonia (NH_3) or nitrate (NO_3^-), which plants can absorb and use to manufacture amino acids and proteins. One of the reservoirs for carbon is calcium carbonate ($CaCO_3$), or limestone, which exists in great beds of sediments beneath the oceans. About 98% of all the carbon in the reservoir exists as carbonates, most of which is limestone.

The second part of the geochemical segment is called the **exchangeable pool,** which consists of chemical forms of elements available to and directly usable by organisms. For example, part of the exchangeable pool for nitrogen is ammonia (NH_3), which is formed mainly by certain bacteria and cyanobacteria from nitrogen gas. The exchangeable pool for carbon includes carbon dioxide and bicarbonate (HCO_3^-), both of which can form spontaneously from dissolved calcium carbonate.

The steps in the biogeochemical cycle for each element are unique. This is not surprising given that each element's role in biological systems, as well as the

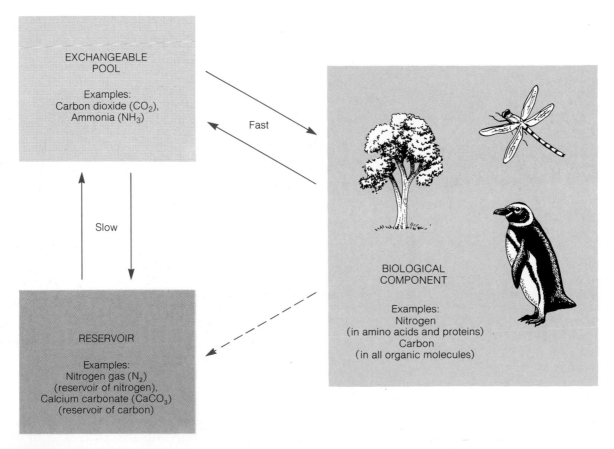

FIGURE 32–7
Biogeochemical Cycles. In the biogeochemical cycle for a given element, the biological component is composed of the living organisms. Animals obtain elements from the food they eat; plants get theirs from an exchangeable pool. When organisms die or excrete organic wastes, the organic matter is degraded and returned to the exchangeable pool, or, as in past geological ages, becomes stored in a fossil reservoir. A slow cycling of elements also takes place between the reservoir and the exchangeable pool.

geochemical forms it can take, are different from those of any other element. The various cycles are often related, though, because different elements frequently combine with one another at various points in their cycles. For example, the carbon and oxygen cycles overlap in their geochemical segments because inorganic forms of carbon exist primarily as molecular oxides (CO_2, HCO_3^-, $CaCO_3$, etc.). The vast majority of biological compounds also contain both carbon and oxygen (sugars, amino acids, and so on).

As we examine the biogeochemical cycles for two very important elements, carbon and nitrogen, two major principles of ecology should become apparent: (1) Nutrients are continuously recycled among the various components of the biological and geochemical segments of the biosphere, and (2) Different species of organisms are inextricably linked to one another by the elements they share. Let us begin with the carbon cycle.

The Carbon Cycle

As you know, carbon is a constituent of all organic compounds. Next to hydrogen, carbon is the most abundant element in living systems.

The carbon cycle has two subcycles: a short-term cycle and a long-term cycle. Both of these subcycles are tied to one another by the exchangeable pool of atmospheric carbon dioxide (Figure 32–8). In the

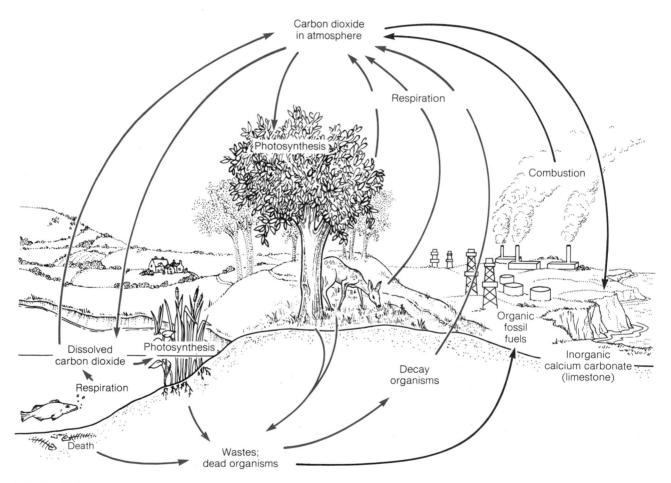

FIGURE 32–8
The Carbon Cycle. In the short-term carbon cycle (colored arrows), carbon dioxide in the exchangeable pool enters the biological component during photosynthesis and is released back to the pool during respiration. The long-term cycle (black arrows) includes reservoirs of carbon such as fossil fuels and limestone. The carbon in these reservoirs becomes available to organisms through combustion and weathering processes that release carbon dioxide.

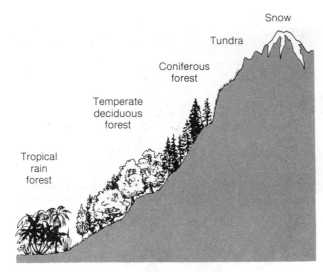

FIGURE 33-6
Altitude and Biomes. The variation in biomes that one sees by ascending a mountain is similar to the variation observed in traveling from low to high latitudes.

mountain to its peak is similar to that seen when traveling (at the same elevation) from low to high latitudes.

In the pages that follow we shall briefly survey the terrestrial biomes. While our emphasis will be on those found in North America, much of what we shall be describing applies equally well to similar biomes found on other continents. Further, as you shall see, terrestrial biomes are usually classified according to the type of vegetation rather than the types of animals found there. The reason is simple: plants are easier to find than animals, which may hide or run away from observers. Furthermore, the type of vegetation present has a major influence on which kinds of animals can exist. Plants tend to determine the animals. That is, animals depend directly or indirectly on plants for food and often shelter. For example, meadowlarks build nests of grass in open grasslands and they feed on the seeds and insects native to these habitats. You are unlikely to find meadowlarks in a mountainous forest, where the appropriate nesting materials and food sources are not available.

Tropical Forests

Broadleaf evergreen tropical forests are the dominant vegetation in areas where temperatures and annual precipitation are relatively high (Figure 33-7). In the United States this type of biome exists only at the southern tip of Florida. In Central and South America, Africa, and southern Asia, however, large areas are covered with tropical forests.

Tropical forests have the highest productivity of any natural terrestrial biome, even though their soils are generally shallow and nutrient-poor. The reason for this paradoxically high productivity has to do with the way minerals are recycled in the tropics. The warm moist conditions are ideal for bacteria and fungi, so fallen leaves decompose quickly on the forest floor, releasing their nutrients into the soil. The nutrients are taken up quickly by the plant roots. This rapid cycling of minerals continues as long as there is a dense covering of trees and vines to continually contribute plant litter and shade the soil.

When a tropical area is cleared for agriculture, however, the soil deteriorates rapidly. Without a continual supply of fallen leaves and other litter, the soil loses most of its minerals in 3 to 5 years. In addition, erosion and the blazing tropical sun transform the soil into a hard crust, which is unsuitable for crop plants. Unfortunately, this problem is becoming more and more common as an expanding human population has increasingly exploited the tropics for lumber and agriculture.

FIGURE 33-7 ➤
Tropical Rainforest. (a) The tropical rainforest houses an incredible variety of animal and plant life. (b) An example of the interesting and abundant vegetation is the plant called "Guarana" by Amazon natives, which contains a high concentration of caffeine and is used to make a cola-like drink. (c) Many species of animals take exotic forms such as the elaborately spined grasshopper shown here. (d) Many species of beautifully colored butterflies abound. (e) One of these spider monkeys is drinking from a Balsa flower.

(a)

(b)

(c)

(d)

(e)

FIGURE 33–8
Grasslands. (a) Until the 1800s most of the land east of the Rocky Mountains was dominated by grasslands such as these. (b) The grasslands supported thousands of grazing animals including huge herds of bison. (c) Today much of this land has been cultivated for grain crops such as wheat, oats, and corn.

(a)

(b)

(c)

Grasslands

From the base of the Rocky Mountains to the deciduous forests of the Mississippi Valley, western North America is dominated by grassland (Figure 33–8). Known as the Great Plains and prairies, this region's native vegetation consists mainly of perennial bunch grasses such as gramagrass and bluegrass. Most of this once vast grassland has now been converted to agriculture, however. Today, annual grasses such as corn, wheat, and rye cover most of the great plains.

Deciduous Forests

Most of the eastern United States from the Mississippi Valley region to the Atlantic is covered by deciduous forest (Figure 33–9). Here, summers are warm and moist, and winters are cold. This type of ecosystem is limited on the north by the severe cold and the short growing season, and to the west by inadequate rainfall.

The woody plants found in deciduous forests (primarily oak, maple, beech, and basswood) drop their leaves during the autumn when temperatures are low and water is scarce. Understory plants are few in number during summer when the tree foliage severely reduces the amount of light reaching the forest floor. However, during a brief period in spring, when sunlight is abundant at ground level, a variety of perennial herbs spring up.

Animals too abound in deciduous forests. These include many small birds that eat seeds and insects, and mammals such as mice, chipmunks, squirrels, raccoons, deer, weasels, and foxes.

FIGURE 33–9
Deciduous Forest. The appearance of deciduous forests is closely linked to seasonal changes: (a) spring brings renewed growth to the buds; (b) foliage abounds during the summer; (c) the cold days in autumn bring on a blaze of color. These woods are home to many animals, such as (d) the white-tailed deer, (e) the red-tailed hawk, and (f) the red-cheeked salamander, which lives in moist, shady recesses.

(a)

FIGURE 33–10
Coniferous Forest. (a) Coniferous forests cover vast areas of the North American continent. (b and c)
In California and the Northwest, trees of immense size are common. The coniferous forest supports a
variety of animals including (d) numerous species of squirrels and (e) the grizzly bear. (f) In the spring,
carpets of wild flowers add color to the meadows.

Coniferous Forests

Two more or less distinct areas of coniferous forest cover sections of North America (Figure 33–10). One is the pine forests in the southern Mississippi Valley region, extending eastward throughout much of the southeast. Its northern and western limits probably are set by cooler temperatures and lower precipitation. Some biologists feel that frequent fires in the southern coniferous forests help to exclude other vegetation. The second and much larger area of coniferous forest covers much of Canada and the northwestern United States, extending southward into the mountain ranges of the west. It includes the redwood forest along the California and Oregon coastline, the spruce and fir forests of the Rocky Mountains, and the pine and fir forests of the Sierra Nevada and Cascade mountain ranges. The northern limits of these coniferous forests are defined by the intense winter cold and short growing season, where forest gives way to tundra. Dryness limits the forests in the south, where they are bordered by scrub (chaparral or brushland) and desert vegetation.

A great variety of animals is found in coniferous forests. These include herbivores, such as the northern flying squirrel, woodrat, deer, porcupine, and snowshoe hare; in addition one finds the primarily carnivorous wolverine (perhaps the most ferocious animal for its size), the marten, omnivorous bears, mice, grouse, and several jays and thrushes.

(b)

(c)

(d)

(e)

(f)

Chaparral and Brushland

From coastal southern California throughout the Great Basin (between the Sierra Nevada and Rocky Mountain ranges), and southeastward through much of Texas, rainfall is scarce and temperatures are often extreme. Here conditions are too harsh for forests to grow, but more moisture is available than in the deserts. A number of small trees, bushes, and grasses adapted to this region, including coastal sage scrub and Great Basin sagebrush (Figure 33–11). As protection against extremes in temperature and water loss, many of these plants have small leaves with protected stomates. Often the leaf surface is gray or whitish and reflects light, or is covered with a tangled mat of hairs that restricts air movement—and therefore evaporation—over the surface of the plant. Many of these plants also have extensive root systems that help them take advantage of available water. Some of the shrubs lose some or all of their leaves during dry periods, such as the most widespread species in this area, Great Basin sagebrush.

The brushlands support a variety of browsing and seed-eating animals: deer, pronghorn, cottontail rabbits, pocket mice, kangaroo rats, and many others. Most of the birds here are seed eaters as well.

(a)

(b)

(c)

(d)

FIGURE 33–11
Chaparral and Brushland. (a) Chaparral covers the hillsides in much of the Southwest. (b) Fire plays an important ecological role in the chaparral community, burning thousands of acres each year.
(c) The chaparral plants exhibit a number of fire-related adaptations such as root crown sprouting.
(d) The pronghorn, which is neither a true antelope nor a true goat, browses on brushland vegetation. It is the fastest North American mammal, sprinting at speeds of up to 50 miles per hour.

FIGURE 33–12

Arctic Tundra. (a) The arctic tundra is characterized by a dense carpet of low-growing vegetation, which continues for great distances in parts of Alaska and Canada. (b) Numerous delicate wild flowers bloom quickly and fade in the short growing season. (c) Herds of caribou roam the tundra. (d) Dall rams are easy to spot as they graze on the mountainsides in McKinley National Park.

(a)

(b)

(c)

(d)

Tundra

At the northern edges of Alaska and Canada, and also in part of western Canada, is the arctic tundra (Figure 33–12). Summers in the tundra are short-lived but the sun shines almost around-the-clock. All of the plants in this region are dormant for about 9 to 10 months of the year because of freezing temperatures. Not surprisingly, the tundra has the second lowest productivity of all terrestrial biomes.

Another limit to life in the arctic tundra is the **permafrost,** a layer of subsoil a few feet below the surface which remains frozen year round. Plant roots and animal burrows are restricted to the soil above

the permafrost. The permafrost also prevents water from seeping away into underground water tables. Thus, much of the arctic tundra is boggy in summer even though little rain falls.

The tundra is treeless, and resembles grassland in its vegetation. In addition to grasses, however, tundra includes many herbs and low-growing shrubs. A lichen known as reindeer moss also is abundant, forming an important food source for the muskox, reindeer, caribou, and lemmings.

The area above the timberline of tall mountains is known as alpine tundra. Here the vegetation is similar to that of the arctic tundra—herbs and shrubs survive, but trees cannot.

(a)

Desert

Large areas in the southwestern United States are covered by desert (Figure 33–13). Here precipitation is scarce, and temperatures range from well below freezing on winter nights to over 100°F every day for several weeks during summer. Because little water is available, desert plants have mechanisms to conserve whatever water they absorb (see Essay 15–1, pp. 248–249). Characteristic plants include the cacti, agave, yuccas, creosote bush, and around oases, the fan palm. Many of these plants have small leaves and a thick waxy covering on the plant surfaces. The stomates of desert plants are usually few in number, and they may open at night rather than in daytime. Some plants also have a mat of tangled hairs over the leaves. However, such adaptations for conserving

water also restrict photosynthesis. Thus, deserts are the least productive of all terrestrial biomes.

When rain does fall, desert plants, especially the annuals, respond quickly. One tiny desert annual in the mustard family completes its life history (from seed germination to fruit production) in just four weeks. When water is not available, these plants can exist as dormant seeds, sometimes for many years.

The adaptations of animals to desert life are just as interesting as those of plants. Many small animals, such as pocket mice, kangaroo rats, and antelope ground squirrels burrow underground, where humidity is higher and temperatures less extreme. Many of these burrowing animals drink little or no water. Rather, they obtain most of their water from the food they eat and the water they produce metabolically.

FIGURE 33–13

Desert. Deserts are characterized primarily by a lack of rainfall. (a) As harsh as this environment is, a number of plants and animals have adapted to life in this ecosystem. (b) While chasing its next meal, a cougar scampers up a saguaro cactus, oblivious to the thorny surface. (c) The kangaroo rat is a nocturnal animal, making the most of the cooler night temperatures to forage. During the hot days, it stays in its burrow. The kangaroo rat obtains most of its water from the food it eats. (d) Capricious spring rains may provide enough moisture to enable annuals and other plants to bloom briefly. However, the annuals will survive only a few weeks and the perennials must rely on numerous water-saving adaptations such as (e) fleshy stems and/or leaves.

MARINE ECOSYSTEMS

At first glance, the oceans of the world, which cover 71% of the earth's surface, may seem to be a more or less homogeneous mass of saltwater. However, they include a number of ecosystems which are as distinct and different from one another as those on land. The two main types of saltwater ecosystems are the coastal system and pelagic system.

Coastal

The most biologically productive region of the sea is the strip of ocean lying over the continental shelf. This is where the continent slopes off from the adjacent dry land into the sea (Figure 33–14). Beyond the shelf, the land drops off rather sharply. The average width of the continental shelf worldwide is about 42 miles, but it differs considerably from place to place. Off the coasts of young mountain ranges, such as the Pacific coastline of the Americas, the shelf is narrow, averaging about 10 miles; in the Arctic, it extends up to 750 miles.

The mineral nutrients that seaweed and other saltwater plants need are more abundant in coastal areas than in the open ocean. In these shallower waters, minerals are continually replenished by dissolved nutrients draining off the land.

While coastal zones comprise only about 10% of the total area of the oceans, they contribute over 85% of the ocean's total productivity. In the shallowest part of the coastal zone, enough light penetrates to support fairly dense plant growth, mainly green, brown, and red algae. The green algae are limited to waters only a few meters deep because most red and blue light, the wavelengths effective in exciting their photosynthetic pigments (chlorophylls and carotenoids), are filtered out at greater depths. The red and brown algae, however, can live 100 meters and more below the ocean's surface. These plants contain accessory pigments which absorb green light (green light penetrates to great depths) and transfer the energy to chlorophyll molecules for photosynthesis. In the deepest shelf waters, all visible light is screened out and no plants can grow.

The unusual demands of living at the water's edge have led to adaptations quite different from those of land animals and plants (Figure 33–15). Along a sandy

FIGURE 33–14

Continental Shelf. Diagram of a typical continental shelf as found on the Atlantic coast at the mouth of the Hudson River off Long Island and New York. The basic profile of a continental shelf is consistent worldwide, although the slopes, depths, and distances vary with locality.

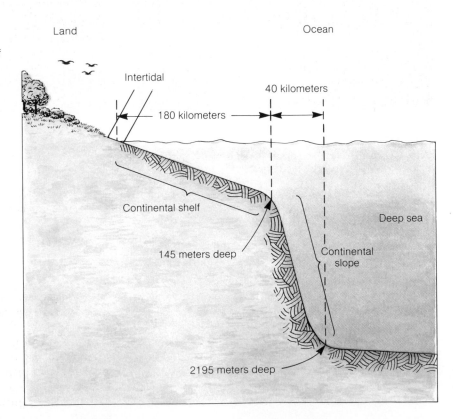

FIGURE 33–15
Coastal—Intertidal. (a) Wave action and repeated exposure to the atmosphere are the two major factors that determine the composition of the coastal—intertidal community. (b) The western coast of the United States exhibits one of the richest populations of coastal—intertidal organisms. (c) The pelican is once again a frequent sight along California shores, having bounced back from near extinction. (d) The delicate anemones attached to this stand of barnacles are from Massachusetts.

beach, many arthropods, molluscs, and worms simply avoid the waves by burrowing into the sand, perhaps projecting a feeding device upwards to filter plankton from the surging water. Many shore birds have long bills which they use to find these buried morsels. On open rocky coasts, organisms of all kinds are found in abundance. They are adapted to the physically punishing waves by virtue of tough shells and streamlined designs; some have flexible bodies that yield to the surging water. Others escape the brunt of the waves by living in crevices or staying on the protected side of a rock.

Most coastal creatures also have protection against drying out. High and low tides alternate twice daily in most areas of the world, exposing the intertidal plants and animals to air at low tides. To conserve their body water when exposed, many animals, such as mussels and barnacles, retreat into their shells. Intertidal algae have a water-retaining slime on their surface that helps prevent excessive water loss.

(a)

Pelagic

Beyond the continental shelf lies the open oceans, which ecologists refer to as the pelagic zone. Mineral nutrients are generally scarce, and this limits the density of organisms to low levels. Single-celled phytoplankton form the base of the often complex food webs in the pelagic. These primary producers are found primarily in the upper 5 meters of the seas, where light is most abundant. Tiny zooplankton feed on the phytoplankton, and both are consumed by tiny shrimp and other crustaceans, which in turn are eaten by small fish and other small animals. Occupying the upper levels of the pelagic food chain are the larger carnivorous fishes and mammals (Figure 33–16).

Recently, deep-water research submarines have discovered some very unusual communities of organisms living on the ocean floor. Often living thousands of feet below the ocean surface, where there is no light to support plant life, these communities are always found near deep-sea vents—cracks in the ocean floor that release hot gases, such as hydrogen sulfide and other reduced compounds, into the overlying water. Certain chemosynthetic bacteria harvest the energy of these reduced inorganic gases by oxidizing them, then apply this energy to manufacture their own organic compounds. These chemoautotrophic bacteria are the food source for other exotic animals, such as the tube worm (Figure 33–17). Thus these ocean floor communities are one of the few whose energy support system derives from the bowels of the earth, not the sun.

FIGURE 33–16
Pelagic. The great oceans of the world make up the pelagic zone. Pelagic organisms include mammals, such as (a) the California gray whale and (b) the dolphin; and many invertebrates, such as (c) the squid and (d) this tiny copepod.

(b)

(c)

(d)

FIGURE 33–17
A Deepwater Ecosystem. The organisms shown here live 2500 meters below the surface of the Pacific Ocean, where there is no sunlight. Chemosynthetic bacteria form the base of the food web, supporting tube worms, crabs, shrimp, bivalves, and others. The tube worms shown here can reach almost 9 meters in length and several centimeters in diameter.

(a)

FRESHWATER ECOSYSTEMS

Freshwater ecosystems include lakes, rivers, brooks, and even small ponds. In each of these environments, the organisms that normally live there are adapted to the special conditions of that particular body of water (Figure 33–18). For example, in a fast-moving mountain stream, the animals must be either strong swimmers, such as trout, or they must remain protected from the current by staying in calmer pools near the water's edge, by attaching firmly to a substrate, or by burrowing into the substratum. In relatively motionless lakes and ponds, on the other hand, weak swimmers and free-floating creatures abound: mosquito larvae, freshwater shrimp, tadpoles, and many, many others. Some kinds of plants and animals occur only in natural depressions that collect water during the rainy season. Another special freshwater ecosystem is found in hot springs, where some cyanobacteria are known to grow at temperatures as high as 75°C.

The producers at the base of freshwater food webs consist largely of unicellular and filamentous forms of green algae. A few flowering plants, such as duckweed and water lilies, live floating or submerged in lakes and ponds. The creatures that feed on these producers include tiny animallike protists (amoeba, *Paramecium,* euglenoids) and animals which feed by filtering the water, such as clams, sponges, fairy shrimp, and water fleas. The larvae of mosquitos, dragonflies, caddis flies, and other insects are also common. Herbivorous fish feed on the primary producers, while bass and trout prey on the insects. Nearby, carnivorous birds such as herons and osprey feed on fish.

FIGURE 33–18
Freshwater. The organisms in freshwater ecosystems vary
according to the special conditions in each particular body of
water. (a) Ponds are alive with many species of small microorganisms such as (b) amphipods and (c) diatoms. (d) Larger
animals such as frogs and birds are usually present. (e) In the
swifter waters of rivers and streams, fish such as the brook
trout must be strong swimmers to fight the fast-moving current.

(a)   (b)

FIGURE 33–19

Human Influences on Ecosystems. (a) Humans have altered the landscape of many areas of the earth, creating new ''biomes''—the urban biome, the agricultural biome, the garbage dump biome, etc. These altered ecosystems present new challenges to organisms, but many are not able to adapt to the drastically altered environments. (b) Manufacturing plants such as this steel mill produce useful materials for humans. However, we must pay a high price for these goods—our air, water, and soil become polluted, mineral resources are used up, and energy is consumed at a phenomenal rate.

HUMAN INFLUENCES ON ECOSYSTEMS

The biomes described in this chapter are natural systems of plants and animals; they were not contrived by humans and they would persist in our absence. Nevertheless, human activities have changed the structure and characteristics of many natural ecosystems, and the rate and degree of such changes are increasing every year (Figure 33–19). Agriculture, mining, clearcutting, and dam construction have taken a heavy toll on natural communities that have been the product of millions of years of evolution. Such human encroachments have already caused the extinction of many plants and animals, and placed thousands more on the endangered species list. Furthermore, the effects of many of these practices go far beyond the immediate and obvious. It is estimated, for example, that perhaps 40% of the increase in carbon dioxide in the atmosphere is a direct result of deforestation.

Part of the difficulty in managing our impact on the environment is that often each individual act contributes so little to the whole problem that we tend to overlook it. We curse the smog and the bumper-to-bumper traffic as we enter a freeway, not thinking about blaming ourselves for adding to the congestion and air pollution. Most of us now recognize the problems our species has created for environmental quality, but feel helpless in controlling them.

Another part of the problem is that, traditionally, we have been inclined to view our activities as distinct from natural events, and therefore subject to different rules. We are now beginning to see that while we may be able to bend the rules in our favor, the rules still persist. The energy from all of our food comes from the sun, and is fed into the system by a single process: the light reactions of photosynthesis. Regardless of what dreamers and space technologists may predict for the future, we still grow plants in the ground and water them, or we don't eat. We depend on rain and snow for our water, as do all other terrestrial creatures. No matter how sophisticated technology becomes, our survival will always hinge on an uninterrupted supply of natural resources. There are no substitutes for food, water, and clean air, but even these most basic of requirements can no longer be taken for granted. Moreover, some of our twentieth-century enterprises have already created an ecological debt from which we cannot escape. This debt will be paid in full in the decades ahead, but just how it will be paid is still in doubt.

SUMMARY

The progressive colonization and replacement of species in new or cleared environments is referred to as ecological succession. The first organisms to occupy an area are called pioneer species. In general these organisms have relatively short lifespans and a rapid rate of reproduction. Gradually the pioneers modify their environment, making it more suitable for larger, slower growing species. The culmination of the successional process is the climax community, a relatively stable assemblage of populations in equilibrium with each other and the environment.

The two most important environmental factors that govern the composition of a particular climax community are temperature and moisture. Because different terrestrial regions on earth vary considerably in their climates, there is variation in the types of communities, or biomes, that exist. This variation is particularly evident as one moves either in latitude or altitude.

One way that biomes differ from one another is in the number of species they contain. Biomes in tropical regions tend to contain numerous species, each one represented by relatively few individuals. With increasing latitude, fewer species are found, but each one is likely to be widely represented. On a smaller geographic scale ecotones, regions between two different communities, usually contain more different species than either community alone. The terrestrial biomes discussed in this chapter include tropical forest, deciduous forest, grassland, coniferous forest, chaparral–brushland, tundra, and desert.

Aquatic ecosystems also have distinctive characteristics. For example, the high nutrient availability in the saltwater coastal zone results in greater productivity than in the nutrient-poor pelagic zone. Freshwater ecosystems occur in numerous areas. Here too the organisms differ depending on the special conditions that characterize the particular pond or stream.

STUDY QUESTIONS

1. Distinguish between primary and secondary succession. What general features characterize the pioneering species in both types of succession?

2. How does the diversity within climax communities change as one proceeds from the equator to northern latitudes?

3. What is an ecotone? How does its species composition differ from areas adjacent to it?

4. What are the two most important environmental factors that govern the range of a particular biome?

5. What happens to the soil in a tropical area when native vegetation is cleared away?

6. List the major biomes found in North America and briefly characterize their vegetation.

7. Why are coastal marine ecosystems much more productive than pelagic ecosystems?

SUGGESTED READINGS

Barbour, M. G., J. H. Burk, and W. D. Pitts. *Terrestrial Plant Ecology.* Menlo Park, CA: Benjamin/Cummings, 1980. There are several chapters in this general plant ecology text which deal with the vegetation types discussed here.

Horn, H. S. "Forest Succession." *Scientific American* 232 (1975):90–98. This article provides a detailed look at forest succession in the woods of New Jersey.

Jannasch, H. W., and C. O. Wilson. "Microbial Life in the Deep Sea." *Scientific American* 236(1977):42–52. This article describes exploration of the deep sea environment utilizing the research submersible "Alvin."

Wright, H. E., Jr. "Landscape Development, Forest Fires and Wilderness Management." *Science* 186(1974):487–495. This review article describes why conifer forests depend on fire for long-term stability.

APPENDIX A
The Metric System

Unit	Measure	Symbol	English equivalent
Linear measure			
1 kilometer	= 1000 meters	km	0.62137 mile
1 meter		m	39.37 inches
1 decimeter	= 1/10 meter	dm	
1 centimeter	= 1/100 meter	cm	
1 millimeter	= 1/1000 meter	mm	
1 micrometer (or micron)	= 1/1,000,000 meter	μm (or μ)	
1 nanometer	= 1/1,000,000,000 meter	nm	
1 angstrom	= 1/10,000,000,000 meter	Å	
Measures of capacity (for fluids and gases)			
1 liter		l or L	1.0567 U.S. liquid quarts
1 milliliter	= 1/1000 liter	ml or mL	
	= volume of 1 g of water at standard temperature and pressure (stp)		
Measures of volume			
1 cubic meter		m³	
1 cubic decimeter	= 1/1000 cubic meter = 1 liter (l)	dm³	
1 cubic centimeter	= 1/1,000,000 cubic meter = 1 milliliter (ml)	cm³ or cc	
1 cubic millimeter	= 1/1,000,000,000 cubic meter	mm³	
Measures of mass			
1 kilogram	= 1000 grams	kg	2.2046 pounds
1 gram		g	15.432 grains
1 milligram	= 1/1000 gram	mg	approx. 0.01 grain

APPENDIX B
Acids, Bases, and pH

Water molecules have a slight tendency to dissociate into hydrogen ions (H^+) and hydroxyl ions (OH^-):

$$H_2O \rightleftharpoons H^+ + OH^-$$

The dissociation is so slight that for every H^+ or OH^- ion in pure water, there are roughly 550 million undissociated water molecules. Many other substances having dissociable H^+ or OH^- groups exhibit greater tendencies to dissociate, particularly when they dissolve in water. Their presence can thus alter the balance between H^+ and OH^- ions in solution. Substances that increase the H^+ concentration (or lower the OH^- concentration) are called **acids.** For example, hydrochloric acid (HCl) in water dissociates into H^+ ions and chloride ions (Cl^-), thereby raising the H^+ concentration.

$$HCl \rightleftharpoons H^+ + Cl^-$$

Other substances called **bases** tend to lower the H^+ concentration (or raise the OH^- concentration) of water. Ammonia (NH_3) is a base that lowers the concentration of H^+ ions by combining with them to form ammonium ions (NH_4^+):

$$NH_3 + H^+ \rightleftharpoons NH_4^+$$

Sodium hydroxide (NaOH) is also a base because it releases OH^- ions when it dissociates:

$$NaOH \rightleftharpoons Na^+ + OH^-$$

The OH^- ions can combine with free H^+ ions to form water, so their presence also acts to lower the H^+ concentration.

Any solution in which the concentrations of H^+ and OH^- ions are equal is said to be **neutral.** Pure water is neutral because the dissociation of water molecules yields an equal number of H^+ and OH^- ions. Solutions containing an excess of H^+ are **acidic** and those containing an excess of OH^- ions are **basic,** or **alkaline.**

A convenient way to express the relative acidity or alkalinity of a solution is by its pH. The **pH** of a solution is a measure of its H^+ concentration, where:

$$pH = -\log_{10} \text{ of the } H^+ \text{ concentration}$$

The pH scale spans from 0 to 14, where neutral solutions have a pH of 7 (see figure). Because pH is related to the negative logarithm of the H^+ concentration, each incremental drop in pH value represents a tenfold increase in H^+ concentration. A pH 6 solution has a tenfold higher H^+ concentration than a pH 7 (neutral) solution; a pH 5 solution has a hundredfold higher H^+ concentration than a pH 7 solution; and so on.

Alkaline solutions have pH values above 7. Their H^+ concentrations are lower than that of pure water (pH 7). The more alkaline a solution is, the higher its pH will be.

Most organisms maintain the pH of their cells in the range of 6.0 (slightly acidic) to 8.0 (slightly alkaline). This is an extremely important activity because enzymes are very sensitive to small changes in pH. If the cellular pH strays too far in either direction on the pH scale, enzymes lose their catalytic activities and cellular metabolism comes to a halt.

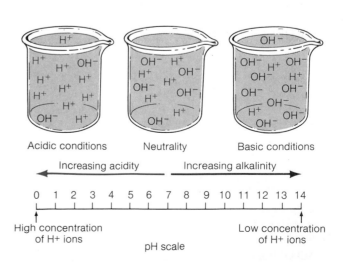

Glossary

Abiotic   Referring to nonliving material or physical factors.

Absorption spectrum   A plot of absorbance of light versus wavelength.

Acetylcholine   One of the neural transmitters active within synapses.

Acetylcholinesterase   The enzyme responsible for the breakdown of acetylcholine to choline and acetate within the synapse.

Acid   Any substance that can release hydrogen ions (H^+).

Acidic   Referring to a substance with a pH less than 7. The hydrogen ion (H^+) concentration is greater than the hydroxyl ion (OH^-) concentration. *See* Alkaline.

Acoelomate   Lacking a central body cavity. *See* Coelom.

Acromegaly   A condition characterized by enlarged hands, feet, and facial tissues which is caused by an oversupply of growth hormone after the attainment of maturity.

ACTH   Adrenocorticotropic hormone, which is produced by the pituitary gland. Stimulates hormone production in the adrenal cortex.

Actin   Contractile protein making up the thin filaments of myofibrils.

Action potential   The rapid and localized depolarization of a neuron or muscle cell membrane caused by ion fluxes.

Action spectrum   A plot of the effectiveness of light in initiating a particular response versus the wavelength.

Activation energy   The energy required to initiate a chemical reaction.

Active site   That portion of an enzyme molecule that binds to the substrates during a catalyzed reaction.

Active transport   The energy-dependent movement of a substance across a cellular membrane against a concentration gradient.

Adaptation   A change in a species that results in an increased ability to survive and reproduce in a particular environment.

Adaptive radiation   The evolution of several species, well-adapted to distinct habitats and ways of life, from a single unspecialized species.

Adenosine triphosphate (ATP)   Nucleotide containing adenine, ribose, and three phosphate groups. Energy for work is released when a phosphate group is removed, forming ADP (adenosine diphosphate).

Adhesion   The attachment of molecules in a solid, liquid, or gaseous state to the surface of a different substance. *See* Cohesion.

Adrenal glands   A pair of glands, located near the upper portion of the kidneys, which secretes over 50 corticosteroids, epinephrine (adrenaline), and norepinephrine.

Adrenaline   *See* Epinephrine.

Adrenocorticotropic hormone   *See* ACTH.

Adventitious roots   Roots arising from a leaf or stem.

Aerobes   Organisms requiring free oxygen for survival.

Aerobic   Requiring or in the presence of oxygen. *See* Anaerobic.

Aerobic respiration   *See* Cellular respiration.

Afterbirth   Fetal membranes and placenta which are discharged from the uterus following childbirth.

Alcohol   An organic compound containing one or more —OH groups.

Alcoholic fermentation   The conversion of sugar to ethyl alcohol in the absence of oxygen.

Aldosterone   A hormone released from the adrenal glands which promotes reabsorption of sodium ions from the nephrons.

Alkaline   Referring to a substance with a basic pH (greater than 7). The hydroxyl ion (OH^-) concentration exceeds the concentration of hydrogen ions (H^+). *See* Acidic.

Allele   Alternate form of a gene that affects a given trait differently.

Allopatric speciation   The evolution of new species after a period of geographic isolation.

Alternation of generations   In the life cycle of a plant, the alternation of the sporophytic or diploid phase with the gametophytic or haploid phase.

Altruism   An individual's self-sacrificing behavior which benefits others in the group.

Alveoli   (Singular, alveolus.) Tiny air sacs located at the terminal ends of the bronchioles in the lungs where gas exchange actually takes place.

Amino acids   Organic molecules containing an amino group ($-NH_2$) and a carboxyl group ($-COOH$); the "building blocks" of protein.

Ammonification   The decomposition of nitrogen-containing compounds by decay microorganisms, generating ammonia.

Amniocentesis   Medical procedure in which a sample of amniotic fluid is withdrawn and fetal cells found therein are cultured and tested for biochemical and chromosomal abnormalities.

Amnion   Innermost membrane surrounding the developing embryo that is filled with amniotic fluid. Present in mammals, reptiles, and birds.

Amniote egg   Shelled egg laid by reptiles and birds, which contains water and nutrients for the developing embryo.

Amniotic cavity   A fluid-filled cavity which surrounds a developing embryo, protecting it from physical injury.

Amphetamines   A group of drugs which act as stimulants to the central nervous system.

Amphibian   Any member of the vertebrate class Amphibia, including frogs, salamanders and their relatives.

Amylase   An enzyme which breaks down starch and glycogen.

Anaerobic   Occurring in the absence of free oxygen. *See* Aerobic.

Analgesic   A substance reducing an individual's sensitivity to pain.

Anaphase   The stage of mitosis during which the chromatids separate from one another and move to opposite poles of the cell. In meiosis, the stage in which homologous chromosomes (anaphase I) or chromatids (anaphase II) move apart.

Androgens   Male sex hormones.

Anemia   Medical condition characterized by a deficiency of red blood cells or hemoglobin and, therefore, lowered oxygen-carrying capacity.

Angiosperm   A flowering plant whose seeds are enclosed in a fruit (mature ovary).

Annelids   Segmented worms. Includes earthworms, leeches, and marine worms.

Anterior   Relating to the front or forward part of a body or organ. *See* Posterior.

Anterior pituitary   Referring to that portion of the pituitary gland located toward the front of the organism; produces several important hormones.

Anther   In flowers, the uppermost part of the stamen where pollen is produced.

Antheridium   The male sex organ of algae, lower vascular plants, and fungi; contains the sperm.

Anthropoid   Resembling humans; group of primates including monkeys and apes.

Antibiotic   Organic substance that inhibits the growth of microorganisms.

Antibody   A protein produced by lymphocytes in response to the presence of specific antigens; major weapon of the immune system.

Anticodon   The sequence of three nucleotides on a tRNA molecule that binds to a complementary codon on mRNA during protein synthesis.

Antidiuretic hormone (ADH)   A hormone which is produced by nerve cells of the hypothalamus and stored in the posterior lobe of the pituitary. Increases the permeability of the collecting tubules in the kidney to water, thus increasing water reabsorption. Also called vasopressin.

Antigen   Any foreign material which stimulates the immune system to produce antibodies against it.

Antiserum   Serum containing antibodies against some specific antigen.

Anus   The terminal opening of the digestive tract.

Aorta   Major artery that carries blood away from the heart.

Apical meristem   Tissue at the tip of a shoot or root that is composed of meristematic (dividing) cells.

Archegonium   The female sex organ of many algae and plants; contains the egg.

Arteriole   Vessels that direct blood from arteries to capillaries.

Artery   Vessel that carries blood away from the heart.

Arthropods   Phylum of invertebrates having exoskeletons and jointed appendages; includes centipedes, crabs, barnacles, spiders, and insects.

Asexual reproduction   Propagation by fission, budding, sporulation, or other means that does not involve the production or union of gametes.

Asexual spore   Reproductive cell produced by mitotic cell division in some plants and fungi.

Atherosclerosis   Narrowing and hardening of arteries caused by the accumulation of lipid deposits.

Atom   The smallest unit of an element that still retains the chemical properties of that element; composed of protons, neutrons, and electrons.

Atomic nucleus   The central region of an atom containing protons and neutrons.

Atomic number   The number of protons contained within the nucleus of an atom.

ATP   *See* Adenosine triphosphate.

Atrioventricular node   Small mass of special muscle fibers located in the heart wall just above the ventricles; initiates a wave of impulses that triggers the ventricles to contract.

Atrium   Chamber of the heart which pumps the blood received from either the body or lungs into the ventricle(s).

Autoimmune diseases   Diseases characterized by the production of antibodies against the body's own proteins.

Autonomic nervous system   Motor portion of the vertebrate nervous system which is normally under involuntary control; innervates glands, smooth muscles, and cardiac muscles.

Autosome   Any of an organism's chromosomes other than the sex chromosomes.

Autotrophic   Referring to organisms that are able to synthesize their own food from inorganic materials (e.g., green plants and some bacteria).

Auxin   A natural or synthetic plant hormone that affects growth and other aspects of plant development.

Axon   Elongated portion of a nerve cell (neuron) which conducts impulses from the cell body to the axonal branches.

Back-crossing   The process of crossing an F_1 hybrid with one of its parents or with an organism of the same genetic constitution as its parents.

Bacteriophage   Any virus that attacks bacterial cells.

Bark   Those tissues of a woody stem or root that occur exterior to the vascular cambium.

Basal body   Cylindrical organelle found at the base of a cilium or flagellum in cells; contains 9 triads of microtubules.

Base   Any substance capable of releasing hydroxyl ions (OH^-), or combining with hydrogen ions (H^+).

Basic   *See* Alkaline.

B cells   Type of lymphocyte which matures within the bone marrow; responsible for antibody production.

Beriberi   Disease caused by a deficiency of thiamin in the diet and characterized by muscle deterioration, paralysis, and loss of mental acuity.

Bilateral symmetry   A body plan in which the right and left sides are approximate mirror images of one another.

Bile   Alkaline secretion of the liver, consisting of bile salts, pigments, and cholesterol; aids in the digestion of fats.

Binomial system   The system of nomenclature in which an organism is given two names, the first indicating genus and the second indicating species.

Biogeochemical cycles   The pathway of chemical elements through the biological and geological components of an ecosystem.

Bioluminescence   The emission of light by living organisms.

Biomass   The total dry weight of living material.

Biome   A persistent association of plants, animals, and other organisms covering a wide geographic area.

Biosphere   The part of earth in which living organisms exist.

Biotic   Referring to life or living organisms.

Bivalent   The chromosomal unit composed of joined homologous chromosomes during meiosis.

Blade   The part of a leaf that is broad and flat. Also, in some algae, the flat, expanded, leaflike part of the thallus.

Blastocoel   The fluid-filled cavity surrounded by undifferentiated cells of the blastula at an early stage in embryo development.

Blastocyst   The stage of embryo development characterized by a structure resembling a hollow ball.

Bone marrow   Inner, soft portion of long bones where blood cells are produced.

Bowman's capsule   The cup-shaped portion of a nephron where blood filtration occurs.

Bronchioles   Small branches of the bronchi that terminate in alveoli of the lungs.

Bronchus   One of two branches of the trachea that direct air into the lungs. (Plural, bronchi.)

Bryophytes   Nonvascular plants which include mosses, liverworts, and hornworts.

Budding   A form of asexual reproduction in which the parent produces an outgrowth that separates to become an independent organism.

Calorie   The amount of heat energy required to raise the temperature of one gram of water $1^\circ C$.

Cambium   A lateral meristem in woody plants that produces secondary xylem and phloem (vascular cambium) or cork (cork cambium).

Capillaries   Smallest vessels of the circulatory system where gases, nutrients, and wastes are exchanged between blood and the body tissues.

Capsule   (1) In bacteria, the slimy exterior covering around the entire cell. (2) The spore case found in mosses and liverworts.

Carbohydrates   Organic compounds (e.g., sugar, starch, and cellulose) composed of carbon, hydrogen, and oxygen atoms in a 1:2:1 ratio.

Carbon cycle   The movement of carbon through the biological and physical components of an ecosystem.

Carboxyl group   —COOH, a functional group characteristic of organic acids.

Carcinogenic   Cancer-causing.

Cardiac muscle   Type of muscle comprising the walls of the heart.

Carnivores   Animals which eat other animals.

Carotene   Yellow pigment located in chloroplasts; precursor of vitamin A.

Carotenoids   The yellow, orange, or sometimes red pigments found in plant plastids; includes the carotenes.

Carpel   The floral organ which bears the ovules. The pistil is composed of one or more carpels.

Carrying capacity   The maximum size of a population that can be supported by a particular environment.

Cartilage   Tough, flexible connective tissue of the skeletal system of vertebrates.

Casparian strip   A bandlike region of wax in the primary walls of the root endodermis; prevents the passage of water and solutes through the endodermal walls.

Catalyst   A substance that enhances the rate of a reaction but is not used up by the reaction (e.g., an enzyme).

Catastrophism   The geological theory that the earth was shaped by sudden and violent events over a short period of time.

Cell   The basic structural unit of all organisms.

Cell body   In nerve cells, the portion of the cell containing the nucleus and from which the dendrites and axon extend.

Cell membrane   *See* Plasma membrane.

Cell plate   In plants, a sheet of Golgi vesicles that forms during early telophase at the equator of a dividing cell.

Cell theory   The idea that all organisms are composed of one or more cells.

Cellular respiration   The breakdown of simple organic molecules, such as glucose, into carbon dioxide and water. The accompanying release of energy is coupled to the formation of ATP.

Cellulose A rigid polysaccharide composed of glucose units; the major component of the cell wall in most plants.

Cell wall In plants, the rigid structure composed largely of cellulose and other polysaccharides; located external to the plasma membrane.

Central nervous system (CNS) In higher animals, the brain and spinal cord.

Centriole Cytoplasmic organelle present in the cells of animals and some lower plants; associated with the mitotic spindle.

Centromere A dense region of a chromosome where spindle fibers attach during mitosis and meiosis.

Cephalization Evolutionary trend in which ganglia and sense organs became concentrated at the anterior end of animals' bodies.

Cerebellum That portion of the vertebrate hindbrain responsible for the control of equilibrium and movement.

Cerebral cortex Gray matter which forms the outer layer of the cerebrum.

Cerebrum The largest area of the vertebrate brain, consisting of two hemispheres connected by the corpus callosum. Responsible for conscious thought and the processing of sensory information.

Cervix The ring of muscular tissue at the opening of the uterus.

Chemical evolution The notion that spontaneous chemical reactions were the basis for the origin of life.

Chemoreceptor Cell specialized for the detection of chemical stimuli, such as smell and taste receptors.

Chemosynthetic Referring to bacteria which utilize the energy derived from the oxidation of certain inorganic compounds to produce food (organic) compounds.

Chiasma The area where physical crossing over of genetic material occurs between chromatids of a homologous pair of chromosomes.

Chitin A tough polysaccharide found in the exoskeletons of arthropods and the cell walls of certain fungi.

Chloroplast A membrane-bound organelle in the cells of plants; the site of photosynthesis.

Chondrichthyes Class of vertebrates including the cartilaginous fishes, sharks, rays, skates, and their relatives.

Chordates Group of animals, members of phylum Chordata, which possess a notochord, a dorsal, hollow nerve cord, and pharyngeal gill slits at some time during their development.

Chromatid One of the two strands of a duplicated chromosome joined by the centromere.

Chromoplast A plastid that contains pigments other than chlorophyll.

Chromosomes Elongated structures in the nucleus, visible during mitosis and meiosis, which are composed of nucleic acid and protein. Carriers of hereditary information.

Chromosome theory of heredity The theory that chromosomes are the vehicles of hereditary factors.

Cilia (Singular, cilium.) Threadlike projections on the surface of some cells which aid in locomotion or serve to move materials past the cell.

Climax community The terminal, stable community of an ecological succession.

Clitoris In female mammals, an erectile tissue situated above the vaginal opening and urethra that is sexually sensitive.

Closed circulatory system System in which the blood is contained within a series of vessels. *See* Open circulatory system.

Cnidarians A phylum of invertebrates which includes sea anemones, corals, jellyfish, etc.

Cochlea Coiled cavity of the inner ear divided into three canals; involved in the conversion of sound vibrations into nervous impulses and, thus, hearing.

Codon Sequence of three nucleotides on mRNA that specifies a particular amino acid in protein synthesis.

Coelom Body cavity of roundworms and higher animals in which the internal organs are suspended.

Coenzyme Compound which plays a vital role in specific enzyme-catalyzed reactions.

Cohesion Tendency of particles or molecules of a given substance to stick together. *See* Adhesion.

Coitus Sexual intercourse.

Coleoptile The protective, sheath-like structure covering the young shoot in grass seedlings.

Collecting ducts Terminal structures within the kidney which collect urine from several nephrons.

Colon The large intestine.

Commensalism A symbiotic relationship in which one organism enjoys a benefit without causing disadvantage to the other.

Community All the organisms living together and interacting with one another in a common environment.

Companion cell A small, specialized parenchyma cell closely associated with the sieve tube cells of the phloem in flowering plants.

Competition The struggle between two or more organisms for life's necessities (e.g., food, water, light, and minerals) when they are in limited supply.

Competitive exclusion principle The ecological principle which states that two species that share the same niche cannot coexist.

Compound A substance composed of atoms of two or more different elements bonded to one another in defined proportions.

Compound microscope A magnifying instrument composed of a series of mounted lenses capable of greater magnification than a simple microscope.

Cone cells Light-receiving cells of the vertebrate eye involved in color perception and the discrimination of detail.

Conifer A cone-bearing plant.

Conspecifics Different individuals of the same species.

Contraception Preventing pregnancy intentionally.

Convergent evolution The origin along separate and unrelated evolutionary lines of similar adaptations.

Cork In woody plants, the protective, secondary tissue that is formed by the cork cambium; prevents water loss. The cells are nonliving at maturity and have waxy walls.

Cork cambium The cambium from which cork develops.

Cornea The transparent outer coat of the eye.

Corpus callosum A mass of white nerve fibers connecting the left and right sides of higher mammals' brains.

Corpus luteum The ovarian structure forming from the egg follicle after it has ruptured. It produces estrogens and progesterone.

Cortex (1) Primary tissues of stems and roots, consisting primarily of parenchyma cells; located between the epidermis and primary phloem (or endodermis in roots). (2) The outer region of an organ, such as a kidney or adrenal gland.

Corticosteroids A group of steroid hormones produced by the adrenal cortex which are involved in carbohydrate and protein metabolism and the regulation of body fluid ion levels.

Cotyledon Seed leaf of an embryo; functions to store or absorb food.

Covalent bond A physical force that holds two atoms together when electrons are shared.

Cristae The foldings of the inner membrane of mitochondria; contain the electron transport components.

Crop Enlarged segment of the digestive system between the esophagus and gizzard, for temporary storage of food in birds and certain invertebrates.

Crossing over The exchange of corresponding chromatid segments of a homologous chromosome pair during meiosis.

Cross-pollination The process in which pollen is transferred from the anther of a flower in one plant to the stigma of a flower in another plant.

Cuticle The waxy layer covering the outer wall of epidermal cells; the soft exoskeleton of arthropods.

Cyanobacteria A major subgroup of kingdom Monera, including those species which use chlorophyll *a* in photosynthesis. Formerly called blue-green algae.

Cyclic AMP Cyclic adenosine monophosphate. A "second messenger" molecule produced in target cells in response to a given hormone which activates certain latent enzymes in the cell.

Cytochromes Electron transport molecules composed of protein and iron, which function in respiration and photosynthesis.

Cytokinesis The division of a cell (eukaryotic) which follows the division of its nucleus.

Cytokinins A class of plant hormones that enhance cell division and inhibit leaf senescence.

Cytoplasm All the protoplasm of the cell with the exception of the nucleus.

Cytoplasmic streaming The circular movement of cytoplasmic components within the cell involving the action of microfilaments.

Cytoskeleton The intracellular scaffolding of microfilaments and microtubules that gives shape to cells.

Cytosol Referring to the soluble portion of the cytoplasm which bathes the cell organelles.

Dark reactions Photosynthetic reactions which do not directly depend on light, during which glucose is synthesized from carbon dioxide with the use of ATP and NADPH$_2$ from the light reactions.

Day-neutral plant A plant which does not require a specific photoperiod to flower.

Deciduous (1) Referring to plant parts (e.g., leaves, fruits, or flower organs) that fall off at the end of the growing period. (2) Broad-leafed plants that lose all of their leaves each year at the end of the growing season, as opposed to evergreens, which always retain some leaves.

Decomposers Organisms that recycle organic matter by breaking it down into smaller units; important links in every food chain.

Dehydration synthesis A linking reaction in which a water molecule is removed from the reactants. The reverse of hydrolysis.

Dendrites Nerve cell extensions which conduct impulses toward the neuron cell body.

Denitrification The reduction of nitrates in the soil to gaseous nitrogen; carried out by certain bacteria.

Deoxyribonucleic acid (DNA) The nucleic acid that carries the genetic information of the cell; contains the bases adenine, guanine, cytosine, and thymine.

Deoxyribose The 5-carbon sugar found in DNA. It has one less oxygen than ribose.

Development The organized sequence of events involving growth and differentiation of an organism or its parts.

Diabetes mellitus Condition characterized by high blood levels of glucose and excess sugar secreted in the urine; caused by insufficient production or inhibition of action of insulin.

Diastole The relaxation phase of the heartbeat cycle.

Diastolic pressure The minimum pressure in the arteries occurring during heart relaxation and influenced by the elasticity of artery walls. *See* Systolic pressure.

Diatoms Members of Division Chrysophyta, Kingdom Protista; single-celled, photosynthetic, with silica-impregnated cell walls.

Dicots Flowering plants (angiosperms) that have two cotyledons or seed leaves. *See* Monocots.

Differentiation The physiological or morphological

modifications of a cell, tissue, or organ that occur during development and result in specialization.

Diffusion The net movement of a substance from a region of high concentration to one of low concentration due to the random motion of the molecules. The end result is diffusion equilibrium—the uniform distribution of the molecules or particles throughout the available area.

Digestion The process of transforming complex polymers into their simple subunits, usually by the action of enzymes.

Diploid Refers to the presence of two homologous sets of chromosomes; the $2n$ number of chromosomes. *See* Haploid.

Directional selection The process by which the gene pool of a population or species changes in a common direction in response to environmental change through natural selection.

Disaccharide A carbohydrate consisting of two simple sugars.

Disruptive selection The process by which the gene pool of a population or a species segregates into two or more groups differing in adaptive traits corresponding to some selective elements in the environment.

DNA *See* Deoxyribonucleic acid.

Dominant An allele that expresses itself phenotypically and masks the effect of the partner allele; also refers to the trait controlled by such an allele.

Dormancy A period of lowered physiological activity in various plant organs (e.g., in seeds, bulbs, and buds).

Dorsal Relating to the back of an organism or body part. *See* Ventral.

Double circulation Circulation of blood involving two passages through the heart, first of deoxygenated blood from the body, then of oxygenated blood from the heart.

Ecdysone The hormone which initiates molting in insects.

Echinoderms A phylum of invertebrates including starfish, sea urchins, and sea cucumbers.

Echolocation The emission of sound waves from an organism, such as a bat, which bounce off objects, allowing the organism to determine its position.

Ecological niche The place and functional role of an organism within an ecosystem which is described by such factors as its relationship with other organisms, its use of resources, and the range of environmental conditions that are optimal for its successful existence.

Ecological succession The sequence of kinds of vegetation either occurring on a disturbed site or beginning on bare ground or rock and leading to a stable vegetation type.

Ecology The study of the interactions among organisms, between organisms and their physical environment, and of the distribution and abundance of organisms which result from these interactions.

Ecosystem An interacting system consisting of living organisms together with their nonliving environment.

Ecotone The marginal area between two adjacent ecological communities.

Ectoderm The outermost of the three germ layers of an animal embryo giving rise to the skin, nerve tissue, etc.

Ectotherm A "cold-blooded" animal that derives most of its body heat from the environment. *See* Endotherm.

Egg A nonflagellate (nonmotile) female gamete.

Ejaculation The forceful release of semen from a penis.

Electrical potential Capacity to do electrical work; the difference in charge between two sides of a membrane due to concentration of positive and/or negative ions.

Electrocardiograph (EKG) Instrument which records the electrical impulses occurring during contraction of the heart; tool for detecting abnormalities.

Electroencephalogram (EEG) A recording of the electrical brain wave patterns of an individual by means of electrodes touching the scalp.

Electromagnetic spectrum Encompasses all types of energy waves, from short X-rays, to visible light, to long radio waves.

Electron Negatively charged subatomic particle. Mass is $1/1837$ that of the proton.

Electron carrier A specialized molecule that can easily gain or lose an electron and functions to transfer electrons from one compound to another.

Electron microscope A microscope which forms the image of very small objects or portions on a photographic plate by means of a beam of electrons (rather than light).

Electron transport system A chain of electron transport components, located in the mitochondria, down which electrons are passed with the concurrent production of ATP.

Element Substance composed of one kind of atom.

Embryo (1) The young sporophyte plant in a seed or archegonium before the beginning of the rapid growth period. (2) In animals, the immature, developing organism.

Embryonic induction An interaction between adjacent tissues in an embryo, in which one tissue induces the other to develop into a specific organ or tissue.

Embryo sac The mature female gametophyte in a flowering plant. Contains the egg nucleus and the polar nuclei.

Endergonic reaction A reaction requiring energy. The products of the reaction contain more energy than the reactants. *See* Exergonic reaction.

Endocrine gland A ductless gland producing hormones which are released directly into the circulatory system. *See* Exocrine gland.

Endoderm The innermost of the three germ layers of an animal embryo; gives rise to the lining of various internal organ systems.

Endometrium The inner lining of the mammalian uterus which is sloughed off during menstruation.

Endoplasmic reticulum (ER) An extensive network of membranes forming channels and compartments in the cytoplasm of a cell. Rough endoplasmic reticulum refers to that associated with ribosomes, as opposed to smooth endoplasmic reticulum.

Endoskeleton Supporting framework located within an organism's body. *See* Exoskeleton.

Endosperm The usually triploid, food-storage tissue found in the seeds of flowering plants.

Endosymbiosis Living within a larger cell, more or less independently and without harm to the larger cell.

Endothelium The layer of thin flat cells that lines blood and lymph vessels.

Endotherm A "warm-blooded" animal that derives most of its body heat metabolically. *See* Ectotherm.

Enkephalins Naturally occurring substances in the brain which possess painkilling properties.

Enzyme A complex protein, synthesized in a living cell, that functions as an organic catalyst, speeding the rate of a chemical reaction.

Epidermis The outermost layer of cells of an organism.

Epinephrine A hormone produced by the adrenal gland which serves to prepare an organism for an emergency by increasing blood sugar, blood pressure, heart rate, etc.; also a synaptic transmitter substance. Also called adrenaline.

Epithelium Cells or tissue which line exterior body surfaces or internal cavities and organs.

Equilibrium A state of balance where no net change is occurring in a system.

Erythrocytes Red blood cells. Shaped like biconcave disks and containing hemoglobin; produced in the bone marrow.

Esophagus Muscular tube through which food materials move from the oral cavity to the stomach.

Estrus The period of time when a female mammal is sexually receptive to the male.

Ethology The study of animal behavior in natural settings.

Ethylene Gaseous hydrocarbon that, at very low levels, affects plant growth and development, particularly fruit ripening.

Eukaryotic Referring to cells that possess membrane-bound nuclei and organelles, as opposed to prokaryotic.

Evolution In biology, refers to the progressive changes in organisms over time.

Excretion The elimination of toxic waste products from the body of an organism.

Exergonic reaction A reaction during which energy is released. The products of the reaction contain less energy than the reactants. *See* Endergonic reaction.

Exocrine gland A gland which releases its secretions into a duct. *See* Endocrine gland.

Exocytosis The process by which cell vesicles fuse with the plasma membrane, releasing their contents outside the cell.

Exoskeleton Rigid skeleton covering the outer part of the invertebrate body; characteristic of arthropods. *See* Endoskeleton.

Facilitated diffusion The movement of a substance down its concentration gradient at a rate faster than is possible with simple diffusion.

Facultative anaerobes Microorganisms which can function in either aerobic or anaerobic conditions. *See* Aerobes and Obligate anaerobes.

FAD Flavin adenine dinucleotide, a hydrogen acceptor molecule involved in cellular chemical reactions.

Fallopian tubes The slender tubes that extend from ovaries to uterus; also called oviducts.

Fatty acids Organic acids that form part of fat and phospholipid molecules.

Fermentation A series of reactions occurring in the absence of free oxygen, during which pyruvic acid is converted to alcohol, lactic acid, or other products.

Fertilization In sexual reproduction, the fusion of two gametes to form a zygote.

Fetus The unborn/unhatched vertebrate past the initial stages of development. In humans, the embryo is termed a fetus at about two months.

F_1 generation The first (filial) generation after a cross. F_2 and F_3 are the second and third generations, respectively.

Fibrin Protein molecule which participates in blood clot formation through crosslinkage with other molecules.

Fibrinogen Soluble plasma protein which is enzymatically converted to fibrin.

Filament (1) Part of a stamen; the stalk that supports the anther. (2) In certain algae, thread-like row of cells.

Filter feeding Mechanism for feeding in which water is strained or filtered through a system of feather-like gills which remove suspended food particles. Used by certain aquatic organisms.

Fission An asexual means of reproduction in which a single-celled organism divides into two.

Flagellum A long whiplike protrusion found on such cells as sperm, zoospores, and certain protists that makes locomotion possible. Similar in internal structure to cilia, but longer.

Florigen A plant hormone, as yet not isolated, which induces flowering.

Follicle Cells surrounding and nourishing the oocyte in the ovary.

Food chain, food web The sequence or pattern of organisms through which materials are cycled and energy flows; depiction of the feeding relationships of organisms.

Forebrain The anterior anatomical region of the vertebrate brain consisting primarily of the thalamus, hypothalamus, and cerebrum.

Fossil Plant or animal parts (or impressions of them) preserved in the earth's crust, thus giving information about life in past geological ages.

Fraternal twins Twins conceived from two eggs and two sperms.

Fruit In angiosperms, the mature, ripened ovary or group of ovaries that contain the seeds; also includes any adjacent parts that have fused with the ovary.

FSH Follicle-stimulating hormone produced by the pituitary gland which stimulates the maturation of an ovarian follicle.

Fungi Kingdom of heterotrophic mostly filamentous organisms which obtain their nutrients by absorption.

Gall bladder Storage organ for bile.

Gametangia Any cell or organ which produces gametes.

Gamete A haploid reproductive cell that may fuse with another gamete to produce a zygote.

Gametogenesis The mitotic and meiotic cell divisions and subsequent development leading to formation of mature eggs or sperm.

Gametophyte The haploid or $1n$ organism in a plant life cycle; produces gametes.

Ganglion A cluster of nerve cell bodies.

Gastric juice The mixture of hydrochloric acid and digestive enzymes produced by the glandular cells of the stomach.

Gastrointestinal tract The two-ended, tubular gut of animals involved in the digestion and absorption of food materials.

Gastrulation Stage in embryonic development when the three germ layers (endoderm, mesoderm, and ectoderm) are formed.

Gene The fundamental hereditary unit that occurs in a linear arrangement with other genes on the chromosome.

Gene pool All of the genes of a given population of organisms.

Generator potential The initial membrane depolarization in a sensory receptor cell caused by an appropriate stimulus.

Genetic drift The change of population gene frequencies due to chance events occurring in relatively small populations.

Genetic engineering The purposeful manipulation of genes, usually using recombinant DNA techniques.

Genetics The branch of biology dealing with heredity.

Genotype An organism's genetic makeup, either latent or expressed, as opposed to the phenotype; all of an individual's genes.

Genus Taxonomic group containing one or more similar species.

Germination The beginnning or resumption of growth in an embryo or spore.

Gibberellin A plant hormone that influences many phases of development, including seed germination, dormancy, cell elongation, and flowering.

Gills (1) In aquatic animals, structures extending outward from the body which are adapted for gas exchange. (2) Plates located on the underside of the cap in certain fungi (e.g., Basidiomycetes, the gill fungi).

Gizzard Muscular sac of the digestive tract in birds and some invertebrates used for grinding food.

Gland Cell or organ producing substances that are secreted.

Glomerulus Rounded mass of capillaries associated with Bowman's capsule of a nephron.

Glucagon Hormone produced by cells of the pancreas which acts to raise blood sugar levels through the initiation of glycogen digestion in the liver.

Glucocorticoids A group of corticosteroid hormones produced by the adrenal cortex; regulate carbohydrate, fat, and protein metabolism.

Glucose A simple sugar with the formula $C_6H_{12}O_6$.

Glycerol A three-carbon alcohol; a component of triglycerides and phospholipids.

Glycogen A food-storage carbohydrate in animals and many fungi; a polymer of glucose.

Glycolysis The early phase of respiration in which glucose is converted to pyruvic acid with the release of a small amount of useful energy.

Goiter Condition characterized by an enlargement of the thyroid gland; usually caused by lowered thyroxine production due to the absence of sufficient iodine in the diet.

Golgi body A cellular organelle that is composed of flat disc-shaped sacs. Chiefly involved with synthesis and packaging of substances for secretion.

Gonads The gamete-producing organs; ovaries and testes.

Graafian follicle In humans, the mature stage of follicle development from which the oocyte is released.

Grana Stacks of thylakoids appearing in the chloroplast. They contain the chlorophyll and carotenoid pigments and are the site of the light reactions of photosynthesis.

Granulocytes Circulating, phagocytic white blood cells.

Gravitropism A growth movement in plants that occurs in response to gravity.

Gray matter Central portion of brain and spinal cord. Appears gray because it consists mostly of nonmyelinated nerve cell bodies.

Gross productivity The total amount of organic material produced in an ecosystem over a period of time. *See* Net productivity.

Growth hormone Hormone which stimulates growth. Produced by the pituitary.

Guard cells Specialized epidermal cells that contain chloroplasts and surround a pore or stomate. Turgor of guard cells regulates stomatal opening.

Gymnosperms Vascular plants with seeds that are not enclosed in an ovary (e.g., the conifers).

Habitat The physical environment occupied by a group of organisms.

Habituation   The diminution and eventual loss of a behavioral response to a particular stimulus due to lack of reinforcement.

Haploid   A term that indicates the presence of only one chromosome of each pair. A single set of chromosomes in a cell. *See* Diploid.

Hardy–Weinberg law   The law of inheritance, expressed mathematically, which states that allelic and/or genotypic frequencies remain unchanged in a population from generation to generation in the absence of factors which change gene frequencies, such as mutation, selection, genetic drift, etc.

Heartwood   The nonliving inner layers of wood in which water conduction has ceased; usually darker in color than the surrounding sapwood.

Heme group   The iron-containing portion of certain protein molecules, such as hemoglobin.

Hemichordates   Members of Phylum Hemichordata, a small group of wormlike animals with pharyngeal gill slits and a dorsal nerve cord.

Hemoglobin   The oxygen-carrying protein in blood, composed of four polypeptide chains, each with a heme-iron group.

Hemophilia   Sex-linked, inherited disease which is characterized by the failure of blood to clot, resulting in excessive bleeding.

Herbaceous   Referring to any nonwoody plant.

Herbivores   Animals which eat plants or plant parts.

Hermaphroditic   Referring to an individual organism which is able to produce both male and female gametes.

Heterotrophic   Referring to organisms that are unable to manufacture their own food and must, therefore, obtain complex organic material from external sources.

Heterozygous   Having two alleles for contrasting characters at corresponding positions on homologous chromosomes. *See* Homozygous.

Hibernation   A prolonged, seasonal period of rest during which the metabolic rate and body temperature of a vertebrate is greatly reduced.

Hindbrain   The posterior anatomical region of the vertebrate brain composed of the medulla and cerebellum.

Homeostasis   The maintenance of a constant internal environment in an organism.

Homeothermic   Maintaining a constant warm body temperature, as in birds and mammals.

Homologous chromosomes   Chromosomes which pair up during the first stage of meiosis and are then segregated into separate daughter cells. Contain identical or dissimilar alleles.

Homozygous   Having the same allele at corresponding positions on homologous chromosomes. *See* Heterozygous.

Hormone   A chemical that is produced in minute quantities in one part of a plant or animal and has a physiological effect in another part of the plant or animal; a chemical messenger.

Host   A living organism upon which a parasitic organism subsists.

Human chorionic gonadotropin (HCG)   A hormone produced by the placenta which functions to maintain the corpus luteum. The resulting elevated progesterone levels serve to prevent the next menstrual cycle and thus maintain pregnancy.

H–Y antigen   Substance produced by young male embryos which initiates the development of the testes.

Hybrid   The offspring from parents of differing phenotypes.

Hybridization   The process of producing offspring from a cross between genetically dissimilar parents.

Hybridoma   A hybrid cell, formed from the fusion of a cancerous cell with another noncancerous cell.

Hydrocarbon   An organic compound consisting only of carbon and hydrogen.

Hydrogen bond   A weak bond linking hydrogen in a polar molecule to another polar molecule or part of the same molecule; an important factor in determining the structure of water, proteins, DNA, membranes, and cellulose.

Hydrolysis   A decomposition process in which a complex compound is broken down into simpler compounds with the addition of water.

Hydrophilic   Referring to substances or parts of substances which readily associate with water.

Hydrophobic   Referring to substances or parts of substances which repel water.

Hyperthyroidism   A condition characterized by nervousness, high blood pressure, and loss of weight, caused by excessive thyroxine production.

Hypertonic   Referring to a solution having a higher solute concentration than the one to which it is being compared. *See* Hypotonic.

Hypervitaminosis   Condition brought about by the overconsumption of fat-soluble vitamins which are stored in the body.

Hyphae   The threadlike filaments that compose the body of filamentous fungi.

Hypoglycemia   Condition characterized by low level of blood glucose brought about by too much insulin.

Hypothalamus   Known as the "seat of the emotions," this part of the vertebrate forebrain serves to control the internal physiological and behavioral state of the body, regulating such things as hunger, pain, and body temperature; produces releasing hormones.

Hypothesis   A tentative explanation of observed events or known facts.

Hypothyroidism   A condition characterized by general lethargy, dry skin, loss of hair, and intolerance to cold, resulting from insufficient thyroxine.

Hypotonic   Referring to a solution having a lower solute concentration than the one to which it is being compared. *See* Hypertonic.

Immunity   The state of being able to resist infection by a foreign agent to which an organism has been previously exposed.

Immunization   Means of bestowing immunity against a given disease agent through prior exposure. *See* Vaccination.

Imprinting   A type of learning which only occurs during a limited period of time shortly after birth in certain species.

Incomplete dominance   The condition that produces a phenotype in heterozygotes that is intermediate between the phenotypes of the two homozygotes.

Inert   Nonreactive. Referring to elements whose atoms have full outer electron shells.

Infant hypothyroidism   *See* Cretinism.

Inferior vena cava   Major vein carrying blood from the lower portion of a mammal's body to the right atrium of the heart.

Inhibiting hormones   Peptide hormones released from the hypothalamus which inhibit the release of certain tropic hormones from the pituitary.

Innate   When referring to behavior, that which is genetically determined and not learned.

Inner cell mass   The cluster of cells contained within the blastocyst which eventually develop into the embryo proper.

Insight learning   Learning in which an organism uses information obtained from past experience to solve a novel problem or meet a new situation. Occurs only in higher organisms.

Instinctive behavior   A genetically-determined, complex behavioral pattern which is independent of practice or past experience.

Insulin   Hormone produced by the pancreas which brings about a decrease in blood glucose levels by stimulating glucose absorption by cells and the synthesis of glycogen in the liver.

Interneuron   Type of neuron found in the brain and spinal cord which connects sensory and motor neurons.

Internode   That portion of the stem situated between any two successive nodes.

Interphase   Cell stage between nuclear divisions.

Invertebrate   Any animal which lacks a backbone or spinal column.

Ionic bond   A bond formed by electrical attraction between ions of opposite electrical charge.

Ions   Atoms or molecules that carry a unit electrical charge or multiple units.

Islets of Langerhans   Specialized cells in the pancreas which secrete glucagon and insulin.

Isotonic   Referring to a solution having the same solute concentration as one to which it is being compared.

Isotope   An atom that exhibits an altered mass number from other atoms of the same element.

Juvenile hormone   Hormone which maintains the juvenile larval stage in insect development.

Kidneys   Vertebrate organs located behind the stomach and liver which are involved in waste removal and the maintenance of water and salt balance.

Kilocalorie (Kcal)   Unit of heat measurement equal to 1000 calories.

Kinetin   A synthetic compound that promotes cell division in plants. Related to cytokinins in chemical structure.

Krebs cycle   The series of chemical reactions occurring in the mitochondria that results in the breakdown of pyruvic acid to carbon dioxide, hydrogen, and electrons.

Kwashiorkor   A disease in children caused by a diet deficient in protein. Symptoms include stunted growth, edema, brain damage, and even death.

Labia majora   The two outer folds of skin at the opening of the vagina.

Labia minora   The two inner folds of skin at the opening of the vagina.

Labor   The process of childbirth, especially the muscular contractions of the uterus and abdomen.

Lactic acid fermentation   The conversion of sugar to lactic acid in the absence of oxygen.

Lactose   A disaccharide of glucose and galactose, with the chemical formula $C_{12}H_{22}O_{11}$, found in milk.

Lancelets   Small group of nonvertebrate chordates; small fishlike animals of shallow coastal waters.

Large intestine   Muscular tube of the digestive tract through which undigested materials pass and where some substances, particularly water, are reabsorbed.

Larva   An immature form of an animal which has a different physical appearance than the adult (e.g., a tadpole is the larval form of a frog).

Larynx   The upper part of the windpipe which contains the vocal cords. Also called the "Adam's apple."

Lateral   Situated on one side of an organ.

Lateral bud   A bud located in the leaf axil.

Leaf   A lateral plant organ developing superficially from the tissues of the shoot apex; usually composed of a blade and petiole, with some variations; major photosynthetic organ in most plants.

Learning   Adjusting one's behavior because of experience.

Lens   A clear biconvex body between the pupil and fluid contents of the eye, which focuses light on the retina.

Leucoplast   A colorless plastid.

Leukocytes   White blood cells produced in the bone marrow and present in the blood and lymph fluid; function in the defense against foreign matter.

Lichen   Organism which consists of an alga and a fungus in a symbiotic relationship.

Light reactions   Photosynthetic reactions that require light and result in the synthesis of ATP and $NADPH_2$, with the release of oxygen.

Limbic system   A diffuse neural network in the vertebrate brain having a role in pain sensitivity and the control of emotions.

Limiting factors   Environmental factors which serve to limit population size.

Linkage   The tendency of a particular group of genes to be inherited together because they are located on the same chromosome. The group of traits of genes so linked.

Lipase   An enzyme which breaks down fats into their component glycerol and fatty acid molecules.

Lipid   General term for fatlike compounds (e.g., oils, fats, and certain steroids).

Long-day plant   A plant which initiates flowering only when the photoperiod includes dark periods shorter than some critical length.

Lower vascular plants   Vascular plants which lack seeds. Includes ferns.

Luteinizing hormone (LH)   A hormone produced by the pituitary gland. Thought to be responsible for initiating ovulation and corpus luteum formation.

Lymphatic system   Transport system consisting of lymph capillaries, veins, ducts, and lymph nodes. Functions to conduct lymph fluid.

Lymph nodes   Small oval nodules of connective tissue located along lymph vessels, which filter out debris.

Lymphocytes   Any of several types of white blood cells involved in the production of antibodies involved in the immune response.

Lysosomes   Membrane-bound cell organelles which contain digestive enzymes.

Macrominerals   Minerals required by animals in relatively large amounts: calcium, phosphorus, magnesium, sodium, potassium, and chloride. *See* Microminerals.

Macromolecules   Large organic molecules that usually possess a rather complex structure (e.g., proteins, DNA, and RNA).

Macronutrients   Those elements required in relatively large amounts for the growth of plants. *See* Micronutrients.

Macrophages   Relatively large, phagocytotic white blood cells usually found in extracellular tissue fluid.

Malnutrition   The condition caused by the consumption of an imbalanced diet which may be high enough in Calories but is lacking in certain vitamins and/or organic precursors.

Maltose   A disaccharide of two glucose molecules, with the chemical formula $C_{12}H_{22}O_{11}$; a product of starch digestion.

Mammals   Members of the vertebrate class Mammalia, including homeotherms with hair who nurse their young.

Marsupials   Group of mammals with no placenta; the young complete development in an external pouch containing teats.

Mass number   The total number of both protons and neutrons of an atom.

Mechanoreceptors   Cells specialized for detecting pressure, position, touch, or pian.

Medulla   (1) That portion of the vertebrate brain located where the spinal cord enters the brain; controls basic bodily functions such as heart rate. (2) The interior part of an organ, such as a kidney or adrenal gland.

Medusa   Cnidarian body form which is umbrella-shaped and free-floating. *See* Polyp.

Meiosis   A sequence of divisions in which a diploid nucleus ($2n$) produces haploid nuclei (n); divisions that halve the chromosome number of a cell.

Membrane potential   The difference in electric charge across a membrane owing to concentration gradients of positively and/or negatively charged ions.

Memory cells   Derivatives of activated B cells which produce surface receptors that recognize specific antigens; immunity-conferring cells.

Menopause   The time in a woman's life when menstruation ceases, usually occurring between 45 and 52 years of age.

Menstruation   Periodic discharge of the uterine lining which occurs in primates. The start of the menstrual cycle.

Meristem   Region of undifferentiated tissue in a plant; gives rise to new tissue through frequent cell division throughout the life of the plant.

Mesoderm   Middle of the three germ layers of an animal embryo giving rise to the major body systems such as the cirulatory system, muscular system, etc.

Mesophyll   Leaf tissue located between the upper and lower epidermis and composed of parenchyma cells containing chloroplasts.

Messenger RNA (mRNA)   A macromolecule synthesized with cellular DNA acting as a template; contains the genetic instructions to guide the synthesis of a polypeptide.

Metabolism   All of the chemical reactions which occur in the living cells of an organism, including synthesis and degradation reactions.

Metamorphosis   The abrupt transformation between two stages in the life cycle of an organism which involves major structural reorganization.

Metaphase   A stage in mitosis or meiosis when the chromosomes are aligned on the equatorial plane of the cell.

Microfilaments   Thin contractile fibers within the cell composed of protein (actin); involved in cytoplasmic streaming.

Microminerals   Minerals required by animals in relatively small amounts: iron, copper, manganese, zinc, molybdenum, iodine, and chromium. *See* Macrominerals.

Micronutrients   Those elements required in relatively small amounts for the growth of plants. *See* Macronutrients.

Microtubules Small tubelike structures located in the cells of eukaryotic organisms and composed of protein filaments.

Microvilli Microscopic projections of the villi which serve to further increase surface area.

Middle lamella The intercellular wall layer; cements together the primary cell walls of adjoining cells.

Mineral An inorganic natural substance; in nutrition, any of several elemental ions required in the diet.

Mineralocorticoids A group of corticoid hormones produced by the adrenal cortex which regulate the levels of mineral ions in extracellular body fluids.

Mitochondria Organelles found in the cytoplasm of eukaryotic cells; site of cellular respiration.

Mitosis A nuclear division in which identical chromatids are segregated.

Molecule Two or more atoms existing in combined form.

Mollusks A phylum of invertebrates including snails, slugs, clams, mussels, scallops, oysters, squids, and octopuses.

Molting The periodic shedding by an organism of its outer covering, such as the exoskeleton of arthropods or the feathers of birds.

Monera Kingdom of single-celled or filamentous organisms whose cells lack a membrane-bound nucleus and organelles.

Monocots Flowering plants that have a single cotyledon (seed leaf). *See* Dicots.

Monosaccharide A simple sugar containing two to several carbon atoms, such as glucose and ribose.

Monotremes Mammals which lay eggs but produce milk to feed their young. The duck-billed platypus and spiny anteater are the only two surviving species.

Morphogenesis The processes resulting in the physiological and morphological differentiation of an organism.

Motor neuron Type of neuron which carries impulses from the central nervous system to a gland or muscle.

Motor unit The combination of all muscle fibers which contract in response to a single neuron's stimulus.

Mucus Polysaccharide mixture produced by glandular, mucosal cells of the digestive and respiratory tracts, which functions to protect and lubricate.

Muscle fiber Skeletal muscle cell.

Mutagen An agent, which may be chemical or some form of radiation, bringing about mutations in living organisms.

Mutant An organism which shows an abrupt unpredictable hereditary change in phenotype.

Mutation A change in the chemical composition of a gene that is reflected as a change in some inheritable characteristic.

Mutualism The situation in which two or more organisms live together, the association being of benefit to all involved.

Mycellium The mass of hyphae that compose a fungal body.

Mycorrhiza(e) A symbiotic association between fungal hyphae and plant roots. Such a relationship can enhance the ability of the plant to take up mineral nutrients from the soil.

Myelin sheath The sheath surrounding many axons, consisting of many layers of Schwann cell membranes.

Myofibrils Contractile subunits extending the length of the muscle fiber.

Myosin Contractile protein making up the thick filaments of myofibrils.

NAD Nicotinamide adenine dinucleotide, a hydrogen acceptor molecule involved in electron transfer reactions, such as those of respiration.

Natural selection A key idea in the theory of organic evolution—individuals in a population vary, and those having characteristics most compatible with their environment survive and multiply more effectively.

Nematocyst Specialized stinging or entangling cell of cnidarians used to obtain food or for protection.

Nematodes Members of Class Nematoda, Phylum Aschelminthes, including hookworms, pinworms, *Trichinella,* and many other parasites.

Nephron The functional unit of the kidney.

Nepotism Behavior which benefits close relatives.

Nerve A group of motor and/or sensory fibers embedded in connective tissue and bound together in a unit.

Nerve impulse A momentary change in the electrical potential of a nerve cell membrane which is conducted down the length of the neuron.

Net productivity The amount of organic material that accumulates in a plant or ecosystem. *See* Gross productivity.

Neuromuscular junction The junction of the terminal ends of a motor neuron with fibers of a skeletal muscle.

Neuron Nerve cell consisting of an elongated axon, cell body, dendrites, and axonal branches.

Neutron An uncharged particle found in the nucleus of an atom; its mass is about the same as that of a proton.

Niche *See* Ecological niche.

Nitrate reduction The conversion of nitrate (NO_3^-) to ammonia (NH_3).

Nitrification Conversion of ammonia to nitrate.

Nitrogen cycle The process by which nitrogen is recycled; involves the fixation of nitrogen, usually by biological processes, its incorporation into organic molecules, and its recycling of molecular nitrogen to the atmosphere.

Nitrogen fixation Conversion of molecular nitrogen into a form available to plants; the process is usually carried out by microorganisms.

Node Region where one or more leaves join a plant stem.

Nodules Enlarged areas on the roots of certain plants containing nitrogen-fixing microorganisms.

Nondisjunction Failure to disjoin; error in mitosis or meiosis in which a chromosome or daughter chromosome moves to the same pole with its homologue (during meiosis) or with its sister chromatid (during mitosis or the second meiotic division).

Nonpolar covalent bond A chemical bond in which there is a fairly equal sharing of electrons.

Nonvascular plants Plants which lack vascular tissue (xylem and phloem). Includes algae and bryophytes.

Notochord The supporting axis of the body in lower chordates and vertebrate embryos.

Nuclear envelope Double membrane surrounding the nucleus of a cell.

Nucleic acids Polymers of nucleotides, such as DNA and RNA.

Nucleotide A subunit of nucleic acids which is composed of a 5-carbon sugar (deoxyribose or ribose), a base, and a phosphate group.

Nucleus (1) In eukaryotic cells, a membrane-bound body that contains the hereditary units of the cell. (2) Central part of an atom.

Obligate anaerobes Microorganisms which are able to survive only in an environment devoid of oxygen.

Omnivores Animals which consume both plant and animal matter.

Oocyte Cell produced by mitosis from oogonia.

Oogenesis The formation of the female gamete or ovum.

Oogonia Diploid cells of the ovary which ultimately give rise to the female gamete or ovum.

Ootid A haploid cell formed as the result of meiotic divisions of an oocyte. Matures into an ovum.

Open circulatory system System in which the blood is allowed to circulate freely within the tissues rather than being confined to vessels. *See* Closed circulatory system.

Operon Group of genes including an operator gene and associated structural genes which are involved in the control of a specific biochemical pathway in prokaryotes.

Oral cavity The first portion of the digestive tract in animals which is involved in the initial processing of food materials.

Organ Tissues organized into a structure serving a particular function.

Organelle Membrane-bounded structural unit within the cytoplasm of a eukaryotic cell (e.g., chloroplast, mitochondrion, and endoplasmic reticulum).

Organic acid An organic compound containing a carboxyl ($-COOH$) group.

Osmoconformer Referring to an organism whose internal osmotic state conforms to that of the external environment.

Osmoregulation The adjustment of the balance of water and ions for the purpose of maintaining a constant internal environment despite fluctuations in the external environment.

Osmosis Movement of water through a membrane in response to a concentration gradient.

Osteichthyes Vertebrate class which includes ray-finned and lobe-finned bony fishes.

Oval window The membrane between the middle ear and inner ear, upon which rests the stirrup.

Ovary (1) In flowers, the enlarged base of the pistil that develops into the fruit; contains the ovules. (2) In animals, the reproductive organ of females.

Oviduct The tube through which eggs pass on their way to the uterus or outside of the body.

Ovulation In mammals, the release of an oocyte from the ovary into a fallopian tube.

Ovule Structure containing the female gametophyte (including egg) in seed plants. The fertilized ovule develops into a seed.

Ovum The female gamete or egg cell.

Oxidation An energy-yielding chemical reaction in which the compound being oxidized loses one or more electrons. Coupled with reduction reactions. *See* Reduction.

Oxytocin A hormone produced by nerve cells of the hypothalamus and stored in the posterior lobe of the pituitary; initiates milk flow and plays a role in labor.

Ozone layer Region of the upper atmosphere (about 10 km to 50 km above the earth's surface) where ozone (O_3) is concentrated.

Pacemaker Same as sinoatrial node.

Paleontology The field of science concerned with the study of life in past geological ages as revealed primarily by fossils.

Pancreas Secretory organ of vertebrates situated near the stomach that produces certain digestive enzymes and hormones.

Panspermia A theory of the origin of life which states that the first life on earth arrived by some means from elsewhere in the universe.

Parasite A heterotroph that derives its food from another living organism. *See* Saprophyte.

Parenchyma A type of cell that is living at maturity, typically thin-walled, unspecialized, and retains the capacity for renewed cell division. The cells have photosynthetic and storage functions. The term may also refer to the tissue composed of such cells.

Passive transport The movement of materials across a membrane without the expenditure of energy, in response to a concentration gradient.

Pathogen Any organism having the ability to cause disease in another organism.

Pedigree A recorded ancestry or family tree.

Pelvis (1) The central cavity of a kidney. (2) Bowl-shaped portion of the vertebrate skeleton between the spinal column and the leg bones.

Pepsin A type of protein-digesting enzyme (a protease).

Peptide Two or more amino acids linked together by peptide bonds. The size division distinguishing peptides from polypeptides is arbitrary.

Peptide bond A bond formed between the acid group (—COOH) of one amino acid and the basic amino group (—NH$_2$) of another amino acid.

Peptide hormones A class of water-soluble hormones consisting of from three to over 200 amino acids.

Peripheral nervous system (PNS) All neurons outside the brain and spinal cord.

Peristalsis Slow rhythmic contractions which move wavelike along walls of a tubular structure such as the digestive tract, or an oviduct.

Permeable Describing materials through which diffusion may occur; usually refers to membranes.

Petiole The stalklike portion of a leaf which supports the blade.

pH Symbol indicating the relative acidity or alkalinity of a solution; the negative logarithm of the hydrogen ion concentration. pH 7 indicates a neutral solution; higher pH values, a basic solution; lower pH values, an acidic solution.

Phage *See* Bacteriophage.

Phagocyte A type of white blood cell which engulfs bacteria, viruses, and debris by phagocytosis.

Phagocytosis A type of endocytosis involving the envelopment of undissolved particles.

Pharyngeal gill slits Embryonic structures in chordates which develop into gills in fishes or into other structures in higher vertebrates.

Pharynx Back portion of the oral cavity which leads into the esophagus and trachea.

Phenotype Appearance of a physiological condition of an organism that results from its inheritance and environment.

Pheromones Chemical messenger substances released by an individual member of a society which influence the behavior or morphological development of other society members.

Phloem A vascular tissue in plants that functions in the long-distance transport of sugar throughout the plant.

Phospholipids Important components of cell membranes. Composed of a glycerol unit attached to two fatty acids and a phosphate-organic group.

Photoautotrophs Organisms which produce their own food materials utilizing the energy of sunlight. Photosynthesizers.

Photomorphogenesis Plant development which is controlled by light.

Photoperiodism Time-measuring in organisms; the response of plants or animals to the day or night length.

Photoreceptors Light-sensitive cells; in humans, the rods and cones in the retina of the eye.

Photosynthesis Process by which light energy is converted to chemical energy; the light-requiring production of organic molecules from inorganic compounds.

Photosystem Photosynthetic units in the chloroplasts consisting of pigment molecules, an electron acceptor, and an electron donor. Involved in the light reactions of photosynthesis.

Phototropism The growth of a plant part toward or away from a source of light.

Phylogeny The evolutionary history of development of a group of organisms, especially in relation to that of other groups.

Phytochrome A plant pigment that exists in two interconvertible forms: P_R and P_{FR}. The latter form is thought to regulate many aspects of plant growth and development.

Phytoplankton *See* Plankton.

Pigment A colored organic compound that has the ability to absorb certain wavelengths of light (e.g., chlorophyll).

Pistil In flowers, the female reproductive organ composed of the stigma, style, and ovary.

Pit A thin place or cavity in the plant cell wall; no secondary cell wall forms at a pit.

Pith The tissue found in the center of a stem, interior to the vascular tissue; consists primarily of parenchyma cells.

Pituitary gland Small gland, located under the hypothalamus in the vertebrate brain, which produces several hormones, some acting on other endocrine glands and some acting on nonendocrine tissue; called the "master gland."

Placenta In mammals, a complex tissue where the circulatory systems of the mother and fetus come in close contact, allowing the exchange of nutrients and waste products.

Placentals Mammals which possess a placenta.

Placoderms The earliest known jawed fishes, now extinct.

Plankton Aquatic organisms, usually microscopic in size, floating freely or having a weak ability to swim. Phytoplankton (plants) and zooplankton (animals) form the base of numerous food chains.

Plasma The liquid portion of the blood.

Plasma cells Derivatives of activated B cells that synthesize and secrete circulating antibodies.

Plasma membrane Delimiting membrane of all cells; regulates passage of materials into and out of cells.

Plasmid Small extrachromosomal piece of DNA in *E. coli* or other bacteria.

Plasmodesmata Very small strands of cytoplasm that penetrate through pores of plant cell walls and connect the living protoplasts of adjacent cells.

Plasmolysis A condition in which the cytoplasm of a cell has shrunk away from the cell wall as a result of water loss.

Plastid A cell organelle bounded by a double membrane, located in the cytoplasm, and involved in food manufacture or storage.

Platelets Cell fragments in the blood functioning in the clotting process.

Pleiotropy Referring to the capacity of a gene to affect more than one characteristic exhibited by an organism.

Polar bodies Small, nonfunctional cells produced during the meiotic divisions of an oocyte.

Polar covalent bond A chemical bond in which there is an unequal sharing of electrons causing one end of the molecule to be slightly positively charged and one end to be slightly negatively charged.

Pollen grain Male gametophyte of seed plants.

Pollination Transfer of pollen from the anther to a receptive female flower part, usually the stigma.

Polygenic inheritance The situation where a particular inherited characteristic is controlled by more than one gene. The effects are often additive.

Polymer A large molecule made of similar subunits (e.g., a protein is a polymer of amino acid units).

Polyp One of two basic body forms in cnidarians, being erect and basally attached to the substrate. *See* Medusa.

Polypeptide A polymer, or long chain, of amino acids.

Polyploid Referring to a cell, tissue, or plant that has more than the diploid ($2n$) number of chromosomes.

Polysaccharide Polymer consisting of many sugar units.

Pons A mass of nerve fibers in the vertebrate brain connecting the medulla with higher brain centers.

Population A group of interbreeding organisms of the same species which inhabit a particular geographic area at a particular time.

Porifera The phylum which includes sponges.

Posterior Relating to the rear part of a body or organ. *See* Anterior.

Prebiotic Before life existed.

Predation The killing and consumption of living organisms by other organisms.

Prey A living organism which is killed and consumed by a predator organism.

Primary cell wall Wall layer composed largely of cellulose, deposited initially in growing plant cells. In some tissues it is displaced outward by the production of a chemically distinct secondary wall after the cell reaches its final size.

Primary consumer *See* Herbivore.

Primary growth Growth that originates from root or shoot apical meristems, as opposed to secondary growth, which originates from a cambium.

Primary oocyte The animal egg before the first meiotic division.

Primary producers Organisms at the base of a food chain or web which produce their own food from inorganic materials.

Primary productivity *See* Net productivity.

Primary spermatocyte The diploid cell that gives rise to four spermatids by meiosis.

Primary succession Initial vegetative changes which occur on uninhabited, barren areas. *See* Secondary succession.

Primary tissues Tissues originating from cell divisions within the apical meristems of the root and shoot. *See* Secondary tissues.

Procambium Meristematic tissue giving rise to the vascular cambium which is responsible for lateral growth.

Progesterone Sex hormone produced in the female by the corpus luteum which functions to prepare the uterus for the reception of a fertilized egg; present in small amounts in the male.

Prokaryotic Referring to organisms that lack membranes around various cell organelles (the bacteria and cyanobacteria).

Prolactin Hormone produced by the pituitary gland which stimulates the production of milk in the breasts.

Prophase First stage in nuclear division, in which the chromosomes thicken and shorten.

Prosimians Group of nonanthropoid primates, including lemurs, tarsiers, and others.

Protein An important functional and structural component of living cells, composed of polypeptide chains which in turn are made up of linear sequences of amino acids.

Prothallium The independent gametophyte of ferns and certain other lower vascular plants. It originates from the germination of a spore and carries the sex organs.

Protista Kingdom of single-celled eukaryotic organisms of diverse structures and life styles.

Proton Subatomic particle with a positive charge. Protons and neutrons form the nucleus of an atom.

Protoplasm The metabolically active contents of cells.

Pseudopods Temporary extensions of cytoplasm involved in locomotion and feeding in amoeboid organisms.

Puberty In humans, that time in development, usually between the ages of 11 and 13, when sexual development occurs and secondary sexual characteristics become evident.

Punctuated equilibrium The idea that periods of rapid evolution alternate with periods of slow change.

Pupa The developmental stage between the larva and adult in insects which undergo metamorphosis.

Pupil The clear area in the center of the iris at the front of an eye.

Purebreeding Organisms which, through inbreeding, produce progeny of the same phenotype.

Radial symmetry A body plan in which any plane drawn through the central longitudinal axis will divide essentially identical halves.

Ray-finned fishes A group of bony fishes with non-fleshy fins, and with gills and swim bladder but no lungs; includes most living fishes.

Recessive The trait or gene that is not expressed in a heterozygote, generally indicated by a lowercase letter. Compare with dominant.

Recombination In progeny, the appearance of new gene combinations differing from those of the parents as a result of meiosis and sexual reproduction.

Rectum Terminal muscular portion of the digestive tract.

Reduction A chemical reaction in which the compound being reduced gains one or more electrons. Coupled with oxidation reactions. *See* Oxidation.

Reductionism The idea that all of life's processes are based upon physical and chemical events.

Reflex arc A simple neural pathway involving at least one sensory neuron, interneuron, and motor neuron; responsible for rapid, unconscious action.

Releasing hormones A group of small peptide hormones produced by the hypothalamus which stimulate hormone release from the anterior lobe of the pituitary.

Replication Copying or reproducing; in biology, most commonly used with reference to molecules such as DNA.

Reptiles Members of the vertebrate class Reptilia, having dry scaly skin and internal fertilization, and laying amniotic eggs.

Reservoir Component of a biogeochemical cycle in which the elements are not readily available to organisms.

Respiration *See* Cellular respiration.

Resting potential The difference in electrical potential across the plasma membrane when a neuron is in an unexcited state; caused by an unequal distribution of ions across the plasma membrane.

Reticular formation A system of neural fibers in the vertebrate brain which regulates levels of awareness and sleep/awake cycles.

Retina The photosensitive portion of the vertebrate eye.

Rhizoid A filamentous or hairlike appendage that penetrates the substrate from the underside of some plants and fungi; functions in anchorage and absorption.

Rhizome An underground stem, usually horizontal, containing storage tissue.

Rhodopsin The visual pigment present in the rod cells of vertebrates.

Ribonucleic acid (RNA) A single-stranded nucleic acid similar in structure to DNA except the sugar ribose replaces deoxyribose and the base uracil replaces thymine.

Ribose A five-carbon sugar occurring in RNA ($C_5H_{10}O_5$).

Ribosomal RNA (rRNA) The RNA which occurs in ribosomes, transcribed from DNA.

Ribosomes Small cellular structures composed of RNA and protein. The site of protein synthesis. They may be free in the cytoplasm or bound to the endoplasmic reticulum.

RNA *See* Ribonucleic acid.

RNA polymerase The enzyme which binds to DNA and polymerizes nucleotides into RNA.

Rod cells Extremely sensitive light receptor cells of the vertebrate eye involved in the perception of dim light (i.e., night vision).

Root In vascular plants, that part which usually grows downward into the soil and serves in anchorage and absorption.

Root cap The dome-shaped protective covering of cells at the tips of many roots.

Root hairs Small extensions of the epidermal cells of a root.

Root nodules Small bulbous enlargements on the roots of leguminous plants containing the bacterium *Rhizobium*, which utilizes atmospheric nitrogen (N_2) to produce ammonia (NH_3).

Rough ER Endoplasmic reticulum which has ribosomes adhering to its surface.

Runner A horizontal creeping stem that initiates root and shoot systems at nodes, as in strawberry plants.

Salivary glands Glands located in the mouth region which secrete saliva.

Saprophyte A heterotrophic organism which feeds on dead material, obtaining organic compounds through absorption.

Sapwood The light-colored outer layer of secondary xylem that is active in water transport. Compare with heartwood.

Sarcolemma The plasma membrane of a muscle fiber.

Sarcomere Functional unit of muscle myofibrils composed of thick myosin filaments and thinner actin filaments.

Sarcoplasm The cytoplasm of a muscle fiber.

Sarcoplasmic reticulum (SR) System of sacs associated with the transverse tubules in a sarcomere, specialized for storing calcium ions; probably homologous with the endoplasmic reticulum.

Schistosomiasis Parasitic disease of the blood, liver, lungs, and other organs caused by the flatworm *Schistosoma*.

Schwann cells Accessory cells which wrap around an axon, forming a sheath.

Sclera The outermost protective layer of the wall of an eyeball.

Scrotum The sac containing the testes in mammals.

Secondary cell wall The final layer or layers of thickening of a plant cell wall; laid down by the protoplast upon the primary wall. *See* Primary cell wall.

Secondary consumer *See* Carnivore.

Secondary growth Growth that results from the action of a cambium.

Secondary oocyte In animals, a meiotic forerunner to the egg that derives from the primary oocyte.

Secondary phloem Phloem which arises from the vascular cambium.

Secondary sex characteristics Physical attributes not related to reproduction which are acquired at puberty, such as a lower voice in males.

Secondary spermatocyte In animals, a meiotic forerunner to a sperm that derives from the primary spermatocyte.

Secondary succession The series of vegetative changes which occur in areas previously inhabited by living organisms. *See* Primary succession.

Secondary tissue In plants, a tissue made up of cells that were produced by a cambium, not by the apical meristem. *See* Primary tissues.

Secondary xylem Xylem which arises from the vascular cambium.

Secretion *See* Exocytosis.

Seed In gymnosperms and angiosperms, the mature ovule consisting of an outer covering called the seed coat, a nutritive tissue made up of endosperm or cotyledons, and an embryo.

Seed coat The outer covering of a seed which is usually thin and papery.

Selection *See* Natural selection.

Selective breeding Breeding program in which individuals of a select phenotype are bred.

Self-pollination (selfing) Transfer of pollen from the anther to a stigma on the same plant.

Semicircular canals Inner ear structures involved in the detection of balance, equilibrium, and acceleration.

Seminiferous tubules The site of sperm production in the male testes.

Sensitivity Ability to respond to stimuli.

Sensory neuron Type of neuron which carries impulses to the central nervous system from various parts of the body.

Sepals In most flowers, the outermost appendages, usually green and leaflike in appearance.

Serum That portion of the blood plasma excluding the blood-clotting proteins.

Sex chromosomes The chromosomes which determine the sex of an individual, generally designated as X and Y chromosomes in mammals.

Sex hormone Any hormones, such as testosterone or estrogens, which influence the development or function of primary or secondary sex characteristics.

Sex-linked Referring to traits whose genes are located on the sex chromosomes. In humans, hemophilia and color-blindness are examples of sex-linked traits.

Sexual intercourse Coitus; insertion of the penis into the vagina.

Sexually transmitted disease (STD) Any infectious disease transmitted primarily through sexual contact.

Sexual reproduction In biology, a type of reproduction involving the union of gametes.

Shoot The stem and leaves of a higher plant.

Short-day plant A plant which initiates flowering only when the photoperiod includes dark periods longer than some critical length.

Sieve plate The perforated endwall between two adjacent cells in a sieve tube of phloem tissue.

Sieve tube Tube in phloem formed by many cells arranged end-to-end and separated by sieve plates.

Sieve tube members Individual sieve tube cells which collectively form the sieve tube.

Single circulation Circulation of blood which involves only one passage through the heart; characteristic of fish.

Sinoatrial node Small mass of special muscle cells located in the upper wall of the right atrium; initiates an electrical impulse causing the atria to contract. Also known as the pacemaker.

Small intestine In animals, the muscular tube which carries out most of the digestion and absorption of food materials.

Smooth ER Endoplasmic reticulum which lacks ribosomes.

Smooth muscle Type of muscle lining internal organs and under involuntary control; does not appear striated.

Social behavior Behavioral interactions among individuals of the same species.

Societies Groups of individuals of one species of animal living together with well-ordered behavioral patterns.

Solute Substance dissolved in a liquid medium.

Solution A liquid containing dissolved substances.

Solvent The liquid medium of a solution.

Somatic cells All those cells of an organism other than the reproductive cells.

Somatic nervous system That portion of the vertebrate nervous system which is normally under voluntary control.

Somatotropin *See* Growth hormone.

Special creation The idea that all life was created by a supernatural Creator.

Speciation The formation of new species from previously existing species.

Species The basic unit of classification; a kind of organism; a group of individuals that interbreed or are potentially capable of interbreeding freely in nature.

Sperm A male gamete or sex cell, usually motile and smaller than the stationary egg.

Spermatid One of four haploid products of the meiotic division of a spermatocyte. Differentiates to become a sperm.

Spermatocytes Enlarged cells formed mitotically from spermatogonia; undergo meiosis to form spermatids.

Spermatogenesis The formation of the male gametes or sperm.

Spermatogonia Diploid cells which ultimately give rise to the male gametes (sperm) by meiosis.

Spinal cord That portion of the vertebrate nervous system enclosed within the vertebral column. Consists of a large bundle of nerve fibers.

Spindle A football-shaped set of microtubules that functions in chromosome movement in a cell during mitosis and meiosis.

Sporangium A special structure or case within which spores are produced.

Spore A reproductive cell.

Sporophyte The diploid or $2n$ phase of a plant's sexual life cycle.

Stamen In flowers, the male reproductive organ that produces pollen. It usually consists of a long, slender filament with an anther at its tip.

Starch A food-storage carbohydrate in plants; a polymer of glucose.

Stem Most often the part of a plant that grows upward and bears leaves.

Steroid hormones Fat-soluble hormones derived from cholesterol.

Steroids Lipids which are composed of several carbon rings (e.g., cholesterol).

Stigma In flowers, the uppermost tip of the pistil where pollen is deposited during pollination.

Stimulus Part of an organism's environment which causes it to respond in some manner (e.g., light, sound, temperature change).

Stomach Digestive tract organ which functions to churn and enzymatically digest food materials.

Stomate The pore in plant epidermis, mostly in leaves, through which gases pass; controlled by guard cells.

Stroma The ground substance that surrounds the chlorophyll-containing grana within the chloroplast.

Style In a pistil, the often thin and elongate appendage between the ovary and stigma.

Substrate (1) Substance upon which an enzyme acts. (2) Solid foundation which something lies on or is attached to.

Succession See Ecological succession.

Sucrose Common table sugar; a disaccharide with the formula $C_{12}H_{22}O_{11}$.

Superior vena cava Major vein carrying blood from the upper portion of a mammal's body to the right atrium of the heart. See Inferior vena cava.

Swim bladder Organ present in many fish species which functions to maintain the fish at particular levels in the water as a result of the gas volume in the bladder.

Symbiosis The close association of two unrelated organisms with little or no ill effects to either.

Sympatric speciation The formation of a new species occurring in the same geographic area as its ancestor, i.e., within normal breeding range.

Synapse The junction between a neuron and an adjacent cell across which an impulse is transmitted.

Synapsis The process in which homologous chromosomes line up in close association with one another, making crossing over possible.

Synaptic vesicles Small membrane-bound sacs from which neural transmitters are released.

Systematics See Taxonomy.

Systole The ventricular contraction phase of the heartbeat cycle.

Systolic pressure The maximum pressure created in the arteries as a result of the ventricular contraction. See Diastolic pressure.

T cell Type of lymphocyte reaching maturity in the thymus gland; regulates antibody production and defends against foreign cells.

Telophase The final stage of mitosis or meiosis, during which the anaphase movement ceases, the chromosomes become diffuse, and nuclear envelopes form around the daughter nuclei.

Territoriality An animal's behavior based upon defining and defending a specific local area for its own use.

Testcross Crossing an individual with the dominant phenotype to one showing the recessive phenotype in order to determine by the progeny whether the dominant individual was homozygous or heterozygous.

Testes (testicles) The male reproductive organs which are responsible for sperm production and are the source of testosterone.

Testosterone Male sex hormone produced by the testes and involved in the development and maintenance of secondary sex characteristics as well as sperm production; present in small amounts in the female.

Tetanus The state of sustained muscle contraction resulting from continual stimulation.

Tetraploid Having four times ($4n$) the haploid number of chromosomes.

Thalamus The relay center of the vertebrate brain, located in the forebrain region.

Thallus A plant body that lacks vascular tissue, such as in thallophytes.

Thylakoids Internal chloroplast membranes, sac-like in appearance, which comprise the grana.

Thyroid gland Endocrine gland located near the larynx in the neck which produces the hormone thyroxine.

Thyrotropin A polypeptide tropic hormone produced by the anterior pituitary which stimulates thyroxine production in the thyroid gland.

Thyrotropin releasing hormone (TRH) Hormone produced by the hypothalamus which stimulates the anterior pituitary to release thyrotropin.

Thyroxine Hormone produced by the thyroid gland which serves primarily to stimulate metabolism.

Tissue Group of cells which together comprise a structural and functional unit.

Tongue A muscular organ in the oral cavity which helps to position and move food. Also functions in speech in humans.

Toxin A poisonous metabolic product of a living organism.

Trachea Cartilaginous tubelike structure lined with ciliated cells which connects the larynx with the bronchi.

Tracheid In plants, an elongate, tapered xylem cell with pitted walls; functions in support and water transport.

Transcription The transfer of information from DNA to RNA; the synthesis of RNA from a DNA template.

Transfer RNA (tRNA) Small RNA molecules which bind to specific amino acids and transport them to the site of protein synthesis. The anticodon of the tRNA is complementary to the codon of the mRNA.

Translation The transfer of information from mRNA to polypeptide; protein synthesis.

Transpiration The loss of water from plant stems and leaves, chiefly through the stomates.

Transverse tubules Membranous tubules that intertwine among the myofibrils of skeletal muscle and carry the stimulatory impulse from the sarcolemma to the sarcoplasmic reticulum.

Trial-and-error learning Learning which is directed by the positive rewards and/or punishment inherent in the organism's environment.

Triglyceride A lipid molecule composed of glycerol and three fatty acid chains.

Triploid Having the 3*n* number of chromosomes. *See also* Tetraploid.

Trophic Level Position within a food chain or web.

Trophoblast The outer cell layer of the blastocyst which develops into a portion of the placenta.

Tropic hormones Hormones produced by the pituitary gland which have their stimulating influence on other endocrine glands.

Tropism The bending movement of a plant part in response to a stimulus such as light (*see* phototropism) or gravity (*see* gravitropism).

Trypsin Protein-digesting enzyme (protease) secreted by the pancreas.

TSH Thyroid-stimulating hormone. Produced by the pituitary gland and acts on the thyroid gland to increase thyroxine release.

Tubal ligation Sterilization method involving the surgical cutting of the oviducts in the female.

Tundra A low, treeless vegetation type growing over permanently frozen subsoil of high elevations and the far north.

Tunicates Members of the chordate subphylum Urochordata. Small sedentary filter-feeding animals whose chordate features appear mostly in their larvae.

Turgor pressure The pressure within a plant cell resulting from water uptake.

Tympanic membrane The eardrum; membrane separating the outer ear from the middle ear which vibrates in response to sound waves.

Umbilical cord Cord containing a vein and two arteries which connects the fetus with the placenta.

Undernutrition The condition of starvation when individuals take in fewer calories than needed.

Unicellular Composed of a single cell.

Urea Soluble nitrogenous waste product excreted by mammals and adult amphibians.

Ureter Tubelike structure that directs urine from the kidney to the urinary bladder.

Urethra The tube connecting the urinary bladder to the exterior; transports urine and, in males, semen.

Uric acid Semisolid nitrogenous waste product excreted by birds, reptiles, insects, and land snails.

Urinary bladder Storage organ for urine.

Urine Watery solution produced in the kidneys; contains urea.

Uterus Muscular organ in mammals in which the embryo develops.

Vaccination The procedure in which an organism is inoculated with a disease agent which has been altered so that it no longer can cause the disease. As a result of the vaccination, however, antibodies are produced, building up the body's defenses against the disease.

Vaccine A suspension of bacteria or viruses designed to be injected into an animal and thereby initiate immune responses; the microbes may be killed pathogenic forms, or living nonpathogenic forms which carry an antigen similar to the pathogen.

Vacuole A large organelle (in mature plant cells) containing a watery solution of stored materials.

Vagina Female organ connecting the uterus to the exterior of the body.

Vascular cambium In plants, a meristematic tissue that produces both secondary xylem and secondary phloem.

Vascular plants Plants which produce vascular tissue, including all seed plants and ferns.

Vascular tissue The xylem and phloem which transports water and mineral nutrients.

Vasectomy Sterilization method in which the vas deferens is surgically cut and tied off in the male.

Vasopressin Hormone released by the hypothalamus which increases water permeability of the distal convoluted tubules and collecting ducts in the kidney. Also called antidiuretic hormone.

Vein Vessel of the circulatory system which carries blood to the heart; contains valves which prevent the blood from flowing backwards.

Ventral Relating to the front or underside of an organism or body part. *See* Dorsal.

Ventricle Chamber of the heart which receives blood from an atrium and pumps it into the arteries.

Venule Blood vessel which delivers blood from the capillaries to a vein.

Vertebral column Series of connected bones protecting the spinal cord and providing support for the body; the backbone.

Vertebrates Animals which possess a backbone.

Vesicle A small, membrane-bound sac within cells.

Vessel element A large water-conducting cell type found in the xylem of plants.

Villi In animals, the small, fingerlike projections of the inner lining of the small intestine.

Virus A nucleic acid-protein particle that can reproduce only within living cells.

Vitalists Adherents to the thesis that a nonphysical living force exists in all living things, one which cannot be explained in purely physical and/or chemical terms.

Vitamins Organic molecules, divergent in structure, that are essential in relatively small concentrations for normal growth and development.

Water potential A measure of the tendency of water to move; affected by pressure and the presence of solutes.

Wavelength The distance between peaks of wavelike motion.

White matter The exterior regions of the brain and spinal cord, composed of myelinated axons.

Xylem A complex plant tissue functioning primarily in water transport; consists of living storage cells, fibers, tracheids, and vessels.

Yolk Stored nutrient material in the egg which nourishes the developing embryo.

Z line The transverse line that delineates one sarcomere from another within a myofibril of skeletal muscle.

Zooplankton *See* Plankton.

Zoospore A motile spore characteristic of some algae, fungi, and protozoans. Motility is the result of flagellum action or amoeboid movement.

Z-scheme A representation of the interrelationships of the energy-capturing light reactions of photosynthesis.

Zygospore The only diploid cell in the life cycle of *Rhizopus* and related fungi; formed by the fusion of nuclei from hyphae of opposite mating types; gives rise to haploid spores by meiosis.

Zygote A diploid cell resulting from the fusion of two haploid gametes.

Acknowledgments

Introduction I–1: Redrawn from "The Search for Extraterrestrial Intelligence," by Carl Sagan and Frank Drake. Copyright © 1975 by Scientific American, Inc. All rights reserved. I–2: Jet Propulsion Laboratory, California Institute of Technology, NASA. I–3 and chapter opener: NASA. I–5: James Mason/Black Star. I–6(a): J. R. Waaland, University of Washington/BPS. I–6(b): L. E. Gilbert, University of Texas at Austin/BPS. I–6(c) and (d): © Lennart Nilsson/*A Child is Born*, Delacorte Press, NY. I–6(e): Frank S. Balthis/Jeroboam. I–6(f): ANIMALS ANIMALS/Stouffer Productions, Ltd. I–6(g): John Gerlach/TOM STACK & ASSOCIATES. I–8(a): P. R. Ehrlich, Stanford University/BPS. I–8(b): R. Humbert, Stanford University/BPS. I–9: Courtesy of Nelson Max, Computer Graphics Group, University of California, Lawrence Livermore Laboratory. I–11: Philip Jon Bailey/Jeroboam.

Chapter 1 1–1 (left): Jerome Wyckoff/Earth Science Photographs. 1–1 (right): Courtesy of Michael Donoghue, San Diego State University. 1–2(a): © Bradley Smith/Gemini Smith, Inc. 1–2(b): George Shoemaker/Meteor Crater Enterprises, Inc. 1–10 (photo): Courtesy of Verne N. Rockcastle, Cornell University. Essay 1–1 (photo): Courtesy of Dale E. Ingmanson, San Diego State University.

Chapter 2 2–4(a) (photo): Courtesy of James Neel and Michael Costello, San Diego State University. 2–4(b) (photo): Courtesy of G. E. Palade, Yale University Medical School. 2–5 (photo): Courtesy of Eva Frei and R. D. Preston, Leeds, England. 2–10: © Martin M. Rotker/PHOTOTAKE. 2–13(b): Courtesy of Richard Weiss, San Diego State University. 2–13(c) and chapter opening art: From *Hemoglobin* by R. E. Dickerson and I. Geis, 1983, Benjamin/Cummings, Menlo Park, CA. Copyright © 1983 Irving Geis. 2–14(a): Hale L. Wedberg, San Diego State University. 2–14(b): L. J. Le Beau, University of Illinois Hospital at the Medical Center, Chicago/BPS.

Chapter 3 3–1(a) (left): P. R. Ehrlich, Stanford University/BPS. 3–1(a) (right): R. Rodewald, University of Virginia/BPS. 3–1(b) (left): M. W. Steer, University of Wisconsin, Madison/BPS. 3–1(b) (right): David L. Rayle, San Diego State University. 3–1(c): H. S. Wessenberg and G. A. Antipa. *J. Protozool.* 17:250–270 (1970). 3–1(d): Courtesy of L. E. Simon, Rutgers University, New Brunswick, NJ. 3–2(a): Hale L. Wedberg, San Diego State University. 3–2(b): P. W. John-

son and J. McN. Sieburth, University of Rhode Island/BPS. 3–2(c): Photo courtesy of Unitron Instruments, Inc., Plainview, NY. 3–3: Courtesy of Carl Zeiss, Inc. 3–4(a): M. Murayama, Murayama Research Laboratory/BPS. 3–4(b) and (c): W. Rosenberg, Iona College/BPS. 3–6(b): J. R. Waaland, University of Washington/BPS. 3–7(b): J. David Robertson, Duke University Medical Center. 3–8(b) and chapter opener: J. J. Cardamone, Jr., University of Pittsburgh/BPS. 3–9: Courtesy of G. Cohen-Bazire, Pasteur Institute. 3–12(b): R. Rodewald, University of Virginia/BPS. 3–12(c): Courtesy of L. Andrew Staehelin, University of Colorado, Boulder. 3–13(c): Courtesy of Keith Porter, University of Colorado, Boulder. 3–14: From *Introduction to the Fine Structure of Plant Cells* by M. C. Ledbetter and K. Porter, 1970, Springer-Verlag, New York, Plate 8.1, p. 132. 3–15(b) and (c): Courtesy of G. E. Palade, Yale University Medical School. 3–17: From "The Ground Substance of the Living Cell," by Keith R. Porter and Jonathan B. Tucker. Copyright © 1981 by Scientific American, Inc. All rights reserved. 3–18: Courtesy of R. O. Hynes, MIT. 3–19(a): L. E. Roth, University of Tennessee; Y. Shigenaka, University of Hiroshima; D. J. Pihlaja, Howe Laboratory of Ophthalmology, Boston/BPS. 3–19(b): P. R. Burton, University of Kansas/BPS. 3–20 (top photo): W. L. Dentler, University of Kansas/BPS. 3–20 (bottom photo): Courtesy of E. de Harven, Rockefeller University. 3–21(b): From *Three-Dimensional Structure of Wood: A Scanning Electron Microscope Study* by B. A. Meylan and G. B. Butterfield, 1972, Chapman and Hall Ltd., Fig. 38, p. 50.

Chapter 4 4–11(b): C. L. Sanders, Batelle-Pacific Northwest Laboratories/BPS. Essay 4–1 (left and middle photos): David L. Rayle, San Diego State University. Essay 4–1 (right photo): Courtesy of Randall W. Davis, Scripps Institution of Oceanography, University of California, San Diego.

Chapter 5 Essay 5–1 (art, p. 84): Redrawn from "The Voodoo Lily," by Bastiaan J. D. Meeuse. Copyright © 1966 by Scientific American, Inc. All rights reserved.

Chapter 6 6–6(a): From *Introduction to the Fine Structure of Plant Cells* by M. C. Ledbetter and K. Porter, 1970, Springer-Verlag, New York, Plate 8.1, p. 132.

Chapter 7 7–2: Courtesy of B. A. Hamkalo and J. B. Rattner, University of California, Irvine. 7–3: Hale L. Wedberg, San Diego State University. 7–4: G. F. Bahr, M.D., Armed Forces Institute of Pathology. 7–5: Courtesy of William T. Jackson, Dartmouth College. 7–6: Courtesy of R. G. Kessel. Appeared in H. W. Beams and R. G. Kessel, *American Scientist* 64:279 (1976). 7–7: B. A. Palevitz and E. H. Newcomb, University of Wisconsin, Madison/BPS. 7–9(a) and (b): Courtesy of James V. Alexander, San Diego State University. 7–9(c): Hale L. Wedberg, San Diego State University. 7–9(d): J. N. A. Lott, McMaster University/BPS. 7–9(e): Courtesy of Boyce Thompson Institute for Plant Research, Ithaca, New York. 7–9(f): Carolina Biological Supply Company. Essay 7–1 (photo): Courtesy of Richard Weiss, San Diego State University.

Chapter 8 8–2: Courtesy of S. Pathak, M.D. Anderson Hospital, Houston, Texas. 8–9: Courtesy of Patricia Oelwein, Model Preschool Center for Handicapped Children, University of Washington.

Chapter 9 9–1: The Bettmann Archive, Inc. 9–12: From A. F. Blakeslee, *J. of Heredity* 5:511 (1914).

Chapter 10 10–3 (left): The Bettmann Archive, Inc. Essay 10–2 (photos): The Bettmann Archive, Inc.

Chapter 11 11–2(b): Courtesy of L. E. Simon, Rutgers University, New Brunswick, NJ. 11–9(a): Courtesy of Blue Sky Records. 11–10: J. Griffith, University of North Carolina Medical School, Chapel Hill. 11–12: Ralph Crane/Time-Life, Inc.

Chapter 12 12–10 (top photo): M. Murayama, Murayama Research Laboratory/BPS. 12–10 (bottom photo): Courtesy of V. Ingram, MIT.

Chapter 13 13–1: Hale L. Wedberg, San Diego State University. 13–2 (photos): R. K. Burnard, Ohio State University/BPS. 13–5 (plant): Courtesy of Anne K. Behnke. 13–5 (fungi): Courtesy of Daniel Stuntz, University of Washington. 13–5 (animal): Courtesy of A. G. Behnke. 13–5 (protist): Courtesy of Gregory A. Antipa, San Francisco State University. Appeared in "The Cytological Sequence of Events in the Binary Fission of *Didinium nasutum* (O.F.M.)," by E. B. Small, G. A. Antipa, and D. S. Marszalek. *Acta Protozool.* 9:275–282 (1972). 13–5 (monera): Courtesy of L. E. Simon, Rutgers University, New Brunswick, NJ. 13–6(a) and (b): Z. Skobe, Forsyth Dental Center/BPS. 13–6(c): J. B. Baseman, University of North Carolina School of Medicine. Appeared in *Infect. Immun.* 17:174–186 (1977). 13–7: Courtesy of Charles C. Brinton, Jr. and Judith Carnahan, University of Pittsburgh. 13–8(a): Courtesy of J. Waterbury, Woods Hole Oceanographic Institution. 13–11: Photo by E. B. Small and D. S. Marszalek. Courtesy of Gregory A. Antipa, San Francisco State University. 13–12: Omikron/

Photo Researchers, Inc. 13–13(a): Courtesy of Gregory A. Antipa, San Francisco State University. 13–14(a): Courtesy of Richard Crawford, University of Bristol. 13–14(b) and (c): Courtesy of James Alexander, San Diego State University. 13–15(a): From J. A. Shemanchuk and H. C. Whisler in *Canada Agriculture,* Winter, 1974. 13–15(b): Courtesy of Michael Donoghue, San Diego State University. 13–15(c): Stephen J. Krasemann/Peter Arnold, Inc. 13–15(d): Courtesy of Daniel Stuntz, University of Washington.

Chapter 14 14–3(a) and (c): J. R. Waaland, University of Washington/BPS. 14–3(b): Carolina Biological Supply Company. 14–3(d): J. N. A. Lott, McMaster University/BPS. 14–4(a): J. R. Waaland, University of Washington/BPS. 14–4(b) and (d): Courtesy of J. Pickett-Heaps, University of Colorado, as published in *The Plant Kingdom: Evolution and Forms* by Samuel E. Rushforth, 1976, Prentice-Hall, Inc., Englewood Cliffs, NJ, Figs. 6–16 and 6–17, p. 83. 14–6(a) and (c): J. R. Waaland, University of Washington/BPS. 14–6(b): J. Merrill, University of Washington/BPS. 14–6(d): W. May/BPS. 14–7(a): Townsend P. Dickinson/Photo Researchers, Inc. 14–7(b): Gregory K. Scott/Photo Researchers, Inc. 14–9(a) and (c): David L. Rayle, San Diego State University. 14–9(b): J. N. A. Lott, McMaster University/BPS. 14–11(a): J. N. A. Lott, McMaster University/BPS. 14–11(b): David L. Rayle, San Diego State University. 14–11(c): R. K. Burnard, Ohio State University/BPS. 14–11(d) L. Egede-Nissen/BPS. 14–13(a): Hale L. Wedberg, San Diego State University. 14–13(b): J. R. Waaland, University of Washington/BPS. 14–16: Courtesy of Michael Donoghue, San Diego State University. 14–18(a): Rod Planck/TOM STACK & ASSOCIATES. 14–18(b): David M. Dennis, Ohio State University. 14–18(c) and (d): J. N. A. Lott, McMaster University/BPS. 14–19(a): C. W. May/BPS. 14–19(b): Michael Hopiak, Laboratory of Ornithology, Cornell University. 14–19(c): Courtesy of Warren P. Stoutamire, University of Akron, Ohio.

Chapter 15 The "green alien" allegory at the beginning of Ch. 15 is adapted from Ch. 1 of *Plant Structure and Function* by Victor A. Greulach, 1973, Macmillan, NY. The chapter in *Plant Structure and Function* was modified from an article by Victor A. Greulach under the pen name of V. A. Eulach published in the October 1956 issue of *Astounding Science Fiction* (now *ANALOG Science Fiction—Science Fact*), copyright 1956 by Smith & Street Publications, Inc., and reprinted in the December 1956 issue of *Science Digest;* current copyright holder is Conde Nast Publications, Inc. 15–2: David L. Rayle and Hale L. Wedberg, San Diego State University. 15–3(b): From *Probing Plant Structure* by J. H. Troughton and L. A. Donaldson, 1972, McGraw-Hill, NY, and A. H. & A. W. Reed, Ltd., Wellington, New Zealand, Plate 29. Courtesy of the Department of Scientific and Industrial Research, Lower Hutt, New Zealand. 15–6(b): J. R.

Waaland, University of Washington/BPS. 15–7: David L. Rayle, San Diego State University. 15–8(b): From *Probing Plant Structure* by J. H. Troughton and L. A. Donaldson, 1972, McGraw-Hill, NY, and A. H. & A. W. Reed, Ltd., Wellington, New Zealand, Plate 60. Courtesy of the Department of Scientific and Industrial Research, Lower Hutt, New Zealand. 15–9(b): Courtesy of R. Anderson and J. Cronshaw, *Planta* 91:173–180 (1970), Springer-Verlag, Berlin. 15–10 (photos): J. R. Waaland, University of Washington/BPS. 15–14(b): From *Three-Dimensional Structure of Wood: A Scanning Electron Microscope Study* by B. A. Meylan and G. B. Butterfield, 1972, Chapman and Hall, Ltd., Fig. 38, p. 50. Essay 15–1 (photos): Hale L. Wedberg, San Diego State University.

Chapter 16   16–2: W. P. Wergin and E. H. Newcomb, University of Wisconsin, Madison/BPS.

Chapter 17   17–2: David L. Rayle, San Diego State University. 17–3: Courtesy of James N. Mills, San Diego State University. 17–7: David L. Rayle, San Diego State University. 17–8: Courtesy of B. O. Phinney and C. Spray, University of California, Los Angeles. With permission from "Chemical Genetics and the Gibberellin Pathway in *Zea mays* L*," by B. O. Phinney and C. Spray. In *Plant Growth Substances,* P. F. Wareing, ed., 1982, Academic Press, London, p. 104. Copyright: Academic Press Inc. (London) Ltd. 17–10: David L. Rayle, San Diego State University. 17–13: Courtesy of Robert L. Hays, Fish and Wildlife Service, Department of the Interior, Fort Collins, Colorado.

Chapter 18   18–2(a): H. W. Pratt/BPS. 18–3(c) and (d): S. K. Webster, Monterey Bay Aquarium/BPS. 18–8(a): C. R. Wyttenbach, University of Kansas/BPS. 18–9(a): S. K. Webster, Monterey Bay Aquarium/BPS. 18–9(b): Courtesy of Barbara A. Broughton. 18–10(a): H. W. Pratt/BPS. 18–11: J. W. Porter, University of Georgia/BPS. 18–12(b): R. K. Burnard, Ohio State University/BPS. 18–13(a): John R. MacGregor/Peter Arnold, Inc. 18–13(b): R. K. Burnard, Ohio State University/BPS. 18–13(c): Courtesy of Dr. Lo-Chai Chen, San Diego State University. 18–13(d) and (e): P. J. Bryant, University of California, Irvine/BPS. 18–14: David Scharf/Peter Arnold, Inc. 18–15 (photo): R. K. Burnard, Ohio State University/BPS. 18–16: Rachel Lamoreux. 18–17(a): C. W. May/BPS. 18–17(b): S. K. Webster, Monterey Bay Aquarium/BPS.

Chapter 19   19–2: C. R. Wyttenbach, University of Kansas/BPS. 19–3(c): S. K. Webster, Monterey Bay Aquarium/BPS. 19–7: Hans Reinhard/Bruce Coleman, Inc. 19–8: A. Kerstitch/TOM STACK & ASOCIATES. 19–10(a): Courtesy of American Museum of Natural History. 19–10(b) and (c): E. D. Brodie, Jr., Adelphi University/BPS. 19–12(a) and (b): J. N. A. Lott, McMaster University/BPS. 19–12(c): Rod Planck/TOM STACK & ASSOCIATES. 19–12(d): R. K. Bur-

nard, Ohio State University/BPS. 19–14: Smithsonian Institution Photo No. 77-8249. 19–15(b): E. D. Brodie, Jr., Adelphi University/BPS. 19–16: Courtesy of American Museum of Natural History. 19–19: Michael Hopiak, Laboratory of Ornithology, Cornell University. 19–21(a): R. K. Burnard, Ohio State University/BPS. 19–21(b): ANIMALS ANIMALS/Stouffer Productions, Ltd. 19–21(c): J. N. A. Lott, McMaster University/BPS. 19–21(d): ANIMALS ANIMALS/Stephen Dalton. 19–22(a): E. D. Brodie, Jr., Adelphi University/BPS. 19–22(b): ANIMALS ANIMALS/Karl Weidmann. 19–22(c): ANIMALS ANIMALS/Irene Vandermolen. 19–22(d): ANIMALS ANIMALS/Miriam Austerman. 19–24(a): Donald C. Johanson, The Institute of Human Origins. 19–25: Courtesy of American Museum of Natural History. 19–26: The Bettmann Archive, Inc.

Chapter 20   Chapter opener: From *Tissues and Organs: A Text-Atlas of Scanning Electron Microscopy* by Richard G. Kessel and Randy H. Kardon, W. H. Freeman & Co., © 1979. 20–8: William Thompson, © 1982. 20–14: Manfred Kage/Peter Arnold, Inc. 20–16: J. G. Hadley, Battelle-Pacific Northwest Laboratories/BPS. Essay 20–1 (photo): Courtesy of Medical Illustration Service, University of Utah College of Medicine.

Chapter 21   Chapter opener: From *Tissues and Organs: A Text-Atlas of Scanning Electron Microscopy* by Richard G. Kessel and Randy H. Kardon. W. H. Freeman & Co., © 1979. 21–1(a): Courtesy of Sara A. Fultz, Stanford University. 21–1(b): From "Enteric Nematodes of Lower Animals: Zoonoses," by Carolyn C. Huntley. In *Medical Microbiology* by Samuel Baron, ed., 1982, Addison-Wesley Publishing Co., Medical Division, Menlo Park, CA, p. 840. 21–1(c): T. W. Ransom/BPS. 21–2: UNICEF photo by FAO. 21–3: UNICEF photo. 21–4(b) and (c): Kim Taylor/Bruce Coleman, Inc.

Chapter 22   22–1: Redrawn from *Life: The Science of Biology* by W. K. Purves and G. H. Orians, 1983, Sinauer Associates, Sunderland, MA, and Willard Grant Press, Boston, MA, p. 575. 22–5: K. E. Muse, Duke University Medical Center/BPS. Essay 22–1 (photo): © Galen Rowell 1983/High & Wild Photography.

Chapter 23   23–5: Rick McIntyre/TOM STACK & ASSOCIATES. 23–9(a): Cristopher Crowley/TOM STACK & ASSOCIATES. 23–9(b): J. N. A. Lott, McMaster University/BPS. 23–10(a): R. Humbert, Stanford University/BPS. 23–10(b): Courtesy of Willow Owens. 23–11: ANIMALS ANIMALS/Breck P. Kent. Essay 23–1 (photo): William Thompson.

Chapter 24   24–6: The Bettmann Archive, Inc. 24–7: Courtesy of William H. Daughaday, Washington University

School of Medicine. From "Clinical Pathological Conference," A. I. Mendeloff and D. E. Smith, eds., *American Journal of Medicine* 20:133 (1956).

Chapter 25   25–6(a): R. Yanagimachi, School of Medicine, University of Hawaii at Manoa/BPS. 25–7(a): Mia Tegner, Scripps Institution of Oceanography. 25–7(b): Gerald P. Schatten and Daniel Mazia, University of California, Berkeley. Both (a) and (b) appeared in "The Program of Fertilization," by David Epel. Copyright © 1977 by Scientific American, Inc. All rights reserved. 25–11(d): Courtesy of K. T. Tosney. Appeared in *Tissue Interactions and Development* by N. K. Wessells, 1977, Benjamin/Cummings, Menlo Park, CA. 25–12, 25–14, and chapter opener: Courtesy of Dr. Roberts Rugh and Landrum B. Shettles. 25–18: William Thompson. (b) and (d) Contraceptives courtesy of Planned Parenthood Association of San Mateo County. (c) Cervical caps courtesy of Women's Health Center, San Francisco General Hospital.

Chapter 26   26–2(b) and chapter opener: Ed Reschke. 26–2(c): E. R. Lewis, Y. Y. Zeevi, and T. E. Everhart, University of California, Berkeley/BPS. 26–3(b): Courtesy of Thomas L. Lentz, Yale University School of Medicine. 26–11(a): P. J. Bryant, University of California, Irvine/BPS. 26–11(b): David Scharf/Peter Arnold, Inc. 26–11(c): Howard Hall/TOM STACK & ASSOCIATES. 26–19: Manfred Kage/Peter Arnold, Inc. 26–20 (photo): Courtesy of Clara Franzini-Armstrong, University of Pennsylvania. Essay 26–1 (photo): Brian Parker/TOM STACK & ASSOCIATES.

Chapter 27   27–12(a): William Thompson, © 1982. 27–13: © Dr. J. A. Hobson and Hoffman-LaRoche Inc., from *Dreamstage* catalog. 27–14: J. N. A. Lott, McMaster University/BPS. Essay 27–1 (photo): Dan McCoy/Black Star.

Chapter 28   28–4: Timothy O'Keefe/TOM STACK & ASSOCIATES. 28–6: The Bettmann Archive, Inc. 28–7: © Zoological Society of San Diego. 28–9 and chapter opener: The Granger Collection, New York. 28–10: The Bettmann Archive, Inc. 28–11: Redrawn from *Botany: A Brief Introduction to Plant Biology* by Rost, et al., copyright © John Wiley & Sons, Inc. Reprinted by permission of John Wiley & Sons, Inc. Essay 28–1 (photos): Courtesy of H. B. D. Kettlewell.

Chapter 29   29–7(a): ANIMALS ANIMALS/Terence A. Gili. 29–7(b): ANIMALS ANIMALS/Frank Roche. 29–7(c): ANIMALS ANIMALS/Alan G. Nelson. 29–10: Galapagos finches courtesy of American Museum of Natural History.

Chapter 30   30–5: Nina Leen/Time-Life, Inc. 30–9: ANIMALS ANIMALS/E. R. Degginger. 30–13(a) and (c): John Gerlach/TOM STACK & ASSOCIATES. 30–13(b): Milton Rand/TOM STACK & ASSOCIATES. 30–14: B. Wilcox, Stanford University/BPS. 30–15: Mark Newman/TOM STACK & ASSOCIATES. 30–16: John Gerlach/TOM STACK & ASSOCIATES. 30–17(a): ANIMALS ANIMALS/Mark. A. Chappell. 30–17(b) and (c) and chapter opener: ANIMALS ANIMALS/Carson Baldwin, Jr. 30–17(d): ANIMALS ANIMALS/Oxford Scientific Films. Essay 30–1 (photos): T. W. Ransom/BPS.

Chapter 31   31–5 (photo): Gary Milburn/TOM STACK & ASSOCIATES. 31–6: Courtesy of the Commonwealth Prickly Pear Board, Australia. 31–9: Norman Owen Tomalin/Bruce Coleman, Inc. 31–11(a): J. N. A. Lott, McMaster University/BPS. 31–12 (photo) and chapter opener: Owen Franken/Stock, Boston. Essay 31–1 (photo): Caron Pepper/TOM STACK & ASSOCIATES.

Chapter 32   32–1(a): Courtesy of Fannie Toldi. 32–1(b): EARTH SCENES/C. C. Lockwood. 32–9: Reproduced by permission of the Director, Institute of Geological Sciences (NERC), England. NERC Copyright reserved/Crown Copyright reserved. 32–12(a): R. Toja, Charles F. Kettering Laboratory. 32–12(b): E. H. Newcomb and S. R. Tandon, University of Wisconsin/BPS. Essay 32–1 (photo): Bob McKeever/TOM STACK & ASSOCIATES.

Chapter 33   33–3(a): Mark Newman/TOM STACK & ASSOCIATES. 33–3(b): ANIMALS ANIMALS/Brian Milne. 33–4: S. K. Webster, Monterey Bay Aquarium/PBS. 33–7(a): Courtesy of Michael Donoghue, San Diego State University. 33–7(b): Courtesy of Jochen Kumerow, San Diego State University. 33–7(c): L. E. Gilbert, University of Texas at Austin/BPS. 33–7(d): P. R. Ehrlich, Stanford University/BPS. 33–7(e): ANIMALS ANIMALS/Michael Fogden. 33–8(a): John M. Kirby/ATOZ IMAGES. 33–8(b): R. Humbert, Stanford University/BPS. 33–8(c): Courtesy of Merrilee Rayle. 33–9(a) and (b): Courtesy of Merrilee Rayle. 33–9(c): J. N. A. Lott, McMaster University/BPS. 33–9(d): L. Egede-Nissen/ BPS. 33–9(e): R. K. Burnard, Ohio State University/BPS. 33–9(f): E. D. Brodie, Jr., Adelphi University/BPS. 33–10(a): J. N. A. Lott, McMaster University/BPS. 33–10(b): David Muench/ATOZ IMAGES. 33–10(c): Courtesy of Merrilee Rayle. 33–10(d): R. K. Burnard, Ohio State University/BPS. 33–10(e): ANIMALS ANIMALS/Stouffer Productions, Ltd. 33–10(f): Courtesy of Merrilee Rayle. 33–11(a): Courtesy of Michael Donoghue, San Diego State University. 33–11(b): Courtesy of Jochen Kumerow, San Diego State University. 33–11(c): Hale L. Wedberg, San Diego State University. 33–11(d): G. C. Kelly/TOM STACK & ASSOCIATES. 33–12(a): Rick McIntyre/TOM STACK & ASSOCIATES. 33–12(b): Bob and Miriam Francis/TOM STACK & ASSOCIATES. 33–12(c): Mark Newman/TOM STACK & ASSOCIATES. 33–12(d): ANIMALS ANIMALS/Brian Milne. 33–13(a): Stewart M. Green/TOM STACK & ASSOCIATES.

33–13(b): ANIMALS ANIMALS/E. R. Degginger. 33–13(c): John Gerlach/TOM STACK & ASSOCIATES. 33–13(d): Courtesy of Michael Donoghue, San Diego State University. 33–13(e): R. Humbert, Stanford University/BPS. 33–15(a): ANIMALS ANIMALS/Adrienne T. Gibson. 33–15(b): ANIMALS ANIMALS/Anne Werthcim. 33–15(c): R. K. Burnard, Ohio State University/BPS. 33–15(d): C. R. Wyttenbach, University of Kansas/BPS. 33–16(a): B. Wilcox, Stanford University/BPS. 33–16(b): R. K. Burnard, Ohio State University/BPS. 33–16(c): S. K. Webster, Monterey Bay Aquarium/BPS. 33–16(d): P. J. Bryant, University of California, Irvine/BPS. 33–17: Robert R. Hessler, Scripps Institution of Oceanography, University of California, San Diego. 33–18(a): Courtesy of Merrilee Rayle. 33–18(b): P. J. Bryant, University of California, Irvine/BPS. 33–18(c): J. R. Waaland, University of Washington/BPS. 33–18(d): J. N. A. Lott, McMaster University/BPS. 33–18(e). ANIMALS ANIMALS/Breck P. Kent. 33–19(a): Don and Pat Valenti/TOM STACK & ASSOCIATES. 33–19(b): Don Rutledge/TOM STACK & ASSOCIATES.

Index

Note: Italicized page numbers indicate figure; t following page number indicates table.

A antigen, 147
ABO blood group, 147, 147t
  inheritance of, *148*
Abortion, 433
  spontaneous, 132
Absorption spectrum, 94, *96*
Abstinence, 430
Acclimatization, 379
Acetic acid, 12
Acetyl group, 80
Acid(s), 615
  acetic, 12
  amino; *see* Amino acid(s)
  citric, 80
  fatty; *see* Fatty acid(s)
  folic, 364
  hydrochloric, in digestion, 367
  lactic, 86, *87*
  linoleic, 359, 362
  nucleic, 28, 359, 362
  organic, 12
  oxaloacetic, 80
  phosphoglyceric (PGA), 100
  pyruvic; *see* Pyruvic acid
  ribonucleic; *see* Ribonucleic acid (RNA)
  uric, 387
Acquired immune deficiency syndrome
        (AIDS), 352–353, 433, 435t
Acromegaly, 404, *404*
Acrosome, 418
ACTH, 402, 403t
Actin, 49, 459
Action potential, 443–445, *445*
Action spectrum, 94, *95*
Activation energy, 74, *75*
Active site of enzyme, 74, *75*
Active transport, 65–66, *65*, *66*
Acupuncture, 481
Adam's apple, 375
Adaptation, xxv, *xxv*
  dark, 450
Adaptive radiation, 524, *526*
Addison's disease, 409
Adenine, 72, *72*, 170, *171*
Adenosine diphosphate (ADP), 28
  in ATP production, 72, *72*
  release of, 73
Adenosine triphosphate (ATP), 28,
        44–45, 72–74
  bonding in, 72–73

conversion of, into cyclic adenosine
        monophosphate, 400
  and evolution of metabolism, 29
  formation of, 82
  as link between respiratory and cellular
        work, *72*
  photosynthesis and, 99
  production of
    in heterotrophs, 70
    in obligate anaerobes, 86
  respiration-linked generation of, 32
  splitting of, 65
  in starch biosynthesis, *73*
  structure of, *72*
Adhesion, water, 60–61, 264
ADP; *see* Adenosine diphosphate
Adrenal cortex, *408*, 408–409
  hormones produced by, 403t
Adrenaline, *401*, 403t, 408
Adrenocorticotropic hormone (ACTH),
        403, 403t
Adrenocorticotropin releasing
        hormone, 406t
Adsorption, 17
Afterbirth, 427, *427*
Agar, 227
Agaricus campestris, 217
Age of Amphibians, 317
Age distribution of populations, 557, *557*
Age of Fishes, 317
Agent Orange, 198, 277
Agnatha, 315
Agriculture, transition to, 568
Agrostis tenuis, 513
AIDS, 352–353, 433, 435t
Air, composition of, xxi, 376
Albinism
  classical, *149*, 176, *176*
  ocular, 159
Alcohol, 483t
  ethyl, 12
  fermentation of, 86, *87*
  formation of, 12
  yeast in production of, 86
Aldosterone, 390
Aleurone layer, 272
Algae, 222–228; *see also* specific types
  blue-green; *see* Cyanobacterium(a)
  brown, 226
  in coastal marine ecosystems, 606

coralline, 227, *228*
  in freshwater ecosystems, 610
  green, 222–225, *224*, *225*
  photosynthesis in, 94
  red, 226–228, *228*
  reproduction in, 117, *118*
Algin, 226
Allele(s), 141
  defined, 139t
  frequency of, 506
    progressive change in, 505
  and law of independent assortment,
        144–145, *145*
  multiple, 147–148
  mutations and, 194–195, 505, 507
  segregation of, during meiosis, 142
  Tay–Sachs, 147
Alligator mississippiensis, 206
Allopatric speciation, 516–518
All-or-none response, 443
Allosaurus, 321
Altitude
  biomes and, 595, *596*
  effects of, on circulatory and
        respiratory systems, 379
  oxygen concentration and, 379
Altruism, 551
Alvarez, Luis, 528–529
Alvarez, Walter, 528–529
Alveoli, *375*, 376
Amanita, 217
Ames test, 198
Amine hormones, 400, *400*
Amino acid(s), 24, *25*
  codons for, 185, 186t
  discovery of, 13
  function of, 24, 26
  and human nutrition, 359
  insulin sequence of, 24, *25*
  nine essential, 359
  in nitrogen cycle, 586
  polymerization of, 17
  types of, 24
Amino acid sequence, 181
  DNA and, 175
  errors in; *see* Mutations
Amino group, 24, *25*
Ammonia
  excretion of, by fishes, 387
  as nitrogenous waste product, 387

Ammonification, 587
Amniocentesis, 134, 426, 428, *429*
Amnion, 422, *422*
Amniote egg of reptiles, 321, *321*
Amniotic cavity, 422, *422*
Amoebas, 214, *214*
　diffusion from, *58*
Amoebocytes in sponges, 290, *290*
Amphetamines, 482, 483t
　dieting and, 361
Amphibians, 317–318, *318*
　Age of, 317
　bone structure of, compared to lobe-
　　finned fish, *319*
　dehydration in, 318
　double circulation in, 337
　respiration in, 318, 377
　temperature regulation in, 394
Amphipods, *611*
Amplitude, 453
Anaerobes, 86
　obligate, 86, 88
Analgesics, 480
Anaphase of mitosis, 114, *115*
Anatomy, comparative, 489
Androgens, 403t, 409
Anemia, sickle cell; *see* Sickle cell anemia
Anemones, sea, *291*
Angiosperms, 235–239
　cross-pollinating, 236
　distinctive features, of, 236
　fertilization and seed development in,
　　237–238
　life cycle of, *236*, 236–241
　major classes of, 238
　structure of, *244*
Animalia, 207, 207t
Animal(s)
　characteristics of, *207*, 288
　communication among, 538–541
　desert adaptations of, 604, *605*
　internal transport systems in, 336–338
　number of species of, 204
　relationships among major phyla of, *289*
　terrestrial, water regulation in,
　　385–386
Animal behavior, 532–555
　genes and, 534–535
　instinctive, 535
　orientation, 541–544
　social, 544–554
　studies in, 533–534
Animal Kingdom, 207, 207t, 208
　invertebrates, 288–309
　lower chordates, 312–314, *312*
　vertebrates, 310–333
Animal societies, 544–545
Anise swallowtail butterfly, stages in life
　of, *305*
Annelids, 300–301
　major characteristics of, 307t

Annuals, 284
Anopheles mosquitoes, 214
Ant(s), mutualism and, 566, *566*
Anteater, spiny, 324
Anterior pituitary gland, 402–405,
　403t, *405*
Anther, 236
Antheridia, 230
Anthropoids, 326, *327*
Antibody(ies), 147, 189, 349, *350*
　composition of, 349
　constant region of, 191, *191*
　monoclonal, 355
　structure of, *191*
　variable region of, 191, *191*
Antibody-mediated responses, 349–352
Anticodon, 183, 187
Antigen(s), 147, 349
　H–Y, 409
　surface, *350*
Antigen–antibody complex, *350*
Antigenic determinant, 349
Antisense strand, 184
Antisera, 351–352
Anus, 370
Apes, 326
Apical meristems, 255, *255*
Appetite suppressants, 361
Arachnida, 301
Archaeopteryx, 322, *322*
Archegonia, 230
Aristotle, 492
　and theory of elements, 4–5
Arrhenius, panspermia theory of, 3–4
Arterioles, 336, 341
Artery(ies), 336
　coronary, blockage of, 344
　elasticity of, 341
　renal, 388, *388*
Arthritis, rheumatoid, 352
Arthrobacter AK 19, 226
Arthropoda, 301–305
Arthropods, 301–305
　characteristics of, 303, 307t
　evolutionary links of, with annelids, 303
　five major classes of, *302*
　nervous system of, *468*, 469
　respiration in, 373
Artificial heart, 345, *345*
Arum lily, 84
Ascaris, *297*
Asexual spores, 117, *118*, 217
Asphyxiation, cause of, 83
Atherosclerosis, 341, 345
　cholesterol and, 24, *24*
Atlantis II deep, 14
Atmosphere
　anaerobic, 70
　formation of, xxi
　prebiotic, 12
　secondary, formation of, xxi

Atom(s), 4–6
　carbon, model of, *5*
　covalent bonding of, 8, *9*
　Dalton's theory of, 5
　ionic bonding of, 8, 10
　structure of, 5–7
　　and bonding ability, 6
Atomic numbers, partial table of, 6t
Atomic theory, 5
ATP; *see* Adenosine triphosphate (ATP)
Atria, human, 338
Atrioventricular node, 339, *339*
AUG codon, 189
Australia, placental mammals in, 525
Australopithecus, 328
Australopithecus afarensis, 328, *329*
　skull and pelvis of, compared to
　　chimpanzee, *330*
Australopithecus africanus, 328, *329*
Australopithecus robustus, 328, *329*
Autoimmune disease, 352
Autonomic nervous system, 469, 471,
　472, 473
　divisions of, 471, *472*, 473
Autosomal linkage, 159–161, 164
　effect of, on inheritance patterns, 159
Autosomes, 159
Autotrophs, 29, 221
Auxin, 274–276
　effects of, on plants, 280t
　plant growth, and, *275*
　synthetic, 277
Avery, Oswald, 168
Axons, 439, *440*

B antigen, 147
B cell activation, 350–351, *351*
B cells, 348, 350
Baboon, social behavior in, 546, *547*
Backbone, 310
Bacteriophages, 213
Bacterium(a), 35, 39–40, 208–210
　cell division in, 110–111, *111*
　chemosynthetic, in deepwater ecosystem,
　　608, *609*
　commercial applications of, 210
　disease-causing, 209–210
　fission in, 111
　penicillin resistance in, selecting for, *512*
　reproduction in, 209
　shapes of, 209, *209*
Barbiturates, 482, 483t
Bark, 257
Barley seed, hormonal messenger systems
　in, 272, *273*
Barnacles, 301
Basal body, 52
　flagellum with, *52*
Basal body temperature method of birth
　control, 430

Base(s), 615
  in DNA, 170, 172, *172*
    replication and, 173–174
  in RNA, 182
  in tRNA, 183
Base deletion mutations, 193, *194*
Base insertion mutation, 193, *194*
Base sequences
  of DNA, 181
  mutations in, 191–195; *see also*
    Mutations
Base substitution mutation, 192–193, *193*
Bats, *325*
  echolocation in, 541, *542*
B-complex vitamins, 362t, 363–364
Beagle, voyage of, 495–496
Bee(s); *see* Honeybee(s)
Beetle, *302*
Behavior
  animal; *see* Animal behavior
  instinctive, in robin, *537*
Belding groung squirrel, 551–552, *552*
Beverly Hills diet, 360
Biceps, 456, *457*
Bile, 368
Biogeochemical cycles, 580–588
  parts of, 582
  steps in, 582–583
Biological species concept, 514
Biology
  nineteenth century landmarks in, *502*
  themes in, xxiv, xxvi
Biomes
  altitude and, 595, *596*
  defined, 594
  factors shaping, 595
  human influence on, 612, *612*
  terrestrial, *594*, 594–605
Biosphere, 556
Biotin, 364
Bipedal locomotion, 326
Bipolar cells, 453
Birds, 322–323
  anatomy of, 322, *322*
  body plan of, 322
  celestial navigation in, 541–544
  passerine, territorial behavior in,
    548–549, *549*
  temperature regulation in, 395
Birth, 427
Birth control, 429–433
Birth control pill
  combination, 431
  mini, 431, 433
  morning-after, 433
  progestin only, 431
Biston betularia, microevolution of, 498
Bivalents, 125, 130
Bivalves, 299, *299*
Bivalvia, 299
Bladder, urinary, 388, *388*

Blade, 245
Blastocoele, 421, *421*
Blastocyst, 421, *421*
Blood, 346–348
  amount of, in human body, 341
  composition of, 346–347, 347t
  filtration of, 388
  oxygen-depleted, 343
Blood cells
  red, *348*
  white, 348, *348*
Blood clots, 344, *347*
Blood clotting, 347
Blood fluke, life cycle of, *295*
Blood plasma, concentration of various
    substances in, 391t
Blood pressure, *340*, 340–341
Blood sugar
  abnormal levels of, 406–407
  insulin and glucagon in regulation
    of, *408*
Blood types, 147–148, 147t, *148*, 509
Blood vessels, 341–346
  structure of, *342*
Blue-green algae; *see* Cyanobacterium(a)
Body plan, in chordates, 310, *311*
Body segmentation, first appearance
    of, 307t
Bond(s)
  ATP, 72–73
  chemical; *see* Chemical bonds
  covalent; *see* Covalent bonds
  hydrogen; *see* Hydrogen bonds
  ionic; *see* Ionic bonds
  peptide, 24, *25*
    formation of, 189
  polar covalent, 60
  in protein, 26, 26t
Bone marrow, 348
Botulism toxin, 88
Boveri, Theodor, 153
Bowman's capsule, 388, *389*
Brain
  chemical activities of, 480–482
  in *Homo erectus,* 330
  in *Homo sapiens,* 330
  human, 328, 473, 474–479
    limbic system of, 477
    reticular formation of, 477–479, *478*
    topographic view of, *475*
  mammalian, 328
  vertebrate, 473
Brain waves, 478, *479*
Bread mold, 217
  life cycle of, *218*
Breathing
  force-pump methods of, 377
  mechanics of, 377, 380
  regulation of, 380
Breathing mechanisms, 377–380
Bristlecone pines, *285*

Bronchi, primary, *375*, 376
Bronchioles, *375*, 376
Brown algae, 226
Brushland, 602, *602*
Bryophytes, 228–230, *229*
  origins of, 230
Budding, 117, *118*
  of fungi, 217
Buds, lateral, 253
Butterfly, anise swallowtail, stages in
    life, of, *305*
Bypass surgery, 345

C_3 photosynthesis, 102–103, 104
C_4 plants, 101, *101*, 104
  dark reaction of, *104*
Cactus(i), adaptations of, 248, *249*,
    604, *605*
Caffeine, 483t
Calendar method of birth control, 430
Calories, 358
  required by humans, 358
Calvin, Melvin, 101, 102, 103
Calvin cycle, 100–101
  carbon dioxide fixation step of, *100*
  temperature and, 107
Cambium
  cork, 257
  vascular, *256*, 256–257
Cambrian explosion, 527
Camptosaurus, 321
Canals, semicircular, 454
Cancer
  environmental pollutants and, 198
  HeLa, 120–121
  liposomes and, 67–68, *68*
  molecular basis of, 200
  new strategies against, 364–365
  oncogenes and, 200
  vaginal, DES and, 436
Candida albicans, 435t
Canis familiaris, 514
Canis latrans, 514
Capillary(ies), 336, 341
  exchange of gases and nutrients in,
    343, *343*
  lymph, 346
  nutrients carried through, 343
Capillary action of water, 60–61
Capsule(s)
  cell, 41
  moss, 230
Carbohydrates, 18–21
  manufacture of, in photosynthesis, 71
  respiration of, 70
  source of, in plants, 267
  storage of, 20
  subgroup of, 18
Carbon
  atomic number and mass of, 6t
  compounds of, 11–12

in ecosystem, 580–581
model atom of, *5*
reservoir of, 584
Carbon cycle, 583–584
  long-term, 584
  short-term, 584
Carbon dioxide, 6
  atmospheric concentration of
    and fossil fuels, 584, *585*
    and photosynthesis, 107
  atmospheric, exchangeable pool of,
    583–584
  concentration in blood, 377, 380
  formation of in cellular respiration, 76
  movement of, in human respiration,
    376–377
  in photosynthesis, 93
  in respiration, 372
  structure of, *9*
  yearly consumption of, by plants, 93
Carbon dioxide fixation, 100, *100*, 102
Carbon-14, 100, 102
  radioactive isotope of, *102*
Carboniferous period, 317
Carbon monoxide, 83
Carbon reduction, 100
Carboxyl group, 12, 24, *25*
Carcinogens, 198
Cardiac muscle, 456, *457*
Carnivores, 559
Carotenoids in photosynthesis, 94
Carrier protein, 63–64, 74
Carrying capacity, population growth
    and, 558
Casparian strip, 251, 262, *262*
Catalysis
  activation energy and, *75*
  defined, 28
  enzyme, *75*
  process of, 74, *75*
Catastrophism, 493
Celestial navigation, 541–544, *542*
Cell(s)
  active transport in, 74
  animal, *35*
    generalized, *42*
  B cells, 348
  blood; *see* Blood cells
  cone, 453
  discovery of, 34, 37
  eukaryotic, 43–46, 54t
  guard, 246, *247*
  HeLa, in medical research, 120–121, *120*
  hypertonic, 62–63, 382, *383*
  hypotonic, 63, 382–383, *383*
  isotonic, 63, 383
  locomotion of, 51–53
  membranes of, 40, *41*, 56–57
  nerve; *see* Neuron(s)
  nucleus of, 43–44
  organelles of, 43, 44–53, 54t

origin of, 28–32
plant, *35*
  generalized, *43*
prokaryotic, 40–42
rod, 450–451
Schwann, 440, *440*
sensory, 449
size of, limiting factors, 38–40, *40*
somatic, 124
structure of, 34–55
surface area of, related to volume, 39, *39*
T cells, 348
target, 398
types of, *35*
water relations of, 56–69
Cell culture in biological research,
    120–121
Cell cycle, defined, 116, *117*
Cell division, 110
  in bacteria, 110–111, *111*
  in eukaryotes, 111–116
Cell elongation
  gravitropism and, 276
  zone of, *255*, 256
Cell lineage, theory of, 37
Cell-mediated immune responses, 352
Cell plate, 114, *115*, 116
Cell theory, 37
Cellulose, structure of, 20, *21*
Cell wall, 53, *53*
  in water transport, 262
Centipedes, 301, *302*
Central nervous system (CNS), 469,
    473–474
Centrioles, 53
Centromere, 113, *113*
Cephalization, 469
Cephalochordata, 312
Cephalopods, 299–300
Cerebellum, 474
Cerebral cortex, 475
Cerebral hemispheres, 475
Cerebrospinal fluid, 473
Cerebrum, 475, 477
  association areas of, 475, *476*, 477
  motor and sensory functions of, 475
Cervical caps, 431, *432*
Cervix, 413
Chaetopleuro apiculata, 298
Chaffinch, experiments with, 535
Chaparral, 602, *602*
Characteristics of life, xxii, *xxiii*
Chargaff, Erwin, 170
Chase, Martha, 168
Chemical bonds, 6–11
  defined, 6
Chemical evolution; *see* Evolution,
    chemical
Chemoreceptors, 448, 449t
Chiasmata, 130
Chilopoda, 301

Chitin, 216
  in arthropod exoskeleton, 303
Chiton, *298*
Chlorella, 224, *224*
  in photosynthesis experiment, 102, 103
Chlorophyll
  light-harvesting, 97–98
  in photosynthesis, 94
Chlorophyll *a*, 93, *95*, 97, 98, 99
  trap, 98
Chlorophyll *b*, 94, 98
Chlorophyta, 222–225, *224, 225*
Chloroplast, *43*, 45
Chlorpromazine, 483t
Cholesterol, 22, *23*, 24
  and atherosclerosis, 24, *24*
Chondrichthyes, 315, 316–317
Chondrus crispus, 228
Chordates, 310–333
  body plan of, 310, *311*
  characteristics and origin of, 310–314
  lower; *see* Lower chordates
Chorion, 422, *422*
Choroid, 450
Chromatid, 113, *113*
Chromatids, 125
  crossing over of, 130, *131*
  homologous, attachment of, 130
Chromatography, paper, 102, 103
Chromoplasts, 45, *46*
  photosynthetic pigments in, 96, *96*
Chromosome(s), *43, 44*
  centromere of, 113, *113*
  chromatids of, 113, *113*
  in fruit flies, *156*
  homologous, 124–125
  human, 123
  human male, *125*
  independent assortment of,
    130–131, *132*
  number of, in various organisms, 111
  sex; *see* Sex chromosomes
  structure of, 112
  X, 153
  Y, 153
Chromosome theory of heredity, 153
Chymotrypsin, 369, 369t
Cigarette smoking and respiratory
    passages, 376
Cilia, 51–53
  in human respiratory system, 375
Ciliates, 214–215, *215*
Circulation
  in arthropods, 303
  closed, *337*
  double, 336, *337*
  function of, 535t
  in mollusks, 297
  human, 338–348, *342*
  open, *336*
  open versus closed, 336

in reptiles, 321
single, 336, *337*
vertebrate, main types of, *337*
Citric acid, 80
Clams, 299
Class, 206
Classification, hierarchy of, 206
Classification system(s)
five-kingdom, 207, 207t
microorganisms and, 204–220
two-kingdom, 206, *207*
Cleavage, 421, *421*
Climax communities, 591–593
Clitoris, 413
Clostridium, 88
Cnidaria, 291–292
anatomy of, *291*
forms of, *291*
major characteristics of, 307t
nervous system of, 467, *468*
Coastal zones, 606, *606*
Coastal–intertidal zones, *607*
Cocaine, 483t
Cochlea, 454
hair cells in, 454, *455*
Cocklebur, time measurement in, 282–283
Code, genetic; *see* Genetic code
Codon(s), 183, 184–185
mRNA, of amino acids, 186t
start, 187
stop, 185, 186t, 189
Coelom
first appearance of, 307t
true, 296
Coenzyme(s), 362
Coenzyme A, 80
Cohesion, water, 264
Coitus, 419–420
Coitus interruptus, 431
Coleoptiles, 274
Collar cells in sponges, 290, *290*
Colon, 369–370
Color blindness, red-green, 157–158
Color perception, 453
Commensalism, 565–566
Communication
animal, 538–541
by chemicals (odors), 541
by sound, 540–541
by visual display, 538, 540
Community(ies), 590–611; *see also* specific
communities
climax, 591–593
defined, 590
Competition, population size and,
563–565, *564*
Competitive exclusion principle, 564
Compound microscopes, 30
Compounds
defined, 6
organic, 18–21

Concentration gradient, 58
Conception, 419
Conditioning, behavioral, 537
Condom, 431, *432*
Conductivity of neuron, 438
Cone(s), 233
pollen, 234, *235*
seed, 234, *235*
Cone cells, 453
Coniferous forest, 600, *600*, *601*
Conifers, 233–234
Conspecifics, 544
Consumers, 573
primary, 573
secondary, 576
Continental shelf *606*
Contraceptive methods, 429–433, *432*
effectiveness and various side effects
of, 430t
Contractile vacuole, 49
Convergent evolution, 525
Copernicus, 493
Coral, 227, 291–292
Coral reef, 292, 593, *593*
Coralline algae, 227, *228*
Cord, nerve, 310, *311*
Cork cambium, 257
Cornea, 450
Corpus callosum, 475
Corpus luteum, 415, *415*
Cortex, root, 250, *251*
Corticosteroids, 409
Cortisone, 409
Cotylosaurs, 320, *320*
Covalent bonds, 8, *9*
nonpolar, 8
polar, 8
Cowper's glands, 418
Cranial nerves, 469
Creation; *see* Special creation
Cretaceous cataclysm, 528–529
Cretaceous extinction, 524, 528–529
Cretinism, 406
Crick, Francis, 169–170, 184, 193
directed panspermia theory of, 3
Cro-Magnon people, *331*
Crop
an annelids, 300, *301*
in birds, 365
in earthworms, 365
Cross
dihybrid, 143–146, *144*, 159, *160*
monohybrid, 137–142, 139, *139*
Mendel's explanation of, 141
Crossing over, 130, *131*
Crossover and gene recombination,
160, *160*
Cross-pollination, 84, 240
Crustacea, 301
Cuticle, 231
leaf, 246

in roundworms, 296, *297*
Cuvier, Georges, 493
Cyanide, 82–83
Cyanobacterium(a), *42*, *210*, 210–211
ecological role of, 211
temperature and, 107
Cycads, *233*
Cycle(s)
biogeochemical, 580–588
carbon; *see* Carbon cycle
nitrogen, 585–588, *586*
Cyclic adenosine monophosphate (cyclic
AMP), 400, *401*
Cytochrome *c*, comparative biochemistry
of, *492*
Cytokinesis, 114, *115*, 116
Cytokinins, 278–279, *279*
effect of, on plants, 280t
Cytoplasm, 44
Cytoplasmic streaming, 49
Cytosine, 170, *171*
Cytoskeleton, 49, *50*, 51

2,4-D, 277
Dalton, John, atomic theory, 5
Dark reactions of photosynthesis, 97,
100–101, 104
in C$_4$ plants, 104
overview of, 97
Darwin, Charles, 205, 492, *497*, 504, 509,
514, 523
and phototropism, 274
Daylength, flowering and; *see* Flower
induction
Day-neutral plants, 282
Decibels, 453
Deciduous forests, 598, *599*
Decomposers, 576–577, 597
Dehydration synthesis, 19
Dendrites, 439, *440*
Denitrification, 587
Deoxyribonucleic acid (DNA), 28, 167–180
amino acid sequence and, 175
bacterial, 110–111, *111*
bases of, 172, *172*
computer-generated model of, *xxv*
contrasted with RNA, 182
eukaryotic, replication of, *174*
and experiments with bacteria, 167–168
and experiments with virus, 168–169
information storage in, 174–176
mutations of, 191–195; *see also*
Mutations
nucleotides of, 170, *171*
and protein synthesis, 44
recombinant, xxviii, 176–179, *178*, 210
legal aspects of, 179
replication of, 110, *173*, 173–174
structure of, 169–173
transfer of, in bacteria, 209, *209*

Deoxyribose, 18, *19*
in DNA, 170
Depolarization, membrane, 444, 445, 449
DES, 433
Desert, 604, 604, *605*
Development, xxii, *xxiii*
of human embryo, *xxiii*, 420–425, *424*
of human fetus, *xxiii*, *424*, 425–427, *426*
Diabetes mellitus, 149t, 407
Dialysis, 392–393
Diaphragm, 377, *378*,
contraceptive, 430t, 431, *432*
Diastole, 340
Diatoms, 216, *216*, *611*
2,4-dichlorophenoxyacetic acid (2,4-D), 277
Dicots, 238
leaf blades of, 245, *245*
seed of, *239*, *272*
structure of, *244*
vascular bundles in, 253, *254*
Dicotyledons, 238
Dictyostelium discoideum, 213; see also
Slime mold
life cycle of, *213*
Diet(s)
Beverly Hills, 360
high-carbohydrate, low-protein, 360
high-protein, low-carbohydrate, 360
Diet aids, 361
Diet pills, starch-blocking, 361
Diethylstilbestrol (DES), 433, 436
Differential permeability, 57
Differentiation, 256, 423
Diffusion, *57*, 57–58
facilitated, 63–65, *64*
simple, versus active transport, *65*
Digestion, 88
in annelids, 300
intracellular, 49
in mollusks, 299
in nematodes, 296
relationship of, to respiration, 88, *89*
in starfish, 306
Digestive enzymes, 369, 369t
Digestive system, 335t, 357–371
blind sac, first appearance of, 307t
human, 366–370, *367*
tubular, 364–365, *366*
Dihybrid cross, 143–146, *144*, 159
gene linkage and, *160*
Dinosaurs, 314, 320, *321*
extinction of, 528
Dioxin, 198, 277
Dipeptide, 189
Diplodocus, 314, *321*
Diploid cells, 125
Diplopoda, 301
Diplovertebron, 318
Disaccharides, 19, *19*
Disruptive selection, 513
Distal convoluted tubule, 389, *389*

DNA; *see* Deoxyribonucleic acid
DNAase, 112
DNA viruses, 213
Dominance, incomplete, *146*, 146–177
defined, 139t
in snapdragons, *146*
Dominance–recessiveness, defined, 139t
Dopamine, 480
Doris, 496–497
Double helix, 170–173; *see also* Deoxy-
ribonucleic acid (DNA), structure of
Down syndrome, 132, *133*, 134, 428
estimated rate of occurrence of, with
different maternal ages, *134*
Dreams, 478
Drift, genetic, 507–509
Drosophila melanogaster, 154, *156*, 205;
see also Fruit fly
Drugs, effects of, on nervous system,
482, 483t
Dryopthicus, 328, *329*
Dwarfism, 404, *404*
Dwarfs, pituitary, 404

Ear, human, 453–454, *454*
Eardrum, 453
Earth
formation of, *xx*, xxi
photograph of, *ixx*
prebiotic
conditions on, 12–13, xviii, *xx*, xxi
gases on, 13–14, *xx*, xxi
Earthworm, 300–301, *301*
Ecdysone, 399, *399*
Echinoderms, *306*, 306–307, 307t, 311
Echolocation, 541
in bats, *542*
Ecological niche, 563, *565*
Ecological succession, 590–591
of Lake Michgan, 590–591, *592*
Ecology
defined, 556
of ecosystems, 572–589
of populations, 556–569
Ecosystem(s), 590–613
aquatic, 606–611
productivity of, 579, *579*, 580t
biogeochemical cycles in, 580–588
defined, 572
ecology of, 572–589
energy flow through, 577–580, *580*
freshwater, *610*, 610–611, *611*
productivity of, 580t
gross and net productivities of, 578–579
human influences on, 612, *612*
major types, 590–613
marine, 606–609
nutrient flow in, 580–588
nutritive relationships in, 573–577
off-road vehicles and, 574–575
productivity of, 578–580, *579*, 580t

saltwater, 606–609
productivity of, 580t
terrestrial, 594–605
productivity of, 579, 580t
Ecotones, 593, *593*
Ectoderm, *422*, 423
Ectotherms, 391, 394–395, *394*
Egg
amniote, of reptiles, 321, *321*
human, *40*, 413
Ejaculation, 419, 420
Electrocardiogram (EKG), 339, *340*
Electroencephalogram (EEG), 478, *479*
Electron(s), 5
respiratory transport of, 76
Electron microscope
scanning, 38
transmission, *37*, 37–38
Electron microscopy, 37–38
Electron transport chain
respiratory, *83*
photosynthetic, 98, *99*
Element(s), 4–6
Elephantiasis, 296
Embryo
environments of, 387t
human development of, *xxiii*,
420–425, *424*
implantation of, 415
preventing implantation or devel-
opment of, 433
seed, 271
Embryo sac, 237
Embryology
comparative, 489, *490*
comparative vertebrate, *490*
Embryonic induction, 423, *423*
Endergonic reaction, 73, *73*
Endocrine glands, 403t, 402–410
Endocrine system, *402*, 402–410
function of, 335t
Endocytosis, 66–67, *67*
Endoderm, *422*, 423
Endometrium, 413
development of, 415–416
Endoplasmic reticulum, 45, *46*, 48
function of, 48
rough, *46*, 48
smooth, *46*, 48
Endorphins, 480
Endoskeleton
in echinoderms, 306
first appearance of, 307
Endosperm, seed, 271
Endosperm cell, 238
Endosymbiosis, 47
Endothelium, arterial, 341
Endotherms, *395*, 395–396
Energy
activation, 74, *75*
ATP, model of, 73

exergonic reaction and, 73
flow of, through ecosystems, 577–580, *580*
gross productivity of, 578
kinetic, 74
in photosynthesis, 94
recovered during glucose respiration, 82t
respiration and, 70–91
transformation of, 44–45
Energy foods, 358
Energy metabolism, 71–76
Energy-coupling mechanisms, 73
Englemann, T. W., 94
Enkephalins, 480
Environment
cancer and, 198
versus inheritance, 149–150
mutations and, 193–194
as selecting agent, 509
Enzymes
active site region of, 74, *75*
catalysis and, *75*
digestive, 369, 369t
function of, 28, 74
in lactose utilization, 196
structure of, 74
temperature and, 74, *76*
Enzyme induction, 196
Eohippus, xxiv, 488, *489*
Epidermis, leaf, 246
Epididymis, 417
Epinephrine, 408; *see also* Adrenaline
Equus, evolution of, *xxiv*, 488, *489*
Erythrocytes, 347, 348, *348*
Escherichia coli (*E. coli*), 196, 210
in colon, 369–370
interferon production and, 344
mutualism and, 566
in recombinant DNA research, 177, *178*
Esophagus, 366, *367*
Estrogens, 403t, 409
menstrual cycle and, 416
Estrus, 413
Estuary, productivity of, 580t
Ethology, 533
Ethyl alcohol, 12
Ethylene, 279–280
effect on plants, 280t
Euglena, 206, *207*
classification problems of, 213
Eukaryotes
cell division in, 111–116
classification of, 213
DNA replication in, 174
transcriptional control in, 196–197, 198
Eukaryotic cells, 43–46
components of, 54t
Evolution
agents of, 505–523
biochemical evidence for, 491

chemical, 3–4, 12, 17–18
occurrence of today, 13–14
and origin of cells, 28–32, *30*
convergent, 525
in cephalopods and vertebrates, 299–300
early theories of, 492
evidence for, 488–491
of horse, *xxiv*, 488, *489*
human, 326–332
Lamarck's theory of, 493–494
Lamarckian versus Darwinian views of, *501*
among mammals, *324*
modern theories of, 504–531, *514*
molecular biology evidence for, 491
patterns in, *524*
photosynthesis and, 70
of reproduction, 412
of reptiles, *320*
respiration and, 70
social behavior and, 545–548
as theme in biology, xxiv–xxvi
theory of, xxiv–xxvi, 205, 488–531
trends in, 523–527
vertebrate, *314*
Exchangeable pool of biogeochemical cycles, 582, *582*
Excitability of neuron, 438
Excretion, 386–391
defined, 387
Excretory system
function of, 335t
human, *388*
Exergonic reaction, 73, *73*
in water formation, 74
Exhalation, 377, *378*, 380
Exocytosis, 66
Exoskeleton
arthropod, 303
first appearance of, 307t
Extinction
Cretaceous, 524
Permian, 523
Extraterrestrial life, xvii
Eye(s)
embryonic formation of, *423*
human, 449–453, *451*
invertebrate, *451*

Facilitated diffusion, 63–65, *64*
FADH$_2$, 80, 82
formation of, 78
Fallopian tubes, 413
Family planning, natural, 430; *see also* Birth control
Fats, 21–22; *see also* Triglycerides
animal, 22
digestion of, 88
Fatty acid(s), 21
digestion of, 88

saturated, 21
unsaturated, 21
Feathers, *323*
as evolutionary adaptation, 323
Fermentation
alcoholic, 86, *87*
defined, 70
of glucose, 83, 86, *87*, 88
lactic acid, 86, *87*
Ferns, 231, *231*
leaves of, 245, *245*
life cycle of, *232*
Fertility awareness methods of contraception, 430
Fertilization, 123
in angiosperms, 237–238
in humans, 412–413
Fertilizers, nitrogen fixation and, 587–588
Fetus, human
development of, *xxiii*, *424*, 425–427
at nine months, *426*
at 28 weeks, *426*
Fiber
digestion and, 369
muscle, 457, *458*
nerve, 469
Fibrin, 347
Fibrinogen, 347
Fight-or-flight reaction, 409
Filter feeding, 299
in tunicates, 312
Finches, Darwin's, 496, 517–518, *519*, *520*
Fishes, 315–317
Age of, 317
bony, *315*, 317, 384–385
osmoregulation in, *385*
cartilaginous, 315, 384; *see also* Chondrichthyes
freshwater, 317, 384
gas exchange in, 373, *374*
heart in, 336
jawless, 315
lobe-finned, 317
ray-finned, 317, *317*
Fission, 111
Flagellum(a), 41, *41*, 51–53
with basal body, *52*
Flatworms, *293*, 293–296
major characteristics of, 307t
Flavin adenine dinucleotide (FAD), 78
Florigen, 283–284, *284*
Flower(s)
development of, 236
structure of, *236*
Flower induction
florigen and, 283–284, *284*
photoperiodism and, 282–284
Fluid mosaic model of membranes, 56, *57*
Fluke(s), 294, 296
blood, life cycle of, *295*
Folic acid, 364

Follicle
  Graafian, 414, *415*
  primary, 414, 415
Follicle-stimulating hormone (FSH), 416
Food chain, 573
Food vacuole, 49
Food web, *577*
  freshwater, 610
  in pelagic zones, 608
Forebrain, 474
Forest(s)
  coniferous, 600, *600*, *601*
  deciduous, 598, *599*
  tropical, 596, *597*
Fossil fuels
  and carbon dioxide concentration, 584
  combustion of, 93
Fossils, 488
Fox, Sidney, 28
Fragmentation of fungi, 217
Franklin, Rosalind, 170
Fraternal twins, 427
Freshwater ecosystems, *610*, 610–611, *611*
  organisms of, *611*
Fructose, 18, *19*
Fruit, 239
  ethylene and ripening of, 279–280
  as seed dispersers, 239, *239*
Fruit fly, 154, 205
  chromosomes in, 156, *156*
  demonstration of sex-linked trait in, 154,
    156–157
  inheritance of eye color in,
    155–157, *157*
Fucus, 227
Fungi, 207, 207t, 216–219
  as decomposers, 217
  mycorrhizal, 267
  reproduction in, 117, *118*, 217

Galactosemia, 149t
Galapagos finches, 518, *519*, *520*
Galapagos Islands
  Darwin's voyage to, *495*, 496
  map of, *519*
Galileo, 493
Gametangium, 219
Gametes, formation of, 123
Gametogenesis, 412
Gametophytes, 221
  of angiosperms, 237
  in moss reproduction, *229*
Ganglion, motor, *468*
Ganglion cells, 453
Gap$_1$ (G$_1$), 116, *117*
Gap$_2$ (G$_2$), 116, *117*
Garrod, Archibad, 175
Gas exchange, 372–380
Gastropods, 297–298, *298*
Gastrovascular cavity, 364, *365*
Gastrulation, 422, *422*

Gause, G. F., 564
Geiidium, 227, 228
Gene(s), 111
  behavior and, 534–535
  discovery of location of, 153
  homologous, 130
  linked, recombination of, 159–160, 164
  Mendel's discovery of, 136–137
  mutations in, 191–195
  new combinations of, 130
  promotor, 196, *197*
  rearrangements of, 189, 191
  regulator, 196, *197*
Gene expression, regulation of, 195–199
Gene flow, 507, *508*
Gene linkage, 153
Gene pool, 504
Generator potential, 449, *450*
Genetic code, 174–175, 181, 184–187
  exceptions to, 185–187
  first experiment to elucidate, *187*
  tRNA and, 183
  universal nature of, 3
Genetic counseling, 176
Genetic drift, 507–509
  in Eskimo populations, *510*
Genetic engineering, 177
Genetic information, nature of, 175
Genetic variability, 128–131
Genetics
  common terms in, 139t
  laws of inheritance and, 136–146
  Mendelian, 136–152
    extended, 146–150
Genital warts, 435t
Genotype, 141
  defined, 139t
Genus, 205
Germination, photodormancy and, 280
Gibberellin, 272, 274, 276–278
  effect of, on plants, 280t
Gigantism, 403
Gill(s)
  in gas exchange, 373, *373*
  structure of, *374*
Gill pouches, 310
Gill slits, pharyngeal, 310, *311*
Ginkgo biloba, 233
Gizzard
  in annelids, 300, *301*
  in birds, 365
  in earthworms, 365
Gland(s)
  adrenal, *408*, 408–409
  anterior pituitary, 402–405, 403t, *405*
  Cowper's, 418
  endocrine, 403t, 402–410
  pituitary, 402
  posterior pituitary, 403t, 405, *405*
  salivary, 366, *367*
  thyroid, 406

Global 2000 Report to the President,
    568, 569
Glomerulus, 388, *389*
Glucagon, 403t
  in blood sugar regulation, *408*
  functions of, 407
Glucocorticoids, 403t, 409
Glucose, 18, *19*; *see also* Sugar
  fermentation of, 83, 86, *87*, 88
  incorporation of, into starch
    molecule, *73*
  respiration of, 76–83
    energy recovered during, 82t
    overview of, 76, *77*, 78
  in sucrose synthesis, 105
  in synthesis of cell constituents, 105
  uses of, 105–106, *105*
Glycogen, *20*
  digestion of, 88, *89*
  storage of, 20
Glycolysis, 76, 78, 79
  in absence of oxygen, 86, 87
  balance sheet for, 78t
Goiter, 406
Golgi bodies, *48*, 48–49, 66
Gonadotropic hormones, 402, 403t
Gonadotropin-releasing hormones, 406t
Gonads, 409–410
  hormones produced by, 403t
Gonorrhea, 433, 435t
  vaccines against, 351
Graafian follicle, 414, *415*
Gradualism versus punctuated equilibrium,
    523, *524*
Grana, 96, *96*
Granulocytes, 348
Grasslands, 598, *598*
Gravitropism, 276, *276*
Great Barrier Reef, 292, 593, *593*
Green algae, 222–225, *224*, *225*
Greenhouse effect, 585
Griffith, Frederick, 167
Growth, xxii
  exponential, 557–558, *558*
Growth hormone, 403t, 403–404
  synthesis of, 178
Growth hormone inhibiting hormone, 406t
Growth hormone releasing hormone, 406t
Growth rings, annual, *257*, 258
Guanine, 170, *171*
Guard cells in leaves, 246, *247*
Gymnosperms, *233*, 233–235
  vascular cambium in, 256–257

Habituation, 535, 537
Hair
  human, *27*
  protein structure of, 26
Hair cells in cochlea, 454, *455*
Haldane, J. B. S., 12
Haploid, 125

Hardy–Weinberg law, 505, 506
Hearing, 453–454
Heart
    amphibian, 336–337, *337*
    bird, 336–338, *339*
    fish, 336, *337*
    four-chambered, *337*, 338
    human, *338*, 338–341
    mammalian, 336–338, *337*
    rate of blood flow in, 341
    reptilian, 337–338, *337*
    three-chambered, 337, *337*
    two-chambered, 336, *337*
    vertebrate, evolution of, 336–338
Heart attacks, incidence of, 344
Heartbeat, anatomy of, *339*
Heart disease, 344–345
Heart murmur, 339
Heart transplant, 345
Heart valves, *338*, 339
Heartwood, 258
HeLa cells, 120–121, *120*
Heme group, 376, *377*
Hemichordates, 311, *311*
Hemodialysis, 392–393
Hemoglobin, *377*
    function of, 376–377
    mutation of, 192
Hemophilia, 158–159
    and European royalty, 162–163
Henle, loop of, 389, *389*
Henslow, John, 495
Herbicides, 277
    in Vietnam, 277
Herbivores, 559, 573
Hermaphroditism, 224–225, 293–294
Herpes, 211, 212, 435t
    genital, 212
        vaccines against, 351
    oral, 212
Herpes Resource Center, 212
Herpes simplex, 212
Hershey–Chase experiment, *170*
Heterokaryotes, 216
Heterotrophs, 29, 221, 357
    dependency of, on photosynthesis, 71
    energy production in, 70
    feeding strategies of, *358*
    and food chain, 573
Heterozygous, defined, 139t
Hibernation, 396
Hindbrain, 474
Histones, 196–197
Holdfast, 226
Homeostasis, 382
Homeotherms, 323, 326, 395
Hominids, evolution of, *329*
Homo erectus, 329, 330, *331*
Homo habilis, 328, *329*
Homologous chromosomes, 124–125
Homo sapiens, 326, 330–331

Homo sapiens neanderthalensis, 329, 331, *331*
Homo sapiens sapiens, 329, 331, *331*
Homozygous, defined, 139t
Honeybees
    behavior of, genetic bases for, *536*
    colonies of, *553*, 553–554
    dance of, 532–533, *533*
    von Frisch's experiments with, 532–533
Hooke, Robert, 34, 36
Hookworms, 296
Hormone(s), 398–401
    adrenocorticotropic (ACTH), 402, 403t
    defined, 274
    flower-inducing, 283–284, *284*
    follicle-stimulating, 416
    gonadotropic, 402, 403t
    growth, 178, 403t, 403–404
    inhibiting, 406, 406t
    insect, 398–399
    juvenile, 398, 399
    levels of, in bloodstream during
        pregnancy, *422*
    luteinizing, 416
    parathyroid, 403t
    plant, effects of, 280t
    releasing, 406, 406t
    sex, 403t, 409
    steroid, action of, *400*
    steroid, amine, and peptide, 400, *400*
    tropic, 402, 403t
    and vascular plant development,
        274–280
    vertebrate, 399–400
    water-soluble, 400
Hormone–receptor, complex, 398
Horse
    classification of, 206
    evolutionary tree of, *xxiv*, *489*
Human(s)
    brain of, 473, 474–479
    brain size of, 328
    caloric requirements of, 358
    chromosomes in, 123
    evolution of, 326–332
    hearing in, 453–454
    heart of, *338*, 338–341
    influence of, on ecosystems, 612, *612*
    kidneys of, 387–391, *392*
    meiosis in, *124*
    mineral requirements of, 364, 364t
    nutritional requirements of, 357–364
    sex-linked traits in, 157–159
    skeleton of, 456, *456*
    vision in, 449–453
    vitamins required by, 362, 362t
    water regulation in, 386
Human chorionic gonadotropin
        (HCG), 421
Human population, 567–570
Huntington's chorea, 480

Hutton, James, 493
H–Y antigen, 409
Hybridization, *522*, 522–523
Hybridoma, 355, *355*
Hybrids, 138
    defined, 139t
Hydra
    digestion in, 364, *365*
    nervous system of, 467, *468*
    reproduction in, *118*, 119
Hydrocarbons, 11–12
    formation of, 12
Hydrochloric acid in digestion, 367
Hydrogen
    atomic number and mass of, 6t
    covalent bonding of, 8, *8*
    molecular, 6
    in prebiotic atmosphere, 12
    in water formation, 74
Hydrogen bonds, 11, *11*
Hydrogenation, 22
Hydrolysis, 19
    in starch or glycogen digestion, 88, *89*
Hydroponics, 266, *266*
Hydroxyl group, 12
Hyperthyroidism, 406
Hypertonicity, 62–63, 382, *383*
Hypervitaminosis, 363
Hyphae, 214
Hypoglycemia, 407
Hypothalamus, 405–406, 474
    hormones produced by, 406, 406t
    role of, in water balance, 391
Hypothesis, defined, xxvii
Hypothyroidism, 406
Hypotonicity, 63, 382–383, *383*

Ichthyosaurs, 320, *320*
Ichthyosis, 159
Identical twins, 131, 427
Imbibition, 61
Immune response
    primary, *351*
    secondary, 351, *351*
Immune system, 349–353
    function of, 335t
Immunity, 351
Immutability of species, 492
Implantation, embryo, 421, *421*
    prevention of, 433
Imprinting, 537–538, *538*
Impulse, neural, 438, 441–445
Independent assortment
    law of, 144–146
    meiosis and, *145*
Induction
    embryonic, 423, *423*
    enzyme, 196
Industrial melanism, 498
Infanticide, 550–551
Inferior vena cava, 346

Inhalation, 377, *378*
Inheritance
    effect of autosomal linkage on, 159
    versus environment, 149–150
    laws of, 136–146
        extended, 146–150
    polygenic, 149, *150*
    of sex, 153–154, *154*
Insect, 304
    body plan of, 304, *304*
    developmental stages of, 304
    eye of, *303*
    hormones in, 398–399
    pollination and, 240–241
    specialization among, 304
Insect societies, 544
Insight learning, 538, *539*
Instinctual learning, 535
Insulin, 403t
    amino acid sequence of, *25*
    in blood sugar regulation, *408*
    roles of, 406–407
    synthesis of, 178
Integumentary system, function of, 335t
Intercourse, sexual, 419–420
Interferon, 354–355
    synthesis of, 178
Interneurons, 439
Internodes of stem, 253
Interphase of cell cycle, 116
Intestine
    large, 369–370
    small, 367–368, *368*
        enzymes produced by, 369t
Intrauterine devices (IUDs), 430t, *432*, 433
Invertebrates, 288–309
    characteristics of, 307t
    evolutionary relationships among, *289*
    eyes of, *451*
    hormones in, 398
    marine
        osmoconformers and osmoregulators
            among, *383*
        salt regulation in, 384–385
        water regulation in, 383–384
    nervous systems of, 467–469
Ion pump, 246
Ionic bonds, 8, 10, *10*
Ionization, defined, 10
Ions, defined, 10
Iris (eye), 450
Islets of Langerhans, 406
Isolating mechanisms, 515–516, 516t
Isotonicity, 63, 383
Isotopes, 102, *102*
IUDs, 430t, *432, 433*

Jacob, Francois, 196
Jarvik-7, *345*
Java man, 330
Jelly fish, *291*, 291–292

nematocysts in, *293*
Johanson, Donald, 328

Kangaroo rat, *605*
    water use by, *386*
Karyotypes, abnormal, 155
Kelp, 226, *227*
Keratin, 27, *27*
Ketosis, 360
Kidney, 387–391, *392*
Kidney failure, 392–393
Kinetic energy, 74
Kinetin, plant growth and, 279
Kingdoms, 206, 207, 207t; *see also*
        specific Kingdoms
Klinefelter syndrome, 155
Krebs, Sir Hans, 78
Krebs cycle, 76, 77, 78, 80, *81*
Kwashiorkor, 359, *359*

Labia majora, 413
Labia minora, 413
Labor in childbirth, 427
    stages of, 427, *427*
Lac operon, 196, *197*
Lactic acid fermentation, 86, 87
Lactose, 19
    enzymes in utilization of, 196
Lamarck, Jean Baptiste, *493*, 493–494, 500
Laminaria, 227
Lampreys, 316, *316*, 562, *562*
Lancelets, *312*, 313–314
Langerhans, islets of, 406
Large intestine, 369–370
Larynx, 375, *375*
Leaf(ves), 244–250
    anatomy of, 246, *247*
    desert adaptations of, 248, *249*
    photosynthesis and, 246
    shapes of, 245, *245*
Leakey, Louis, 328
Leakey, Mary, 328
Learning
    conditioning, 537
    defined, 535
    habituation, 535, 537
    imprinting, 537–538, *538*
    insight, 538, *539*
    trial-and-error, 537
    types of, 535, 537–538
Lemmings, 560–561, *560*
Lemur, 326, *327*
Lens of human eye, 450
Leucoplasts, 45
Leukocytes, 347, 348, *348*
Lichens, 230, 267
    mutualism and, 566–567, *567*
Life
    characteristics of, xxii, *xxiii*
    conditions necessary for, 12
    defining, xxii–xxiii

dependency of, on photosynthesis, 292
elements of, 4t, 4–6
extraterrestrial, xvii
origins of, 2–16, 17–33
    major events in, *30*
    theories of, 2–4, *4*
origin of cells and, 28–32
processes characterizing, xxii, *xxiii*
Light
    far-red, effect of, on germination,
        280, 281t
    in photosynthesis, 94
    and photosynthetic pigments, 94–96
    pigments in absorption of, 94
    plant development and, 280–284
    red, effect of, on germination, 280, 281t
    ultraviolet, 94
    visible, 94
Light intensity as limiting factor in
        photosynthesis, 106, *106*
Light microscopy, *36*
    invention of, 34
    limitations of, 37
Light reactions, *97*, 97–100
    overview of, 97
    purpose of, 99
Light saturation point, 106
Lily, arum, 84
Limbic system, 477, *477*
Linkage, autosomal, 159–161, 164
Linkage group, 154
Linnaeus, Carolus, 204
Lipid(s), 21–24
    structure of, 21
Liposomes, 67
    anticancer, 67–68, *68*
Liverwort, *229*
Lizards, 320, *320*
Lobe-finned fishes, *315*, 317
Locomotion
    in arthropods, 303
    bipedal, 326
    in nematodes, 296
Long-day plants, 282, *283*
Loop of Henle, 389, *389*
Lorenz, Konrad, 537, *538*
Lower chordates, 310–333
    species of, 312–313
LSD, 483t
Lungs, *373*
    capacity of, 380
    evolution of, 374
    human, 375–380
Luteinizing hormone (LH), 416
Lyell, Charles, 496
Lymph, 346
Lymph capillaries, 346
Lymph nodes, 346
Lymph veins, 346
Lymphatic system, human, 346, *346*
Lymphocytes, 348

Lysine, 359
Lysosomes, 49

Macrocystis, 226
Macronutrients, soil, 265, 265t
Macrophages, 348, *349*
Malaria
   acute, coincident with sickle cell
      anemia, *513*
   vaccines against, 351
Malnutrition, 358
Malthus, Thomas, 496
Maltose, 19
Mammal(s), 323–326
   body coverings of, 323
   brain size of, 328
   evolutionary relationships among, *324*
   forelimbs of, comparative anatomy
      of, *490*
   homeothermic nature of, 326
   marine, 385
   mating in, 413
   mating behavior in, 550–551
   nervous system in, 326
   placental, 525, *526*
   reproduction in, 326, 327
   temperature regulation in, 395
   vision in, 327
Mammary glands, 326
Mantle of mollusks, 297
Marchantia, 229
Marijuana, 483t
Marine ecosystems, 606–609
   coastal, *606,* 606–607
   coastal–intertidal, *607*
   deepwater, 608, *609*
   pelagic zone, 608
Marine kelp, 226
Marine lampreys, 562, *562*
Marine polychaetes, 300–301
Marsupials, 324, 527
   adaptive radiation in, *526*
Mass number
   defined, 5
   parial table of, 6t
Mating behavior in mammals, 550–551
Mating rituals, 540, *540*
Mating types in molds, 219
Mechanoreceptors, 448, 449t
Medulla, 474, *474*
Medusa, 291, *291*
Megaspores, 236–237
Meiosis, 123–135, *126, 127*
   anaphase I of, 128
   anaphase II of, 128
   compared to mitosis, *129*
   errors in, 132, *133,* 134
   first meiotic division of, 125–128
   gamete formation and, 412

and genetic law of segregation, 142
and genetic variability, 128, 130–131
in humans, *124*
metaphase I of, 126
metaphase II of, 128
overview of, *126*
prophase I of, 125–126
prophase II of, 128
second meiotic division of, 128
segregation of alleles during, *142*
stages of, 125–128, *127*
telophase I of, 128
telophase II of, 128
Meiotic nondisjunction, 132, *133,* 154, 155
Melanism, industrial, 498
Membrane(s)
   cell, 40, *41*
   fluid mosaic model of, 56, *57*
   function of, 56
   fusion of, 67
   invagination of, 47
   plasma, 40, *41*
   postsynaptic, 446
   presynaptic, 446
   structure of, 56–57
   tympanic, 453
Membrane potential, 443
Memory cells, 350
Menarche, 413
Mendel, Gregor, 136, *137,* 175, 505
Mendelian genetics, 136–152
Menopause, 413
Menstrual cycle, 413–416, *416*
   contraceptive methods based on,
      429–431
   hormonal regulation of, 415
Menstruation, 413
Meristems, 255, *255*
   apical, 255, *255*
   lateral, 256–257
Mescaline, 483t
Mesoderm, *422,* 423
Mesophyll cells, 246, *247*
Messenger RNA (mRNA), 183, *183*
   editing, 189, *190*
   transcription of, 184, *185*
   translation of, 187, *188,* 189
Messner, Reinhold, 379
Metabolic pathways, evolution of, 29
Metabolism, xxii, 358
   evolution of, 29
   heat as product of, 84
   site of, 39
Metals, shiny, 6, *7*
Metaphase of mitosis, 114, *115*
Methionine, 359
Metric system, 614t
Met-tRNA, 187
Microbes, nitrogen-fixing, 585, *587*
Microfibrils, 20, *21*

Microfilaments, 49, *50,* 51
Microfilaria, 296
Micronutrients, soil, in vascular plants,
   265, 266
Microorganisms, classification systems and,
   204–220
Microscopes
   compound, 34
   images of, compared, *38*
   electron, 37, 37–38
   light, *36*
      invention of, 34
      limitations of, 37
Microspores, 236
Microtine rodents, 560
Microtubules, 50, *50*
Microvilli, 368, *368*
Midbrain, 474
Milk, production of, in mammals, 323
Miller, Stanley, 13, *13*
Millipedes, 301, *302*
Mineral nutrition in vascular plants,
   265–266
Mineralocorticoids, 403t, 409
Minerals
   in human diet, 364
   major roles of, 364t
Miscarriage, 132
Mitochondrion, *43*
   compartments of, *80*
   Krebs cycle in, 78
   respiration and, 44–45, *45*
Mitosis, 110–122
   anaphase of, 114, *114, 115*
   compared to meiosis, *129*
   metaphase of, 114, *114, 115*
   prophase of, 113, *113, 115*
   stages of, *113–114, 115*
   telophase, 114, *114, 115*
Mold(s)
   bread, 217, *218*
   slime, life cycle of, *213*
Molecule(s)
   defined, 6
   polymerization and, 17–18
Mollusca, 296–301
   classes of, 297; *see also* specific classes
Mollusks, 296–301
   characteristics of, 307t
   anatomy of, 297
   evolutionary advances in, 297
   major structures of, *298*
Molting, arthropod, 303
Monera, 207t, 208–211
Mongolism, 134; *see also* Down syndrome
Moniliasis, 435t
Monoclonal antibodies, 355, *355*
Monocots, 238
   leaf blades of, 245, *245*
   seed of, *239, 272*

vascular bundles in, 253, *254*
Monocotyledons, 238
Monod, Jacques, 196
Monohybrid, 138, 139t
Monohybrid cross, 137–142, *139*
    Mendel's explanation of, 141
Monosaccharides, 18, *19*
Monotremes, 324
Morgan, Thomas Hunt, 154, *156*, 205
Mosses, sexual life of, 228, *229*, 230
Moth, peppered, microevolution of, 498
Motility, 438
Motor ganglion, *468*
Motor neuron, 439
Motor unit, 461
Mountain sickness, 379
mRNA; *See* Messenger RNA
mRNA editing, 189, *190*
mRNA–ribosome complex, 183
Mucosa of stomach, 367, *367*
Mucus
    in human respiratory system, 375
    in stomach, 367
Mucus method of birth control, 430
Muscle(s)
    antagonistic, 456
    arthropod, 455–456
    cardiac, 456, *457*
    coordination of, 438
    nervous sytem and, 454–465
    relaxation of, 461
    skeletal; *see* Skeletal muscle
    smooth, 457, *457*
Muscle contraction
    biochemical events in, 459, 461
    control of, 461, 464
    electrical events in, 459
    sliding filament hypothesis, of, *460*, 461
Muscle fatigue, *465*
Muscle fiber, 457, *458*
Muscle twitch, 464, *465*
Muscular dystrophy, Duchenne, 159
Muscular system, function of, 335t
Mutagens, 194, 198
Mutant, 191
Mutations, 505, 507
    base deletion, 193, 194
    base insertion, 193, *194*
    base substitution, 192–193, *193*
    DNA, 191–195
    during replication, 193
    environment and, 193–194
    estimated rates of, 507
    and genetic change, 195
    as source of new alleles, 194–195
    types and causes of, 192–194
Mutualism, 566
Myasthenia gravis, 352
Mycorrhizae, 267
Myelin sheath, 440, *441*

Myofibrils, 457, *458*
Myoglobin, *27*
Myosin, 49

NAD⁺, 78
    extramitochondrial regeneration of, 86
NADH, 80, 82
    formation of, 78, 80, *81*
NADPH, light reaction and, 99
Nasal cavity, 375, *375*
Natural selection, xxv, 495–500, 509–514
    directional, 511–513, *512*
    disruptive, 513–514
    key components of theory of, 499–500
    modes of, 509–513, *511*
    stabilizing, 509, 511
Nature versus nurture, 149–150
Navigation, celestial, 541–544, *542*
Neanderthal people, 331, *331*
Neisseria gonorrhoeae, 435t
Nematocysts, 291–292, *293*
Nematoda, 296, *297*
Nematodes
    evolutionary advances in, 296
    as human parasites, 296
    major characteristics of, 307t
Nephron, 388–389, *389*
    processes occurring in, *390*
Nepotism in ground squirrels, 551–553
Nereocystis, 227
Nerve(s), 469
    cranial, 469
    optic, 453
    spinal, 469
    structure of, *471*
Nerve cells; *see* Neuron(s)
Nerve cord, dorsal hollow, 310, *311*
Nerve fibers, 469
Nerve impulse, propagation of, 445–446
Nerve net, 467, *468*
Nervous system(s), 467–484
    arthropod, *468*, 469
    autonomic; *see* Autonomic
        nervous system
    central, 469, 473, 474
    Cnidaria, 467, *468*
    effect of drugs on, 482, 483t
    functions of, 335t, 438
    Hydra, 467, *468*
    of invertebrates, 467–469
    in mammals, 326
    muscle response and, 454–465
    peripheral; *see* Peripheral
        nervous system
    Planaria, 467, *468*
    respiration and, 380
    somatic, 469, 470–471
    vertebrate, 469–474
Net productivity of plants, 578

Neuromuscular junction, 459
Neuron(s), 438–448
    all-or-none response of, 443
    basic properties of, 438
    communication between, 446–448, *447*
    and information transmittal, 441–445
    motor, 439
        muscle fibers and, 461
    postganglionic, 471, *472*
    preganglionic, 471, *472*
    sensory, 439
    sodium–potassium pump of, 443, *444*
    structure of, 439–440, *440*
    types of, 439
Neuropeptides, 480
Neurotransmitters, 446–447, *446*, 480
Neutrons, 5
Niacin, 364
Niche, ecological, 563, *565*
Nicholas II, 162, *162*
Nicotine, 483t
Nitrate reduction, 586
Nitrification, 585
Nitrogen
    atomic number and mass of, 6t
    covalent bonding in, *9*
    in ecosystem, 580–581
    molecular, 585
Nitrogen cycle, 585–588, *586*
Nitrogen fixation, 585, *586, 587*
    fertilizers and, 587–588
Nitrogenous waste, 343
    produced by various animal groups, 387t
Node(s)
    atrioventricular, 339, *339*
    lymph, 346
    sinoatrial, 339, *339*
    stem, 253
Nodules, root, 585, *587*
Nondisjunction, meiotic, 132, *133*, 154, 155
Nonhistone proteins, 197, 199
Norepinephrine, 403t, 408, 482
Notochord, 310, *311*
Novocaine, 483t
Nuclear envelope, *43*, 44
Nucleases, 369, 369t
Nucleic acids, 28
Nucleolus, *43*, 44
Nucleotide, 28
    DNA, 170, 171
    RNA, 181
    sequence of, 174
    structure of, *171*
Nucleotide alphabet, 174
Nucleus
    atomic, 5
    cellular, *43*, 43–44, *44*
Numbers, atomic, 6t
Nutrients
    essential, in vascular plants, 265t

flow of, in ecosystems, 580–588
functions of, 357
Nutrition
human requirements in, 357–364
mineral, in vascular plants, 265–266

Obelia, life cycle of, *292*
Obligate anaerobes, 86, 88
Obligatory parasites, 211
Octopuses, 299–300, *300*
Ocular albinism, 159
Odors, communication by, 541
Oedogonium, 224–225, *225*
Oil(s), 21–22
Omnivores, 559
Oocytes
primary, 413
secondary, 414, *415*
Oogenesis, 413–415, *415*
Oogonia, 413
Ootid, 415
Oparin–Haldane proposal, 12–13
Operator gene, 196, *197*
Opiates, natural, 480, 482
Opium, *482*, 483t
Optic nerve, 453
Oral cavity, 366, *367*
Organelles, 43, 44
origin of, 47
Organic acid, 12
Organic compounds, carbohydrate, 18–21
Organic precursors, 358–359
Organs
in chordates, 310
as evolutionary advance, 293
first appearance of, 307t
transplants of, rejection of, 352
Organ systems, review of, 334, 335t, *335*
Orgasm, 419
Orientation behavior, 541–544
Osculum in sponges, 290, *290*
Osmoconformers, *383*, 383–384
Osmoreceptors in hypothalamus, 391
Osmoregulation, 382–386
in bony fishes, *385*
Osmoregulator, *383*, 384
Osmosis, 58–59, 62, 63
Osmotic equilibrium, 62
Osmotic status, 383
Osteichythyes, 315, *315*, 317; *see also*
Fishes, bony
Oval window, 453
Ovaries, 413
Ovett, Steve, 462
Ovulation, 413, *415*
preventing, 431, 433
Ovulation method of birth control, 430
Ovum, 415
Oxaloacetic acid, 80
Oxidation, 13–14

Oxygen
atomic number and mass of, 6t
buildup of, 70
as by-product of photosynthesis, *71*
carried in capillaries, 343
covalent bonding of, 8, *9*
in ecosystem, 580–581
glycolysis in absence of, 86, *87*
movement of, in human respiration,
376–377
production of, consequences of, 30, 32
in respiration, 372
transport of, 348
in water formation, 74
Oxygen concentration, 376, 380
at high altitude, 379
Oxygen deficit, 86
responses to, 86
Oxytocin, 403t, 404, 405
Ozone layer, 31–32

P_{660}, 280, 281, 282
P_{730}, 280, 281, 282
Pacemaker, 339, *339*
Palisade layer of mesophyll, 246
Palm, sago, *233*
Pancreas, 368, 406–407
enzymes produced by, 369t
hormones produced by, 403t
Pancreatic amylase, 369, 369t
Pancreatic lipase, 369, 369t
Pangaea, Supercontinent of, 523, 524,
525, *525*
Panspermia, Arrhenius' theory of, 3–4
Pantothenic acid, 364
Paramecium, 215, *215*
competition among, 564
Parasite, 217, 357, *358*
nematodes as, 296
obligatory, 211
Parasitism, 559, 566
Parasympathetic division of autonomic
nervous system, 471
Parathyroid hormone, 403t
Parenchyma, defined, 250
Parkinson's disease, 480
Passerine birds, territorial behavior in,
548–549, *549*
Pasteur, Louis, xxvi
Peas
Mendel's, 136, 137, *138*
traits of, 140, *140*
Pectin, 53
Pedigree, analysis of, 158, *158*
Pelagic zone, 608
Pelvis
of *A. afarensis* compared to chimp, *330*
and human posture, *330*
Penicillium, 217
Penis, 416, *417*, 419

Pepsin, 367
Peptidases, 369, 369t
Peptide bond, 24, *25*
formation of, 189
Peptide hormones, 400, *400*
Perennials, 285
Pericycle, 251, *251*
Peripheral nervous system, 469–473
divisions of, 469, *470*
function of, 469
Peristalsis, 366, *367*
Peritoneal dialysis, 392
Permafrost, 603
Permeability
differential, 57
selective, 40
Permian extinction, 527
Petals, 236
Petiole, 245, *245*
PGA (phosphoglyceric acid), 100
PGAL (phosphoglyceraldehyde), 100
pH, 615, *615*
Phaeophyta, 226
Phages, 213
Phagocytes, 348, 349
Phagocytosis, 67, 214, *214*
Pharynx, 366, *367*, 375, *375*
Phenotype, 141
defined, 139t
distribution of, 141
recombinant, crossover and, *161*
Phenotype ratio, probability of, 142
Phenylalanine, 359
codon for, *187*
high levels of, 148
Phenylketonuria (PKU), 148
Pheromones, 541
Phloem, 246, *247*, 252
food-conducting cell of, *253*
sap movement in, 267, 268t
secondary, 257
of stem, 253
Phosphate group, 22, *23*, 72, *72*
of DNA, 170
transfer of, 73
Phosphoglyceraldehyde (PGAL), 100
Phosphoglyceric acid (PGA), 100
Phospholipids, 22, 56
structure of, *23*
in water, *23*
Photoautotrophs, 221
Photodormancy, discovery of phytochrome
and, 280–281
Photomorphogenesis, 281–282
Photoperiod, 282
Photoperiodism, flower induction and,
282–283
Photoreception, 449–453
defined, 449
Photoreceptors, 448, 449t

Photosynthesis, 45, 92–108
  action spectrum of, 94, *95*
  C_3, 102–103
  carbon dioxide in, 93
  carbon dioxide concentration and, 107
  and carbohydrate production, 71
  chemical energy production during, 578
  chlorophylls in, 94
  dark reactions of, 100–104
  defined, 92
  emergence of, 29–30
  energy in, 94
  and formation of ozone layer, 32
  leaf's relationship to, 246
  light and dark reactions of, 97–104
    overview of, *97*
  light intensity and, 106–107
  limiting factors of, 106–107
  origin of, 70
  overview of, 97, *97*
  relationship of, to respiration, 70, *71*
  temperature as limiting factor in, 106–107
  visible light in, 94
  water availability and, 107
Photosynthetic organisms as primary producers, 573
Photosystem, 97–98, *98*
Photosystem I (PS I), 98, *99*
Photosystem II (PS II), 98, *99*
Phototaxis, 449
Phototropism, *274*, 274–276
Phylum, 206
Phytochrome
  forms of, 280–281, 282
  photodormancy and, 280–281
  photogenesis and, 281–282
  seedling development and, *281*, 281–282
Phytoplankton in pelagic zones, 608
Pigments
  and light absorption, 94
  photosynthetic, in chloroplasts, 96, *96*
Pill, birth control, 431, 433
Pine(s)
  bristlecone, *285*
  life cycle of, *234*, 234–235
  lodgepole, *233*
Pine cones, 234, *235*
Pinworms, 296
Pistil, 236, *236*
Pith of stem, 253, *254*
Pithecanthropus erectus, 330
Pituitary gland, 402
PKU, 148
Placenta, 326, 425, *425*
Placentals, 324, *325*, 326
Placodermi, 315, 316
Planarians, *293*
  nervous system of, 467, *468*

Plant(s); *see also* Vascular plant(s)
  aging in, 284–285
  alternation of generations in, 221–222
  annuals, 284
  C_4, 101, *101*, 104, *104*
  cell of, *35*
    generalized, *43*
  classification characteristics of, *207*
  desert, strategies of, 248–249
  evolutionary trends among, 242t
  flowering; *see* Angiosperms
  gross productivity of, 578
  major groups of, 222, *223*
  net productivity of, 578
  nonvascular, 222–230, *223*
  number of species of, 204
  perennials, 285
  photoperiodic, time measurement in, 282–283
  sexual life cycle of, 221–222, *222*
  structural adaptations of, to land, 242t
  turgor pressure and plasmolysis in, *62*
  vascular; *see* Vascular plants
  woody, 598
Plant Kingdom, 207, 207t, *208*, 221–242
  evolutionary trends in, 242t
Plantae, 207, 207t, 221–242; *see also* Plants(s); Vascular plants
Plasma, 346
  blood, concentration of various substance in, 391t
Plasma cells, 350
Plasma membrane, 40, 41
Plasmablasts, 350
Plasmids, 177
Plasmodesmata, 262, *262*
Plasmodium, 214
Plasmolysis in plant cells, *62*
Plastids, 44, 45
Platelets, 347
Platyhelminthes, 293–296
Platypus, duck-billed, 324
Pleiotropy, 148, 149t
Plesiosaurs, 320, *320*
Pleural cavity, 377, *378*
Polar body, 414, *415*
Polarity, 8
  in water molecule, *10*
Pollen grains, 235
Pollen tubes, 235
Pollination, 235
  cross, 84, 240
  mechanisms of, *240*, 240–241
Polychaetes, marine, 300
Polygamy, 550
Polygenic inheritance, 149, *150*
Polymerase, RNA, 184, *185*
Polymerization, 17–18
  of amino acids, 17
  dehydration synthesis, 19

Polymers, 17
Polyp, 291, *291*
Polypeptides, 24, *25*
  translation of mRNA to, 187, *188, 189*
Polyploidy, 520–523, *521*
Polysaccharides, 19–21, *20, 21*
  Golgi bodies and, 49
  storage, *20*, 74
Poly U, 185
Pons, 474, *474*
Population(s)
  age distribution of, 557, *557*
  characteristics of, 556–557
  doubling times for, 569t
  ecology of, 556–569
  human, 567–570
  interactions among, 559, 562–567
  isolation of, 514
  natural selection and, 512
  size of, 556–557
Population growth, 467–468, *468*
  carrying capacity and, 558, *558*
  control of, 570
  doubling time of, 568
  dynamics of, 557–558
  future prospects for, 568–570
Population size
  competition and, 563–565, *564*
  regulation of, 558–559
Pore cells in sponges, 290, *290*
Porifera, 290–291, 307t
Porphyra, 227, *228*
Postelsia, 227
Posterior pituitary gland, 403t, 405, *405*
Postganglionic neuron, 471, *472*
Postmating isolating mechanisms, 515
Postsynaptic membrane, 446
Post-translational modifications, 189
Potassium pump, 246
Potato, reproduction in, 119
Predation, *562*
  defined, 559
Predators, 559
Predator–prey systems, 559, *559*
Preganglionic neuron, 471, *472*
Pregnancy, 420–428
  DES and, 436
  hormone levels in maternal bloodstream during, *422*
  tests to confirm, 421
Premating isolating mechanisms, 515
Pressure, water potential and, 59
Pressure flow model, sugar transport and, 269
Presynaptic membrane, 446
Primary succession, 591
Primates, 326–332
  characteristics of, 326–328
  examples of, *327*
Principle of edges, 592–593

Pritikin diet, 360
Procambium, 257
Productivity
    ecosystem, 578–580, *579*, 580t
    of tropical forests, 596
Progesterone, 403t, 409
    in menstrual cycle, 416
Prokaryotes
    in Kingdom Monera, 208
    transcription control in, 195–196
Prokaryotic cells, 40–42
Prolactin, 403t, 404
Prolactin inhibiting hormone, 406t
Prolactin releasing hormone, 406t
Promotor gene, 196, *197*
Prophase of mitosis, 113, *115*
Prosimians, 326
Prostigmine, 482
Protein(s), 24–28
    accuracy in synthesis of, 183
    amino acid sequence of, 181
    as amino acid source, 359
    carrier, 63–64
    catalysis function of, 28
    composition of, 24
    deficiency of, 359, *359*
    enzymes, 28
    fibrous, 26, *27*
    formation of
        translational level of, 195
        transcriptional level of, 195
    function of, 27–28
    and genetic expression, 195
    globular, 26, *27*
    histones, 196–197
    nonhistone, 197, 199
    primary structure of, 26, 26t
    protective function of, 27
    quaternary structure of, 26t, 27
    regulatory, 175
    secondary structure of, 26, 26t
    storage function of, 27
    structure of, 26t, 26–27, *27*
    synthesis, 44, 48, 184–189, *188*
    tertiary structure of, 26t, 26–27
    transport, 175
    transport function of, 27–28
Proteinoids, 28, *28*
Protista, 207, 207t, 231–216
Protons, 5
Protoplasm, 39, 42
Proximal convoluted tubule, 388–389, *389*
Pseudocoelom, 296
    first appearance of, 307t
Pseudopods, 214, *214*
PSI, 98, *99*
Pterosaurs, 320
Puberty, hormonal changes during, 409
Punctuated equilibrium, 527
    versus gradualism, 523, *524*
Punnett square, 506

Pupil, 450
Purebred, defined, 139t
Purebreeding strains, 137
Pyruvic acid, 76, 86
    formation of, 78
    splitting of, 80

Queen Victoria, 158, 162

Radial symmetry in echinoderms, 306
Radiation
    adaptive, 524, 526
    solar
        fates of, 577, *578*
        spectrum of, 94, *94*
Radiolarians, 214, *214*
Ramapithecus, 328, *329*
Rapid eye movement sleep (REM), 478, *479*
Rasputin, 162–163, *163*
Rays, 384
Reabsorption, selective, 389
Reactants, 74, *75*
Reaction(s)
    with and without catalysts, *75*
    dark; *see* Dark reactions
    endergonic 73, *73*
    exergonic, 73, *73*, 74
    light; *see* Light reactions
Receptor potential, 449
Recombinant DNA, 176–179, 210
Rectum, 370
Red algae, 226–228, *228*
Red-green color blindness, 157–158
Reductionism, xxvi
Redwoods, 600, *601*
Reflex arc, 439, *439*, 446
Regulator gene, 196, *197*
Reinforcement, 537
Remoras, commensalism of, 566, *566*
Renal arteries, 388, *388*
Renal veins, 388, *388*
Replication, DNA, 173–174, *173*, *174*
    DNA, compared to RNA transcription, 184
    mutations and, 193
Replication bubbles, 174
Reproduction, xxii
    in angiosperms, 236–241
    in annelids, 300
    asexual, 110–122, *118*
    bacterial, 209
    in cyanobacteria, 211
    in fungi, 217
    in green algae, 224–225, *225*
    human, 412–437
    in mammals, 326, 327
    in plants, 221–222, *222*
    sexual, 123–124
Reproductive isolating mechanisms,
    515–516

Reproductive system(s), human
    female, 413–416, *414*
    male, 416–418, *417*
Reptiles, *318*, 318–323
    adaptive radiation of, *526*
    circulation in, 321, 337–338
    evolution of major groups of, *320*
    land adaptations of, 320–321
    reproduction in, 321
    respiration in, 321
Respiration
    aerobic, 318
    in amphibians, 318
    of carbohydrates, 70, 76–83
    defined, 70, 76
    energy and, 70–91
    evolution of, 30, 32
    of glucose, 76–83
    mitochondria and, 44–45, *45*
    the mobilization of polysaccharides and
        fats for, 88
    overview of, 76, 77, 78
    relationship of, to photosynthesis, 70,
        71, 93
    in reptiles, 321
    in yeasts, 86
Respiratory electron transport, 76
Respiratory electron transport chain, 80,
    82–83, *83*
Respiratory structures
    evaginated, 272–273, *273*
    invaginated, 272–273, *273*
Respiratory system
    function of, 335t
    human, *375*, 375–377
    types of, 372–374, *373*
Resting potential, 443, *444*
Reticular formation, 477–479, *478*
Retina, 450, *452*
Revolution, industrial, 568
Rhesus monkey, *327*
Rheumatic fever, 352
Rheumatoid arthritis, 352
Rhizobium, *587*, 588
    nitrogen fixation and, 585
Rhizoids, 228
Rhizome, 232
Rhizopus, life cycle of, *218*
Rhodophyta, 226–228
Rhythm method of birth control, 430
Ribonucleic acid (RNA), 28, 181–184
    contrasted with DNA molecule, 182
    messenger (mRNA), 183, *183*
    ribosomal (rRNA), 183, *183*
    structure of, 182, *182*
    transfer (tRNA), 183, *184*
    transcription of, 184
    types of, 182–183
Ribonucleic acid polymerase, 184, *185*
Ribose, 72, *72*, 182
Ribosomal RNA, 183, *183*

Ribosomes, 48, 184
Ribulose bisphosphate (RuBP), 100
Rickets, 362t, 363, *363*
RNA; *See* Ribonucleic acid
Rod cells, 450–451
Rodents, mictrotine, 560
Root(s), 250–252
    anatomy of, 250–252, *251*
    embryonic, *272*
    gravitropism and, 276
    lateral, 244
    micropathways of water movement
        in, *261*
    primary, 244
    roles of, 250
    surface of, 250
    vascular cylinder of, 250, *251*
    in water transport, 261
Root cap, 250, *251*
Root hairs, 250, *251*
Root nodules, 585, *587*
Root system, 244, 250, *250*
Roundworms, 296, *297*
rRNA, 183, *183*
RuBP (ribulose bisphosphate), 100

Saccharomyces cerevisiae, 217
Sagebrush, Great Basin, 602, *602*
Sago palm, *233*
Salivary amylase, 366
Salivary glands, 366, *367*
    enzyme production by, 369t
Salmonella, Ames test and, 198
Sap, movement of, 267; *see also* Sugar
    transport
Saprophytes, 357, *358*
Sapwood, 258
Sarcodina, 214
Sarcolemma, 459
Sarcomeres, 459
Sarcoplasmic reticulum, 459
Saturability, 63
Schistosoma, 294, 295
Schistosomiasis, 294, *295*, 296
    vaccine against, 351
Schleiden, Matthias, 37
Schwann, Theodor, 37
Schwann cells, 440, *440*
Science, society and, xxvii–xxviii
Scientific method, xxvi–xxvii
Sclera, 450
Sclerenchyma, 253, *254*
Scrotum, 416, *417*
Scurvy, 362
Sea anemones, *291*
    movement in, 455, *455*
Seed, 233
    barley, hormonal messenger system in,
        272, *273*

development of, in angiosperms,
    237–238
    dicot, *272*
    dormancy of, 272, *273*
    germination of, 271–273, *272*
    longevity of, 270, 270t
    monocot, *272*
    parts of, 271
    structure of, *239*, 271–272, *272*
Seed coat, 238, 271
Seed cones, 234, *235*
Seedling, phototropism and, 274
Seedling development, phytochrome and,
    281, 281–282
Segmentation, body, first appearance
    of, 307t
Segregation, law of, 140–141, 142
Selection, natural; *see* Natural selection
Selective breeding, 136, 489, 491, *491*
Selective permeability, 28–29
Selective reabsorption, 389
Selfish gene hypothesis, 548–554
Semen, 417
    ejaculation of, 420
Semicircular canals, 454
Seminiferous tubules, 418
Semispecies, 515
Sense strand, 184
Sensitivity, xxii, *xxiii*
Sensory cells, activation of, 449
Sensory neuron, 439
Sensory reception, 448–454
Sensory receptors, *448*, 448–449, 449t
Sepals, 236
Serine, tRNA and, *184*
Serotonin, 482
Sex, inheritance of, 153–154, *154*
Sex chromosomes, 153, *154*
    human, abnormalities of, 155
Sex determination, 153–154, *154*
Sex hormones, 403t, 409
Sex-linked traits, 154
    in fruit flies, 154, 156–157
    in humans, 157–159
    number of, 158–159
Sexual intercourse, 419–420
Sexual reproduction, 123–124
Sexually transmitted diseases,
    433–434, 435t
Shells, atomic, 6, 7
Shiny metals, 6, 7
Shoot apex, 253
Shoot system, 244
Short-day plants, 282, *282*
Sickle cell anemia, 192, *192*, 512–513
    coincident with acute malaria, 513
Sieve tube, 252, *253*
Sinoatrial node, 339, *339*
Skeletal muscle(s), 456, *457*
    contraction of, 459, *460*, 461, 465
    structure of, 457, *458*, 459, *459*

Skeletal system, function of, 335t
Skeleton, human, 456, *456*
Skull
    of *A. afarensis, 303*
    in vertebrates, 310
Sleep, 478
    rapid eye movement, *479*
Sliding filament hypothesis of muscle con-
    traction, *460*, 461
Slime mold
    classification problems of, 213
    life cycle of, *213*
Small intestine, 367–368, *368*
    enzymes produced by, 369t
Smooth muscle, 457, *457*
Social behavior, 544–554
    altruism, 551–554
    of baboons, 546, *547*
    of California sea lions, 550
    evolution and, 545, 548
    of ground squirrels, 551–553
    of honeybees, 553–554, *553*
    of hoofed mammals, 550, *551*
    mating behavior in mammals, 550–551
        infanticide, 550–551
        polygamy, 550
    nepotism, 552
    of passerine birds, 548–549, *549*
    territoriality, 548–549
Societies, animal, 544–545
Sociobiology, defined, 544
Sodium–potassium pumps, 443, *444*
Soil macronutrients, 265, 265t
Soil micronutrients in vascular plants,
    265, 266
    effect of, on water potential, 59
Solute composition, 382
Solute concentration, 382, 384
    water potential and, 59
Solutes, 61
Solvent
    function of, 17
    universal, 61
Somatic cells, 124
Somatic nervous system, 469, 470–471
Somatotropin, 403t, 403–404
Special creation, 2–3, 492, 493
Speciation, 514–523
    allopatric, 516–518
    defined, 514
    by geographic isolation, 517, *518*
    sympatric, 518–523
Species
    biological concept of, 514
    competition among, 563–565, *564*
    concept of, 514–515
    defined, 205
    diversity of
        climate and, 592
        in higher latitudes, 592, *592*
    immutability of, 492

number of, 523
problems with the concept of, 514, 515, *515*
trophic level of, 573
Specific heat of water, 60
Spectrum(a)
absorption, 94, *96*
action, 94, *95*
Sperm
barriers to movement of, 431
human, *419*
production of, 418
Spermatids, 418, *418*
Spermatocytes, 418, *418*
Spermatogenesis, 418, *418*
Spermatogonia, 418, *418*
Spermicidal foam, 431
Sphagnum, 229
Sphincter, 366
Sphygmomanometer, 341
Spicules in sponges, 290, *290*
Spider monkey, *327*, *597*
Spinal cord, *472*
Spinal nerves, 469
Spindle apparatus, 51, 113
Spindle fibers, 113
Spirogyra, 223–224, *224*, 610
Spirulina, 210
Sponge(s)
anatomy of, *290*
major characteristics of, 307t
species of, 290
Spongin, 290
Spontaneous abortions, 132
Sporangia, 232
Spores, asexual, 117, *118*, 217
Sporophytes, 222
in fern reproduction, *232*
of pines, 234
Sporozoans, 214
Spring wood, *257*, 258
Squids, 299–300, *609*
Stamens, 236, *236*, 237
Starch
accumulation of, 105
biosynthesis, of, ATP in, *73*
digestion of, 88, *89*
in human diet, 20
Starch-blocking diet pills, 361
Starfish, 306, *307*, *607*
Start codon, 187
Stem(s), 253, *254*
anatomy of, 253, *254*
components of, 253
gravitropism and, 276
secondary growth in, 257
vascular cambium in, 257
Sterilization, 433, *434*
Steroid(s), 22, 24
Steroid hormone(s), 400, *400*, *401*
Stickleback, mating ritual of, *540*

Stigma, 236
Stimulus, threshold, 441
Stipe, 226
Stirrup, 453
Stomach, 367, *367*
enzymes produced by, 369t
mucus in, 367
Stomates, 246, *247*
defined, 106
opening and closing of, 246, *247*, 250
Stop codons, 185, 186t, 189
Stretch receptors, 448
Stroma, 96, *96*
Style, 236, *236*
Succession
ecological, 590–591
primary, 591
Sucrose, 19, 105
Sugar; *see also* Glucose
blood
abnormal levels of, 406–407
insulin and glucagon in regulation of, *408*
simple, 18
transport of, in plants, 246
uses of, 104–106, *105*
Sugar sinks, 267
Sugar sources, 267
Sugar transport
aphids in experiments on, 268
pressure flow model and, 269, *269*
in vascular plants, 267–269
Summer wood, *257*, 258
Sunlight
range of, 94, *94*
as source of organisms' energy, 577
Superior vena cava, 346
Surgery, bypass, 345
Sutton, Walter S., 153
Symbiosis, 565–566
Symmetry
bilateral, 293, 307t
radial, 306, 307t
Sympathetic division of autonomic nervous system, 471
Sympatric speciation, 518–523
Synapse(s), 446
clearing of, 447–448
communication across, 446–448, *447*
Synaptic cleft, 446, *447*
Synaptic vesicles, 446
Synthesis, protein; *see* Protein, synthesis of
Syphilis, 433, 435t
vaccines against, 351
Systematics, 204
Systole, 340

2,4,5-T, 277
T cells, 348, 350, 352
Tadpoles, 318
Tapeworms, 296

Taproot system, 250, *250*
Target cells, 398
Taxonomy, 204
Tay–Sachs disease, 147, 149t
Telophase of mitosis, 114, *115*
Temperature
biomes and, 595
as limiting factor in photosynthesis, 106–107
Temperature regulation, 391, 393–396
ectothermic, 391, 394–395
endothermic, 395–396
Template, DNA, 173
Terrestrial biomes; *see also* specific biomes
descriptions of, 594–605
map of, *594–595*
Territoriality, 548–549
Testes, 416, *417*
Testicles, 416, *417*
Testosterone, 409
Tetanus, *465*
Tetanus toxin, 88
Thalamus, 474
Thallus, 222
Thecodonts, 320
Therapsids, 320, *320*, 323, 324
Thermoreceptors, 448, 449t
Thiamin, deficiency of, 362
Threonine, 359
Threshold stimulus, 441
Thylakoids, 96, *96*
Thymine, 170, *171*
Thyroid, hormone produced by, 403t
Thyroid gland, 406
Thyrotropin, 402, 403t
Thyrotropin releasing hormone, 406, 406t
Thyroxine, 403t, 406
feedback regulation of, 406, *407*
overproduction of, 406
Time measurement in photoperiodic plants, 282–283
Tools, *Homo sapiens*' use of, 330
Tortoise, Galapagos, *319*
Trace elements, 265t, 266
Tracheae, *374*, 375, *375*
Tracheids, 252, *252*
Traits
inheritability of, xxv–xxvi
sex-linked; *see* Sex-linked traits
Transcription, 184, *185*
histones and, 196–197
RNA, compared to DNA replication, 184
Transcriptional control
in eukaryotes, 196–197, 199
in prokaryotes, 195–196
Transcriptional level of protein formation, 195
Transfer RNA (tRNA), 183, *184*
and genetic code, 183
serine and, *184*
Transforming factor, 168, *168*

Translation, 187, *188*, 189
Transmission electron microscope, *37*, 37–38
Transmitters, amino acid, 480
Transpiration, 261, 261t
Transplant
  heart, 345
  organ, rejection of, 352
Transport
  active, 65–66, *66*
    versus simple diffusion, *65*
  vesicular, 66–68
Transport systems, animal, 336–338; *see also* specific systems
Trap chlorophyll *a*, 98
Treponema pallidum, 435t
Trial-and-error learning, 537
Triceps, 456, *457*
Trichinella, 358
Trichinella spiralis, 296
2,4,5-trichlorophenoxyacetic acid (2,4,5-T), 277
Trichomonas vaginalis, 435t
Trichomoniasis, 435t
Triglycerides, 21–22, *22*
Tripeptide, 189
Triplet code, 181, 186t
Trisomy-21, 134; *see also* Down syndrome
tRNA, 183, *184*
Trophic level of food chain, 573, *576,* 580
Trophoblast, 421, *421*
Tropic hormones, 402, 403t
Tropical communities, 592
Tropical forest, 396, 596, *597*
Tryon, R.C., 534
Trypsin, 369, 369t
Tryptophan, 24, 359
Tubal ligation, 433, *434*
Tubules, transverse, 459
Tubulins, 51
Tundra, 603, *603*
Tunicates, 312, *312*
Turgor pressure, 62, *62*
Turner syndrome, 155
Turtles, 320, *320*
Twins, 427
Twitch contraction, 464
Tympanic membrane, 453

UGA codon, 186
Ulcer, 367
Ultraviolet light, 94
Ultraviolet radiation, 31
Umbilical cord, 425
Undernutrition, 358
Uniformitarianism, theory of, 493
Uracil, 182, *182*
Urea, 343, 387
Ureter, 388, *388*
Uric acid, 387
Urinary bladder, 388, *388*

Urine, 388
  concentration of various substances in, 391t
  formation of, 389–391
Urochordata, 312
Urogenital system, 417
Uterus, 413

Vaccines, 350
Vacuole, *43*, 49
  contractile, 49
  food, 49
Vagina, 413
  acidity of, 420
  cancer of, and DES, 436
Vaginal sponge, 431
Valonia, 39, *40*
Valves
  heart, *338*, 339
  one-way, in veins, *343*
Van Helmont, Jean Baptiste, 92
Van Leeuwenhoek, Anton, 34
Vaporization, latent heat of, 60
Vascular bundles, 253, *254*
Vascular cambium, *256*, 256–257
Vascular cylinder of root, 250, *251*
Vascular plants, 230–241
  auxin and, 274–276
  carbohydrate source in, 267–268
  development of
    and growth regulation, 271–286
    hormones and, 274–280
    light and, 280–284
  essential nutrients in, 265t
  evaporation-driven water potential gradient in, 264, *264*
  fibrous root system in, 250, *250*
  gravitropism and, 276
  growth of, 255–258
    cytokinins and, 278–279
    gibberellin and, 278
    kinetin and, 279
    primary, *255*, 255–256
    secondary, *256*, 256–258
  leaves of, 244–250
  lower, *231*, 231–233
  major organs of, 244
  phototropism and, 274–276
  roots of, 250–252
  seed-producing, 233–241
  soil macronutrients in, 265, 265t
  soil micronutrients in, 265t, 266
  stems of, 253, *254*
  structure of, 244–254
  structure and growth of, 243–259
  sugar transport in, 267–269
  taproot system in, 250, *250*
  transpiration in, 261
  water functions in, 260
  water transport in, 260–266
  xylem and phloem of, 252

Vascular tissues, 222
Vas deferens, 417
Vasectomy, 433, *434*
Vasopressin, 391, 403t, 405
VD National Hotline, 212
Vein(s), 336
  lymph, 346
  one-way valves in, *343*
  renal, 388, *388*
Vena cava, *342*, 346
Venereal diseases, 433–434, 435t
Venereal warts, 435t
Ventricles of human heart, 339
Venules, 336
Vertebral column, 310
Vertebrata, 314
Vertebrate(s), 314–326
  ancestry of, *313*
  brain of, 473
  classes of, *314*
  comparative embryology of, *490*
  evolution of heart in, 336–338
  evolutionary relationships among, *314*
  hormones in, 399–400
  marine, 384–385
    salt regulation in, 384–385
    water regulation in, 384–385
  nervous system of, 469–474
  organ systems of, 334, 335t, *335*
Vesicular transport, 66–68
Vessel elements of xylem, 252, *252*
Villi, 368, *368*
Virchow, Rudolph, 37
Viruses, 211–213
  DNA, 213
  DNA experiments and, 168–169
  structure of, *169*
Visible light, 94
Vision, 450
  color, 157–158
  mammalian, 327
Visual displays, 538, 540
Vitalists, xxvi
Vitamin(s), 362t, 362–364
  B-complex, 362t, 363–364
  defined, 362
  fat-soluble, 362–363
  require by humans, 362, 362t
  water-soluble, 363–364
Volvox, 224
Von Frisch, 532
Vulva, 413

Wallace, Alfred Russel, 499, *499*
Warts, genital, 435t
Water, 6, 60–61
  adherence of, 60
  adhesion and cohesion of, 60–61, 264
  availability of, and photosynthesis, 107
  capillary action of, 60–61
  characteristics of, 60–61

competition for, in desert, 248
conduction of, in xylem, 252, *252*
conservation of, by kangaroo rat, *386*
covalent bonding in, *9*
effect of solutes on, *59*
expansion of, 61
formation of, 74, 78
functions of, in plants, 260
human use of, 386
importance of, 60–61
latent heat of evaporation of, 60
movement of, into and out of cells,
    58–59, 62–63
polarity in molecule of, *10*
regulation of
    in freshwater animals, 384
    in marine invertebrates, 383–384
    in marine vertebrates, 384–385
    in pig, *385*
    in terrestrial animals, 385–386
specific heat of, 60
transport of, in plants, 246; *see also*
    Water transport
Water balance, 391
    vacuoles in, 49
Water potential, 58–59

effect of pressure on, *59*
Water transport
    mechanism of, 263–265
    micropathways of
        in leaf, 263, *263*
        in root, *261*
    pathway of, 261–263
    in vascular plants, 260–266
    in xylem and phloem, 267, 268t
Water vascular system of echinoderms, 306
Watson, James, 169–170
Weight, ideal, 361
Weight loss, 360–361
Went, Frits, 275
Wheat, bread, hybridization and polyploidy
    of, *522*
White blood cells, 348, *348*
White matter of spinal cord, 473
Wilkins, Maurice, 170
Wood, 257–258, *257*
Woody plants, 598
Worms, segmented, 300–301

X chromosome, *125*, 153–154, *154*
XO, 155
XXX, 155

XXY, 155
XYY, 155
Xylem, 246, *247*, 252
    root, in water transport, 261
    sap movement in, 267, 268t
    secondary, 257
    in stem, 253
    water conduction in, 252, *252*

Y chromosome, *125*, 153–154, *154*
Yeast(s), 216
    in alcohol production, xxvi, 86
    brewer's, 217
    budding, 117, *118*
    reproduction in, 117, *118*
    respiration in, 86
Yeast infections, 433, 435t
Yolk sac cavity, 422, *422*

Z line, 459, *459*
Zone of cell differentiation, *255*, 256
Zone of cell elongation, *255*, 256
Zoospores, 225
Z-scheme, 98, *99*
Zygospore, 219
Zygote, 123, 420